Springer Atmospheric Sciences

More information about this series at http://www.springer.com/series/10176

Weihong Qian

Temporal Climatology and Anomalous Weather Analysis

Springer

Weihong Qian
Department of Atmospheric and Oceanic Sciences
School of Physics, Peking University
Beijing
China

ISSN 2194-5217 ISSN 2194-5225 (electronic)
Springer Atmospheric Sciences
ISBN 978-981-10-9928-1 ISBN 978-981-10-3641-5 (eBook)
DOI 10.1007/978-981-10-3641-5

© Springer Nature Singapore Pte Ltd. 2017
Softcover reprint of the hardcover 1st edition 2017
This work is subject to copyright. All rights are reserved by the Publisher, whether the whole or part of the material is concerned, specifically the rights of translation, reprinting, reuse of illustrations, recitation, broadcasting, reproduction on microfilms or in any other physical way, and transmission or information storage and retrieval, electronic adaptation, computer software, or by similar or dissimilar methodology now known or hereafter developed.
The use of general descriptive names, registered names, trademarks, service marks, etc. in this publication does not imply, even in the absence of a specific statement, that such names are exempt from the relevant protective laws and regulations and therefore free for general use.
The publisher, the authors and the editors are safe to assume that the advice and information in this book are believed to be true and accurate at the date of publication. Neither the publisher nor the authors or the editors give a warranty, express or implied, with respect to the material contained herein or for any errors or omissions that may have been made. The publisher remains neutral with regard to jurisdictional claims in published maps and institutional affiliations.

Printed on acid-free paper

This Springer imprint is published by Springer Nature
The registered company is Springer Nature Singapore Pte Ltd.
The registered company address is: 152 Beach Road, #21-01/04 Gateway East, Singapore 189721, Singapore

Preface

Weather extremes such as heat waves, cold surges, heavy rains, severe drought and forest wildfire, unusual storm tracks, and tornadoes frequently hit humankind. Climatic anomalies like the Arctic warming and sea ice decline have become a hot topic of climate change in recent years. Can we find early signals or methods to predict these weather extremes? And how can we explain these climatic anomalies? Some questions have been raised by meteorological researchers and weather forecasters. Since more data from satellite observations and mathematical model simulations have become accessible, we feel that atmospheric motion is more complex and full of chaos. We worry about if we are still standing on a mound while more data are filled like a high mountain. We need to stand on a higher platform to inspect the general circulation and anomalous atmospheric motion. Finally, we can physically achieve to build a connection between the surface weather extremes and atmospheric variable anomalies.

To quantitatively describe an extreme weather event, we first need a common reference in measuring the event's intensity, coverage, and duration. The common reference is a precise climatic clock (or climatic cycle), which is divided into 365-day $\times$ 24-hour = 8760 terms in a year. This climatic cycle or temporal climatology should be identified and studied, but does not need to be predicted. On the contrary, the hourly and daily departures from the temporal climatology need to be predicted. However, researchers nowadays directly apply reanalysis products and model simulations to study the historical weather episodes, and forecasters are still employing the traditional synoptic chart drawn by current observations and medium-range model outputs to predict the future weather extremes. This book demonstrates that the temporal climatology is useful to understand the climatic state and improve the forecast of weather extremes.

As a teacher at Peking University, I think we need an innovative concept and method to re-identify: What is weather? What are climate and general circulation? What are their differences? These three questions have been confusing teachers and students. We hope to cultivate the new generation of researchers and forecasters using this new concept and method. To achieve this goal, I have published a book titled "Principles of Medium to Extended Range Weather Forecasts" in Chinese in 2012.

In that book, all cases of weather extremes have been published in Chinese journals by analyzing the NCEP/NCAR reanalysis products with the four-term decomposition, namely zonally averaged climatology, its zonal departure, the zonally averaged anomaly, and its zonal departure. In this book, the cases have been published by 20 papers in international journals by use of the ERA-Interim product using the two-term decomposition, including the temporal climatology and anomaly. Ten chapters supported by the National Natural Science Foundation of China (41375073) and the two patents of China (201110061438.7 & ZL201210134358.4) are presented to readers as follows.

This book tells the readers a story from the universal dynamics of the Newton's single particles to Laplace's all particles in the Universe. Dynamics for each system, such as the Milky Way galaxy, the Solar and Earth systems, and a tropical cyclone with their similarity are described by decomposing total field into a basic state and anomaly in Chap. 1, "Weather and Climate." Wind is the oriented motion of air particles. The general circulation is a climatic representation of wind state. Therefore, multi-scale general circulations with diurnal and seasonal cycles are summarized in Chap. 2, "Wind and General Circulation." In this chapter, we will also introduce two new meridional mean circulations, the Arctic cell and the Antarctic cell, besides the Hadley, Ferrel, and Polar cells. Spatial patterns of climatic state are responses of the solar forcing and surface condition, so the four meridional cells in each hemisphere, the zonal Walker cells in boreal summer, and the atmospheric active centers are illustrated in Chap. 3, "Spatial Patterns and Time Scales of Climatic State," including annual, semiannual, and biweekly cycles. Monsoon has been an archaic topic since BD 23–22 centuries. The study of global monsoon started in the 1950s but the keyword of "global monsoon" describing dry–wet alternation was not proposed until the late 1990s. There are lots of monsoon definitions from regional to planetary scales and from precipitation variation to wind reversal. Therefore, the literature for monsoon is abundant. Chapter 4 has not only summarized previous monsoon results but also given a reasonable definition for the global monsoon domain with dry–wet alternation and low-level prevailing wind reversal associated with the cross-equatorial flow. Based on the global monsoon definition, division of the global monsoon, outer edge active zones of the global monsoon, and millennium monsoon dry–wet modes are also summarized in Chap. 4, "Global Monsoon."

The first four chapters introduce the temporal climatology of the atmospheric motion. Chapter 5, "Waves and Vortices," is a transition from temporal climatology to anomaly and their applications. A total atmospheric variable can be decomposed into temporal climatic state and anomaly. Spatial structures of climatic state show wave features with propagations while the anomalous component shows transient vortices with different spatiotemporal scales. As an example, the climatic states and anomalies of geopotential height, temperature, and wind at 0000 UTC 2 July 1998 are plotted from the troposphere to the stratosphere to illustrate those waves and vortices as well as their connections with surface climate and extremes in Chap. 5.

To let readers and forecasters understand the connection between anomalous vortices aloft and surface weather extremes, Chapter 6, "Regional Convective

Events," introduces the application of regional heavy rains and tornado-producing systems while Chap. 7, "Heat Waves and Cold Events," illustrates the applications on heat waves and cold surges. The intensity of a tropical cyclone (TC) can be well evaluated following its center by the vertical profile of height and temperature anomalies. The explanation and prediction of most TC unusual tracks are still difficult from direct observations and complex model outputs. Chapter 8, "Unusual Tropical Cyclones," introduces an optimal-level dynamical model, showing its application in the prediction of most unusual TC tracks which affected China in the past 30 years. This model is also used in the explanation of the left turning of Hurricane Sandy (2012) and the binary interaction of two typhoons in the Northwest Pacific. Chapter 9, "Circulation and Climatic Anomalies," gives the explanation of the Arctic warming and sea ice decline associated with the anomalies of meridional Polar and Arctic cells from seasonal, interannual, and interdecadal scales. The intraseasonal activity of the South Asian high and the western Pacific subtropical high is also illustrated in Chap. 9 to describe their connection with anomalous weather patterns in Eurasia-Pacific region. Finally, standing low-frequency oscillations and moving intraseasonal oscillations are described in Chap. 10, "Low-frequency Oscillations." There is no special chapter to describe the regional drought events and forest wildfires because they are usually accompanied with heat waves or warm anomalies.

Shortly, several important innovations are demonstrated in the book. It presents an approach, decomposing the atmospheric variable into temporal climatology and anomaly. It helps students, teachers, researchers, and forecasters understand weather, climate, and general circulation as well as their anomalies. It confirms the fourth meridional circulation (the Antarctic cell and the Arctic cell) in each hemisphere. The Arctic cell is directly linked to the warming in the high latitudes. It proposes the anomaly-based synoptic chart that can locate weather extremes. It develops an optimal-level dynamical model which can predict most unusual TC tracks. These points are the background to develop an anomalous nonhydrostatic atmospheric model in the future.

Observations of the Milky Way galaxy system, the Earth system, and even a TC show the feature of chaos. This book introduces a method to identify the large-scale structure (climatic state) from observations and to extract vortex signals to indicate weather extremes. Although observations are chaos and headache for forecasters, the decomposed climatic chart and anomalous chart can illustrate many beautiful spatial structures and temporal evolution from basic atmospheric variables.

When we enjoy these beautiful pictures from anomalous synoptic charts, forecasters need to remember their duty to send information of weather extremes to the public. I had been working as a forecaster at Yancheng station for 9 years. Whenever there were weather extremes, I usually collected data and then published papers to Chinese journals. In the afternoon of 23 June 2016, a tornado hit the villages of Yancheng and caused 99 fatalities. This is a new case that we have studied using the basic method introduced in this book.

In 1976, all faculties in the meteorological group of Peking University wrote a textbook titled "Weather Analysis and Forecast" for undergraduate students. Forty years later, the book, "Temporal Climatology and Anomalous Weather Analysis" is written for graduate students as an introductory course. The course of "Medium to Extended Range Weather Forecast" has employed this book draft in the first term of 2016.

Beijing, China
October 2016

Weihong Qian

Acknowledgements

This book is a gift of my 60-year-old birthday contributed by all my students who have joined my group since 2000. Therefore, I should thank them for working so hard during their student ages. I would like to thank a number of people who have contributed much in my 60 years. I hope my story can have a favorable impression for the next generation.

I thank my parents, Longzhen Qian and Defeng Wu, for giving me a good start in life. They supported me to stay with my grandma, Renzhen Shi, whose home was near the Anfeng downtown of Dongtai City, for the first 8 years when my parents migrated to the Yellow Sea Beach and cultured new farmland. My grandma talked and sang to me about the Sun at day and the moon and stars at evening during my childhood. During the second 8 years, I entered a primary school named "Dark-home and Salt-place School" near the beach. Many farmers suffered from hunger and cold during that period. I had a happy period with my two younger brothers, Weijiang and Weibo, because my father often asked us to help him to catch fish or seafood in the beach. My first time meeting a typhoon with strong wind and storm surge was during that period. My happiest days were those when I visited my grandma in Chinese spring festival.

When I was 16, I lived with my grandma and my adoptive father, Longgen. I had been studying at the Anfeng Middle School for 2.5 years. The ratio of students from rural villages and the town was about one to one. I borrowed novel books from other students and read at nights beside a kerosene lamp. During that time, the kerosene was planned to be used in each family. Exchange between kerosene and egg was a usual way. Wu, a store boss, always gave me an additional spoon of kerosene. He said to others that the boy might be one of talents to know astronomy and geography in the future. I was bashful but happy to have additional kerosene.

I was an educated youth as many students went back to countryside for farming after graduating from middle school. I met a tornado passing near my home and destroying neighboring houses. I dreamed to become a teacher in a primary school or a worker in a cotton factory in Anfeng town but all were difficult to realize. Ten years after the China's Cultural Revolution, the enrollment system was reformed by

President Xiaoping Deng. Most students graduated in the last 10 years from countryside were able to attend the college entrance examination. However, only a few passed the examination which made me so sad. Huiren Zhang, my chemistry teacher, helped me fill the volunteer information to study meteorology at Nanjing Meteorological Institute. I did not know it was my career started, just because of this volunteer sheet.

This institute is far away from the downtown of Nanjing, so most students studied hard at campus even in weekends. I read the books "Introduction to Typhoon" and "China Heavy Rains" during vacations. Many descriptions in the books were based on experiences similar to the traditional Chinese medicine. I liked the lecture "Dynamic Meteorology" very much, so I covered it by a sheet and drew a picture of a beautiful rotating typhoon on it. Professor Yongqing Peng saw the colored picture and asked me to meet him after class. Hence, Prof. Peng was my first advisor of meteorological study.

My dream was to become a teacher like Prof. Peng, but I had no background. As a single son in my adoptive father's family, I chose to return to Yancheng in Jiangsu. I worked as a forecaster at Dongtai Weather Station. Sometimes, only one forecaster and one observer worked at the station, so I had much free time to plant vegetables and to submit Chinese papers to journals and even wrote a book "Introductory to Planetary Earth Dynamics" after reading the terrestrial globe. The chief engineer of Jiangsu Meteorological Bureau, Shizhong Wang, asked me to work at his office and allowed me to have a personal computer early in 1985. It was difficult to get a job at Nanjing City due to the limitation of permanent residence. Two years later, Wang arranged me to work at Yancheng Weather Station, so that I could get an engineer position. I worked hard as a forecaster and wrote papers after weather extremes occurred. I felt ill and stayed at my wife's hospital of Dongtai City for about one year.

I acquired an award from China Meteorological Society in 1990 when I was in Xi'an City. At the society conference, I met Dr. Jianping Huang, who was assisting Prof. Jifan Chou to recruit a new Ph.D. student. The condition was whether the student could do research by himself. I replied to Prof. Chou and said I did not have a master degree. Then, Chou wrote to me: You have got 20 master degrees based on your 20 papers. During the heavy flood in summer of 1991, when I went to Lanzhou University, I stopped at Nanjing Railway Station because the railway had been damaged by the heavy rains. On the other hand, no message was sent to me that my daughter was ill and my bed at home was submerged by water. After the flood disaster, my wife and daughter returned to Dongtai and lived with my parents-in-law, Yizhang Cui and Yaying Feng.

I stayed at Lanzhou University only for the first term. During most time of my 3-year Ph.D. program, I studied in Beijing and met many experts in the astronomy and geography fields. At any fields, Prof. Chou always discussed with me and commented to me. During my Ph.D. period, I published my first book "Introductory to Planetary Earth Dynamics" in 1994. The book was supported by Dr. Han Sun, because the story of the Earth evolution made him unable to sleep for three days and nights.

My family came to Beijing when I had my postdoctoral at Peking University. My advisor was Prof. Shaowu Wang, who told me to combine Chou's dynamical idea and his historical data to study climatology. I did not know the difference between weather and climate. My idea still stopped at weather extremes, so I published a book titled "Convection and Structure in the Atmosphere" in 1996. I learned so much after Wang who asked me to organize a series of lectures which invited more than ten experts in the spring term of 1995. Using these lectures, I wrote a teaching material "Climate Dynamics" and taught graduate students since 1996. It was a good chance for me because many young teachers went abroad. One old teacher asked me to study monsoon and another asked me to teach the course of Synoptic Meteorology. Before I got the professor position in 2000, I was very busy studying the East Asian monsoon, El Niño-Southern Oscillation (ENSO) prediction, and China historical climate change, teaching the course of Global Climate System for graduate students and Synoptic Meteorology for undergraduate students, attending two national 973 projects led by the two academicians, Congbin Fu and Guoxiong Wu.

I have experienced several busy years. One was in 2008 before the Beijing Olympic Game. My group attended the extended-range forecast at National Climate Center. I asked one of my students to decompose the daily-mean 500-hPa height field into 100 components using the spherical-harmonic expansion. My idea was to statistically predict each component in the future 10–20 days. The student worked hard, but he failed and left me. He might think it was not science. Before each important event in China, seniors are invited to be advisors because they have experience. Usually, these advisors do not directly give any detailed predictions. Their experience is difficult to teach the new forecasters by words. So, it is like a miserable thing for all.

This event made me realize that I cannot mislead my students again. After a deep thinking, an approach of four-term decomposition, namely zonally averaged climatology, its zonal departure, the zonally averaged anomaly, and its zonal departure, was proposed. My students and I had published more than ten papers in 2012 on Chinese journals, studying the China weather extremes such as winter freezing rains, summer heavy rains, droughts, heat waves, and cold surges. Now we know that to study weather extremes, it is better to use transient anomalies rather than zonally averaged anomaly or its zonal departure.

Now the book is dedicated to readers, but I will not forget all the students who have made contribution. In Chap. 1 "Weather and Climate," the climatic charts were drawn by Ning Jiang. In Chap. 2, "Wind and General Circulation," many climatic charts were plotted by Kaijun Wu and Ning Jiang. Chapter 3, "Spatial Patterns and Time Scales of Climatic State," was contributed by Haoyuan Liang. Chapters 4 and 5, "Global Monsoon" and "Waves and Vortices," were mainly contributed by Ning Jiang. Outer edge active zones of the global monsoon and millennium monsoon dry–wet modes were contributed by Dr. Bo Lu and Prof. Yafen Zhu. Xiaolong Shan and Ning Jiang made their efforts to contribute for studies on heavy rains while Jeremy Cheuk-Hin Leung and Weimeng Luo made contribution on tornado-producing systems in Chap. 6, "Regional Convective

Events." All figures in Chap. 7, "Heat Waves and Cold Events," were plotted by Yun Chen and Tiantian Yu. In Chap. 8, "Unusual Tropical Cyclones," Xiaolong Shan developed the optimal-level dynamical model and many figures were drawn by Jing Huang, Guangwen Zhang, and Weimeng Luo. In Chap. 9, "Circulation and Climatic Anomalies," figures were plotted by Kaijun Wu, Jian Shi, and Haoyuan Liang. Jeremy Cheuk-Hin Leung contributed Chap. 10, "Low-frequency Oscillations." Finally, the book was edited by Jeremy Cheuk-Hin Leung and Jian Shi. This book could not be accomplished without the efforts from previous students. Those students graduated 3–4 years ago. They are Jin Li, Man Jiang, Ting Ding, Zongjie Zhang, Jiaolan Fu, Shuaiqi Tang, Cai Yao, and Xu Tang.

I thank the researchers in advance who will work on the following four studies using the anomaly-based approach. One is to study climatic anomalies in the Southern Hemisphere including the increasing trend of the Antarctic sea ice against the declining trend of the Arctic sea ice. Second is to study the strong convective events such as tornado-producing anomalous systems using observations of high spatiotemporal resolution. Third is to study the air–sea interaction using the maximum height anomaly (MHA) at the upper troposphere and the maximum subsurface sea temperature anomaly (MSTA) at the thermocline level. Last but important one is to develop anomalous nonhydrostatic models to predict weather extremes.

I thank many professional friends. For monsoon study, Prof. Song Yang from Sun Yat-sen University is my co-author. Professor Deliang Chen from Goteborg University and Prof. Qi Hu from Nebraska University are my faithful supporters. In recent three years, Dr. Jun Du helps me to publish papers on topics from heavy rainfall to unusual-track tropical cyclones. Many friends have given me constructive suggestions during my career. They are Minghua Zhang, Guangjun Zhang, Runhua Yang, Yi Deng, Zhongwei Yan, and Congwen Zhu. Some weather stations such as Guangxi, Nantong, and Shanghai as well as the Innovation Institute of Disaster Prevention and Reduction at Inner Mongolia provided my students' allowance after I stopped attending national 973 projects for the last 10 years.

This book is not only dedicated to readers but also to my family, especially my wife, Hengping Cui, my children, Huimin Qian and Feng Han, my grandson, Alex Han (Shuai Qian), whose love and sacrifice are essential to its completion. I feel happy when my two-year-old grandson, Alex, joined my Group Skype Meeting on my shoulder.

October 2016 Weihong Qian

Contents

1 Weather and Climate 1
1.1 Universal Dynamics of a System 1
1.2 Similarity and Evolution 4
1.3 Dynamical Definitions of Climate and Anomalies 13
1.4 Descriptions of Weather and Climate 15
1.5 Weather and Climatic Anomalies as Well as Their Predictions 17
1.6 Decomposition of Temporal Climatology and Anomaly 21
References 30

2 Wind and General Circulation 31
2.1 Diurnal-Cycle General Circulation 31
2.2 Seasonal-Cycle General Circulation 33
2.3 History of Trade Circulation 35
2.4 Modern General Circulation 39
2.5 The Arctic Cell and the Four-Cell Model 41
2.6 Eddy Heat Flux and Eddy Momentum Flux 51
2.7 Global Eight Meridional Cells 55
2.8 Hemispheric Height and Vorticity 59
2.9 An Aqua Control Experiment 61
References 66

3 Spatial Patterns and Time Scales of Climatic State 69
3.1 Seasonal Evolution of Zonally Averaged Variables 70
3.2 Zonally Averaged Vertical–Meridional Structure 77
3.3 Zonal Asymmetric Vertical Structure 79
3.4 Horizontal Patterns of Climatic Variable Departure 86
3.5 Seasonal Variation of Climatic Variable Departure 93
3.6 Temporal Decomposition of Climatic Field and Annual Cycle Systems 101

3.7 Semiannual and Biweekly Cycle Systems. 106
3.8 Identifying Three Period Bands 121
3.9 Annual and Semiannual Asymmetric Systems 126
References. 151

4 Global Monsoon 153
4.1 Wind Seasonality and Symmetric Dry–Wet Alteration 153
4.2 Cross-Equatorial Flow and Asymmetric Seasonal Dry–Wet Alteration 158
4.3 Identifying Global Monsoon Domain 165
4.4 Global Seasonal Precipitation Division 177
4.5 Seasonal Circulation Systems 186
4.6 Global Monsoon Division and Trend 191
4.7 Outer Edge Active Zones of the Global Monsoon 193
4.8 East Asian Monsoon Outer Edge Active Zone 196
4.9 Millennium Monsoon Dry–Wet Modes. 200
References. 209

5 Waves and Vortices. 215
5.1 Lower Stratosphere Decomposition. 215
5.2 Upper Tropospheric Decomposition 222
5.3 Lower Tropospheric Decomposition 225
5.4 Other Level and Vertical Decompositions. 231
5.5 Vertical Sections of Total and Anomalous Variables 231
5.6 Regional Total- and Anomaly-Based Synoptic Charts 239
5.7 Examples of Model Prediction 244
References. 249

6 Regional Convective Events 251
6.1 Site and Regional Rainfall Events in China 251
6.2 Comparison Between Total and Anomalous Fields. 257
6.3 Normalized Anomaly and Raw Anomaly 269
6.4 Heavy Rains and Anomalous Circulation Patterns 280
6.4.1 Three Anomalous Circulation Patterns in South China 281
6.4.2 Two Anomalous Circulation Patterns along the Yangtze River Valley 289
6.4.3 Anomalous Vortex Patterns in North China and Northeast China 292
6.5 Moist Divergence and Moist Vorticity 298
6.6 Nonconservative Thermodynamic Parameters. 316
6.7 Tornado-Producing Anomalous Systems in the Central US 326
6.8 Tornado-Producing Anomalous Systems in the Eastern China 334

6.9 Model Prediction Ability 339
References 343

7 Heat Waves and Cold Events 347
7.1 Definitions of Heat Waves and Cold Events 347
7.2 Summer Moving-Type Heat Waves 352
7.3 Heat Waves in Model Predictions 365
7.4 Spring and Autumn Low Temperature Events 369
7.5 Positive Height Anomalies and Warm Events 386
7.6 Negative Height Anomalies and Cold Events 398
7.7 Summer Standing-Type Heat Events 407
References 420

8 Unusual Tropical Cyclones 425
8.1 Separating the Typhoon Vortex from Its Surrounding Flow 425
8.2 Evaluating Typhoon Intensification 437
8.3 The Generalized Beta-Advection Model (GBAM) 445
8.4 GBAM Predictions on Typhoon Megi (2010) 450
8.5 More Applications of the GBAM 456
8.6 Hurricane Sandy (2012) and Left-Turning Prediction 474
8.7 Reexamining the Binary Interaction of Tropical Cyclones 492
8.8 Ideal Binary Interaction of Tropical Cyclones 504
References 517

9 Circulation and Climatic Anomalies 523
9.1 Meridional Circulation Anomalies and Arctic Sea-Ice Events 523
9.2 Seasonal Variation of the Arctic and Polar Cells 530
9.3 Meridional Cell Anomalies and Climatic Anomalous Patterns 535
9.4 Annual Mean Meridional Cell Trends and the Arctic Climatic Anomalies 551
9.5 Boreal Winter–Spring Arctic Amplification and Meridional Cells 562
9.6 Regional Relationship Between Arctic Sea Ice and Atmospheric Anomalies 571
9.7 The South Asian High and Eurasian Summer Climatic Anomalies 585
9.8 The Western Pacific Subtropical High and Precipitation Pattern 604
References 618

10 Low-Frequency Oscillations 625
10.1 Maximum Anomalies of Atmospheric Variables 625
10.2 Low-Frequency Standing Oscillations in the Northern Hemisphere 629

10.3 Low-Frequency Standing Oscillations in the Southern Hemisphere 636
10.4 Low-Frequency Standing Oscillations in the Tropics 639
10.5 The Madden-Julian Oscillation and Height Anomalies 644
10.6 Eastward and Westward Propagations of Tropical Oscillations 663
References 674

Data Sources 677

References 681

Index 683

Chapter 1
Weather and Climate

Weather and climate in our planet are basic environments for human beings and directly rely on a larger environment in our solar system. However, difference between weather and climate is sometimes confusing. We always hear and read some terms like "climate change" , "weather extremes," and "climatic anomalies" but many cannot clearly tell what they are. Thus, the definition of weather and climate needs to be physically given first. In this chapter, we first have a glance at our environments and their evolutions from the universe to the solar and the Earth systems. Then all observational atmospheric variables from the surface up to the stratosphere are decomposed into two components: temporal climatology and anomaly. The climatic state at a certain site should be an equilibrium forced by the regular solar radiation and the surface condition without the influence of other disturbances. Therefore, the climatic state needs to be identified while the anomaly needs to be predicted. Hence, the weather forecast is the superposition of the climatic state and the anomaly, while the climate projection is the superposition of the climatic state and a relatively long-period (low-frequency) anomaly.

1.1 Universal Dynamics of a System

Weather and climate are consequences of atmospheric motion in the Earth. The Newton's laws of motion and fluid dynamics (or mechanisms of continuous media) are the basic principles which lead to the development of atmospheric dynamics. Sir Isaac Newton (1642–1727) was an English scientist and mathematician. He made major contributions in mathematics, theoretical and experimental physics, achieving a remarkable synthesis of the work of his predecessors on the laws of motion, especially the law of universal gravitation.

Newton's laws of motion are three physical laws that form the basis for classical mechanics. They describe the relationship between an object's motion and forces

© Springer Nature Singapore Pte Ltd. 2017
W. Qian, *Temporal Climatology and Anomalous Weather Analysis*,
Springer Atmospheric Sciences, DOI 10.1007/978-981-10-3641-5_1

acting on it. They have been expressed in several different ways over nearly three centuries, and can be summarized as follows:

1. First law: If there is no net force on an object, then its velocity is constant. The object is either at rest (if its velocity is equal to zero), or it moves with constant speed in a single direction.
2. Second law: The acceleration of an object is parallel and directly proportional to the net force F acting on the object, and is inversely proportional to the mass m of the object, i.e., $F = ma$.
3. Third law: When a first object exerts a force F_1 on a second object, the second object simultaneously exerts a force $F_2 = -F_1$ on the first object. This means that F_1 and F_2 are equal in magnitude but opposite in direction.

Newton's laws can be applied to objects which are idealized as single mass point, in the sense that the size and shape of the object can be neglected in order to focus on its motion more easily. This can be done when the object is small compared to the distances involved, or the deformation and rotation of the object are of no importance. In this way, even a planet can be idealized as a point for the analysis of its orbital motion around a star. Although Newtonian mechanics has been superseded by relativistic mechanics, it is correct with negligible error when the speeds involved are much slower than the speed of light.

In keeping with his grand vision, Newton proposed just one law for gravity—a law that worked for every scrap of matter in the Universe, for moons and planets as well as for apples and the Earth. By combining this law with his general laws of motion, Newton was able to demonstrate mathematically that a single planet would move around the Sun in an elliptical orbit, just as Kepler claimed each of the planets did.

Newton's discoveries became the basis for detailed and comprehensive mechanic studies of force and motion. The detailed character of the Newtonian laws was such that once this majestic clockwork had been set in motion and then its future development was, in principle, entirely predictable. This property of Newtonian mechanics is called determinism. It had an enormously important implication: Given an accurate description of the character, the position and velocity of every particle in the Universe at some particular moments (i.e., the *initial condition* of the Universe), and the forces that operated between those particles, the subsequent development of the Universe could be predicted with as accurate as desired.

In 1814, the French scientist Pierre Simon Laplace (1749–1827) published what is usually known as the first articulation of causal or scientific determinism: "*We may regard the present state of the Universe as the effect of its past and the cause of its future. An intellect which at a certain moment would know all forces that set nature in motion, and all positions of all items of which nature is composed, if this intellect was also vast enough to submit these data to analysis, it would embrace in a single formula the movements of the greatest bodies of the Universe and those of the tiniest atom; for such an intellect nothing would be uncertain and the future just like the past would be present before its eyes*" (A Philosophical Essay on Probabilities).

This intellect is often referred to as *Laplace's demon* and sometimes *Laplace's Superman*. According to determinism, if someone knows the precise location and momentum of every atom in the Universe, their past and future values for any given time are determined; they can be calculated from the laws of classical mechanics.

We can try to write the Laplace's causal determinism into a formula. The universe consists of many kinds of galaxies and solar systems. Each galaxy constitutes a subsystem. The system is always composed of matters. We assume that there are sufficient numbers of R variables which can describe the systems or subsystems. These R variables construct a system $S(t)$ at a specific moment t,

$$S(t) = (d_1, d_2, \ldots, d_R) \tag{1.1.1}$$

where $(d_1, d_2, \ldots, d_R)$ are the variables such as temperature (T), density (ρ), and velocity (v) that change with time t.

A system has its defined space range. Its size is determined by a boundary, by which the system is separated into internal and external parts. The descriptions of the so-called system dynamics are not complete without three aspects: (1) The system is influenced by external forcing; (2) There is a restriction mechanism, namely internal dynamics; (3) We should be able to easily describe the system's change or tendency. An ideal system dynamics can be regarded as a prediction model

$$\frac{\partial S(t)}{\partial T} + (N + L + D)S(t) = F(t) \tag{1.1.2}$$

where the $F(t)$ is the sum of forces including periodic-cycle forcing $\sum_{i=1}^{I} \widetilde{F}_i(t)$ and random forcing $\sum_{j=1}^{J} \widetilde{F}'_j(t)$, i.e., $F(t) = \sum_{i=1}^{I} \widetilde{F}_j(t) + \sum_{j=1}^{J} F'_j(t)$, acting on the system. N denotes the complexity or nonlinear relationship that controls the internal status of the system. Other two, L and D, describe the linear relationship and dissipative effect within the system, respectively. Together, $(N + L + D)$ describes the internal dynamics of the system.

Based on the Laplace's causal determinism, we can predict the future state of the universal system as well as other systems such as the Milky Way galaxy, the solar system, the Earth system, the atmospheric system, and even a tropical cyclone and a tornado. For these predictions, the basic conditions include that (1) the initial state $S(t_0)$ can be quantitatively measured, (2) all periodic-cycle forcing $\widetilde{F}_i(t)$ and random forcing $F'_j(t)$ are known first, (3) the internal dynamics $(N + L + D)$ can be written in mathematical form. Similar to the modern numerical models or climate models, this set of nonlinear equations can be integrated using computer. From the deterministic view, the predictability is limited by the above three conditions.

1.2 Similarity and Evolution

Although there are multi-scale systems in the Universe, the similarity among their rotational structures could be identified. The Milky Way galaxy is a rotation system in the Universe. On our Earth, a typical rotation system is tropical cyclone (TC) or typhoon in the Northwest (NW) Pacific or hurricane in the North Atlantic. Figure 1.1 shows two similar systems in the Earth and the Universe, respectively. For a super TC, a central eye surrounding by four spiral rain (cloud) bands can be clearly observed from satellite images. It is well-known that the TC system moves under the environmental flow in the troposphere. At point A in Fig. 1.1a, the motion of a small air particle can be separated into two orthogonal velocity components: along the tangential (***t***) and normal (***n***) directions. The particle at point A has a tendency of motion toward the TC eye under the two velocity components along a spiral cloud band (red-dashed arrow). Four spiral cloud bands are identified in the typhoon in the southeast of Taiwan. Similarly, the Milky Way system is a spiral galaxy consisting of over 400 billion stars mainly concentrated along the four spiral high-density star bands (the red-dashed arrows in Fig. 1.1b). We expect that all stars including the Sun system along the four spiral star bands will move toward the Galactic Center according to the similarity.

The similar horizontal disk structure is compared between the super TC and the Milky Way galaxy. During the developing stage of a super TC, we can find that many strong convective cloud clusters move inward along spiral cloud bands. The Milky Way galaxy is similar to the super TC in the developing stage. We can infer the evolution of the Milky Way galaxy after all star clusters entering the Galactic Center. At the last moment, the Galactic Center will become a "massive ball" in space because many star clusters have entered within the Galactic Center. The final disk structure of the Milky Way galaxy is similar to the Saturn and Saturn's rings.

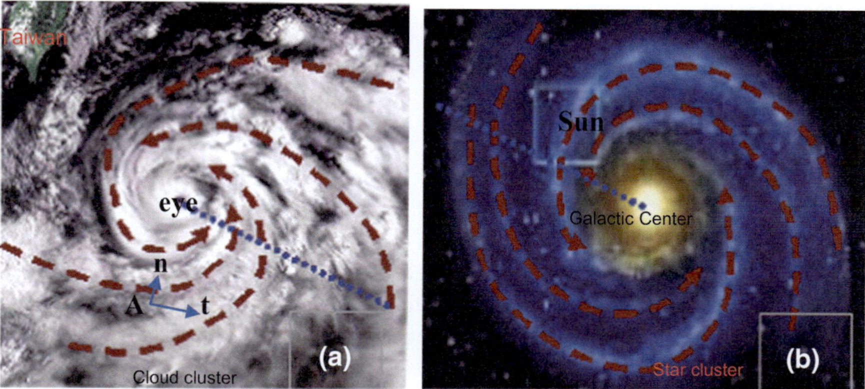

Fig. 1.1 **a** A typhoon to southeast of Taiwan and **b** the Milky Way galaxy. *Four red-dashed arrows* indicate cloud bands in **a** and star bands in **b**. Modified from Qian (2011a)

If we regard the atmosphere as a universe, we can find most "atmospheric galaxies" such as TCs formed along the inter-tropical convergence zone (ITCZ) (Fig. 1.2). All TCs are initiated from a small disturbance (wind convergence) and then developed into typhoon even super typhoon through attracting nearby air particles. During the TC developing process into a super TC, four spiral cloud bands finally become clear with time. They usually die out when entering high latitudes or making landfall. During this vanishing process, the four spiral cloud bands reduce as three, two to one, and finally dissipate. Therefore, a typhoon is a phenomenon showing that lots of dry and wet airs are organized and packed (four spiral cloud bands) together in a limited spell and space by a mechanism. After it dissipates, some of dry and wet airs are reorganized into other new typhoons or cyclones. In the atmospheric universe, the Mei-yu front from eastern China to Japan during boreal summer and polar fronts in high latitudes are large-scale environment to form extra-tropical cyclones in the atmosphere. So in boreal summer, China is often influenced by tropical cyclones initiated from tropics, Jianghuai cyclones in the leeside of Tibetan Plateau and extra-tropical cyclones in middle latitudes.

Several cyclones can generate and develop along a planetary-scale front. In Fig. 1.2 four extra-tropical cyclones are referred to as a cyclone family. In East Asia, a cyclone family often forms along the Mei-yu front during boreal summer. All cyclones move east-northeastward. The oldest cyclone is an occluded cyclone (L4) in the east of Japan. An explosive cyclone (L3) moves eastward to the Japan Sea. A mature cyclone (L2) generates at the lower Yangtze River and a disturbance (L1) just forms in the upper Yangtze River. Following the life-cycle process of a cyclone, we can find that it originates from a small disturbance in the upper Yangtze

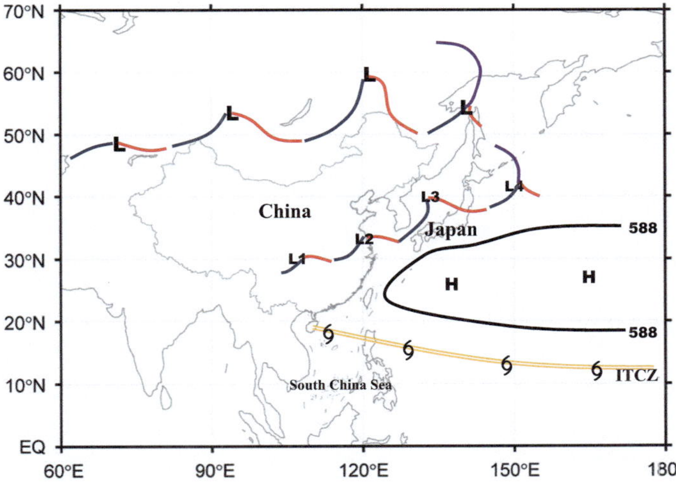

Fig. 1.2 A schematic representation of planetary-scale (ITCZ, Mei-yu front and polar front) and small-scale (typhoon 6 and cyclone L) weather systems in East Asia. *Blue*, *red*, and *purple lines* are cold, warm, and occluded fronts, respectively.

River (L1), generates in the lower Yangtze River (L2), deepens over the Japan Sea (L3), occludes in the east of Japan (L4), and finally dissipates in the northern Pacific. During these stages of its life-cycle process, a rain band forms along the initial disturbance with a stationary front, two rain bands along the cold and warm fronts can be observed in the mature and explosive stages, then three rain bands are associated with the cold and warm fronts in the southern part and the occluded front in northern part of the occluded cyclone, respectively. Therefore, the number of rain bands can indicate the intensity of a cyclone. The strongest cyclone is super typhoon with four spiral rain (cloud) bands as shown in Fig. 1.1a.

The above description shows that the atmosphere is like a complex universe with motions of dry and wet air. The evolutions of atmospheric and universal motions are different mainly in their temporal and spatial scales. Different systems have their spatial structures at different stages of their life cycle. For example, the Saturn has a prominent ring system that consists of nine continuous main rings and three discontinuous arcs, which are mostly composed of ice particles with a smaller amount of rocky debris and dust (Fig. 1.3). Sixty-two moons are known along Saturn's orbit, of which fifty-three are officially named. This does not include the hundreds of moonlets comprising the rings. The Titan is the largest moon in the Saturn system. It means that the last stage of the Milky Way system is similarly like the Saturn system, having its central body and encircled rings.

Now we come to the evolution of the Solar system and the formation of its surrounding planets. Figure 1.4 illustrates the developing stage of the Solar system and its disk. The four star-cloud spiral lines are assumed to exist surrounding the Solar Core in that stage. There should be four rings surrounded the Sun when the Sun formed after all the star-cloud clusters having entered the Solar Core.

At that time, 99% masses of the Solar system had converged to the Sun and four encircled rings with 98% angular momentum of the solar system surrounded the

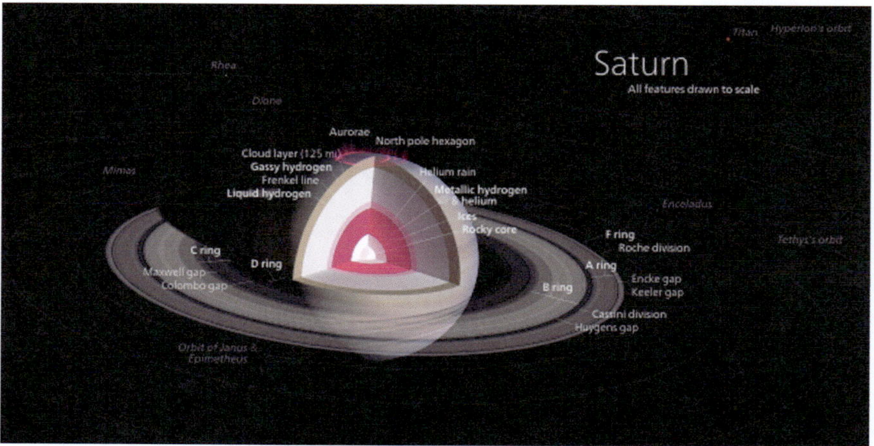

Fig. 1.3 The Saturn and Saturn's encircled rings. *Source* https://zh.wikipedia.org/wiki/File:Saturn_diagram.svg

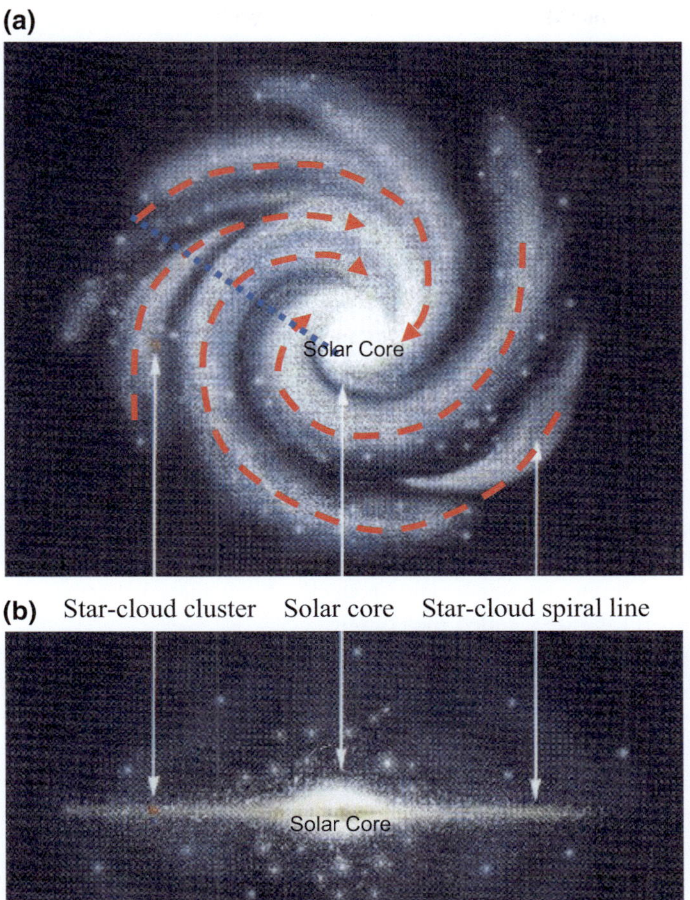

Fig. 1.4 The developing stage of **a** the solar system and **b** its disk. Modified from http://sciencevault.net/ibphysics/astrophysics/pics/milkyway.gif

Sun. The rings had different material components depending on their distances to the Sun. Dense and rocky compositions are mostly concentrated near the core while light and gaseous compositions are far away from the Sun. Masses have higher orbital speeds near the Sun than those far away as shown in Fig. 1.5. We assume that there were four waves or fluctuations surrounding the Sun and forming four shear circle bands centered the Sun. Each circle band is consisted of materials which can be seen as fluid-like motion. Therefore, small-scale vortices and their cores were generated with the same rotation direction at the same shear circle band. The two adjacent shear bands formed opposite rotation directions of vortices and cores. These cores were embryos of planets. Within the same shear band, smaller embryos were attracted by larger ones or captured as satellites of larger embryos. In shear band "A," one embryo attracted all other smaller embryos and became a planet named as the Mercury. In its outside band, an opposite shear band formed the Venus with an opposite rotation direction to the Mercury. Shear band "C" is relatively wider, so two planets (the Earth and the Mars), as well as several satellites (such as the Earth's moon) finally formed. The rotation direction of the Earth and the Mars as well as their satellites is opposite to that of the Venus but the same as the Mercury. The Asteroid belt with lots of embryos formed in shear circle band "D". Similar to shear band "C", two planets the Jupiter and the Saturn with their satellites generated in the shear band "E" in the same direction of rotation. The Uranus and the Neptune formed in the same way in shear bands "F" and "G". Beyond shear "G", there should have other shear bands and vortices to form with different rotations. But their surrounding periods to the Sun are longer than those of the above planets. Therefore, we explain that the four smaller inner planets (i.e., Mercury, Venus, Earth, and Mars) are terrestrial planets, being primarily composed of rock and metal. The four outer planets are giant planets, being substantially more massive than terrestrial planets. The two largest planets, Jupiter and Saturn, are gas giants, being composed mainly of hydrogen and helium. The two outermost planets, Uranus and Neptune, are ice giants, being composed mostly of substances with relatively high melting points compared with hydrogen and helium, called ices, such as water, ammonia, and methane. All planets have their orbits lying within a nearly flat disk called the ecliptic.

It is clear that different rotation directions of the eight planets in the Solar system can be explained under the evolution of four shear bands. In the atmosphere, there are many cases showing that there are four spiral bands when their intensity reaches the strongest level. For example, only over a 24-h period (midnight-midnight central daylight time, CDT) of 27 April 2011, 199 tornadoes occurred in the southern USA (Knupp et al. 2014). Four convective bands (black dashed lines) of reflectivity factor surround the tornado (Enhanced Fujita EF-4 scale damage) in Fig. 1.6a and the mesoscale vortex (MV) in Fig. 1.6b, respectively. These convective bands can be seen as intense quasi-linear convective systems (QLCS) while the four convective bands construct a convective system.

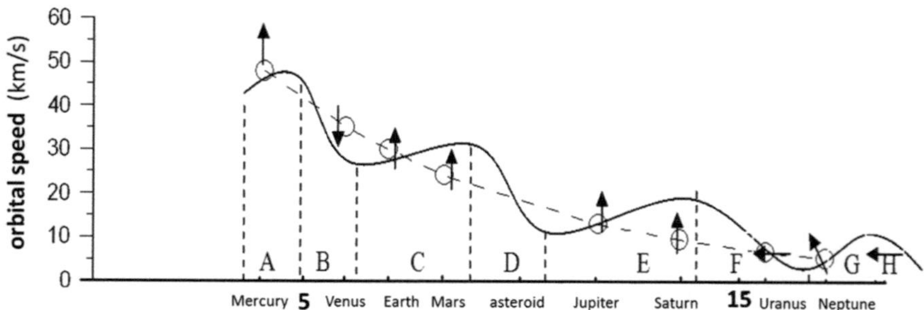

Fig. 1.5 The orbital speed (km/s) and distance ($x = 3\ln(10R)$) of planets relative to the Sun. $R = 1$ (1 astronomical unit, 150,000,000 km) is the distance between the Sun and the Earth. *Arrows* denote different directions of planet rotation. Modified from Qian (2011a)

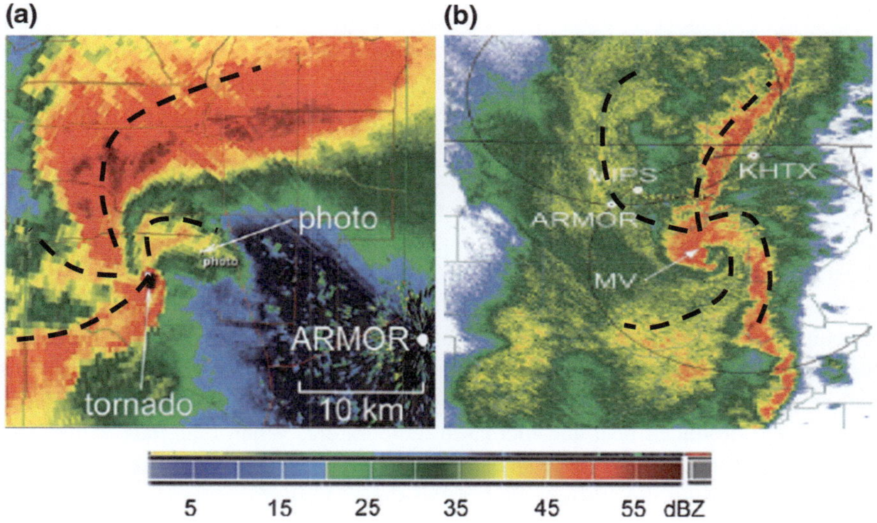

Fig. 1.6 Equivalent reflectivity factor (*shading*, 2.5 dBZ interval) of **a** the Hackleburg tornado at 2125 UTC 27 April 2011 and **b** an image from the Columbus, MS at 0735 UTC 27 April 2011. The *black dashed lines* indicate the convective bands. Modified from Knupp et al. (2014)

These tornado-producing systems (Fig. 1.6), TCs (Fig. 1.1a) and extratropical cyclones (Fig. 1.2) have different spatial scales and life periods. We can observe all dry and wet air masses within a developing system and find that most air masses are contracted into a system surrounding their spiral bands. At the same time, most air masses are left a system when the system is expanded and some of them are reorganized by other new systems. Two opposite processes including one system contraction and another system expansion can be commonly observed in the atmosphere. Similarly, two opposite processes of particle contraction and particle expansion could also be commonly observed in the Universe. If we focus an air mass or a star, it is rapidly far away from us so it cannot be used to explain the cosmic expansion.

According to radiometric dating and other sources of evidence, the Earth formed about 4.54 billion years ago. The Earth gravitationally interacts with other objects in space, especially the Sun and the Moon. In a complete cycle orbiting around the Sun, the Earth rotates around its own axis 365.26 times, creating 365.26 solar days in one sidereal year. The Earth's axis of rotation is tilted 23.4° away from the perpendicular of its orbital plane, producing seasonal variations on the planet's surface with a period of one tropical year (365.26 solar days). The Moon is the Earth's only permanent natural satellite. Its gravitational interaction with the Earth causes ocean tides, stabilizes the orientation of Earth's rotational axis, and gradually slows the Earth's rotational rate.

The Earth's lithosphere is divided into several rigid tectonic plates that migrate across the surface over periods of millions of years. 71% of the Earth's surface is covered with water, with the remainder consisting of continents and islands that form lakes and other sources of water. These water sources also contribute to the hydrosphere. Its polar regions are mostly covered with ice, including the Antarctic ice sheet and the sea ice of the Arctic ice pack. Its interior remains active with a solid iron inner core, a liquid outer core that generates the magnetic field, and a convective mantle that drives plate tectonics.

According to the above description, the Earth and other planets are formed at the same time after the Sun's body reached the maximum when all star clusters along the four main spiral bands had attracted and converged into to the Solar Core. From the appearance of the Earth embryo to its planet formation, we refer to this process as the astronomical evolution spell while after that it is referred to as the geographical evolution process. During the astronomical evolution spell, the Earth was initially formed from a vortex and then to an embryo when lots of smaller embryos had been attracted by the Earth embryo. These smaller embryos hit the Earth embryo and produced heat and increased temperature. When materials melted at a higher temperature, dense materials such as iron sank to the inner core while light materials floated on the surface. Part of the shear angular momentum in band "C" (Fig. 1.5) transferred into rotated angular momentum of the Earth embryo. The rotation rate of the Earth embryo increased when the smaller embryos hit the Earth embryo. At the last stage, there were three embryos including the Earth embryo, the Moon embryo, and a nameless embryo (Fig. 1.7). The gravitational force between the Earth embryo and the nameless embryo was larger than the centrifugal force, so the latter was attracted and hit into the Earth embryo. This is the last largest event as the end of astronomical evolution process. At this moment, the Earth formed with the fastest rotation rate, the largest body, and the highest temperature with a melted surface. At the last moment, the gravitational force between the Earth embryo and the Moon embryo was equal to the centrifugal force so the Moon finally became a natural satellite of the Earth.

Since the geographical evolution process, the Earth's most significant feature is the declining surface temperature after the nameless embryo hit its surface. During one orbit around the Sun, the earlier Earth rotated around its own axis about 700 times, creating 700 solar days a year, so the shape of the earlier Earth was largely flatter than that today. The geopotential height of the Earth surface is higher in the two poles with two natural diamond mine fields. Following the surface cooling,

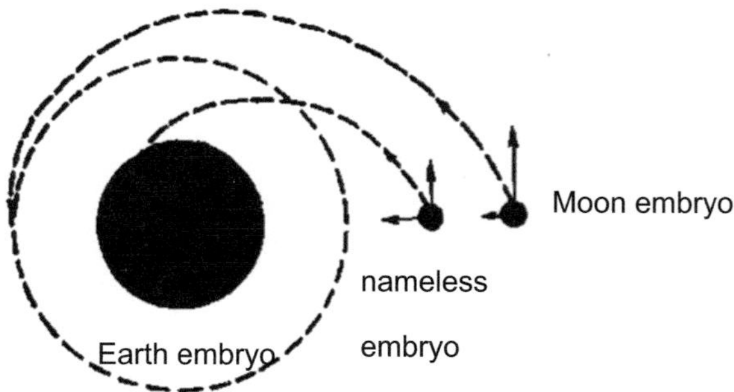

Fig. 1.7 The formation of the Earth embryo and the Moon embryo. Modified from Qian (1994, 1996)

several geologic layers formed from the inner-outer core, mantle, and crust. Out of the Earth surface, the hydrosphere and the atmosphere are also formed. The hydrosphere was an original ocean basin located within the tropical zone, which separated two large plates of continents in the two hemispheres. In earlier stage, the inner-outer core and the outer crust can be seen as solid layers while the mantle was a fluid layer. It is well-known that Earth's rotational rate is slowing down. If we have a reference at the core and only consider the core-mantle system, the slowing rotation causes the mantle to have an eastward and equator-ward motion. This motion of mantle drives the split and drift of the two original continents (Fig. 1.8). The original continent in the Southern Hemisphere (SH) was split into five plates while the original continent in the Northern Hemisphere (NH) was split into four large plates. The two natural diamond mine fields are located in South Africa (P_S) and Siberia (P_N) in Fig. 1.8.

In the SH, the five plates had been split and shifted northeastward over the mantle fluid layer but their shapes were different. These differences of plates change the motion of mantle fluid and shift the ITCZ away from the rotated equator as well. All the plates and islands in the two hemispheres cannot move cross the ITCZ when the mantle fluid had an eastward motion relative to the core. It implies that the accumulated extents of continents and islands are equal in the two hemispheres separated by the ITCZ of the mantle motion. The distributions of minerals, living beings, even resources, and disasters can be explained from the geographical evolution process.

The initial distributions of the split continents are shown in Fig. 1.8 while the current distributions are plotted in Fig. 1.9. The heavy black-dotted line indicates the ITCZ of mantle motion, which extends from the Mediterranean Sea in the northernmost to the Fiji Islands in the southernmost. The largest oil field in the

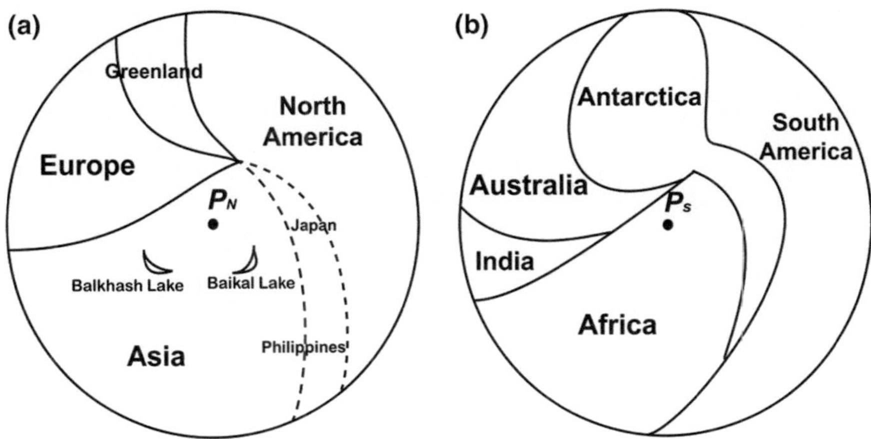

Fig. 1.8 The two original continents in **a** the Northern Hemisphere and **b** the Southern Hemisphere with different plates centered at the two poles of the Earth. Modified from Qian (1994, 1996)

world is located in the northernmost of the ITCZ, namely the Middle East, which was a paleo-ocean place but merged by the African continent from the SH and the Eurasian continent from the NH. In the SH, the South American plate, the African plate, and the Australian plate as well as their islands drifted northeastward as indicated by the yellow-dashed arrows in Fig. 1.9. They left three basins surrounding the Antarctic plate. As indicated by Qian (1992, 1994), all continental plates had drifted northeastward and returned southwestward to the south of the ITCZ many times while southeastward and northwestward to the north of the ITCZ. The three mid-oceanic ridges (yellow-solid arrows) were once connected with the three continental plates. It means that the three mid-oceanic ridges were the traces left when the three continental plates returned southwestward to the south of the ITCZ. The three red solid arrows indicate that the three islands can be traced back to their continents. The two spheres of the core and the mantle, as an enclosed system, have angular momentum exchange so there are relative motions between the surface mantle fluid and continental plates. Figure 1.9 recorded the history of geological evolution process and left lots of traces such as mid-oceanic ridges, oceanic basins, seas, and trenches.

The main goal of this book is to explore those weather extremes. It is well-known that the temporal climatic state is the background to reveal and predict weather

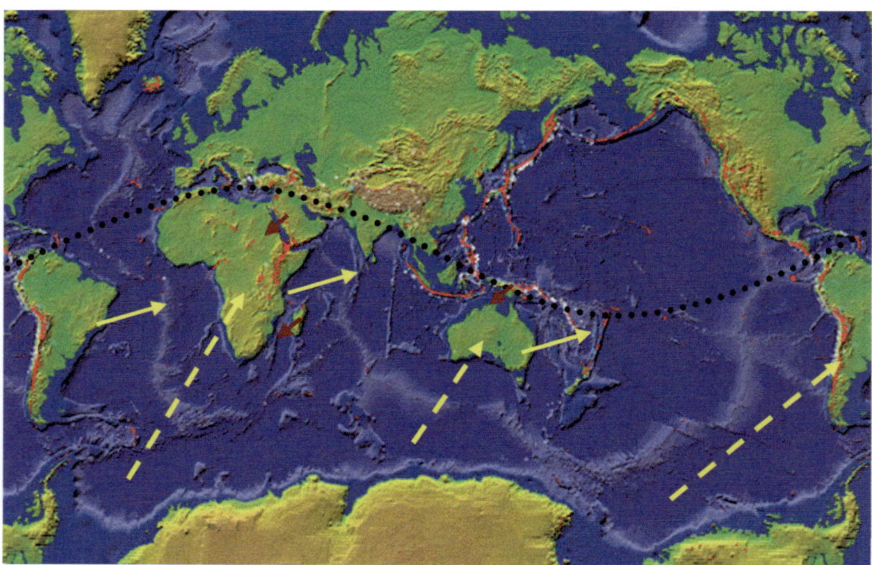

Fig. 1.9 The current continents in the world. The *heavy black-dotted line* indicates the ITCZ of the mantle fluid motion. The *three yellow- solid arrows* indicate that the three plates were once connected with the three mid-oceanic ridges. The *three yellow-dashed arrows* indicate that the three plates were shifted from the three oceanic basins. The *three red solid arrows* indicate that the three islands can be traced back to their continents. The *red dots* indicate earthquakes occurred in the history. Modified from Qian (2011a)

extremes. Furthermore, the temporal climatic state relies on the formation of sea-land distributions and mountain lift. When simulating the paleo-climate features, one needs to first correctly know the geographical evolution process of the Earth.

1.3 Dynamical Definitions of Climate and Anomalies

We turn to study the atmospheric system. Assuming that the climate system is controlled by external forcing and internal dynamics of linear and dissipation factors, we have the following equation according to Eq. 1.1.2

$$\frac{\partial \widetilde{S}(t)}{\partial t} = \sum_{i=1}^{I} \widetilde{F}_t(t) - (L+D)\widetilde{S}(t). \quad (1.3.1)$$

The total state $S(t)$ can be decomposed into climatic state $\widetilde{S}(t)$ and anomaly $S'(t)$.

$$S(t) = \widetilde{S}(t) + S'(t) \quad (1.3.2)$$

Thus, the climate change $\frac{\partial \widetilde{S}(t)}{\partial t}$ determined by Eq. 1.3.1 is caused by both periodic-cycle forcing such as diurnal and seasonal cycles of solar radiation$\widetilde{F}(t)_{\text{diurnal}} + \widetilde{F}(t)_{\text{seasonal}}$ and surface conditions $-(L+D)\widetilde{S}(t)$. For example, the diurnal and seasonal cycles of solar radiation forcing, $\widetilde{F}(t)_{\text{diurnal}}$ and $\widetilde{F}(t)_{\text{seasonal}}$, have a maximal value at noon (12:00 a.m.) and on June 22 but the local surface maximal temperature is delayed at 2:00–3:00 p.m. on calendar day and in late July in the Northern Hemisphere, respectively. The delayed hours and days are caused by the internal linear-dissipative processes. The solution of climatic state $\widetilde{S}(t)$ can be obtained from Eq. 1.3.1, which describes the dynamical definition of climate.

The climatic (or weather) anomaly is determined by nonlinear and anomalous forcing factors as follow:

$$\frac{\partial S'(t)}{\partial t} = -N\left[\widetilde{S}(t) + S'(t)\right] - (L+D)S'(t) + \sum_{k=1}^{K} F'_k(t) \quad (1.3.3)$$

The climatic (or weather) anomaly (Eq. 1.3.3) composes of four parts:

$$\frac{\partial S'_1(t)}{\partial t} = -N\widetilde{S}(t) \quad (1.3.4)$$

$$\frac{\partial S'_2(t)}{\partial t} = -NS'(t) \quad (1.3.5)$$

$$\frac{\partial S_3'(t)}{\partial t} = \sum_{k=1}^{K} F_k'(t) \tag{1.3.6}$$

$$\frac{\partial S_4'(t)}{\partial t} = (L + D)S'(t) \tag{1.3.7}$$

The first part $S_1'(t)$ is caused by climatic cycles that have been decomposed into many orthogonal components using spherical harmonic expansion in current climate models. Lots of orthogonal components derived from mathematical decomposition have no physical meaning, so the nonlinear interaction $-N\widetilde{S}(t)$ of these components produces new climatic (or weather) anomalies and "false alarms." In the atmospheric system, adjacent anomalous systems such as anomalous anticyclones and anomalous cyclones have nonlinear interaction $-NS'(t)$ and produce the second part $S_2'(t)$ of anomalies. The third part $S_3'(t)$ of anomalies is a delayed response to anomalous external forcing $F_k'(t)$ such as irregular solar activity, volcanic activity, urban developing, El Niño/La Niña events, and long-term atmospheric component impact, etc. The fourth part of anomalies $S_4'(t)$ is a process of linear and dissipation, so it can be traced back and predicted for the future. This process has been widely applied in revealing weather extremes and anomalous climatic events as in many examples in this book and other publications (Qian 2012, 2013, 2014b).

This analysis of decomposition indicates that the future field of the system is composed of a certainty part $S_{\mathrm{Ce}}(t)$ and a random part $S_{\mathrm{Ra}}(t)$ i.e.,

$$\frac{\partial S(t)}{\partial t} = \frac{\partial S_{\mathrm{Ce}}(t)}{\partial t} + \frac{\partial S_{\mathrm{Ra}}(t)}{\partial t}, \tag{1.3.8}$$

where

$$\frac{\partial S_{\mathrm{Ce}}(t)}{\partial t} = \sum_{i=1}^{I} \widetilde{F}_i(t) - (L + D)\widetilde{S}(t) - (L + D)S'(t), \tag{1.3.9}$$

and

$$\frac{\partial S_{\mathrm{Ra}}(t)}{\partial t} = \sum_{k=1}^{K} F_k'(t) - NS'(t). \tag{1.3.10}$$

The deterministic part of climate or weather can be predicted, which only depends on periodic forcing $\sum_{i=1}^{I} \widetilde{F}_i(t)$ and the linear and dissipation processes of decomposed climatic component and anomaly $(L + D)\widetilde{S}(t) - (L + D)S'(t)$. It is difficult to obtain the random part if random forcing is not known and nonlinear model does not perform well.

1.4 Descriptions of Weather and Climate

What is weather, what is climate, and what are their differences? We hear about weather and climate all the time. Most of us check the local weather forecast to make our plan. Climate and climate change are certainly a "hot" topic nowadays in the news. There is, however, still a lot of confusion over the difference between the two.

Weather is naturally what you see in the atmosphere at any particular moment of a day. Most people think of weather in terms of temperature, humidity, precipitation, cloudiness, brightness, visibility, wind, and atmospheric pressure. For example, it may be 5 °C at noon with drizzling rain and cloud, or it could be −5 °C at night with heavy snow, or it is sunny in the afternoon with calm wind. Therefore, weather is what we actually get (Herbertson 1901). In most places, weather can change from minute to minute, from hour to hour, and from day to day.

Some scientists define climate as the average weather over a particular region and a period or interval, usually taken over 30 years. When scientists talk about climate, they are looking at the averages of precipitation, temperature, humidity, sunshine, wind velocity, phenomena such as fog, frost, hail storms, and other measures of the weather that occur over a long period in a particular place. The climate record also includes extreme values such as record high (low) temperatures or record amounts of rainfall.

Some scientists define climate change including climate average value and extreme values of weather varying from year to year, from decade to decade, and from century to century. For example, after looking at the rain gauge data, lake and reservoir levels, and satellite data, scientists can tell if an area was drier than average during a particular summer. If it continues to be drier than normal over the course of many summers, then it is likely to indicate a change in climate.

Globally, climate and climate change are considered as the amount of solar energy reaching the Earth, which determines the forcing from the Sun. Serbian geophysicist, Milutin Milankovitch, proposed the theory that the collective effects of changes in the Earth's climate rely on variations of eccentricity, axial tilt, and precession of the Earth's orbit (Fig. 1.10). These variations determine climatic patterns on the Earth through orbital forcing. The Earth's axis completes one full cycle of precession in approximately 19–24 kyr. The angle between the Earth's rotational axis and the normal to the plane of its orbit (obliquity) oscillates between 22.1 and 24.5° on a 41 kyr cycle. It is currently 23.44° and is decreasing. The elliptical orbit rotates more slowly. Its variation in eccentricity is about 95–125 and 400 kyr cycles. These cycles cause variations in the amount of solar energy reaching the Earth, which explains the advance and retreat of the polar ice caps (Fig. 1.11). These cycles lead the irregular oscillation of solar forcing along 65°N in the boreal summer and along 65°S in the austral summer. Many scientists believe these cycles play a role in the Earth's climate change. In the last 1000 kyr, the glaciation has experienced 13 hot and cold stages.

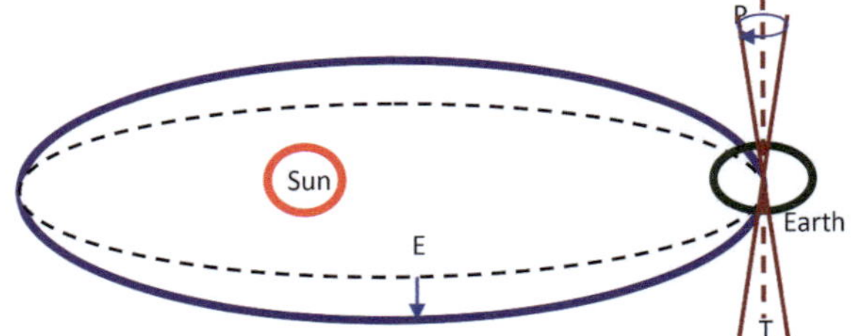

Fig. 1.10 Schematic representation of Milankovitch cycles. Letters "E" for eccentricity, "T" for axial tilt, and "P" for precession of the Earth's orbit. Modified from http://www.southwestclimatechange.org/files/cc/figures/milankovitch.jpg

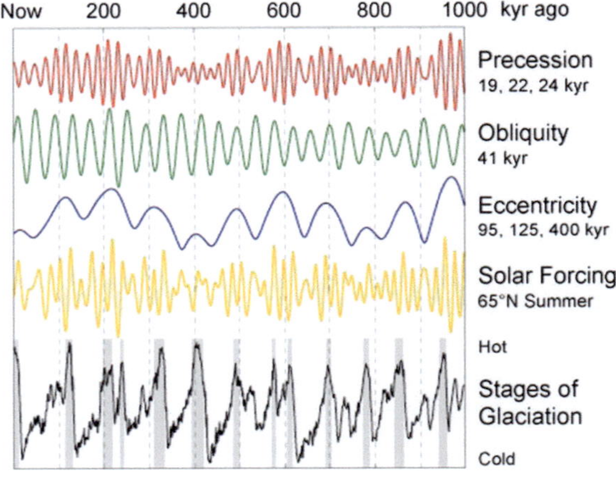

Fig. 1.11 Milankovitch cycles over the past 1000 kyr. *Source* http://it.snhu.edu/naturalsciences/science252/chapter3/lectures/climate/images/MilankovitchIce.gif

The earliest known classification of climate, devised by the Greeks, simply divided each hemisphere into three zones, i.e., the "summerless," "intermediate," and "winterless," which are now known as the Frigid, Temperate, and Torrid Zones, only accounting for the latitudinal differences in solar effect. The Greek word, "klima," means "inclination" of incoming solar ray.

Solar insolation on the surface of the Earth is the amount of sunlight received. The amount of heat energy received at any location on the globe depends on the Sun angle. The angle at which sunlight strikes the Earth varies by location, minute of hour, time of day, and season due to the Earth's orbit around the Sun and the Earth's rotation around its tilted axis. The effect of solar insolation is determined by several factors including the inclination of incoming solar ray, atmospheric conditions, the Earth's tilt, and solar radiation intensity caused by the solar activity and the distance between the Earth and the Sun.

The Earth's surface curve and the angle of solar insolation affect the distance that a beam of sunlight travels through the Earth's atmosphere before it reaches the Earth's surface. A beam of sunlight penetrates a shorter distance through the atmosphere at the equator than the incoming sunlight arriving at higher latitudes. As incoming sunlight travels through more of the atmosphere, larger amount of sunlight is reflected or scattered, thus the insolation at the Earth's surface is reduced. For a beam of sunlight, it is spread over a greater area in polar region than that in lower latitudes.

Climatologically, the surface temperature changes in seasonal and diurnal scales are largely controlled and balanced by the changes in the amount of energy received and redistributed. Climate change includes the climatic state changes from diurnal cycle to seasonal cycle and long-term cycles. From these cycles, we can say, climatic state changes from minute to minute, hour to hour, day to day, year to year, decade to decade, and century to century, only depending on the motion of the Earth relative to the Sun and the balance of energy under the surface condition. Weather is the sum of climatic state and anomaly from minute to minute, hour to hour, and day to day. Therefore, a modern climatic clock needs at least two pointers to indicate the climate change from hour to hour and from day to day.

As an example, Table 1.1 lists the rates of air temperature changes in different timescales at a specific site of the Earth surface. The maximal rate of temperature change is the diurnal cycle (1.0×10 °C/h) and the second is the seasonal cycle (6.8×10^{-3} °C/h). The rates of temperature changes in 11-, 22-, and 200-year cycles are about 1.0×10^{-6}, 5.0×10^{-7}, and 5.0×10^{-8} °C/h, respectively. The inter-annual change rate of temperature is about 1.4×10^{-4} °C/h, but it is not periodic.

1.5 Weather and Climatic Anomalies as Well as Their Predictions

Weather is what we get naturally in the atmosphere at a moment. Our body can sense how hot or cold and how dry or wet it is. There are many devices to measure the weather quantitatively. These weather elements include temperature, precipitation, wind, humidity, visibility, sunshine, and some phenomena like rain, snow, haze, fog, dust, etc. Any value (W) of measured weather element at position p and at time t can be decomposed into two components: the climatic state $(\widetilde{C}(p,t))$ and weather anomaly $(W'(p,t))$, i.e.,

$$W(p,t) = \widetilde{C}(p,t) + W'(p,t). \tag{1.5.1}$$

Table 1.1 Rates of site air temperature changes in different timescales

Time-scale	Diurnal	Seasonal	11-year	22-year	200-year
Amplitude (°C)	10.0	60	0.1	0.1	0.13
Rate (°C//h)	1.0×10^0	6.8×10^{-3}	1.0×10^{-6}	5.0×10^{-7}	5.0×10^{-8}

Source Qian (2011b)

In order to get the weather anomaly $W'(p,t)$, we must first know the climatic state $\widetilde{C}(p,t)$. Climatic state at a certain site should be an equilibrium state forced by the regular solar radiation and the surface condition without any influences of other disturbances. As Herbertson (1901) put it: "Climate is what on an average we may expect." It means that we do not need to predict a climatic state at any time but it can be estimated by an average.

Up to now, we have not got the true climatic state. It is assumed that we have a site observation of temperature ($T_{d,y}(p,t)$) at a certain moment (t) such as at 0000 UTC on a calendar date (d) for years (y) from 1 to Y. The Y is at least 30 years or more than 100 years. We can estimate the climatic state $\widetilde{T}_d(p,t)$ indicated by temperature at a specific site or position p and at a specific moment (t) as follow:

$$\widetilde{T}_d(p,t) = \sum_{y=1}^{Y} T_{d,y}(p,t)/Y. \tag{1.5.2}$$

Using observations from other hours at the same calendar day, we can estimate climatic states of other hours to form a diurnal cycle of climatic temperature. After getting the climatic state of temperature, the weather anomaly can be obtained by

$$T'_{d,y}(p,t) = T_{d,y}(p,t) - \widetilde{T}_d(p,t). \tag{1.5.3}$$

By this way, we can obtain precipitation anomaly, pressure anomaly, surface wind anomaly, and other elements' anomalies after we know the climatology of all these variables. So weather anomaly is a deviation relative to the climatic state. Usually, absolute climatic temperature is locally larger than temperature anomalies while climatic precipitation is often smaller than precipitation extremes.

From Eq. 1.5.2, it is known that climatic state is the average of weather during a long period. At a specific moment of a calendar day, the climatic state is statistics of weather, and varies with time in the diurnal cycle and seasonal cycle. Thus, weather at a specific moment of a calendar day is the sum of its climatic state and weather anomaly.

Climatic state is what we expect without weather anomalies and climatic anomalies. A perfect climatic state is where its standard deviation of weather anomalies is very small. If there is no deviation of weather anomalies, the weather and the climatic state are both the same. Weather anomaly changes frequently from minute to minute, hour to hour, and day to day. Positive (or negative) weather anomaly may persist for minutes, hours, or days and then it dissipates. If positive (or negative) weather anomaly persists for more than months, years, decades, and centuries, we refer to it as climatic anomaly, anomalous climatic event, or extreme climatic event.

Weather forecast is the application of science and technology to predict the future state of the atmosphere. Human beings have attempted to predict the weather informally for millennia, and formally since the nineteenth century. On the first day of my synoptic lecture, I asked why my students chose to study this course. One of students said: I will tell the daily weather for the following year to my parents and grandparents at the Chinese spring festival when I go back home, so they do not need to watch the television show of weather forecast each evening. Weather forecasts are made by collecting quantitative data about the current atmosphere on a given place and using scientific understanding of atmospheric processes to project how the atmosphere will evolve at that place. Before a major event, such as the Olympic Games or the National Games in China, seniors are invited to be advisors of weather forecast because they have rich experience. Usually, these advisors are not directly giving any detailed predictions. Their experience is difficult to teach new forecasters. So weather forecasts are not as simple as they are thought and expected.

There are a variety of users of weather forecasts. Weather warnings are important information because they can protect lives and properties. Forecasts of temperature and precipitation are important to agriculture, and therefore to traders within commodity markets. Temperature forecasts are used by utility companies to estimate demand over following days. On a daily basis, people use weather forecasts to determine what to wear on a given day. Since the outdoor activities can be severely curtailed by heavy rain, snow, and wind chill, weather forecasts can help when planning activities, so as to stay away from bad weathers.

Weather forecast can be divided as now-casting, short-term, medium-range, and extended-range spells. The weather forecast within the next six hours is often referred to as now-casting. Over this time range, it is possible to forecast smaller-scale and strong phenomena, such as individual showers and thunderstorms with reasonable accuracy, which are too small to be resolved by a computer model. A man given the latest radar, satellite, and observational data will be able to make a better analysis of the small-scale features and so will be able to make a more accurate forecast for the following few minutes or hours, according to the current moving speed and direction of small-scale system.

The short-term weather forecast focuses from the next six hours to three days. The products of numerical weather model can be used starting from the next six hours. Synoptic (or weather) chart analysis is a traditional method used in the short-term forecast. On the traditional daily synoptic chart, there are some moving large-scale waves, which are called Rossby waves, in the middle troposphere. These Rossby waves or ridges and troughs are very important for the propagation of mid-latitude weather systems. The surface cold and warm fronts and the distributions of weather phenomena are located within a weather system. The moving speed and direction of the ridges and troughs can be determined from their previous positions in the past several six-hour intervals. This motion can be used to estimate the position of ridge and trough in the next six hours to three days.

For the forecasts of medium to extended ranges, their time lengths cover 4–9 days and 10–30 days, respectively. Predictions of all the three spells (short-term, medium, and extended ranges) rely mainly on the outputs of numerical weather prediction (NWP) model. Currently, the European Centre for Medium-Range Weather Forecasts (ECMWF) model can generate a 15-day forecast product, while the US

Global Forecast System model regulated by the Environmental Modeling Center can generate a 16-day forecast product. Exceeding 16 days, we cannot obtain information from these operational models.

It can be told from Eq. 1.5.1 that the climatic state can be estimated rather than forecast. In order to forecast the weather tomorrow, we must first know and trace the weather anomalies today and yesterday, respectively. Currently, the forecasts of temperature, precipitation, and other phenomena of elements, from either NWP models or traditional synoptic chart, are not decomposed into climatic state and weather anomaly. Climatic state or climatology should be identified instead of predicted. Weather anomaly needs be predicted by using objective methods.

A climatic prediction $C(p,t)$ at a position p for the future time t has two parts including the climatic state $\widetilde{C}(p,t)$ and climatic anomaly $C'(p,t)$, namely

$$C(p,t) = \widetilde{C}(p,t) + C'(p,t) \tag{1.5.4}$$

Climatic anomaly is a persistent positive (or negative) weather anomaly for more than months, years, decades, and centuries. As the climatic state should be identified first, climatic anomaly needs be predicted by objective methods. Climatic state is changing with time due to various periodic forcing through physical processes. Climatic state can be seen as natural resource while the climatic anomalies are causes of natural disasters. Climatic predictions are generally the products most eagerly sought for longer-term decisions and early warning of potential hazards. Climatic predictions can be produced on global, regional, or local scales.

The current methods of weather and climatic predictions do not decompose the output into climatic state, weather anomaly, and climatic anomaly. The process of climatic predictions from 30 days up to two years on the global scale requires huge amounts of computer resources along with a specialized knowledge. Under this condition, there are only few centers around the world that are able to produce climatic predictions. These centers are known as World Meteorological Organization (WMO) Global Producing Centers of Long-Range Forecasts (GPCLRFs). One of the strongest signals of climatic anomaly is the surface sea temperature anomalies in the equatorial eastern Pacific, namely El Niño/La Niña events. The WMO produces El Niño/La Niña updates on a quasi-regular basis (approximately once in three months) through a collaborative effort between WMO and the International Research Institute (IRI) for Climate and Society. The ability to predict the status of the El Niño/La Niña phenomenon is particularly important, due to its impact on global climate conditions through atmospheric circulation anomalies.

Weather prediction and climatic prediction are mostly dependent on numerical models. The basic idea of numerical weather prediction is to sample the total state of the atmosphere at a given time and use the equations of fluid dynamics and thermodynamics to estimate the total state of the atmosphere at a given time in the future. The main inputs from country-based weather services are the surface observations from automated weather stations at ground level over land and from weather buoys over sea. The WMO acts to standardize the instrumentation, the observing practices, and the timing of these observations worldwide. Station reports either hourly or every six hours. Sites launch radiosondes, which rise through the depth of the troposphere and well into the stratosphere. Data from weather satellites are used in areas where traditional data sources are not available.

Models are initialized using observed data, which include both climatic state and weather anomaly. The irregularly spaced observations are processed by data assimilation and objective analysis methods, which perform quality control and obtain atmospheric variable values at locations. The data are then used in the model as the starting point for a forecast. Commonly, the equations used to predict the known physics and dynamics of the atmosphere are called primitive equations. These equations are initialized from the analysis data (climatic state and anomaly) and the rates of change are determined. The rates of change (atmospheric climatic state change and anomaly) predict the atmospheric field for a short time into the future. The equations are then applied to this new atmospheric field to estimate the new rates of change, and these new rates of change to predict the atmospheric field at a future time point. This time stepping procedure is continually repeated until the solution reaches the desired forecast time.

Essentially, a model is a large number of computer programs that produce meteorological information for future time points at given locations and altitudes. The model is a set of the primitive equations used to predict the future field of the atmosphere. These equations—along with the ideal gas law—can simulate the evolutions of air density, pressure, and potential temperature scalar fields and the velocity vector field. The equations used are nonlinear partial differential equations which are impossible to solve exactly through analytical methods, but can be approximated by numerical methods. Different models use different solving methods: some global models use spectral methods in the horizontal dimensions and finite-difference methods in the vertical dimension, while regional models and other global models usually use finite-difference methods in all three dimensions. No matter what numerical methods are used in the models, they use the observed data that include the atmospheric climatic state and anomalies. The climatic state is also decomposed into spectral components in global spectral models or calculated as grid values in finite-difference models. Thus, spectral expansion or finite difference of climatic state will produce errors, false weather systems, or false alarms of extreme weather events.

1.6 Decomposition of Temporal Climatology and Anomaly

Traditionally, the climatic state is defined, to a large extent, as the average conditions over a time interval, such as monthly, seasonal, and annual averages (Starr and White 1954; Peixoto and Oort 1992). If the time interval is taken as a month (or a season), the climatic seasonal cycle is divided to 12 (or 4) solar terms.

In ancient China, a relatively complete calendar, namely "Primordial Calendar" , was established during the Western Han Dynasty by Lou Xia-Hong and Deng Ping, who summarized many historical climatic documents of previous dynasties. The astronomical positions for the 24 solar terms had been determined on the division of the ecliptic place, which depends on the relative location of the Earth to the Sun and separate 360° of Sun's longitude into 24 equal segments, with each of solar terms by 15° Sun's longitude. Thus, there are two solar terms each month. At the Spring Equinox, the Sun's longitude is 0°. Such astronomical 24 solar terms reflect the periodic cycle as the Earth orbits the Sun, and thus, the date in the Gregorian calendar of each solar term is almost fixed. For example, the solar terms in the first half year are mainly on the 6th or 21st day of the month, whereas those in the latter half year are mainly on the 8th or 23rd day from February to July, with a bias smaller than 2 days. Each term has its Chinese name, which represent the climatic seasonal cycle. For example, "spring begins" is the first solar term, usually from 5 to 20 February on the calendar, and marks the beginning of the spring season and the steady rise of temperatures. The "cold dew" is the 17th solar term, beginning around on October 8 of the calendar and lasting for 15 days, which indicates the beginning of decrease in temperatures with increasing likelihood from dew to frost that can damage crops. Accordingly, each term is corresponding to a growing stage of certain crops in certain regions in China, and is a reference for agricultural activities, as well as judges whether the current weather is normal or abnormal. This calendar, which has been built in 104 BC, is like a climatic clock with 24 intervals for a year.

The idea of ancient Chinese 24 solar terms is to separate the climatic seasonal cycle from the anomalous weather or the anomalous climate in a half-month interval. This idea can also be used to decompose any meteorological field into climatology and anomaly with an hourly interval on each of 365 solar (or calendar) days. The climatology is quasi-stationary and reflects an equilibrium state between solar radiation and local geographical features, while the anomaly is a departure or deviation from the equilibrium state varying at any moment. The climatic cycle in a year can be divided into 365 days × 24 hours = 8760 terms.

An atmospheric total or full field $S_{\{d,y\}}(\lambda, \varphi, p, t)$ such as geopotential height H, temperature T, specific humidity q, and winds (u, v) at diurnal time t (24 h a day) on calendar date d in year y at a spatial point of longitude λ, latitude φ, and pressure level p, can be decomposed into a climatic field $\widetilde{S}_d(\lambda, \varphi, p, t)$ and an anomalous field $S'_{\{d,y\}}(\lambda, \varphi, p, t)$ following Qian et al. (2014a):

$$S_{\{d,y\}}(\lambda, \varphi, p, t) = \widetilde{S}_d(\lambda, \varphi, p, t) + S'_{\{d,y\}}(\lambda, \varphi, p, t) \tag{1.6.1}$$

The climatic field is estimated by averaging over 30 years (1981–2010) based on any one of reanalysis datasets on calendar date d and at diurnal time t,

$$\widetilde{S}_d(\lambda, \varphi, p, t) = \sum_{y=1981}^{2010} S_{\{d,y\}}(\lambda, \varphi, p, t)/30, \tag{1.6.2}$$

where y ranges from 1981 to 2010. For example, when d is set to 1 May, $\widetilde{S}_d(\lambda, \varphi, p, t)$ is the average of $S_{\{d,y\}}(\lambda, \varphi, p, t)$ from 1 May 1981, 1 May 1982, …, 1 May 2010 for the entire 30 years. It is assumed that the positive and negative anomalies of meteorological variables at a specific grid point and a given calendar time cancel each other during the 30-year period to approximate the quasi-static climatic state. The climatic state (or climatology) estimated by Eq. 1.6.2 varies from hour to hour and day to day.

The temporal climatic field remains the diurnal cycle and seasonal cycle of temperature, wind, and height. Near the surface or the lower troposphere, there are not only large-scale climatic systems (such as an aloft subtropical high in the free atmosphere) caused by large-scale sea-land contrast but also small-scale climate systems such as local sea breeze and local land breeze (diurnal cycle). There are two examples. The strength of summer low-level jet near the 850 hPa in southern China has a climatologically diurnal cycle with the maximum strength at the late night around 0200 Beijing Time (BJT; Zhou and Wu 1979), so the rain over nighttime is stronger than that over daytime (Yu et al. 2007). Other facts show that the summer monsoon onset in the South China Sea (SCS) has a rapid climatic change from pentads 27 to 28 (e.g., Qian and Yang 2000; Qian and Jiang 2015).

In the following paragraphs we give some updated analyses. Figure 1.12a, b show the seasonal cycle of climatic wind at 0200 and 2000 BJT at 850 hPa along 110°E. The SCS summer monsoon onset transforms from easterly to westerly winds as indicated by the first black dashed line (1) in pentads 27–28. The second black dashed line (2) is in pentad 54 (late September). On that day or pentad, the north part of the red-dashed line is northeasterly wind while its south part is southwesterly wind, so there is a monsoon convergence line (or monsoon trough) in the central SCS. Therefore, the monsoon convergence line will be masked when we take monthly-mean climatic wind. Figure 1.12c illustrates the climatic wind difference between the 0200 BJT and the 2000 BJT at 850 hPa along 110°E. It clearly shows a diurnal cycle of climatic wind. In summer, southwesterly velocity at 0200 BJT can increase over 3–4 m s^{-1} in southern China comparing to that at 2000 BJT on the same calendar day. Such features of diurnal cycle will be lost if we take a daily-mean climatic wind.

In pentad 27, the summer monsoon is not onset in the SCS because its daily-mean precipitation rate is low (less than 4 mm/day) and the subtropical high at the lower troposphere is in the central SCS (Fig. 1.13a). In pentad 28 (Fig. 1.13b), however, a precipitation center (above 8 mm/day) climatologically appears at the central SCS because the subtropical high has moved eastward with wind direction reversal from easterly to westerly. The strengthened southwesterly wind and increased precipitation rate are centered in the SCS (Fig. 1.13c). If we take running mean of 4 (time points per day) by 90 (days of three months) by 30 (years) consecutive variable fields, the amplitude of diurnal cycle and seasonal cycle will be largely reduced and then the anomaly will be largely enlarged. It is clear that the daily climatology in pentad 27 is rather different from that in pentad 28.

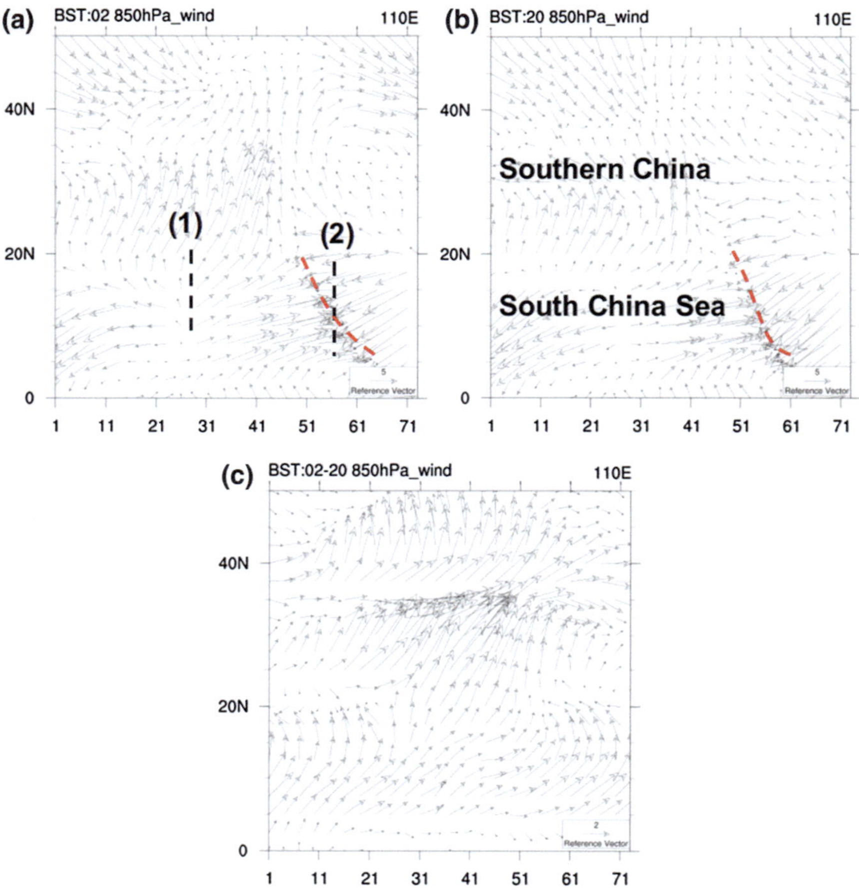

Fig. 1.12 Climatic winds (m s^{-1}) with season (1–73 pentads) crossing Central China (110°E) at 850 hPa at **a** 0200 and **b** 2000 BJT, and **c** the climatic wind difference by the 0200 BJT minus the 2000 BJT at 850 hPa along 110°E, using the ECMWF Reanalysis Interim (ERA-Interim) product. Southern China covers 20°N–35°N and the South China Sea bands over 0–20°N. The *dashed lines (1) and (2)* indicate in pentads 27 and 54 while the *red-dashed line* is the monsoon trough in the SCS

In the middle summer, a strengthened southwesterly wind at 850 hPa in pentad 40 is clearer than that in pentad 39 accompanying increased precipitation rate over the Shandong Peninsula in China (Fig. 1.14a). This situation changes rapidly in Fig. 1.14b in which a strengthened northeasterly wind in pentad 42 has decreased precipitation rate when compared with that in pentad 41.

In the central SCS, a rapid withdraw of subtropical high ridge is observed from pentads 27 to 28. In the same place, a rapid deepening of the monsoon trough is found from pentads 54 to 55 with increased daily-mean precipitation rate (Fig. 1.15). However, decreased daily-mean precipitation rate is clearly observed in the East China and the subtropical NW Pacific due to that there are two strengthened anticyclones (Fig. 1.15c).

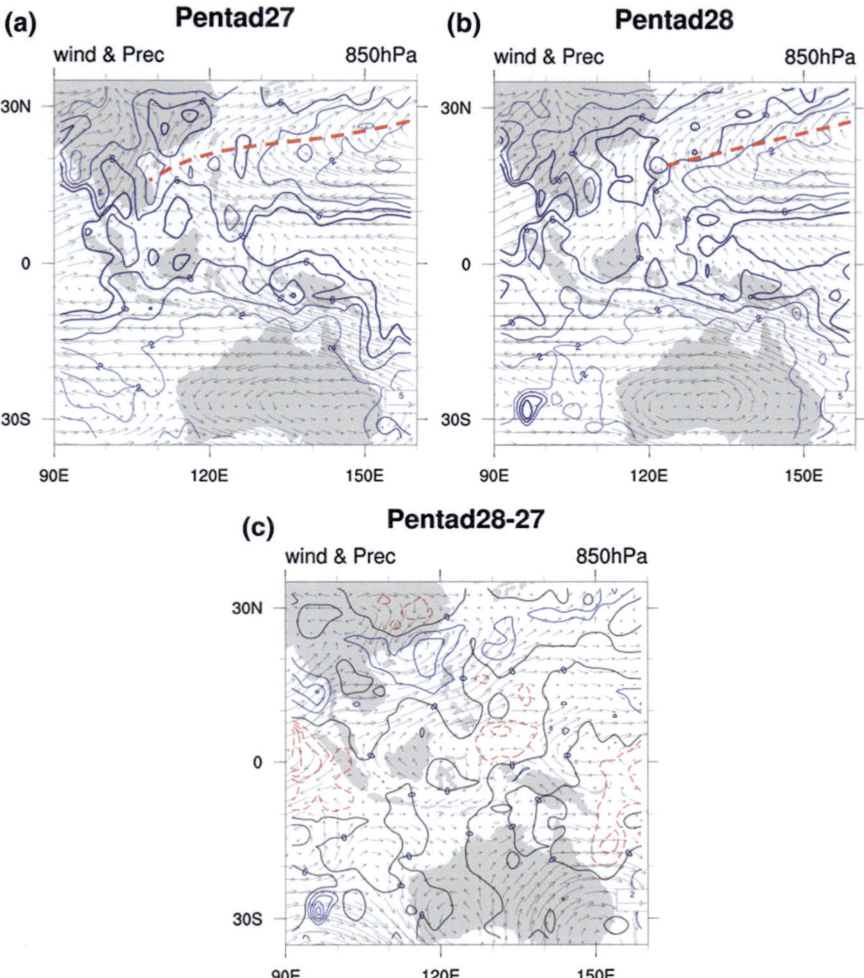

Fig. 1.13 Climatic wind (*arrow*, m s^{-1}) at 850 hPa and precipitation rate (*blue line and thin-red-dashed line*, 2 mm/day interval) in pentads **a** 27 and **b** 28 as well as **c** their difference by the later minus the former. The *thick-red-dashed line* is the ridge line of the subtropical high in the NW Pacific. The wind and precipitation rate are used from the ERA-Interim product and the CMAP dataset, respectively

The daily-mean differences of climatic variables are also large in many regions. Large difference of climatic wind directions in the SCS can be observed in Fig. 1.16a, b. The differences of climatic wind and specific humidity centered in the SCS, eastern China, and the Japan Sea can be found in Fig. 1.16c. In the SCS and eastern China, southwesterly velocity and northeasterly velocity strengthen 1–2 m/s. A center of climate specific humidity differences can be found in the Japan Sea. The deepened climatic cyclone in the SCS is more obvious in Fig. 1.17c by the latter (Fig. 1.17b on 21 May) minus the former (Fig. 1.17a on 20 May). The

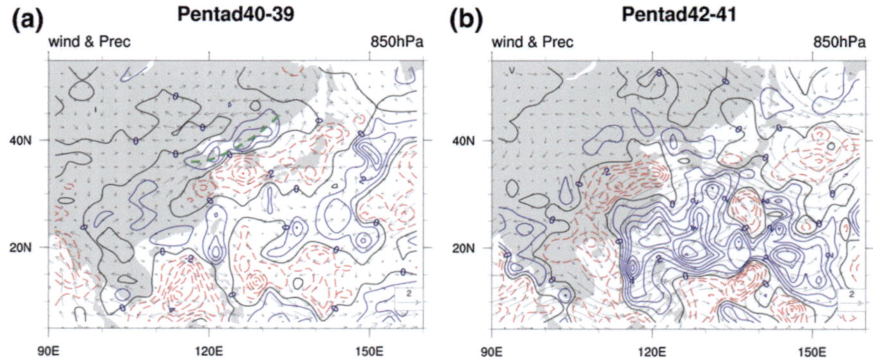

Fig. 1.14 Same as in Fig. 1.13c except by **a** pentads 40 − 39 and **b** pentads 42 − 41. The *thick dashed green line* is the axis of daily-mean precipitation rate

climatology with seasonal and diurnal cycles is critical to diagnose and predict the location and intensity of heavy rains and tropical cyclones.

In the study of atmospheric circulation, the monthly mean or seasonal mean of zonally averaged variables are usually used to examine the structures of meridional and vertical sections from the troposphere to stratosphere. Indeed, zonally averaged atmospheric circulations described climatic cycles also vary with a diurnal cycle and seasonal cycle. If our planet is a water ball (aqua planet) without the sea-land distribution, the major form of atmospheric motion should be meridional circulations or cells driven by solar radiation on the rotating Earth and varied with their strengths and locations, with a diurnal cycle and seasonal cycle in climatology. However, the irregular sea-land distribution leads some departures of observational meridional cells related to the water ball cells. Even so, meridional cells (or climatic features) are still identified from zonally averaged variables.

$$\widetilde{S}_d(\varphi, p, t) = \sum_{\lambda=1}^{M} \widetilde{S}_d(\lambda, \varphi, p, t)/M \tag{1.6.3}$$

Here, M is the total grids along a zonal circle.

The Hadley cell and the Ferrel cell can be obtained by the zonally averaged component. Thus, the climatic component can be spatially separated as zonally averaged climatic part and climatic zonal departure, which varies with longitudes

$$\widetilde{S}_d(\lambda, \varphi, p, t) = \left[\widetilde{S}_d(\varphi, p, t)\right] + S_d^*(\lambda, \varphi, p, t) \tag{1.6.4}$$

The climatic zonal departure or asymmetric system is partly caused by the sea-land topographical distribution and sea-land thermal contrast, which can be used to indicate the climatic monsoon circulation varying with season.

In practice, the westerly index proposed by Rossby et al. (1939) is often used in the medium-range weather forecast. The westerly index describes westerly wind

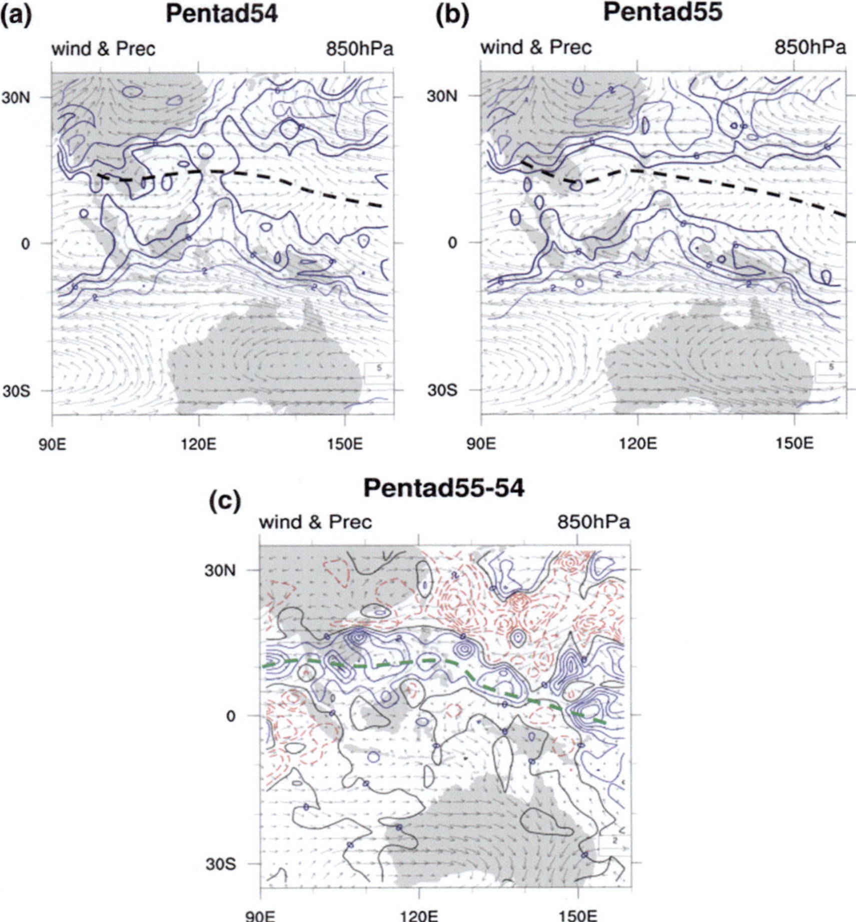

Fig. 1.15 Same as in Fig. 1.13 except in **a** pentad 54, **b** pentad 55, and **c** their difference by the later minus the former. The *thick-black dashed line* in **a** and **b** is the monsoon trough while the *thick-green dashed line* in **c** is the axis of daily-mean precipitation rate

anomaly or geopotential height anomaly in the middle and high latitudes. It can be written as

$$S'_{d,y}(\varphi, p, t) = \sum_{\lambda=1}^{M} S'_{d,y}(\lambda, \varphi, p, t)/M \tag{1.6.5}$$

This temporal zonally averaged anomalous part can be used to describe climatic change with time and latitudes (Qian and Liang 2012). If there is only positive temperature anomaly in the Western Hemisphere (WH) but without any anomaly in the Eastern Hemisphere (EH), the zonally averaged temperature anomaly is positive

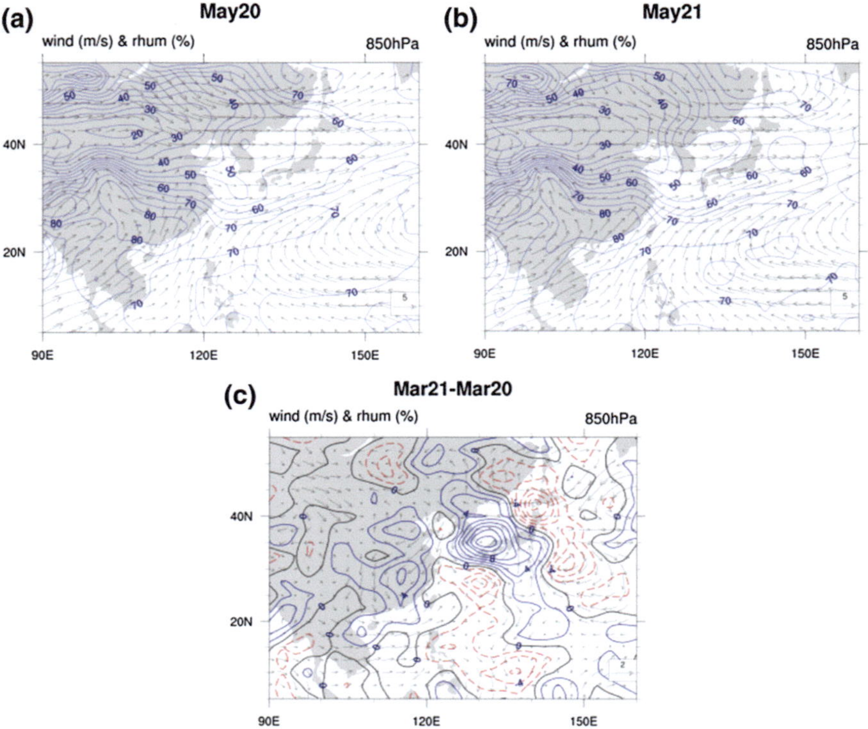

Fig. 1.16 Same as in Fig. 1.13 except wind and specific humidity at 850 hPa on **a** May 20, **b** May 21, and **c** their difference of the latter minus the former

but it is weaker than that in the WH. If two opposite temperature anomalies are in the EH and WH, respectively, the zonally averaged temperature anomaly is zero. Thus, the temporal zonally averaged anomalous part has no physical meaning, unless there is a forcing along any circle of latitude.

Similarly, a temporal anomaly $S'_y(\lambda,\varphi,p,t)$ can be divided as temporal zonally averaged anomaly and temporal departure:

$$S'_{d,y}(\lambda,\varphi,p,t) = \left[S'_{d,y}(\varphi,p,t)\right] + S'^{*}_{d,y}(\lambda,\varphi,p,t) \tag{1.6.6}$$

The temporal zonally averaged anomaly is smaller than temporal departure. It means that

$$S'_{d,y}(\lambda,\varphi,p,t) \approx S'^{*}_{d,y}(\lambda,\varphi,p,t) \tag{1.6.7}$$

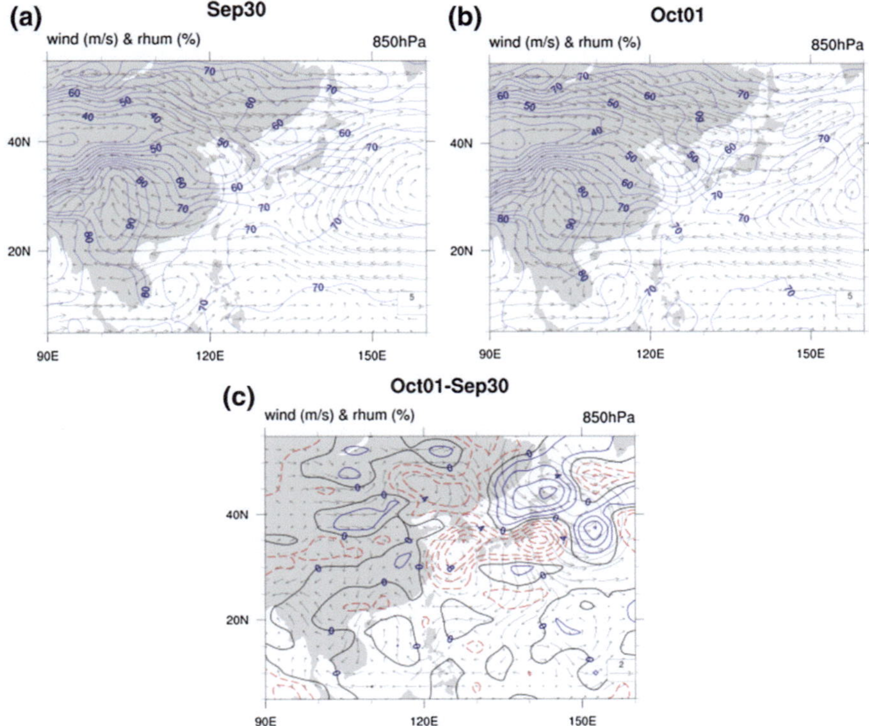

Fig. 1.17 Same as in Fig. 1.16 except on **a** 30 September, **b** 1 October, and **c** their difference by the latter minus the former

Questions

1. The spatial scales and life durations of natural systems are different. What are the similarities of different natural systems' evolutions?
2. What is the dynamical definition of climate? Why does the climatic state need to be identified while the anomaly needs to be predicted?
3. Why does the traditional synoptic chart contain information of weather extremes but cannot correctly locate weather extremes?
4. Why do current complex numerical models produce climate shift and false weather extremes?
5. An airport forecaster observed that 10,000 passengers take flights to depart the airport at 0800 UTC. How about their spatial domain or expansion speed of all these passengers after 1, 2, …, 10, even 24 h?

References

Herbertson AJ (1901) Outlines of Physiography: An Introduction to the Study of the Earth. Edward Arnold, 312 pp

Knupp KR, Murphy TA, Coleman TA, … Klockow K (2014) Meteorological overview of the devastating 27 April 2011 tornado outbreak. Bull Amer Meteorol Soc 95: 1041–1062

Peixóto JP, Oort AH (1992) Physics of Climate. Amer. Inst. Physics, New York, pp 520

Qian WH (1992) Distribution of continents and islands over the Southern Hemisphere and the long-term deceleration in speed of the Earth's rotation. Scientia Geologica Sinica 10(3): 305–308

Qian WH (1994) Dynamical Introduction of the Planetary Earth. Meteorol. Press, Beijing, pp 213

Qian WH (1996) The motion of the Earth interior liquid and global tectonics. Earth Sci Frontiers 3: 152–160

Qian WH, Yang S (2000) Onset of the regional monsoon over Southeast Asia. Meteorol Atmos Phys 75: 29–38

Qian WH (2011a) Who have driven the climate change? Beijing, Scientific Press, pp 210

Qian WH (2011b) Climate change of multiple timescales. Adv Meterol Sci Tech 3(2): 36–40

Qian WH (2012) Principles of medium to Extended Range Weather Forecasts. Scientific Press, Beijing, China, pp 410

Qian WH (2013) Atlas of Anomalous Circulations associated with Regional Heavy Rainfall in China. China Meteorological Press. Beijing, China, pp 227

Qian WH, Jiang N (2015) The global monsoon definition using the difference of local minimum and maximum pentad precipitation rates associated with cross-equatorial flow reversal. Theor Appl Climatol doi:10.1007/s00704-015-1448-4

Qian WH, Liang HY (2012) Propagation of planetary-scale zonal mean wind anomalies and polar oscillations. Chinese Science Bulletin 57(20): 2606–2614

Qian WH, Shan XL, Liang HY (2014a) Typhoon Turning Atlas. World Scientific, pp 441

Qian WH, Shan XL, Liang HY, Huang J, Leung CH (2014b) A generalized beta advection model to improve unusual typhoon track prediction by decomposing total flow into climatic and anomalous flows. J Geophys Res Atmos 119: 1097–1117

Rossby CG, Collaborators (1939) Relation between Variation in the Intensity of the Zonal Circulation of the Atmosphere and the Displacements of the Semipermanent Centers of Action. J Marine Res 2: 38–55

Starr VP, White RM (1954) Balance requirements of the general circulation. Air Force Cambridge Research Center, Geophys Res Papers

Yu RC, Zhou TJ, Xiong AY, Zhu YJ, Li JM (2007) Diurnal variations of summer precipitation over contiguous China. Geophys Res Lett 34: L01704, doi:10.1029/2006GL028129

Zhou CC, Wu PP (1979) The role of low-level jet in a continuous rainstorm (in Chinese). Plateau and Mountain Meteorol Res 33(4): 30–34

Chapter 2
Wind and General Circulation

Wind is the directional motion of air particles. The general circulation is a representation of climatic state of wind, which is a linear and delayed response of periodic solar radiation forcing. The timescale of general circulation is from diurnal cycle to seasonal cycle, while its spatial scale ranges from a small peninsula to continent and basin. The study of general circulation has a long history since Halley (1686) and Hadley (1735) to Maury (1855) and Ferrel (1856), from the single-cell model, the two-cell model, and to the three-cell model. In 1921, V. Bjerknes proposed the four-cell model based on his theoretical speculation. Using four sets of reanalysis products and a climate model simulation, the four-cell model has been confirmed through zonally averaged calculation from basic atmospheric variables. The fourth cell located over the two polar areas are, respectively, named as the Arctic cell in the Northern Hemisphere and the Antarctic cell in the Southern Hemisphere. The Hadley, Ferrel, Polar, and Arctic/Antarctic cells exist in each hemisphere but their intensities vary from day to day and from month to month. The result of an aqua control experiment further identifies the existence of eight cells in the global atmosphere.

2.1 Diurnal-Cycle General Circulation

Traditionally, the general circulation of the atmosphere includes the totality of motions that characterizes the global-scale atmospheric flow. It is to identify average structures of the wind field along with other meteorological variables including temperature, humidity, pressure, and phenomena such as precipitation. Thus, the general circulation can be considered to consist of the flow averaged over a period of time sufficiently long for months and seasons, which can remove the random variations associated with individual weather systems. We here believe that

© Springer Nature Singapore Pte Ltd. 2017

W. Qian, *Temporal Climatology and Anomalous Weather Analysis*,
Springer Atmospheric Sciences, DOI 10.1007/978-981-10-3641-5_2

the general circulation is climatic winds or prevailing winds with various spatiotemporal scale subcomponents caused by regular forcing from solar radiation and surface conditions.

Land and sea breezes are such climatic winds with local spatial-scale component caused by regular diurnal-cycle forcing. Without large-scale climatic wind and wind anomalies, a sea breeze or onshore breeze is a gentle wind that develops over bodies of water near land due to differences in air pressure created by their different heat capacities. It is a common occurrence along coasts during the morning as solar radiation heats the land more quickly than the water. Land breezes or offshore breezes have a reverse effect, caused by land cooling more quickly than water in the evening. Sea breezes dissipate and wind blows from the land toward the sea. Offshore wind refers more generally to any wind over open water.

The cause is that the sea has a greater heat capacity than land, so the sea surface warms up slower than the land's surface during the daytime. As the temperature of the land surface rises, the land heats the air above it by conduction (sensible heat flux). Warm air is less dense and so it expands, decreasing the pressure over the land near the coast. The air above the sea has a relatively higher pressure, causing air near the coast to flow toward the lower pressure over land. The strength of the sea breeze is directly proportional to the temperature difference between the land and the sea.

Atmospheric moisture rises at a place of wind convergence near the surface. Sea-breeze fronts are such convergence systems created by sea breezes. Cold air from the sea meets warm air from the land and creates a boundary like a shallow cold front. If the air is humid and unstable, cumulonimbus clouds and even showers can be formed along the mesoscale front.

At night, sea breezes usually change to land breezes, due to reversal of the sea–land temperature gradient. During night, the land cools off faster than the ocean due to differences in their heat capacity, which forces the dying of the daytime sea breeze as the temperature of the land approaches that of the ocean. If the land becomes cooler than the adjacent sea surface temperature, the air pressure over the water will be lower than that over the land, setting up land breezes blowing from the land to the sea, as long as the environmental surface wind pattern is not strong enough to oppose it. If there is sufficient moisture and the atmosphere is unstable, land breezes can cause showers, or even thunderstorms, over the water. Land breezes die once the land warms up again the next morning.

The reversal wind directions on the surface and rising–sinking flows between land and sea crossing the coastline form two opposite overturning circulations from day and night. The intensity of this local closed circulation can be quantitatively described by mass stream function, which is locally associated with the solar radiation and the surface sea–land condition. The intensity of circulation should be delayed to the solar radiation in the diurnal timescale. Its surface zonal wind $\tilde{u}(t)$ crossing the coastline can be, according to Eq. (1.3.1), written as

$$\frac{\partial \tilde{u}(t)}{\partial t} = \tilde{\Theta}(t) - (L + D)\tilde{u}(t), \tag{2.1.1}$$

where the term $\tilde{\Theta}(t)$ is the solar forcing and the term $-(L + D)\tilde{u}(t)$ is the delayed factor depending upon the linear relationship and dissipative effect $-(L + D)$. If the solar radiation forcing $\tilde{\Theta}(t)$ approaches its maximum at noon, the maximum wind $\tilde{u}(t)$ and other weather phenomena such as shower should appear in the afternoon. The diurnal cycle of this local circulation and the surface wind and regular weather is a climatic phenomenon which does not need to be predicted but should be identified. This reversal circulation caused by sea–land heating contrast with diurnal cycle can also be referred as daily sea breeze and daily land breeze during day and night, respectively.

From Eq. (1.5.1), the local wind is written as

$$u(p, t) = \tilde{u}(p, t) + u'(p, t). \tag{2.1.2}$$

It is clear that only wind anomalies $u'(p, t)$, which may be caused by anomalous cyclone and anomalous anticyclone, need to be predicted. And, mountain and valley breezes are also local climatic circulation with a diurnal cycle.

2.2 Seasonal-Cycle General Circulation

Solar insolation has an obvious seasonal cycle besides the diurnal cycle. Therefore, apart from sea and land breezes as shown in Fig. 2.1, the general circulation should also have a seasonal cycle. This seasonal-cycle general circulation is directly associated with the seasonal change in solar radiation over adjacent continent and ocean. The reason is that the land responds to the seasonal changes in solar radiation much more quickly than the ocean does. Therefore, this seasonal-cycle general circulation of continent–ocean contrasts is a large-scale sea breeze. The direct response to the solar radiation is land rather than ocean, which occur when the temperature on land is significantly warmer in summer or cooler in winter than the temperature of the ocean. These temperature imbalances happen because oceans and land absorb heat in different ways. Over oceans, the air temperature remains relatively stable for two reasons: water has a relatively high heat capacity and because both conduction and convection will equilibrate water temperature in the vertical direction (up to 50 m). In contrast, dirt, sand, and rocks have lower heat capacities over land and they can only transmit heat into the earth by conduction and not by convection. Therefore, the bodies of water stay at a more even temperature, while land temperatures are more variable.

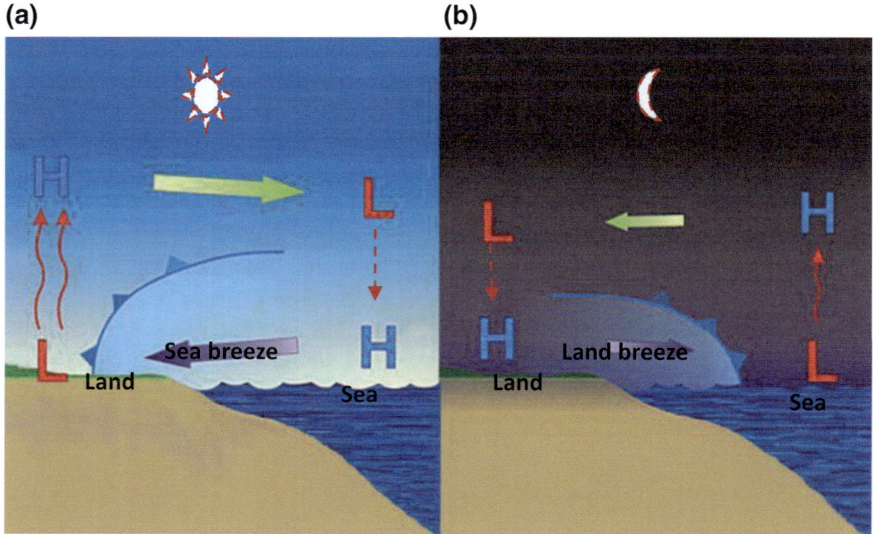

Fig. 2.1 A schematic representation of **a** daily sea breeze and **b** daily land breeze. Modified from https://earthsciencecenter.files.wordpress.com/2010/11/landsea.jpeg

During summer, sunlight heats the surfaces of both land and oceans, but land temperatures rise more quickly. As the land's surface becomes warmer, the air above it expands and an area of low pressure develops. Meanwhile, the ocean remains at a lower temperature than the land, and the air above it retains a higher pressure. This difference in pressure causes seasonal sea breeze to blow from ocean to land, bringing moist air inland. This moist air rises to a higher altitude over land and then it flows back toward the ocean, completing a closed circulation. However, when air rises, and while it is still over the land, the air cools. This decreases the air's ability to hold water, and this brings precipitation over the land (Fig. 2.2a).

In winter, the cycle is reversed. Land cools faster than ocean and the air over the land has higher pressure than air over the ocean. This causes the air over the land to flow to the ocean, namely seasonal land breeze. When humid air rises over the ocean, it cools. This causes precipitation over the oceans. Cool air then flows toward the land to complete a closed circulation (Fig. 2.2b).

Similar rainfall occurs when moist ocean air is lifted upwards by mountains, surface heating, with convergence at the surface and divergence aloft. It should be noted that this circulation has a seasonal cycle but its spatial scale diversifies from a small island to peninsular and continental scales. It can appear in any latitudes even near the two polar areas on the Earth. In many places, this circulation is not reality of monsoon particularly in high latitudes. One can easily obtain the seasonal-cycle general circulation and its normal weather procedure caused by the land–ocean contrast locally.

(a)

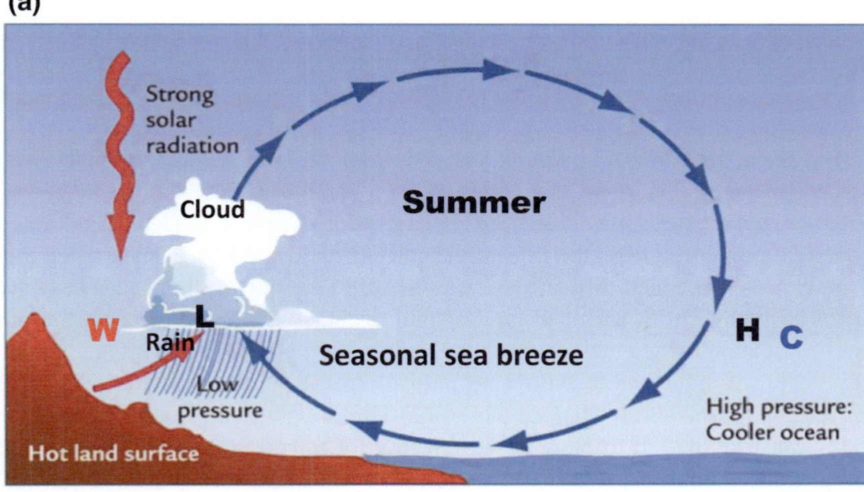

(b)

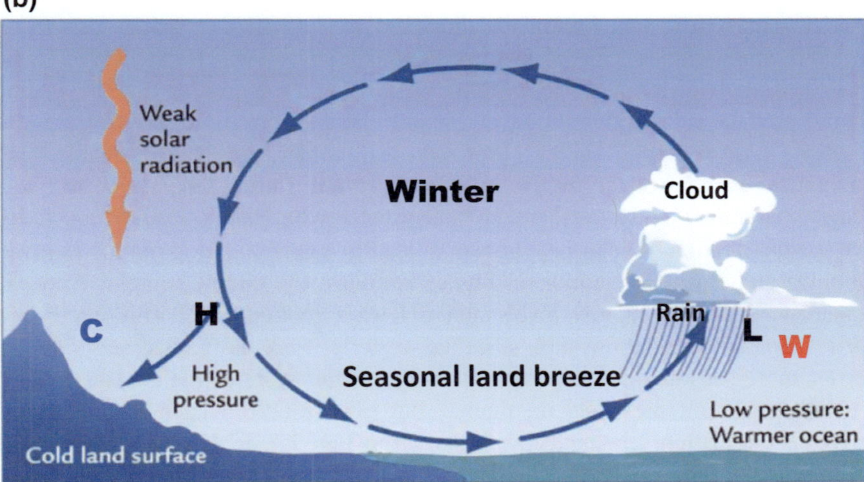

Fig. 2.2 **a** Intense solar heating of the land causes an in-land-up circulation (seasonal sea breeze) of moist air from the ocean (modified from http://www.atmosedu.com/Geol390/matter/WC6.jpg) and **b** weak solar radiation allows the land to cool off and creates a down- and -out circulation (seasonal land breeze) of cold dry air from the land. Modified from http://www.atmosedu.com/Geol390/matter/WC7.jpg

2.3 History of Trade Circulation

The diurnal-cycle and seasonal-cycle circulations occupy two timescales. On the other hand, one finds that the prevailing wind direction does not change with season in many places over the Earth. The trade winds are the prevailing pattern of easterly surface winds found in the tropics, in the lower troposphere near the Earth's equator. The trade winds blow predominantly from the northeast in the Northern Hemisphere (NH) and from the southeast in the Southern Hemisphere (SH). Historically, the trade winds help captains of sailing ships to cross the world's oceans for centuries, and enable the expansion of European empire into America and the trade routes established across the Atlantic and Pacific oceans.

The term trade wind originally derived from the late Middle English word "trade," meaning "path" or "track" (Braham et al. 2001). The Portuguese recognized the importance of the trade winds in navigation over both the North and South Atlantic Ocean as early as the fifteenth century (Muelder 2007). They learned that in order to reach South Africa, they needed to go far out in the ocean, head for Brazil at around 30°S, and then go eastward again. In the Pacific Ocean, the full wind circulation, which includes both the trade wind (easterlies) and higher latitude westerlies, was unknown to Europeans until Andres de Urdaneta's voyage in 1565 (Hayes 2001). The ancient captain of a sailing ship sought a course along which the winds can be expected to blow in the direction of travel. It is clear that the trade wind has been applied in sailing across an oceanic basin for many centuries.

Except some captains collected enough information to create prevailing wind and current charts, meteorologists attempted to explain the spatial pattern or structure of circulation. One of the most famous scientists was Hadley (1735), who explained the trade winds and prevailing westerly winds by noting that the heating should produce a direct meridional cell in each hemisphere (Lorenz 1967). Prior to Hadley's time, Edmund Halley (1686), an astronomer, presented a detailed and methodical account of the trade winds as observed in three separate oceans, and sought a common cause for them. In concordance with Halley, Hadley concluded that the distribution of solar heating would lead to a general rising motion in lower latitudes and a sinking motion in higher latitudes, the circuit completed by the equatorward motion at low levels and poleward motion aloft. This circulation carries more angular momentum, since the westerly winds aloft are stronger. It also carries more sensible heat and internal energy, if the stratification is stable.

Hadley himself presented no figures, but Lorenz (1967) later introduced the Hadley's circulation schematically as shown in Fig. 2.3. In Hadley's picture, the horizontal transports needed to satisfy the balance requirements that are accomplished by the simplest possible mechanism: A meridional circulation where the uniform poleward current at one elevation carries a different amount of each transported property from the uniform equatorward current at another elevation. Near the surface, northeast trade winds and southeast trade winds are clearly pictured in the tropical regions of the two hemispheres. The surface northwesterly winds in the NH and the surface southwesterly winds in the SH show an unstable observation that they originate from the polar highs. Therefore, this picture has not entirely described the spatial dynamical relation between surface winds and meridional circulations. In the NH, the surface northwesterly winds proposed by Hadley began to cast some doubt due to growing evidence that the prevailing westerly winds in the NH tended to blow from somewhat southwesterly rather than northwesterly winds.

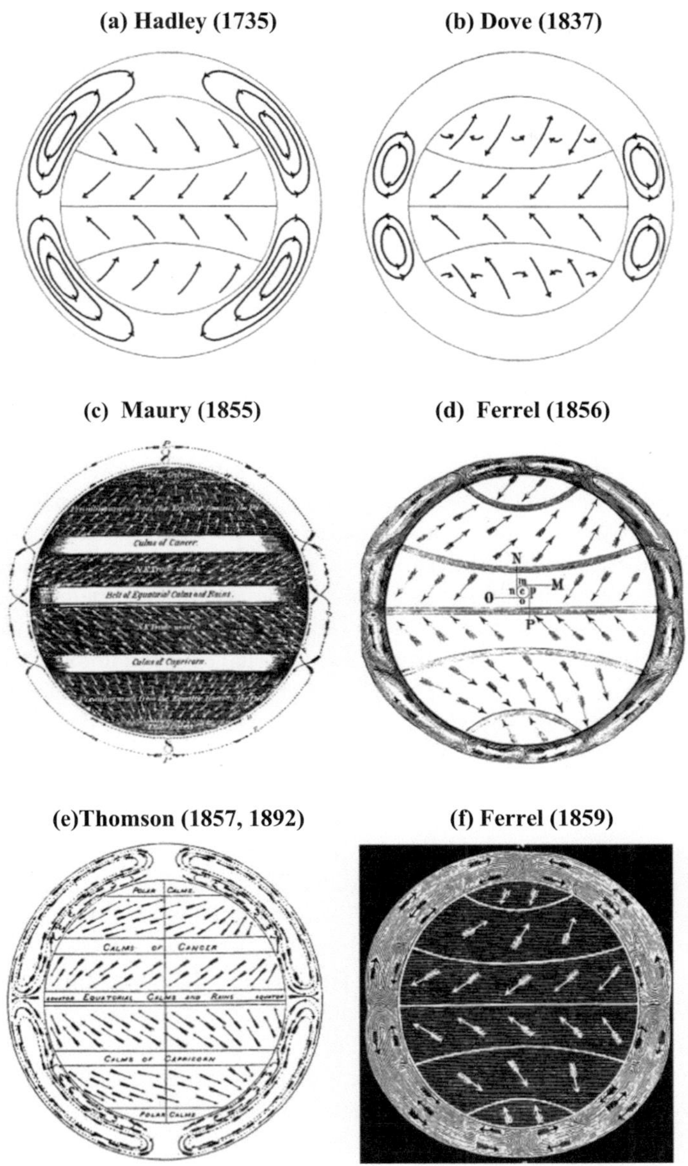

Fig. 2.3 Schematic representations of the general circulation model of the atmosphere as envisioned by **a** Hadley (1735), **b** Dove (1837), **c** Maury (1855), **d** Ferrel (1856), **e** Thomson (1857, 1892), and **f** Ferrel (1859). Sourced from Lorenz (1967)

After Hadley's work in 1735, similar explanation was also proposed by other savants during more than half century because the scientific community was generally unaware of this work. Figure 2.3a pictures a single circulation model from the equator to the high latitudes. Large-scale eddies such as cyclones are absent in this picture because eddies influence the zonally averaged motion by transporting some property from one latitude band to another. One hundred years later, Dove (1837) modified the previous model (Fig. 2.3b). He made two modifications on Hadley's explanation. One is that southwesterly winds and northeasterly winds vary with moving eddies in middle latitudes. Another is that the single circulation shrinks within tropics. It implies that a vertical circulation exists in meridional section in the tropics while transient southwesterly winds and northeasterly winds with moving eddies are horizontally active in middle latitudes. In his southwesterly winds and northeasterly winds, the effect of Earth rotation contributed by Coriolis in 1835 is visible.

Matthew Maury, a naval officer, was one of the most diligent compliers of weather observations. He charted of the winds over the oceans and began to consider the reduction in the normal sailing times between distant points. About 120 years after Hadley, Maury (1855) proposed a model shown in Fig. 2.3c. It followed Dove's circulation and trade winds in the tropics and proposed a second meridional circulation in the vertical direction over the mid-high latitudes. The new cell is an opposite current beside Dove's circulation. The effect of the Earth rotation was well considered in his scheme where trade winds were below the tropical meridional circulation and southwesterly winds were under the second circulation. Therefore, there were two cells: A direct cell like Hadley's within the tropics and an indirect cell in higher latitudes. This picture correctly showed that the two direct cells in the tropics rose in the doldrums along the equator while the direct cell and indirect cell combined to form a sinking flow in the horse latitudes. However, the indirect cell had not been explained.

Maury's book is extremely readable. It became rather popular and instrumental in initiating some of the more rational theories which were to follow. A school teacher, William Ferrel, was one of those readers. He first learned that normal pressure was not uniform over the Earth's surface, but higher in the horse latitudes, and lower in the doldrums and especially in the two polar regions. He disagreed with some of Maury's ideas, particularly the crossing of the meridional currents. In the following year, Ferrel came forth with a theory of his own and proposed a spatial circulation pattern as shown in Fig. 2.3d. The circulation which he envisioned was somewhat like Maury's, except that there were now three cells in each hemisphere, which he felt were demanded by the observations. Unlike Maury, however, Ferrel believed that he could present a complete explanation for the three cells. Ferrel's great contribution was the introduction of a "new" force, the meridional component of the Coriolis force acting on winds. He definitely believed that proper consideration of this new force would account for the previously unexplained features not only of the general circulation but also of cyclones and smaller disturbances as well. Ferrel was the first person who presented to the meteorological world a correct account of the Coriolis force, a quantitative description of the geostrophic wind, and a partial explanation for its occurrence.

In the spatial circulation pattern of Ferrel's model, the third cell was respectively located in two polar regions. This should be a direct cell and its surface winds should be northeasterly flow due to the Earth's rotation or the Coriolis effect. Two surface convergence zones were, respectively, located along the equator and in high latitude while a surface high zone was in the horse latitudes. Ferrel (1859) revised his meridional cells by two closed currents vertically located in two layers but the indirect cell limited in the lower layer of middle latitudes (Fig. 2.3f). The revised pattern may be influenced by Thomson (1857)'s model (Fig. 2.3e), which can also explain the two northeasterly zones, respectively, in the tropics and two polar regions while a southwesterly zone in NH middle latitudes. Ferrel's original version has been popular till nowadays, but his revision has not been accepted.

2.4 Modern General Circulation

In the Ferrel era, the general circulation was identified using mainly theoretical speculation, with less observations. In that era, eddies were only considered by Dove (1837) in high latitudes of his circulation picture but they did not exist in the Ferrel (1856)'s model. Since the early twentieth century, many observations in high latitudes can be obtained. A school group led by V. Bjerknes particularly focused on the study of observed cyclones in NH middle latitudes. Then, J. Bjerknes (1919), the son of V. Bjerknes, identified observational cyclones as waves on the polar frontal surface. These cyclones, as some disturbances growing upon a preexisting flow pattern, required a consideration of the preexisting pattern itself. This preexisting pattern is the general circulation. In the discussion of the general circulation, V. Bjerknes (1921) noted the difficulties in deducing a zonally symmetric circulation from a pure theory, and based his picture upon the combination of theory and observation. His theoretical result is illustrated in Fig. 2.4a. Then, he noted that this circulation appeared to be unstable, and that cyclones should develop, ultimately leading to the distribution of asymmetric circulation, as shown in Fig. 2.4b. For the symmetric scheme, he described that on each hemisphere there are four circulations, running as toothed wheels, the circulations of the trades, the circulation of the temperate zone, the circulation of the easterly polar zone, and the circulation of the theoretically introduced westerly polar zone. The first and the third of these circulations represent thermodynamically direct cells, in which the motion is maintained by heat energy. But the temperate cell is an indirect one, by which kinetic energy is transformed into heat, and this is also the case for the westerly polar zone. In his symmetric scheme (Fig. 2.4a), the fourth cell is the smallest one, so its westerly polar zone was not plotted.

The scheme of symmetric circulations gives two zones with descending motion, where very limited precipitation should be expected, namely a zone around the pole, and a zone near subtropical calms; and further two zones with ascending motion and abundant precipitation, one along the equator and one along the polar front, situated on the polar side of it. Finally, a third theoretical zone of precipitation should be at the pole itself (Fig. 2.4a).

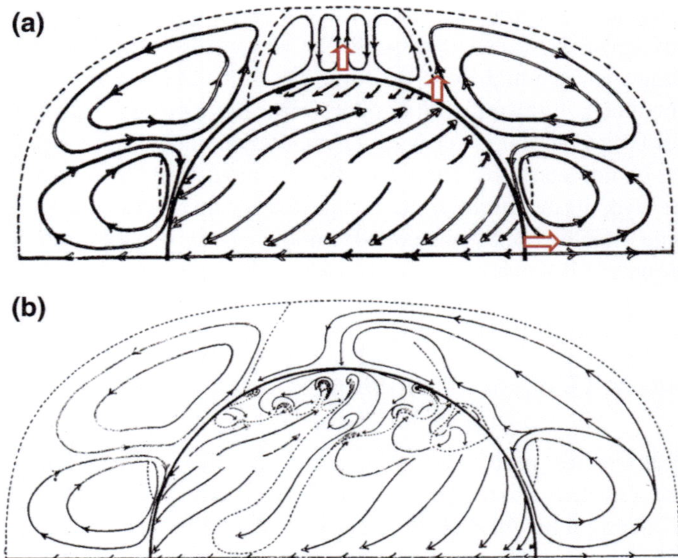

Fig. 2.4 **a** Symmetrical and **b** asymmetrical circulation schemes proposed by Bjerknes (1921) and sourced from Lorenz (1967). In **a** three zones of ascending motion with abundant precipitation are indicated by the *red arrow*. The *shading* in **b** is rain areas located in cyclone family. Sourced and modified from Lorenz (1967)

The first direct cell in the tropics stably exists in symmetric and asymmetric schemes. However, many transient eddies are active in mid-high latitudes and make other three meridional cells unstable (Fig. 2.4b). Cyclones and anticyclones are introduced in proper places as essential links in the mechanism of circulation, drawing the supply of cold air from the polar circulation and the supply of warm air from the inversely going temperate circulation. We now understand that the symmetric cells occupy their climatic features while the asymmetric scheme shows the transient systems as horizontal vortices and vertical stream lines. It is clear that Fig. 2.4a is a climatic chart and Fig. 2.4b is a total-based field chart. In Fig. 2.4b, two cyclone families have been plotted in mid-high latitudes.

In the last part of Bjerknes (1921), he gave a detailed description of the two circulations in the polar region. The two will be mixed and the stronger one will gain. It showed that the two polar circulations were combined into one, which has the direction of the thermodynamic tendency. But the effect will to some extent be checked by the tendency toward the formation of the polar cyclone. He said that "we have introduced it to remind the reader of this checking effect." We all know that the polar cyclone does not exist in daily synoptic chart because the diurnal and seasonal general circulations and transient eddies caused by sea–land contrast coexist.

The fourth cell in the polar region is seldom mentioned in the literature. One reason is that it is not important to daily weather and climate in mid-low latitudes.

Another is the lack of observations to prove its existence. Beyond 80°N, the historical radiosonde observations are sparse with about five stations from the beginning of record through 1987 (Kahl et al. 1992). To examine the intensity and extent of meridional cells, stream line and stream function (or mass stream function) were used to indicate the existence of meridional cells in previous studies. Some calculation covered the 60 latitudes (Peixoto and Oort 1992). Reanalysis products are available since the late twentieth century (Kalnay et al. 1996). Using the NCEP/NCAR reanalysis product, Li (2001) calculated the zonally averaged vertical velocity in the pressure-latitude section. As indicated by Bjerknes (1921), there are three ascending flows in the NH along the equatorial zone, the polar frontal zone and near the poles. Near the North Pole, the center of ascending motion is located round 850 hPa while the center near the South Pole is at upper troposphere in Li's (2001) atlas.

2.5 The Arctic Cell and the Four-Cell Model[1]

The traditional three-cell model was first proposed by Ferrel (1856) and has been widely accepted in the past century. The model describes that the Hadley cell ranges from the equator to about 30°N latitude, the Polar cell covers from about 60°N to the poles, and the Ferrel cell exists between them in the NH. Ferrel believed that the three meridional cells are equally crossing about 30 latitudes in each hemisphere. However, the Polar cell, in theory, cannot cross latitudes as many as the Hadley cell does in the lower latitudes because a larger decrease in the rotational radius will cause a larger increase in westerly velocity aloft according to the conservation of angular momentum (Ahrens 2012).

Recently, Qian et al. (2015a) proposed a fourth meridional cell, named the Arctic cell, based on the analyses of meridional-vertical section flow and meridional-mass stream function. The Arctic cell was found existing in the troposphere to the north of 80°N, and its intensity is very weak compared with other three cells. Due to the sparse long-term atmospheric radiosonde observations in the Arctic troposphere (Kahl et al. 1992), researches on the Arctic cell by observational datasets, climate model simulations as well as visualized methods are few. Therefore, more analyses based on different datasets and simulations are required to prove the existence of the Arctic cell.

Figure 2.5 shows annual mean zonally averaged meridional cells depicted by vertical sections of flow and zonal velocities averaged over 30 years (1981–2010) using four different reanalysis products and the CESM1-CAM5 model simulation. The meridional flow is simply drawn using the monthly mean wind components $v(\varphi, p, t)$ and $\omega(\varphi, p, t)$. Five centers of meridional cells covering between 30°S and

[1]Modified article originally published in *Qian et al.* (2015a). *The Arctic and Polar cells act on the Arctic sea ice variation. Tellus A, 67: 27692.* Published with kind permission of © 2015 W. Qian et al. All Rights Reserved.

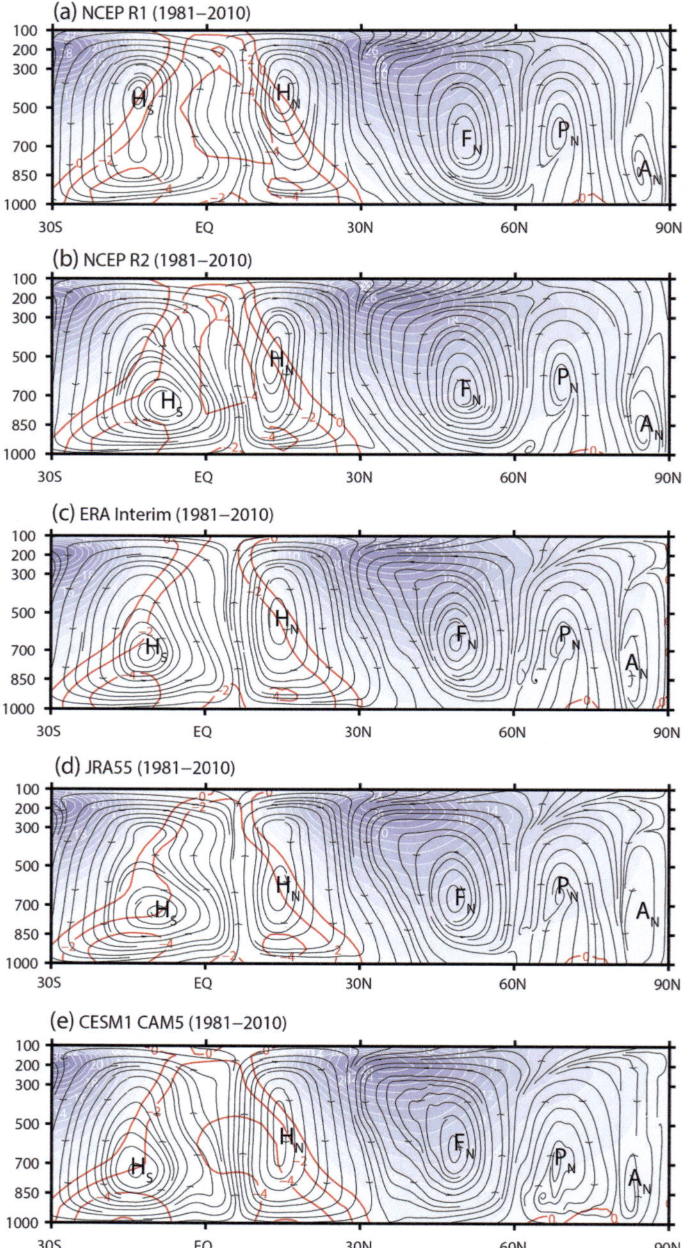
(a) NCEP R1 (1981–2010)
(b) NCEP R2 (1981–2010)
(c) ERA Interim (1981–2010)
(d) JRA55 (1981–2010)
(e) CESM1 CAM5 (1981–2010)
100
200
300
500
700
850
1000
30S
EQ
30N
60N
90N
H_S
H_N
F_N
P_N
A_N

◀ **Fig. 2.5** Five zonally averaged meridional cells described by the pressure-latitude section of flow and westerly (*shading*, 2 m s^{-1} interval) or easterly (*red line*, 2 m s^{-1} interval) averaged over 30 years (1981–2010) using the five reanalysis products: **a** NCEP R1, **b** NCEP R2, **c** ERA-Interim, **d** JRA-55, and **e** CESM1-CAM5. Symbols of H_N, F_N, P_N, and A_N indicate the Hadley, Ferrel, Polar, and Arctic cells in the Northern Hemisphere, while H_S indicates the Hadley cell in the Southern Hemisphere. Sourced from Wu (2016)

90°N are clearly obtained from each of the datasets in Fig. 2.5. The two Hadley cells are well defined from around 30°S to 30°N with an ascending flow in the tropics and two descending flows in the two subtropical zones. The northeasterly trade wind in the tropical NH and the southeasterly trade wind in the tropical SH are commonly observed in the tropical lower troposphere from five datasets. The flow in the NH is characterized by westerly jet with the maximum zonal velocity (a core) at about 35°N near the tropopause. A shallow layer northeasterly wind is centered below the Polar cell while a southwesterly wind is located below the Arctic cell near the polar region. In the NH, four meridional cells can be identified from the four reanalysis products and a model simulation. The fourth cell named by the Arctic cell is located beyond 80°N with its center at 850 hPa (Qian et al. 2015a).

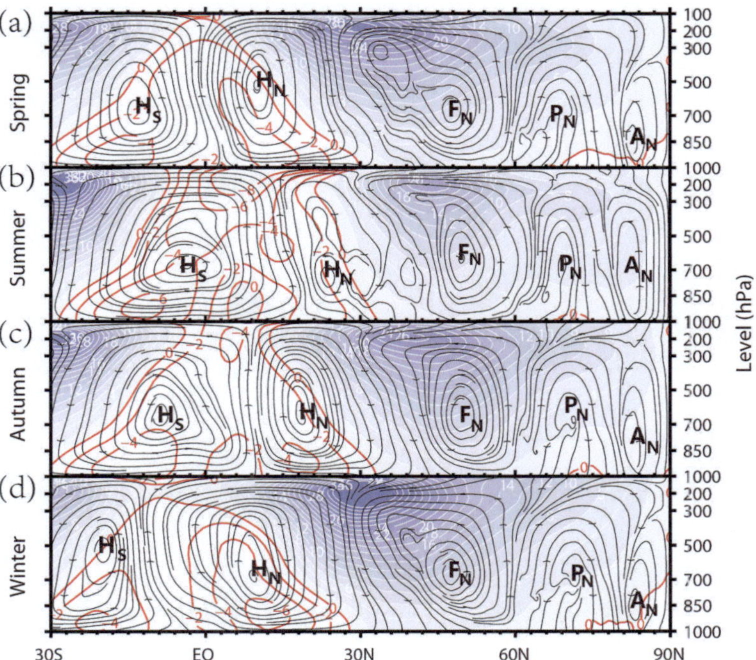

Fig. 2.6 Same as in Fig. 2.5 except for seasonal-mean flow and westerly (*shading*, 2 m s^{-1} interval) or easterly (*red line*, 2 m s^{-1} interval) averaged over 30 years (1981–2010) using the ERA-Interim product in the boreal **a** spring, **b** summer, **c** autumn, and **d** winter. Sourced from Qian et al. (2015a) © 2015 W. Qian et al., with permission

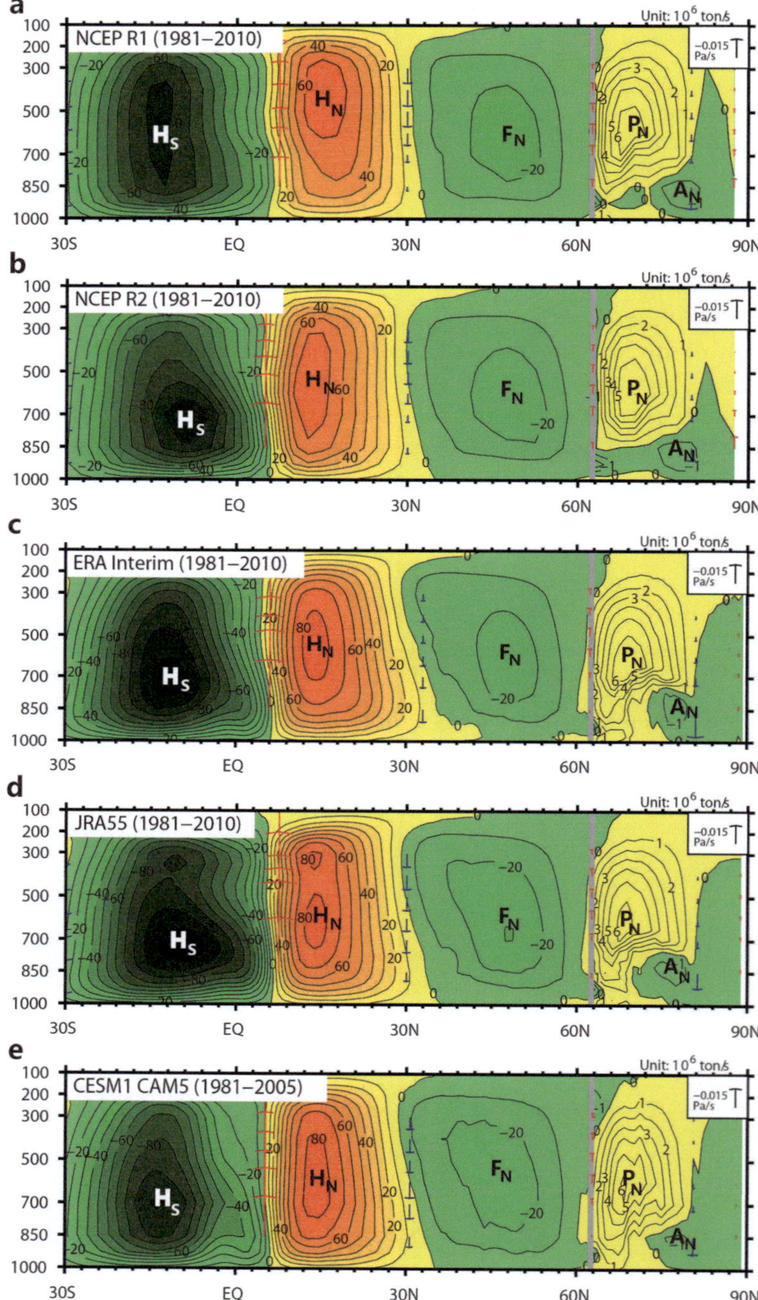

Fig. 2.7 Same as in Fig. 2.5 except for annual mean MSFs (*yellow-red shading* for positive and *green shading* for negative). The MSF interval is 25 × 10^6 tons^{-1} between 30°S–30°N, 5 × 10^6 ton s^{-1} between 30°N–62.25°N and 1 × 10^6 ton s^{-1} between 62.25°N–90°N. *Red* and *blue arrows* indicate ascending and descending pressure velocities (−0.015 Pa s^{-1}) respectively. Sourced from Wu (2016)

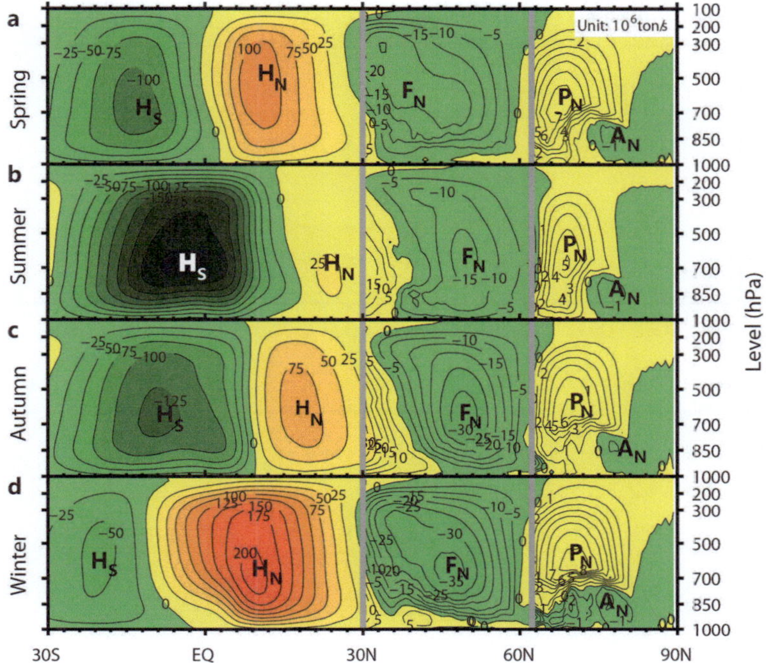

Fig. 2.8 Same as in Fig. 2.7 except of seasonal-mean MSFs using the ERA-Interim product in the boreal **a** spring, **b** summer, **c** autumn, and **d** winter. The MSF interval is 25×10^6 ton s^{-1} between 30°S–30°N, 5×10^6 ton s^{-1} between 30°N–62.25°N and 1×10^6 ton s^{-1} between 62.25°N–90°N

To examine the seasonal features, Fig. 2.6 shows vertical sections of climatic seasonal-mean flow and westerly (or easterly) averaged over 30 years (1981–2010) using the ERA-Interim product in the four boreal seasons. It is noted that the central locations of the four cells vary with different seasons. The centers of two Hadley cells have the largest seasonal migration. In the boreal spring, the Ferrel cell even splits into two in the NH, probably because the Ferrel cell is an indirect circulation caused mainly by eddies. The Polar and Arctic cells are always limited in the high latitudes between 60°N and 90°N. The easterly flow is stably observed in the tropical troposphere associated with the Hadley cell, but the westerly flow in the polar region is highly unstable because there are strong interactions between the meridional cells and the synoptic eddies, as well as the influences from the topographic, frictional and diabatic processes. The same result can also be illustrated from other reanalysis products and model simulations.

As we mentioned, while the flow clearly depicts the spatial structures and basic patterns of the four meridional cells in the NH, the mass stream function (MSF) can quantitatively illustrate the intensity of each cell. Five zonally averaged meridional cells illustrated by the vertical section of climatic annual mean MSFs averaged over 30 years (1981–2010) are shown in Fig. 2.7 using the four reanalysis products and one model simulation. Four meridional cells are commonly observed in the NH,

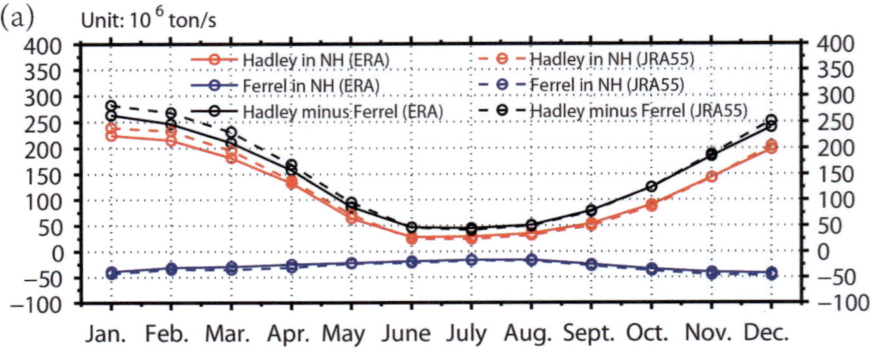

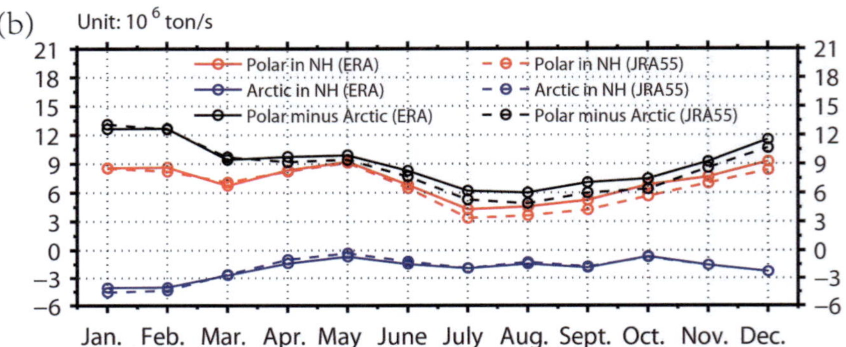

Fig. 2.9 Monthly mean intensities of the four cells in the Northern Hemisphere. The MSF intensities (*red* and *blue* lines, 10^6 ton s^{-1}) of **a** the Hadley and Ferrel cells and their differences (*black line*, 10^6 ton s^{-1}) based on ERA-Interim and JRA-55 reanalysis products; **b** same as **a** but for the Polar and Arctic cells. Sourced from Qian et al. (2015a) © 2015 W. Qian et al., with permission

with comparable central intensities and locations from all five datasets. The intensity ratios of the Ferrel, Polar, and Arctic cells to the Hadley cell are approximately 1:3, 1:13, and 1:80, respectively, while the intensity ratio of the Arctic cell to the Polar cell is approximately 1:6 (Qian et al. 2015a). The Arctic cell exists next to the Polar cell as a weak opposite circulation and is centered at 850 hPa and 80°N.

Five zonally averaged meridional cells illustrated by the vertical section of climatic seasonal-mean MSFs averaged over 30 years (1981–2010) are shown in Fig. 2.8 using the ERA-Interim product in the four boreal seasons. The central intensities and locations of two Hadley cells, as expected, vary with seasons. The Hadley cell in the NH has the minimal (maximal) MSF value 25×10^6 ton s^{-1} (200×10^6 ton s^{-1}) and the northernmost (southernmost) latitude in the boreal summer (winter). The central MSF value gradually reduces from the Ferrel cell to

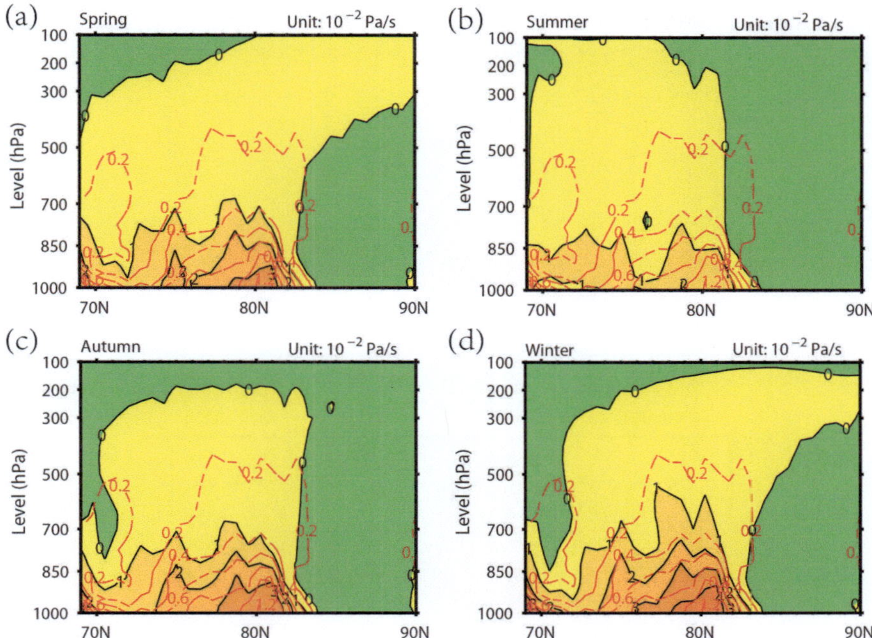

Fig. 2.10 Climatic seasonal-mean vertical velocity. Vertical and meridional sections of zonally averaged descending (*yellow*) and ascending (*green*) pressure velocities (*solid line*, 1×10^{-2} Pa s^{-1} interval) and standard deviation (*red-dashed line*, 0.2×10^{-2} Pa s^{-1} interval) between 69°N and 90°N and from 100 to 1000 hPa in the boreal **a** spring, **b** summer, **c** autumn, and **d** winter using a 30-year average (1981–2010) of the ERA-Interim product. Sourced from Qian et al. (2015a) © 2015 W. Qian et al., with permission

the Polar and Arctic cells, so the MSF interval is drawn as 25×10^6 ton s^{-1} between 30°S–30°N, 5×10^6 ton s^{-1} between 30°N–62.25°N and 1×10^6 ton s^{-1} between 62.25°N–90°N. It is noted that the Arctic cell is influenced by the Polar cell in terms of intensities and locations. The central location of the Arctic cell is generally lower than that of the Polar cell. In the boreal winter, when the intensity of the Polar cell reaches its maximum, however, the central location of the Arctic cell is entirely located beneath the Polar cell (Fig. 2.8d). In contrast, the Arctic cell stands alongside the Polar cell in summer when the Polar cell is relatively weak (Fig. 2.8b). Similar distributions of MSFs can also be observed from other reanalysis products and model simulations.

Two pairs of meridional cells in the NH are well defined using the two 4D-VAR ERA-Interim and JRA-55 reanalysis products in Fig. 2.9. The first pair is the Hadley and Ferrel cells, whose monthly mean intensities have a climatic seasonal cycle and opposite in phase (Fig. 2.9a). The MSF difference between the Hadley

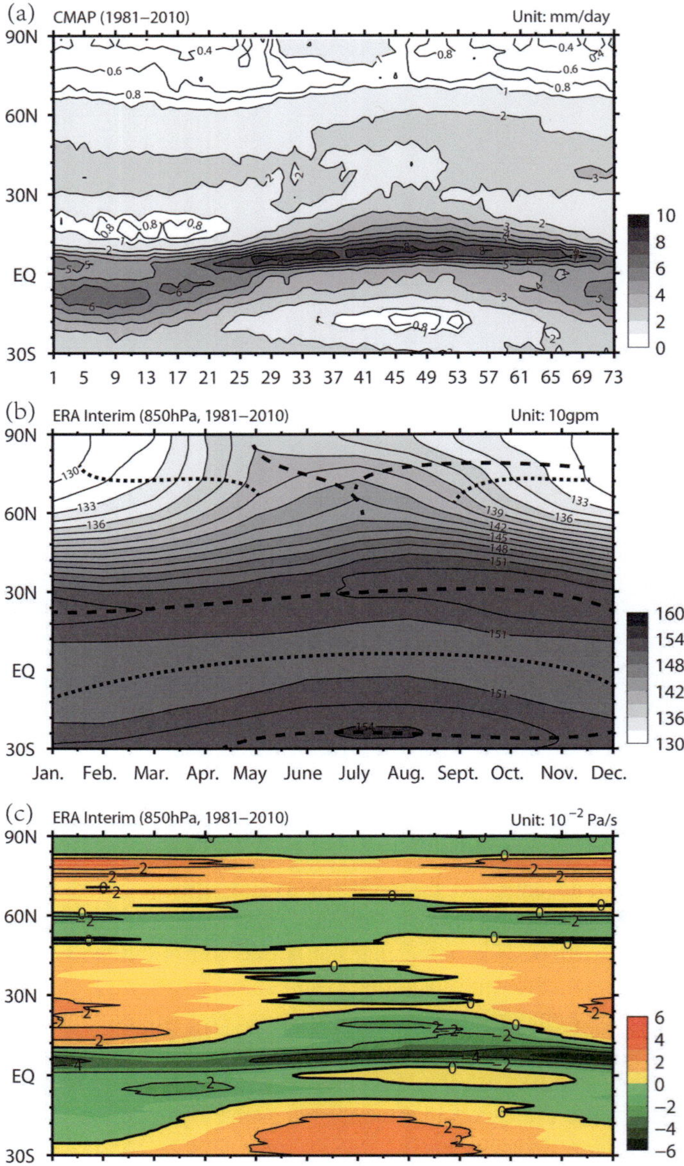
(a) CMAP (1981–2010)
Unit: mm/day
90N
60N
30N
EQ
30S
1 5 9 13 17 21 25 29 33 37 41 45 49 53 57 61 65 69 73
10 8 6 4 2 0
(b) ERA Interim (850hPa, 1981–2010)
Unit: 10gpm
160 154 148 142 136 130
Jan. Feb. Mar. Apr. May June July Aug. Sept. Oct. Nov. Dec.
(c) ERA Interim (850hPa, 1981–2010)
Unit: 10^{-2} Pa/s
6 4 2 0 −2 −4 −6
Jan. Feb. Mar. Apr. May June July Aug. Sept. Oct. Nov. Dec.

◀ **Fig. 2.11** Time-meridional (30°S–90°N) sections of **a** climatic pentad (1–73) precipitation (*gray shading*, 1 mm d^{-1} interval) using the enhanced CMAP dataset, **b** monthly mean geopotential height (*gray shading*, 30 gpm interval), and **c** vertical pressure velocity (*yellow-red* for positive and *green* for negative, 1×10^{-2} Pa s^{-1} interval) at 850 hPa using the ERA-Interim product averaged over 30-year (1981–2010). **a** The limited precipitation rate (<1 mm d^{-1}) along the subpolar zone separates two higher precipitation (>1 mm d^{-1}) zones respectively in the Arctic and middle latitudes. **b** Three height ridges (*dashed line*) separate a low-pressure center in the Arctic and two height troughs (*dotted line*) respectively in the equator and middle latitudes. **c** Two major descending zones (*red shading*) in the subpolar and the subtropical regions separate three ascending zones (*green shading*) in the Arctic, middle latitudes and tropical regions, respectively. Sourced from Qian et al. (2015a) © 2015 W. Qian et al., with permission

and Ferrel cells is maximum in January and minimum in July. Similarly, the intensity difference between the second pair (the Polar and Arctic cells) is also maximum in January and minimum in July (Fig. 2.9b).

In the symmetrical circulation scheme proposed by Bjerknes (1921), there should be a descending flow in the subpolar zone (70°–80°N) and an ascending flow in the polar region (Fig. 2.4a). Figure 2.10 shows the intensities of the descending and ascending flow branches in the NH high latitudes. The descending zone is concentrated from 69°N to 82°N while the ascending flow zone is located beyond 83°N. The strongest descending flow is climatologically observed in the boreal winter (Fig. 2.10d).

Two descending zones with limited precipitation and three ascending zones with plentiful precipitation speculated by Bjerknes (1921) in his scheme of symmetric circulations are also confirmed in Fig. 2.11. In the Arctic, higher precipitation rate is mainly concentrated in summer from pentads 29 to 45. During this period, the precipitation rate is more than 1 mm day^{-1} in the Arctic, while in the subpolar there is a dry zone with the precipitation rate less than 1 mm day^{-1} (Fig. 2.11a). In the Arctic, there is a year-round low center of climatic geopotential height at 850 hPa, whereas there is a high-pressure zone in the subpolar region (Fig. 2.11b). The low center in geopotential height and the higher precipitation rate in the Arctic are closely linked to an ascending flow region at 850 hPa (Fig. 2.11c). In the traditional three-cell model, the Polar cell ranges from 60°N to 90°N. In that case, the limited precipitation rate, the high-pressure system and the descending flow are expected to exist climatologically in the Arctic, which is inconsistent with the new results. Figure 2.11 gives additional evidences challenging the traditional three-cell model.

Theoretically, there should be southwesterly wind below the center of Arctic cell. Figure 2.12 shows climatic winds at 2 m level and surface air temperature in the NH high latitudes. On an average, there is southerly wind beyond 80°N. It means that the regional-mean southerly is climatologically larger than that of northerly. But as illustrated in Sect. 2.2, seasonal-cycle general circulation caused by sea–land contrast has a large deviation relative to the zonally averaged winds. Seasonally, there are three pairs of southerly and northerly centers. The strongest pair is caused by Greenland and the Greenland Sea. Other two pairs are associated with the two continents and seas in the high latitudes. In boreal summer, the surface air temperature is relatively higher beyond 80°N with southwesterly winds.

In a study of the frontal development, Bergeron (1928) found the necessity to introduce a model of the general circulation, so he proposed a three-cell meridional circulation, somewhat similar to the one which Ferrel introduced. It has been widely reproduced and it marked the beginning of the general acceptance of the three-cell model. The middle-latitude indirect cell has come to be known as the Ferrel cell since then. Figure 2.13 illustrates the traditional three-cell model and the new four-cell model. Their difference appears only in the polar region beyond 60°N, where an additional cell exists over the Arctic. Their climate effects should be completely different in the polar region.

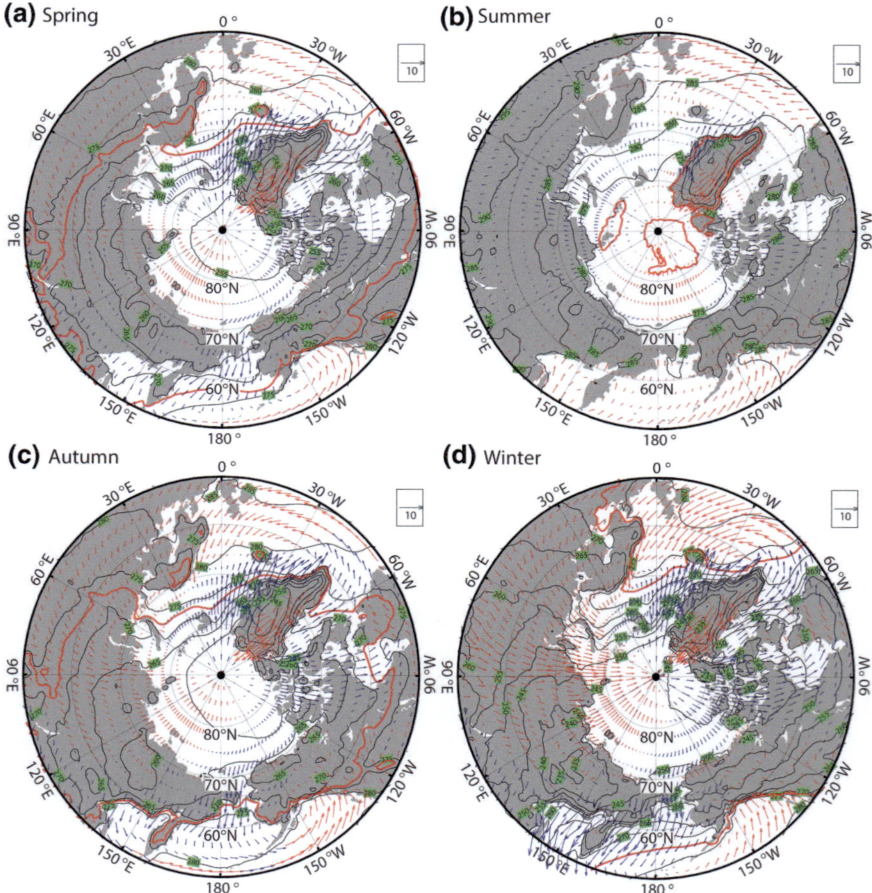

Fig. 2.12 Climatic winds (m/s) at 2 m level and surface air temperature (K, the *red line* is 273 K) of **a** spring, **b** Summer, **c** Autumn, and **d** Winter in the NH polar region using a 30-year average (1981–2010) of the ERA-Interim product. *Gray shading* denotes land area. Sourced from Qian et al. (2016) © 2016 John Wiley & Sons, Inc., with permission

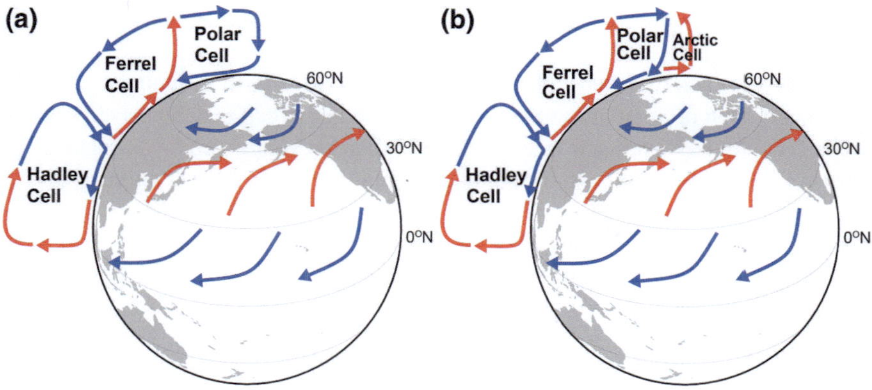

Fig. 2.13 **a** The traditional three-cell model and **b** the new four-cell model

2.6 Eddy Heat Flux and Eddy Momentum Flux[2]

In the NH, there should be two direct cells and two indirect cells along the meridional direction. The two indirect cells are associated with eddies. We here illustrate the eddy heat flux and eddy momentum flux. Monthly mean wind vector $\vec{V}_m(\lambda,\varphi,p,t)=\vec{i}u_m(\lambda,\varphi,p,t)+\vec{j}v_m(\lambda,\varphi,p,t)+\vec{k}\omega_m(\lambda,\varphi,p,t)$ consists of three components: zonal wind velocity u, meridional wind velocity v, and vertical pressure velocity ω (positive downward) or z-vertical velocity $\vec{k}\omega_m(\lambda,\varphi,p,t)$ (positive upward). The zonally averaged monthly mean climatic wind vector $\vec{V}_m(\varphi,p)$ at a specific month of a year t is calculated by

$$\left[\tilde{\vec{V}}_m(\varphi,p)\right]=\sum_{t=1}^{N}\sum_{\lambda=1}^{K}\vec{V}_m(\lambda,\varphi,p,t)/(K\times N), \tag{2.6.1}$$

where N = 30 years from 1981 to 2010, K is the total grid number along a zonal circle, and m represents the month. Similarly, the zonally averaged seasonal-mean and annual mean climatic wind vectors can also be calculated from the monthly mean climatic wind vector. The zonally averaged deviation of climatic monthly mean wind vector is

$$\tilde{\vec{V}}^*_m(\lambda,\varphi,p)=\tilde{\vec{V}}_m(\lambda,\varphi,p)-\left[\tilde{\vec{V}}_m(\varphi,p)\right]. \tag{2.6.2}$$

[2]Modified article originally published in *Qian et al.* (2015a). *The Arctic and Polar cells act on the Arctic sea ice variation. Tellus A, 67: 27692.* Published with kind permission of © 2015 W. Qian et al. All Rights Reserved.

The two components, $\tilde{u}_m^*$ and $\tilde{v}_m^*$, can produce the stationary meridional eddy transport $[\tilde{v}_m^* \tilde{u}_m^*]$.

It is well-known that the Hadley cell is thermally driven and its boundary is set by the conservation of angular momentum, while the Ferrel cell is eddy driven (Holton 2004). The zonally averaged momentum and thermodynamic energy equations for quasi-geostrophic motions on the middle latitude β plane can be written as

$$\frac{\partial[\tilde{u}]}{\partial t} - f_0[\tilde{v}] = -\frac{\partial[\tilde{u}^* \tilde{v}^*]}{\partial y} + [\tilde{X}], \tag{2.6.3}$$

$$\frac{\partial[\tilde{T}]}{\partial t} + N^2 H R^{-1}[\tilde{w}] = -\frac{\partial[\tilde{v}^* \tilde{T}^*]}{\partial y} + \frac{[\tilde{J}]}{c_p}, \tag{2.6.4}$$

where $[\tilde{u}]$, $[\tilde{v}]$ and $[\tilde{w}]$ are the zonally averaged climatic westerly, southerly and z-vertical components of velocity, $[\tilde{T}]$ is the zonally averaged climatic temperature, $[\tilde{v}_m^* \tilde{T}_m^*]$ and $[\tilde{v}^* \tilde{u}^*]$ are the zonally averaged stationary eddy heat fluxes and eddy momentum fluxes, $[\tilde{X}]$ and $[\tilde{J}]$ are the zonally averaged climatic turbulent drag and diabatic heating rate, N is the buoyancy frequency defined by $N^2 \equiv \frac{R}{H}\left(\frac{\kappa T_0}{H} + \frac{\mathrm{d}T_0}{\mathrm{d}z}\right)$, in which H is scale height, R is the gas constant of dry air, and c_p is the specific heat of constant pressure.

If $[\tilde{u}]$ and $[\tilde{T}]$ are steady, namely $\frac{\partial[\tilde{u}]}{\partial t} = \frac{\partial[\tilde{T}]}{\partial t} = 0$, the zonally averaged climatic direct circulation ($[\tilde{v}]_d$ and $[\tilde{w}]_d$) is mainly forced by the zonally averaged climatic turbulent drag and diabatic heating rate

$$-f_0[\tilde{v}]_d = [\tilde{X}], \tag{2.6.5}$$

$$N^2 H R^{-1}[\tilde{w}]_d = \frac{[\tilde{J}]}{c_p}, \tag{2.6.6}$$

while the indirect circulation ($[\tilde{v}]_i$ and $[\tilde{w}]_i$) is mainly driven by the zonally averaged stationary eddy heat fluxes and momentum fluxes

$$N^2 H R^{-1}[\tilde{w}]_i = -\frac{\partial[\tilde{v}^* \tilde{T}^*]}{\partial y} \tag{2.6.7}$$

$$f_0[\tilde{v}]_i = \frac{\partial[\tilde{u}^* \tilde{v}^*]}{\partial y}. \tag{2.6.8}$$

The two terms, $[\tilde{v}^* \tilde{T}^*]$ and $[\tilde{u}^* \tilde{v}^*]$, can be calculated using reanalysis products while $[\tilde{X}]$ and $[\tilde{J}]$ are difficult to obtain directly. The four terms together contribute to form the total circulation.

Using four reanalysis products and one model simulation, we have revealed that there are four meridional cells in the NH from individual monthly mean, climatic seasonal-mean flows, and MSFs (Qian et al. 2015a). The dynamical explanations and additional evidence will be given here to support the existence of the Arctic cell. It is noted that the Hadley, Ferrel and Polar cells have complex structures. They may split, shift, and merge in a complicated manner. Near the surface, topographic, frictional, and diabatic processes also affect these circulations.

The number of meridional cells existing in the NH troposphere can be illustrated by MSFs. The tropical Hadley cell is dynamically and thermally driven by the zonally averaged turbulent drag force $[\overline{X}]$ and diabatic heating rate $[\tilde{J}]$ while the Ferrel cell is believed to be contributed mainly by the zonally averaged stationary eddy heat flux $[\tilde{v}^*\tilde{T}^*]$ and eddy momentum flux $[\tilde{u}^*\tilde{v}^*]$ (Holton 2004). Figure 2.14 shows the northward eddy heat flux $[\tilde{v}^*\tilde{T}^*]$ calculated from the ERA-Interim product in October, November, December, and January. As expected, positive and strong centers of the eddy heat flux exist in the middle latitudes (30°N–60°N) while weak and negative centers are found in the tropics. From the negative weak center (H) to the positive strong center (F), $-\frac{\partial[\tilde{v}^*\tilde{T}^*]}{\partial y}<0$, i.e. $[\tilde{w}]_i<0$, so there is a descending branch (blue dashed arrow) between them. From the positive strong center (F) to a negative weak center (P), there is an ascending branch (red-dashed arrow) because $-\frac{\partial[\tilde{v}^*\tilde{T}^*]}{\partial y}>0$, i.e. $[\tilde{w}]_i>0$. Thus, if the Hadley cell forms in the

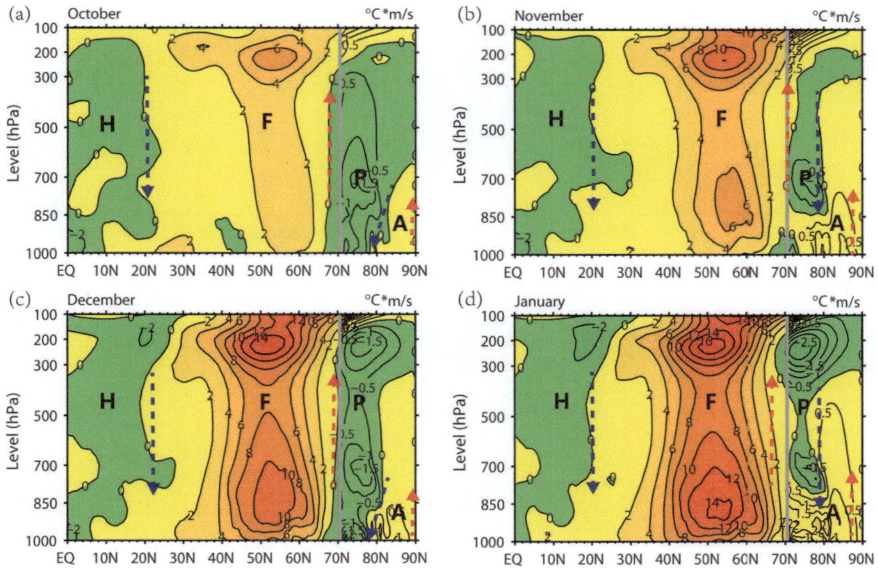

Fig. 2.14 Northward eddy heat flux ($[\tilde{v}^*\tilde{T}^*]$, *yellow-red shading* for positive and *green* for negative, 2 °C m s^{-1} interval in mid-low latitudes and 0.5 °C m s^{-1} interval in the Arctic) calculated from a 30-year (1981–2010) average of ERA-Interim product in the NH in **a** October, **b** November, **c** December, and **d** January. Letters "H", "F", "P" and "A" are central locations of eddy circulation near the Hadley, Ferrel, Polar and Arctic cells respectively. Sourced from Qian et al. (2015a) © 2015 W. Qian et al., with permission

tropics, the reversed and closed loop next to it should be the Ferrel cell in the middle latitudes. It can be identified that the Ferrel cell is mainly contributed by the stationary eddy heat fluxes.

In previous studies, no detailed comparison between the high and mid-low latitudes stationary eddy heat fluxes was made. As shown by Fig. 2.14, there are two pairs of negative and positive centers of eddy heat flux respectively in mid-low latitudes and high latitudes. The first pair is indicated by "H" and "F" while the second pair is denoted by "P" and "A," although the latter is weaker and smaller than the former. In Fig. 2.14, these four centers of eddy flux are near the centers of the Hadley, Ferrel, Polar, and Arctic cells.

Similarly, Fig. 2.15 shows the northward stationary eddy momentum flux $[\tilde{v}^*\tilde{u}^*]$ in October, November, December, and January in the NH. From an ascending flow (red-dashed arrow) to a descending flow (blue dashed arrow), $\frac{\partial[\tilde{u}^*\tilde{v}^*]}{\partial y} > 0$, i.e. $[\tilde{v}]_i > 0$, so there is a southerly wind component (white dashed arrow). Otherwise, there should be a northerly component. There are also two pairs of cells described by the eddy momentum flux, respectively, in mid-low and high latitudes. The first pair indicated by "H" and "F" is, respectively, near the centers of the Hadley cell

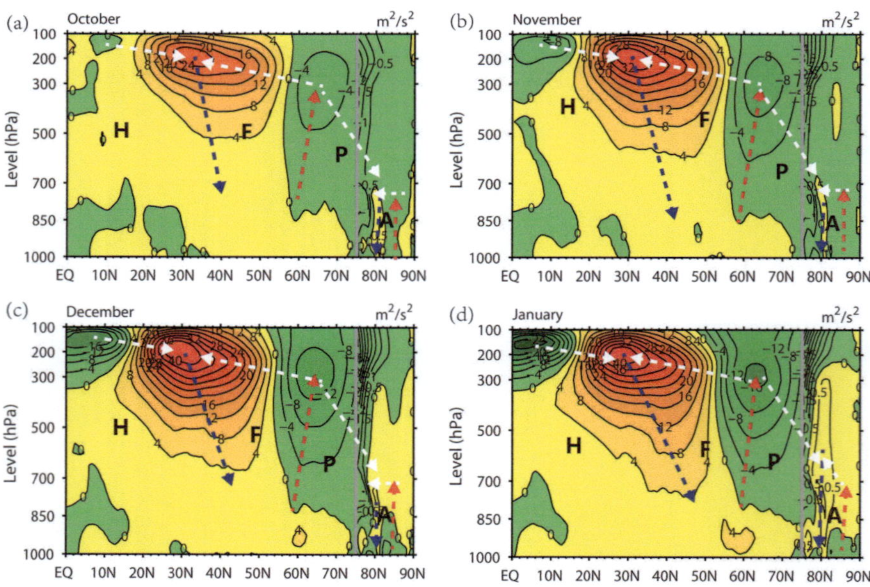

Fig. 2.15 Same as in Fig. 2.14 except for the northward eddy momentum flux $[\tilde{v}^*\tilde{u}^*]$ (*yellow-red shading* for positive and *green* for negative, 4 m^2 s^{-2} interval in mid-low latitudes and 0.5 m^2 s^{-2} interval in the Arctic) calculated from a 30-year (1981–2010) average of ERA-Interim product in the NH in **a** October, **b** November, **c** December, and **d** January. Sourced from Qian et al. (2015a) © 2015 W. Qian et al., with permission

and the Ferrel cell, while the second pair denoted by "P" and "A" is near the centers of the Polar cell and the Arctic cell. Similar distributions as shown in Figs. 2.14 and 2.15 can also be observed from the other three reanalysis products and the model simulation.

2.7 Global Eight Meridional Cells

Theoretically, there should be eight meridional cells globally. Using the four reanalysis products and a model simulation, Fig. 2.16 shows zonally averaged flow and westerly velocity climatologically during 1981–2010. For the westerly, all data show that there is only one center at the upper troposphere near 35°N while two centers exist, respectively, in 30°S and 50°S. It shows that the subtropical high-level jet stream and the polar frontal jet stream can be separated from the annual mean climatology in the SH, but they are not for the NH. The easterly is weak and unsteady below the Polar cell in the NH but is relative strong in the high latitudes of SH. We particularly concern the meridional cells in the SH. The two NCEP reanalysis products show that there are also four meridional cells in the SH. The fourth cell over the Antarctic has been referred to as the Antarctic cell (Qian et al. 2015b). The center of the Antarctic cell is located in the upper troposphere while the center of the Arctic cell is in the lower troposphere (Fig. 2.16a, b). But the Antarctic cell is not so clear from the analysis of other two products and model simulation (Fig. 2.16c–e).

The MSF analysis from all the five datasets shows that the intensities of the traditional three (Hadley, Ferrel and Polar) cells in the SH are stronger than those in the NH (Fig. 2.17). That may be the reason why the intensity of the Antarctic cell is so weak. This is because the Polar cell in the SH is largely stronger than the Antarctic cell. The intensity ratio of the Antarctic cell to the Polar cell in the SH is approximately 1:12 (Qian et al. 2015b).

Under the Antarctic cell, there should be northerly with convergence. Figure 2.18 shows climatic winds at 2 m level and surface air temperature in the SH high latitudes. The northerly is averagely larger than that of southerly wind beyond 80°S. The Antarctic is a continent surrounded by several peninsulas so the seasonal-cycle sea–land heating contrast forms zonal climatic deviation. Therefore, several pairs of northerly and southerly contrast exist over the Antarctic. The sea ice extent has seasonal variability with temperature.

Although we have not found a closed Antarctic cell from its annual mean climatology (Fig. 2.16c), it can be observed in an individual month. Figure 2.19 gives a profile of meridional cells in September 2004 in the SH. The four cells including the Antarctic cell are clearly observed in the SH.

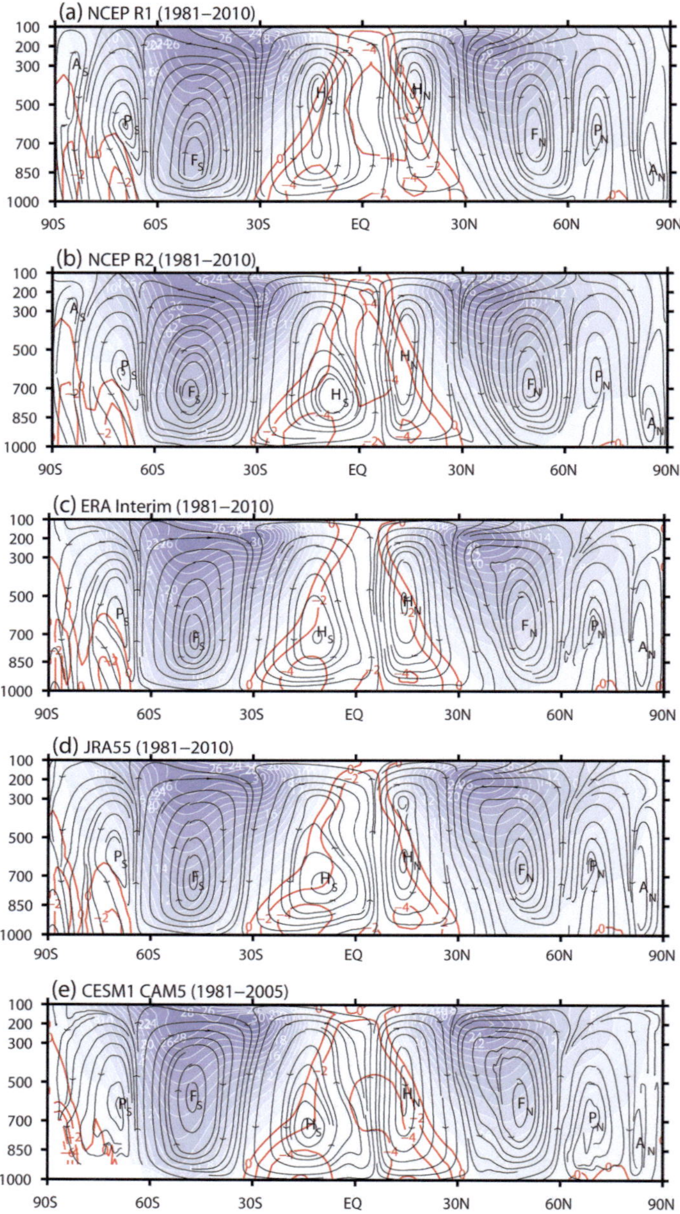

Fig. 2.16 Same as in Fig. 2.5 except flow and zonal velocities between the two poles globally

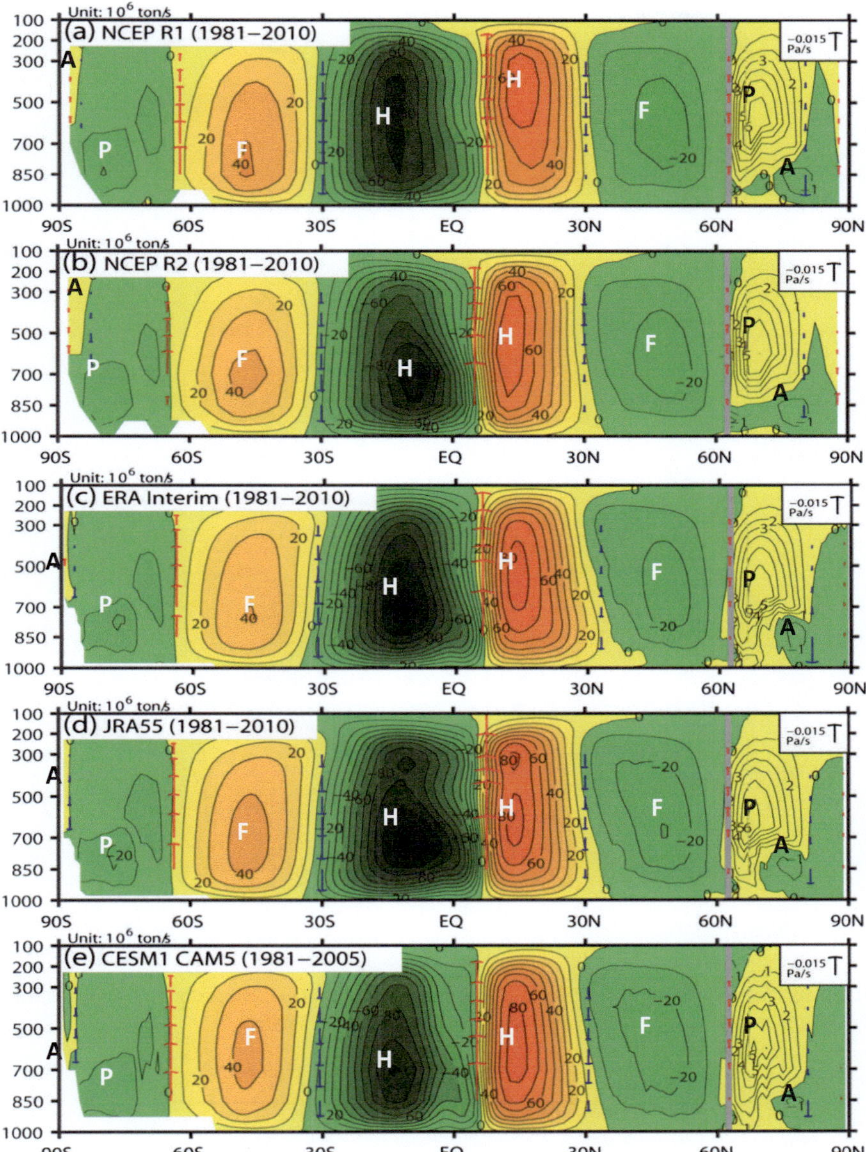

Fig. 2.17 Same as in Fig. 2.7 except between the two poles globally. Symbols "H", "F", "P" and "A" indicate the central locations of the Hadley, Ferrel, Polar and Arctic (Antarctic) cells. The MSF interval is 10×10^6 ton s^{-1} between 90°S–60°N while it is reduced to 1×10^6 ton s^{-1} between 62.5°N–90°N. *Red* and *blue arrows* indicate ascending and descending pressure velocities (-0.015 Pa s^{-1}), respectively

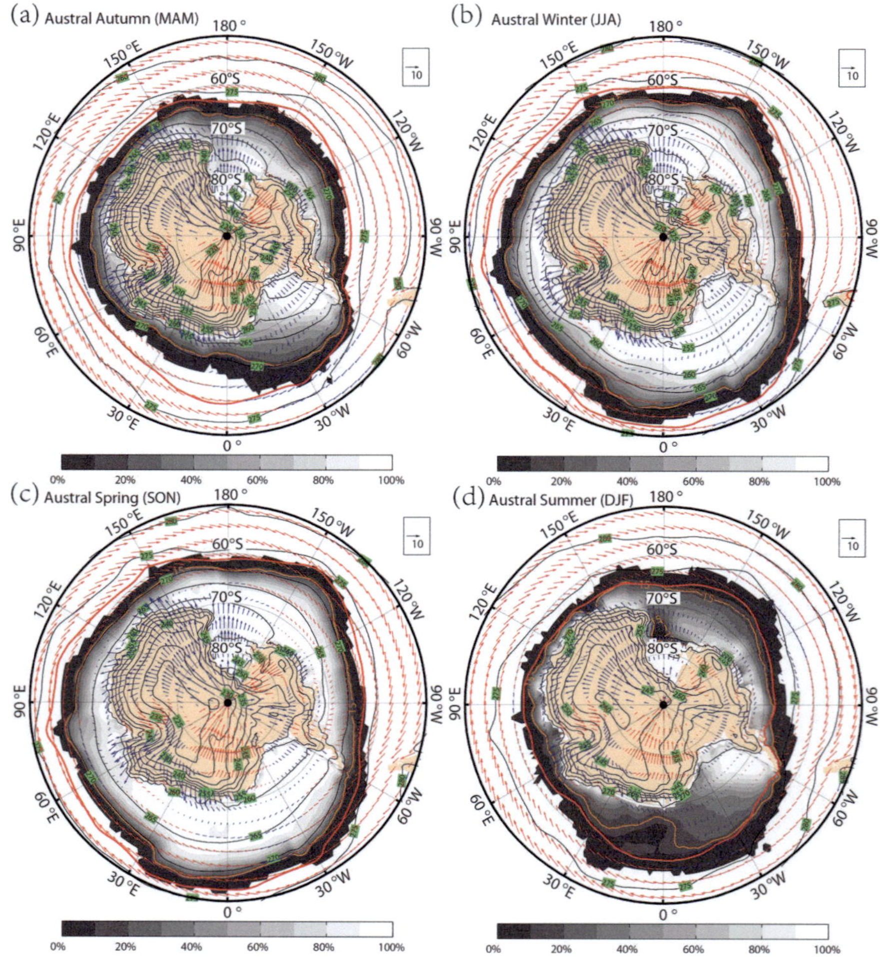

Fig. 2.18 Same as in Fig. 2.12 except climatic winds at 2 m level and surface air temperature in the SH polar region. The *shading* covers the maximum sea ice extent

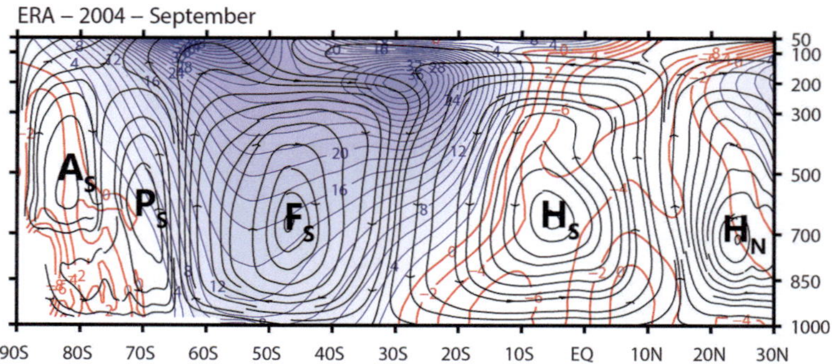

Fig. 2.19 Same as in Fig. 2.5 except flow and zonal velocities in September 2004

2.8 Hemispheric Height and Vorticity

Are the four cells certainly existing in each hemisphere? If yes, they should be visible not only in wind field but also in zonally averaged pressure (geopotential height) and vorticity fields. Even at an isobaric level, we should also find their basic zonal features. We take the climatic annual mean height and vertical pressure velocity at 700 hPa and avoid the Antarctic terrain. In the SH, a zone with higher height gradient is concentrated in the middle latitudes, while in the Antarctic and tropics heights are relatively homogeneous (Fig. 2.20). This implies that there is strong westerly zone in the middle latitudes over the SH. Three zones of ascending velocity (blue shading) indicated by dashed line, dash-dotted line, and dotted line are clearly observed in the tropics, high latitude, and in the polar area. The zones of higher height gradient and ascending pressure velocity at 700 hPa should be symmetric in the NH (Fig. 2.20a). The asymmetric distribution is influenced by the sea–land contrast in the NH. Many troughs of height and individual ascending areas are thus exhibited. A fan-type area without the influence of sea–land contrast in the central North Pacific is taken to examine their zonal features. Three zones of ascending velocity can be comparable in the central North Pacific and the central South Pacific. Although the ascending area in the two polar areas is very small, it is the evidence to indicate the Arctic cell in the NH and the Antarctic cell in the SH.

On daily synoptic chart, there are always two low vortices near two polar regions and two longitudinal high zones in the subtropics, especially in the middle-upper troposphere. Figure 2.21 shows the climatic distribution of vorticity at 300 hPa in the NH in four seasons. In spring, positive (or negative) vorticity indicates that there

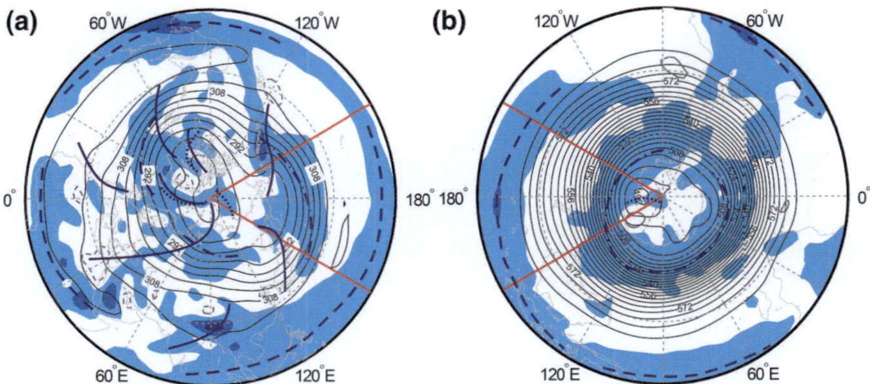

Fig. 2.20 Climatic annual mean geopotential height (contour, 10 gpm) and vertical pressure velocity (0.05 Pa s^{-1}) at 700 hPa in **a** the NH and **b** the SH. *Shading grade* (0.05 Pa s^{-1} interval) and *dashed lines* denote the ascending and descending velocities, respectively. The heavy *blue line* denotes the height trough at 700 hPa. Heavy *blue dashed*, *doted* and *dashed-dot lines* are axes of ascending velocity respectively in lower, high and middle latitudes. Two *red lines* cover the central North (South) Pacific between 150°E and 150°W. Sourced from Qian et al. (2015b) © Springer-Verlag Wien 2015, with permission

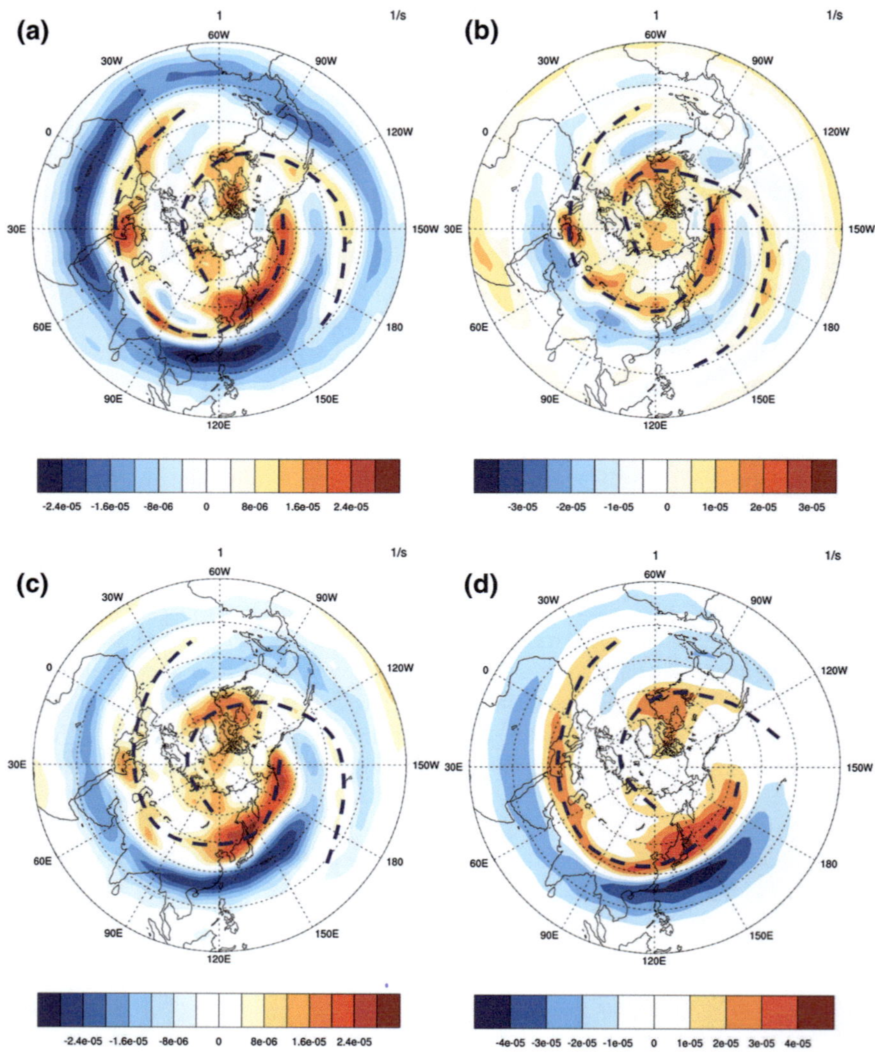

Fig. 2.21 Climatic vorticity (*shading*, 1×10^{-5}/s) at 300 hPa in the NH in boreal **a** spring, **b** summer, **c** autumn, and **d** winter. Heavy *dashed line* indicates the bands of positive vorticity

are two spiral bands extending from the north-polar region to middle and lower latitudes. The two spiral bands become weaker in summer, autumn and winter. The negative vorticity indicating the subtropical high at 300 hPa is the strongest in spring (Fig. 2.21a). For the two special days on March 4 and June 18, positive vorticity shows that there are about four spiral bands extending from the north-polar region to middle and lower latitudes (Fig. 2.22).

Differences are found in the SH, where there are three climatic spiral bands of positive vorticity, extending from the south-polar region to middle and lower

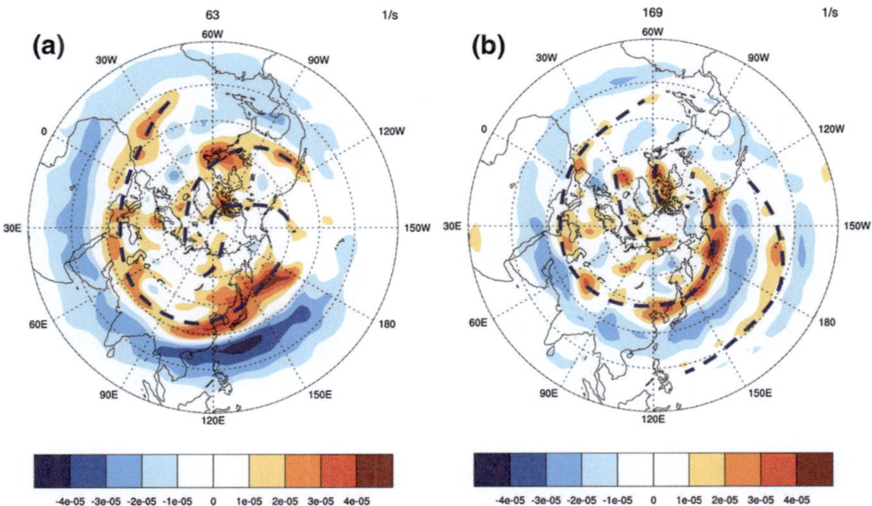

Fig. 2.22 Climatic vorticity (1×10^{-5}/s interval) at 300 hPa in the NH on **a** 4 March and **b** 18 June. Heavy *dashed line* denotes the bands of positive vorticity

latitudes in four seasons (Fig. 2.23). For the two special days on March 4 and June 18, positive vorticity shows that there are four spiral bands (Fig. 2.24).

It is noted that spiral vorticity bands, extending from the two polar regions to middle and lower latitudes, exist in the two hemispheres. But their strengths and bands in the two hemispheres change with days and seasons. These structures of vorticity rely on topography, thermal contracts, and temporal periods as well. The spiral vorticity structures in the two hemispheres are similar to those of a super typhoon in Fig. 1.1a, the Milky Way galaxy in Fig. 1.1b and the mesoscale vortex (MV) in Fig. 1.6.

We also found that the spiral convective bands exist in extra-tropical cyclones. Figure 2.25a shows the surface pressure superimposed on combined reflectivity factor of radar echo at 0000 UTC 20 July 2016. Four spiral convective bands are clearly indicated by the combined reflectivity factor of radar echo at 0000 UTC 20 July 2016. This extra-tropical cyclone caused 130 deaths and 110 missed. The four spiral convective bands of reflectivity factor (Fig. 2.25a) are consistent with the four shears of anomalous flow at 700 hPa (Fig. 2.25b).

2.9 An Aqua Control Experiment

The topography distribution is asymmetric in the Earth's two polar areas. We can always find the Arctic cell in the NH but the Antarctic cell is not always observed. So, are the two cells associated with the topography? We here give the results of an aqua control experiment of CCSM4 model. It is the fourth version of the

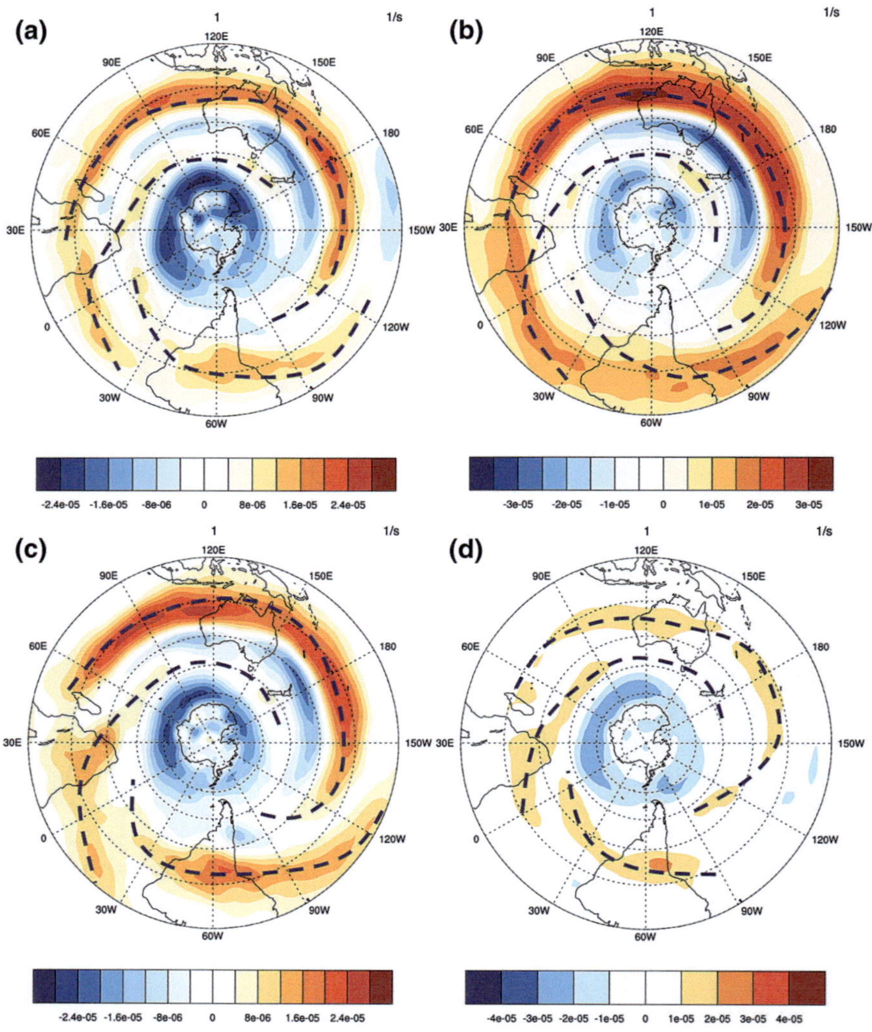

Fig. 2.23 Same as in Fig. 2.21 except the climatic vorticity at 300 hPa in the SH in boreal **a** spring, **b** summer, **c** autumn, and **d** winter. Heavy *dashed line* indicates the bands of positive vorticity

Community Climate System Model (CCSM) consisting of atmosphere, land, ocean, and sea ice components that are linked through a coupler that exchanges state information and fluxes between the components (Gent et al. 2011). An aqua control experiment removed land and sea ice components and only remained ocean. The data of the aqua control experiment is derived from website www.earthsystemgrid.org/dataset/cmip5.output1.NCAR.CCSM4.aquaControl.mon.atmos.Amon.r1i1p1.html#variablesTab.

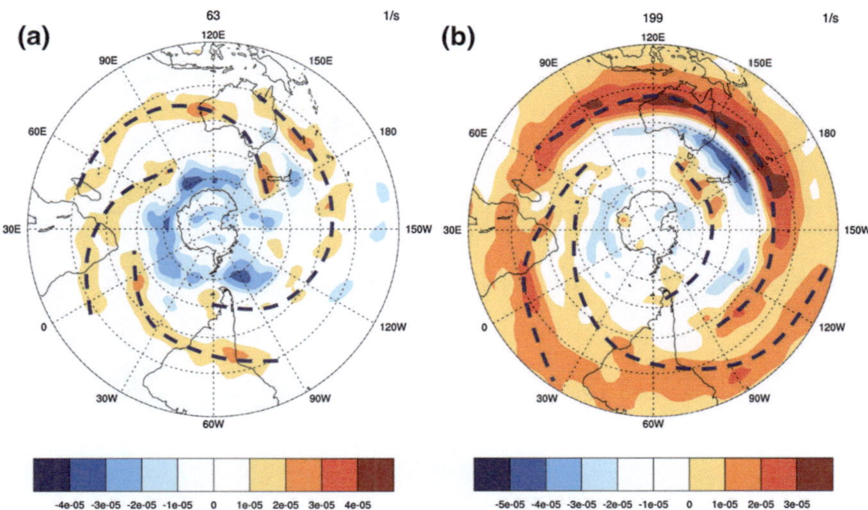

Fig. 2.24 Same as in Fig. 2.22 except the climatic vorticity at 300 hPa in the SH on **a** 4 March, and **b** 18 July. Heavy *dashed line* indicates the bands of positive vorticity

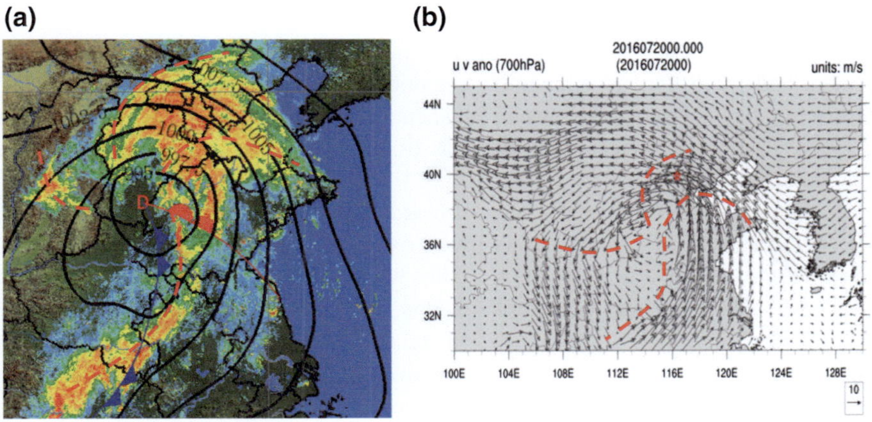

Fig. 2.25 Four spiral convective bands in **a** and shears in **b** indicated by *red-dashed lines*. **a** Surface pressure (*solid contour*, 2.5 hPa interval) superimposed on combined reflectivity factor of radar echo (Data from CMA MICAPS4.0) and **b** wind anomalies (m s^{-1}) at 700 hPa and 0000 UTC 20 July 2016

The zonally averaged flow and zonal velocities by the five-year mean are given in Fig. 2.26. The meridional cells are clearly found in the SH with the Antarctic cell at the middle-upper troposphere. In the NH, the fourth circulation is not closed but the ascending flow is observed beyond 80°N. Five ascending zones and

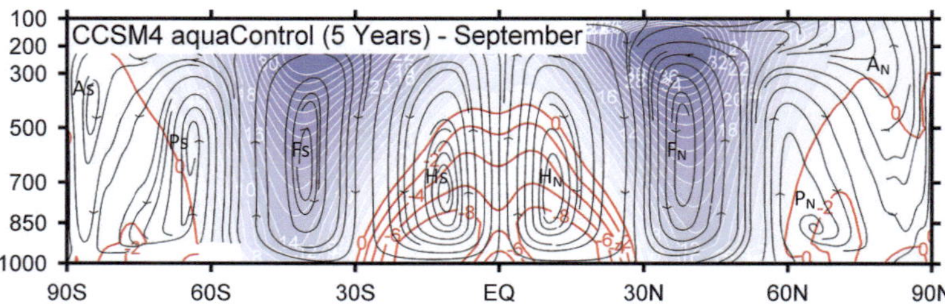

Fig. 2.26 Same as in Fig. 2.5 except for the result of an aqua control experiment in September

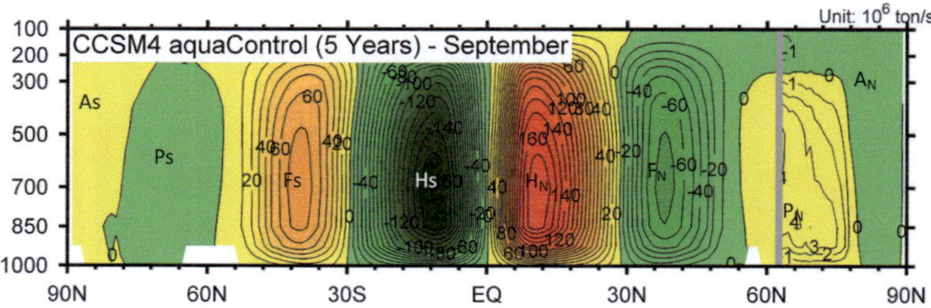

Fig. 2.27 Same as in Fig. 2.7 except for the result of an aqua control experiment in September

four descending zones are well simulated in the aqua control experiment but the centers and intensities of circulation are not symmetric in the two hemispheres. This may be caused by eddies. The spatial distribution of MSF is shown in Fig. 2.27. Although there are eight areas of MSF, their intensities are really not symmetric.

For the aqua control experiment, northward eddy heat flux is shown in Fig. 2.28. The maximum of eddy heat flux is found over the indirect Ferrel cell. Its center shifts northward to 65°N in Fig. 2.28b when comparing with reanalysis in Fig. 2.14b. The intensities of northward eddy heat flux are not stable to indicate the Polar and Arctic cells. This situation is also found in the northward eddy momentum flux (Fig. 2.29). Two positive areas of the northward eddy momentum flux are clearly observed in Fig. 2.29c, d, but not in Fig. 2.29a, b. This indicates that the eddies are also frequently produced in the aqua experiment. The intensity of eddies is not steady through the model's forcing and the boundary condition are determined in symmetry in the two hemispheres. This difference indicates that it is difficult to obtain a climatic state from the aqua model.

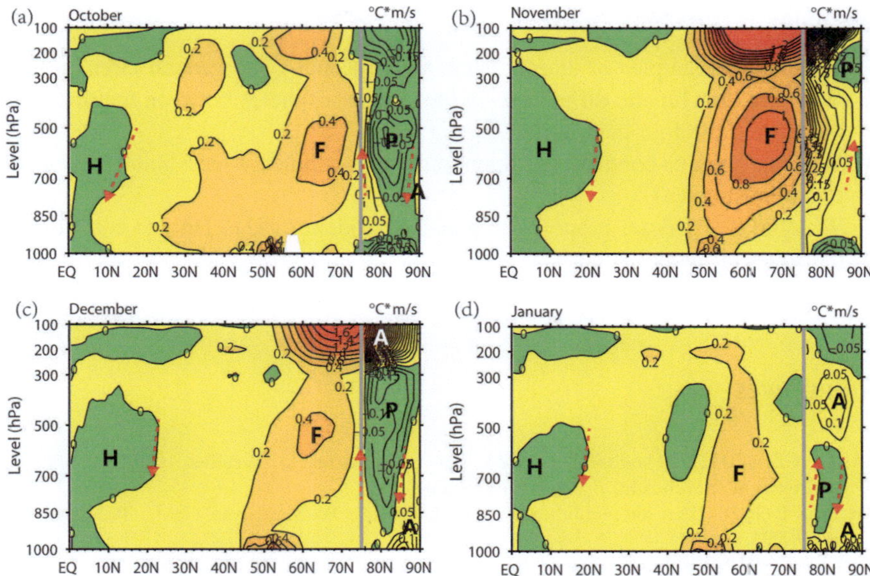

Fig. 2.28 Same as Fig. 2.14 except the result (northward eddy heat flux) of an aqua control experiment in **a** October, **b** November, **c** December, and **d** January

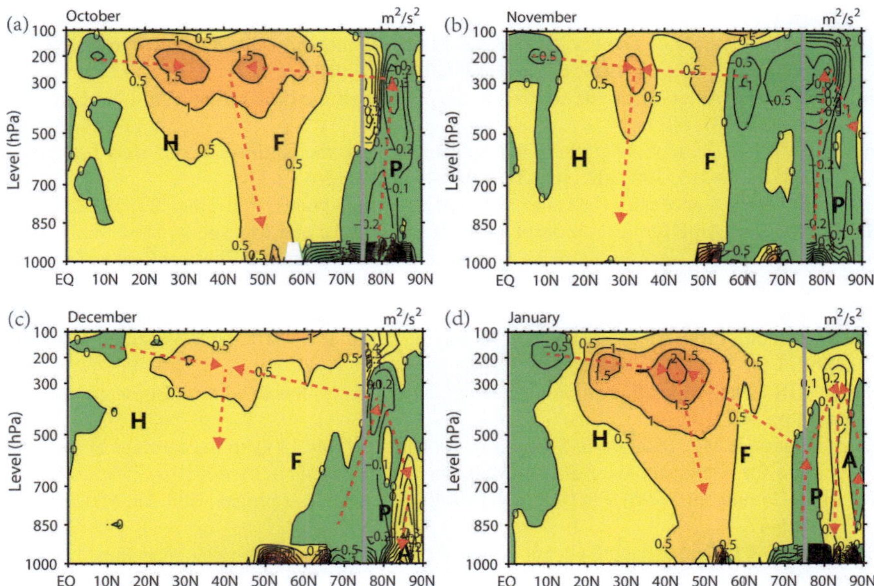

Fig. 2.29 Same as Fig. 2.15 except the result (northward eddy momentum flux) of an aqua control experiment in **a** October, **b** November, **c** December, and **d** January

Questions

1. Give some examples of diurnal-cycle and seasonal-cycle general circulation.
2. What are the climatic differences in the two hemispheres between the three-cell model and the four-cell model?
3. Why can an aqua control experiment produce asymmetric meridional cells in the two hemispheres?
4. Why did V. Bjerknes propose the four-cell model in 1921? How to confirm that the four-cell model is correct?

References

Ahrens CD (2012) Meteorology Today: An Introduction to Weather, Climate, and the Environment (Seventh Ed.). Thomson Learning, Inc

Bergeron T (1928) Uber die dreidimensional verknupfende Wetteranalyse. Geofys Publ 5(6):111, pp(59,75)

Bjerknes J (1919) On the structure of moving cyclones. Geofys Publ 1(2): 1–8pp (74)

Bjerknes V (1921) On the dynamics of the circular vortex with applications to the atmosphere and atmospheric vortex and wave motions. Geofys Publ 2(4): 1–88

Braham CG, Pearsons E, Posner DM, Maas GS, Goodman R (2001) Random House Webster's College Dictionary (second ed.). Random House. p. 1385. ISBN 0-375-42560-8

Dove HW (1837) Meteorologische Untersuchungen. Berlin, Sandersche Buchhandlung, 334pp. (61)

Gent PR, Danabasoglu G, Donner LJ, et al. (2011) The community climate system model version 4. J. Climate, 24, 4973–4991. doi: 10.1175/2011JCLI4083.1

Ferrel AJ (1856) An essay on the winds and the currents of the ocean. Nashville J Medicine and Surgery 11: 287–301

Ferrel AJ (1859) The motions of fluids and solids relative to the Earth's surface. Math Monthly 1: 140–147, 201–216, 300–307, 366–372, 397–406. (66, 136)

Hadley G (1735) Concerning the cause of the general trade winds. Phil Trans 29: 58–62.

Halley E (1686) An Historical Account of the Trade Winds, and Monsoons, Observable in the Seas between and Near the Tropicks, with an Attempt to Assign the Physical Cause of the Said Winds. Philosophical Transaction, 16,172–175. doi:10.1098/rstl.1686.0028

Hayes D (2001) Historical atlas of the North Pacific Ocean: Maps of Discovery and Scientific Exploration, 1500–2000. Douglas & McIntyre. p. 18. ISBN 978-1-55054-865-5. Retrieved 2009-11-08

Holton BJR (2004) An introduction to dynamic meteorology, fourth edition. Elsevier Academic Press, pp 535

Kahl JD, Serreze MC, Shiotani S, Skony SM, Schnell RC (1992) In situ meteorological sounding archives for Arctic studies. Bull Am Meteorol Soc 73: 1824–1830. doi:10.1175/1520-0477

Kalnay E, Co-authors (1996) The NCEP/NCAR 40-year reanalysis project. Bull Am Meteorol Soc 77: 437–471

Li JP (2001) Climatic Atlas of Global Atmospheric Circulation, Meteorological Press, Beijing, pp 279

Lorenz EN (1967) The nature and theory of the general circulation of the atmosphere. World Meteorol. Organization. pp 158

Maury MF (1855) The physical geography of the sea. New York, Harper, pp 287 (63)

Muelder HR (2007) Years of This Land - A Geographical History of the United States. Read Books. p. 38. ISBN 978-1-4067-7740-6. Retrieved 2009-11-09

Peixoto JP, Oort AH (1992) Physics of Climate, Springer, Am. Inst. Phys. Press, New York

Qian WH, Wu KJ, Chen DL (2015a) The Arctic and Polar cells act on the Arctic sea ice variation. Tellus A 67: 27692

Qian WH, Wu KJ, Leung JCH, Shi J (2016) Long-term trends of the Polar and Arctic cells influence the Arctic climate since 1989. J Geophys Res Atmos doi: 10.1002/2015JD024252

Qian WH, Wu KJ, Liang HY (2015b) Arctic and Antarctic cells in the troposphere. Theor Appl Climatol. doi: 10.1007/s00704-015-1485-z

Thomson J (1857) Grand currents of atmospheric circulation. British Assoc. Meeting, Dublin, (65,115,136)

Thomson J (1892) On the grand currents of atmospheric circulation. Phil Trans Roy Soc A, 183:653–684

Wu K (2016) The research on the mean meridional circulations in the Arctic and their climatic influences. Dissertation. Peking University

Chapter 3
Spatial Patterns and Time Scales of Climatic State

Solar radiation is the essential driving of the atmospheric motion. The spatial distribution and temporal evolution of climatic state is also determined by various surface conditions. The zonally averaged climatic state and its longitudinal deviation have a delayed response to the solar radiation for about one month. For the seasonal march of maximum zonally averaged precipitation, the southernmost, transition and northernmost time points are pentads 8, 28 and 44, respectively. The intensity and location of the eight meridional cells also experience seasonal cycle. Besides meridional cells, the Walker circulation is the result of zonal sea–land contrast and surface temperature difference. Zonal cell in the Pacific and other two zonal cells in the Atlantic and Indian Ocean only exist in the equatorial zone and the subtropical Northern Hemisphere (NH). All centers of atmospheric action described by pressure (height), air temperature, and winds are the spatial response. Changes in intensity and location of atmospheric active centers form seasonal variations of different atmospheric variables. Temporally, climatic field can be decomposed into annual cycle, semiannual cycle and biweekly cycle, which are the atmospheric responses to the solar radiation and surface conditions. The three components of different timescales and their systems have different spatial scales, but they have the same vertical structure. Usually, the largest center of climatic height component is located at the upper troposphere (around 200–300 hPa). For each climatic component, a warm air column and a cold air column are, respectively, located below and above a positive center of height component, which satisfy the hydrostatic equilibrium. All these climatic systems can be climatically traced, and do not need to be predicted. This chapter demonstrates the spatial patterns and temporal evolutions of climatic state.

© Springer Nature Singapore Pte Ltd. 2017

W. Qian, *Temporal Climatology and Anomalous Weather Analysis*,
Springer Atmospheric Sciences, DOI 10.1007/978-981-10-3641-5_3

3.1 Seasonal Evolution of Zonally Averaged Variables

We first examine the zonally averaged seasonal evolution of precipitation rate. Figure 3.1a shows the seasonal cycle of zonally averaged climatic precipitation obtained from the CMAP (Xie and Arkin 1992). A precipitation threshold of 4 mm/day was used to determine the onset of the summer monsoon in the central South China Sea (SCS) (Qian and Yang 2000) and to identify the northernmost monsoon location in Eastern Asia (Hu and Qian 2007). The 4 mm/day precipitation threshold extends to 20°S in pentad 8 (early February) and to 20°N around pentad 44 (early August). Therefore, there are two dry–wet transition zones from 6°N to 18°N in the NH and its south from 2°N to 19°S (Fig. 3.1b). Between 2°N and 6°N, there is a wet zone around the equatorial circle with the precipitation always larger than 4 mm/day over the whole year. Two dry–wet transition zones of monsoon are clearly observed in this illustration. Another significant feature is the seasonal transitions between the two hemispheres in middle May and the middle November indicated, respectively, by dotted lines. Middle May and middle November are also the seasons of the SCS summer monsoon onset and retreat, respectively.

Seasonal standard deviations (SDs) of zonally averaged climatic variables reflect the variations of climatic cycle. Figure 3.1c is the seasonal SDs of zonally averaged climatic precipitation. In each hemisphere, there are four relatively large peaks of SDs for climatic precipitation near latitudes 10, 35, 60, and 85. The largest peak is located at 10–15°N and the second largest peak is at 10–15°S. These eight peaks should be in accordance with the seasonal shifts of ascending branch. Thus, the changes in locations, intensity, and widths of ascending branch should produce zonally averaged dry–wet variations from regular seasonal to irregular intra-seasonal and inter-annual and inter-decadal timescales. Eight peaks of precipitation difference are clearly observed in Fig. 3.1c.

In this chapter, atmospheric variables are from the NCEP/NCAR reanalysis 1. The air temperature is a basic variable that can be used to represent climatic state. It is known that air temperature decreases with altitude and latitudes in the troposphere. Figure 3.2 shows the zonally averaged climatic air temperature at eight isobaric surfaces from stratosphere to troposphere. At 10 hPa, a warm center of temperature (−22.5 °C) first appears in winter (November and December) over the South Pole. This warm center then seasonally propagates northward and arrives in the equator in late April and to the North Pole in early July. The cold center is first observed in the South Pole in early July (−80 °C) and then seasonally propagates northward to the North Pole (−70 °C) in the late November.

At 50 hPa, the tropical zone between 30°N and 30°S is colder than its southern and northern zones. In the middle-high latitudes of the NH, a warm center experiences a seasonal cycle that first appears in the latitude of 50°N in January and then propagates northward to the North Pole in early July. While in the middle-high latitudes of the SH, a warm center also experiences a seasonal cycle that first appears in the South Pole in winter (December and January) and then propagates

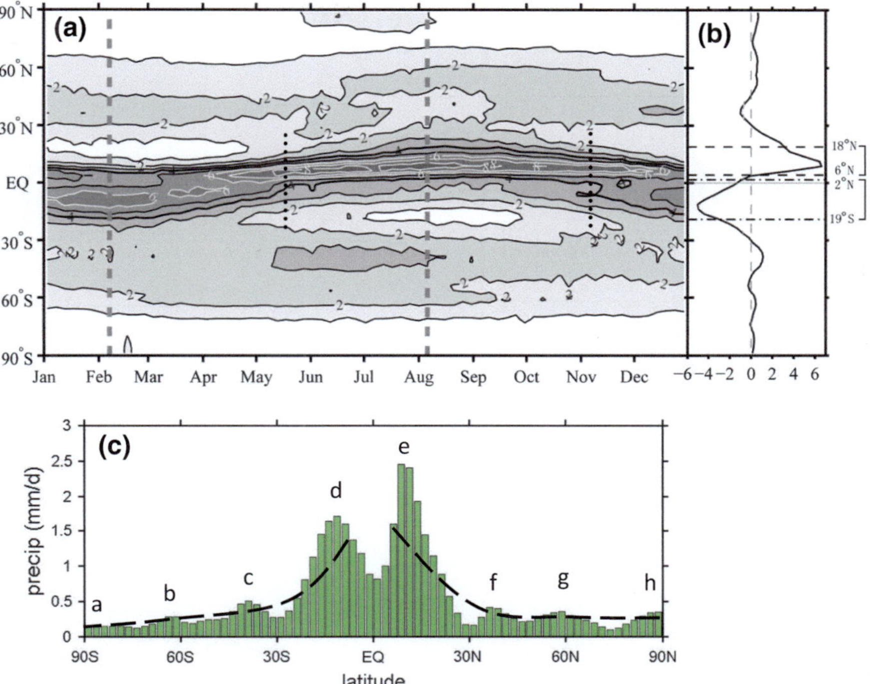

Fig. 3.1 **a** The zonally averaged climatic precipitation (*black and white solid lines*, 1 mm/day interval) from 90°S to 90°N and from January to December. The *heavy solid line* indicates 4 mm/day. The *heavy grey-dashed lines* in early February (pentad 8) and in early August (pentad 44) indicate that the 4 mm/day precipitation reaches the southernmost and northernmost locations. The *dotted lines* denote the seasonal transitions between two hemispheres. **b** The curve is the difference of precipitations between pentads 44 and 8. **c** Seasonal SDs of zonally averaged climatic precipitation (mm/day) with two *heavy dashed lines* are fitted with the climatic precipitation in the two hemispheres. Letters "a"–"h" in **c** denote eight relative large peaks of climatic precipitation variation

northward to around 45°S in early July. Similar seasonal cycle of temperature is also observed at 100 hPa.

The temperature at the upper troposphere of 300 hPa is somewhat different from that at the stratosphere. Temperature in the tropical zone between 30°N and 30°S is warmer than its southern and northern zones. The coldest temperature (lower than −57.5 °C) in the NH is found in January and the coldest temperature (lower than −62.5 °C) in the Southern Hemisphere (SH) appears in August. The warmest season in the middle latitude of NH is in July.

Seasonal temperature cycle is obvious in the NH and at the middle-lower troposphere such as at 500, 850, 925, and 1000 hPa. The highest temperature in the middle latitude of NH is found in July. At 500 hPa above the tropical zone, the warmest center (−5 °C) occurs in April and May. But the warm centers at 850, 925, and 1000 hPa are located in the subtropical NH from June to September. From

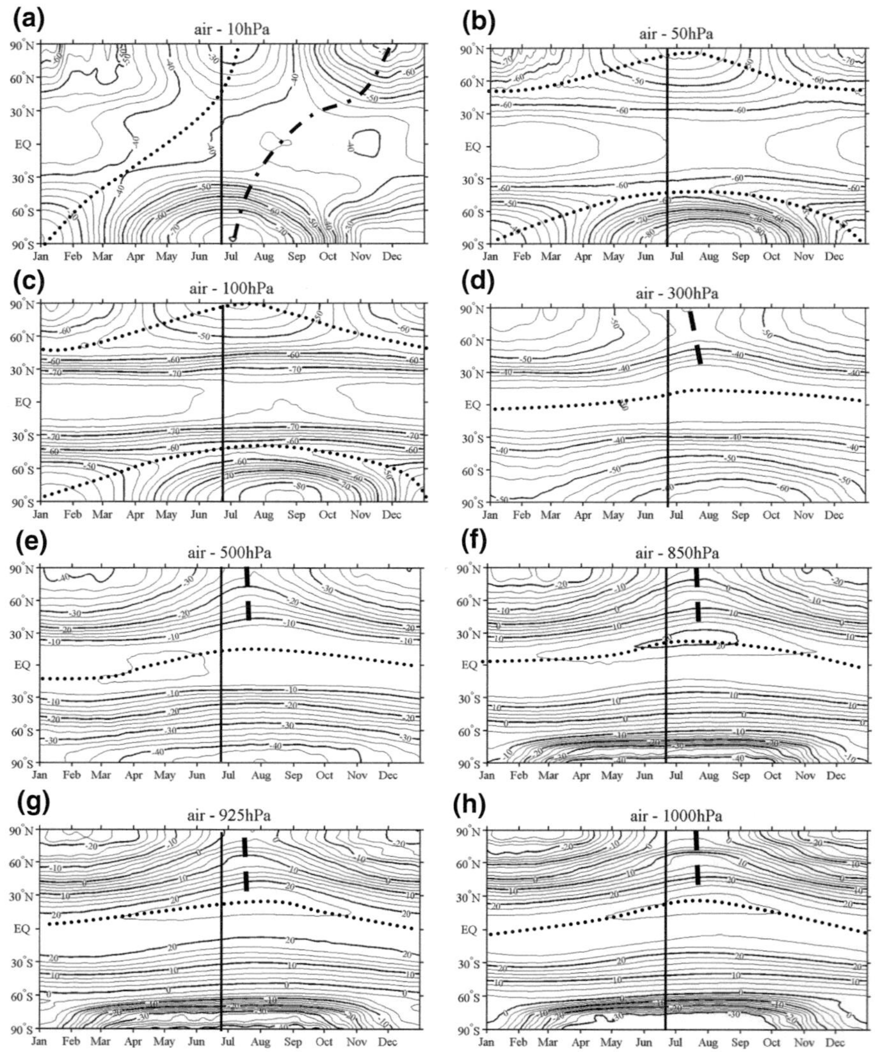

Fig. 3.2 The zonally averaged climatic air temperature (2.5 °C interval) from 90°S to 90°N and from January to December at **a** 10 hPa, **b** 50 hPa, **c** 100 hPa, **d** 300 hPa, **e** 500 hPa, **f** 850 hPa, **g** 925 hPa, and **h** 1000 hPa. The *dotted lines* denote the seasonal cycles of warm centers and the *dot-dashed line* indicates the seasonal cycle of the cold centers in **a**. The *heavy dashed line* connects the warmest time points in the NH. The *solid line* denotes 22 June, when the solar rays radiate directly at the Tropic of Cancer

500 hPa to the surface, the highest temperature in the NH mid-high latitudes is observed in late July, which lags the day with the strongest solar radiation on 22 June for about one month. The lagged relation can be explained using Eq. (1.3.1) with an atmospheric delayed operator $-(L+D)$.

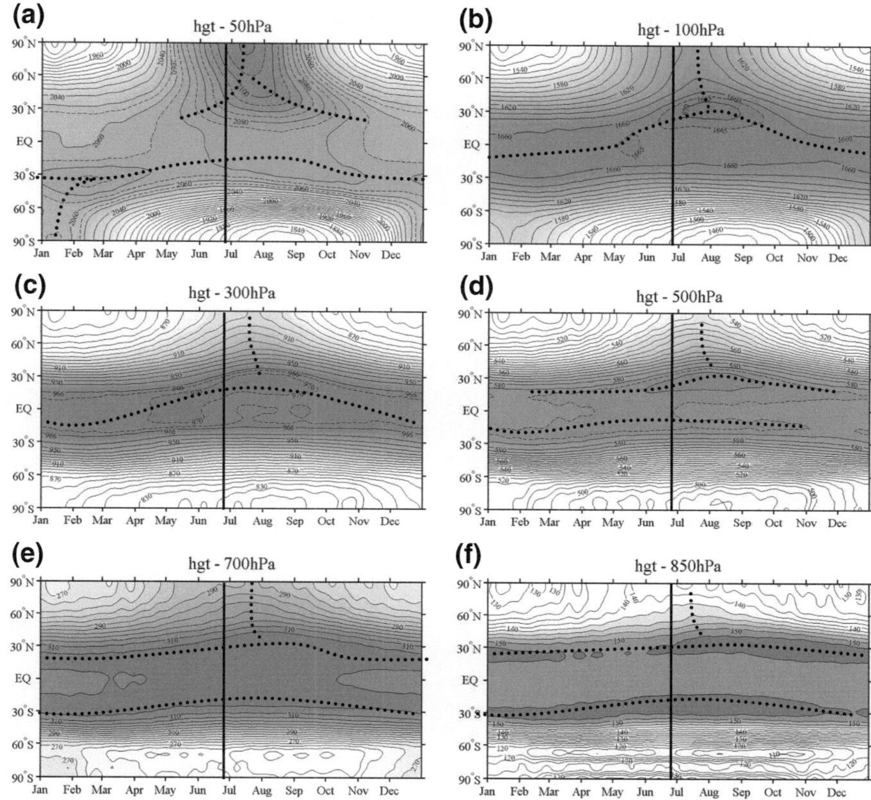

Fig. 3.3 The zonally averaged climatic geopotential height (5 × 10 and 10 × 10 gpm intervals) from 90°S to 90°N and from January to December at **a** 50 hPa, **b** 100 hPa, **c** 300 hPa, **d** 500 hPa, **e** 700 hPa, and **f** 850 hPa. The *dotted lines* denote the seasonal cycles of height centers and ridges. The *solid line* denotes 22 June, when the solar rays radiate directly at the Tropic of Cancer

The zonally averaged climatic geopotential height also shows obvious seasonal cycle, varying with latitudes (Fig. 3.3). At 50 hPa, a high center is basically located in the SH subtropical area around the year. In the middle-high latitudes of NH, the higher value of height appears in the boreal summer (JJA) and centers in July, while the lower value of height is in the boreal winter and centers in January. At 100 hPa, the higher value of height experiences a seasonal cycle with its center in the south of the equator in the boreal winter and spring and then shifts to the NH subtropical area in summer. The highest value of height in late June to early September is contributed mainly by the South Asian High (SAH). The subtropical high is also isolated in the NH at 300 hPa but its center shifts from SH in boreal spring (April to May) to NH in summer (June to September). Two subtropical highs with varying intensity in two hemispheres shift with seasonal cycle at 500, 700, and 850 hPa. The NH subtropical high reaches its peak intensity in boreal summer at 500 hPa but it is weak in boreal spring (March–May) at 850 hPa. From the stratosphere to the troposphere, the maximum geopotential height in the NH mid-high latitudes is in late July, lagging the day with the strongest solar radiation on 22 June for about one month.

The water vapor content is high in the low troposphere. The zonally averaged climatic specific humidity shows a climatic seasonal cycle along each of latitudinal zones (Fig. 3.4). At 500 hPa, a moist center has a seasonal cycle near the equator with its maximum in early August. Another moist center with a seasonal cycle is located in the subtropical NH with its maximum in late July. The maximum value of specific humidity in the SH appears in early February. The maximum values of specific humidity in two hemispheres lag the solar radiation on the Tropic of Cancer (22 June) and on the Tropic of Capricorn (December 22). Three centers of specific humidity can be clearly observed at 925 hPa. Two of them are close to the equator. The specific humidity center in the subtropical NH is rapidly formed in late May and observed at 500, 700, 850, and 925 hPa.

Seasonal cycles of zonally averaged westerly are complicated at every isobaric level and latitude as shown in Fig. 3.5. At 10 hPa, there are transitions between westerly and easterly in the middle-high latitudes over the two hemispheres. But easterly prevails in the tropical zone with two centers in February (−30 m/s) and July (−25 m/s), respectively. In the middle latitudes of SH, the westerly can reach the strongest 80 m/s in early August but the westerly in the NH can only reach 35 m/s in winter (December and January).

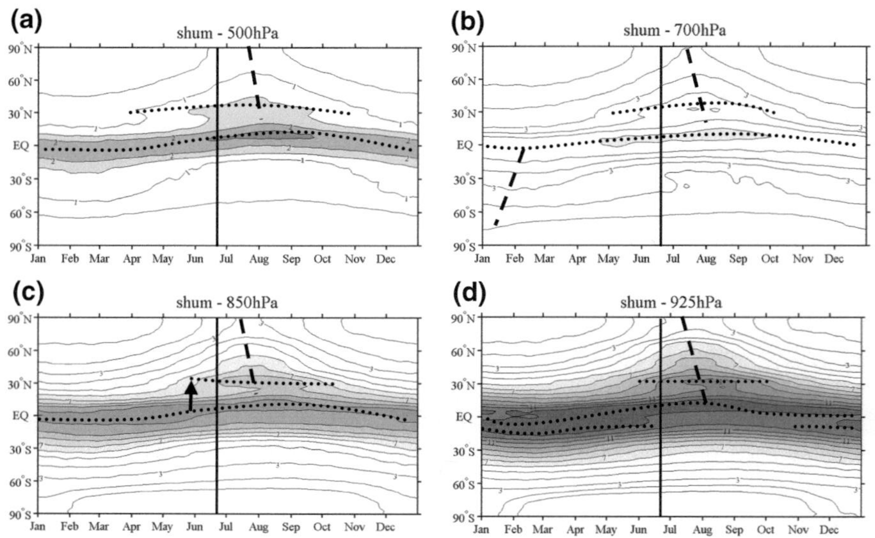

Fig. 3.4 The zonally averaged climatic specific humidity (0.5 and 1 g/kg intervals) from latitudes 90°S to 90°N and from January to December at **a** 500 hPa, **b** 700 hPa, **c** 850 hPa, and **d** 925 hPa. The *dotted lines* denote the seasonal cycle of specific humidity and the *dashed lines* indicate the maximum of high specific humidity. The *solid line* denotes 22 June, when the solar rays radiate directly at the Tropic of Cancer

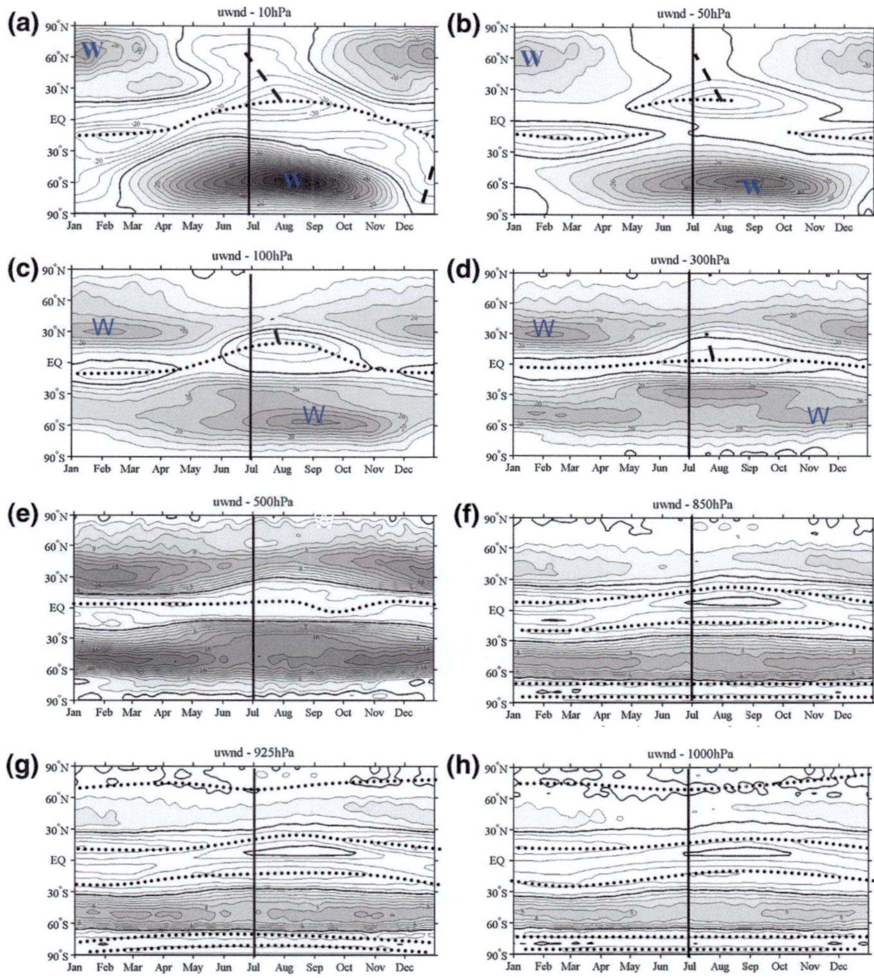

Fig. 3.5 The zonally averaged climatic westerly (*shading*, 2 and 5 m/s intervals) from 90°S to 90° N and from January to December at **a** 10 hPa, **b** 50 hPa, **c** 100 hPa, **d** 300 hPa, **e** 500 hPa, **f** 850 hPa, **g** 925 hPa and **h** 1000 hPa. The *dotted lines* denote the seasonal cycles of easterly and the dashed lines indicate the maximum of easterly. The *solid line* denotes 22 June, when the solar rays radiate directly at the Tropic of Cancer

The intensities of westerly and easterly are weak at 50 and 100 hPa when comparing with those at 10 hPa. But the transition between westerly and easterly exists in all latitudes. The center of westerly is located at 60°N in January and at 60°S in August.

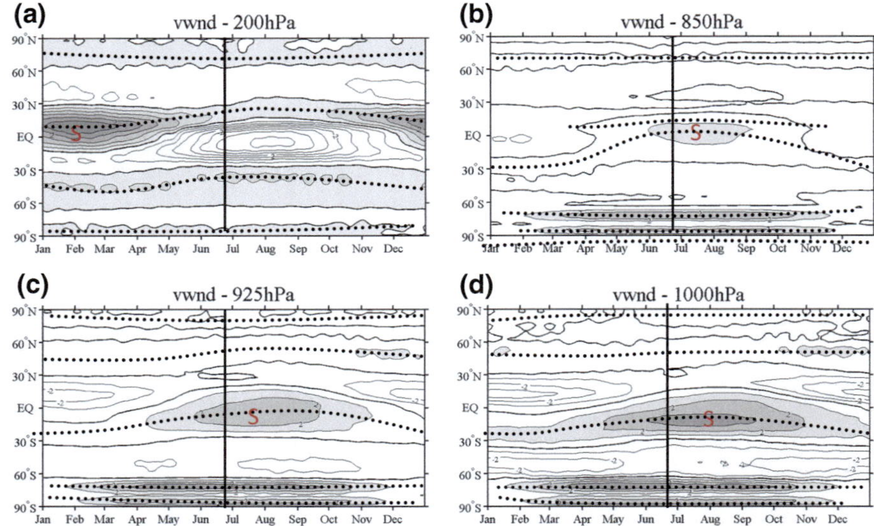

Fig. 3.6 The zonally averaged climatic southerly (*shading*, 0.5 and 1 m/s intervals) from 90°S to 90°N and from January to December at **a** 700 hPa, **b** 850 hPa, **c** 925 hPa, and **d** 1000 hPa. The *dotted lines* denote the seasonal cycles of southerly. The *solid line* denotes 22 June, when the solar rays radiate directly at the Tropic of Cancer

At 300 and 500 hPa, easterlies are concentrated near the equator and westerlies prevail over middle-high latitudes in the two hemispheres over the whole year. At the lower troposphere (850, 925, and 1000 hPa), two centers of easterlies indicate the trades in the subtropical area. To the north of the equator, there is a center of westerly in the lower troposphere from late June to September, which indicates the southwesterly monsoon flows. In the high latitudes of SH, there are two centers of easterly winds. The center of easterly in boreal summer from 10 to 300 hPa lags the day with strongest solar radiation on 22 June for about one month.

The transition between southerly and northerly and their strength changes are important to exchange atmospheric mass and water vapor between the two hemispheres. Along the equator, the southerly at 200 hPa prevails in the boreal winter half year (December–April) and the northerly is found in the boreal summer half year (Fig. 3.6). In the lower troposphere at 850, 925, and 1000 hPa, the cross-equatorial southerly clearly appears in the boreal summer half year (April–October), with the highest value in July. In the boreal summer, there is another southerly center at 850 hPa in the subtropical NH, besides the center on the equator. Southerly and northerly winds are opposite between the upper and lower troposphere in summer and winter. In the SH high latitudes, there are two pairs of southerly and northerly centers. In the boreal summer, the southerly center in the lower troposphere (1000–850 hPa) and the northerly center at the upper troposphere (200 hPa) lag the day with the strongest solar radiation on 22 June for about one month.

3.2 Zonally Averaged Vertical–Meridional Structure

For the seasonal evolution of zonally averaged precipitation, four time points, including pentads 8, 28, 44 and 58, have been noted in Figs. 3.7, 3.8, 3.9, and 3.10. We will give comprehensive vertical distributions of zonally averaged climatic variables including precipitation rate, specific humidity, temperature, height, and winds in this section. Figure 3.7 shows the climatic state of meridional–vertical cells, zonal wind speed, geopotential height and temperature in pentad 8. In the NH, there are four meridional circulations: the Hadley cell, the Ferrel cell, the Polar cell, and the Arctic cell. This is during the boreal winter, so the ascending branches of the two Hadley cells meet near 15°S. Easterlies are prevailing in the tropical zone and westerlies are in the middle latitudes with the wind speed of 44 m/s at 200 hPa and 30°N. The ridge line of subtropical high is obliquely separated upward by easterly in the tropical zone and westerly in the middle latitudes. Here, the upper flow of the Ferrel cell is well defined by the northwesterly in the upper troposphere and the southwesterly in the lower troposphere. The Polar cell shows the northeasterly near 75°N in the lower troposphere and southwesterly at high altitude. In the Arctic area, a closed cell is referred to as the Arctic cell (Qian et al. 2015a, b). In the SH, the traditional three cells are well displayed in Fig. 3.7. Ascending branches are well accompanied with larger precipitations, except that the precipitation near

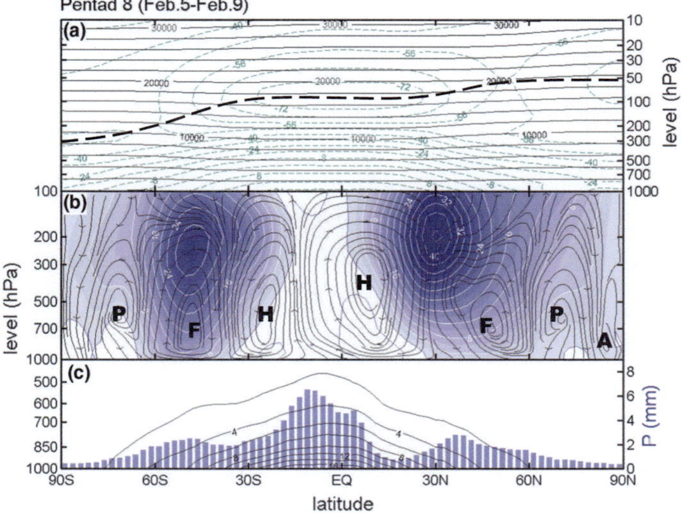

Fig. 3.7 Zonally averaged distributions of climatic variables including **a** height (*black contour*, 200 gpm interval) and temperature (*blue-dashed contour*, 8 °C interval), **b** meridional–vertical cells and zonal wind speed (*shading* with white solid line for westerly, 4 m/s interval) in pentad 8 (5–9 February), and **c** precipitation (bar, mm/day) and specific humidity (*contour*, 2 g/kg interval). Symbols "H", "F", "P" and "A" in **b** indicate the Hadley cell, Ferrel cell, Polar cell and Arctic (Antarctic) cell. *Heavy dashed lines* in **a** denotes the minimum temperature

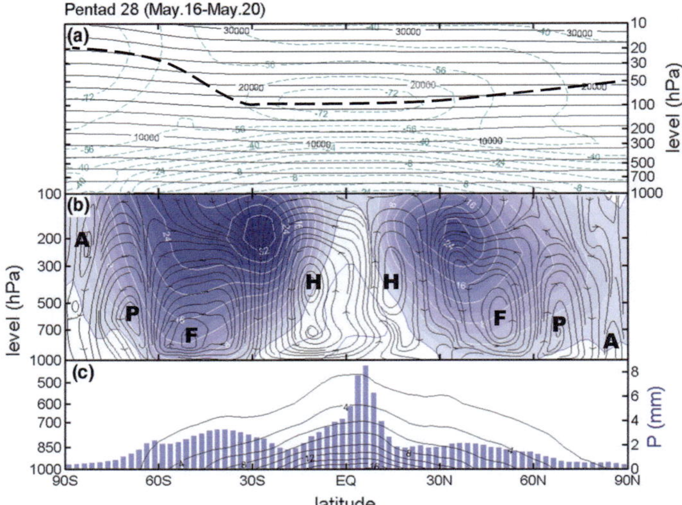

Fig. 3.8 Same as in Fig. 3.7 except in pentad 28 (16–20 May)

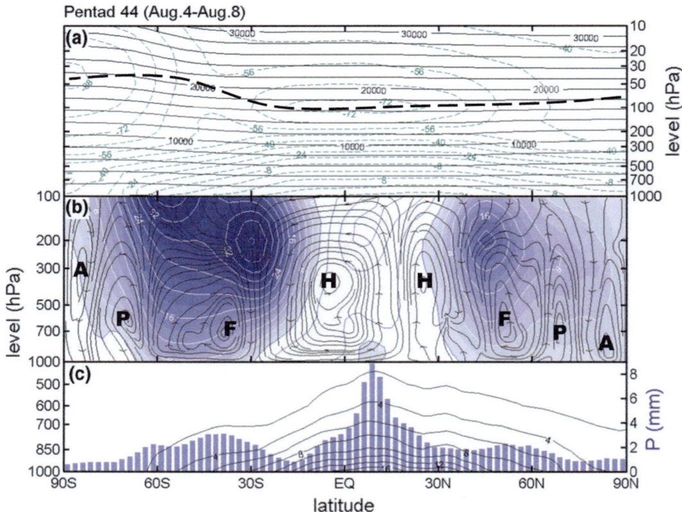

Fig. 3.9 Same as in Fig. 3.7 except in pentad 44 (4–8 August)

35°N is contributed by the subtropical frontal precipitation locally. The lapse rate of zonally averaged temperature is obvious while the zonally averaged height has no distinguished differences from latitudes.

Eight meridional cells are clearly shown in pentad 28 in Fig. 3.8, when the summer monsoon is climatologically onset in the South China Sea (Qian and Yang

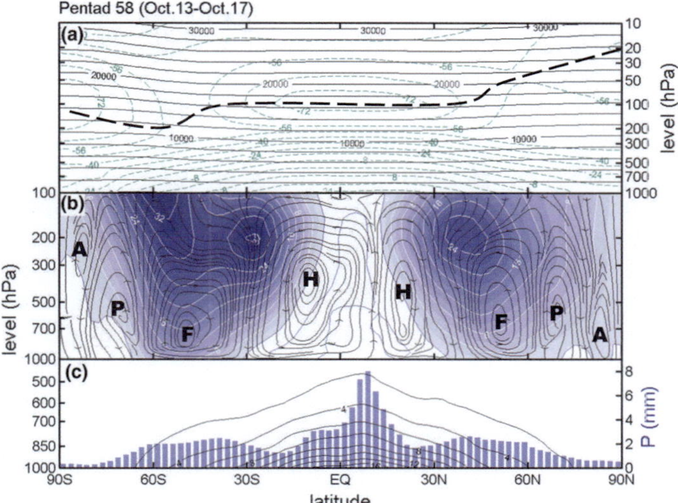

Fig. 3.10 Same as in Fig. 3.7 except on pentad 58 (13–17 October)

2000; Qian and Lee 2000). This is also the time for seasonal transition between the two hemispheres. The Arctic cell is lowly centered at 850 hPa while the Antarctic cell is highly centered between 200 and 300 hPa. The precipitation peaks at 5–10°N, where the ascending branches of the two Hadley cells are combined. In the SH, the Hadley cell has two centers separated in the upper and lower troposphere. The eight cells are well defined by their flows, such as the low-level northeasterly for the North Polar cell and low-level southeasterly for the South Polar cell.

Seasonally, pentad 44 is the time when the atmospheric variables reach their largest values in the NH. Figure 3.9 shows the zonally averaged climatic precipitation, height, temperature, meridional–vertical cells and westerly wind speed on pentad 44. Pentad 44 is during the boreal summer, when the summer monsoon reaches its northernmost location, so the ascending branch of the two Hadley cells combines near 20°N. The eight cells are also climatologically well defined by their flows. Another transition time from the boreal summer to winter is in pentad 58 (Fig. 3.10). Eight cells are also well defined by their flows in this pentad. The precipitation peaks at the 5–10°N with a strong ascending branch.

3.3 Zonal Asymmetric Vertical Structure

While the Hadley, Ferrel, Polar, and Arctic cells are major vehicles in global heat and water vapor transport, they do not work alone. These meridional circulations are the consequence of the fact that incidence of solar radiation is the highest at the "heat equator," and decreases as the latitude increases, reaching its minimum at the poles. Disparities in temperature also drive a set of longitudinal circulation cells, which are known as the zonal overturning circulations influenced by zonal sea–land contrast.

Longitudinal circulation forms because water in oceans has a higher specific heat capacity than land, with less temperature changes. Even at mesoscales, this effect is noticeable. It is the reason for sea breeze in the day and land breeze at night. This is also a climatic state of local circulation with diurnal cycle, which has been described in Chap. 2.

On a larger scale, warm air rises over the equatorial, continental, and western Pacific Ocean regions. When it reaches the tropopause, air flows eastward or westward depending on its location, and subsides in other regions. Figure 3.11 shows the vertical–zonal climatic departure of temperature, height, wind, and precipitation along the equatorial circle at various pentads. In all seasons, warm and cool airs are opposite in the western and eastern Pacific in the troposphere. Obviously, deep and warm air exists in the troposphere over the western Pacific. On the other hand, shallow and cool air exists in the lower troposphere over the eastern Pacific, but they are opposite in the upper troposphere. Height differences are not obvious along the equatorial zone and in the troposphere.

A typical zonal circulation describes that the air rises over the warm western Pacific to the upper troposphere and descends over the cool eastern Pacific, so westerly in the upper troposphere and easterly in the lower troposphere are observed. This cell in the equatorial Pacific is of such importance that it has been named as the "Walker circulation" after Sir Gilbert Walker, a director of British observatories in India in early twentieth century, who sought a means of predicting when the monsoon would fail. While he was never successful in doing so, his work led him to the discovery of an indisputable link between periodic pressure variations in the Indian Ocean and the Pacific, which was termed the "Southern Oscillation."

The Walker cell over Pacific basin is only observed in the boreal summer such as pentad 44 (Fig. 3.11b). In pentad 8 (Fig. 3.11a), there is no typical overturning circulation in the equatorial Pacific basin, but two typical overturning circulations are observed in the equatorial Indian and Atlantic basin, respectively. These two overturning circulations are referred to as the Walker-like cell. In pentad 44, three typical overturning circulations are well defined over the three equatorial oceanic basins, but the overturning circulation over the Indian Ocean has a reversed and closed loop next to the Walker cell.

More precipitation is found in the western Pacific with warm and rising flow. In the equatorial Atlantic, warm-rising flow is also observed over the western part and the cool subsiding flow is over its eastern part, so there is more precipitation in the equatorial western Atlantic Ocean. Along the equatorial Indian Ocean, precipitation well corresponds with cross-equatorial southwesterly flow in the lower troposphere, although the ascending flow is weaker than that in western Pacific and western Atlantic.

The maximum precipitation rate in pentad 44 along the equatorial zone is shown in Fig. 3.11b. In this pentad, four positive peaks are clearly indicated by "A", "B", "C",

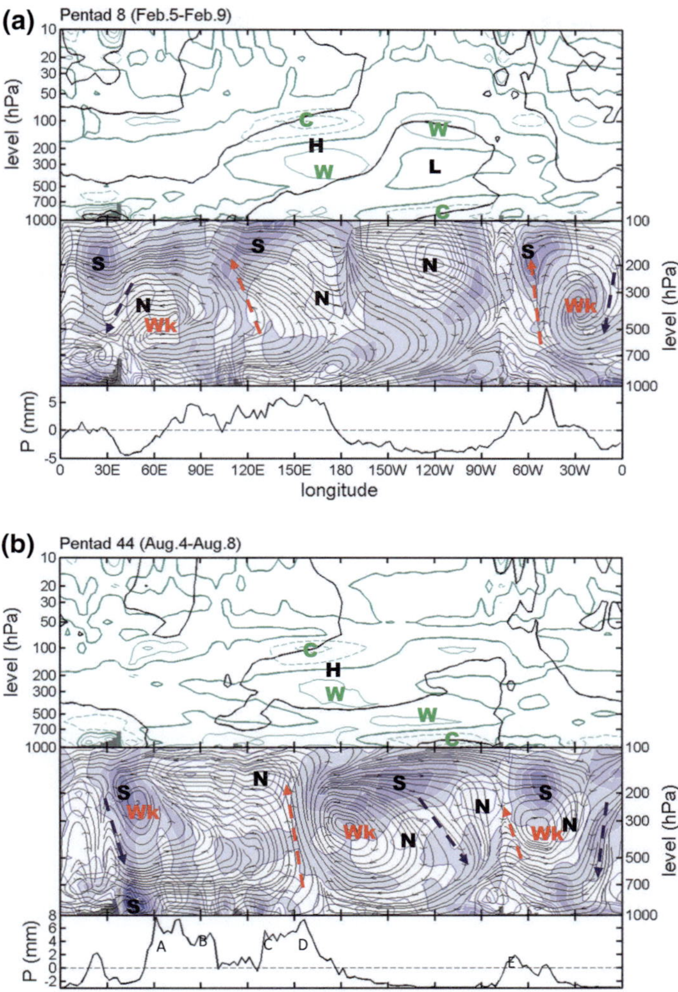

Fig. 3.11 Vertical sections of climatic departure along the equatorial circle in pentads **a** 8 and **b** 44. The *upper panel* shows the temperature departure (*green solid contour* for positive and *green dashed contour* for negative, 1 °C interval) and height departure (*black contour*, 4 × 10 gpm interval); the *middle panel* shows the longitudinal departure of circulation (*black bar* for topography) and meridional wind departure (*shading* for southerly, 4 m/s interval); the *lower panel* is the precipitation rate departure (mm/day). Letters "H/L" denote the center of height departure; letters "W/C" denote the center of temperature departure; letters "S/N" indicate the centers of the southerly and northerly, respectively; letter "Wk" indicates the Walker cell and the Walker-like cell with a overturning circulation. *Red- and blue-dashed arrows* indicate the ascending and descending flows, respectively

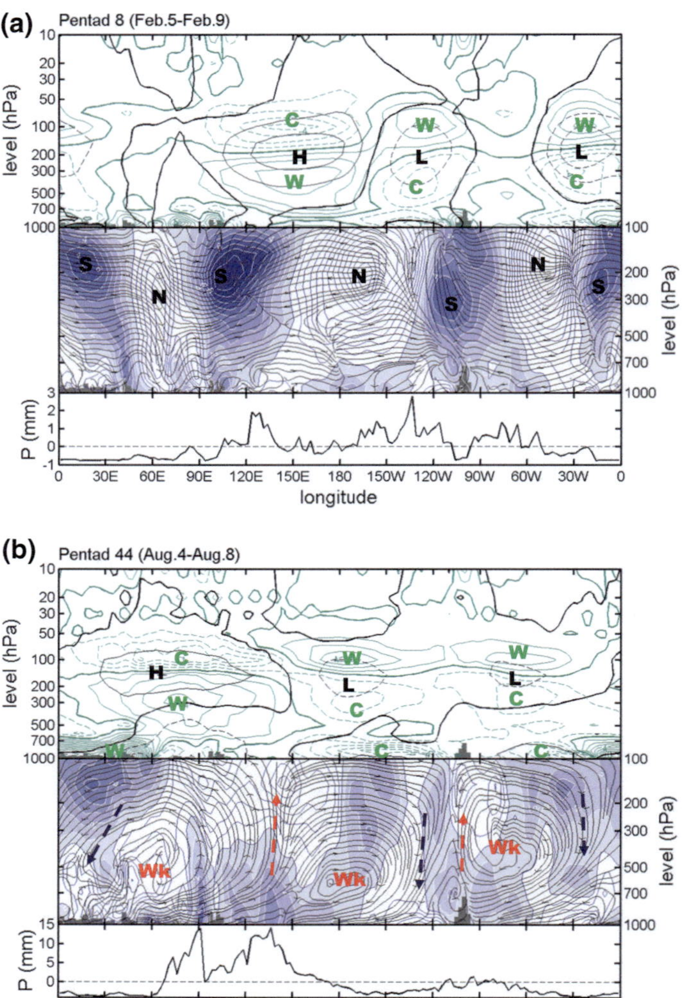

Fig. 3.12 Same as Fig. 3.11 except the vertical section of climatic departure along the 20°N in pentads **a** 8 and **b** 44

and "D" in the Arabian Sea, the Bay of Bengal (BOB), Southeast Asia and western Pacific, respectively. From the equatorial Indian Ocean to the equatorial western Pacific around 150°E, southwesterlies are prevailing in the lower troposphere and northeasterlies are prevailing in the upper troposphere. This cell in the equatorial Indian ocean is opposite in the direction to that in the Pacific and the Atlantic.

The relationship between height and temperature departures can be clearly observed along the 20°N. Temperature departures have opposite signs in the lower stratosphere and troposphere. In pentad 8 (Fig. 3.12a), a positive center of height departure is located in the mid-high troposphere of western Pacific, accompanied with a negative center and a positive center of temperature departure in its upper and lower parts. Opposite configuration is observed in the eastern Pacific and the Atlantic. In winter, there is no closed zonal cell along the 20°N. Peaks of precipitation departure are accompanied with southwesterly or southeasterly in the mid-lower troposphere.

The three Walker (or Walker-like) cells are evidently observed over along the 20°N in pentad 44 (Fig. 3.12b). The vertical pattern with a positive center of height departure at the upper troposphere separating a warm center below and a cold center above is observed in southern Asia, but with a negative center of height departure in the lower troposphere. The opposite configuration is found over the subtropical Pacific and Atlantic. The precipitation departure peaks around the rising flow of each Walker cell.

In boreal winter over the middle latitudes of NH, negative temperature departure and positive height departure (cold-high departure) are observed near the surface in the eastern part of the two continents while cold-low departures are in western part of the two oceans (Fig. 3.13a). The vertical pattern with a positive center of height departure at the upper troposphere separating a warm air column below and a cold air column above is located over the Northwest America. The same configuration is located in the Atlantic but the opposite configuration is in the Pacific. Two deep descending branches extending in the whole troposphere exist in the eastern part of the two continents while two deep ascending branches exist in the eastern part of the two oceanic basins. In the central part of the Eurasian continent, there is another ascending branch. Therefore, three southerly centers and three northerly centers are clearly observed in Fig. 3.13a. Baroclinity is clear from the troposphere to stratosphere. Positive precipitation departures are accompanied well with southwesterly rising flows.

In boreal summer, the surface warm departure is located in the Asian continent and the American continent along the 40°N (Fig. 3.13b). Over the Asian continent, there is warm center in the upper troposphere (around 200 hPa). Two centers of height departure are near northern parts of the SAH and over North America. In this zonal circle, positive precipitation departures are mainly located in the western Pacific and the western Atlantic. The Walker-like cell does not exist over this latitude in boreal summer. Those systems illustrated by height, temperature, and wind departures become complicated with small scale.

The configuration of temperature-height-flow departures with baroclinity is simple and clear from the troposphere to stratosphere in the high latitudes along the 60°N during boreal winter (figures not shown). The cold-low departures extend downward and eastward from western Europe to Northeast Asia, so the East Asian Trough is climatologically observed at the lower troposphere in winter. The North Atlantic Trough is vertically active in boreal winter. Two peaks of major precipitation departure are consistent with the two ascending flow branches. In boreal summer, the configuration becomes complicated. Many systems of temperature-height-flow departures with many peaks of precipitation departure are observed.

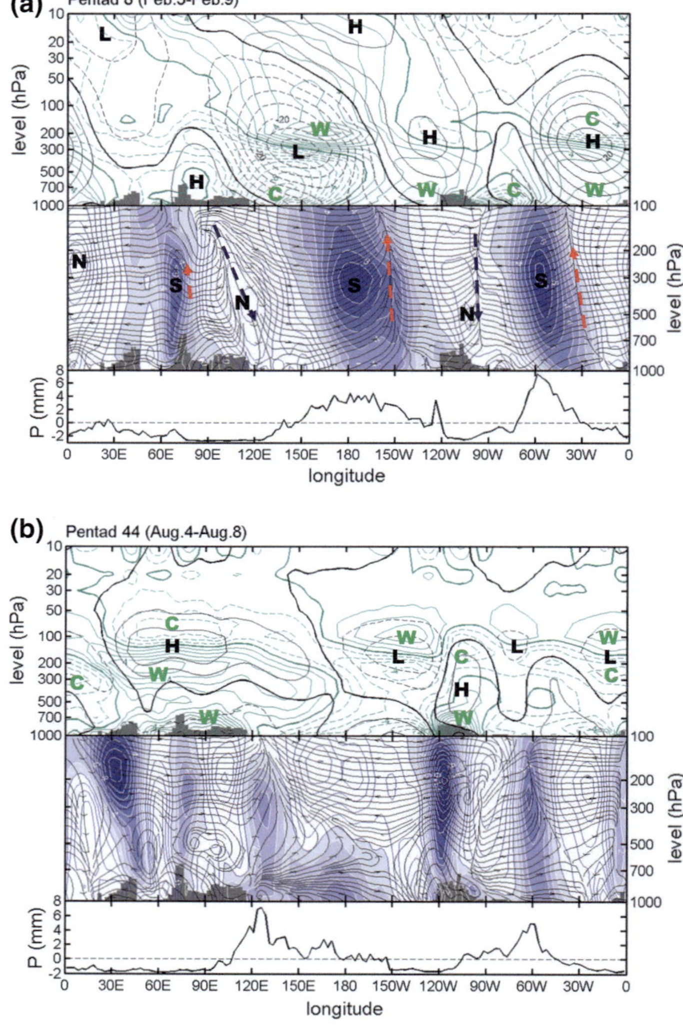

Fig. 3.13 Same as Fig. 3.11 except the vertical section of climatic departure along the 40°N in pentads **a** 8 and **b** 44

In austral summer (pentad 8) and along the SH subtropical area (20°S), three centers of surface warm departure are found in South Africa, Australia, and South America (Fig. 3.14a). Near these places, rising flows are accompanied with positive precipitation departure. Three subsiding flows are found in the subtropical Indian Ocean, the subtropical Southeast Pacific, and the subtropical Southeast Atlantic. It is somewhat different from that in the NH subtropical area, because there are not any basin-scale overturning circulations over the SH subtropical area in pentad 44 (Fig. 3.14b). Flows are waved in the Pacific. Along the 60°S, there are not any

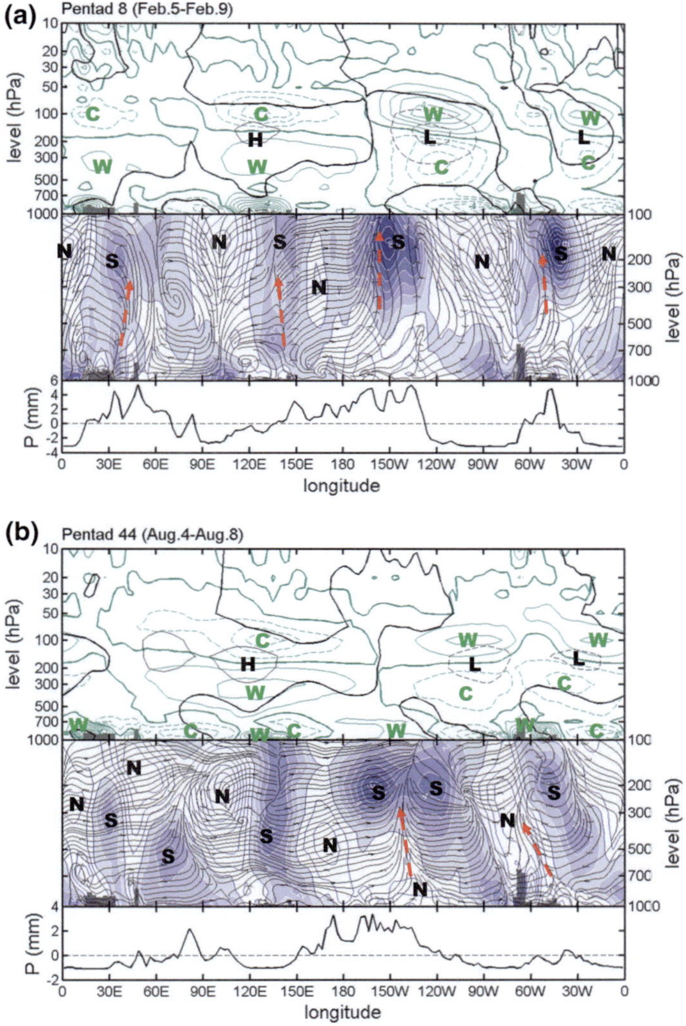

Fig. 3.14 Same as Fig. 3.11 except the vertical section of climatic departure along the 20°S in pentads **a** 8 and **b** 44

basin-scale overturning circulations in winter and summer either, but there are three pairs of southerly and northerly flows changing their directions in the three basins (figures not shown).

The Walker cell in the Pacific basin plays a particularly important role in the Earth's climate. This entirely ocean-based cell is the result of a marked difference in the surface temperature between the western and eastern Pacific. Under ordinary circumstances, the western Pacific are warm and the eastern are cool. The process begins when the strong convective activity over equatorial East Asia and the subsiding cool air off South America's west coast create a wind pattern, which pushes the water westward and piles it up over the western Pacific. Water levels in the western Pacific are about 60 cm higher than those in the eastern Pacific, a difference entirely due to the force of moving air. The closed cell in the Pacific and other two closed cells in the Atlantic and Indian Ocean only exist in the equatorial zone and the NH subtropical area.

3.4 Horizontal Patterns of Climatic Variable Departure

The influence of sea–land contrast on atmospheric variable fields is strong in the lower troposphere than that in the upper troposphere. This influence can be well described by the climatic departure of geopotential height (pressure), temperature, and wind. All centers of atmospheric action change with seasons and altitudes. Monsoon can also be well illustrated by the climatic departure of wind field.

Daily mean and pentad mean climatic departures can be estimated at various isobaric levels. Figure 3.15 shows the estimation of climatic departure indicated by winds at 1000 hPa and the surface level pressure. In pentad 3 (Fig. 3.15a), the Siberian High in the east part of Eurasian continent and the Aleutian Low in the northern part of North Pacific, as two atmospheric active centers, reach their maximum intensity. Therefore, a large gradient of sea level pressure (SLP) exists and northerly prevails in the northwestern Pacific. Other two atmospheric active centers are the Iceland Low in north and the Azores High in south over the North Atlantic. Their SLP gradient forms a prevailing southwesterly band in the middle latitudes of the North Atlantic. Some climatic convergence lines and climatic troughs are indicated by those dashed lines in several basins.

The surface circulation in boreal winter is also characterized in pentad 8 (Fig. 3.15b). Two climatic troughs are deeply located in the Northwest Pacific and the Northwest Atlantic. Other two climatic troughs from the Iceland Low extend northwestward along the Baffin Bay, northeastward in the Norwegian Sea, and the Barents Sea. In the tropical zone, there are convergence zones in the subtropical South Indian Ocean, Southwest Pacific, subtropical Northeast Pacific, and the equatorial Atlantic. Other two climatic troughs indicated by SLP departure are located in middle latitudes of the SH.

It is well known that the pentad 28 is a season of the monsoon onset over the SCS. As shown in Fig. 3.15c, the North Pacific High and the North Atlantic High become two main anticyclonic systems in the NH. The convergence zones are active in the tropics. In the region of Asian marginal seas, four peninsular scale troughs are located in the BOB, the SCS, the East China Sea (ECS), and the Japan Sea.

The atmospheric active centers in the NH become the clearest in pentad 44 (Fig. 3.15d). In Asia and Pacific, the strongest low is centered over Asia and the strongest high is located in the North Pacific. So there is southwesterly monsoon flow in South Asia and southerly monsoon flow in the East Asia. Around the Asian

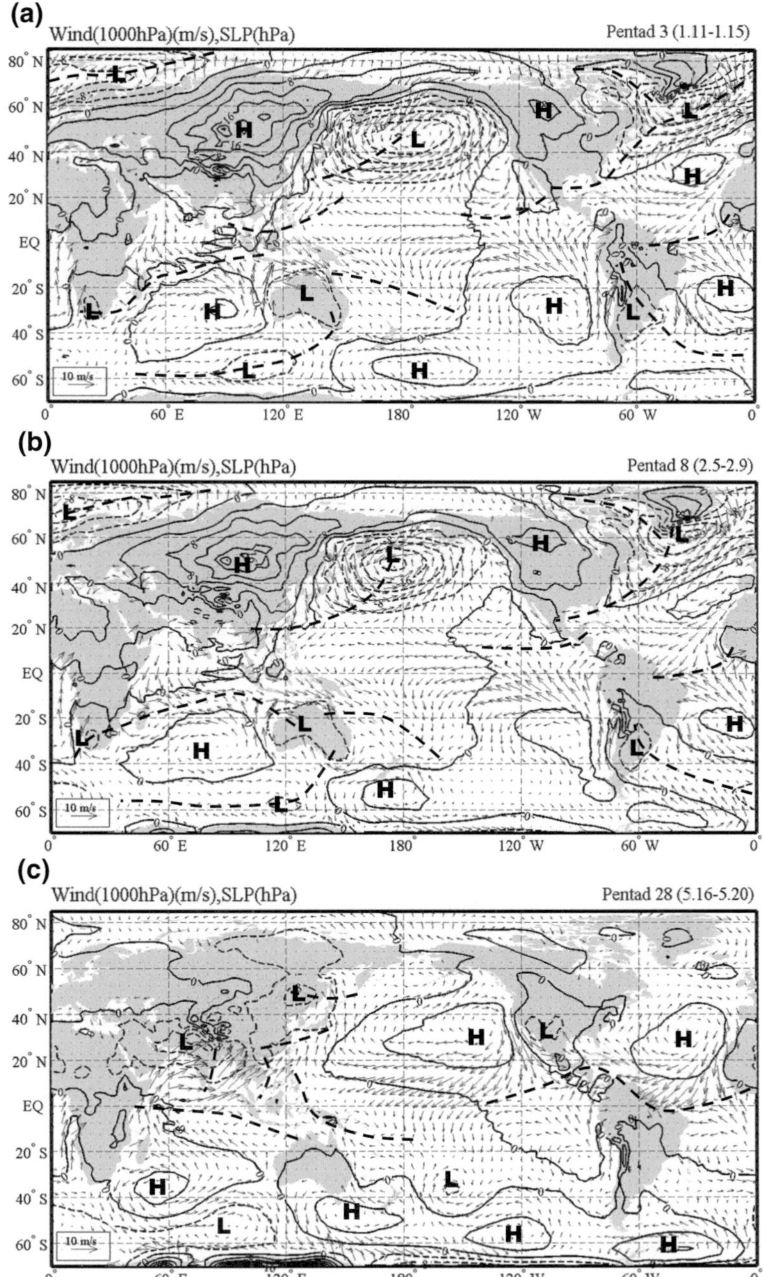

Fig. 3.15 The climatic departure of wind (m/s) at 1000 hPa and surface level pressure (*solid line*, 4 hPa interval) in **a** pentad 3 (11–15 January), **b** pentad 8 (5–9 February), **c** pentad 28 (6–10 May), and **d** pentad 44 (4–8 August). *Heavy dashed lines* are climatic troughs and monsoon troughs. Letters “H/L” indicate high/low centers

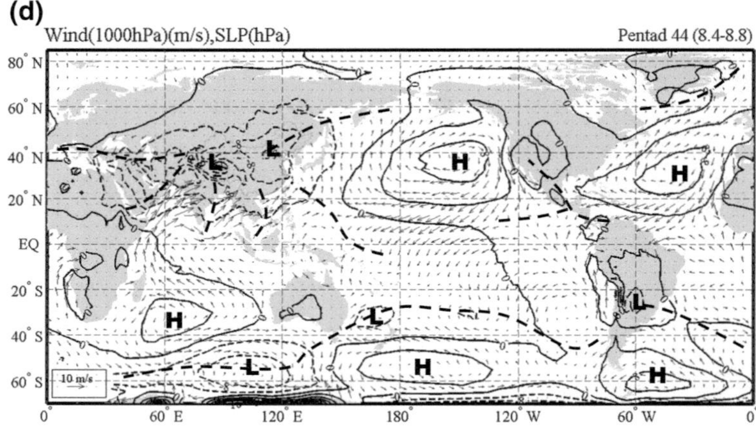

Fig. 3.15 (continued)

continental low, five peninsula troughs are seasonally located in the Mediterranean Sea, the Arabian Sea, the BOB, the SCS, and over the Northeast Asia (Qian and Tang 2010). In the subtropical Northwest Pacific, the ITCZ seasonally reaches the Southeast Asia. In the region of the North Atlantic and North America, the strongest high is the North Atlantic High while three weak troughs are observed in the high-latitude Atlantic and tropical Northeast Pacific region. In the SH, there is a trough extending along the middle latitudes.

In boreal winter (pentad 8, Fig. 3.16a), atmospheric active centers in the NH are obvious at 850 hPa, with three positive centers of height departure in the Asian continent, the Northwest American, and the subtropical Southeast Atlantic, and with two negative centers of height departure in the North Pacific and the Northeast American. Two troughs are situated in east of Japan and the eastern coast of North America. The geopotential height contours are mostly perpendicular to temperature contours to the northwest of the two troughs. Therefore, there are strong cold advections in East Asia and eastern America considering the geostrophic relation. Reversely, warm advections are prevailing in the western coast of America and Europe. In the tropical and SH regions, several warm centers are observed in the continents of Africa, South Asia, South America, and Australia. In the SH middle latitudes, there are four centers of height departure, respectively, in the South Indian Ocean, the South Atlantic, and the South Pacific. Troughs along the equatorial Pacific and Atlantic and in SH subtropical area are well accompanied with major precipitation zones.

Pentads 26 and 28 show a rapid transition before and after the SCS monsoon onset. In pentad 26 (Fig. 3.16b), three troughs are well accompanied with more precipitation in the BOB, from South China to Japan, and along the ITCZ surrounding the subtropical high. Similar relationship among the subtropical high, its surrounding precipitation, and troughs can also be observed in the North Atlantic. In the SH, a zonal trough is well accompanied with more precipitation mostly crossing the three basins.

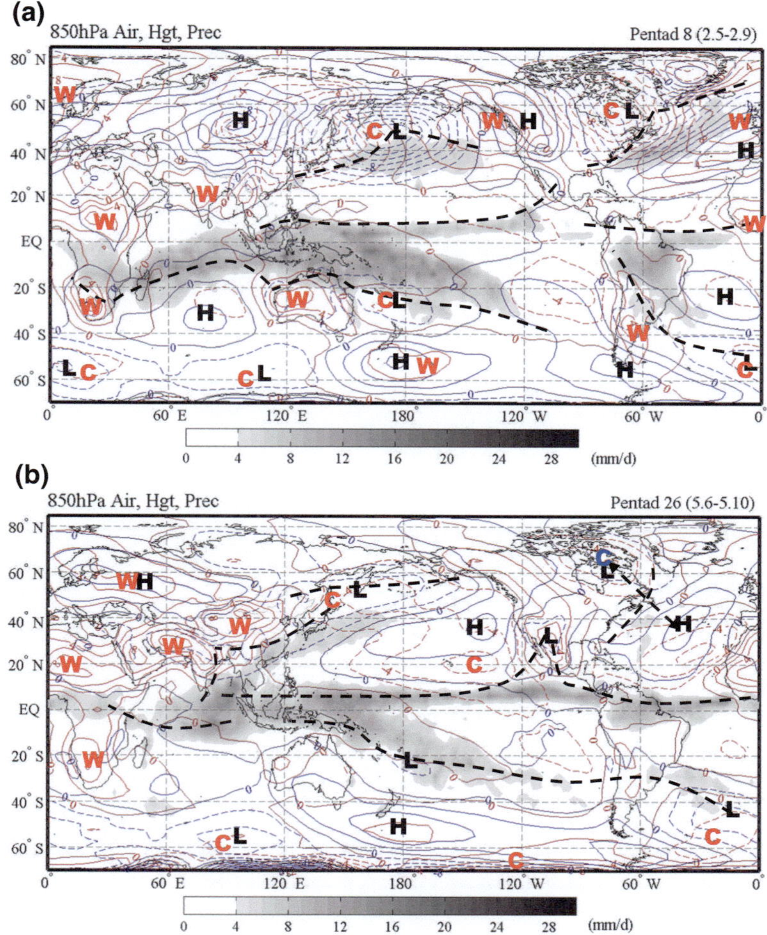

Fig. 3.16 The climatic departure of height (*blue contour*, 20 gpm interval) and temperature (*red contour*, 2 °C interval) at 850 hPa in **a** pentad 8 (5–9 February), **b** pentad 26 (6–10 May), **c** pentad 28 (16–20 May), and **d** pentad 44 (4–8 August). *Heavy dashed lines* are climatic troughs and monsoon troughs. Letters "H/L" and "W/C" indicate high/low and warm/cold centers

In pentad 28 (Fig. 3.16c), precipitation departure larger than 4 mm/day appears in the SCS and the peninsular scale monsoon trough is also formed in the SCS. In pentads 26 and 28, high and low centers with similar intensity exist in the two hemispheres. In the mid-low latitudes of NH, there are three high centers in the western Europe, subtropical North Pacific, and subtropical North Atlantic. In the mid-high latitudes of SH, there are one high center in the South Indian Ocean, two in the South Pacific, and one in the South Atlantic. Warm centers of temperature departure are mainly located in continents of Africa, South Asia, Northwest-North China, and western North America. During this transition season, temperature

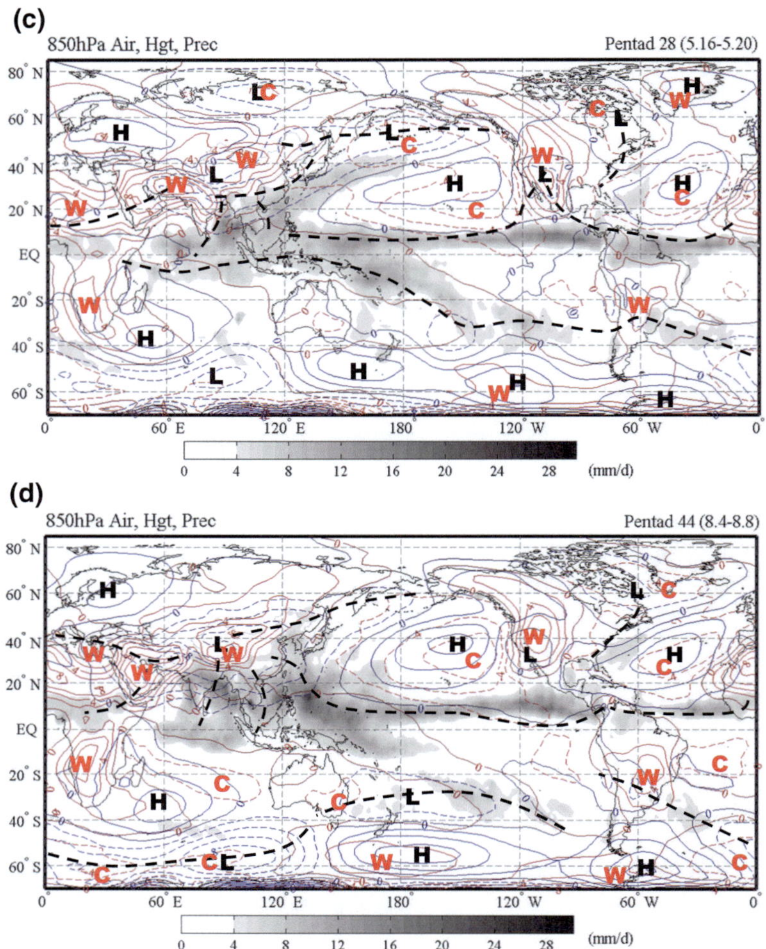

Fig. 3.16 (continued)

departures are relatively lower in Southeast China and eastern America. The subtropical summer monsoon trough forms from East China to Japan and connects to a trough in the northern part of North Pacific. Between the subtropical monsoon troughs in the East Asia and the subtropical high in the North Pacific, southwesterly should be prevailing from the SCS to Japan.

In pentad 44, the largest warm-low center and the largest cold-high center form in the Asian continent and the North Pacific, respectively (Fig. 3.16d). Another pair of warm-low and cold-high centers is located in the western America and the subtropical North Atlantic. Six troughs are surrounding the largest warm-low center in the Asian continent. The Mediterranean Sea trough is relatively dry. Three peninsula troughs in the BOB, the SCS, and the East China Sea have precipitation departure more than 4 mm/day. The ITCZ in the Northwest Pacific extends to the Southeast

China with positive precipitation departure. The subtropical summer monsoon trough and positive precipitation departure belt seasonally advance to the northernmost part in the Northeast Asia in pentad 44. In this pentad, gradients of height and temperature departures are large between the East Asia and Northwest Pacific, so southerly monsoon flow should be the strongest with wet climatic departure. Longitudinal gradients of height and temperature departures are also large in the western coast of North America, but its climate is different from that in East Asia.

Climatic departure indicated by geopotential height and temperature at 500 hPa is rather clear in Fig. 3.17. In the mid-high latitudes (Fig. 3.17a), high and low

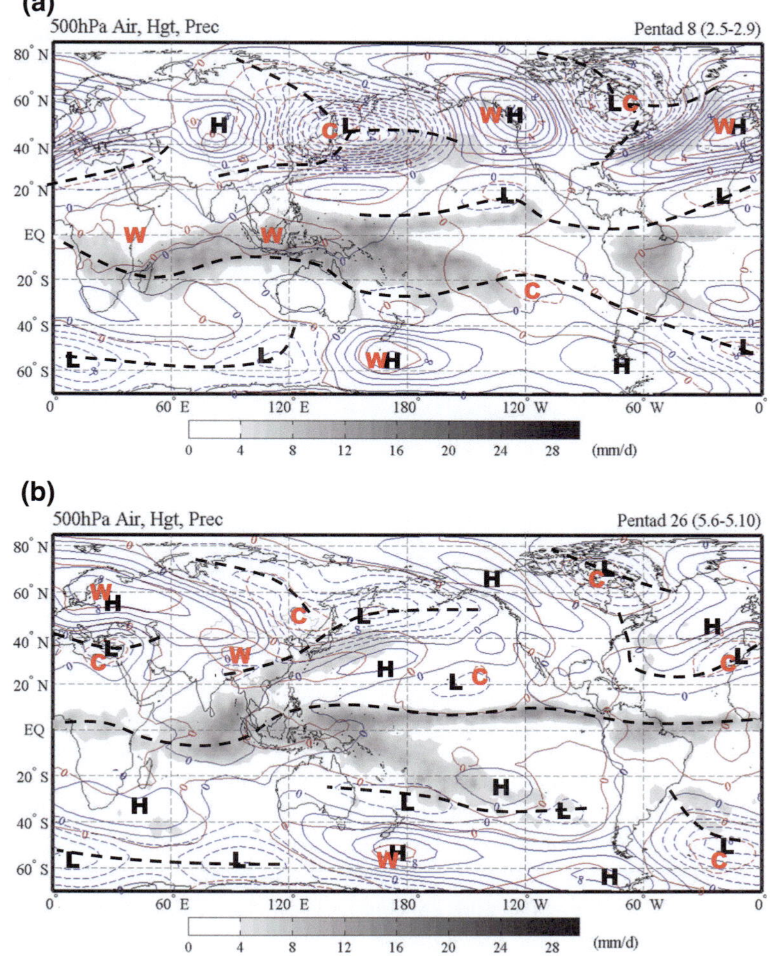

Fig. 3.17 Same as in Fig. 3.15 except of the climatic departure of height and temperature at 500 hPa in **a** pentad 8 (5–9 February), **b** pentad 26 (6–10 May), **c** pentad 28 (16–20 May), and **d** pentad 44 (4–8 August)

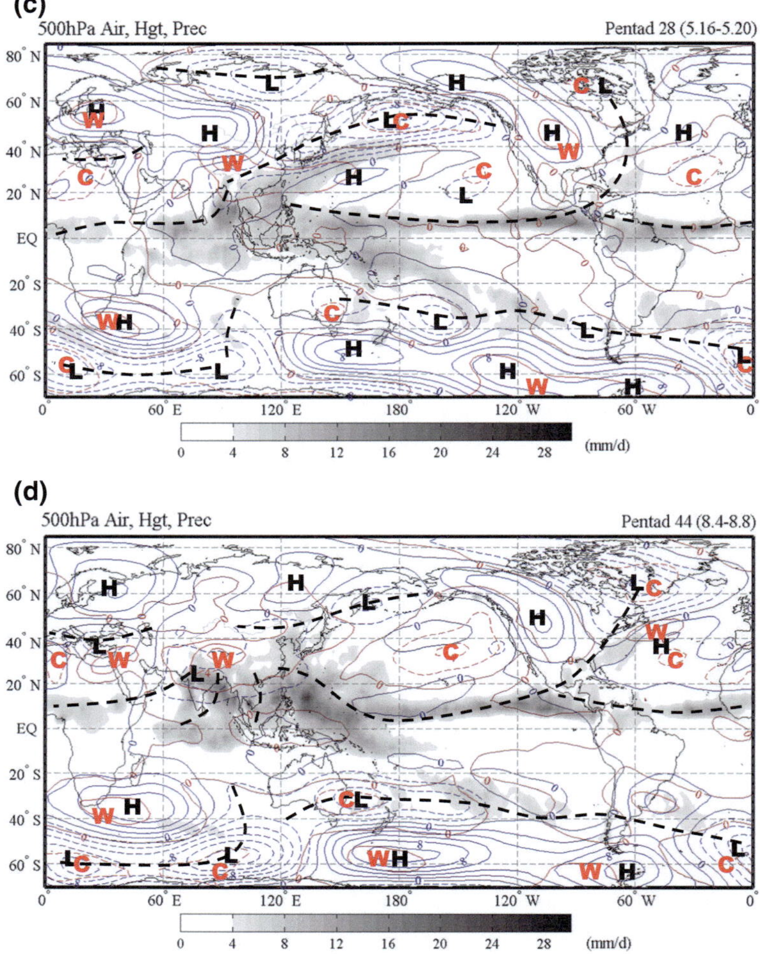

Fig. 3.17 (continued)

centers are, respectively, accompanied with warm and cold centers. In the mid-low latitudes, troughs are accompanied with positive precipitation departure nearby, except in the North Africa and the Mediterranean Sea. In pentad 8, there are two warm-high centers in Northwest North America and the Northeast Atlantic, separated by two cold-low centers near Japan and Northeast America, except another high in Northwest China.

In pentads 26 (Fig. 3.17b) and 28 (Fig. 3.17c), the ITCZs extend along the equatorial circle with positive precipitation departure. In SH, troughs in the South Pacific and the South Atlantic are accompanied with positive precipitation departure. The East Asian subtropical monsoon trough seasonally advances northward after the summer monsoon onset in the SCS.

In pentad 44 (Fig. 3.17d), peninsular scale monsoon troughs in South Asia and East Asia as well as ITCZs and troughs in the Pacific and the Atlantic are well accompanied with positive precipitation departure. High and low centers in the NH, except the low in the Northeast Canada, are weak. In the mid-high latitudes of SH, warm-high and cold-low centers become comparable.

In pentad 8 (Fig. 3.18a), cold and warm centers separated by high and low centers are clear in the mid-high latitudes of the NH at 300 hPa. Pentad 28 (Fig. 3.18b) is the transition season in the two hemispheres, so the intensities of high-low and warm-cold centers are weak. In pentad 44 (Fig. 3.18c), those centers of warm/cold and high/low departures are similar to those in pentad 28 in their intensities. The North Pacific Low (NPL) moves eastward and the SAH moves northwestward from pentad 8 to pentad 28 and pentad 44.

In the stratosphere at 30 hPa, seasonal variations of warm/cold and high/low centers are obvious. In pentad 8 (Fig. 3.19a), there is only a pair of warm/cold and high/low centers in the mid-high latitudes of the NH. In pentad 44 (Fig. 3.19b), only a pair of warm/cold and high/low centers is predominant in the mid-high latitudes of the SH. Warm/cold and high/low centers are not accompanied with each other at their locations. The southerly warm advection exists in Northeast Asia, where it may be the channel to form the warming near the North Pole.

3.5 Seasonal Variation of Climatic Variable Departure

Change in intensity and location of atmospheric active centers forms the seasonal variations of different basic variables. The SDs of climatic daily departures of temperature, height, and winds at different isobars provide information of their seasonal variations. At each spatial point, a time series of variables has 365 daily time points in its seasonal cycle. Some major characteristics are illustrated in this section.

Seasonal SD distributions of temperature departure at three isobaric levels have some regional characteristics. At 300 hPa, large SD centers of temperature departure are mainly distributed in the NH. Three large SD areas are separated in the Asia, central Northeast Pacific, and North Atlantic with their values larger than 2 °C (Fig. 3.20a). At 500 hPa, three large SD areas of temperature departure are still in the NH, but they are centered in Northeast Asia, the Northeast Pacific, and the Northeast Atlantic with central values from 2.5 to 4 °C (Fig. 3.20b). At 850 hPa, the SD centers of temperature departure with its value larger than 6 °C are only found in Northeast Asia in the NH and over the Antarctic seas (Fig. 3.20c). Other SD centers with values of 4 °C appear in the Northeast Pacific, the Northeast Atlantic, and Southwest Asia. These large SD centers of temperature departure imply that there are large amplitudes of seasonal temperature cycle. The large SD centers of temperature departure are mainly concentrated in the NH and the lower troposphere.

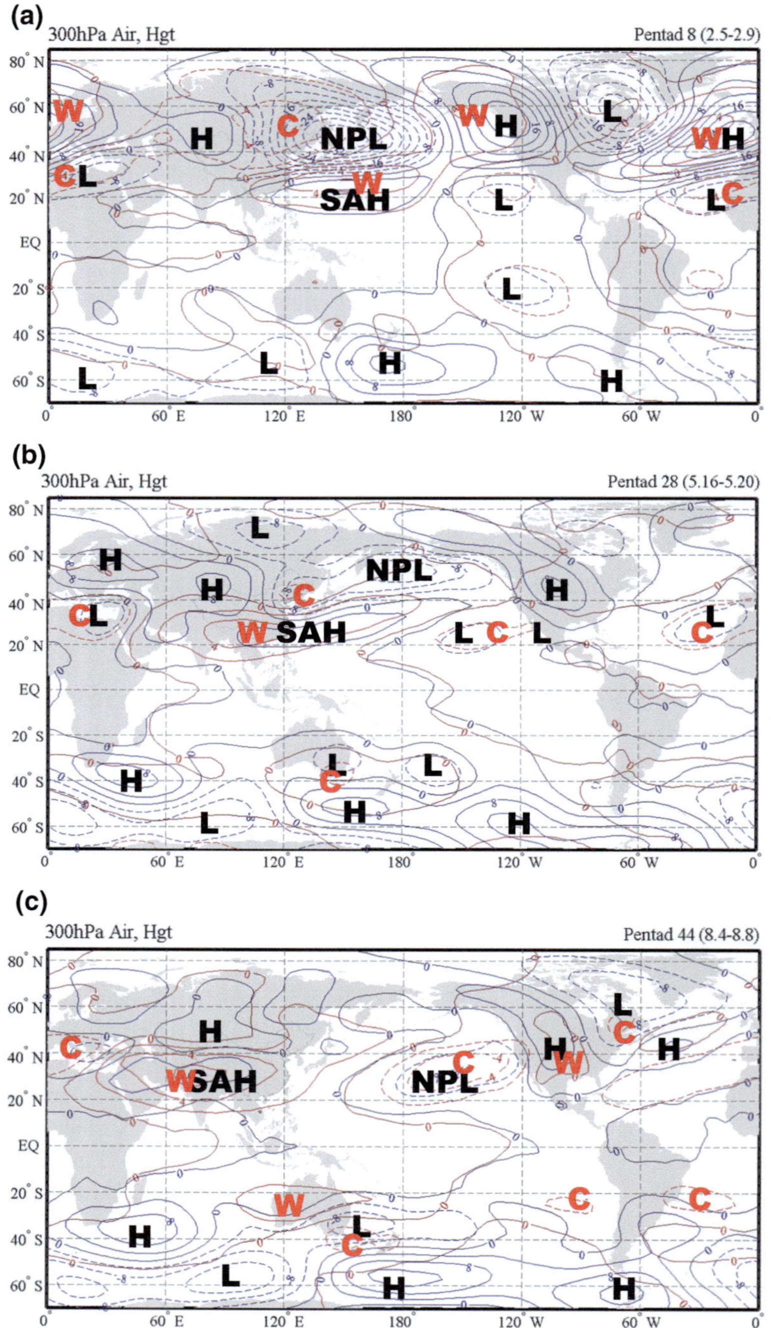

Fig. 3.18 Same as in Fig. 3.15 except of the climatic departure of height (*blue contour*, 40 gpm interval) and temperature (*red contour*, 2 °C interval) at 300 hPa in **a** pentad 8 (5–9 February), **b** pentad 28 (16–20 May), and **c** pentad 44 (4–8 August)

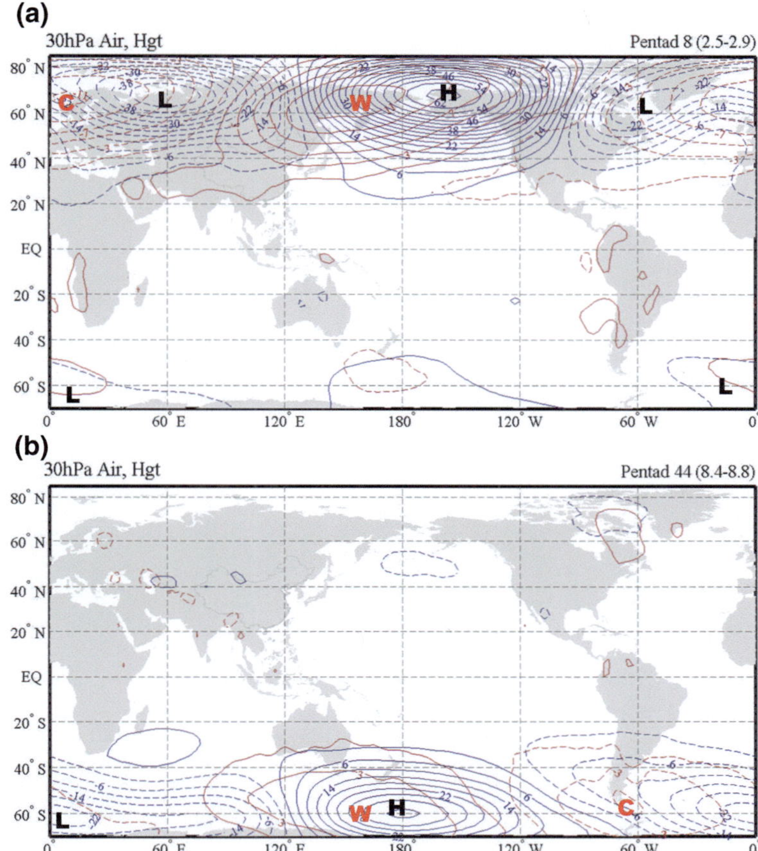

Fig. 3.19 The climatic departure of height (*blue contour*, 40 gpm interval) and temperature (*red contour*, 2 °C interval) at 30 hPa in **a** pentad 8 (5–9 February) and **b** pentad 44 (4–8 August)

At 300 hPa, large SD centers of height departure are mainly observed in Northeast Asia, Northeast Pacific, Northeast America, and Northeast Atlantic (Fig. 3.21a). The largest SD center of height departure with the value larger than 130 gpm is located in the Sea of Okhotsk. The second largest SD center of height departure with the value 100 gpm is observed near the Iceland. Only two SD centers of height departure with their values larger than 40 gpm are found over the Antarctic seas.

At 500 hPa, five large SD areas of height departure are observed in the NH (Fig. 3.21b). Their SD centers are noted in Kazakhstan, the Sea of Okhotsk, the Northeast Pacific, Northeast America, and the Northeast Atlantic. The largest SD center of height departure is located in the north of Japan with the value 80 gpm, which reflects the seasonal shift of the East Asian Trough. The second largest SD center of height departure is near the Iceland with the value 64 gpm. In the SH, there is a large SD zone of height departure between 40°S and 70°S.

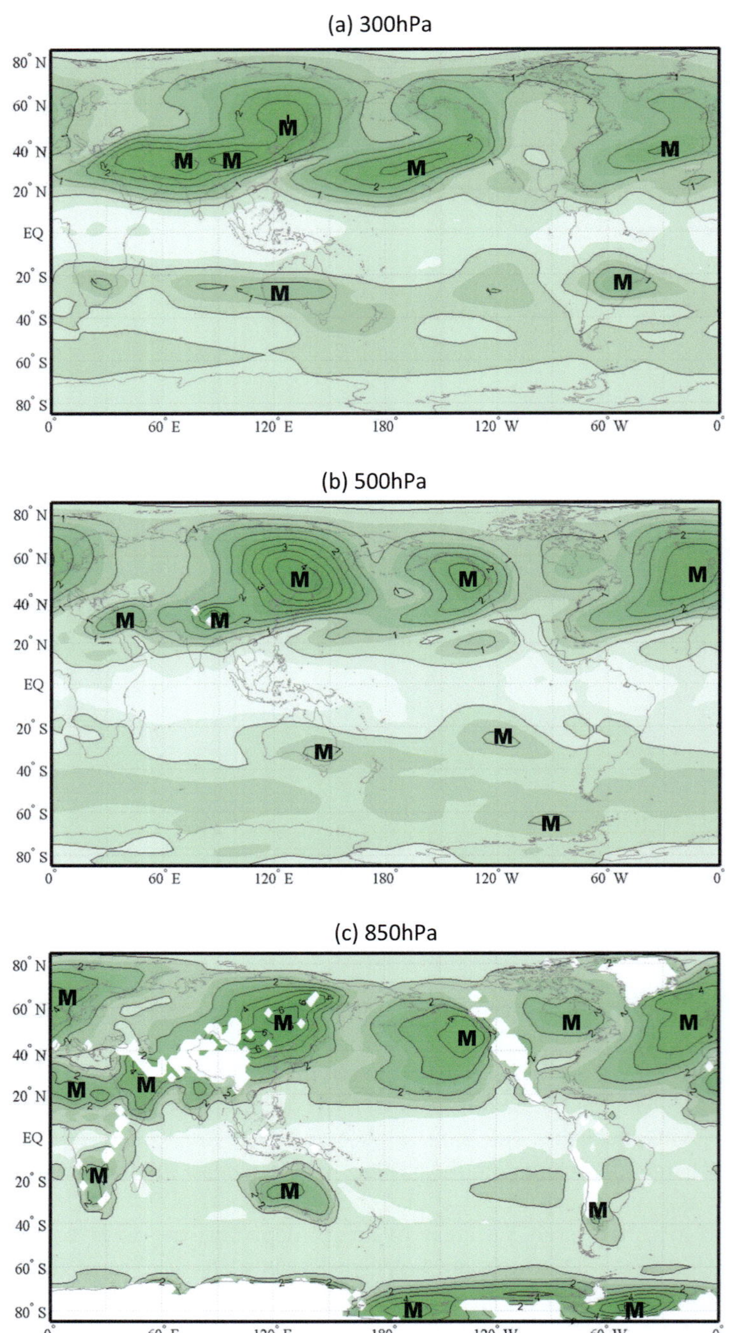

Fig. 3.20 Seasonal SDs of temperature departure at **a** 300 hPa (0.5 °C interval), **b** 500 hPa (0.5 °C interval), and **c** 850 hPa (1.0 °C interval). Letter "M" denotes the maximum center of SD

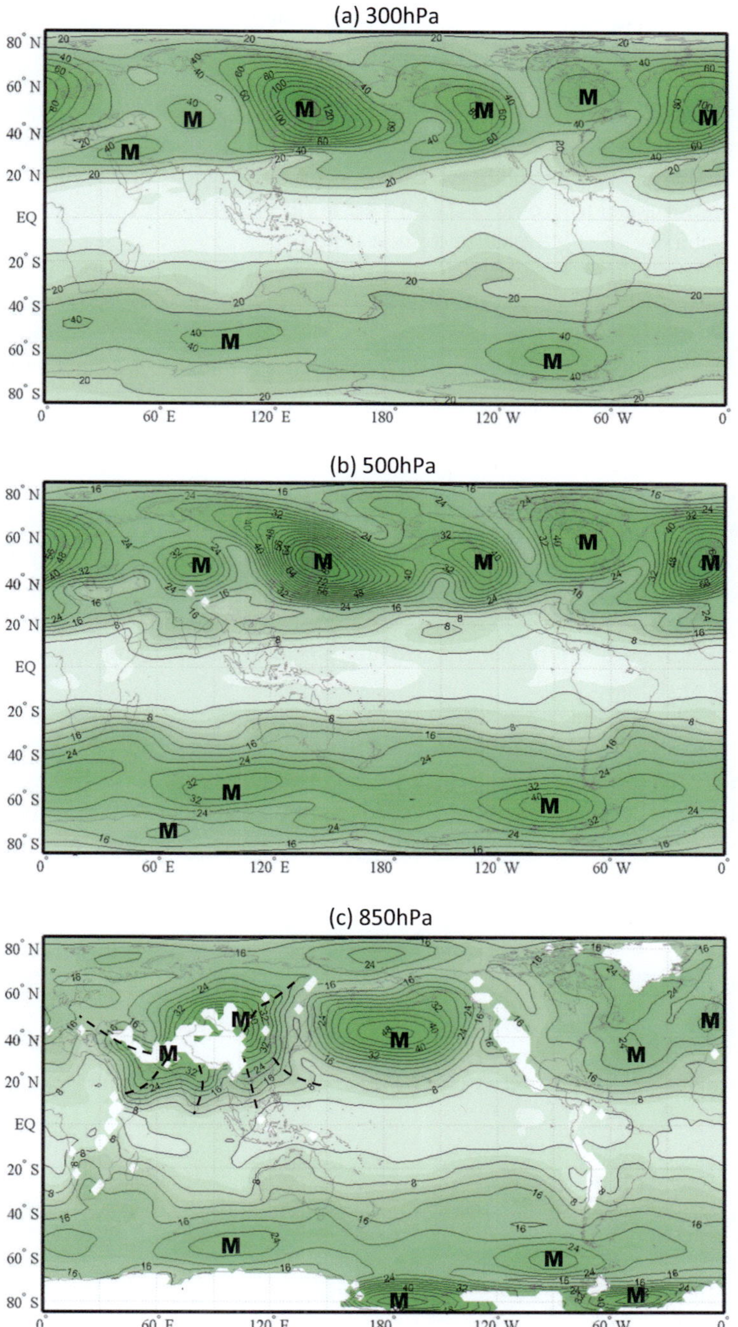

Fig. 3.21 Seasonal SDs of height departure at **a** 300 hPa (10 gpm interval), **b** 500 hPa (4 gpm interval), and **c** 850 hPa (4 gpm interval). The *heavy dashed line* in **c** indicates the active area of peninsular monsoon troughs. Letter "M" denotes the maximum center of SD

At 850 hPa, large SD centers of height departure with values larger than 40 gpm are located in Asia and the middle latitudes of the North Pacific (Fig. 3.21c). It is interesting to note that there are six dashed lines surrounding the Tibetan Plateau from its western to its southern and northeastern sides. They are the activities of peninsular monsoon troughs in the Mediterranean Sea, the Arabian Sea, the BOB, and the SCS. From the Southeast China to the subtropical Northwest Pacific, it reflects the seasonal activity of ITCZ. In the north of Northeast China, it reflects the seasonal activity of subtropical monsoon trough.

In the troposphere, seasonal variation of vertical velocity departure is examined at 500 hPa. Figure 3.22 shows that those large SD areas of seasonal vertical velocity extend from the Mediterranean Sea, Asia, the mid-latitude North Pacific to the Northwest America. It means that these regions have large seasonal variation of convective activity.

In the atmosphere, water vapor is mainly concentrated in the mid-low troposphere. Figure 3.23 shows the seasonal SD distributions of specific humidity departure at 500, 700, and 850 hPa. At 500 hPa, large variation of specific humidity departure is located around the Tibetan Plateau and extends into South Asia, Southeast Asia, and East Asia (Fig. 3.23a). At 700 and 850 hPa, large seasonal variations of specific humidity departure are clearly observed in the global monsoon regions (Fig. 3.23b, c). These regions are from North Africa, South Asia, and East Asia to subtropical North America in the NH as well as from South Africa to Australia in the SH.

The seasonal variations of zonal wind in the troposphere are examined at 300, 500, and 850 hPa. At 300 hPa, the large SD distributions of zonal wind departure are mainly concentrated between 20°N and 40°N (Fig. 3.24a). There are three parallel large SD zones of zonal wind departure. The longest one extends from off-equatorial North Atlantic to North Africa, South Asia, East Asia, and the east of Japan (Northwest Pacific). Along this zone, two SD centers are located in North

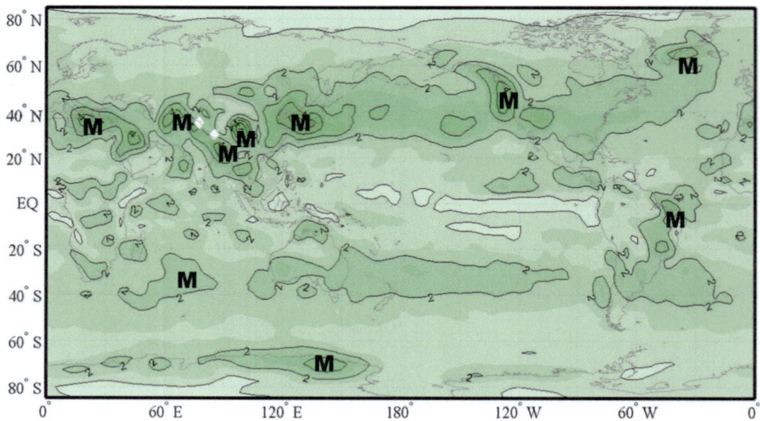

Fig. 3.22 Seasonal SD (0.01 Pa/s) of vertical velocity (omega) departure at 500 hPa

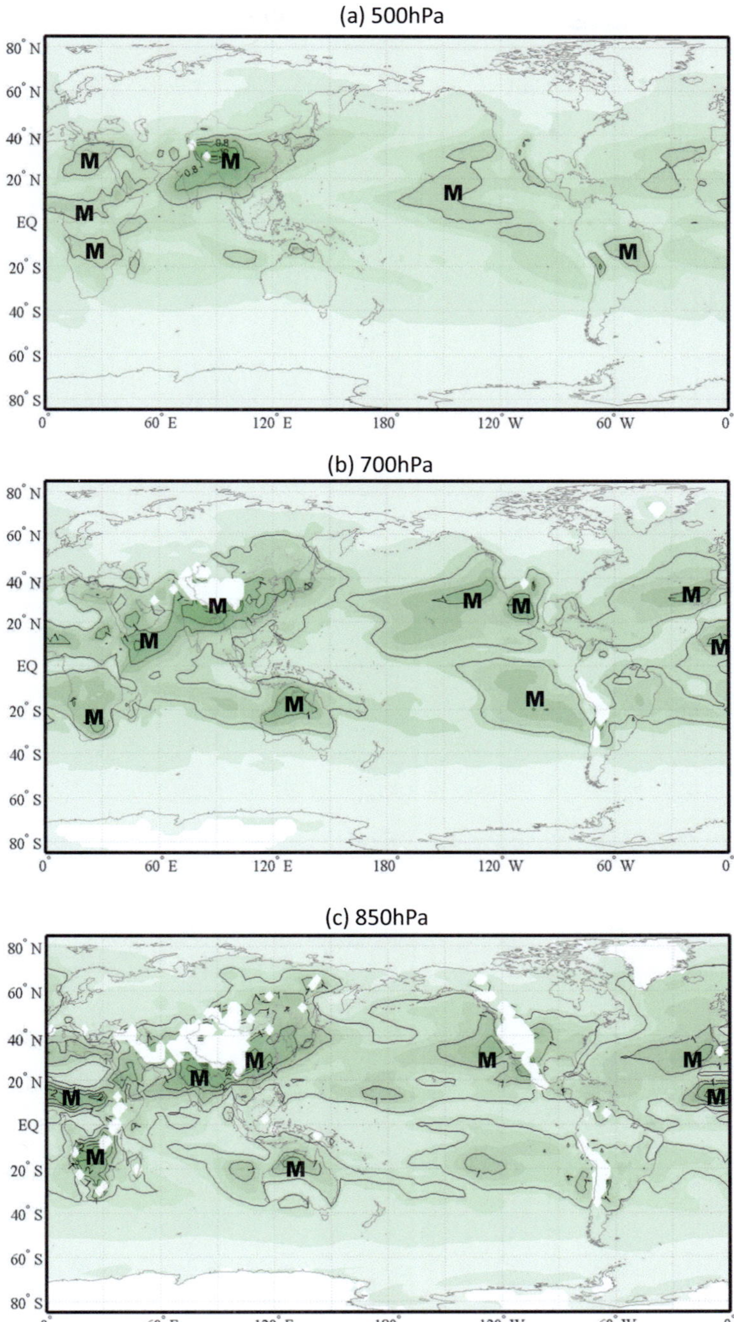

Fig. 3.23 Seasonal SD of specific humidity departure at **a** 300 hPa (0.4 g/kg interval), **b** 500 hPa (0.5 g/kg interval), and **c** 850 hPa (0.5 g/kg interval)

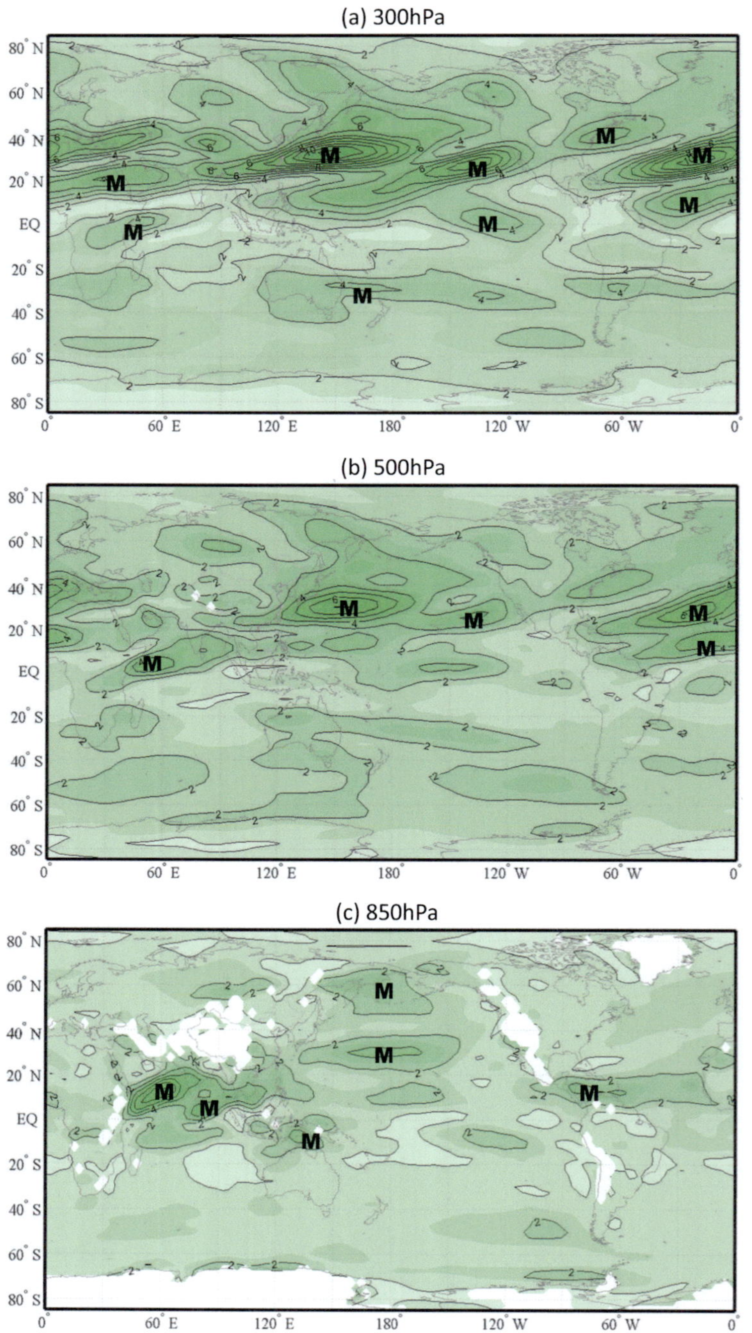

Fig. 3.24 Seasonal SD (1 m/s interval) of zonal wind departure at **a** 300 hPa, **b** 500 hPa, and **c** 850 hPa

Africa and over Japan (12 m/s). Along other large zones, two large SD centers are found in the subtropical Northeast Pacific and the subtropical Northeast Atlantic.

At 500 hPa, there are also three large and parallel SD zones of zonal wind departure between 20°N and 40°N (Fig. 3.24b). The two largest SD centers with values 6 m/s are located in the east of Japan and the subtropical Northeast Atlantic. At 850 hPa, large SD distributions of seasonal zonal wind departure with values larger than 4 m/s are mainly found from the equatorial East Africa to the Arabian Sea and the southern BOB (Fig. 3.24c). In these two SD regions, westerlies and easterlies are prevailing in summer and winter, respectively.

Seasonal SD distributions of meridional wind departure in the troposphere are examined at 300, 500, and 850 hPa. At 300 hPa, large seasonal variations of meridional wind departure are mainly concentrated between 20°N and 70°N (Fig. 3.25a). Near 60°N, there are about six large SD centers of meridional wind departure with values larger than 5 m/s. In the subtropical NH, there are several other large SD centers of meridional wind departure. Between 40°S and 60°S, there is a large SD zone.

At 500 hPa, there are five large SD centers of meridional wind departure with values of 3 m/s centered in Northeast Asia, Northeast Pacific, western America, central North Atlantic, and the northwestern part of North Africa (Fig. 3.25b). A large zone is situated around in middle latitudes (40–60°S) over the SH. At 850 hPa, large SD areas of meridional wind departure with values larger than 2 m/s are mainly centered from the tropical East Africa to the tropical western Indian Ocean, the BOB, the SCS, Northeast Asia, and western North America (Fig. 3.25c). Large centers in the tropical eastern Africa, South Asia, Southeast Asia, and East Asia indicate the transition of monsoon flows between summer and winter. The seasonal transition of Somali cross-equatorial flow is characterized with the largest SD center (5 m/s).

It can be summarized that large seasonal variations of atmospheric variables are mainly concentrated between 20°N and 60°N with several SD centers. The largest departure center of seasonal variations of temperature, height, and winds is located in Northeast Asia. In the mid-low troposphere, large seasonal variations of atmospheric variables such as height at 850 hPa, specific humidity at 500 hPa, and winds at 850 h well indicate the domain of monsoon from Africa to Asia. The global monsoon regions can also be indicated indirectly by the seasonal variation of specific humidity at 700 and 850 hPa.

3.6 Temporal Decomposition of Climatic Field and Annual Cycle Systems

Physically, temporal climatic state is forced by the solar radiation and influenced by the surface conditions. Daily series of atmospheric variables at a spatial point should include their information. Figure 3.26a shows the daily mean climatic height

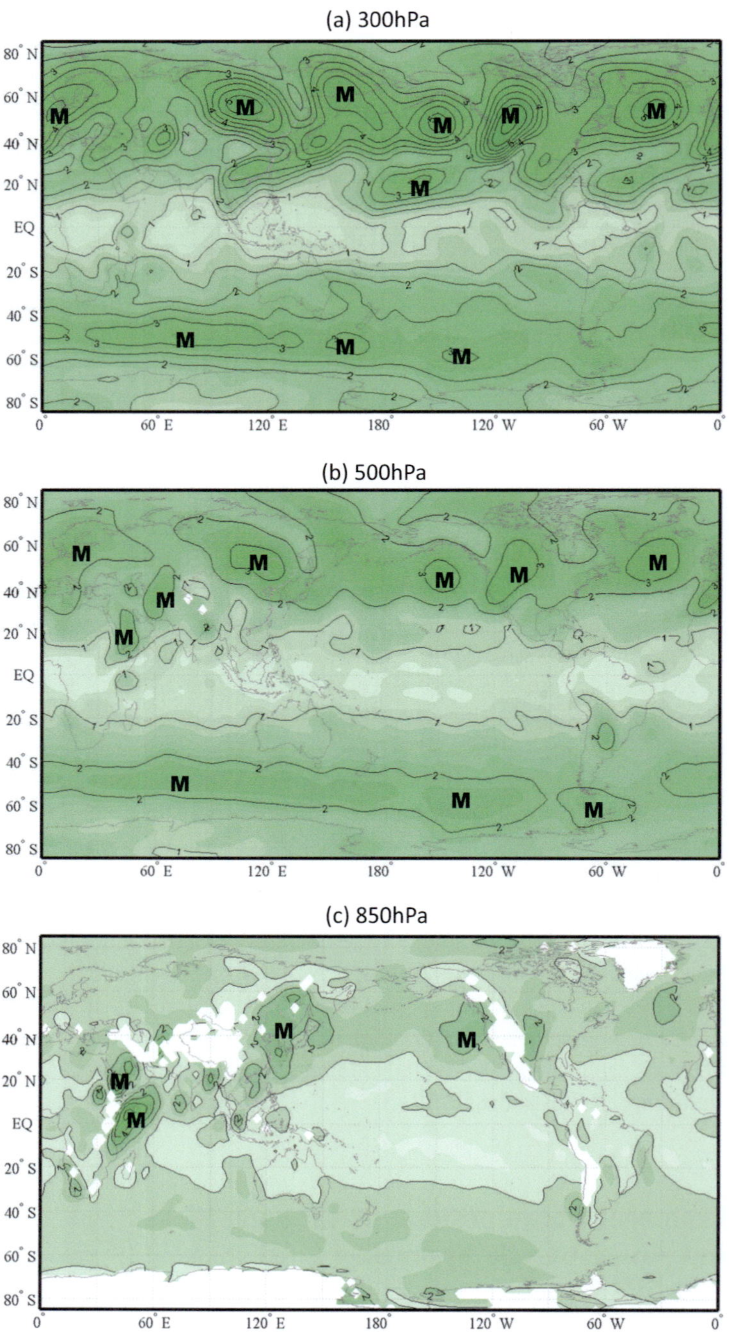

Fig. 3.25 Seasonal SD (m/s) of meridional wind departure at **a** 300 hPa, **b** 500 hPa, and **c** 850 hPa

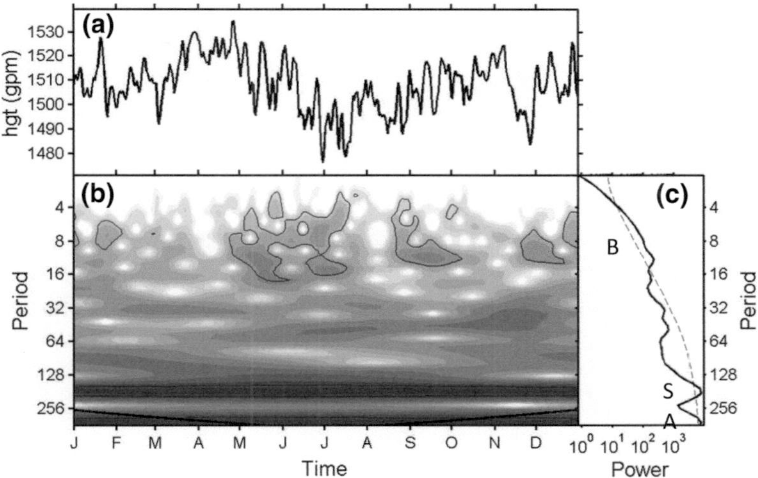

Fig. 3.26 **a** Daily mean climatic height series (gpm) at the grid point (180°, 30°S) at 850 hPa, **b** its wavelet power spectrum (unit of period: days), and **c** Fast Fourier Transformation (FFT) power spectrum (unit of period: days). In **c**: The *dashed line* denotes the 95% confidence level and letters "A", "S", and "B" indicate three peaks of annual, semiannual, and biweekly cycles

series at the grid point (180°, 30°S) of 850 hPa. This series is not as smooth and regular as we would expect. Its wavelet power spectrum illustrated in Fig. 3.26b shows that there are three peaks of periodic bands. These are annual, semiannual, and biweekly cycles, indicated by three peaks of "A", "S", and "B" in Fig. 3.26c. Among the three peaks, the annual and semiannual cycles are directly associated with the solar radiation while the biweekly cycle is the response to the surface conditions. The three cycles should widely exist in the global atmosphere from climatic height, temperature, and winds. Thus, climatic mean field can be temporally decomposed into three components: the annual cycle, semiannual cycle, and biweekly cycle.

$$\tilde{S}_d(\lambda, \varphi, t) = \tilde{S}_{\text{annal}}(\lambda, \varphi, t) + \tilde{S}_{\text{semiannual}}(\lambda, \varphi, t) + \tilde{S}_{\text{biweekly}}(\lambda, \varphi, t) \qquad (3.6.1)$$

In order to characterize the annual and semiannual cycles in an easier way, two trigonometric functions are used to simulate their amplitudes and phases.

$$H_A = H_0 + H_1 \sin\left(2\pi \frac{t}{365} + \beta_1\right) \qquad (3.6.2)$$

$$H_B = H_2 \sin\left(4\pi \frac{t}{365} + \beta_2\right) \qquad (3.6.3)$$

The least-squared method is used to obtain five parameters including H_0, H_1, H_2, β_1, and β_2. Then, a series that only includes annual and semiannual cycles for 365 days can be reconstructed.

The daily mean climatic height series (Fig. 3.26a or Fig. 3.27a) is decomposed into three components as shown in Fig. 3.27b. For the annual cycle component (heavy solid line in Fig. 3.27b), the maximum value is on 25 February and the minimum value is on 27 August. Two maximum values on 17 April and 16 October and two minimum values on 15 January and 17 July appear in the semiannual cycle component (dashed line in Fig. 3.27b). The reconstructed series using annual and semiannual cycle components is also shown in Fig. 3.27a (gray line). The departure between the original series and the reconstructed series is shown in Fig. 3.27b (thin curve line). Biweekly cycle is expected to dominate in this departure series. In this decomposition, H_o = 1507.8 gpm, H_1 = 6.52 gpm, H_2 = 7.86 gpm, $\beta_1 = 0.192\pi$, and $\beta_2 = 1.331\pi$.

The daily change rate of biweekly cycle is larger than that of annual and semiannual cycle (Fig. 3.27b). The difference between biweekly cycle of climatic heights on 26 and 30 June is 34.7 gpm. This difference has reached a magnitude of general vortex formation. Thus, the biweekly cycle of climatic height should be considered in daily weather forecast. A band filter from 1 to 31 days is used to extract the biweekly cycle from height, temperature, and precipitation seasonal series.

According to Eq. (3.6.2), there is the constant term H_0 that is not changing with season. Figure 3.28a shows the height distribution of the constant term H_0 at 850 hPa. There are three high centers in the subtropical NH and four high centers in

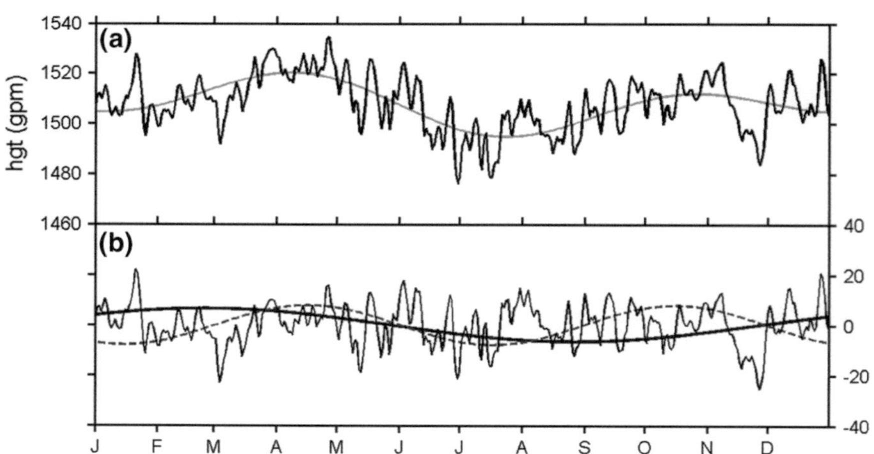

Fig. 3.27 **a** Daily mean climatic height series (*black curve*, gpm) at the grid point (180°, 30°S) and the reconstructed series (*gray line*) using the annul and semiannual cycle components, and **b** three components consisting of the annual cycle (*heavy black line*), semiannual cycle (*dashed line*) and the biweekly cycle (*thin curve line*) which is the difference between the daily mean climatic height and the reconstruction from the first two cycles

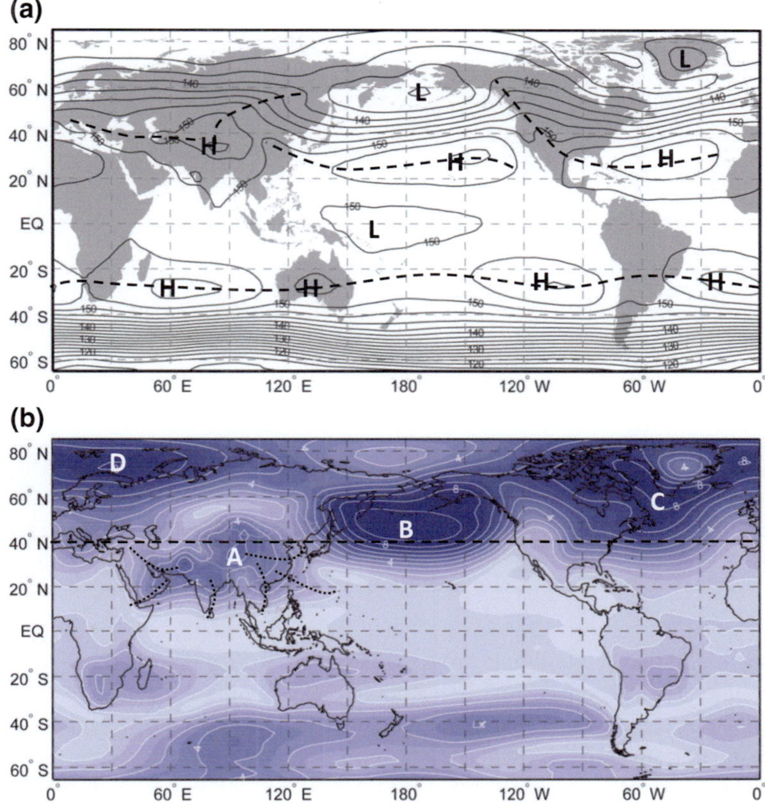

Fig. 3.28 Annual cycle component indicated by climatic height (10 gpm) at 850 hPa: **a** constant term (H_0) and **b** amplitude (H_1). *Dashed lines* are height ridges in **a**. Letters "A", "B", "C", and "D" denote four main centers of amplitude in **b**. *Dotted lines* in **b** indicate the active regions of peninsular scale monsoon troughs surrounding the Asian continent

the subtropical SH. One low center in the equatorial western Pacific, two low centers near the Aleutian Islands, and the Iceland are three stable systems. The three centers resemble the distribution of atmospheric active centers, reflecting a field of annual mean sea surface pressure or the surface geopotential height (e.g. Peixoto and Oort 1992).

Figure 3.28b shows the amplitude H_1 of height departure, which changes with season. Four large high centers are located in the Asian continent, North Europe, the middle latitudes of North Pacific, and Northwest Atlantic. The annual cycle of these active centers with positive/negative heights illustrates the monsoon and climatic annual cycle in Asia and other regions of the NH. Around center "A" in Asia, six dotted lines indicate the active regions of peninsular scale monsoon troughs surrounding the Asian continent (Qian and Tang 2010). The distribution of annual cycle climatic component is similar to that of spatial climatic departure in the lower troposphere.

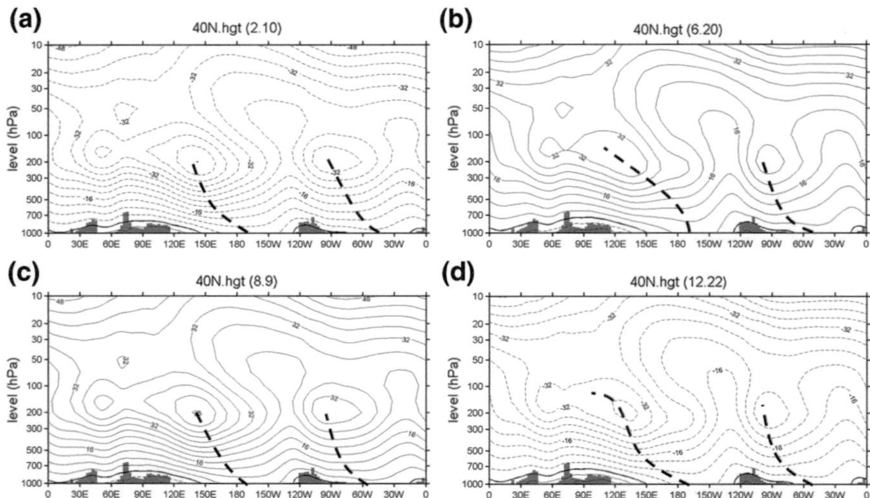

Fig. 3.29 Vertical–zonal cross sections of climatic annual cycle component indicated by height (gpm) on **a** 10 February, **b** 30 June, **c** 9 August, and **d** 22 December along 40°N. *Heavy dashed lines* indicate height ridge or trough

Three high areas of amplitude cross 40°N in Fig. 3.28b. Four vertical sections of height annual cycle component along 40°N on four different dates are shown in Fig. 3.29. It is clearly noted that the large amplitude "A" in Fig. 3.28b is caused by the seasonal transition of heights only near the lower troposphere. Two large amplitude areas "B" and "C" in Fig. 3.28b are the down-eastward extension of height centers. Most large centers of height annual cycle are located around 200 hPa. Thus, Fig. 3.29 shows the annual cycle component of climatic height at 200 hPa, indicated by the constant term (H_0) and amplitude (H_1).

Constant term (H_0) of height at 200 hPa is smooth in the SH. However, there are two troughs near the eastern coasts of Asian and North American continents in the NH (Fig. 3.30a). The maximum height with annual cycle is centered in the northeast Asia, with its central value about 500 gpm (Fig. 3.30b). Two large centers indicated by "A" and "B" in Northeast Asia and Northeast North America imply that positive and negative heights are climatologically annual cycle in the two areas from summer to winter, respectively. Height ridges and troughs are active in summer and winter over these two areas.

3.7 Semiannual and Biweekly Cycle Systems

After extracting the semiannual cycle component, its height amplitude H_2 at 850 hPa is shown in Fig. 3.31. Six large centers are located between 20°N and 80°N in the NH and several centers between 40°S and 60°S in the SH.

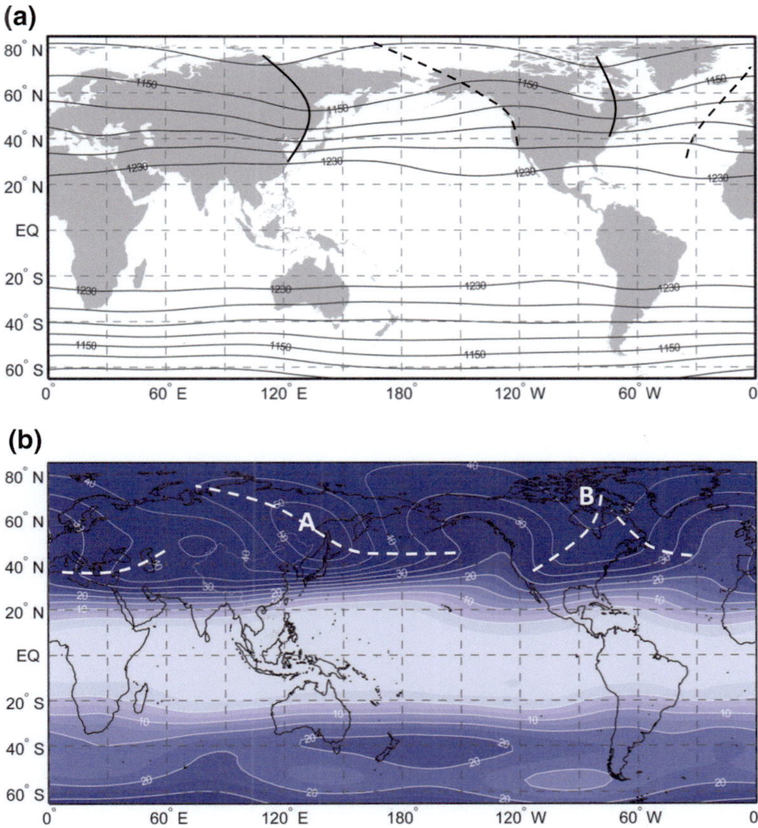

Fig. 3.30 Same as Fig. 3.28, but for annual cycle component of climatic height (10 gpm) at 200 hPa: **a** constant term (H_0) and **b** amplitude (H_1). *Heavy solid line* and *heavy dashed line* indicate the height trough and ridge in **a**. *Heavy dashed lines* indicate height ridge in **b**

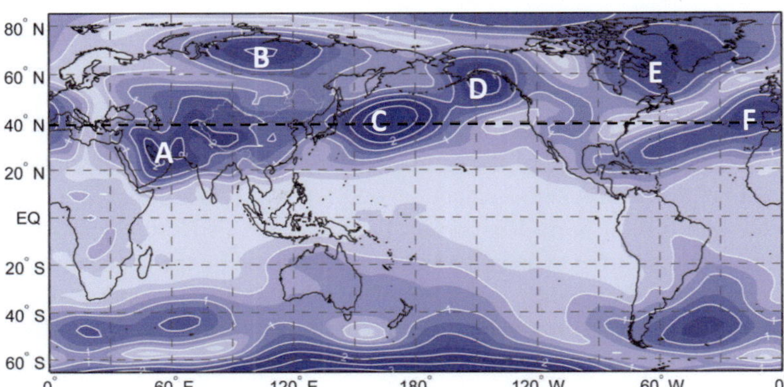

Fig. 3.31 Semiannual cycle component indicated by the height amplitude (1 × 10 gpm interval) at 850 hPa. Letters "A"–"F" indicate six centers with large amplitude

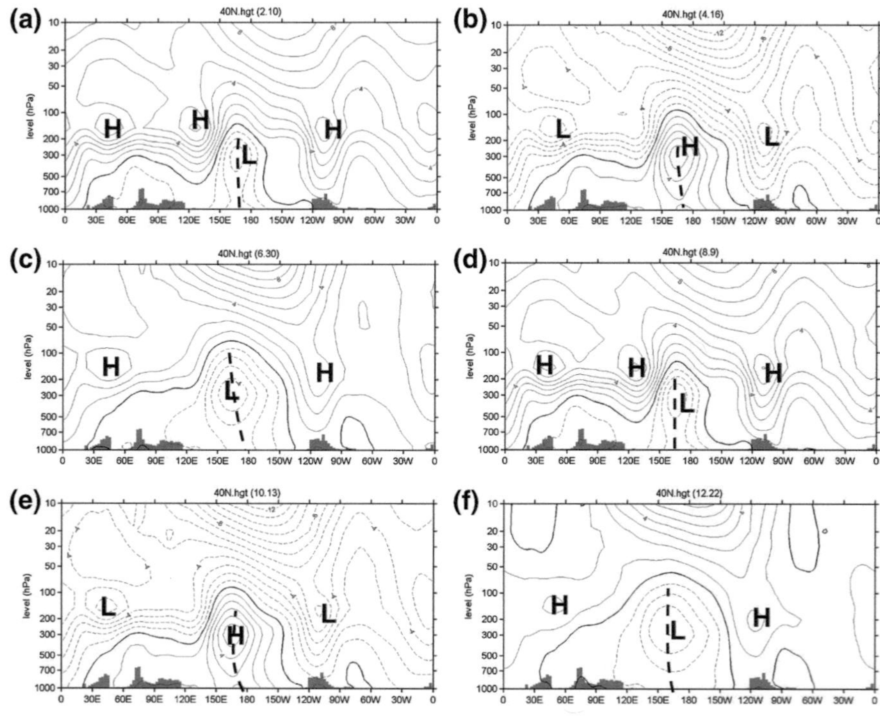

Fig. 3.32 Vertical–zonal cross sections of semiannual cycle component indicated by climatic height (*red-solid contour* for positive and *blue-dashed contour* for negative, 1 × 10 gpm interval) along 40°N on **a** 10 February, **b** 16 April, **c** 30 June, **d** 9 August, **e** 13 October, and **f** 22 December. *Heavy dashed lines* indicate the height trough and ridge while letters "H" and "L" are positive and negative centers

The maximum height center of semiannual cycle is indicted by letter "C" in the Northwest Pacific. Six panels of vertical–zonal section show that this center at 850 hPa in the Northwest Pacific extends downward from 300 hPa (Fig. 3.32). On 16 April and 13 October, heights are positive while on the calendar dates of 30 June and 22 December heights are negative at 40°N and 170°E, respectively. The maximum height amplitude of the semiannual cycle is found in the stratosphere.

Several height centers with semiannual cycle at 300 hPa are shown in Fig. 3.33. Two centers in central north Russia and the northwest of North America indicate that heights from positive to negative values are changed by the semiannual cycle in the two areas. In the middle latitudes of Northwest Pacific and eastern North Atlantic, two centers are, respectively, described the activities of trough and ridge for a semiannual cycle. The amplitude of the semiannual cycle is smaller than that of the annual cycle.

The vertical–zonal cross sections of height and temperature indicated by the daily biweekly cycle activities in winter and summer are shown in Fig. 3.34. It is

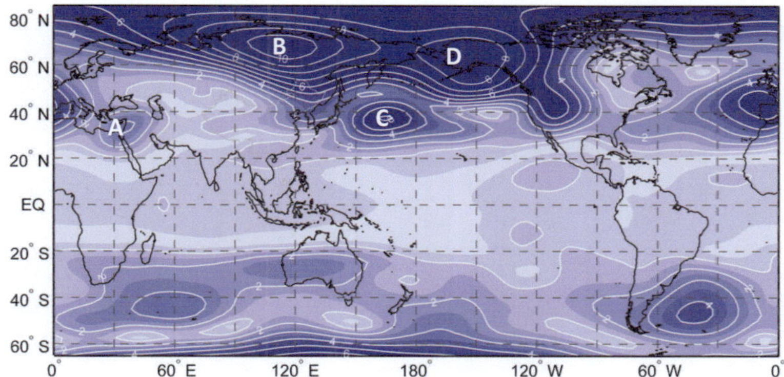

Fig. 3.33 Semiannual cycle component indicated by the height amplitude (1 × 10 gpm) at 300 hPa. Letters "A"–"D" indicate four centers with large amplitude

interesting to note that in the troposphere the maximum centers of positive (negative) height exist near 300 hPa while negative (positive) centers of temperature are, respectively, located at around 200 hPa (600 and 850 hPa). On 7 Feburary along 60°N (Fig. 3.34a), a negative center (−35 gpm) of height exists at 300 hPa and 170°E, while a negative center of temperature is just below 500 hPa, extending downward to the surface and a positive temperature center is above 200 hPa. Two large centers and two small centers of positive height are, respectively, observed at the upper troposphere in 35°E, 95°E, 100°W, and 60°W while four positive centers of temperature are located below in the lower troposphere, extending downward to the surface. On 11 August, a positive center (+45 gpm) of height is located at 280 hPa above 160°E and 60°N. Above and below this positive center of height, negative and positive centers of temperature are observed (Fig. 3.34b). This vertical structure of temperature and height is also observed from seven positive height centers and five negative height centers (Fig. 3.34b). These vertical structures of climatic temperature and height satisfy the hydrostatic equilibrium.

It is interesting to compare these biweekly cycle centers with the temporal anomalous centers. Figure 3.35 shows the vertical–zonal cross sections of the original field, climatic field, and anomaly field of daily mean height and temperature along 60°N on 11 August 2005. The original field shows that the height has some weak ridges and troughs while temperature has some strong ridges and troughs on 11 August 2005 along the 60°N (Fig. 3.35a). These ridges and troughs in the original height and temperature are partly contributed by the climatic height and climatic temperature on August 11 (Fig. 3.35b). After removing the climatic height and temperature, the anomalous vertical structures are shown in Fig. 3.35c.

It is noted that in the troposphere, the maximum centers of positive (negative) height anomalies exist around 250 hPa while negative (positive) centers of temperature anomalies are respectively located over 150 hPa (mid-low troposphere) on 11 August 2005 (Fig. 3.35c). Four positive and four negative centers of height anomalies are clearly located in the upper troposphere along 60°N. The central

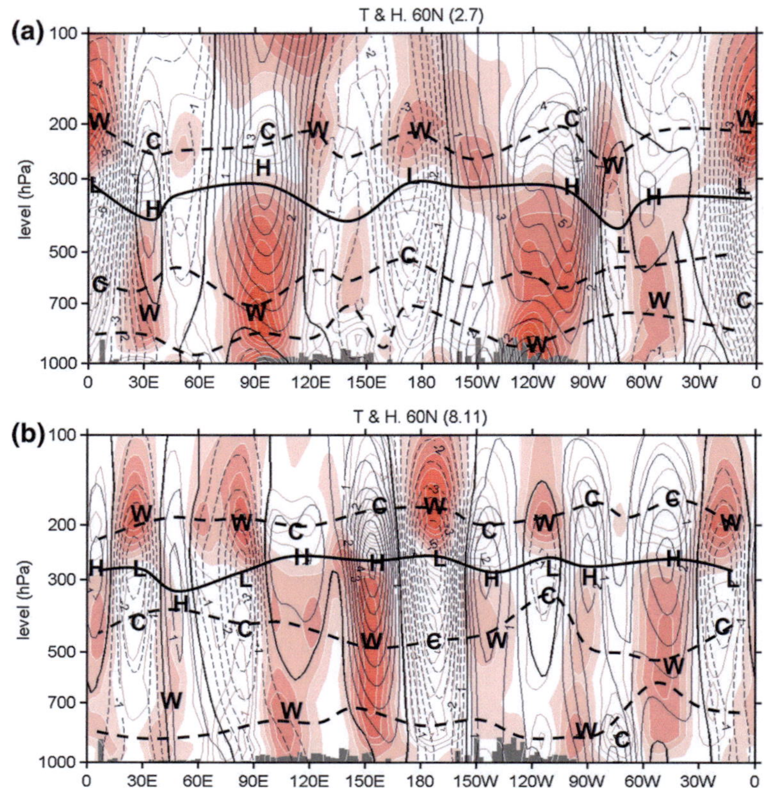

Fig. 3.34 Vertical–zonal cross sections of daily biweekly cycle activities of height (*black-solid contour* for positive and *black-dashed contour* for negative, 1 × 10 gpm interval) and temperature (*shading and white contour*, 0.25 °C interval) along 60°N on **a** 7 February and **b** 11 August. The *heavy solid line* and *heavy dashed line* connect positive/negative centers of height (H/L) and temperature (W/C), respectively

values of positive and negative height anomalies are from +200 to −200 gpm. The temperature ridge (trough) associated with the height ridge (trough) extends downward to the surface from the height anomalous center, which can explain the positive (negative) temperature anomalies at the surface. These ridges and troughs of height and temperature anomalies in the troposphere satisfy the hydrostatic equilibrium relationship. These signals of height and temperature anomalies in the troposphere have been used in the analyses and forecasts of extreme weather events (Qian et al. 2013, 2014; Qian and Zhang 2012; Ding and Qian 2012; Qian and Jiang 2014).

This comparison shows that the vertical structures and spatial scales of height/temperature anomalies and climatic biweekly cycle height/temperature are

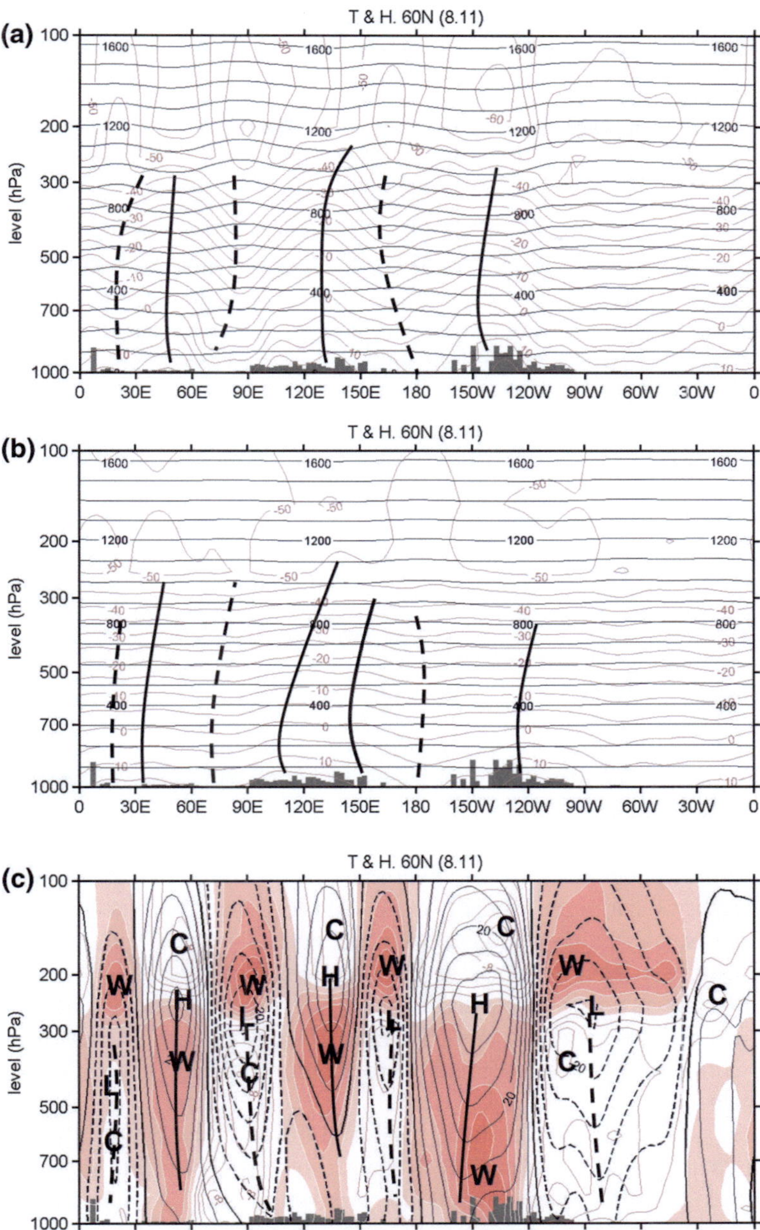
T & H. 60N (8.11)
(a)
(b)
(c)
level (hPa)
0 30E 60E 90E 120E 150E 180 150W 120W 90W 60W 30W 0

◀ **Fig. 3.35** Vertical–zonal cross sections of the **a** original field, **b** climatic field, and **c** anomalous field of daily height and temperature along 60°N on August 11, 2005. The *black contour* indicates height (100 × 10 gpm interval) and the *red contour* denotes temperature (5 °C interval) in **a** and **b**. *Black-solid contour* and *black-dashed contour* are height anomalies (H/L, 5 × 10 gpm interval) while *shading* with *white contour* (*red contour*) indicates positive (negative) temperature anomalies (W/C, 2 °C interval) in **c**. Bars denote topography. *Heavy solid* and *heavy dashed lines* indicate ridge and trough of height, respectively

similar. The former has a central value about 200 gpm while the latter has a central value about 45 gpm. The vertical structure of temperature and height anomalies satisfies the hydrostatic equilibrium relationship. Similarly, biweekly cycle height and temperature also satisfy the following hydrostatic equilibrium relationship

$$\tilde{T} = -\frac{1}{R} \cdot \frac{\tilde{H}}{\Delta \ln p} \tag{3.7.1}$$

and geostrophic wind relationship

$$\tilde{u}_g = -\frac{1}{f}\frac{\Delta \tilde{H}}{\Delta y} \tag{3.7.2}$$

$$\tilde{\zeta} = \frac{1}{f}\nabla^2 \tilde{H} \tag{3.7.3}$$

It means that the biweekly cycle components of temperature $\tilde{T}$, wind $\tilde{u}_g$ and vorticity $\tilde{\zeta}$ can be derived or reconstructed from the biweekly cycle height $\tilde{H}$. The temperature, wind, and vorticity with annual and semiannual cycles also have the same vertical structure satisfying hydrostatic equilibrium and geostrophic wind relationships as that of the biweekly cycle. If the biweekly cycle height component has a center in the upper troposphere, the biweekly cycle wind and vorticity components also have a center at the same altitude while the biweekly cycle temperature component should have two opposite centers above and below the height center. These structure and relationship between biweekly cycle height and temperature objectively exist similar to the vertical synoptic scale systems in the troposphere.

The spatial scales of semiannual cycle and biweekly cycle height systems are smaller than that of the annual cycle height system. According to Fig. 3.34, a biweekly cycle height system at 300 hPa is shown in Fig. 3.36 from December 31 to January 7. The negative center "B" located in Siberia on 31 December moves southeastward to Northeast China on January 3 and further to the central northern Pacific on January 7, while the positive center "A" moves northeastward for about 8 days along the eastern coast of America. These biweekly cycle height systems are like synoptic scale systems, but they are indeed moving climate systems.

The maximum positive and negative centers of biweekly cycle height are basically located at around 300 hPa in the upper troposphere (Fig. 3.34). The signals are

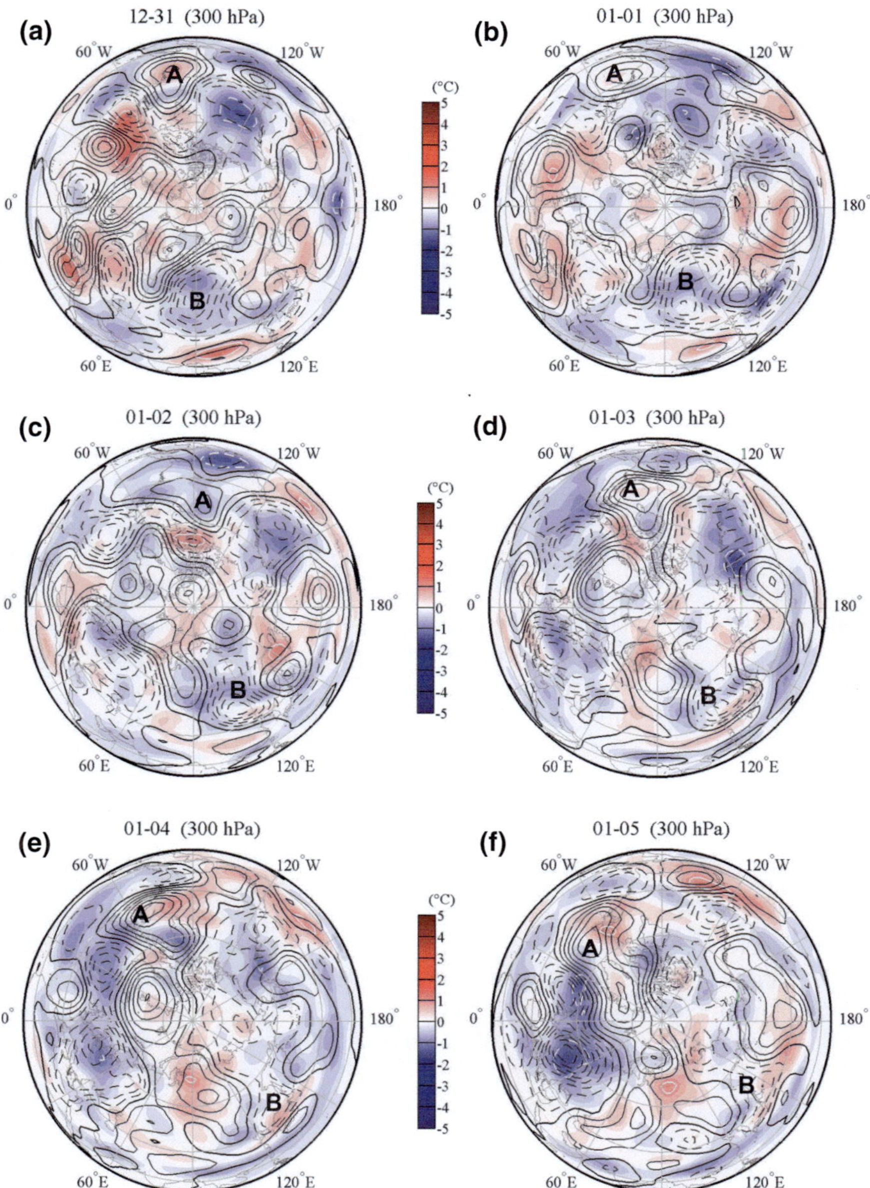

Fig. 3.36 Daily biweekly cycle height (*solid and dashed lines*, 1 × 10 gpm interval) and temperature (*shading*, 1 °C interval) at 300 hPa from 31 December to 7 January in the NH. Letters "A" and "B" indicate the daily centers of positive and negative biweekly cycle height

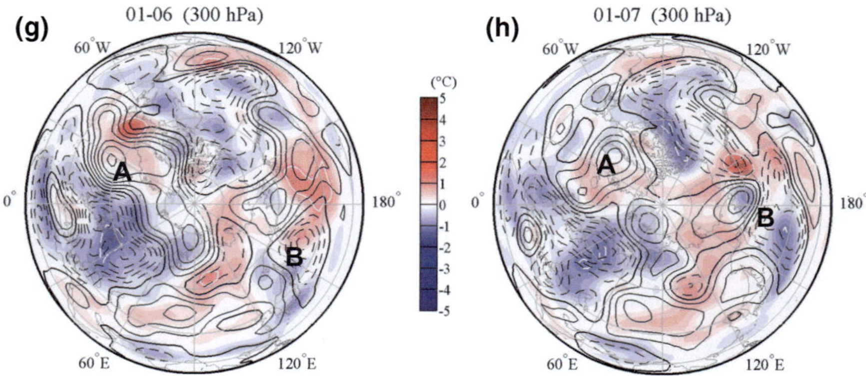

Fig. 3.36 (continued)

stronger in winter and spring than those in summer and autumn. Propagating features can also be found from the daily biweekly cycle activities of height at 300 hPa along different latitudes in the SH (Fig. 3.37). Along 40°S and 60°S, all signals mainly propagate eastward. But along 20°S, signals propagate westward from January to April and eastward from May to October (figures not shown).

In the NH, the propagating directions and speeds as well as spatial scales change with season and latitudes (Fig. 3.38). Along 20°N, the signal is strong and propagates basically eastward from January to May, while the signal is weak and propagates westward from June to October (figures not shown). Along 40°N, all signals basically propagate eastward around the year (Fig. 3.38a). Along 60°N, the signal is strong and propagates westward from January to August and the signal moves eastward from September to December (Fig. 3.38b).

A height system of the daily biweekly cycle can propagate along a zonal circle for months, but positive and negative height centers at a point can only persist 1–2 weeks. These height systems are like synoptic scale systems but they are indeed moving climate systems. The climatic systems of the biweekly cycle need to be identified.

In weather forecast, surface air temperature (SAT) and precipitation at the future time $t+\delta$ of year y at any point p can be predicted by

$$W_y(p,t+\delta) = \bar{W}_0(p) + \tilde{W}(p,t+\delta) + W_y'(p,t+\delta), \quad (3.7.4)$$

where $\bar{W}_0(p)$ is annual mean SAT (or precipitation) at the point p, and its climatic component $\tilde{W}(p,t+\delta)$ is decomposed into annual, semiannual, and biweekly cycles:

$$\tilde{W}(p,t+\delta) = \tilde{W}(p,t+\delta)_{\text{annual}} + \tilde{W}(p,t+\delta)_{\text{semiannual}} + \tilde{W}(p,t+\delta)_{\text{biweekly}} \quad (3.7.5)$$

The largest change rate and the largest horizontal gradient of any variable come from the biweekly cycle.

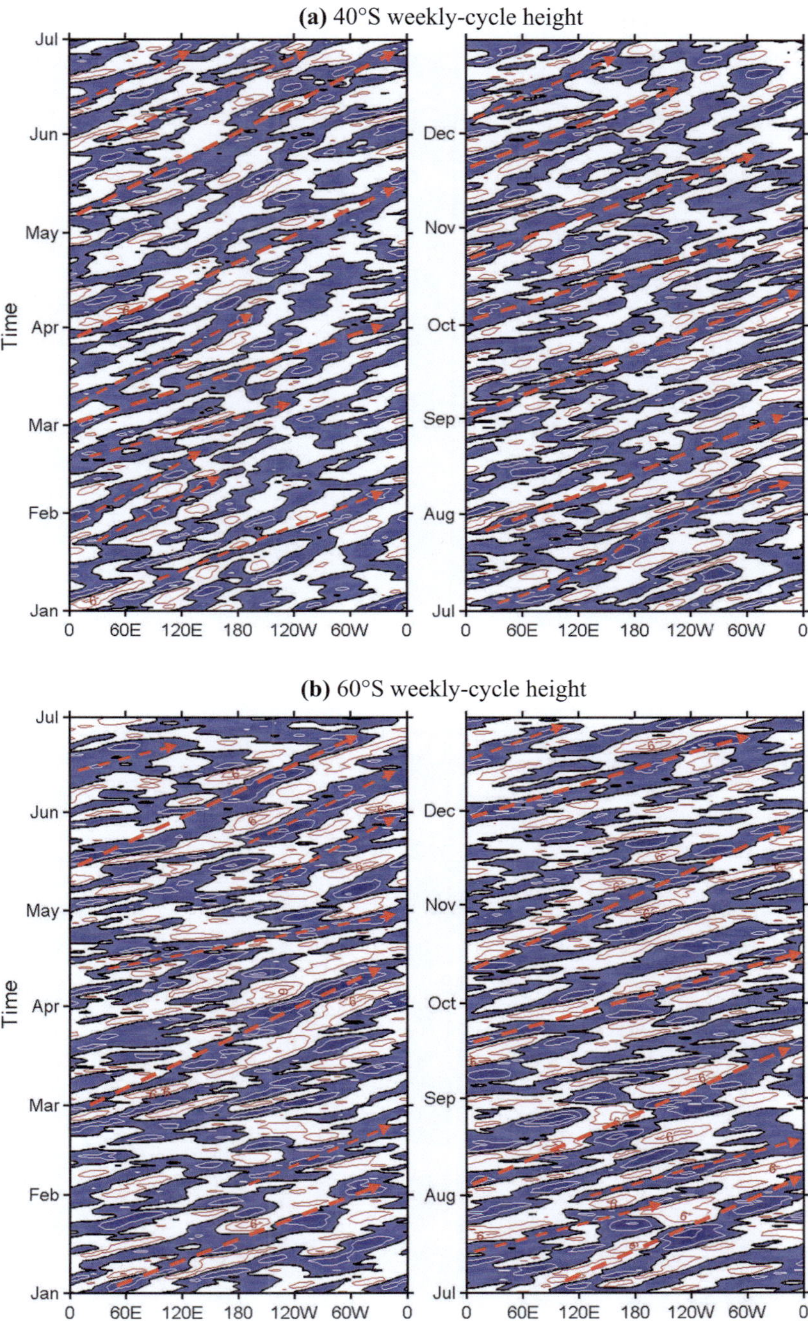

Fig. 3.37 Time-zonal sections of daily biweekly cycle height activities (3 × 10 gpm interval) at 300 hPa along **a** 40°S and **b** 60°S. *Red dashed arrows* indicate propagating directions

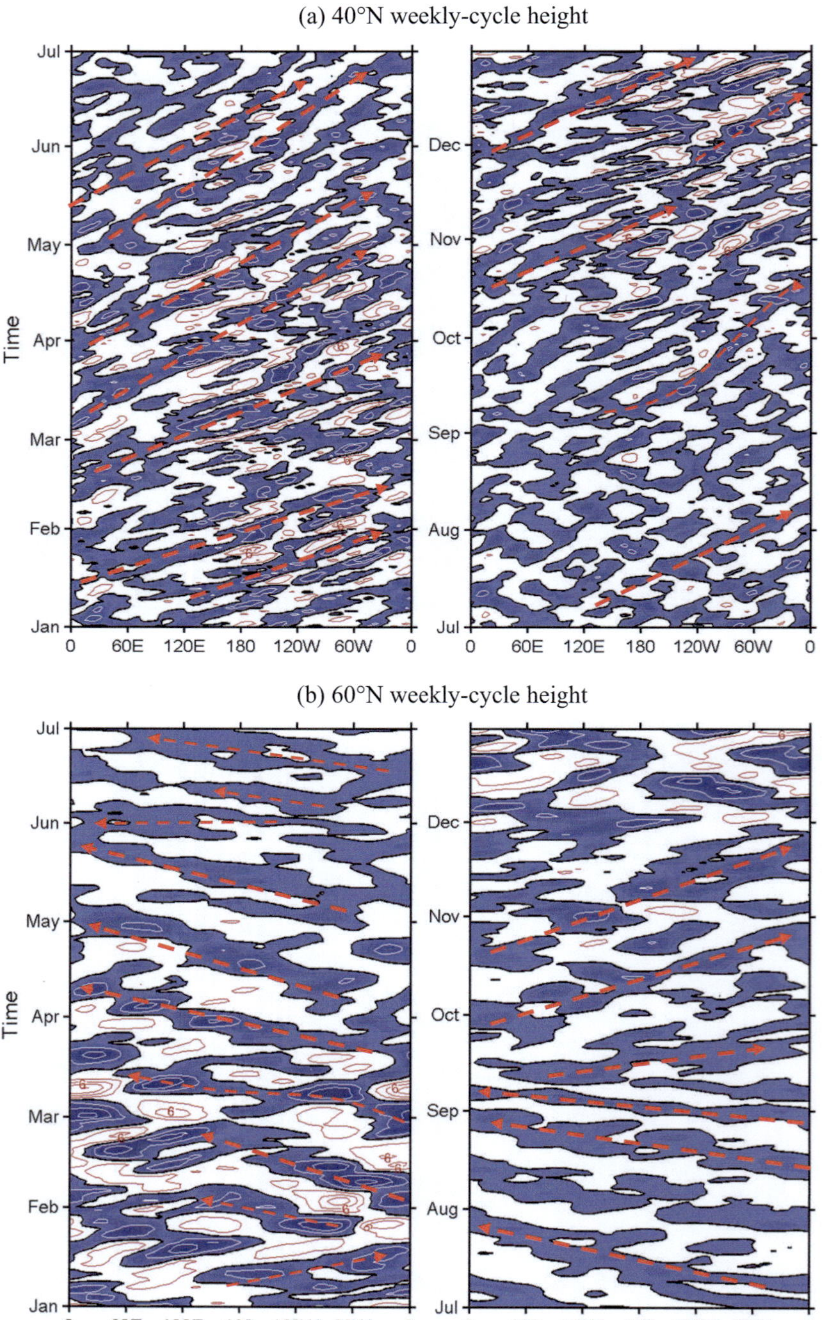

Fig. 3.38 Time-zonal sections of daily biweekly cycle height activities (3 × 10 gpm interval) at 300 hPa along **a** 40°N and **b** 60°N

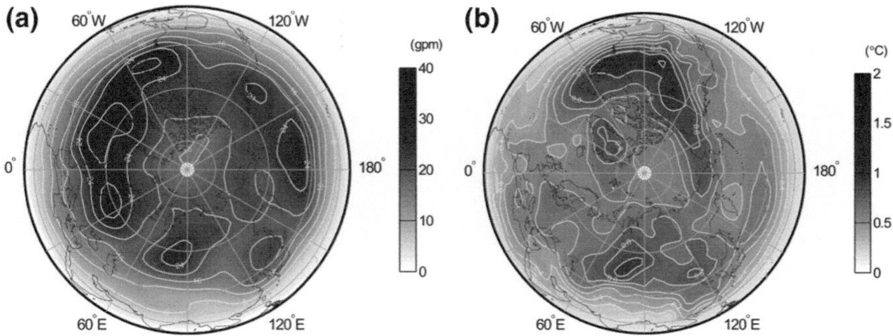

Fig. 3.39 SDs of the biweekly component of climatic **a** height (4 gpm interval) at 300 hPa and **b** temperature (0.1 K interval) at 850 hPa

As an example, the relationship between the climatic biweekly cycle height and regional biweekly cycle SAT and precipitation in China is examined. Figure 3.39 shows the SDs of the biweekly component of climatic height at 300 hPa and temperature at 850 hPa in the NH. Height SD at 300 hPa illustrates seven high centers along the middle latitudes in the NH. Among them, a center is located in Northeast China (Fig. 3.39a). Temperature SD at 850 hPa displays the largest center in the North America and the second largest one in Europe. In Northeast China, there is also a center (Fig. 3.39b). Thus, we examine the relationship of biweekly cycles from the surface climate (SAT and precipitation) and 300 hPa height in Northeast China.

Correlation between biweekly cycle height at 300 hPa averaged over the Northeast China (42.5–52.5°N, 125–135°E) and the biweekly cycle SAT (precipitation) in China is illustrated in Fig. 3.40. Positive correlations for the SAT are found from Northeast China to the upper Yangtze River basin while negative correlations are located in the Xinjiang region over Northwest China. The precipitation shows positive correlation in northern China and negative correlation in southern China with one day lagging.

Two biweekly cycle series of the averaged 300 hPa height, averaged SAT series over 114 stations, and averaged precipitation series over 112 stations covering North-Northeast China (area marked by dotted lines in Fig. 3.40) are shown in Fig. 3.41. The correlation between the height and the SAT series is 0.47 while the correlation between the height and the precipitation series is 0.23 with height leading precipitation one day. Height and SAT have larger amplitudes in spring and autumn while precipitation has relatively large amplitude in summer.

It is found that the correlations of height with SAT and precipitation change with season. Figure 3.42 shows the running correlations of height with SAT and precipitation. For the SAT, positive correlations cover the main part, while negative correlations are observed from late January to February and from late June to August. For the precipitation, two negative correlation spells are from late January to February and from late May to June. The highest correlation for the SAT (0.88)

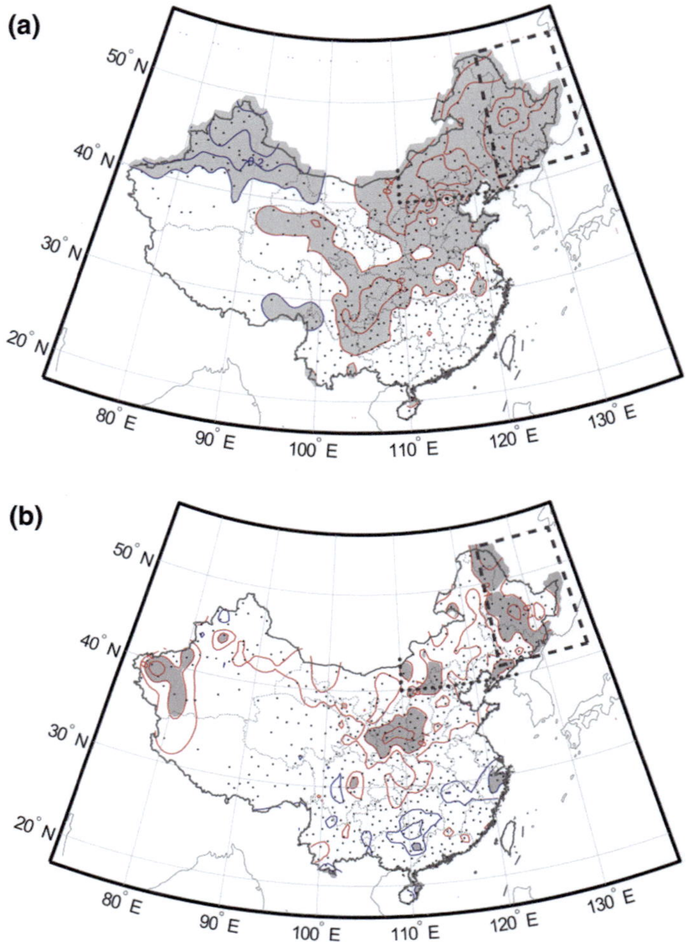

Fig. 3.40 Correlation distributions (0.1 interval) between the biweekly cycle of 300 hPa height averaged over the *dashed area* (42.5–52.5°N, 125–135°E) in Northeast China and the biweekly cycles of **a** the station SAT and **b** precipitation in China for the whole year. *Shading* areas indicate correlation reaches the 95% confidence level. In **b**: Precipitation is 1 day lagging the height biweekly cycle component

appears on October 6 and the highest correlation for the precipitation (0.73) on April 9. The dominating significant correlations demonstrate the usefulness of the biweekly cycle.

Biweekly cycle systems are tilted in vertical section as shown in Fig. 3.43. On May 12 (Fig. 3.43a), a biweekly negative height center is located in North China (120°E, 47°N) and its trough extends southward to Central China (32°N–36°N). On the same day, a strong positive height center is situated in Central China and its ridge extends southward to South China. Independent datasets of SAT and

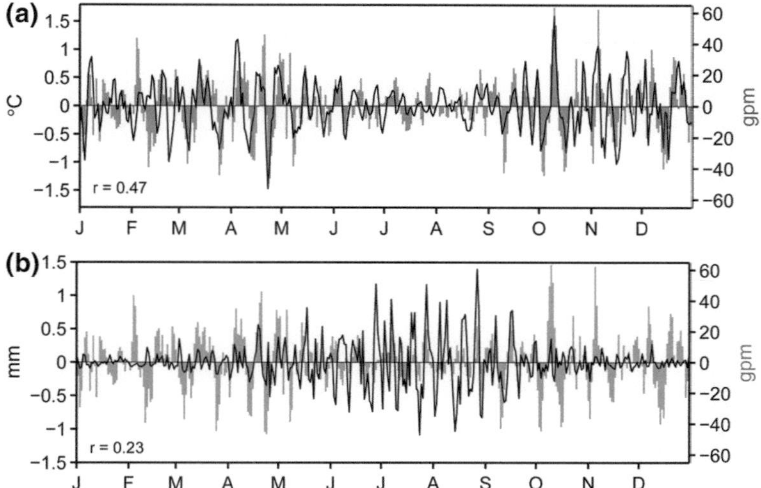

Fig. 3.41 Two biweekly cycle series of climatic height (*bar*) at 300 hPa averaged over the area (42.5–52.5°N, 125–135°E; indicated in Fig. 3.40 by *dotted line*), associated with **a** SAT series (*solid line*) averaged over 114 stations, and **b** precipitation series (*solid line*) averaged over 112 stations

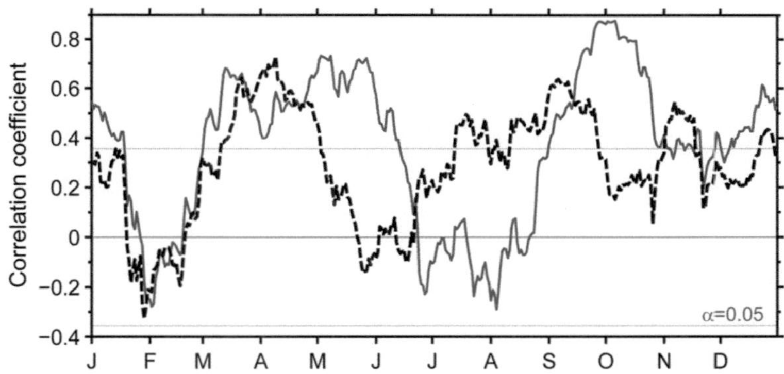

Fig. 3.42 The running correlations indicated by the *solid line* (height vs. SAT) and the *dashed line* (height vs. precipitation)

precipitation in China are used to compare with the NCEP/NCAR reanalysis results. The positive and negative precipitation in biweekly cycle along 120°E can be well explained by the height trough and ridge. The temperature trough and ridge associated with the height trough and ridge can explain the negative and positive SAT in northern and southern China. On July 30 (Fig. 3.43b) along 40°N, the temperature trough and ridge associated with the height trough and ridge can also explain the negative and positive SAT as well as the positive and negative precipitation in biweekly cycle over western and eastern China. However, all these troughs and

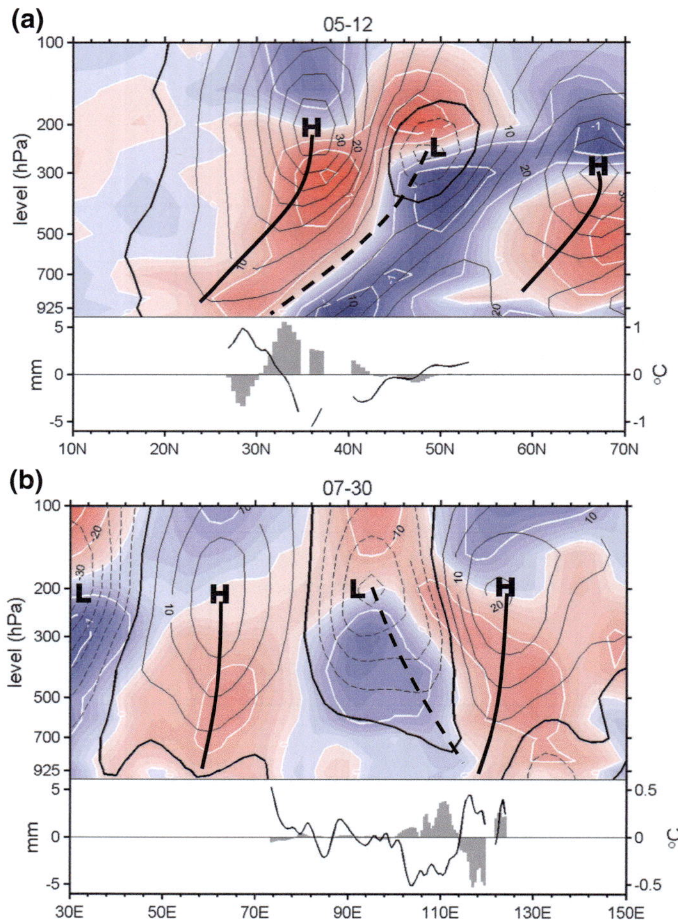

Fig. 3.43 Vertical cross sections (*upper panel*) of height (*solid and dashed contour*, 5 gpm interval) and temperature (*red and blue shadings* for positive and negative respectively, 0.5 K interval) in biweekly cycles along **a** 120°E on 12 May and **b** 40°N on 30 July. In *lower panel* the *solid line* is the biweekly cycle distribution of SAT (°C) and the *bar* indicates the biweekly cycle distribution of precipitation (mm/d) along **a** 120°E on 12 May and **b** 40°N on 30 July. The *heavy solid and dashed lines* are height ridges and troughs, respectively

ridges are tilted vertically related to their height centers around 200 hPa. These ridges and troughs of height and temperature in the troposphere satisfy the hydrostatic equilibrium relationship. The positive (negative) SAT in biweekly cycle is a result of the downward extension of positive (negative) temperature ridge (trough) to the surface, while positive and negative precipitations in biweekly cycle are associated with biweekly height cyclones and anticyclones in the lower troposphere, respectively. It is re-identified by this example that the structure and relationship between height and temperature objectively exist in the troposphere like some synoptic scale systems.

3.8 Identifying Three Period Bands

It is noted that the annual cycle, semiannual cycle and biweekly cycle are the three most significant timescales for atmospheric variables. Percentage distributions of periodic cycle of climatic temperature at 300, 500, and 850 hPa in zonal direction (144 grids) are shown in Fig. 3.44. Their wavelet transform reaches the 0.05 significance level along various latitudes with different periodic cycles. At all three isobaric levels and all latitudes (Fig. 3.44), there are three major bands of periodic cycles: the biweekly cycle (3–14 days), semiannual cycle (150–210 days), and

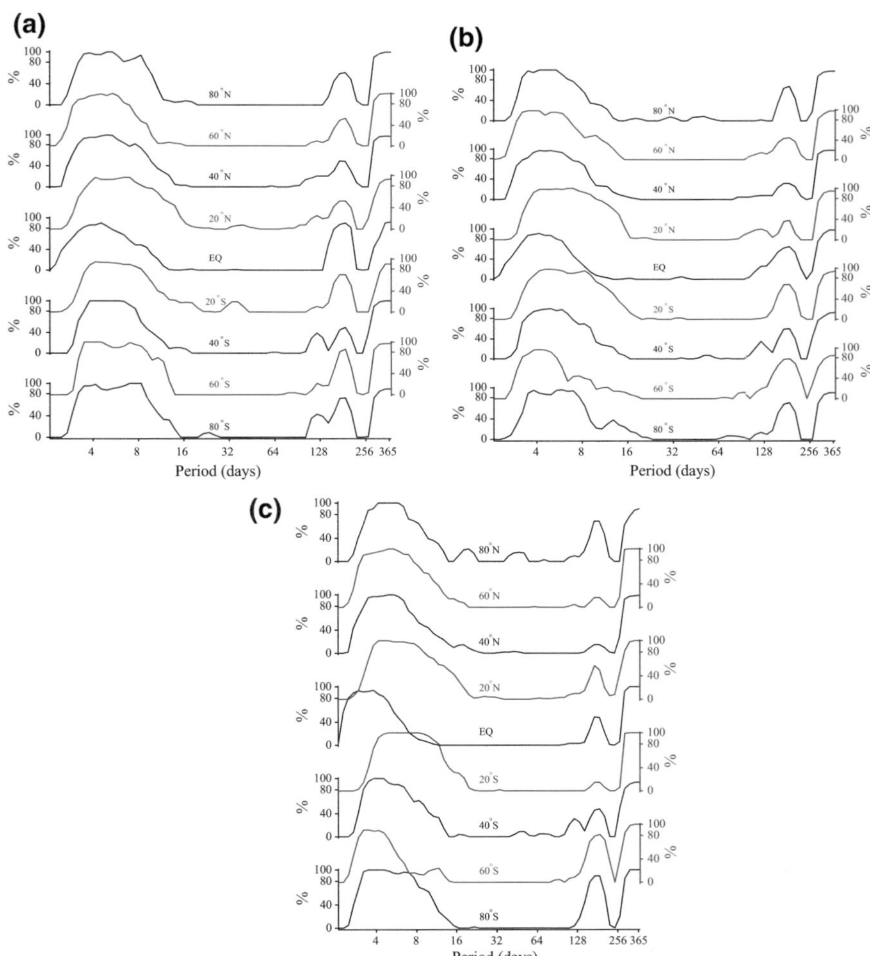

Fig. 3.44 Percentage (%) of periodic cycle for climatic temperature at **a** 300, **b** 500, and **c** 850 hPa in zonal direction. Their wavelet transform reaches the 0.05 significance level along various latitudes with different periodic cycles

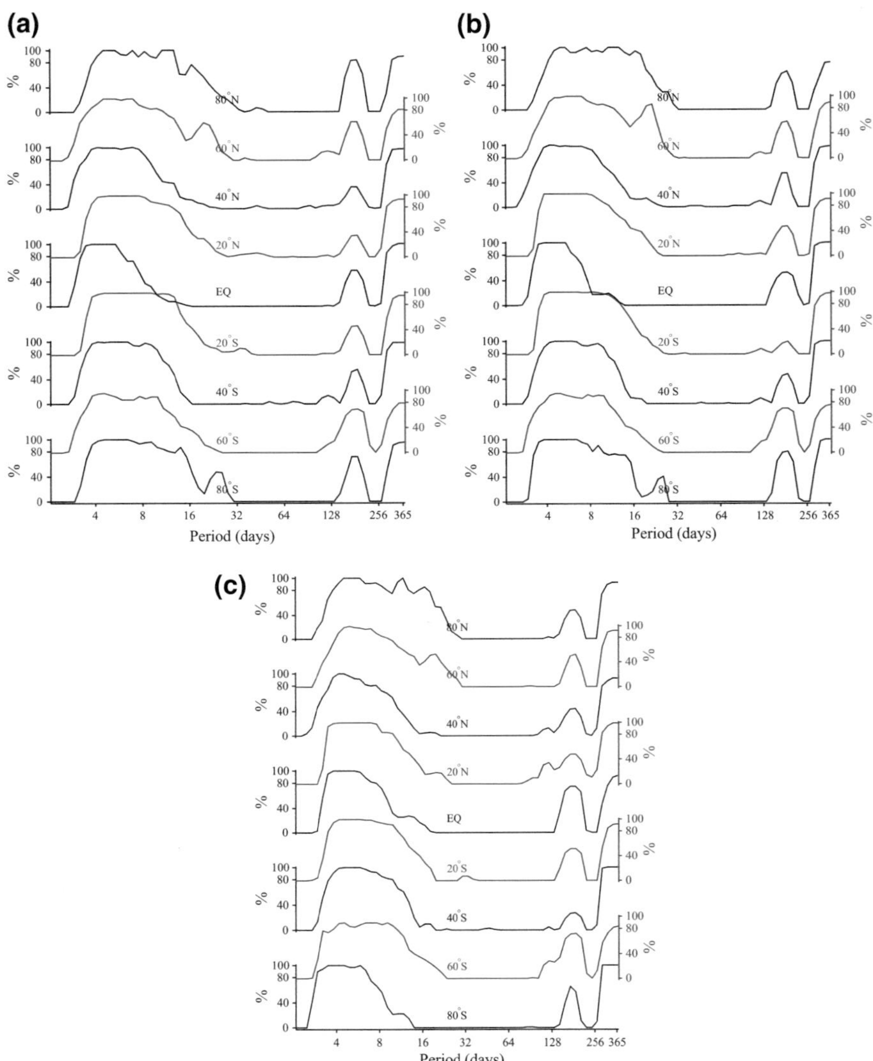

Fig. 3.45 Percentage (%) of periodic cycles for climatic height at **a** 300, **b** 500, and **c** 850 hPa in zonal direction

annual cycle (300–365 days). The three periodic bands not only robustly exist in climatic temperature (Fig. 3.44) but also in climatic height (Fig. 3.45), westerly (Fig. 3.46), and southerly (Fig. 3.47).

Variances of climatic height in terms of total, annual cycle, semiannual cycle, biweekly cycle, as well as the sum of three cycles at 300 hPa along different latitudes in the NH are shown in Fig. 3.48. Along the equator, the largest peak of annual-cycle variance is located in the equatorial eastern Pacific. For the semiannual

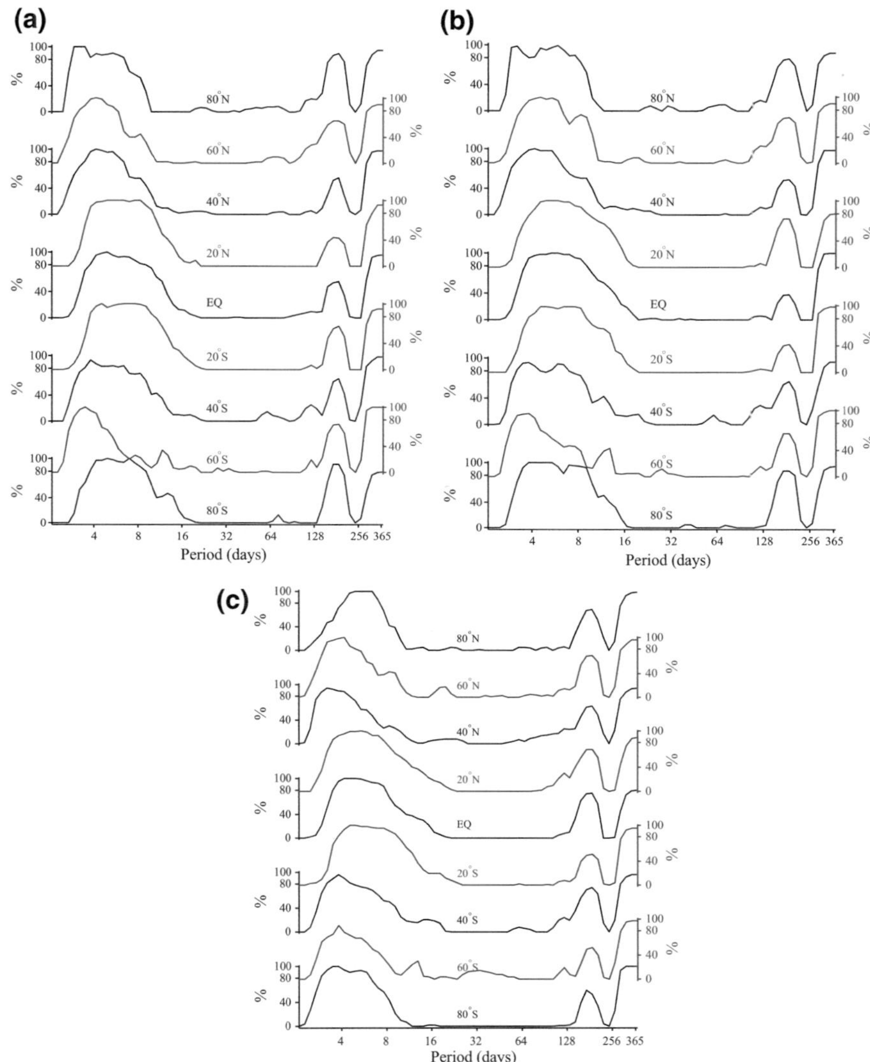

Fig. 3.46 Percentage (%) of periodic cycles for climatic westerly at **a** 300, **b** 500, and **c** 850 hPa in zonal direction

cycle, the largest variance is found in the equatorial central Pacific. The variance of biweekly cycle is smaller compared to that of other two cycles, without any peak along the equator. The reconstructed series of the three components well match the variance of climatic state (Fig. 3.48a). Along 10°N, there are three large peaks of annual cycle variance in the tropical western Pacific, the tropical eastern Pacific, and the central tropical Atlantic, respectively. The three peaks of semiannual cycle variance are found near the dateline, 120°W, and 30°W. The reconstructed series of

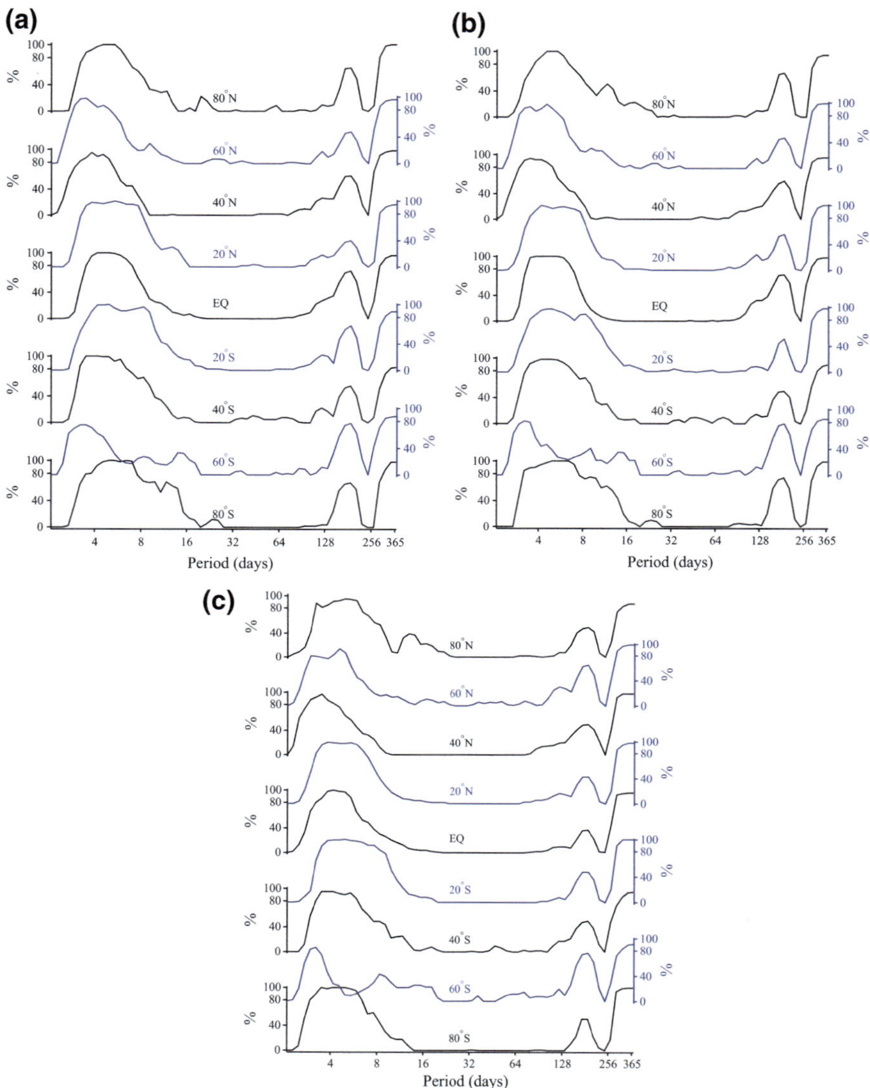

Fig. 3.47 Percentage (%) of periodic cycles for climatic southerly at **a** 300, **b** 500, and **c** 850 hPa in zonal direction

the three components also well match the variance of climatic state (Fig. 3.48b). Similarly, there are four peaks of annual cycle variance along 20°N, four peaks along 30°N, three peaks along 40°N, and four peaks along 50°N, 60°N, 70°N, and 80°N. In 70°N and 80°N, height variances of semiannual cycle and biweekly cycle become larger relative to those of annual cycle.

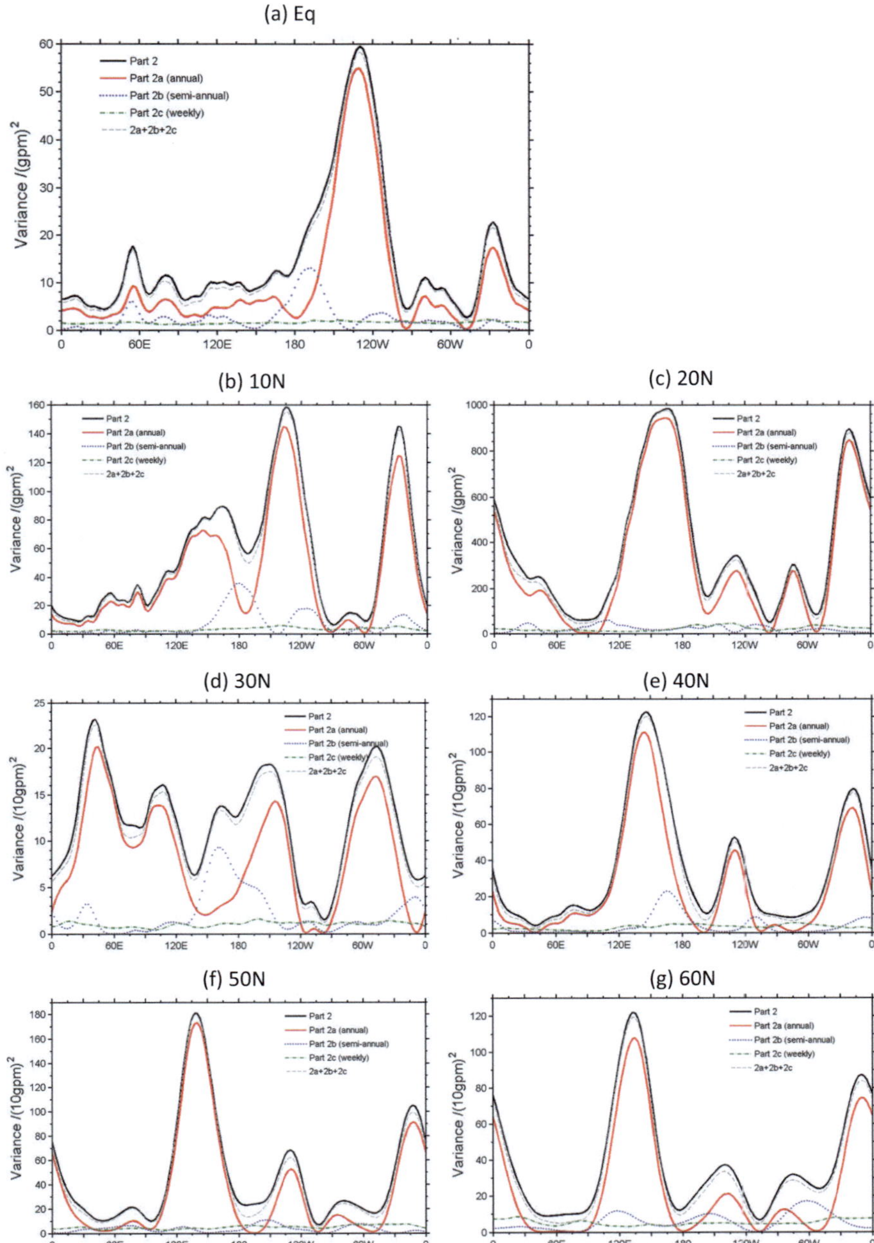

Fig. 3.48 Variances of climatic height in terms of total (*black line*), annual cycle (*red line*), semiannual cycle (*blue dotted line*), biweekly cycle (*green dashed line*, 30-day band), as well as the sum of three cycles (*grey-dashed line*, 2a + 2b + 2c) at 300 hPa along **a** the equator, **b** 10°N, **c** 20°N, **d** 30°N, **e** 40°N, **f** 50°N, **g** 60°N, **h** 70°N, and **i** 80°N

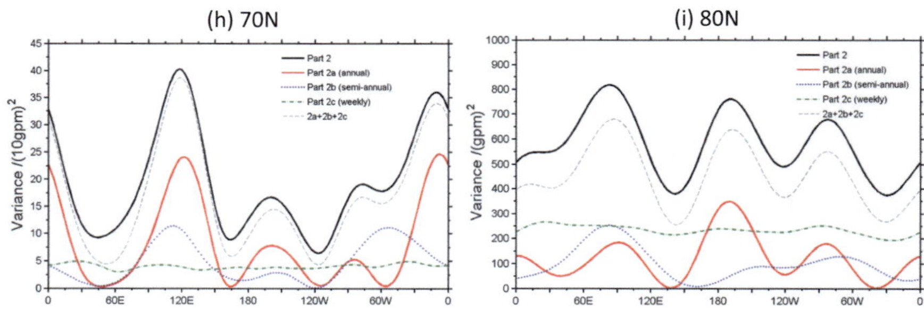

Fig. 3.48 (continued)

Along 10°S, there are four large peaks of annual cycle variance in the tropical eastern Pacific, near 70°W, 30°W, and 30°E. Three peaks of semiannual cycle variance are found near the dateline, 120°W, and 30°W. The variance of climatic height is well constructed by the three components (Fig. 3.49a). Four annual cycle peaks and four semiannual cycle peaks of height variances are observed along 20°S (Fig. 3.49b). Along 30°S, 40°S, and 50°S, the variance of biweekly cycle is larger than that of semiannual cycle. Along 60°S, 70°S, and 80°S, the same variance can be found in the biweekly cycle and semiannual cycle.

Variances of climatic height in terms of total, annual cycle, semiannual cycle, biweekly cycle, as well as the sum of three cycles at 300 hPa along different longitudes are shown in Fig. 3.50. Along 30°E (Fig. 3.50a), there are three peaks of annual-cycle variance, three peaks of semiannual cycle variance, and two peaks of biweekly cycle variance. Along 90°E (Fig. 3.50b), there are three peaks of annual cycle variance, two peaks of semiannual cycle variance, and two peaks of biweekly cycle variance in two mid-high latitudes. The reconstructed series of the three components well match the variance of climatic height. Along 120°E (Fig. 3.50c), the largest variance is the annual cycle. Peaks of height variances in annual cycle, semiannual cycle, and biweekly cycle are comparable along the dateline, 120°W, and 60°W in the mid-high latitudes (Fig. 3.50d–f). However, variances at different longitudes are different.

3.9 Annual and Semiannual Asymmetric Systems

The annual cycle of the climatic zonal departure or asymmetric systems is mainly caused by large-scale continent-ocean contrast. Figure 3.51a shows the constant term H_0 of climatic height at 850 hPa. In the NH, there are two negative height areas from the subtropical Asia to the North Pacific and from the subtropical North America to the North Atlantic, which are separated by two positive height areas from the subtropical Atlantic to the Europe-North Asia and from the subtropical eastern Pacific to the Northwest America. In the SH, positive and negative height

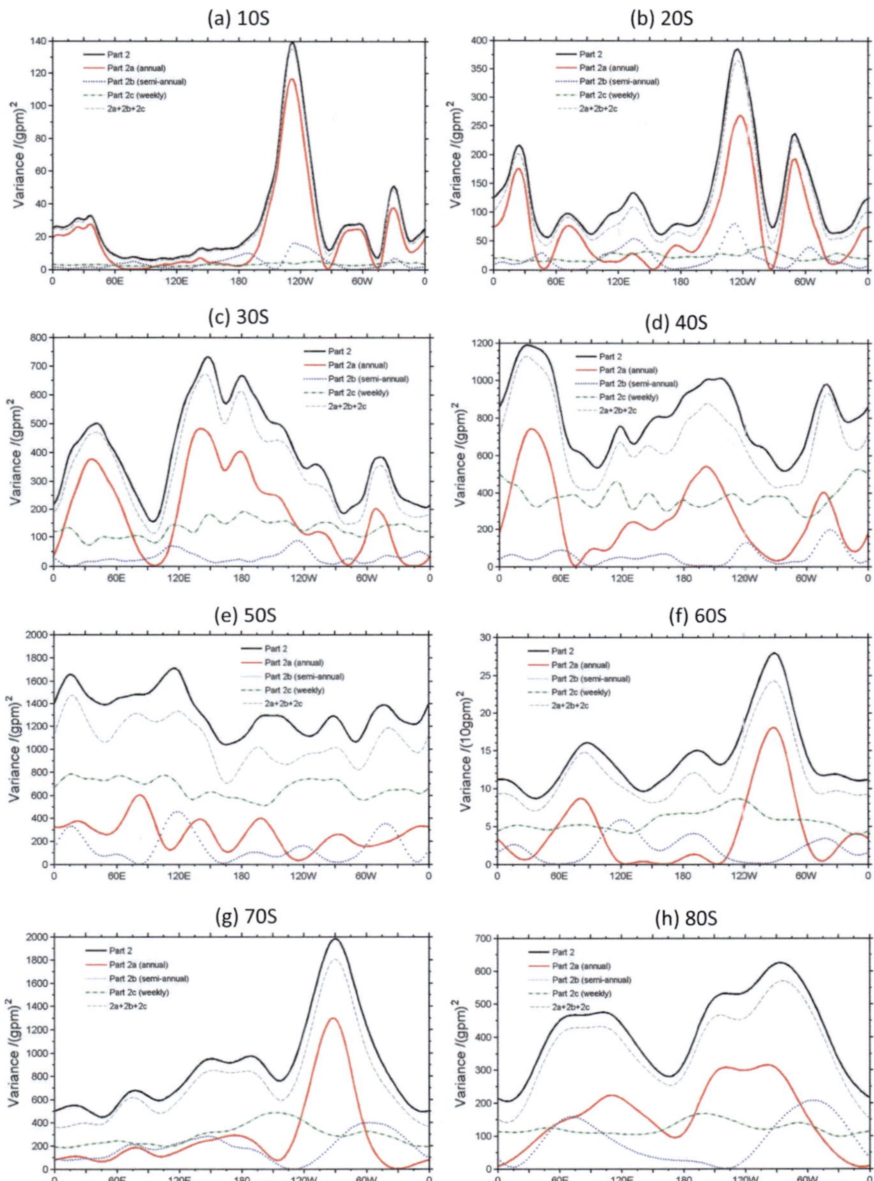

Fig. 3.49 Variances of climatic height in terms of total (*black line*), annual cycle (*red line*), semiannual cycle (*blue dotted line*), biweekly cycle (*green dashed line*, 30-day band), as well as the sum of three cycles (*grey dashed line*, 2a + 2b + 2c) at 300 hPa along **a** 10°S, **b** 20°S, **c** 30°S, **d** 40°S, **e** 50°S, **f** 60°S, **g** 70°S, and **h** 80°S

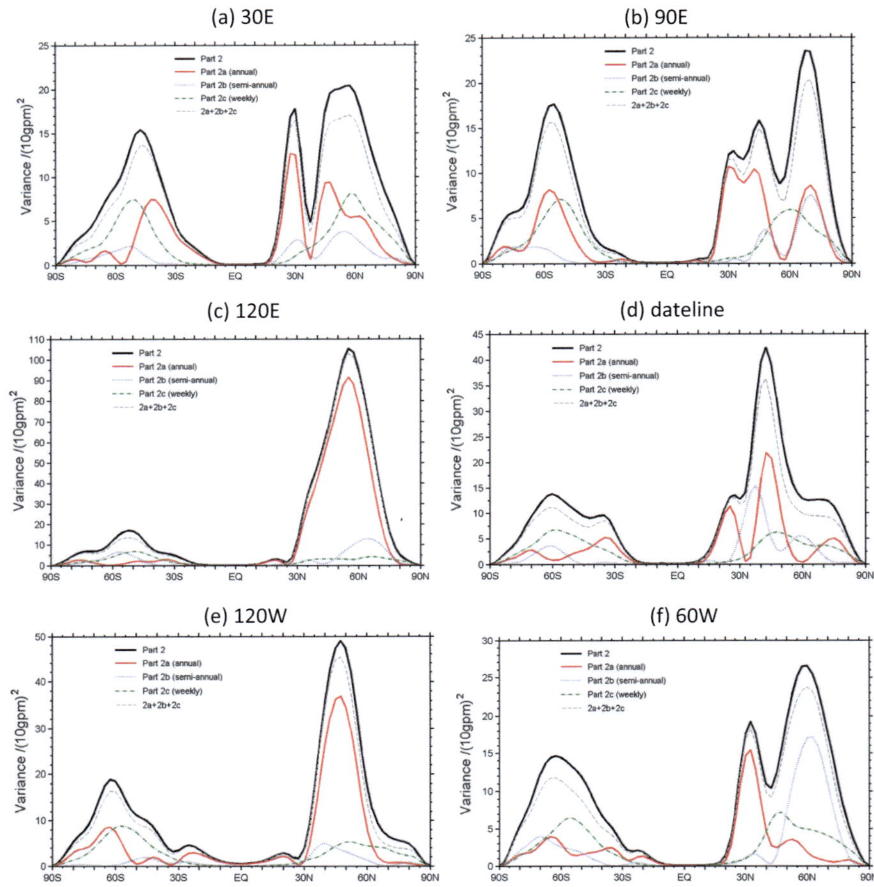

Fig. 3.50 Variances of climatic height in terms of total (*black line*), annual cycle (*red line*), semiannual cycle (*blue dotted line*), biweekly cycle (*green dashed line*, 30-day band), as well as the sum of three cycles (*grey dashed line*, 2a + 2b + 2c) at 300 hPa along **a** 30°E, **b** 90°E, **c** 120°E, **d** 180°E, **e** 120°W, and **f** 60°W

areas are mainly situated in the middle latitudes. Figure 3.51b shows the global height amplitude at 850 hPa. Two large centers are located in the Asian continent and the middle latitudes of North Pacific. In North America and the North Atlantic, there are other three small centers. In the SH, there are five centers in the middle latitudes between 20°S and 60°S. This annual cycle distribution of asymmetric systems is basically similar to that shown in Fig. 3.28.

In Fig. 3.51b, three asymmetric height centers of annual cycle amplitude move westward, eastward, even polarward, and equatorward seasonally. Their tracks with season can be clearly observed from Fig. 3.52. On February 10, the positive center "A" is located in the Siberia region and the negative center "B" is in the region of Aleutian Islands. The two centers are the Siberia High and the Aleutian Low in

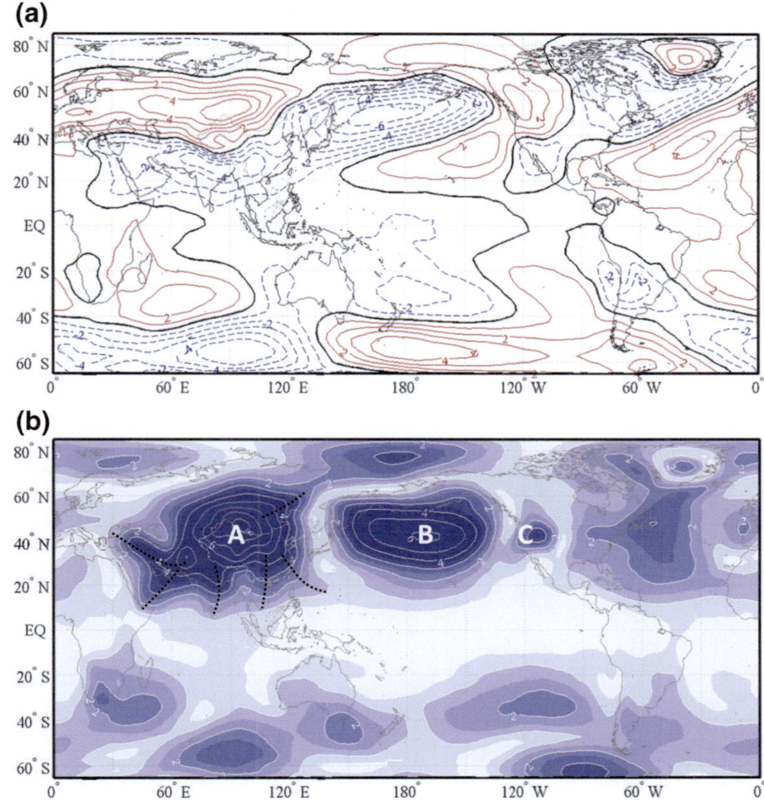

Fig. 3.51 The annual cycle of asymmetric height (10 gpm) at 850 hPa in terms of **a** constant term (H_0) and **b** amplitude (H_1). Letters "A", "B", and "C" denote three centers of amplitude. *Dotted lines* in **b** indicate the active regions of peninsular scale monsoon troughs surrounding the Asian continent

boreal winter. On May 1, the Siberia High moves westward to the Caspian Sea while the Aleutian Low moves southwestward and splits into two centers ("B1" and "B2") respectively in the north of Japan and South Asia. Another positive center "C" moves southwestward from Northwest America on February 10 to the subtropical Northeast Pacific on May 1.

From May 1 to June 20, the Siberia High moves westward to Europe, and the Aleutian Low moves southwestward, developing into the South Asian Low. The high center C moves westward, developing into the central North Pacific High. From February 10 to May 1, high and low systems are altering in the Asian-North Pacific region. Based on this alteration of height fields, geostrophic winds are completely opposite in their directions in East Asia. In winter, it should be northerlies and northeasterlies from East Asia to South Asia. In summer, southwesterlies prevail from South Asia to East Asia. Thus, this is a typical monsoon region from South Asia to East Asia.

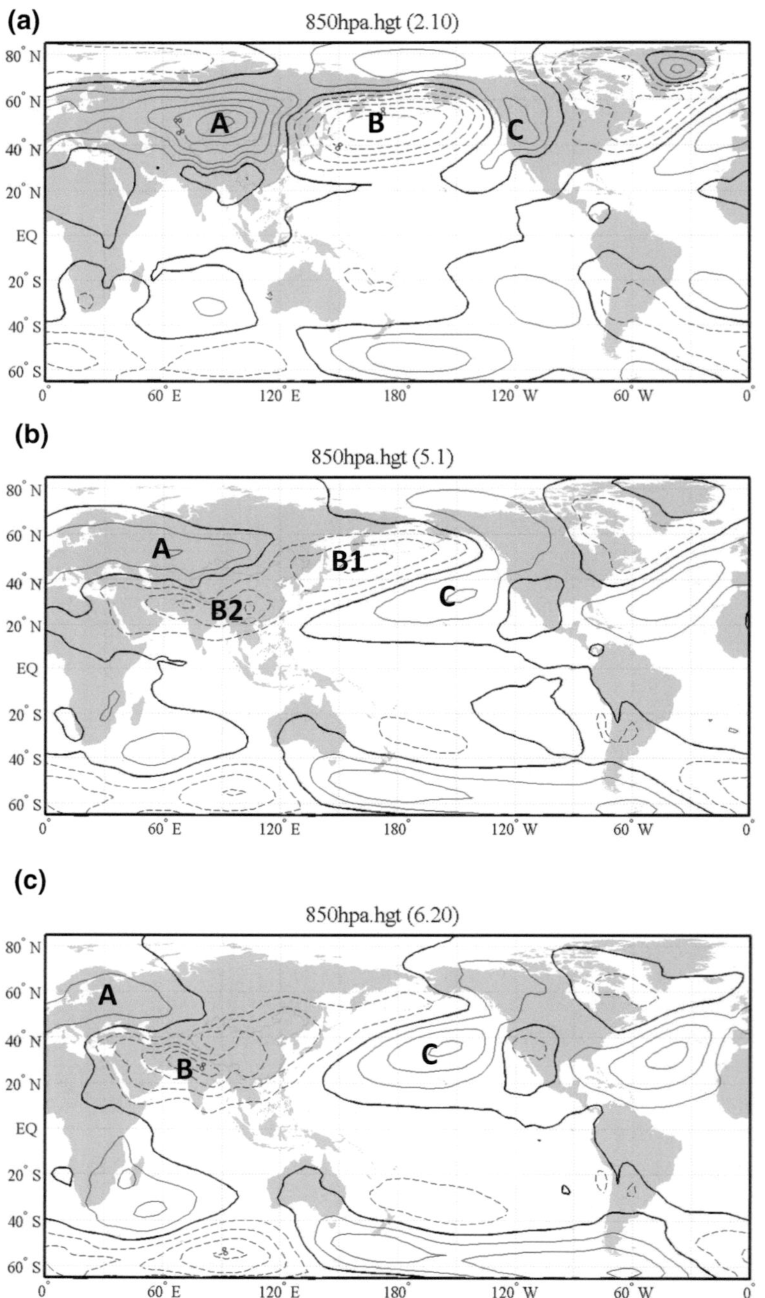

Fig. 3.52 The annual cycle component of asymmetric height (gpm) at 850 hPa on **a** 10 February 10, **b** May 1, **c** June 20, **d** 9 August, **e** September 28, and **f** December 22. Letters "A", "B", and "C" denote the height centers of annual cycle

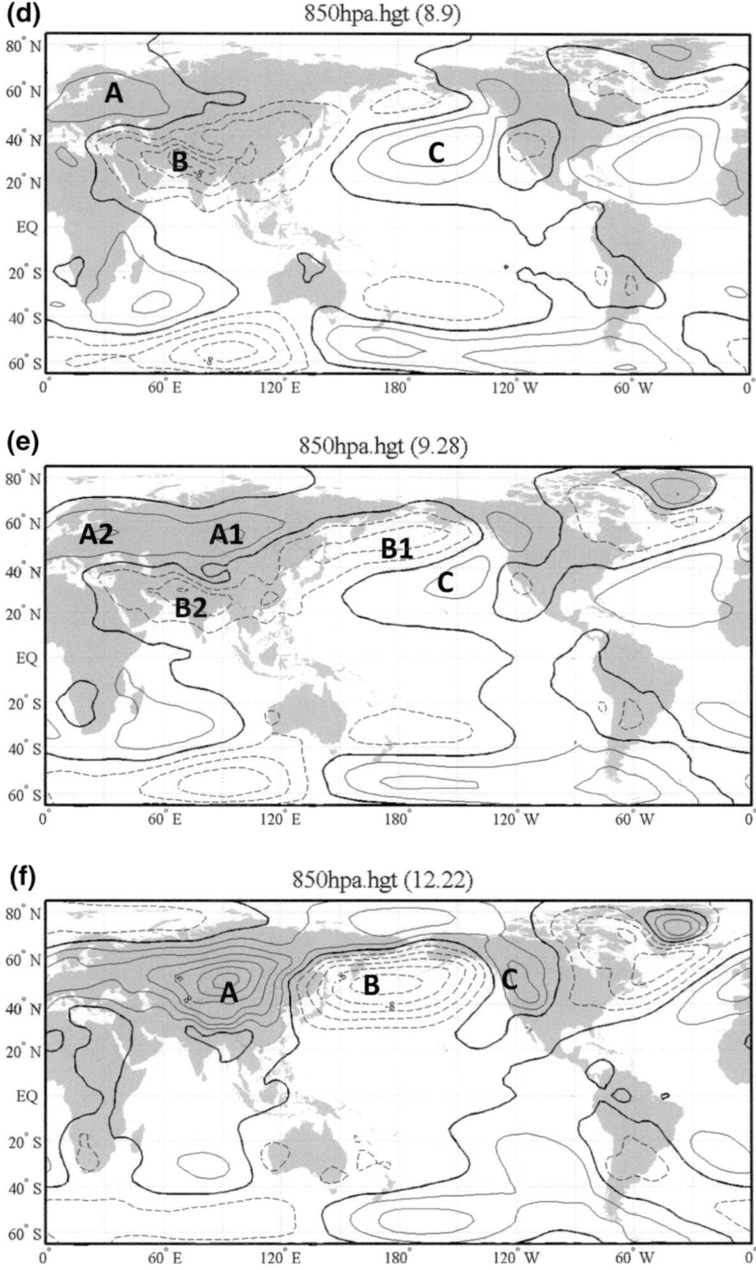

Fig. 3.52 (continued)

From June 20 to August 9, three high-low centers are basically similar in their intensities and locations. But from August 9 to September 28, there are two high centers, respectively, in Europe and Siberia while two low centers, respectively, in South Asia and the region of Aleutian Islands. On December 22, three high-low systems return to the winter situation as on February 10. In the latitudes between 40°S and 60°S, the annual cycle only shows the intensity changes of positive height in the Southern Pacific and negative height from the Southern Atlantic to the Southern Indian Ocean.

At 850 hPa, the annual cycle component of asymmetric height is shown in Fig. 3.52. Three height centers are basically moving along 40°N. Vertical sections of annual cycle asymmetric height are shown in Fig. 3.53. In winter, the Siberia High extends vertically from the surface to about 300 hPa. For the Aleutian Low, its center at 1000 hPa is located in the central North Pacific but its actual center is at 250 hPa and 140°E. For the North American High, its center is near 200 hPa and 130°W. In boreal summer, the Siberia High disappears from the surface but the northern part of the SAH can be observed at 150 hPa and 60°E. The center of Aleutian Low can be observed near the surface. The North American High and the Northeast Pacific High alter with season.

For the annual cycle component of asymmetric height, vertical distributions along 60°N and 60°S on February 10 are compared in Fig. 3.54. In the NH, there are two pairs of positive/negative height centers. In the SH, there is only one pair of positive/negative height centers. Vertical distributions of height along 60°N and 60°S on August 9 are compared in Fig. 3.55. In the NH, there are two pairs of positive/negative height centers, but the positive center in North America is weak (Fig. 3.5a). In the SH, there is only one pair of positive/negative height centers (Fig. 3.55b).

From Figs. 3.53 to 3.54, it is noted that most strong centers of annual-cycle component of asymmetric height are located at around 300 hPa. Figure 3.56 shows the time-zonal sections of the annual cycle component of asymmetric height at 300 hPa along six latitudes. Along 20°N and 30°N, positive and negative centers move westward from boreal winter to summer. In boreal summer, the SAH has two centers near 60°E and 90°E. Along other latitudes in the NH, two pairs of positive and negative centers basically stand with season. In the SH, a pair of positive and negative centers also basically stand with season (Fig. 3.57).

The semiannual cycle amplitude of asymmetric height at 850 hPa is shown in Fig. 3.58. Seven large centers are located between 20°N and 80°N in the NH and four centers are located between 40°S and 60°S in the SH.

In Fig. 3.58, seven large centers move westward and eastward, even polarward and equatorward or disappear seasonally in the NH. Their seasonal tracks can be clearly observed from Fig. 3.59. On February 10 (Fig. 3.59a), positive center "D" is near the Alaska Peninsula and then it moves southwestward to the east of Japan on April 6 (Fig. 3.59b), while the negative center "F" moves southeastward from the Greenland on February 10 to the southeastern North Atlantic on April 6. On August 9 (Fig. 3.59d), seven centers reappear in the NH. In the SH, center "S" moves

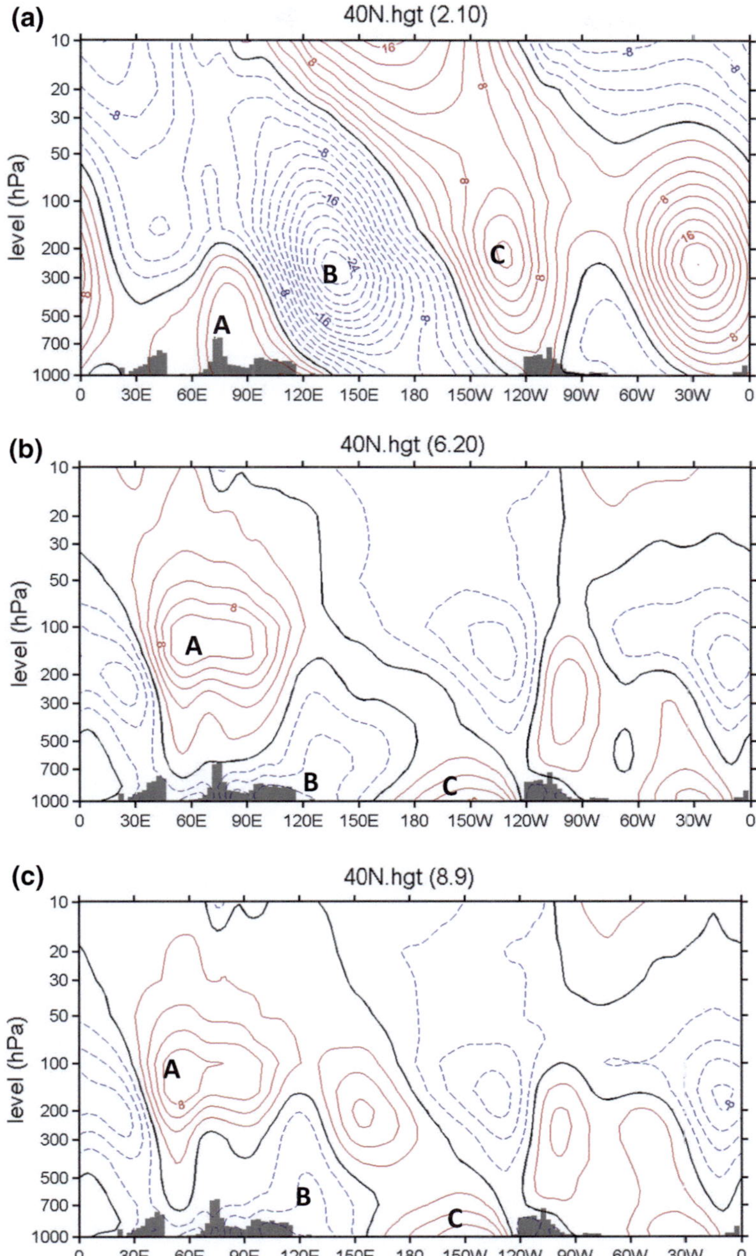

Fig. 3.53 Vertical–zonal sections of the annual cycle component of asymmetric height (gpm) on **a** February 10, **b** June 30, **c** August 9, **d** December 22. Letters "A", "B", and "C" denote the height centers of annual cycle component

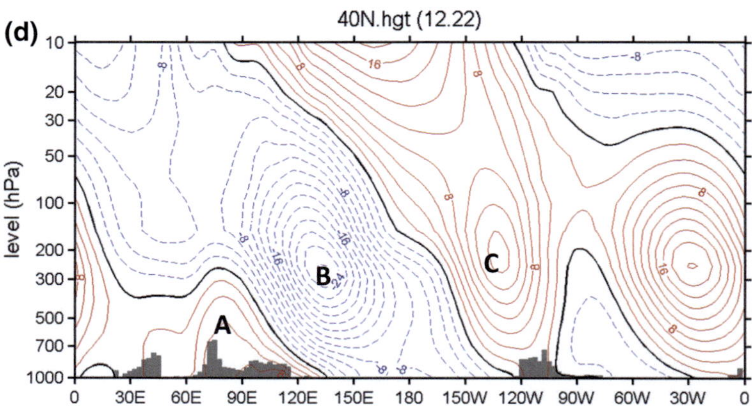

Fig. 3.53 (continued)

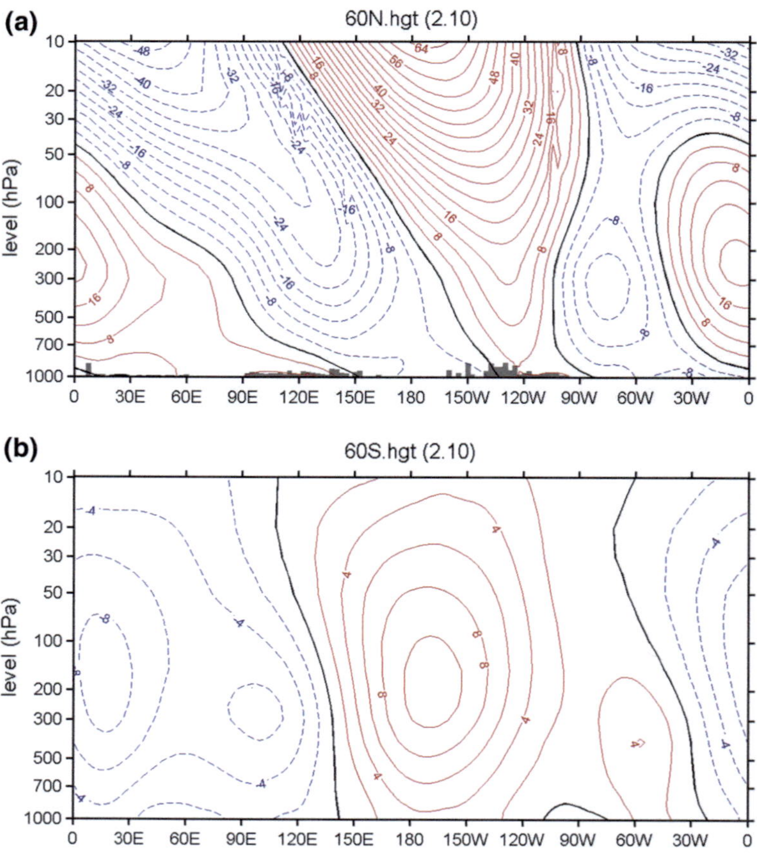

Fig. 3.54 Vertical sections of the annual cycle component of asymmetric height (gpm) along **a** 60°N and **b** 60°S on February 10

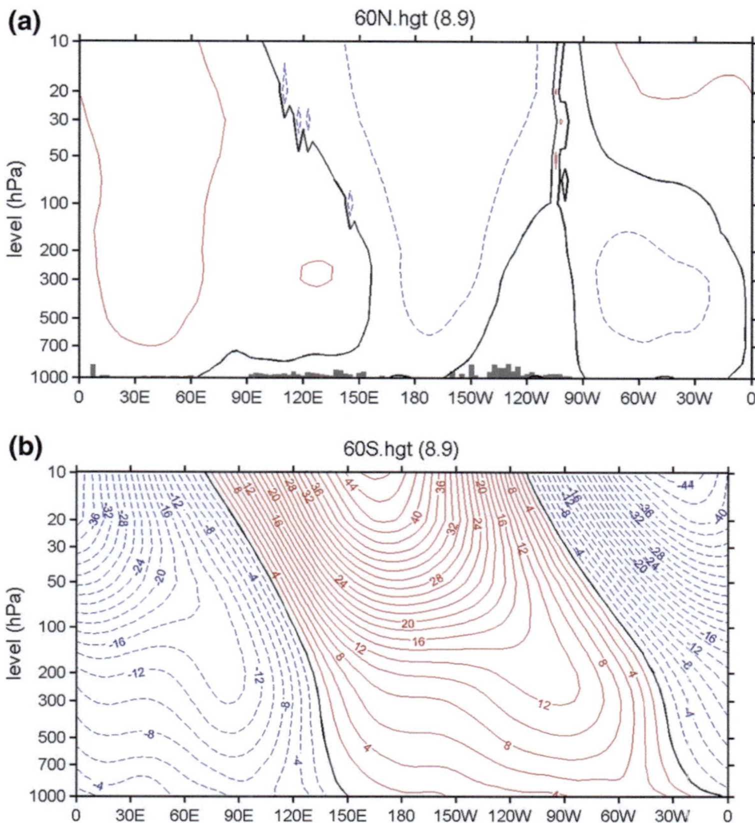

Fig. 3.55 Vertical sections of the annual cycle component of asymmetric height (gpm) along **a** 60°N and **b** 60°S on August 9

seasonally from central South Pacific (160°W, 55°S) to Australia and the southeastern Indian Ocean (130°E, 55°S).

The central position, intensity, and extent of positive and negative asymmetric heights change within the semiannual cycle. It is clearly noted that large centers of positive and negative heights exist near the tropopause and in the upper stratosphere (Fig. 3.60). Along 40°N, a negative center "A" moves eastward from central North Pacific (170°E) at 200 hPa on February 10 to the North America (110°W) at 300 hPa on April 16, and then returns to the North Pacific on 30 June.

For the semiannual cycle component of asymmetric height, vertical sections of height along 80°N, 60°N, and 60°S on February 10 are compared in Fig. 3.61. Along 80°N, a pair of positive and negative height centers increases from the lower troposphere to the upper stratosphere and tilts eastward with altitude. Along 60°N, the positive and negative height centers stands vertically with the positive one from 80°E to 105°W. Along 60°S, a pair of positive and negative height centers with

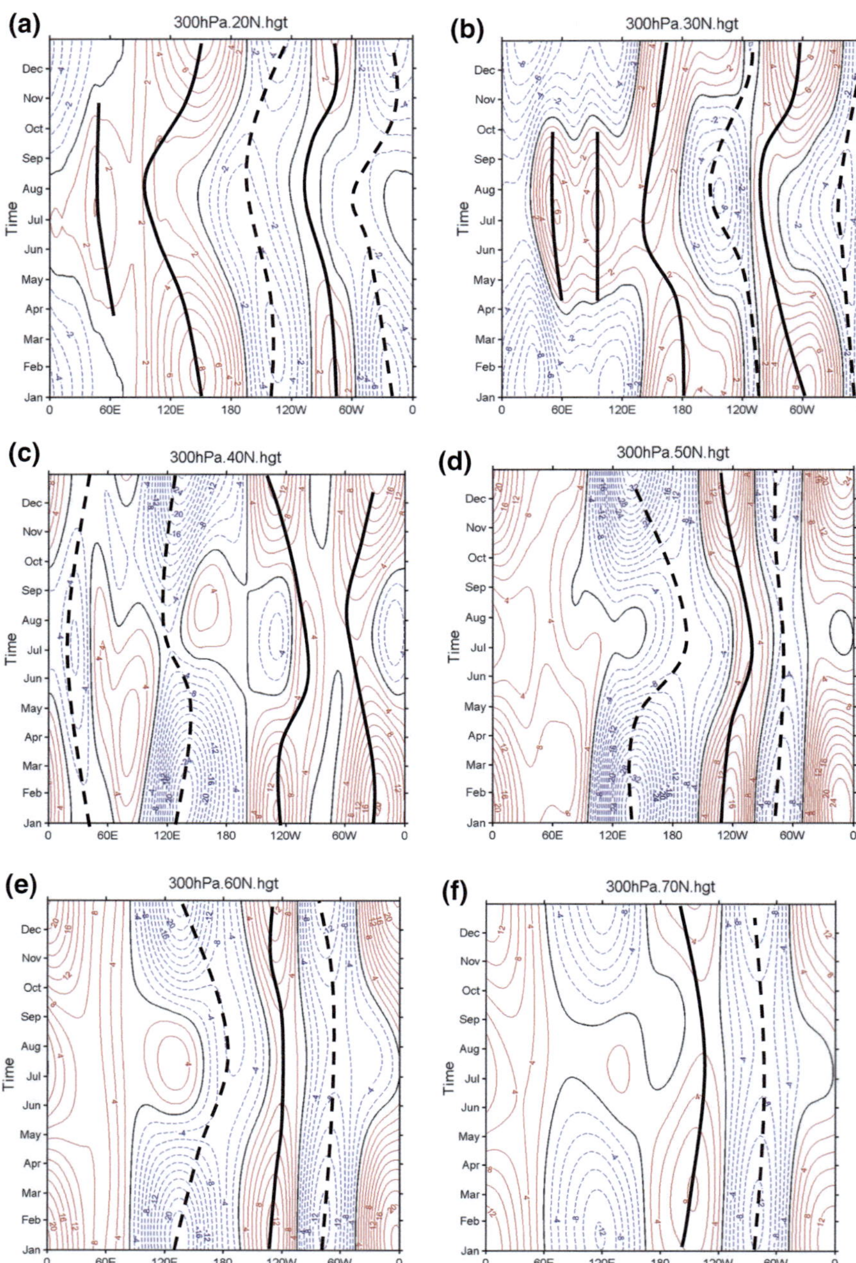

Fig. 3.56 Time-zonal sections of the annual cycle component of asymmetric height (gpm) at 300 hPa along **a** 20°N, **b** 30°N, **c** 40°N, **d** 50°N, **e** 60°N, and **f** 70°N. *Heavy solid and dashed lines* indicate the maximum of positive and negative heights

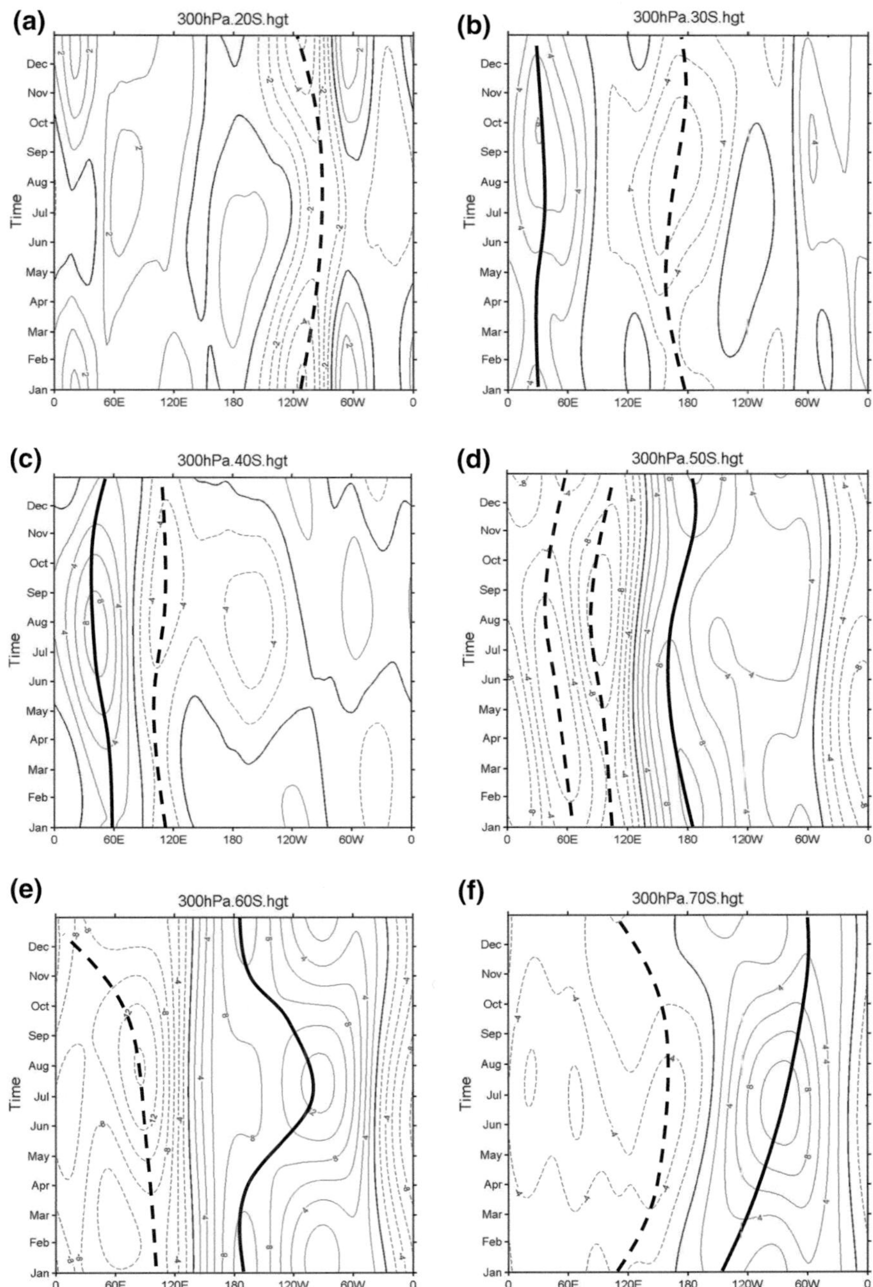

Fig. 3.57 Time-zonal sections of the annual cycle component of asymmetric height (gpm) at 300 hPa along **a** 20°S, **b** 30°S, **c** 40°S, **d** 50°S, **e** 60°S and **f** 70°S

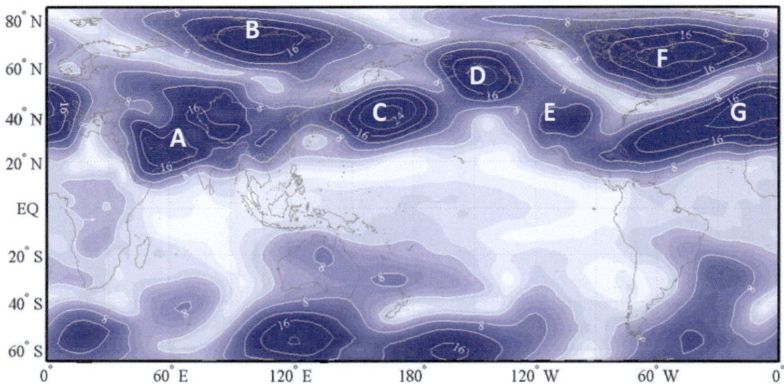

Fig. 3.58 The semiannual cycle amplitude of asymmetric height (gpm) at 850 hPa. Letters "A"–"G" indicate seven centers with large amplitude in the NH

opposite signs is separated in the stratosphere and the troposphere. Similar distributions are found on August 9 along 80°N, 60°N, and 60°S.

As shown in Figs. 3.60 and 3.61, most strong centers are located at around 300 hPa. Figure 3.62 shows the time-zonal sections of the semiannual-cycle component of asymmetric height at 300 hPa along six latitudes. Along 20°N, a positive center moves eastward from North Africa in boreal spring to the central Pacific in July, and further the subtropical Northeast Pacific in November. Along 30°N, a positive center moves eastward from North Africa in boreal spring to the central Pacific in May, and then the North America in August. Along 40°N, a positive center in spring first appears in North China and then moves zonally. This center moves westward to Europe in June and July. It moves eastward to the central North Pacific in April and May, then to North America and the North Atlantic in June and July. Along 50°N, a positive center first appears from Kazakhstan in May, and moves eastward to Northeast Asia in July and the central North Pacific in August. Along 60°N, a positive center moves eastward from Europe in April to the eastern Siberia and the Bering Sea in July. Conversely, along 70°N, a positive center first appears over the Greenland in April and then moves to the west of Alaska in July.

In the SH, the time-zonal sections of the semiannual-cycle component of asymmetric height at 300 hPa along four latitudes are shown in Fig. 3.63. Along 20°S (Fig. 3.63a), a positive center in April first appears in South America and then moves zonally. In August, it moves westward to Northwest Australia. Along 30°S (Fig. 3.63b), two positive centers first appear in the Cape Basin in June and February then move zonally. Finally, they combine together over the South Pacific in November and May, respectively. Along 40°S (Fig. 3.63c), two positive centers first appear in the Southern Indian Ocean in February and August, and then move eastward to the Southwestern Atlantic in August and February. Similar situation is also observed along 50°S (Fig. 3.63d).

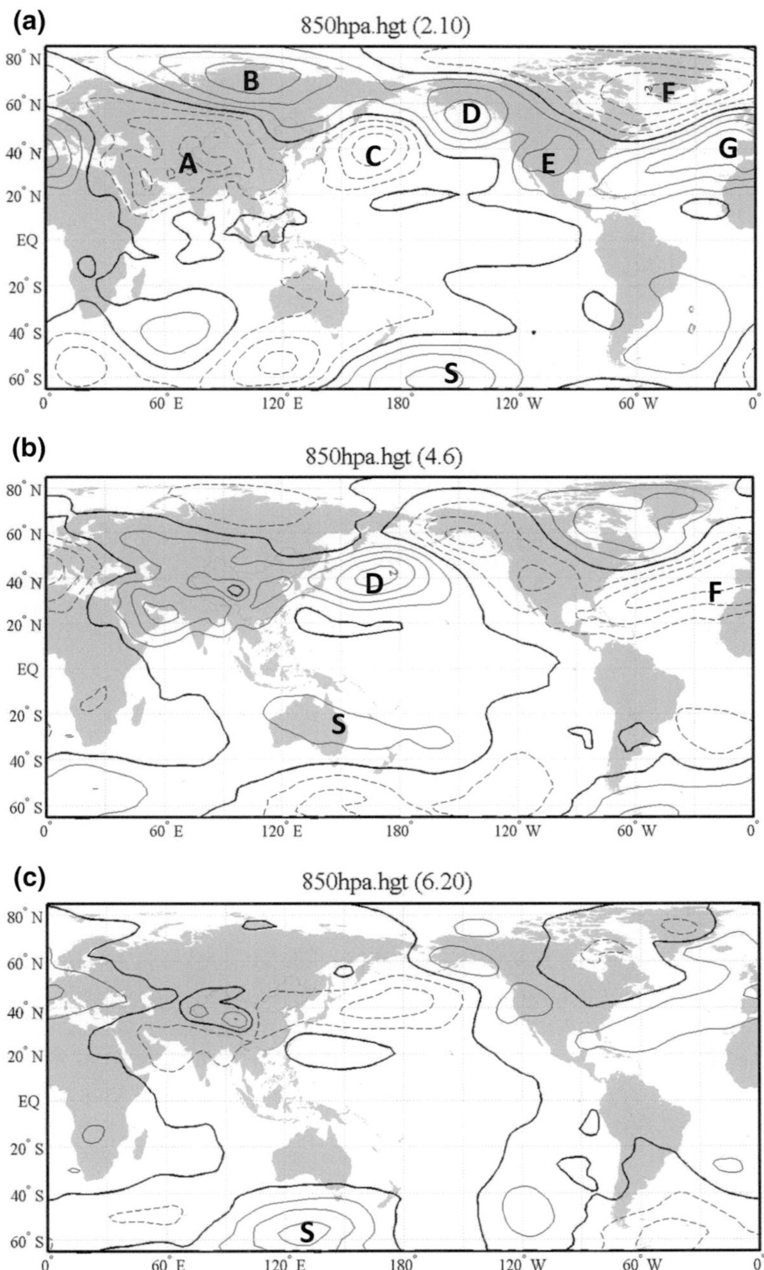

Fig. 3.59 The semiannual cycle component of climatic asymmetric height (gpm) at 850 hPa on **a** February 10, **b** April 6, **c** June 20, **d** August 9, and **e** December 22. Letters "A"–"G" and "S" denote the height centers

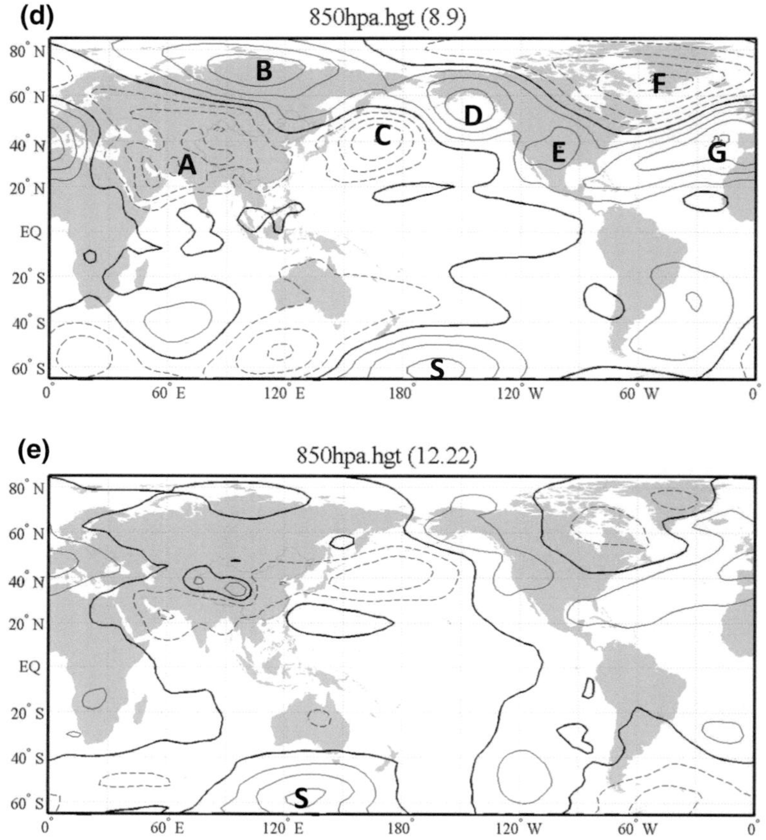

Fig. 3.59 (continued)

Changes in asymmetric systems should be obvious in Asia because of huge topography and sea–land thermodynamical contrast. The asymmetric height at 850 hPa averaged between 120 and 130°E, which crosses the East China Sea and Northeast Asia, is shown in Fig. 3.64. Near 60°N, the annual cycle of asymmetric height indicates that the Siberia High is dominant in the winter half year (Fig. 3.64a). In the mid-low latitudes, the southwest trough and the East Asian Trough are active in East Asia. In the subtropical region, the southwest trough starts to strengthen and move northward from March to early July, and then weakens and moves southward from July to December. Around 45°N, the East Asian Trough experiences an annual cycle with the strongest intensity in late June. The semiannual cycle of asymmetric height is clearly strong in the subtropical and mid-high latitudes (Fig. 3.64b). There are two regions with large amplitudes of semiannual cycle around 35°N and 72°N. It may indicate that the high and low departures of semiannual cycle first appear in the high latitude 72°N then move southward to 35°N after 2 months. The combination of the annual and semiannual components of

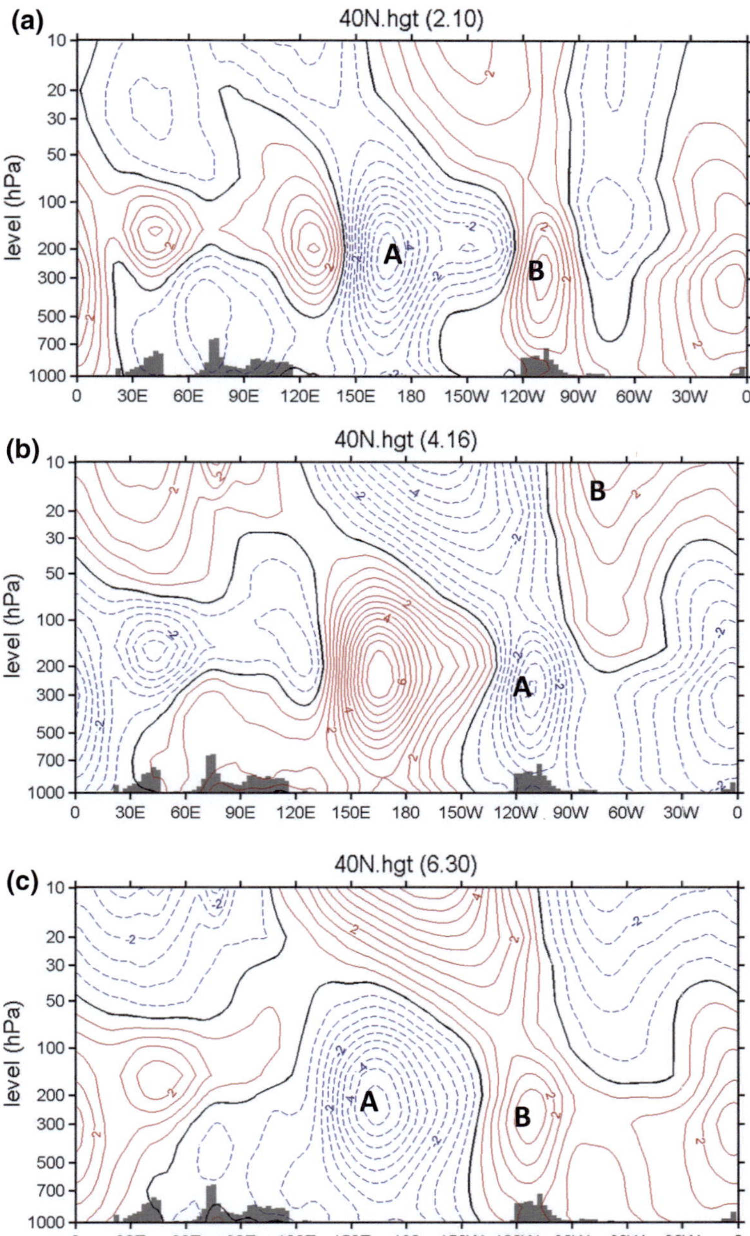

Fig. 3.60 Vertical sections of the semiannual cycle component of asymmetric height (gpm) along 40°N on **a** February 10, **b** April 16, **c** June 30, **d** August 9, **e** October 13, and **f** December 22. Letters "A" and "B" indicate the height centers

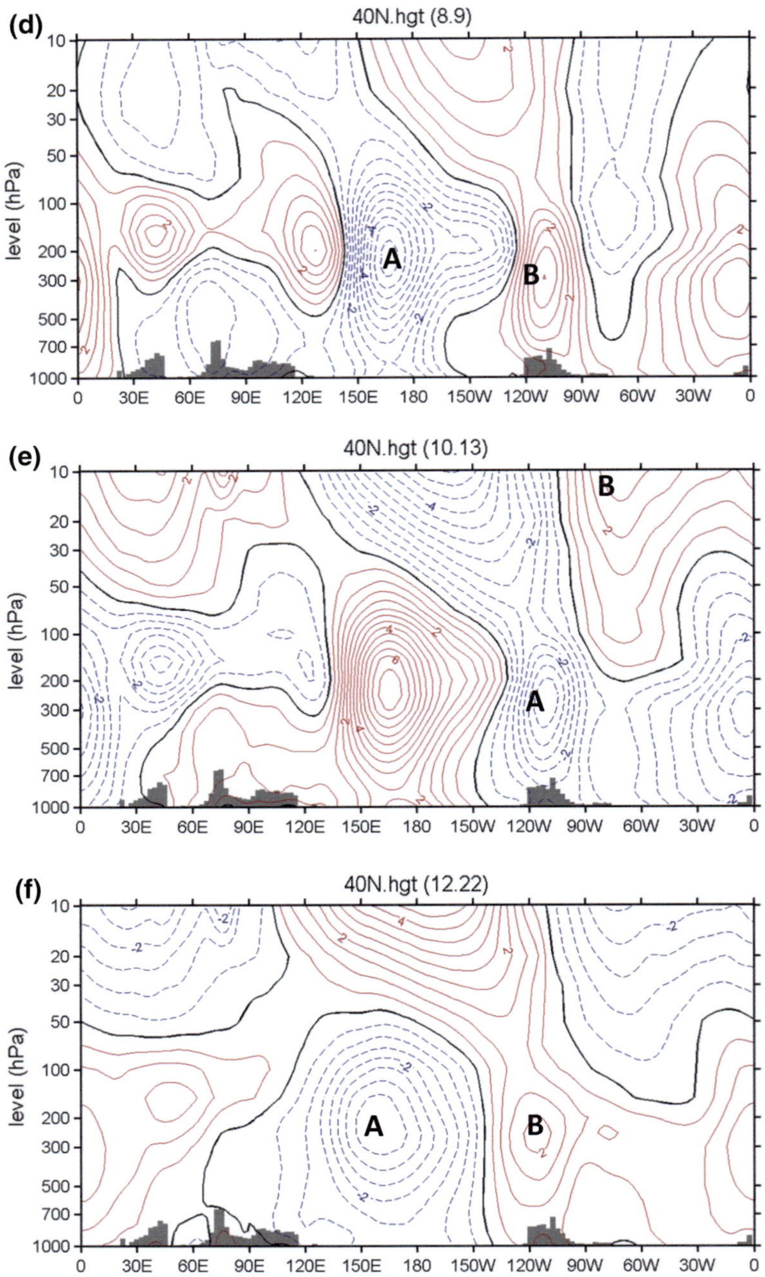

Fig. 3.60 (continued)

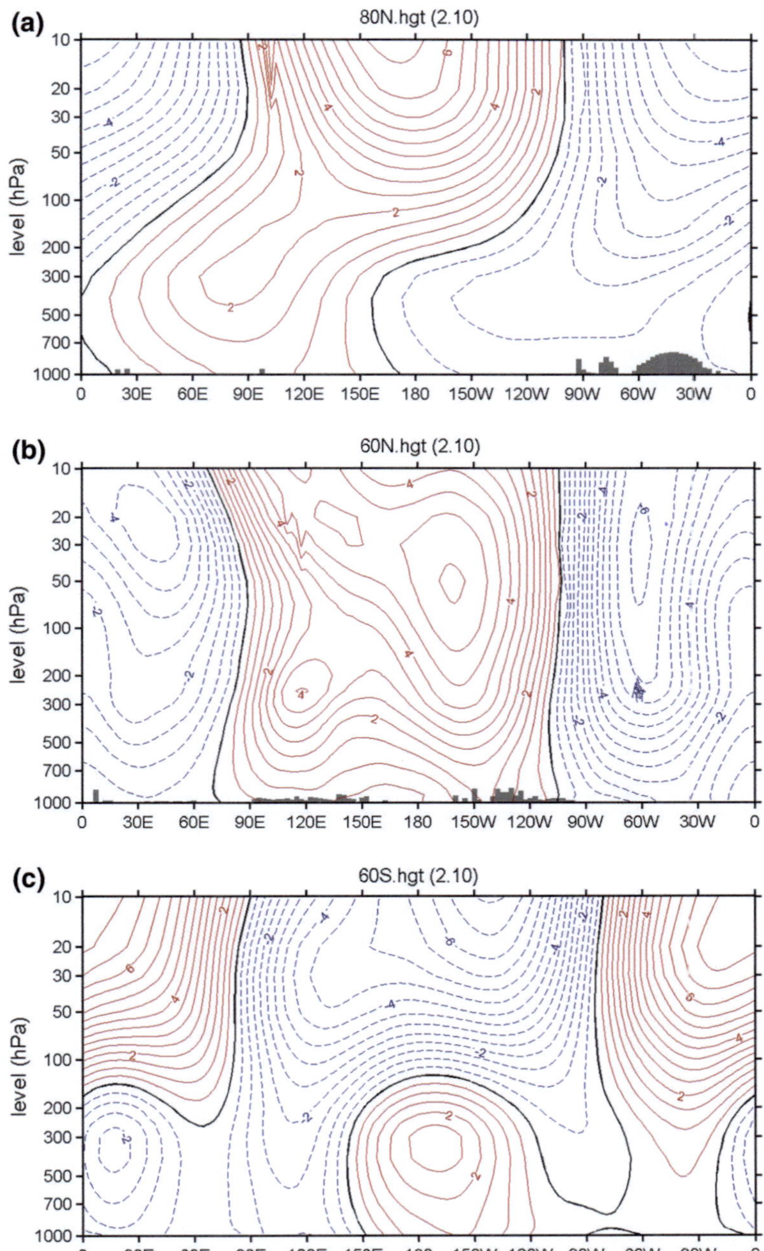

Fig. 3.61 Vertical sections of the semiannual cycle component of asymmetric height (gpm) along **a** 80°N, **b** 60°N, and **c** 60°S on February 10

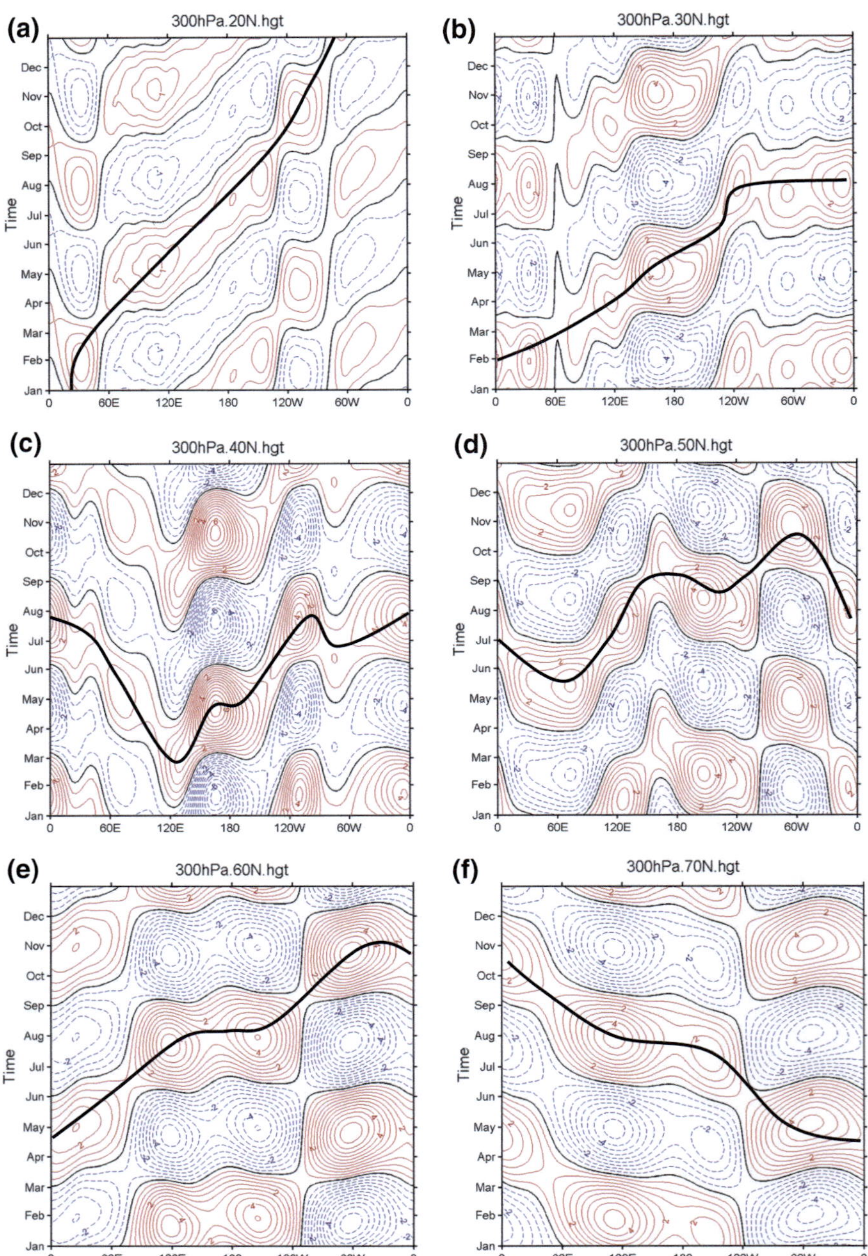

Fig. 3.62 Time-zonal sections of the semiannual cycle component of asymmetric height (gpm) at 300 hPa along **a** 20°N, **b** 30°N, **c** 40°N, **d** 50°N, **e** 60°N, and **f** 70°N. *Heavy solid lines* indicate the maximum of positive height

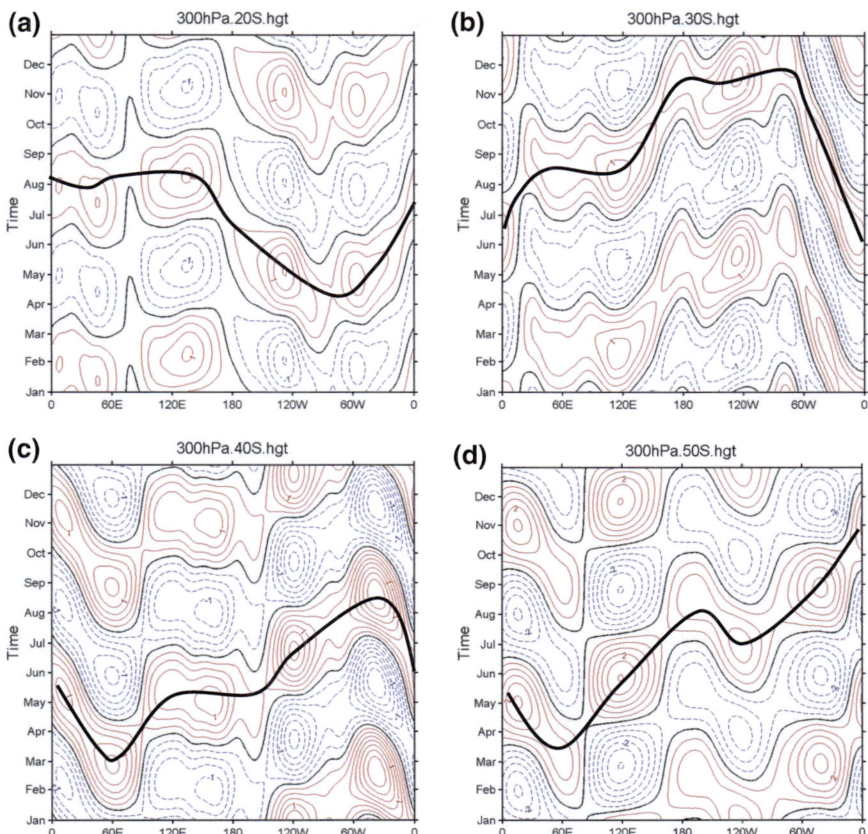

Fig. 3.63 Time-zonal sections of the semiannual cycle component of asymmetric height (gpm) at 300 hPa along **a** 20°S, **b** 30°S, **c** 40°S, and **d** 50°S

asymmetric height is shown in Fig. 3.64c. It is noted that the Siberia High is strong in December, January, and February. In May, the polar vortex frequently appears in Northeast Asia (75°N). Two troughs reach the deepest in East Asia indicated by EAT in July.

The asymmetric height averaged over 120–130°E and banded 1–30 days at 850 hPa is shown in Fig. 3.65. Biweekly cycle signals of climatic height propagate northward from the lower latitudes and southward from mid-high latitudes. Sometimes, propagating signals confront in the East China Sea.

Along the eastern China and central Siberia, the asymmetric height averaged over 110–120°E and the combination of annual cycle and semiannual cycle at 850 hPa are shown in Fig. 3.66. The Siberia High starts to strengthen from August around 60°N, reaches the maximum in January, and extends to 30°N. From

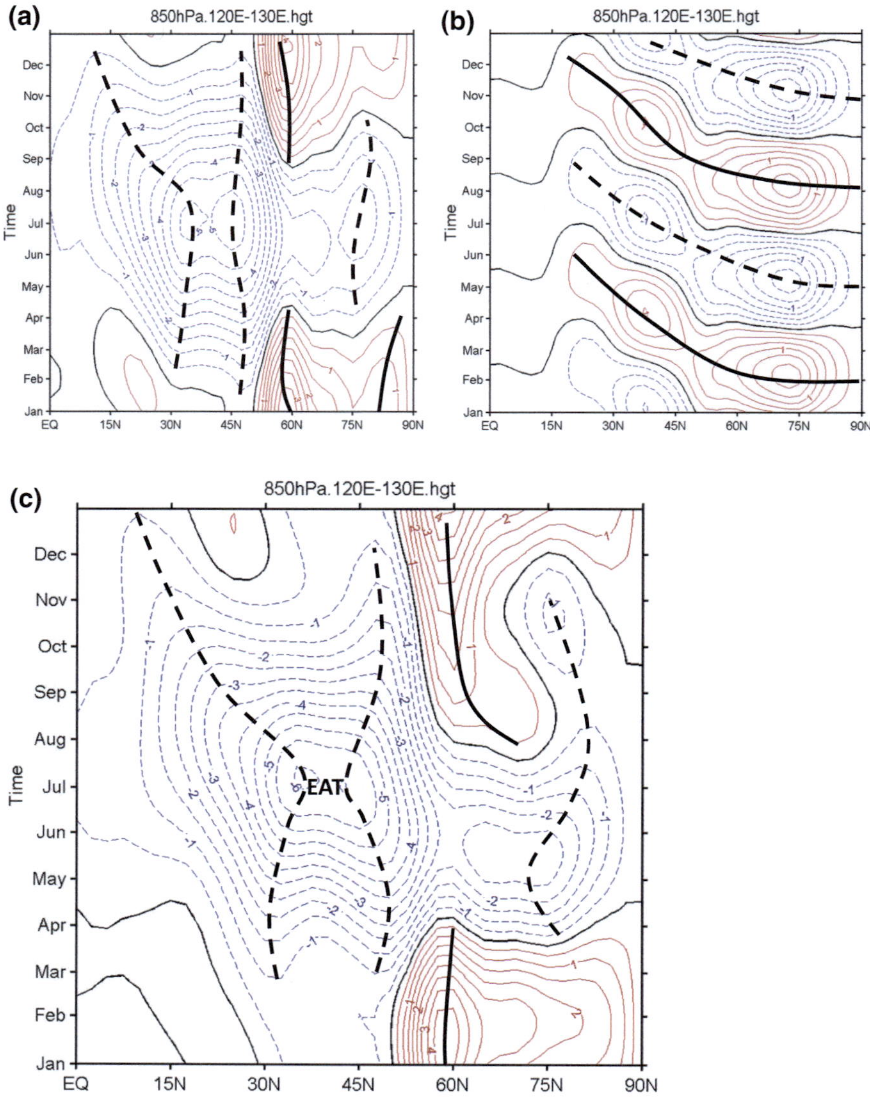

Fig. 3.64 Time latitude sections of asymmetric height (10 gpm) averaged between 120 and 130°E at 850 hPa in terms of **a** annual cycle, **b** semiannual cycle, and **c** the sum of them. *Heavy solid and dashed lines* indicate the axes of positive and negative heights

February, the subtropical trough starts to strengthen in South China, propagates northward slowly to the Yangtze River in June and July, and then retreads southward quickly.

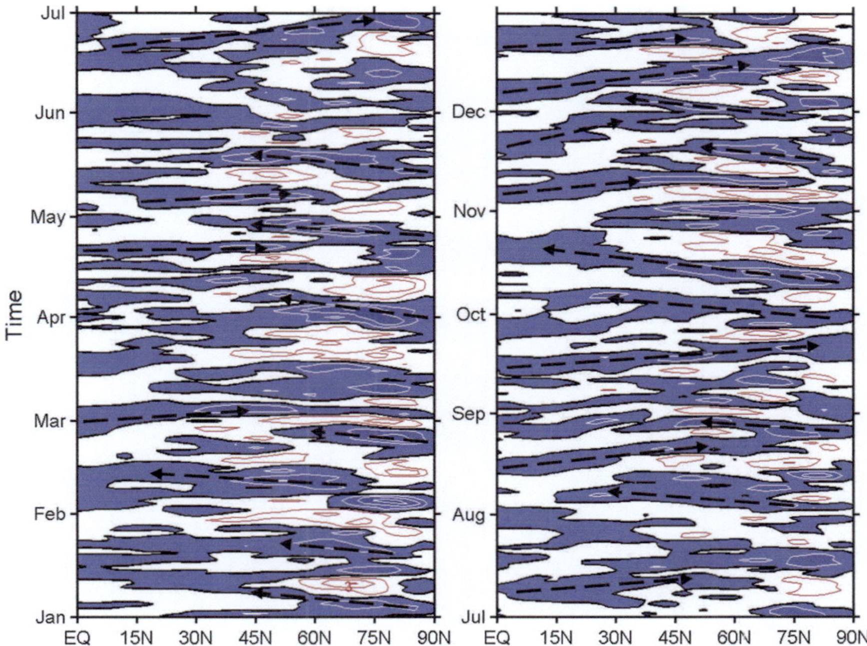

Fig. 3.65 Asymmetric height (10 gpm interval) averaged over 120–130°E and banded 1–30 days at 850 hPa

The asymmetric height averaged over 110–120°E and banded 1–30 days at 850 hPa is shown in Fig. 3.67. Most of the asymmetric height signals propagate southward from mid-high latitudes in winter half year. Sometimes, propagating signals confront in central and South China.

The asymmetric height averaged over 10–20°N at 850 hPa is shown in Fig. 3.68. This area crosses the tropical North Africa, the Arabian Peninsula, the Arabian Sea, the Indian Peninsula, BOB, the Indo-China Peninsula, the SCS, and the subtropical Northwest Pacific. There are three pairs of peninsular scale sea–land distributions. The famous tropical monsoon area is located along this zone. The annual cycle component of asymmetric height shows four active troughs in summer and autumn (Fig. 3.68a). The strongest is the BOB trough. The semiannual cycle component of asymmetric height shows three active centers, respectively, in the Arabian Sea, the BOB, and the SCS (Fig. 3.68b).

The combination of annual and semiannual cycles of asymmetric height is shown in Fig. 3.68c. A trough in the central Africa appears from January to the early May. The trough in the BOB forms in February and strengthens from April, and then approaches the strongest in July. In July, three troughs are clearly observed

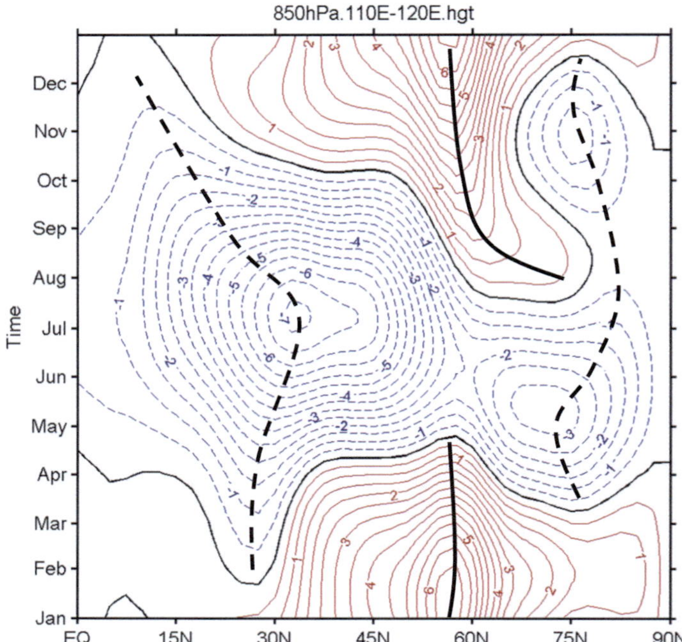

Fig. 3.66 Asymmetric height (10 gpm interval) averaged over 110–120°E and the combination of annual cycle and semiannual cycle at 850 hPa. *Heavy solid and dashed lines* indicate the axes of positive and negative heights

in the Arabian Sea, the BOB, and the SCS. In the subtropical NW Pacific, the tough arising from August is the ITCZ.

At all grids, asymmetric climate consists of annual cycle, semiannual cycle and biweekly cycle. Figure 3.69 shows the asymmetric height averaged over 10–20°N and banded 1–30 days at 850 hPa. From the subtropical North Africa to the subtropical North Indian Ocean, some signals of 1–30 days move eastward mainly from February to May. There are many westward signals of 1–30 days from the subtropical NW Pacific to the subtropical North Indian Ocean around the year.

Questions

1. Why do the strongest values and the northernmost locations of all atmospheric variables lag the strongest solar radiation on June 22 for about one month in the NH?
2. What are the vertical–meridional configurations of the zonally averaged circulation, geopotential height, temperature, and humidity indicating the zonal precipitation climatology?

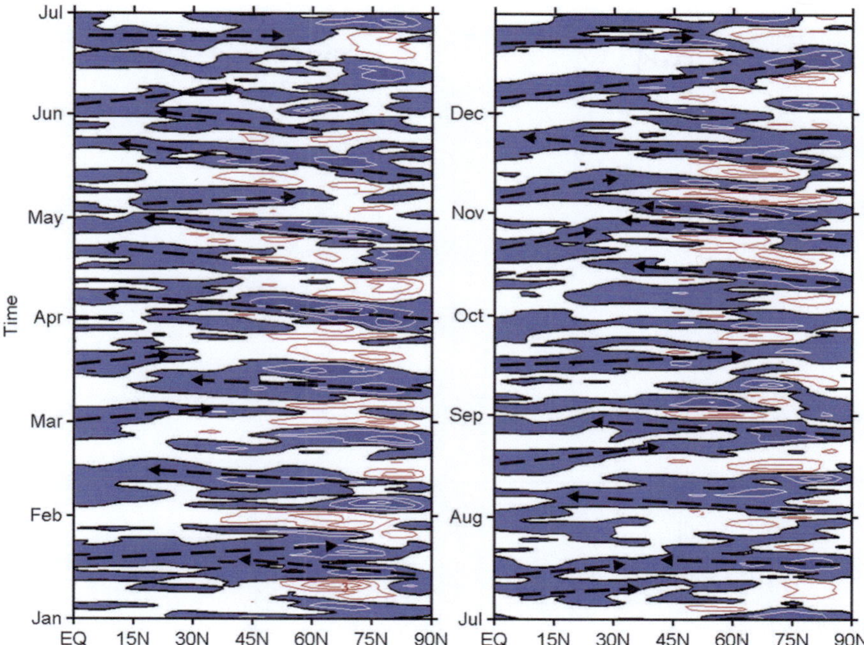

Fig. 3.67 Asymmetric height (10 gpm interval) averaged over 110–120°E and banded 1–30 days at 850 hPa

3 Why are the three Walker (or Walker-like) cells evidently observed between the equatorial area and subtropical NH (near 20°N) during boreal summer?
4 How can we find the atmospheric active centers and climatic troughs in boreal summer from the surface pressure field?
5 The climatic state is forced by the solar radiation and influenced by the surface conditions. How many timescales of climatic cycles should be identified?
7 What are the summer monsoon systems associated with annual and semiannual climatic components?
8 How to trace the biweekly cycle systems and what are their applications?
9 Why do we need to identify climatic components of different timescales in detail? How can we obtain a temporal climatic state of the atmosphere?

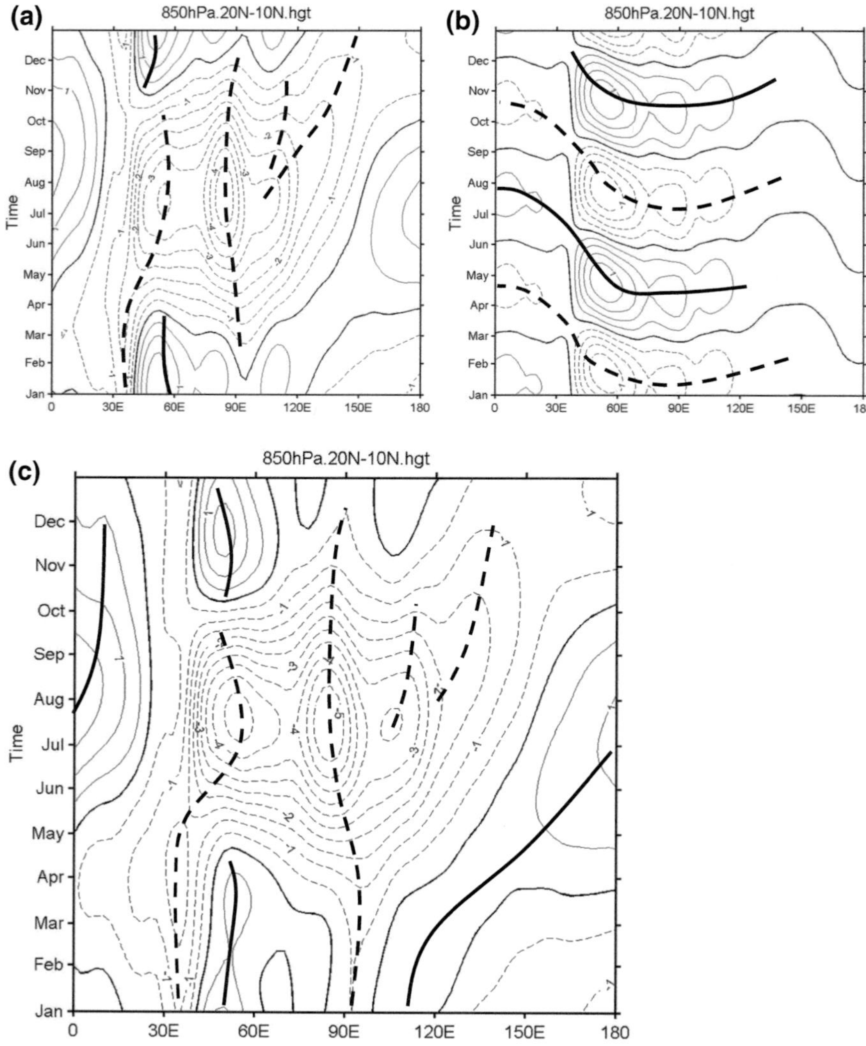

Fig. 3.68 Time-latitude sections of asymmetric height (10 gpm interval) averaged over 10–20°N at 850 hPa in terms of **a** annual cycle, **b** semiannual cycle, and **c** the sum of them. *Heavy solid and dashed lines* indicate the axes of positive and negative heights

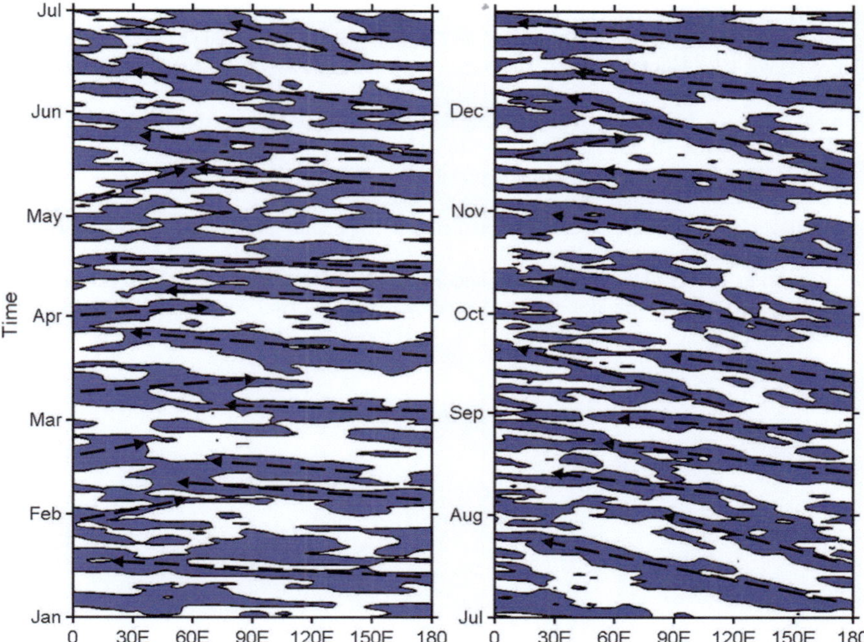

Fig. 3.69 Asymmetric height (10 gpm) averaged over 10–20°N and banded 1–30 days at 850 hPa

References

Ding T, and Qian WH (2012) Statistical characteristics of heat wave precursors in China and model prediction. Chin J Geophys 55: 1472–1486

Hu HR, Qian WH (2007) Identifying the northernmost summer monsoon location in East Asia. Progress in Natural Science 17(7): 812–820

Peixoto JP, Oort AH (1992) Physics of Climate, Springer, Am Inst Phys Press, New York

Qian WH, Jiang M (2014) Early signals of synoptic-scale atmospheric anomalies associated with the summer low temperature events in Northeast China. Meteorol Atmos Phys doi:10.1007/s00703-013-0306-0

Qian WH, Lee D-K (2000) Seasonal march of Asian summer monsoon. Inter J Climatol 20(11): 1371–1386

Qian WH, Li J, Shan XL. 2013. Application of synoptic-scale anomalous winds predicted by medium-range weather forecast models on the regional heavy rainfall in China in 2010. Sci China (Earth Sci) 56: 1059–1070. doi:10.1007/s11430-013-4586-5

Qian WH, Shan XL, Liang HY, Huang J, Leung CH (2014) A generalized beta-advection model to improve unusual typhoon track prediction by decomposing total flow into climatic and anomalous flows. J Geophys Res Atmos 119: doi:10.1002/2013JD020902

Qian WH, Tang SQ (2010) Identifying global monsoon troughs and global atmospheric centers of action on a pentad scale. Atmospheric and Oceanic Science Letters 3(1):1–6

Qian WH, Wu KJ, Chen DL (2015a) The Arctic and Polar cells act on the Arctic sea ice variation. Tellus A 67: 27692

Qian WH, Wu KJ, Liang HY (2015b) Arctic and Antarctic cells in the troposphere. Theor Appl Clim 2015b, DOI 10.1007/s00704-015-1485-z

Qian WH, Yang S (2000) Onset of the regional monsoon over Southeast Asia. Meteorol Atmos Phys 75: 29–38

Qian WH, Zhang ZJ (2012) Precursors to predict low-temperature freezing-rain events in southern China. Chin J Geophys 55: 1501–1512

Xie PP, Arkin PA (1992) Global precipitation: a 17-year monthly analysis based on gauge observatios, satellite estimates, and numerical model output. Bull Amer Meteor Soc 78: 2539–2558

Chapter 4
Global Monsoon

A definition is needed to identify the global monsoon (GM) domain using a monsoon index. Since most previous studies to establish monsoon indices have been limited to specific regions, they have lacked the applicability to universally describe the GM domain. In this chapter, the GM domain is defined by the regions where the difference of two pentad mean precipitation rates exceeds 4 mm day^{-1}, which is also influenced by the low-level prevailing wind reversal associated with the cross-equatorial flow. This definition not only confirms previous results of the classical GM domain from the tropical Africa to Asia–Australia and non-classical monsoon region in the tropical America, but also solves the issue of missing summer monsoon spots. By applying the *k*-means clustering method on the climatological evolution of global precipitation, the global precipitation domain is divided into 22 major dry–wet alterative regions under six categories. Considering the seasonal influences of the low-level cross-equatorial flow in these major dry–wet alternative regions, the GM is objectively divided into 16 major regions. The Earth's surface can be roughly divided into two regions, i.e., monsoon and non-monsoon regions, based on whether the region is influenced by monsoon flow and monsoon rainfall. Thus, there are outer edge active zones of the GM. The East Asian monsoon outer edge active zone has been an old topic since AD 688–742 in ancient China, but a summer monsoon dry–wet index series has only been available since AD 950.

4.1 Wind Seasonality and Symmetric Dry–Wet Alteration

The earliest worldwide monsoon description was given by Khromov (1957), who used the annual frequency of surface wind reversal to categorize three types of monsoon: the first is from East Africa to South Asia and Southeast Asia as well as the equatorial western Africa; the second covers the equatorial Africa, East Asia–Northwest (NW) Pacific, the maritime continents, and other local areas; the third is

© Springer Nature Singapore Pte Ltd. 2017

W. Qian, *Temporal Climatology and Anomalous Weather Analysis*,
Springer Atmospheric Sciences, DOI 10.1007/978-981-10-3641-5_4

located in mid-high latitudes. Later, Ramage (1971) mapped a worldwide monsoon including the first two types of Khromov's description. The great progress on worldwide monsoon was made by Webster (1987) who combined prevailing wind reversal and dry–wet alteration between winter and summer and mapped six monsoon regions over the whole world.

Prior to this century, many works focused on the study of regional-scale monsoon variations, mostly on continents over South Asia or Indian Continent (e.g., Lau and Yang 1997; Webster et al. 1998; Wang and Fan 1999), East Asia (e.g., Tao and Chen 1987; Chen et al. 1991), Australia (e.g., McBride 1987), Africa (e.g., Nicholson and Kim 1997), South America (e.g., Zhou and Lau 1998), and North America (e.g., Adams and Comrie 1997), but very few on the GM. This is because observed precipitation was not available in many places over ocean in the past.

Since satellites have started providing a global observation in late 1970s, some products such as the outgoing long-wave radiation (OLR) and the upper tropospheric water vapor band brightness temperature (BT) can be used in the worldwide monsoon study. Using the OLR and highly reflective clouds, Wang (1994) delineated the global tropical monsoon regimes.

Two terms, global warming and GM, are appearing frequently in scientific literatures and public news in recent decades. The former describes the tendency of global surface air temperature, while the latter focuses on the seasonal change of global major precipitation. Thus, surface air temperature and surface precipitation form two basic climatic elements for human beings. The term of "global monsoon" (GM) first appeared in the works of Sankar-Rao (1966) and Sankar-Rao and Saltzman (1969), and Sankar-Rao (1970), but they did not give any precipitation (or dry–wet change) descriptions except a set of equations of atmospheric motion and the impact of land–sea thermal contrast. The term of GM used in dry–wet alteration from the global scale was first found in the literature of the late twentieth century using BT product (Qian et al. 1998a, b). Globally, six major monsoon regions in North–South Africa, Asia–Australia, and South–North America were revealed by the analyses of OLR (Wang 1994) and BT (Qian 2000) despite the differences in their coverage. Two years later, Trenberth et al. (2000) depicted the GM as the global scale seasonally varying overturning circulation throughout the tropics. Meanwhile, Qian (2000) described the GM by dry–wet alternation in the global scale using monthly mean BT satellite data. To determine the GM, Qian et al. (2002a) used the NCEP/NCAR 850-hPa wind and BT to summarize the GM distribution based on the analysis of both dry–wet alteration and wind reversal associated with the cross-equatorial flow. From the GM (Qian et al. 1998a, b, 2002a; Qian 2000) and some regional monsoons in Asia (Qian and Yang 2000; Qian and Lee 2000), the summer monsoon in the subtropics was defined as the expansions of three deep convective centers or three wet centers in the equatorial region, driving by low-level cross-equatorial flow. The largest deep convective center (or wet center) is located in the equatorial maritime continents while the other two are in the equatorial America and the equatorial Africa.

The geographical distribution of the GMs, including the tropical, subtropical, and temperate-frigid monsoons, was pictured by Li and Zeng (2002) using the normalized seasonality of wind field. Zeng and Lu (2004) depicted the onset and retreat of GM using the normalized precipitable water data. Using the Global Precipitation Climatology Project (GPCP) data, Wang and Ding (2006) as well as Zhang and Wang (2008) illustrated the GM domain, defined by the difference between varied seasonal mean precipitation rates from local summer season in the Northern Hemisphere (NH) and winter season in the Southern Hemisphere (SH). Furthermore, Wang and Ding (2008) as well as Zhang and Wang (2008) demonstrated while Chang et al. (2011) reviewed that the GM represents the dominant mode of the annual variation of the tropical precipitation and low-level winds. Considering the relationship between the GM and general circulation systems, Qian and Tang (2010) compared the GM troughs and global atmospheric active centers in the two calendar pentads 8 and 44.

The existing worldwide monsoon or GM studies can be classified into three categories. The first studies the prevailing wind reversal such as those made by Khromov (1957), Ramage (1971), Trenberth et al. (2000), and Li and Zeng (2003). For the first category, prevailing wind reversals not only exist in the subtropical areas but also in the equatorial zone and mid-high latitudes. Figure 4.1 shows a case. The geographical extent of the global surface monsoons is outlined by the normalized seasonality (Li and Zeng 2002, 2003). The global surface monsoons contain three kinds of monsoons: the tropical, subtropical, and temperate-frigid monsoons, respectively. The extent of the classical tropical monsoons is surrounded by the positions of the intertropical convergence zone (ITCZ) in boreal summer and

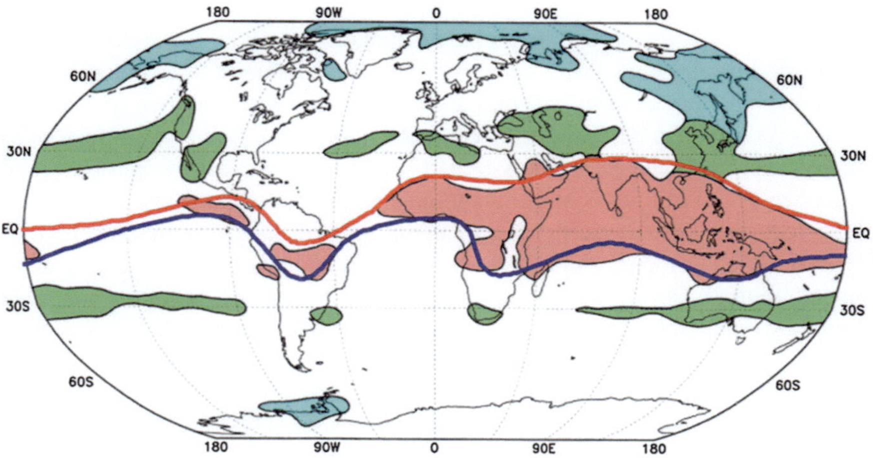

Fig. 4.1 Geographical extent of the global surface monsoons defined by 850 hPa winds using NCEP/NCAR reanalysis product for 1968–1996. The *red, green, and blue areas* indicate the tropical, subtropical, and temperate–frigid monsoon, respectively. The *red and blue thick lines* represent the ITCZ in boreal summer and winter, respectively. Sourced from Li and Zeng (2005)

winter. They used a dynamical normalized seasonality (DNS) index of wind field. The DNS index is given by

$$\delta_{m,n} = \frac{\left\| \overrightarrow{V}_{\mathrm{J}} - \overrightarrow{V}_{m,n} \right\|}{\left\| \overrightarrow{V}_{J-J} \right\|} - 2, \tag{4.1.1}$$

where $\overrightarrow{V}_{\mathrm{J}}$ and $\overrightarrow{V}_{\mathrm{J-J}}$ are the January climatological wind vector and the mean of January and July climatological wind vectors, respectively. $\overrightarrow{V}_{m,n}$ is the monthly mean wind vector in month m and year n. It is indicated that $\delta_{m,n} > 0$ in monsoon area.

The second is the dry–wet alteration, studied by Wang (1994), Qian et al. (1998a, b), Qian (2000), Zeng and Lu (2004), and Wang and Ding (2006). In this category, three datasets including OLR, BT, and precipitable water data collected from satellites can roughly describe the GM domains. The last category combines dry–wet alteration and prevailing wind reversal, such as from the studies of Webster (1987), Qian et al. (2002a), Wang and Ding (2008), Zhang and Wang (2008), and Qian and Tang (2010). As indicated recently by Nguyen et al. (2014), some "spots" could not be detected using local rainfall or zonal wind threshold of local difference between summer and winter. A spot in South China also exists in the study using a rainfall of local precipitation difference between two calendar pentads 8 and 44 (Qian and Tang 2010). The limitation is, indeed, because the studies only put dry–wet alteration and prevailing wind reversal together without explicitly establishing a unified index to picture the GM. The above two calendar seasons or pentads are not suitable to distinguish where the monsoon regions are, because the climatic maximal and minimal precipitation rates are asymmetrically seasonal advance from one place to another.

Figure 4.2a shows the GM precipitation domain defined by the region in which the local annual range (AR) intensity (the local summer (JJA)-minus-winter (DJF) precipitation) exceeds 180 mm and the local summer monsoon precipitation exceeds 35% of annual rainfall (Wang and Ding 2006). Some spots such as in South China, from the East China Sea to south of Japan, and the western South China Sea (SCS) can be found. In the middle latitudes, the Northeast Asia beyond 50°N and the central North America are located in the monsoon regions. Figure 4.2b shows the summer monsoon rainy domains, where the summer 5-month rainfall (MJJAS for the NH, and NDJFM for the SH) is equal to or above 3 mm day^{-1}, and the ratio of summer to annual rainfall is greater than 55% (Zhang and Wang 2008). The coverage of GM domain mapped by Zhang and Wang (2008) is smaller than that mapped by Wang and Ding (2006), but some spots indicated by Nguyen et al. (2014) still exist. Recently, a new proposed GM domain satisfies that (1) the local summer (MJJAS)-minus-winter (NDJFM) precipitation rate exceeds 2 mm day^{-1}, and (2) the local summer precipitation exceeds 55% of the annual total (Wang et al. 2014). This is a set of less strict condition so that the coverage becomes larger than that in Fig. 4.2a, b, particularly in Europe, Northeast Asia, and

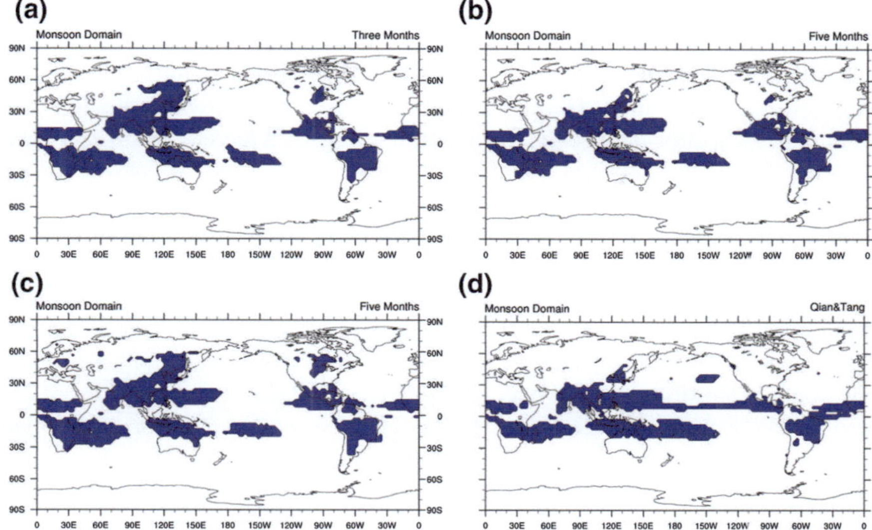

Fig. 4.2 Four global monsoon precipitation domains (*shading*) defined by **a** Wang and Ding (2006), **b** Zhang and Wang (2008), **c** Wang et al. (2014), and **d** Qian and Tang (2010). Sourced from Qian and Jiang (2015) © Springer-Verlag Wien 2015, with permission

North America, but those spots still remain. When the difference of pentad mean precipitation rate between pentads 8 and 44 is used to detect the tropical GM domain (Qian and Tang 2010), six monsoon regions are clearly observed off the equatorial zone, but some spots also exist (Fig. 4.2d). All these definitions cannot eliminate the spots such as in South China where a stationary frontal precipitation is often active from spring to early summer (Zhang and Wang 2008; Zhao et al. 2009; Zhu et al. 2011).

There are different viewpoints to explain the causes of monsoon (Chao and Chen 2001). Globally, the solar radiation and the land–ocean distributions are considered as the two primary driving forces for the formation of the GM (Ramage 1971; Webster 1981, 1987). Two of the most important early studies were done in the late seventeenth by Edmund Halley and in early eighteenth centuries by George Hadley (Webster 1981). Halley attributed the monsoon circulation primarily to differential heating and cooling of the land and the ocean. Different heatings would cause pressure differences in the atmosphere that would be equalized by winds. Hadley modified Halley's monsoon model by considering the Coriolis force due to the rotating Earth. Voyeikov (1879) linked seasonal precipitation variations to the monsoon over East Asia. Wang et al. (2012) indicated that the annual variation of solar radiation is a fundamental driver and a necessary and sufficient condition for the existence of monsoon, while the land–sea thermal contrast is critical for the location and strength of monsoon but it is neither a necessary nor a sufficient condition. In fact, however, wind reversal, besides heating contrast, also involves in monsoon formation. Thermal contrast, due to the variation of solar radiation and the land–sea distributions, causes wind reversal, which changes the locations of the wind convergence in the lower troposphere, and as a result, forms regional dry–wet alteration. Thus, the physical mechanism of monsoon, in short, should be the wind reversal as a result of thermal contrast and hence the dry–wet alteration over the surface. It is clear that the thermal contrast change is a thermodynamical index, the wind direction reversal is a dynamical index, and the dry–wet alteration is a climatological index or a final result.

Although the term of GM domain has been proposed in many previous works using different indices, the debates on whether such a global domain can be objectively determined still remain, as summarized by Wang and Ding (2008) and Wang et al. (2008) and indicated by Nguyen et al. (2014). By an annual precipitation range between local summer and winter, Wang and Ding (2008) depicted a GM domain. However, as Nguyen et al. (2014) indicated, some "spots" could not be detected by their definitions due to the use of a local rainfall or zonal wind threshold with the difference between local summer and winter. This is a critical issue concerned by climatologists and short-term climate forecasters.

4.2 Cross-Equatorial Flow and Asymmetric Seasonal Dry–Wet Alteration[1]

For an ideal GM, one could imagine that if the entire Earth is covered by water and solar rays always radiate on the equatorial zone without seasonal variation, there should be only three large precipitation zones caused by, respectively, a trade flow convergence along the equator and two polar frontal convergences near 60° (Qian and Jiang 2015). The three large precipitation zones also correspond to the convergence of two Hadley cells in tropics and the convergences between the Ferrel and Polar cells in the two middle latitudes. In this situation, dry–wet alteration and wind reversal cannot be observed in the global scale (Fig. 4.3a). The cross-equatorial flow is not observed in this case. In another situation, if seasonal and meridional variations of solar radiation are considered, there should be dry–wet alteration and prevailing wind reversal caused by location shifts of the ITCZ and two polar frontal convergence zones. There are two subtropical dry zones under the two subtropical high zones. In pentad 44 during the boreal summer (Fig. 4.3b), the southwesterly cross-equatorial flows and the northeasterly trades form the ITCZ in the north of equatorial circle while the two polar frontal convergence zones shift north of 60°N in the NH and 60°S in the SH. In pentad 8 during the boreal winter (Fig. 4.3c), however, the northeasterly cross-equatorial flows and the southeasterly trades form the ITCZ in the south of equatorial circle while the two polar frontal convergence zones shift south of 60°N in the NH and 60°S in the SH. The tropical

[1]Modified article originally published in Qian and Jiang (2015). Published with kind permission of © Springer-Verlag Wien 2015. All Rights Reserved.

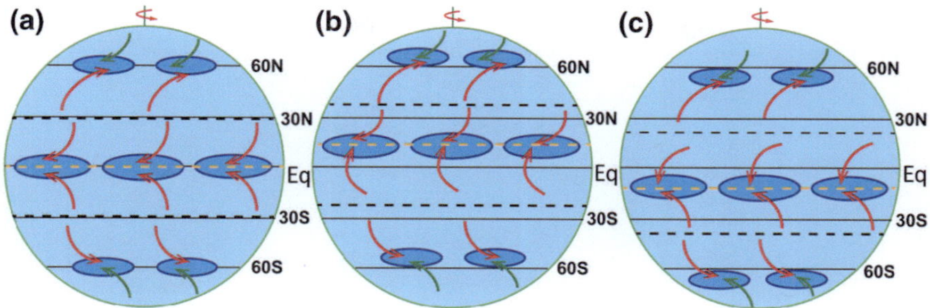

Fig. 4.3 Surface flows and precipitation areas (*shading*) of an aqua Earth, **a** when solar rays always radiate on the equatorial zone without seasonal variation, **b** in boreal summer (pentad 44) when solar rays have seasonal variation and **c** in boreal winter (pentad 8) when solar rays have seasonal variation. The *black-dashed line* denotes the subtropical high ridge line while the *yellow-dashed line* is the ITCZ

GM can be identified from either or both of the dry–wet alteration and wind direction reversal associated with the low-level cross-equatorial flow in the tropics. Other two zones in the middle latitudes also satisfy dry–wet alteration but wind direction reversal is not associated with the low-level cross-equatorial flow. Thus, whether the dry–wet alteration is associated with cross-equatorial flow is a critical condition to define monsoon regions.

Figure 4.3 describes a zonal symmetric circulation and the seasonal shift of precipitation zone if the entire Earth is covered with water. However, the Earth's surface is covered by multi-scale oceans and continents, which form various circulation systems under the solar radiation. Climatologically, the intensity and location changes of these systems are influenced by solar seasonal radiation and sea–land distributions. The ideal two subtropical high zones become five subtropical high centers over three oceans (Fig. 4.4a). Globally, large portion of precipitation falls in oceans (Fig. 4.4a). Over continents, two major large precipitation regions are located in the two tropical rain forests of equatorial South America and equatorial western Africa. Over the Atlantic and the America, the two mid-latitude precipitation centers, c1 and b1, are caused by subtropical highs and mid-latitude westerly flows. The precipitation region d1 is formed at the ITCZ, and the precipitation center a1 is formed between two neighbor SH subtropical highs in the Atlantic and Pacific. The same pattern but with stronger highs and mid-latitude westerly flows are also noted in the mid-low latitudes of Pacific for the four precipitation centers (a2, b2, c2, and d2). Similar pattern exists, with the three precipitation centers (a3, b3, and d3) surrounding the subtropical high over the Indian Ocean. From the three ideal global precipitation zones to the 13 precipitation centers or regions in Fig. 4.4a, regional or local precipitation can be explained by different circulation systems. The three precipitation regions near the equatorial zone (d1, d2, and d3) are caused by the ideal ITCZ. The five precipitation regions b1, b2, and b3 in the SH as well as c1 and c2 in the NH are caused by the mid-latitude westerly flows and the subtropical highs. However, the four

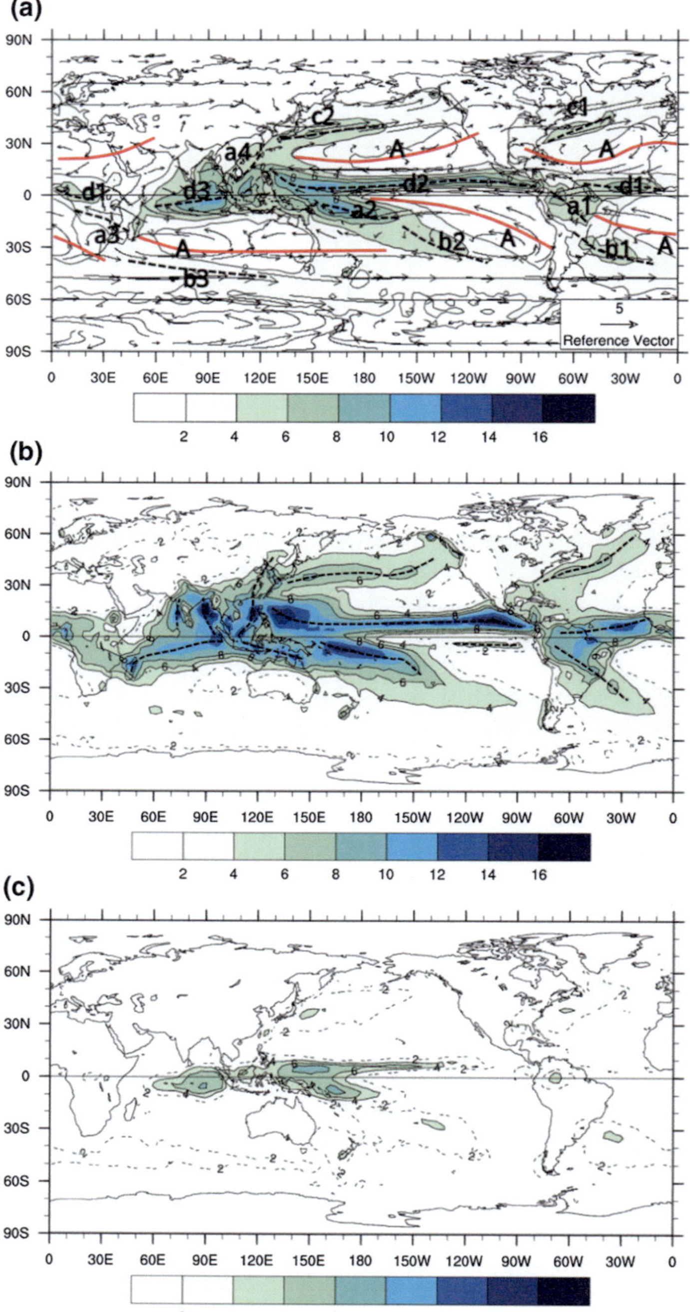
(a)
c2
c1
a4
A
A
d1
d3
d2
d1
a2
a1
a3
A
b2
A
b1
A
b3
5
Reference Vector
(b)
(c)
90N
60N
30N
0
30S
60S
90S
0
30E
60E
90E
120E
150E
180
150W
120W
90W
60W
30W
0
2
4
6
8
10
12
14
16

◀ **Fig. 4.4** Global climatic distribution of **a** annual mean precipitation rate (*shading*, 2 mm day^{-1} interval) and wind arrow (m s^{-1}) at 850 hPa, **b** maximal pentad mean precipitation rate (2 mm day^{-1} interval), and **c** minimal pentad mean precipitation rate (2 mm day^{-1} interval). In **a**, the *thick red line* denotes subtropical high (anticyclone) ridge while the *thick black-dashed line* is along the axis of precipitation rate, letter "A" denotes the location of anticyclone at 850 hPa, letters "a," "b," "c," and "d" indicate precipitation axes. In **b**, the *heavy-dashed lines* denote the axes of precipitation rate. Sourced from Qian and Jiang (2015) © Springer-Verlag Wien 2015, with permission

precipitation centers (a1, a2, a3, and a4) are caused by the large-scale sea–land (topography) contrast and five subtropical highs. It is clear that the GM precipitation is directly associated with the intensity and location of seven precipitation centers (a1, a2, a3, a4, d1, d2, and d3). Climatologically, the location shift of three precipitation regions off the equatorial zone (d1, d2, and d3) is due to the asymmetry of large-scale sea–land distribution in the two hemispheres, while the formation of other ten precipitation centers is mainly attributed to the zonally oriented sea–land distributions and subtropical highs.

To obtain a smoother daily climatology of atmospheric variables, a Fourier analysis is performed at each of all grids for a 365-day or 73 pentad series by Qian and Jiang (2015). The method of extracting the annual and semiannual harmonics is derived from the website (http://www.ncl.ucar.edu/Document/Functions/Contributed/smthClmDayTLL.shtml). A smooth climatic cycle contains only annual and semiannual components, while all harmonics greater than two are set to zero. This method is reasonable in both the tropical and subtropical zones to find the seasonal maximum and minimum precipitation rates. The twice cross-equatorial solar radiation is the main cause.

If we only concern the maximal pentad precipitation rate of the smooth climatic cycle, the large-scale structure is similar to that of annual mean precipitation rate except for the complexity found in the Eastern Hemisphere, particularly over the region between Asia and Australia (Fig. 4.4b). From Asia to NW Pacific, there are five centers of maximal pentad mean precipitation rate, respectively, in the Arabian Sea, the Bay of Bengal (BOB), the SCS, the eastern sea of the Philippines, and the subtropical NW Pacific. In East Asia, the precipitation rate of 4 mm day^{-1} can reach Northeast China (130°E, 50°N). In the subtropical SH, there are four centers of pentad mean precipitation rate, respectively, in the Madagascar Island, the south of equatorial central Indian Ocean, the Arafura Sea, and the Solomon Islands. The most prominent feature shows that these centers of maximal pentad mean precipitation rate are located in off-equatorial zones. As a comparison, the global distribution of minimal pentad mean precipitation rate is shown in Fig. 4.4c, where the center exceeding 4 mm day^{-1} is located over the equatorial eastern Indian Ocean and the equatorial western Pacific. After illustrating the climatic pentad mean precipitation intensity, Fig. 4.5 shows the climatic precipitation durations with precipitation rate exceeding 4 and 6 mm day^{-1} worldwide, respectively. We first examine the duration with precipitation rate exceeding 4 mm day^{-1} in Fig. 4.5a. Equatorial Indian Ocean and equatorial western Pacific are the two regions with the longest precipitation duration (72–73 pentads). It means that there is no climatic dry

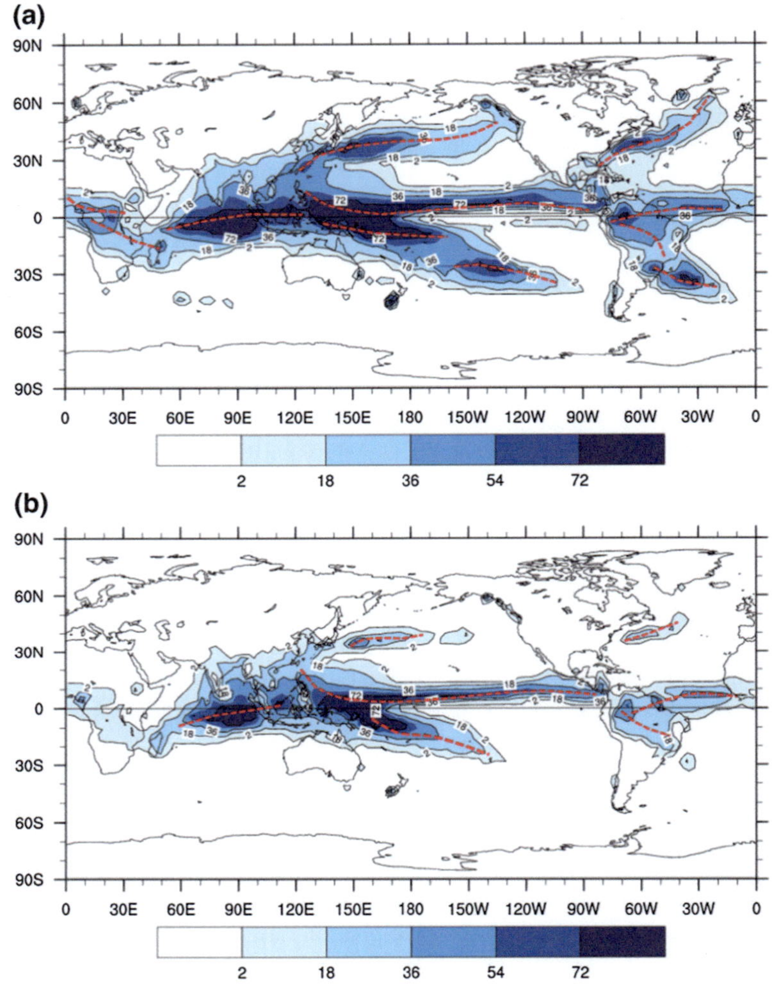

Fig. 4.5 Climatic precipitation durations (pentads) with precipitation rate exceeding, **a** 4 mm day^{-1} and **b** 6 mm day^{-1}. The *red-dashed line* denotes the belt with high duration. Sourced from Qian and Jiang (2015) © Springer-Verlag Wien 2015, with permission

spell or no climatic dry–wet alteration over the whole year. Wet spell centers with durations longer than half a year (36 pentads) can be observed in subtropics of NW Pacific, NW Atlantic, central southern Pacific, and Southwest Atlantic. These centers are mostly located over open oceans. Over continents, two centers are found in equatorial western Africa and tropical South America lasting for more than 36 pentads. In the Asian continent, the duration is less than 36 pentads.

We then compare the duration with precipitation rate exceeding 6 mm day^{-1} in Fig. 4.5b. The distribution is similar to that with 4 mm day^{-1}. Two duration centers of precipitation rate exceeding 6 mm day^{-1} are still found in equatorial Indian

Ocean and equatorial western Pacific. Over continents, the largest area with the precipitation duration longer than 18 pentads (three months) is observed in tropical South America. In the Asian continent, the region with three-month duration is found near the BOB, over the Indo-China Peninsula (ICP), and the coast of South China. If taking account of the precipitation rate larger than only 3 mm day^{-1}, the coverage of the climatic precipitation duration is larger than that in Fig. 4.5a, such as in eastern China and North America. However, Figs. 4.4 and 4.5 cannot tell where a tropical monsoon region is because one of monsoon definitions is the seasonal dry–wet alteration.

Regarding the fact that the summer wet spell is shorter than 1 month in many subtropical regions, Fig. 4.6 shows the climatic difference between maximal and minimal climatic pentad mean precipitation rates, which successfully eliminates the spots mentioned above. Along the equatorial zone, many places do not reach the threshold of 4 mm day^{-1} difference except western Africa, the maritime continents, and tropical South America. These three areas also have relative dry spell as shown in Fig. 4.4b. Comparing to the tropical GM domain of Wang and Ding (2006, 2008), the domain of 4 mm day^{-1} difference covers a larger area, but that of 6 mm day^{-1} difference misses some places such as that in East Asia. Figure 4.6 shows the places that experience dry–wet alteration worldwide but it depends on the thresholds such as 4 or 6 mm day^{-1}. If taking 2 mm day^{-1} as threshold, regions of dry–wet alteration can be observed in middle latitudes as indicated by the thin-dotted lines in Fig. 4.6.

Figure 4.7a, b, respectively, shows the two calendar pentads when the precipitation rate of each grid climatologically reaches the maximum and minimum. But we here only focus on the regions that satisfy the above condition of dry–wet alteration (the difference between two pentad mean precipitation rates exceeding

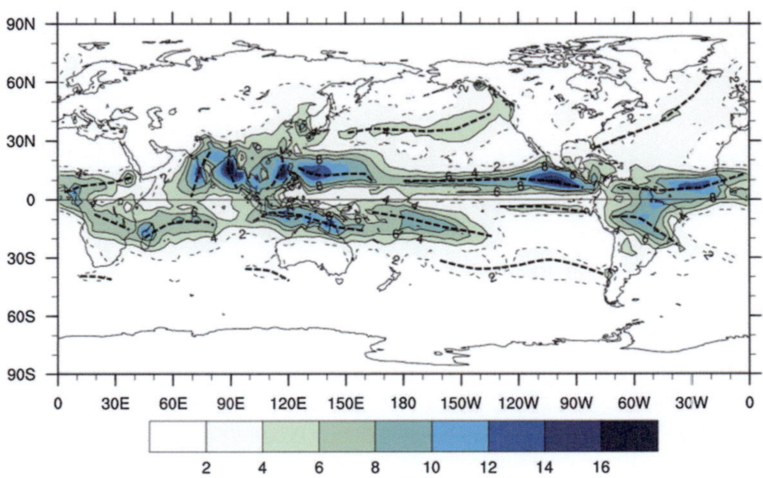

Fig. 4.6 Climatic difference (2 mm day^{-1} interval) of global maximal and minimal pentad mean precipitation rates with the *heavy-dashed lines* denoting the axes of precipitation rate difference. Sourced from Qian and Jiang (2015) © Springer-Verlag Wien 2015, with permission

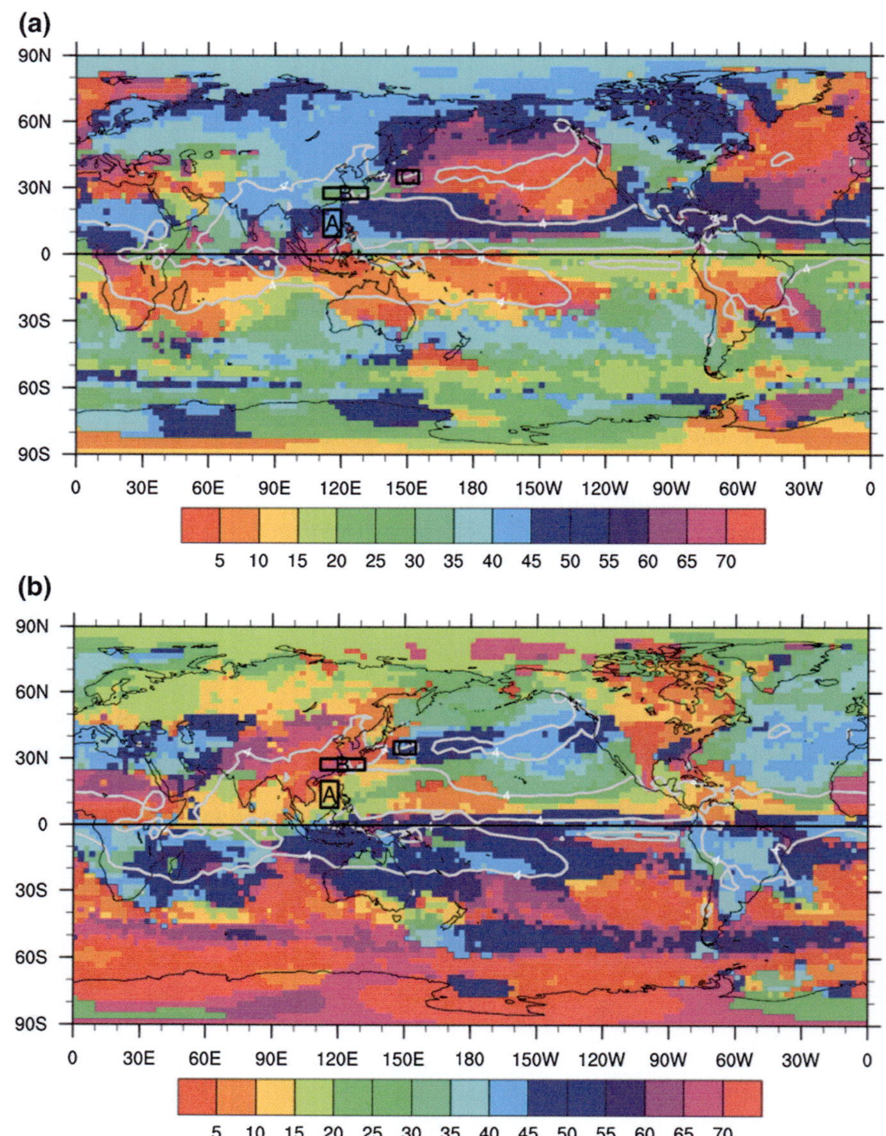

Fig. 4.7 Global distribution of pentads (five pentad interval) when precipitation rate climatologically reaches, **a** maximum and **b** minimum, respectively. *Heavy white line* indicates the regions with the precipitation difference exceeding 4 mm day^{-1} as shown in Fig. 4.6. Three boxes A, B, and C are, respectively, located in the eastern SCS (8–20°N, 112–120°E), from South China to the East China Sea (25–30°N, 112–132°E) and in the eastern part of the NW Pacific monsoon domain (32–38°N, 145–155°E). Sourced from Qian and Jiang (2015) © Springer-Verlag Wien 2015, with permission

4 mm day^{-1} or covered by white lines). In the subtropical SH (Fig. 4.7a), the maximal precipitation pentad is concentrated in the austral summer (from pentads 70 to 15). The complexity is observed in NH and along the equatorial zone. In the equatorial western and eastern Africa, maximal precipitations are from pentads 60 to 70 and from pentads 20 to 30, respectively. In the maritime continents, maximal precipitation pentads are from 60 to 73. The maximal precipitation in the eastern BOB appears in pentads 30–40 while the western BOB in pentads 60–65 and the ICP in pentads 45–55. During pentads 45–55, the maximal precipitation is in the eastern sea of the Philippines along the ITCZ, but that in the eastern SCS is earlier in pentads 40–45. From South China to the East China Sea and south of Japan, the maximal precipitation occurs in the spell of pentads 25–40, so the spot depicted in the previous studies (e.g., Zhang and Wang 2008; Qian and Tang 2010) disappears. These pentads of maximal precipitation rate described here are the same as that illustrated by Zhang and Wang (2008) from their six monsoon regions, but we have shown all the pentads of maximal precipitation rate from the worldwide scope. If the 4 mm day^{-1} precipitation rate is used as one of the thresholds, the duration of wet period at each place can be centered by the maximal precipitation pentad over the total consecutive pentads from Fig. 4.5a. In many regions such as in the equatorial eastern Africa, northern India, North China, and from the Gulf of Mexico to the Caribbean Sea, the duration of wet spell is from 2 to 10 pentads.

The pentad when the precipitation rate reaches its minimum is also complicated as shown in Fig. 4.7b. But its progress is clear as that from North China in pentads 60–70, and Central China in pentads 70–5, to the SCS in pentads 5–20. In SH, minimal precipitations appear in pentads 30–45 over the subtropical African and South American continents. A special region is located in the mid-North Pacific, where the precipitation difference is larger than 4 mm day^{-1} with the maximal precipitation in pentads 70–5 (winter) and the minimal precipitation in pentads 40–50 (autumn). It means that the maximal and minimal precipitation rates are not climatologically centered in pentads 8 and 44 as used by Qian and Tang (2010) or in boreal summer (MJJAS) and austral winter (NDJFM) as used by Zhang and Wang (2008), although it is in the classical monsoon region. This phenomenon suggests that we cannot use a local pentad mean or seasonal mean dry–wet alteration as a threshold to define a monsoon region.

4.3 Identifying Global Monsoon Domain[2]

The cause of the dry–wet alternation and wind reversals has been a controversial question for a long period since Ramage (1971). The wind and precipitation (or dry–wet) changes are dynamically responses to solar seasonal radiation and sea–

[2]Modified article originally published in Qian and Jiang (2015). Published with kind permission of © Springer-Verlag Wien 2015. All Rights Reserved.

land thermal contrast. We here only identify the dry–wet alternation associated with wind reversal. There are several places with the seasonal difference between maximal and minimal precipitation rates exceeding 4 mm day^{-1}, such as in the mid-North Pacific and the south of equatorial eastern Pacific. But it is not enough to indicate where the monsoon regions are. In previous studies, the tropical GM domain was identified, respectively, using dry–wet alteration (Qian et al. 1998a, b; Qian 2000; Wang and Ding 2006) and wind reversal (Trenberth et al. 2000; Li and Zeng 2003) as well as a combination of the above two (Webster 1987; Qian et al. 2002a). To identify the tropical GM domain in detail, Fig. 4.8a shows the 850-hPa wind vector difference between the two pentads when precipitation rates reach the maximum and minimum, respectively. It is noted that the prevailing wind reversal from both velocity and positive angle difference (clockwise) is predominantly located in the tropical Eastern Hemisphere (a heavy-dashed area). This wind reversal region is similar to that described by Khromov (1957) and Ramage (1971). Other significant regions of wind reversals are detected in the middle latitudes from East Asia to the NW Pacific as well as in the North Atlantic and the subtropical South America. Obviously, East Asia and the subtropical South America experience a negative angle change (anticlockwise). The reversal between southwesterly and northeasterly is clearly observed in the north of the equator, but with large velocity difference in the Eastern Hemisphere and small velocity difference in the eastern Pacific. Another reversal between northwesterly and southeasterly is observed in the south of the equator from Africa to Indian Ocean and western Pacific. The cross-equatorial flow can be noted in the East Africa, central Indian Ocean (south of Indian Peninsula), and from the maritime continents to the eastern Pacific. The changes from southwesterly to northwesterly are found over East Asia while those from northwesterly to easterly in the tropical South America (Qian et al. 2002a). In the mid-North Pacific, the reversal is caused by the interchange of westerly and easterly when the subtropical high moves seasonally.

As an example for the SCS monsoon (Fig. 4.7 in the area A), its regional precipitation rate has at least three main maximum peaks (Qian and Yang 2000). The three peaks with the precipitation rate more than 11 mm day^{-1} appear in late June, from July to August, and in September while the minimum precipitation rate less than 2 mm day^{-1} appears in March and a rapid increase of precipitation rate from 2 to 6 mm day^{-1} occurs in middle May (Fig. 4.9a). The dry–wet difference for this rapid transition is 4 mm day^{-1}. The maximum precipitation rate using the first two harmonics peaks in late July while the peak in annual cycle is in early August. Other short-spell peaks are caused by the local small-scale climatic circulation systems. In the area B from South China to the East China Sea (Fig. 4.7), two major peaks of precipitation rate are in June (Fig. 4.9b). The combination (red line) is composed of the annual cycle (green line) and semiannual cycle (yellow line) and its peak is near the original precipitation maximum (blue line). The summer (JJA or MJJAS)-minus-winter (DJF or NDJFM) precipitation in area A (SCS) is larger than 4 mm day^{-1}, but this criterion cannot be used in the area B because its wet period is only about 1 month from late May to early June. The cross-equatorial flow (figures not shown) at 850 hPa in pentad 41 for the wet peak

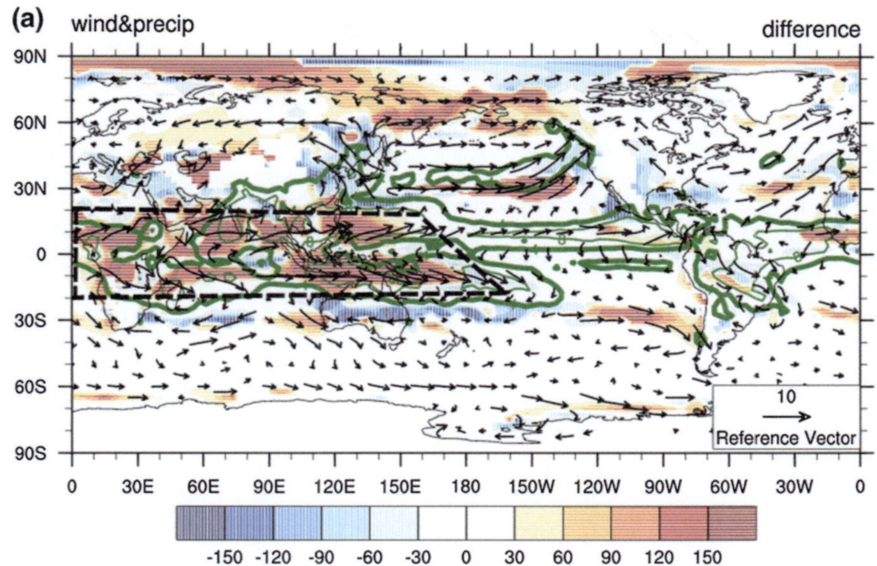
(a)
wind&precip
difference
90N
60N
30N
0
30S
60S
90S
0 30E 60E 90E 120E 150E 180 150W 120W 90W 60W 30W 0
10
Reference Vector
-150 -120 -90 -60 -30 0 30 60 90 120 150

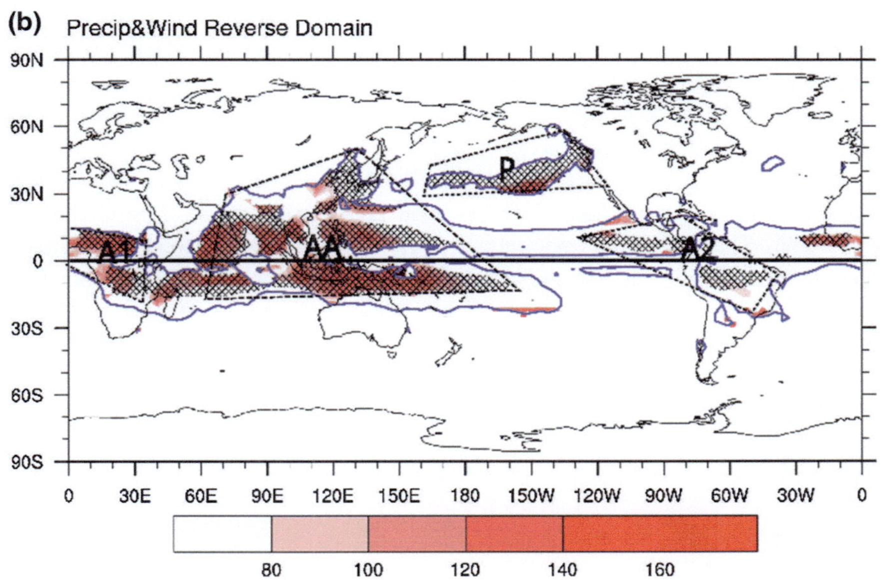
(b)
Precip&Wind Reverse Domain
90N
60N
30N
0
30S
60S
90S
P
A1
AA
A2
0 30E 60E 90E 120E 150E 180 150W 120W 90W 60W 30W 0
80 100 120 140 160

◀ **Fig. 4.8** **a** Coverage exceeding 4 mm day^{-1} of global maximal and minimal pentad mean precipitation rate differences (*thick green line*) and 850 hPa wind vector difference (m s^{-1}) between the two pentads of the extrema of precipitation rates. *Shadings* are reversal angle (30° interval) of the wind vector by the two pentads of maximal precipitation rate and minimal precipitation rate. *Positive (negative) value* indicates the wind vector reversal in clockwise (anticlockwise) direction. **b** The *thick-dashed line* covers regions with the max–min precipitation rate difference exceeding 4 mm day^{-1}, *red shading* denotes alteration angle of wind direction between the two pentads larger than 80°, *gray-crossing lines* denote the wind difference exceeding 5 m s^{-1}. Letters "A1," "AA," "A2," and "P" denote four regions of dry–wet alteration. Sourced from Qian and Jiang (2015) © Springer-Verlag Wien 2015, with permission

(Fig. 4.9a) has influenced the SCS. The cross-equatorial flow at 850 hPa for the wet peak (Fig. 4.9b) approaches from South China to the south of Japan (Fig. 4.9c), and the wind reversal as well as the standardized pentad precipitation rate anomalies are clear from the SCS to South China and the East China Sea (Fig. 4.9d vs. c). In Fig. 4.9b, there are two large precipitation rate differences between climatology (blue line) and combined annual–semiannual cycles (red line) in pentads 34 and 39, respectively. Figure 4.9e, f, respectively, shows the spatial differences of their precipitation rates and 850-hPa winds at the two pentads. It is noted that the differences of precipitation rates are caused by local climatic biweekly circulation systems from South China to the south of Japan. It means that the two opposite distributions of precipitation rate differences are not associated with the cross-equatorial flow at the lower troposphere. Although four harmonic components (cyan line in Fig. 4.9b) resemble the climatology but their spatial differences are caused by local climatic systems (Fig. 4.9e, f).

Another example is area C, taken from the eastern sea of Japan (Fig. 4.7). In Fig. 4.10a, there are two major precipitation peaks in June and from October to November. The first peak (red line) in early June is the monsoon precipitation because it is influenced by the cross-equatorial flow at 850 hPa as shown in Fig. 4.9c. The precipitation rate maximum in pentad 62 and minimum in pentad 45 as well as their 850-hPa flow are shown in Fig. 4.10b, c. It is identified that the westerlies tend to locate to the west/northwest of the convective region. Thus, the wind reversal when the precipitation reaches maximum is not clear in the region of wet–dry alternation over eastern part of the NW Pacific monsoon domain. In pentad 45 (Fig. 4.10c), the cross-equatorial flow can arrive in Northeast China but the dry–wet transition beyond 50°N in Northeast Asia (Fig. 4.6) are influenced by the westerly at 850 hPa (Hu and Qian 2007).

Critical information from dry–wet alteration and prevailing wind reversal is extracted from Fig. 4.8a and presented in Fig. 4.8b. In this distribution, the mid-North Pacific and the south of equatorial eastern Pacific were excluded in the previous studies. According to the work of Wang and Ding (2008), the region in the mid-North Pacific belongs to the oceanic storm track. However, in other studies (Qian and Yang 2000; Qian and Lee 2000; Qian 2000; Qian et al. 2002a), a monsoon region should be directly influenced by the low-level cross-equatorial flow. The dry–wet alternation in the mid-North Pacific is directly influenced by westerly and the subtropical anticyclone, while that in the south of equatorial

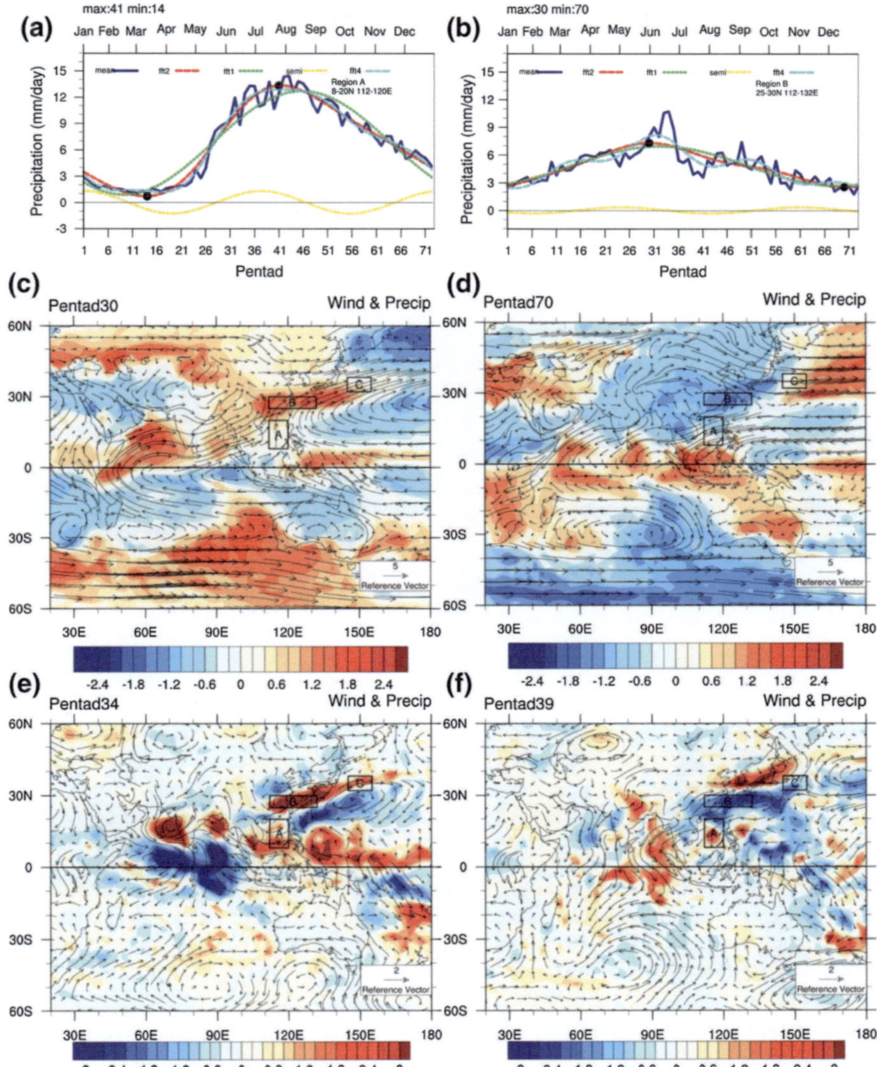

Fig. 4.9 Pentad precipitation rates (mm day^{-1}) in areas, **a** A (SCS) and **b** B (South China and the East China Sea) as well as standardized pentad precipitation rate anomalies relative to the annual average and wind (m s^{-1}) at 850 hPa in pentads, **c** 30 and **d** 70. The differences of precipitation rate (mm day^{-1}) and wind (m s^{-1}) in pentads, **e** 34 and **f** 39 based on original climatology and combined annual–semiannual cycles. In **a** and **b**, original, annual cycle, semiannual cycle, and combined annual–semiannual cycle pentad precipitation rates as well as the first four harmonic components are indicated by *blue, green, yellow, red, and cyan lines*. Four heavy black dots in **a** and **b** along the *red line* denote their peaks in pentads 41 and 30 as well as valleys in pentads 14 and 70, respectively. Sourced from Qian and Jiang (2015) © Springer-Verlag Wien 2015, with permission

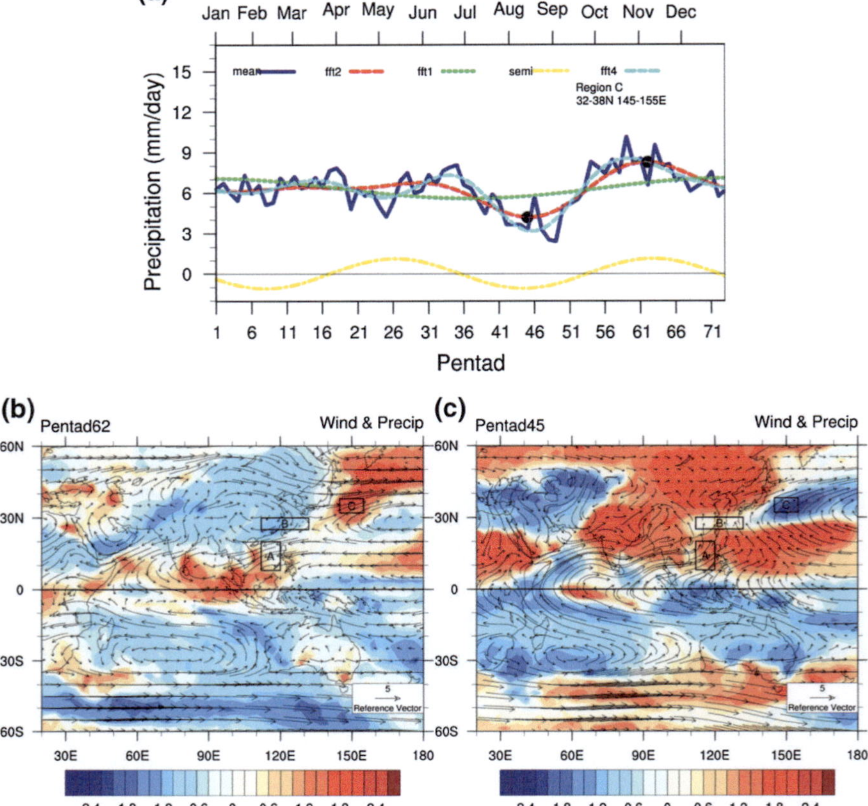

Fig. 4.10 Same as in Fig. 4.9 except in the area "C" (eastern part of the NW Pacific monsoon domain) with *two heavy black dots* in pentads 62 and 45 in **a**. The 850-hPa winds and standardized precipitation rates in pentads 62 **b** and 45 **c**. Sourced from Qian and Jiang (2015) © Springer-Verlag Wien 2015, with permission

eastern Pacific is influenced by southeasterly flow. Therefore, the mid-North Pacific region should not be counted as a monsoon region. After excluding the region in the mid-North Pacific, the tropical GM domain satisfying the dry–wet alteration and prevailing cross-equatorial wind reversal more than 5 m s^{-1} is clear in the tropical South and North Africa, from South Asia to East Asia and the tropical NW Pacific, and from the subtropical Indian Ocean to the maritime continents. It is noted that if we only consider the dry–wet alteration, other two regions in the north of equatorial central–eastern Pacific along the Pacific ITCZ and from the tropical South America to the tropical Atlantic, are also included in the tropical GM domain.

From the above description, four regions with dry–wet alteration can be categorized into two classical monsoon regions, a non-classical monsoon region, and a non-monsoon region, respectively, marked by A1, AA, A2, and P in Fig. 4.8b. Region A1 located in the tropical Africa and region AA situated between Asia and

Table 4.1 Categories and features of global dry–wet alteration regions

Category	Region	Wind angle change >120°	Wind difference >5 m s^{-1}	Cross-equatorial flow influence
Classical monsoon	African monsoon (A1) Asian–Australian monsoon (AA)	Yes Yes	Yes Yes	Yes Yes
Non-classical monsoon	American monsoon (A2)	No	Yes	Yes
Non-monsoon	Mid-North Pacific (P)	Yes	Yes	No

Four regions marked by A1, AA, A2, and P are shown in Fig. 4.8b. Sourced from Qian and Jiang (2015) © Springer-Verlag Wien 2015, with permission

Australia are influenced by the cross-equatorial flow with the wind reversal larger than 120° or wind velocity difference exceeding 5 m s^{-1}. These two regions are consistent with the description of Khromov (1957) and Ramage (1971), so they are referred to as the classical GM domain. Region A2 located in the tropical America is influenced by the cross-equatorial flow with the wind difference exceeding 5 m s^{-1} but the wind reversal is smaller than 120°. This region is not considered in the description of Khromov (1957) and Ramage (1971), so it is referred to as the non-classical monsoon region. Region P located in the mid-North Pacific is not influenced by the cross-equatorial flow, although its wind difference is more than 5 m s^{-1} and the wind reversal is larger than 120°, so it is referred to as the non-monsoon region. The wind reversal appears in early November when the precipitation maximum is not clear in the region of wet–dry alternation over eastern part of the NW Pacific monsoon domain, but its second precipitation maximum in June is influenced by the cross-equatorial flow at 850 hPa. The categories and features from this classification are given in Table 4.1. Region P is caused by the polar front when its location shifts seasonally. Similar location shift of polar frontal precipitation can also be observed in the middle latitudes over the North Atlantic and SH (Fig. 4.6).

Some basic climatic features of the tropical monsoon dry–wet alteration and prevailing wind change can be further identified from Fig. 4.11. Figure 4.11a shows the latitude–pentad section of pentad mean precipitation rate, 850-hPa wind vectors and 850-hPa temperature averaged over the whole latitude circle. An average dry–wet alteration can be identified in Fig. 4.11a by 6 mm day^{-1} precipitation rate with an asymmetric wet spell in the NH from pentads 18 to 71. The southernmost and northernmost locations of the 6 mm day^{-1} line and easterly flow (trade wind, black-dashed line) in the two hemispheres appear in pentads 8 and 44. In each hemisphere, wet spell appears during the period of weak trade wind and poleward wind.

In the meteorological community, the study of South Asian monsoon has a long history. Figure 4.11b shows the latitude–pentad section of the three climatic variables averaged over 60°E and 100°E. A reversal of easterly and westerly features in pentads 24 and 56. The precipitation rate reaches a high flat from pentads 30 to 56. The strongest southwesterly flow peaks in pentad 44, but the temperature peaks in pentad 53.

A unique feature can be identified for the SCS summer monsoon. In previous studies, an abrupt onset of the summer monsoon observed from dry–wet alteration and easterly-westerly reversal is from pentads 27 to 28 (middle of May) in the central SCS (Qian and Yang 2000; Qian and Lee 2000). This feature is also clearly indicated by the 4 mm day^{-1} (or 6 mm day^{-1}) precipitation line and wind reversal from easterly to southwesterly (Fig. 4.11c). Between 20 and 30°N over South China, the southwesterly gradually increases from pentads 1 to 46, but the 6 mm day^{-1} precipitation line starts from pentads 20 to 42. After the onset of central SCS monsoon, southwesterly and 2 mm day^{-1} precipitation line extend northward from the Yangtze River to the Yellow River. Beyond 45°N, the

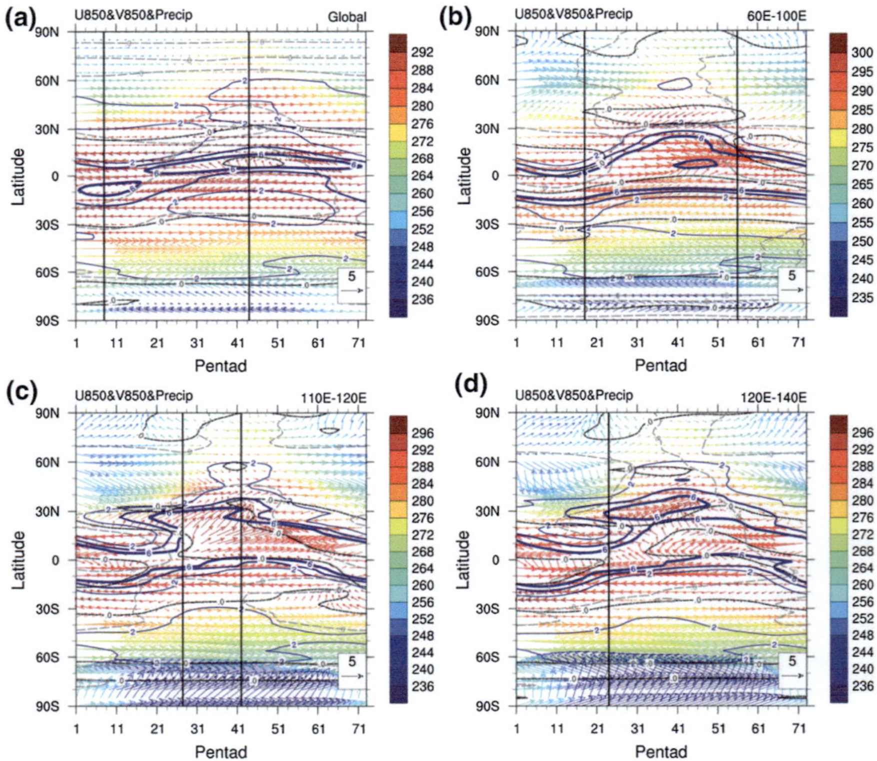

Fig. 4.11 Latitude–pentad sections of climatological pentad mean precipitation (*solid-blue line*, 2 mm day^{-1} interval), 850-hPa wind vectors (m s^{-1}) and 850-hPa temperature (*shading*, 4 K interval) averaged over, **a** the zonal circle, **b** 60–100°E, **c** 110–120°E, and **d** 120–140°E, respectively. *Heavy straight lines* are marked in pentads, **a** 8 & 44, **b** 18 & 56, **c** 27–28 & 42, and **d** 24. The *thin black* (*gray*)*-dashed line* indicates the zero line of zonal (meridional) wind. Sourced from Qian and Jiang (2015) © Springer-Verlag Wien 2015, with permission

2 mm day^{-1} precipitation rate corresponds with the northwesterly, so summer monsoon precipitation should be regarded as a condition of poleward moisture transport influenced by the cross-equatorial flow.

The western Pacific monsoon is also a well-studied topic (e.g., Tanaka 1992; Holland 1995; Ueda et al. 1995; Wang and Lin 2002). Figure 4.11d shows the latitude–pentad cross section of the three climatic variables averaged over 120–140°E. There are two regions covered by the 6 mm day^{-1} precipitation line, which are separated in the subtropical NW Pacific from pentads 24 to 48. Near 30°N, the southwesterly flow and 6 mm day^{-1} line are similar to that shown in Fig. 4.11c. Between 40 and 50°N, the transition from a cold northwesterly to a warm southwesterly starts since the central SCS monsoon onset. The southwesterly flow can reach 50°N and the precipitation falls over Japanese islands and the Korean Peninsula (Qian et al. 2002b).

Several important pentads are highlighted in Fig. 4.11. Their spatial structures of pentad mean precipitation rate and warm-cold flows at 850 hPa are also illustrated in this section. Figure 4.12a shows these structures in pentad 8 as in Fig. 4.3c. An axis of precipitation rate in the south of equator from subtropical Africa to the Southwest Pacific ends up with the northerly flow. It implies that the axis of precipitation rate is also the convergence line in the austral summer. The wind direction changes when it crosses the convergence line. Another axis of precipitation rate in the north of equatorial North Pacific accompanies with a convergence between southeasterly and northeasterly trade flows. Thus, the two convergence zones in the tropical western Pacific are resulted from different flows. The former is termed the monsoon convergence zone while the latter is the trade wind convergence zone. In the tropical Atlantic, two axes of precipitation rate are accompanied well by a monsoon convergence zone and a trade convergence zone. These axes of precipitation rate in the tropics are associated with low-pressure zones and convergence zones, separated by the subtropical high systems. In the mid-North Pacific and mid-North Atlantic, there are two axes of precipitation rate, which are along the storm tracks and resulted from polar frontal activities seasonally in boreal winter.

Pentad 24 is the transition time from the boreal winter half to the summer half of a year, so tropical convergence zone and precipitation rate axis mostly approach the equator similar to that shown in Fig. 4.3a. At this time, the cross-equatorial flow is very weak (Fig. 4.12b). All subtropical anticyclone systems in SH move equatorward while anticyclone systems in NH are enhanced and move westward relative to those in pentad 8. Meanwhile, two axes of frontal precipitation in NW Pacific and NW Atlantic extend westward to South China and eastern America, respectively.

In pentad 27, an obvious feature is that the central SCS is still dry (not covered by 6 or 4 mm day^{-1} line) and the subtropical NW Pacific high ridge is also controlled in the central SCS (Fig. 4.12c). At this time, 6 mm day^{-1} line covers the eastern Arabian Sea, the BOB, and the southern part of SCS, influenced by southwesterly cross-equatorial flow. Meanwhile, 6 mm day^{-1} line has covered over South China while the axis of precipitation rate is a frontal place extending westward. Indeed, the westerly flow is still prevailing over the Indian Peninsula, the ICP, and South China (Fig. 4.12a–c). Therefore, the precipitation in South China is

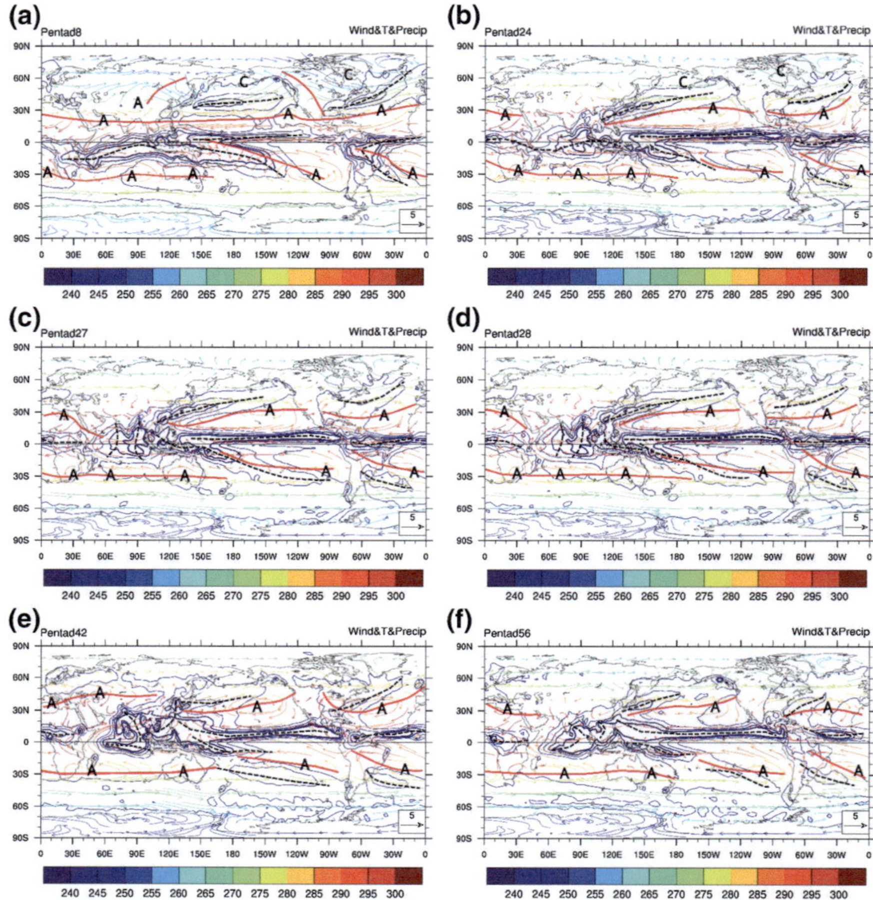

Fig. 4.12 Climatic precipitation rate (*solid-blue contour*, 2 mm day^{-1} interval) and climatic wind at 850 hPa (*arrow*, m s^{-1}) and climatic temperature (*shading*, 5 K interval) in **a** pentad 8, **b** pentad 24, **c** pentad 27, **d** pentad 28, **e** pentad 42, and **f** pentad 56. The *thick red line* denotes subtropical high (anticyclone) ridge while the *thick black-dashed line* is along the axis of precipitation rate. Letters "A" and "C" denote the locations of anticyclone and cyclone, respectively. Sourced from Qian and Jiang (2015) © Springer-Verlag Wien 2015, with permission

not influenced by the cross-equatorial monsoon flow at this moment. An abrupt dry–wet alteration occurs in the central SCS in pentad 28 (Fig. 4.12d). Surrounding the neighboring seas of Asian continent, 6 mm day^{-1} line and cross-equatorial southwesterly arrive at the Arabian Sea, the BOB, the SCS, and South China, because the subtropical high ridge has withdrawn from the SCS. These basic features of the SCS summer monsoon onset have been revealed a decade ago using different datasets (Qian and Yang 2000; Qian and Lee 2000).

In pentad 42, it is similar to that in Fig. 4.3b. Two subtropical highs in North Pacific and North Atlantic have reached the strongest stage and the northernmost location, so the Asian summer monsoon flow also reaches its strongest stage and the northernmost location, particularly in Northeast Asia (Fig. 4.12e). In this pentad, the trade wind convergence zone (or ITCZ) has also reached the strongest in the equatorial Pacific and the equatorial Atlantic. Another alteration time from the boreal summer half to the winter half year is in pentad 56 (Fig. 4.12f), but the tropical precipitation is stronger than the former seasonal alteration in pentad 24 because the 6 mm day^{-1} line still covers the SCS, although the 4 mm day^{-1} line has left South China. The axis of precipitation rate over the tropical Southwest Pacific is now located in the trough between the two anticyclones centered in Australia and the subtropical Southeast Pacific.

Globally and seasonally, the distribution of pentad mean precipitation rate corresponds well with the major large-scale circulation systems forced by seasonal solar radiation and sea–land contrasts as illustrated in Fig. 4.12. In NH Pacific, the longest zonally oriented axis of pentad mean precipitation rate in the north of the equatorial Pacific is caused by the ITCZ. Another axis of pentad mean precipitation rate in mid-latitudes is along the storm track associated with the interaction between the westerly flow and the subtropical North Pacific high with its center moving southwestward from spring to summer. Two precipitation rate axes in the NH Atlantic along the ITCZ and the storm track are similar to that in the NH Pacific. In the SH, two precipitation rate axes in the tropical Southwest Pacific and the tropical Southwest Atlantic are caused by the interactions between two adjacent anticyclones and middle-latitude westerly flow. In the tropical Indian Ocean, a zonally oriented precipitation rate axis is observed along the austral ITCZ. Between Asia and Australia as well as over the tropical African continents, the distribution of pentad mean precipitation rate is complicated. The precipitation rate axis in the south of equatorial Africa is formed in boreal summer and caused by the interactions of northeasterly flow from South Asia with two adjacent subtropical anticyclones in the Indian and Atlantic oceans. Another precipitation rate axis in the north of equatorial Africa is connected with the ITCZ in the equatorial Atlantic. Several centers of pentad mean precipitation rate situated in the neighboring seas of Asian continent are associated with peninsular-scale monsoon troughs (Qian and Tang 2010). The climatic difference exceeding 4 mm day^{-1} between maximal and minimal pentad mean precipitation rates captures the global dry–wet alteration. Although some spots such as from South China to the East China Sea cannot be found from this difference, there are some places with large precipitation difference in middle latitudes such as in the mid-North Pacific and the mid-North Atlantic.

A new index based on the climatic difference between maximal and minimal pentad mean precipitation rates exceeding 4 mm day^{-1} is used to identify the regions of dry–wet alteration (Qian and Tang 2010). Along the equatorial zone, there is a large portion of area without seasonal dry–wet alteration. Thus, dry–wet alteration is mainly concentrated in two subtropical zones. From South Asia to the subtropical NW Pacific, the region of dry–wet alteration is wider than that over other subtropical regions in the NH. In East Asia, the dry–wet alteration even extends to Northeast China and the Korean Peninsula near 45° N, but almost no dry–wet alteration in the Arabian region. There is a very narrow zone of dry–

wet alteration in the south of equatorial eastern Pacific and another dry–wet alteration in the middle latitudes of North Pacific.

The timing of the seasonal dry–wet alteration can be identified in more details at various locations (Qian and Jiang 2015). The most complicated is observed in Asia, but much simpler over ocean. Maximal precipitation rates are observed in pentads 30–40 in the Arabian Sea, 40–45 in the Indian Peninsula, 60–65 in the western BOB, 35–40 in the eastern BOB, 45–55 in the ICP, 40–45 in the central SCS, and 28–35 from South China to Japan. In middle latitudes of North Pacific, maximal precipitation rate mainly appears in boreal winter over pentads 60–73–10. In the subtropical SH, maximal precipitation rate appears over pentads 1–25.

In the work of Qian and Jiang (2015), the cause behind the dry–wet alteration was explored mainly from low-level wind changes or atmospheric circulation systems. Two dry–wet alteration regions in the north of equatorial eastern Pacific and the north of equatorial Atlantic are linked with two hemispheric subtropical anticyclones and associated with the shift of ITCZ with little wind direction change. In the south of equatorial eastern Pacific, there is a small dry–wet alteration region but without wind direction change. In the mid-latitude North Pacific, there is a dry–wet alteration region together with prevailing wind reversal but its wind is not coming from the low-level cross-equatorial flow. The GM domain satisfying dry–wet alteration and prevailing wind reversal associated with the low-level cross-equatorial flow is mainly located in the tropical Eastern Hemisphere from the tropical Africa to tropical western Pacific, including South Asia, East Asia, tropical NW Pacific, and the maritime continents.

After the analysis about the places and causes of the dry–wet alteration, some regional differences of calendar maximal and minimal precipitation rates in two pentads can be well explained. In the subtropical SH, there should be only one circular zone of dry–wet alteration if not considering the sea–land distributions, so persistent dry regions in the subtropical Southeastern Pacific and the subtropical South Atlantic are influenced by the three continents. The breadth of dry–wet alteration regions from South Asia to East Asia and the subtropical NW Pacific and the absence of dry–wet alteration in the Arabian region are influenced by the Tibetan Plateau and forced by the contrast between the Asian continent and the Indian Ocean. Therefore, the global dry–wet alteration domain in Fig. 4.6 is a result influenced by the seasonal change of planetary-scale general circulations in Fig. 4.3 and sea–land multiple-scale circulation in Fig. 2.2.

The global main regions of dry–wet alteration are distinguished by whether it is influenced by the low-level cross-equatorial flow with wind difference exceeding 5 m s^{-1} and wind angle change larger than 120°. The classical monsoon domain satisfying this condition is confirmed from the tropical Africa to the tropical Asia–Australia region. The non-classical monsoon region satisfying the difference of the low-level cross-equatorial flow exceeding 5 m s^{-1} but wind angle change smaller than 120° is located in the tropical America. The non-monsoon region satisfies the wind difference exceeding 5 m s^{-1} and wind angle change larger than 120° but with no influence of the low-level cross-equatorial flow in the mid-North Pacific. Since the late 1990s, different definitions were used to identify the GM domain. Even in the GM domain identified, the time of maximal pentad mean precipitation rate is complicated. How many monsoon regions can be objectively divided over the GM domain is another question.

4.4 Global Seasonal Precipitation Division[3]

The GM domain has been determined utilizing two criteria: the difference of local maximum and minimum pentad mean precipitation rates exceeding 4 mm day^{-1}, and the wind reversal of low-level cross-equatorial flow. We here examine how to objectively divide monsoon regions from the GM domain. The *k*-means clustering algorithm coupled Fourier analysis is introduced here and has been applied by Jiang et al. (2016) to objectively identify the categories and independent regions from the global precipitation climatology.

Supposing the global dry–wet alteration domain consists of *N* grids, our goal is to partition the *N* points into *K* categories or clusters ($1 \leq K \leq N$) by the similarities of their climatic cycle behaviors. Now, we have a smooth climatic time series for each spatial grid. However, due to the high dimensionality of time series, traditional clustering methods for raw data are not valid for time series clustering problem (Balasubramaniyan et al. 2005; Song and Pei 2012). Therefore, dimension reduction of the time series is an important prerequisite for effective realization of time series clustering (Liu and Ni 2009).

For a particular grid, $\tilde{F}_p(\lambda, \varphi)$ $(p = 1, 2, 3, \ldots, 73)$ is a sequence of 73 numbers (pentads). This sequence can be regarded as a set of coordinates of a point in a 73-dimensional space. Fourier analysis is proposed to preprocess the normalized climatic cycle of precipitation. In the harmonic analysis, we construct a set of vectors called a basis such that every grid in the 73-dimensional space can be written as a linear combination of the orthogonal trigonometric functions:

$$\tilde{F}_p(\lambda, \varphi) = \sum_{k=0}^{[73/2]} \left[a_k \cos\left(\frac{2\pi kp}{73}\right) + b_k \sin\left(\frac{2\pi kp}{73}\right) \right] \quad (p = 1, 2, 3, \ldots, n). \quad (4.4.1)$$

Noted that only harmonics smaller than or equal to 2 are kept in the smoothing the climatic cycle. Therefore, the climatic behavior of each grid is featured by five Fourier coefficients (a_0, a_1, a_2, b_1, b_2). Here, the Fourier coefficient a_0 represents the annual mean of the grid. a_1 and b_1 represent the annual cycle, while a_2 and b_2 represent the semiannual cycle.

[3]Modified article originally published in Jiang et al. (2016). Published with kind permission of © Springer-Verlag Berlin Heidelberg 2016. All Rights Reserved.

Here, we are only interested in the phase evolution of precipitation climatology, which reflects the evolution of monsoon. Thus, normalization is done on the time series of each grid's precipitation climatology in order to compare the climatic cycle features in different regions. After normalization, a_0 becomes zero and the climatic cycles of all N grids are now characterized by four coefficients (a_1, a_2, b_1, b_2), which gives the unimodal (annual) and bimodal (semiannual) behaviors of the climatic cycles. Then, a grid's climatic precipitation behavior is located in a 4-dimensional space instead of the original 73-dimensional space, and it is now prepared for the *k*-means clustering. The following steps are to obtain the minimum mean within-cluster distance and the maximum mean between-cluster distance during this procedure.

The *k*-means clustering is a method of vector quantization, which can partition the data space into Voronoi cells (Hartigan and Wong 1979). The *k*-means cluster analysis has been a popular method in the atmospheric and ocean sciences for decades (e.g., Michelangeli et al. 1995; Christiansen 2007; Johnson and Feldstein 2010; Riddle et al. 2013; Johnson 2013). We here use the *k*-means clustering to objectively identify the categories of climatological precipitation.

Generally, given a set of point observations (X_1, X_2, …, X_N), where each observation is a *d*-dimensional real vector, the *k*-means clustering algorithm helps to partition the N points into K categories $C = \{C_1, C_2, \ldots, C_i, \ldots, C_K\}$. One challenge in *k*-means cluster analysis is the determination of the optimal number of clusters (K) (Johnson 2013). Many studies have suggested various useful heuristic methods for determining K (e.g., Michelangeli et al. 1995; Christiansen 2007; Johnson 2013; Riddle et al. 2013), but an objective optimal K has remained elusive (Johnson 2013). Here, the cluster distance variation (CDV) and the total distance variance (TDV), which are referred to as the trajectories clustering algorithm of the *HYSPLIT* model (Draxler et al. 1999), are introduced in this method to determine a reasonable value of K. The two indicators, CDV and TDV, are calculated as follows:

$$\mathrm{CDV}_i = \sum_{X \in C_i} \|X - \mu_i\|^2 \tag{4.4.2}$$

$$\mathrm{TDV}_K = \sum_{i}^{K} \mathrm{CDV}_i, \tag{4.4.3}$$

where μ_i is the central position of points in C_i. Given an initial set of K-means $m_1^{(1)}, m_2^{(1)}, \ldots, m_K^{(1)}$, the algorithm consists of the following steps:

(1) Assignment step: Assign each observation to the cluster whose mean yields the least within-cluster sum of squares (WCSS). Since the sum of squares is the squared Euclidean distance, this is intuitively the "nearest" mean:

$$C_i^{(t)} = \left\{ x_p : \left\| x_p - m_i^{(t)} \right\|^2 \leq \left\| x_p - m_j^{(t)} \right\|^2 \forall_j, \quad 1 \leq j \leq K \right\}, \tag{4.4.4}$$

where each x_p is assigned to exactly one $C_i^{(t)}$, even if it could be assigned to two or more of them.

(2) Update step: Calculate the new means to be the centroids of the observations in the new clusters:

$$m_i^{(t+1)} = \frac{1}{\left| C_i^{(t)} \right|} \sum_{x_j \in C_i^{(t)}} x_j. \tag{4.4.5}$$

Since the arithmetic mean is a least-squares estimator, this also minimizes the WCSS objectively.

The algorithm converges when the assignments no longer change. Various categories are tested from N to 1 using the 4 coefficients at all global dry–wet alteration grids. Initially, each grid is defined to be a category, so TDV is zero, i.e., there are N categories for the N grids. Then various categories are calculated until all the grids (N grids) are clustered in one group. The judgment of convergence depends on whether the TDV (%) approaches a stable stage as K (K = 1, 2, 3, …, $N - 1$) increases

$$\text{Change in TDV}(\%) = \frac{\text{TDV}_K - \text{TDV}_{K+1}}{\text{TDV}_K} \times 100. \tag{4.4.6}$$

The change in *TDV* represents the difference between the clustering results of K and $K + 1$ categories. A small change in *TDV* means that the clustering results become stable. Therefore, 10% TDV change is set as an objective criterion for determining the optimal number of categories. Figure 4.13 shows the relationship between the change in *TDV* and the number of categories K from the global dry–wet alteration domain (1855 grids). It is noted that the change in *TDV* rapidly decreases from 1–2 categories, and slowly decreases from 3–5 categories, and then it is relatively stable beyond six categories (*Change in* TDV <10%). Therefore, the global dry–wet alteration region can be divided into six categories.

Figure 4.14 shows an example of dividing the global dry–wet alteration regions into two categories (CA and CB). The two categories are basically equal in size with a contrast between two hemispheres. Category 1 (red shading) is mainly located in the tropical NH while category 2 (blue shading) covers the tropical SH

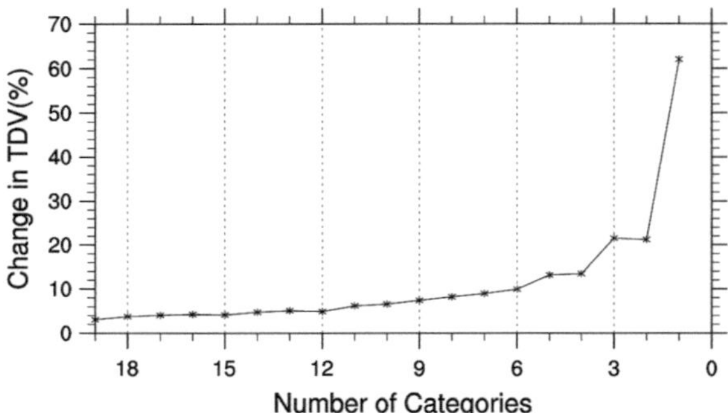

Fig. 4.13 Change in TDV(%) and the number of categories of the global dry–wet alteration domain (1855 grids). Sourced from Jiang et al. (2016) © Springer-Verlag Berlin Heidelberg 2016, with permission

and part of the mid-North Pacific and mid-North Atlantic. The centroids of the two categories, marked with black stars, are located in the SCS in the NH and near the Java Island in the SH, respectively, representing the average evolutions of their clusters.

Figure 4.14b shows the normalized seasonal cycles of the centroids of the two clusters. The period of positive (negative) values of the time series is taken as wet (dry) season, whose peak (valley) is defined according to the maximum (minimum) values of the time series. The onset dates (ODs) and the withdrawal dates (WDs) of the wet season are determined as the first and last days of the positive period. As shown in Fig. 4.14b, the two time series are almost completely anti-symmetric. In the NH (SH), the OD (WD) of the wet season is about 2 months after the spring equinox, and the WD (OD) also lags the Autumnal Equinox for about 2 months. Similar to the OD and WD, the peak and the valley lag behind the solstices. However, the peak (pentad 8) of category CB is 3 pentads in advance of the valley (pentad 11) of category CA, while CA peaks (pentad 44) 2 pentads in advance of the valley (pentad 46) of CB. Their different responses to the solar forcing are mainly due to the difference of land distribution between the two hemispheres. The peak positions match the previous findings that the zonal mean variables, such as temperature, U-wind, and V-wind at 850 hPa, reach the southernmost and northernmost locations in pentad 8 and pentad 44, respectively (Qian et al. 2009; Qian and Tang 2010).

Following the above example, the global dry–wet alteration domain is partitioned into six categories, and their corresponding normalized amplitude series of centroids are obtained and shown with different color areas and curves, respectively, in Fig. 4.15a, b. The six curves in Fig. 4.15b reach the maximum (minimum) in pentads 73 (43), 21(48), 8 (52), 44 (13), 33 (10), and 56 (12), respectively. The six categories in Fig. 4.15 are the subsets of the two categories in Fig. 4.14. Based on their climatic precipitation behaviors, including ODs, WDs, the periods of the wet (dry) seasons, and other characteristics of series, the six categories can be divided into four types: the dominant categories (C3 and C4), the equinoctial

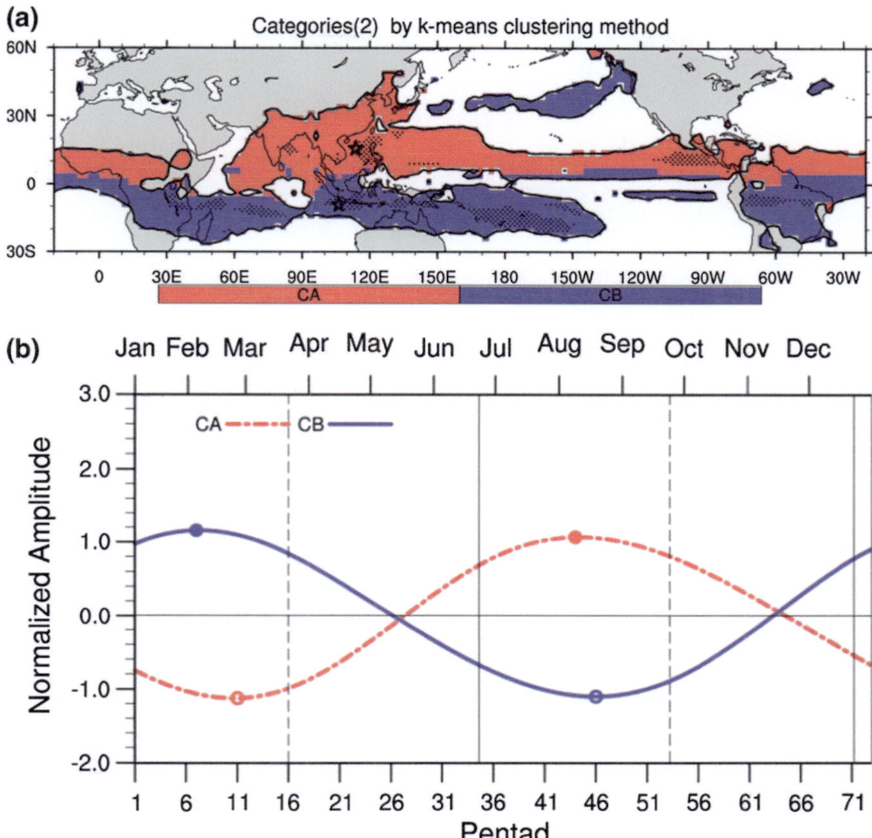

Fig. 4.14 **a** Two categories (CA and CB) of dry–wet alteration and **b** their seasonal cycle (annual cycle plus semiannual cycle) indicated by normalized time series of central (*star*) point precipitation rate. **a** *Shading* points reach the 0.95 high correlations to the central point precipitation rate series. **b** Two curves (CA and CB) peak the *positive* (*negative*) amplitude in pentads 44 (7) and 11 (46). The equinoxes and solstices are indicated by the *vertical-dashed* and *solid lines*, respectively. Sourced from Jiang et al. (2016) © Springer-Verlag Berlin Heidelberg 2016, with permission

asymmetric categories (C2 and C6), the bimodal categories (C5), and the symmetric category (C1).

As shown in Fig. 4.15a, categories C3 and C4 cover the largest area of the global dry–wet alteration domain. They have climatic behaviors (Fig. 4.15b) similar to those of the SH and NH categories (CB and CA) in Fig. 4.14. In Fig. 4.15a, the same category of grids which are continuous in space can be grouped as a region. However, we here mainly focus on the large regions (marked by numbers and letters), which cover more than three grids in both meridional and zonal directions. The consistency of seasonal precipitation rates can be, respectively, observed in Figs. 4.16c–d but the opposite phase evolution is obvious between

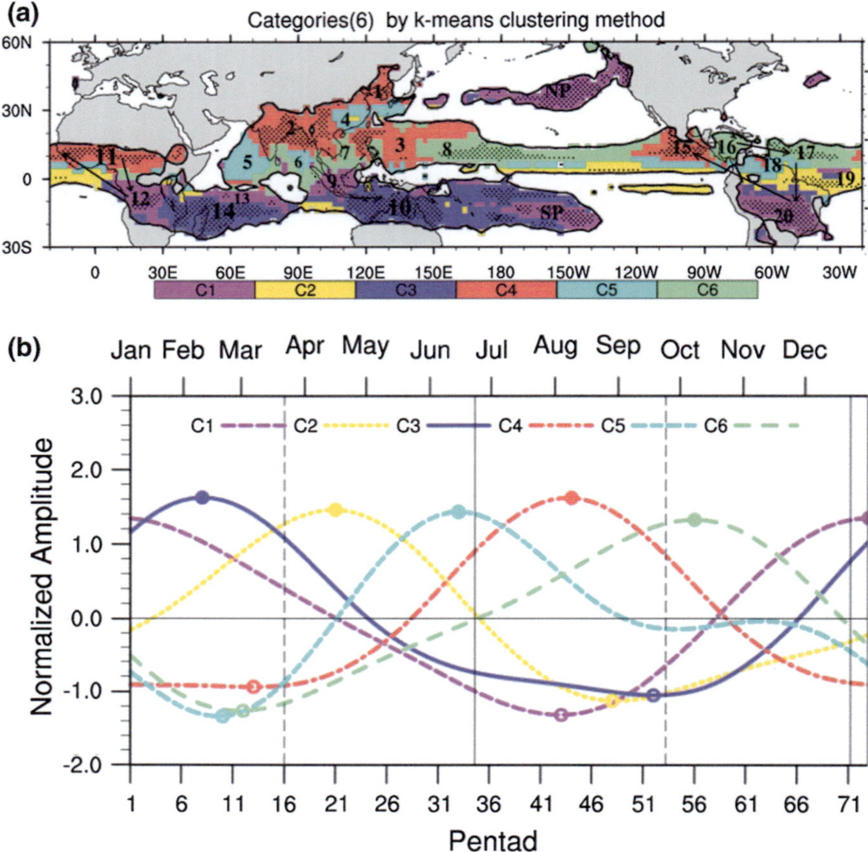

Fig. 4.15 **a** Same as Fig. 4.14 except of six categories of dry–wet alteration. **b** Six curves (C1, C2, C3, C4, C5, and C6) peak the *positive* (*negative*) normalized series of central point precipitation rate in pentads 73 (43), 21 (48), 8 (52), 44 (13), 33 (10), and 56 (12), respectively. Sourced from Jiang et al. (2016) © Springer-Verlag Berlin Heidelberg 2016, with permission

them. Both categories have a short wet summer with a concentrated rainy period in comparison of a "flat" and dry winter, which is in agreement with the typical characteristics of the monsoon regions depicted by Wang and Ding (2008). The peaks of C3 and C4 are about 1.5 months after the maximum solar heating in the two hemispheres (Qian and Tang 2010), while the valleys are near the equinoxes. Since C3 and C4 cover most regions of the dry–wet alteration domain, they are referred to as the dominant categories of the dry–wet alteration between the two hemispheres.

For category C4, regions 1, 2, and 3 cover most areas of dry–wet alteration domain in Asia, surrounding region 4 in South China, which belongs to category C5. There is a yellow spot in region 4, implying the complication of the precipitation evolution in South China. Another two red regions (Fig. 4.15a) with correlation dots are located in the tropical North African continent (region 11) and the tropical North Eastern Pacific (region 15), which are the two typical monsoon regions: the Northwestern African Monsoon (e.g., Gasse and Van Campo 1994) and the North America Monsoon (e.g., Adams and Comrie 1997). The precipitation rates of the five red-shading regions in Fig. 4.15a are illustrated in Fig. 4.16d. Their precipitation rates have similar seasonal evolutions. But the annual mean precipitation rates decrease as latitude increases, except North Africa (region 11), and regions 3 and 11 peak a bit later than other regions. The precipitation rate of region 3 has the biggest amplitude among the five curves in Fig. 4.16d.

In particular, region 3 is located from the SCS to the tropical NW Pacific, in which only the precipitation rate in the SCS part has a high correlation to that in the centroid of category C4. And as depicted in Fig. 4.14a, the SCS as the centroid of category CA is a key area characterizing the abrupt transition from the dry season to

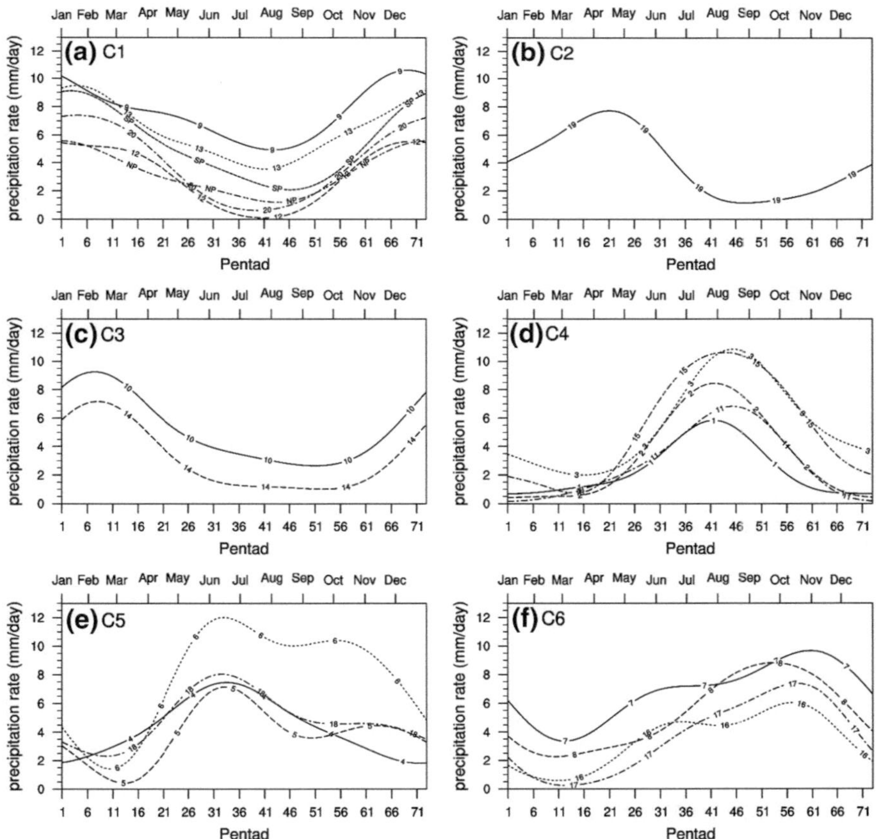

Fig. 4.16 Seasonal climatic precipitation rate (mm day^{-1}) curves of regions in categories. **a** C1, **b** C2, **c** C3, **d** C4, **e** C5, and **f** C6 regions marked by numbers or letters from Fig. 4.15a. The seasonal climatic precipitation rate of each region is averaged over all grids in the corresponding region. Sourced from Jiang et al. (2016) © Springer-Verlag Berlin Heidelberg 2016, with permission

the rainy (wet) season and the subsequent seasonal change in the Asian monsoon region (Xue et al. 2015). From the result of the *k*-means clustering, precipitation rate in the SCS can be implied as a key indicator not only for Asian monsoon onset but also for the NH dry–wet transition. For its importance and the unique features, many works have been done on the SCS summer monsoon (SCSSM). However, there is still not a consensus on the SCS domain yet. For example, several common SCS domains in previous studies are 5°N–20°N 110°E–120°E (Mao and Chan 2005), 10°N–20°N 110°E–120°E (Qian and Yang 2000; Wang et al. 2004), and 5°N–15°N 110°E–120°E (Zhou and Chan 2007). Here, in Fig. 4.15a, a reference domain of the SCSSM region can be determined by the high correlation dots in region 3 and centered in Fig. 4.14a.

For category C3, regions 10 and 14 are located from the Indian Ocean to the South Western Pacific Ocean, having a similar climatic behavior of precipitation rate in Fig. 4.16c. And the precipitation rate in region 10 is about 2 mm day^{-1} higher than that in region 14. The high correlation located in the north of Australia illustrates the typical Australia Monsoon region (Fig. 4.15a). The precipitation rate in the western part of region 10 near the Java Island (a star in the SH in Fig. 4.14a) can be implied as a key indicator for the climatological cycle in the SH.

Another pair of categories, C2 and C6 (Fig. 4.16b, f), the equinoctial asymmetric categories, has a gentle cycle of climatological behaviors. The OD and WD are consistent with the solstices, while the peaks and valleys are, respectively, after and before the equinoxes. The transition from the wet to dry phases is faster than that from the dry to wet phases. The two categories have a half-year wet period with a half-year dry phase without a sharp rainy season, which reflects a flat evolution feature. Therefore, these characteristics can be distinguished from the typical monsoon categories.

The regions which belong to the two categories C2 and C6 are mainly located to the south and the north of the equator. The spatial distributions of C2 and C6 regions are similar to the equinoctial asymmetric mode (the second mode of MV-EOF) defined by Wang and Ding (2008). These regions (7, 8, 16, 17, and 19) peak in the transitional seasons and are depicted as the spring–fall asymmetry, showing one of the fundamental features of the seasonal variation in the tropical circulations (Chang et al. 2005). The seasonal precipitation rates of regions 7 and 16 have double peaks, indicating that the one around November is stronger than that in June (Fig. 4.16f).

The regions of category C5 are characterized by an asymmetric bimodal climatic cycle with an obvious major peak in June and a small peak in November. In Fig. 4.16e, the seasonal precipitation rates of most C5 regions have double peaks except in South China (region 4), which has only one peak in June. Among all the regions in C5, the seasonal precipitation rate of region 6 in the southern part of the BOB is the highest. In Fig. 4.15a, no high correlation regions with dots can be found, which implies that these regions of C5 have their own dispersed distribution besides the asymmetric bimodal climatic cycle (Fig. 4.16e).

Some regions of category C5 are in the equatorial region between C2 and C6. The equatorial rainfall of such regions follows the maximum solar forcing when the sun crosses the equator. One large region 18 is in the north of the South America. Other two regions 5 and 6 are located at the eastern Arabian Sea and the BOB, respectively. Particularly, region 4 in South China, which belongs to C5, is ambiguous in previous works (Wang and Ding 2006; Zhang and Wang 2008; Qian and Tang 2010) as well as mentioned by Nguyen et al. (2014) and Qian and Jiang (2015). From the clustering result, region 4 is surrounded by regions of category C4 (red shadings in Fig. 4.15a). This explains why region 4 usually cannot be determined by a fixed calendar (calendar seasons of summer and winter) difference of rainfall in GM domain.

The SH precipitation categories are not as complicated as those in the NH. The regions of categories C1 and C3 cover almost all the dry–wet alteration domain in the SH. The climatological behavior of the C1 precipitation is similar to that of C3 as shown in Fig. 4.16a, c. But the seasonal distribution of C1 is more symmetric than that of C3 and other categories. The seasonal asymmetric distributions of these regions give rise to a complementary semiannual component (Wang and Ding 2008). The OD and the peak of C1 are both about 1.5 months earlier than those of C3 (Zhang and Wang 2008). The C1 curve reaches its valley when the C4 curve almost gets to its peak. The WD of C1, which is corresponding to the OD of C5, is a bit earlier than that of C3.

Though most regions of category C1 are located in the SH, some regions and spots can also be found in the mid-North Pacific and mid-North Atlantic. These regions can also be found in Fig. 4.14a and the leading MV-EOF mode (Wang and Ding 2008). From Figs. 4.15a and 4.16a, the six marked regions of category C1 are analyzed respectively. Two big regions in purple with correlation dots are located in the North (NP) and South (SP) subtropical Pacific, respectively, and a narrow region (region 13) located to the north of region 14. The seasonal evolutions of precipitation rates of regions NP and 13 are gentle (Fig. 4.16a). Region 9 mainly covers Borneo and Pulau Sumatera around the equator, and the contribution of its semiannual component is obvious in its seasonal precipitation rate (Fig. 4.16a). Two large regions of category C1 are located on South Africa (12) and South America (20) continents, which are investigated as the South African Monsoon and South American Monsoon regions. The climatological behaviors of categories C1 and C3 tell us that the OD and peaks of the South African Monsoon and the South American Monsoon are earlier than those of the Australian Monsoon. Particularly, the rainy season of the two C1 monsoon regions (12 and 20) is gentler than other typical monsoon regions such as C3 and C4 regions.

In summary, regions of categories C3 and C4, the dominant dry–wet alteration regions, are characterized by the typical monsoon precipitation cycle with a shorter and more concentrated rainy period in comparison with a “flat” and dry winter. The regions of categories C2 and C6 act as the transition belt of the dry–wet alteration between the two hemispheres. An asymmetric bimodal climatic cycle can be detected in most of the C5 regions located around the equator. In the off-equatorial regions, the climatological cycle of the South China precipitation (region 4, C5) is unimodal (single maximum and minimum). The semiannual variation in South China is an artifact of the asymmetric seasonal distribution of precipitation between the dry and wet seasons (Wang 1994). The seasonal distribution of category C1 is more symmetric than that of other categories. These C1 regions are mainly located in the subtropical Pacific as well as North Africa and North America.

If we focus on the relationships among these six categories and their corresponding regions, it is interesting to note that the peaks of six categories regularly exist every 2 months while the valleys are concentrated from February to March and from August to September (Fig. 4.15b). On the whole, an early peak is associated with an early OD and an early WD of climatological precipitation. Based on the climatological behaviors of precipitation of the six categories and the adjacency relations among these regions of different categories, cyclic routes of phase propagation are identified in Fig. 4.15a (black arrows). They are composed of northward and southward routes. Over the American and African sectors, the northward routes are similar, which is confirmed by the results of Wang (1994). The categories passed by the northward routes are in the order of C1, C3, C2, C5, and C4; and that by the southward routes are in the order of C4, C5, and C1. The routes are mostly closed loops, but a jump is detected in the southward route of southward propagation over the American sector. Based on the above analyses, the phase routes of southward propagation are different from the northward routes.

The clustering result provides us the information about the climatological precipitation behaviors of each category. However, it is possible that the regions with different climatological precipitations are controlled by the same circulation, since the differences among these regions may be due to the seasonal march of the same circulation system. On the other hand, regions of the same category may also be assigned with different circulation systems. In order to derive a comprehensive division and to understand the phase propagation features, the circulation systems should be investigated as the dynamical aspect of the dry–wet alteration.

4.5 Seasonal Circulation Systems[4]

As the peaks of the six categories regularly exist in two consecutive months, the 850-hPa circulation and the normalized precipitation rates of the six peak pentads are plotted in Fig. 4.17. In each panel, the precipitation domains of categories are extracted from Fig. 4.15a and marked at their corresponding peak pentads. In this section, the circulation systems and the precipitations are analyzed to depict the seasonal cyclic process.

From Fig. 4.15a, category C3, which includes the south Indian Ocean (region 14), Australia and its adjacent Pacific Ocean (region 10), nearly covers the southernmost of the dry–wet alteration domain and their climatological precipitation peaks in pentad 8 (Fig. 4.15b). The peaks of regions 10 and 14 are mainly

[4]Modified article originally published in Jiang et al. (2016). Published with kind permission of © Springer-Verlag Berlin Heidelberg 2016. All Rights Reserved.

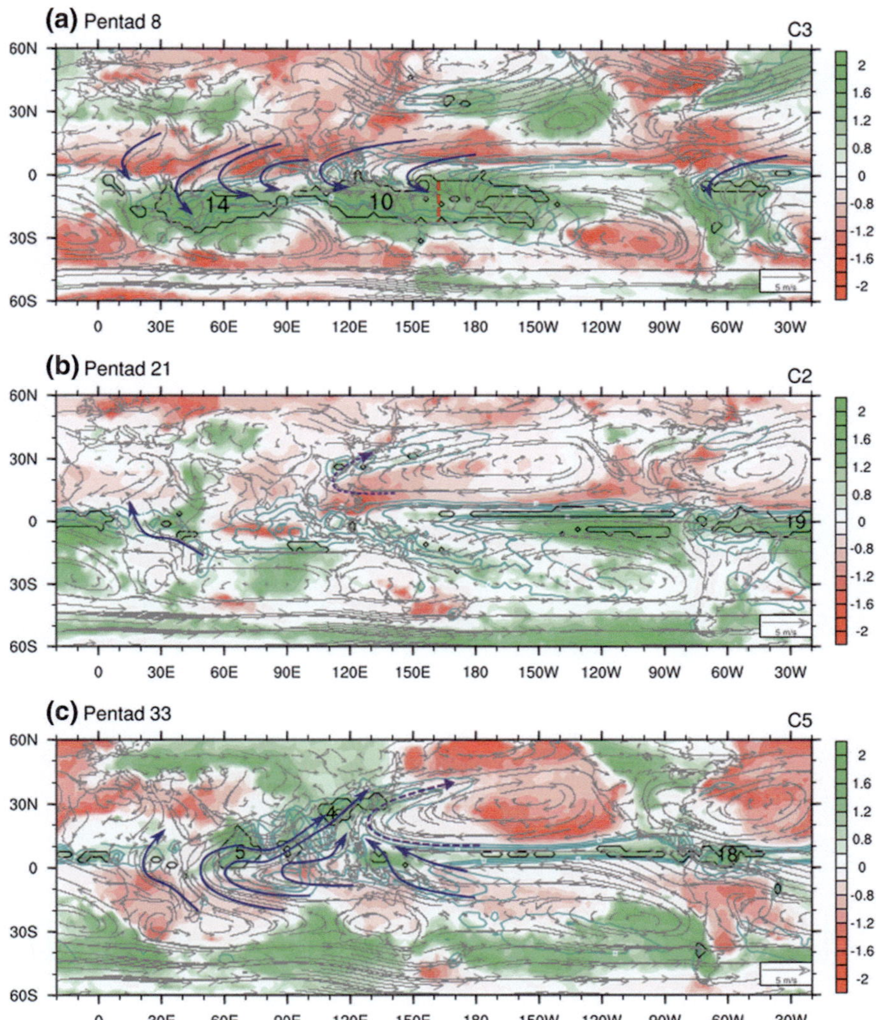

Fig. 4.17 850-hPa climatic flow (m s^{-1}) and normalized precipitation rate (*green shadings* for positive and *red* for negative, 0.4 interval) as well as the precipitation rate (*dark green contour*, 2 mm day^{-1} interval) in **a** pentad 8, **b** pentad 21, **c** pentad 33, **d** pentad 44, **e** pentad 56, and **f** pentad 73. The *solid-blue arrows* indicate the cross-equatorial flows and the *dashed-purple arrows* depict the northward flows surrounding the subtropical highs. The *red-dashed line* in **a** separates region 10 into the western and the eastern parts based on the influence of the cross-equatorial flows, and the eastern part is controlled by the trade winds. These six selected pentads are corresponding to the peaks of the six categories in Fig. 4.15. The regions marked by *numbers* or *letters* and covered by the *heavy black line* are also depicted in their peak pentads. Sourced from Jiang et al. (2016) © Springer-Verlag Berlin Heidelberg 2016, with permission

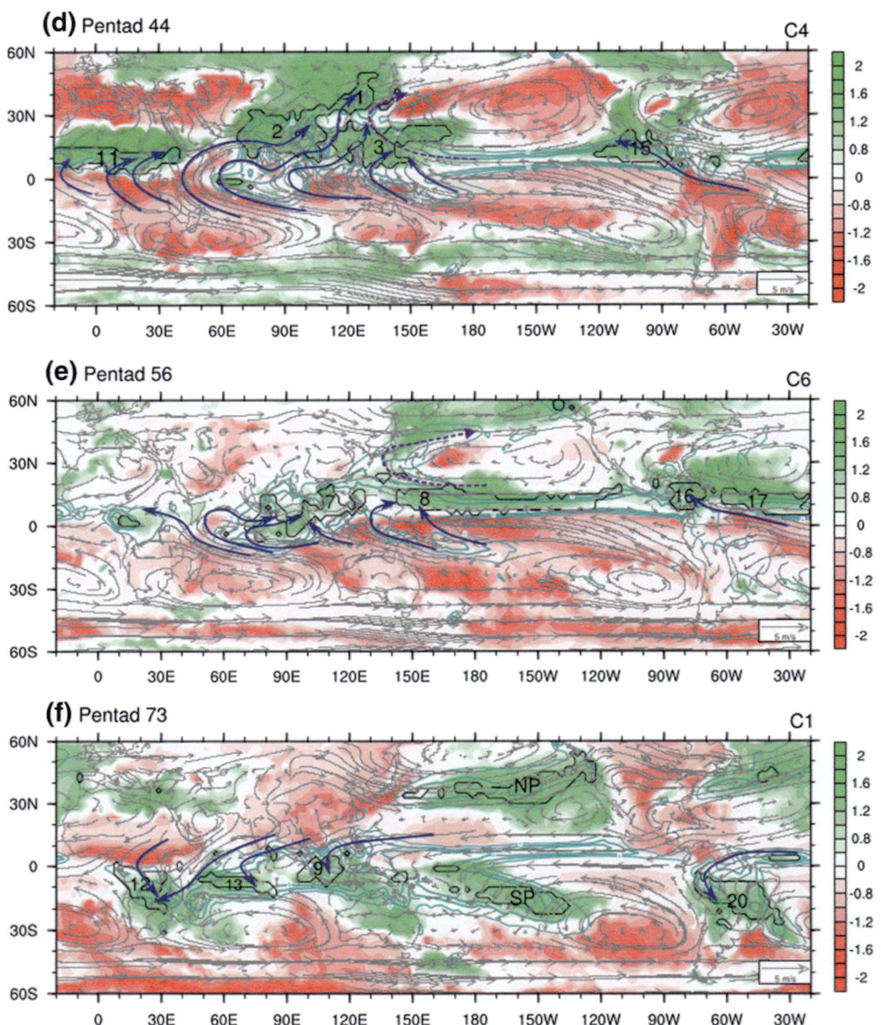

Fig. 4.17 (continued)

characterized by the pronounced southward cross-equatorial flows (Fig. 4.17a). The rain belts associated with the cross-equatorial flows expand to southernmost between the south Indian Ocean and the southwest Pacific Ocean. In particular, the precipitation in the eastern part of region 10 is mainly associated with the Southern Pacific convergence zone (SPCZ), which is oriented diagonally, stretching northwest–southeast (Vincent 1994) from near the Solomon Islands to southern French Polynesia (Salinger et al. 2014; Wiel et al. 2015) and separated by a red-dashed line in Fig. 4.17a. The SPCZ is a major feature in the precipitation distribution over the SH tropical Pacific during austral summer (Wiel et al. 2015). Besides, a narrow belt located in the north of South America, on the northern edge of the core monsoon region (region 20), also reaches its precipitation rate peak in this pentad. In pentad 8, the precipitation rates of C1 regions have already begun to decrease. The decay phases of the South American Monsoon System (SAMS) and the South African Monsoon System (SAfMS) continue through the boreal autumn as convections gradually retreat northward to the equator.

Pentad 21 is a transition point from the boreal winter half to the summer half, so the tropical convergence zones and precipitation rate axes nearly approach to the equator. The precipitation of category C2 reaches its peak phase in this pentad, and the OD of category C5 and the WDs of categories C1 and C3 are also around this time point. During pentad 21, rain belts shrink to the equator and the southward cross-equatorial flows weaken compared to pentad 8 (Fig. 4.17b vs. a). Thus, most regions of category C2 are around the equator, and their precipitation peak phases are due to the convergence of the trade winds. A spot of category C2 can be detected in South China where the precipitation peak of the spot is due to the southwesterly (dotted purple arrow) at the edge of the subtropical anticyclone extending from the NW Pacific to the SCS. The anticyclone systems in the NH are enhanced and move westward from pentad 8 to 21, while the subtropical anticyclone systems in the SH move equatorward. Meanwhile, two axes of frontal precipitation in the NW Pacific and NW Atlantic extend westward to South China and eastern America, respectively.

The regions of category C4 (red shadings in Fig. 4.15a) mainly cover the major monsoon regions in the NH. The ODs of these monsoon regions in the NH are around pentad 28 (Fig. 4.15b). Five pentads later, the regions of category C5 reach their peaks in pentad 33 during the development phase of the C4 precipitation cycle. In the tropical Pacific, some narrow regions of category C5 located near the equator are associated with the ITCZ. Other three major regions cover the eastern Arabian Sea, the southern BOB, and from South China to the south of Japan, respectively. After the boreal summer monsoon onset, the cross-equatorial flows from the SH strengthen and become a dominant feature.

In pentad 33 (Fig. 4.17c), the Somali cross-equatorial flow is the strongest. The subtropical anticyclone on the NW Pacific enhances and moves eastward. Its western edge (dotted purple arrow) controls the southeastern part of Japan instead of South China in pentad 21. The precipitation of South China is now associated with the cross-equatorial flows and cold fronts. The cross-equatorial flows from the western Pacific and the New Guinea converge with the southeasterly of the subtropical anticyclone, bringing rainfall to the subtropical NW Pacific. In Africa, the strong northward cross-equatorial flows are found over the East African coast and they curve to change directions over the Arabian Sea before reaching India. However, the cross-equatorial flows over the Atlantic Ocean are not clear, so the tropical rain belts around America and Africa are mainly associated with the trade winds.

In pentad 44 (Fig. 4.17d), the precipitation rates of the NH monsoon regions (C4) reach their peaks. The rain belts associated with the cross-equatorial flows reach the northernmost. Compared to pentad 33, the subtropical anticyclone in the North Pacific, again, enhances and moves further eastward. The branches of the cross-equatorial flows between the Indian Ocean and the western Pacific converge over the Asian continent and reach the northernmost to combine with the westerlies (region 1). Though the precipitation rate of South China begins to decrease from pentad 33, its precipitation in pentad 44 is also controlled by the cross-equatorial flows. However, the precipitation peak of the eastern part of region 3 is due to the southeasterly of the subtropical anticyclone. In this pentad, the northward cross-equatorial flow over equatorial North America is not as strong as that in other seasons and other continents. The strong cross-equatorial flows from the Indian Ocean influence the off-equatorial North Africa. In the Atlantic Ocean, relatively weak northward cross-equatorial flows can be detected.

After the peak of category C4, the NH monsoons begin to retreat. In pentad 56 (Fig. 4.17e), C6 regions reach their precipitation peaks. Though the precipitation of category C4 enters the decline stage, the NH monsoon regions are still in the wet period. The rain belts associated with the cross-equatorial flows in the NH shrink equatorward, but the Somalia cross-equatorial flows are still apparent. The C6 regions are restricted in the northern tropics, whose precipitation peak of region 7 is due to the convergence of the cross-equatorial flows from the Somalia and the north of Australia. The biggest region of category C6 is located in the tropical Pacific (region 8), where its western edge is associated with the retreated cross-equatorial flows from the Southeast Pacific. The precipitation of region 8 is mainly due to the convergence of the trade winds. The region 16 located in the tropical America is also influenced by the weak and retreated cross-equatorial flows. The rain belt of region 17 is corresponding to the convergence of the trade winds.

Comparing to pentad 56, the cross-equatorial flows have reversed southward in pentad 73 (Fig. 4.17f). The rain belts also expand southward associated with the southward cross-equatorial flows. The phases of the cross-equatorial flows, as well as the rain belts' immigration, are not consistent in different regions in the SH. In pentad 73, the cross-equatorial flows over South Africa and South America have already reached the southernmost while those in the Asia–Australia monsoon regions just turn southward and reach their northern edges (regions 9 and 13). Meanwhile, the South African monsoon and the South American monsoon have already entered the mature stage. It is noted that the cross-equatorial flows are monsoonal types in the Eastern Hemisphere while they are characterized by the trade winds in the Western Hemisphere (Gao and Xue 2006). Therefore, the cross-equatorial flows in the Americas are not as clear as those in the Asia–Australia monsoon regions. Two regions (NP and SP) of category C1 are situated in the northern and southern subtropical Pacific Ocean, respectively. Similarly, other two regions are located in the North and South Atlantic, respectively. These four precipitation zones that tilt eastward and poleward are in association with the oceanic storm tracks in the ocean (Wang and Ding 2008). And these regions should not be distinguished as monsoon because the wind is not steady (Wang and Ding 2008; Zhang and Wang 2008). Associated with the southward immigration of the cross-equatorial flows, the rain belt between the Indian Ocean and the Southwest Pacific expands to the southernmost in pentad 8, resulting in the mature stage of the Australian Monsoon. A complete dry–wet alteration cycle is depicted with the cross-equatorial flows and other circulation systems.

4.6 Global Monsoon Division and Trend[5]

The traditional concept of monsoon refers to certain areas that exhibit significant seasonal variation in circulation and precipitation, and this regional concept has been extended to the global scale (An et al. 2015). Based on the previous studies (Webster 1987; Trenberth et al. 2000; Qian et al. 2002a; Li and Zeng 2003; Wang and Ding 2008; Zhang and Wang 2008; Qian and Tang 2010) and a recent review (An et al. 2015) of GM, the surface climate of GM is characterized by two points: (a) a seasonal reversal of prevailing wind direction and (b) a seasonal alteration of dry and wet conditions. Most studies of GM focused on the prevailing wind reversal (Khromov 1957; Ramage 1971; Trenberth et al. 2000; Li and Zeng 2003), the dry–wet alteration (Wang 1994; Qian et al. 1998a, b; Qian 2000; Zeng and Lu 2004; Wang and Ding 2006) and combination of the wind reversal and dry–wet alteration (Webster 1987; Qian et al. 2002a; Wang and Ding 2008; Zhang and Wang 2008; Qian and Tang 2010).

According to the characteristics of precipitation categories, some monsoon regions, such as regions of categories C3 and C4, can be recognized, which agrees with the results of previous studies (e.g., Wang and Ding 2008; Zhang and Wang 2008). However, in order to identify the monsoon regions comprehensively, circulation cannot be neglected. Different monsoon regions are influenced by various circulation systems, but both global and regional monsoons are closely related to the cross-equatorial flow, which plays a key role in mass, moisture, and energy transport between the hemispheres (Wang and Li 1982; Zeng and Li 2002; Li and Li 2014). Many analyses have confirmed the contribution of the cross-equatorial flow toward local monsoon, such as monsoon over Asia (Wang and Xue 2003; Xie 2013), India (Kumar et al. 1999), SCS (Gao and Xue 2006), the NW Pacific (Lin and Chou 2014), Australia (Godfred-Spenning and Reason 2002), America (Wang and Fu 2002), and Africa (Benjamin and Serge 2003). By the cross-equatorial flow, the regional monsoon can be explained in a global scale. The GM is defined as the expansions of three deep convective centers in the equatorial region driven by the low-level cross-equatorial flow (Qian et al. 1998a, b, 2002a; Qian 2000). The relationship between monsoon and the cross-equatorial flow has also been deeply analyzed globally by Zeng and Li (2002). Furthermore, some special monsoon regions, such as the American monsoon regions (without clear wind reversal) and the South China region (a precipitation peak in the boreal winter), can be depicted by considering the influence of the cross-equatorial flow (Qian and Jiang 2015). Hence, not only the precipitation characteristics but also the influence of the cross-equatorial flow is selected to define monsoon regions.

After considering the influence of cross-equatorial flows on the dry–wet seasonal transition of six categories, 16 regions of the GM are highlighted in Fig. 4.18. The major monsoon regions are basically identical to the previous works (Wang 1994;

[5]Modified article originally published in Jiang et al. (2016). Published with kind permission of © Springer-Verlag Berlin Heidelberg 2016. All Rights Reserved.

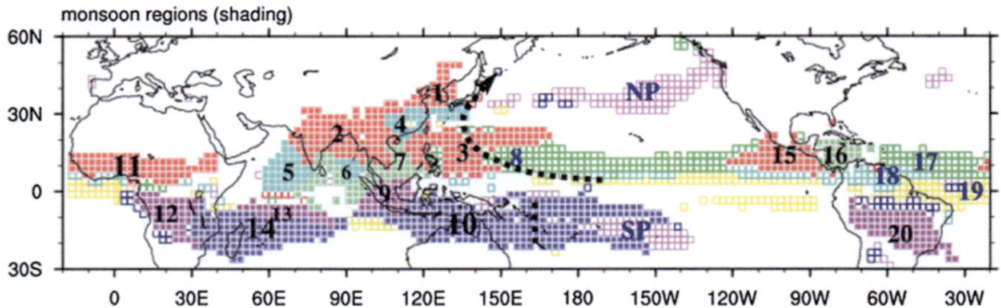

Fig. 4.18 Derived GM regions and non-monsoon regions. The GM regions influenced by the cross-equator flows are shaded in *full-color grid boxes* marked by the *black numbers* as Fig. 4.15a while the other regions are indicated by the *hollow-color grids* marked by the *blue numbers* or *letters*. The *dashed-black arrow* is the same as the *dashed-purple arrow* in Fig. 4.17d. The *dotted-black line* in region 10 is the same as the *dashed-red line* in Fig. 4.17a. The *dashed arrow* and the *dashed line* are used to separate regions 3 and 10 into two parts, respectively. The eastern parts of regions 3 and 10 are mainly influenced by the trade winds. Sourced from Jiang et al. (2016) © Springer-Verlag Berlin Heidelberg 2016, with permission

Wang and Ding 2006, 2008; Qian et al. 2002a; Qian and Tang 2010; Qian and Jiang 2015). Four regions (11, 12, 13, and 14) are located in the Africa and the Southwest Indian Ocean. The Asia–Australia monsoon includes 9 regions. Based on the analysis above, the eastern parts of regions 3 and 10 are not influenced by the cross-equatorial flows but by the trade winds and the SPCZ, respectively. Therefore, only the western parts of these two regions are defined as monsoon regions. Three regions (15, 16, and 20) cover Americas. These 16 monsoon regions belong to five categories (C1, C3, C4, C5, and C6). Only three categories can be found in the Africa and its adjacent ocean (C1, C3, and C4) or Americas (C1, C4, and C6). Five categories exist in the Asia–Australia monsoon regions, implying the complexities of the Asia–Australia monsoon system.

Long-term precipitation trends of 16 monsoon regions in rainy periods from 1981 to 2010 are depicted in Fig. 4.19. Consistent with Wang and Ding (2006), Zhou et al. (2008), and Lin et al. (2014), increasing trends are observed in most of the derived monsoon regions. Among the Asia–Australia monsoon regions, no obvious trends are observed in South Asia (region 2) and Australia (region 10), but all the other regions experience increasing trends especially for the East Asia (regions 1, 4, and 7; Yao and Zhao 2015). Increasing trends are also found in the South America (region 20), while the trends in the North (Central) America are not apparent (Barrett and Longoria 2013). Only four regions (regions 11, 12, 13, and 14) with significant decreasing trends are found in the Africa and its adjacent Indian Ocean. The trend in North Africa is consistent with the reanalysis datasets but contrary to the GPCP dataset (Wang et al. 2012; Lin et al. 2014). In this study, using the CMAP dataset, the biggest increasing and decreasing trends are found in South China (region 4), and South Africa (region 12), respectively. During the rainy seasons of 1981–2010, decreasing trends are concentrated in Africa and its adjacent Indian Ocean, while increasing trends of monsoon precipitation are found in East Asia, especially in South China with the biggest increasing trend.

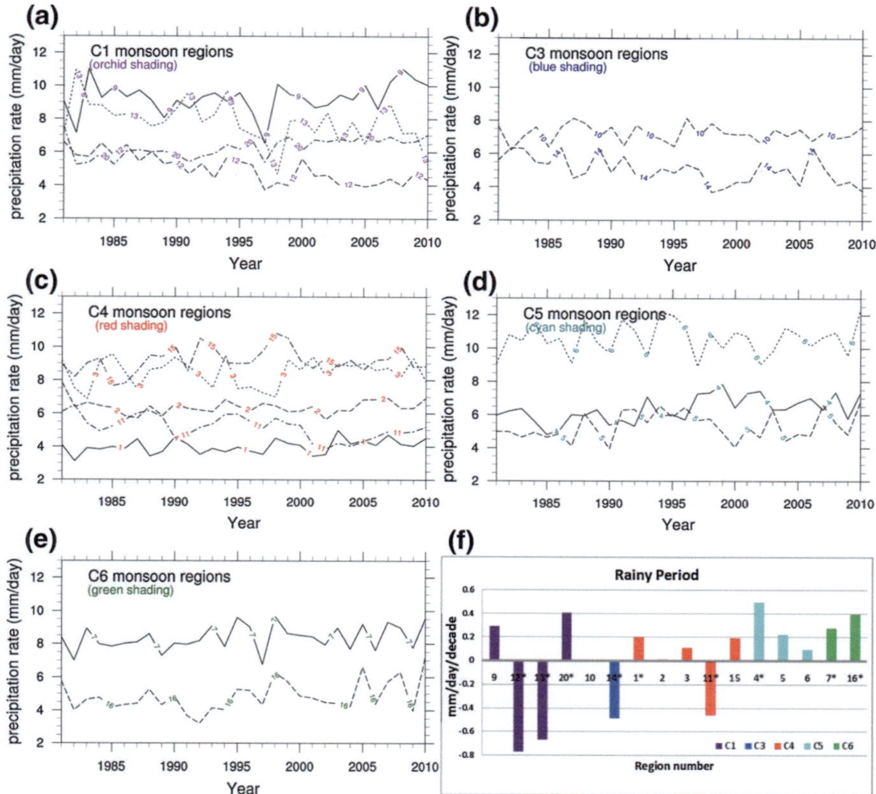

Fig. 4.19 The mean series of regional-averaged precipitation rate (mm day^{-1}) during rainy seasons from 1981 to 2010 in the regions of the five categories from **a** to **e**. In **f**, the regional trend values (mm/day/decade) of the five categories are depicted, and the symbol star marks the regional precipitation trends reaching the 90% confidence level. Sourced from Jiang et al. (2016) © Springer-Verlag Berlin Heidelberg 2016, with permission

4.7 Outer Edge Active Zones of the Global Monsoon

Climate worldwide can be roughly divided into two regions, i.e., monsoon and non-monsoon regions, based on whether the region is influenced by monsoon flow and monsoon rainfall. The low-level monsoon flow is seasonally active in the tropical and some subtropical regions. In mid-high latitudes, low-level flow is prevailing by the westerly climatologically. There are two circular zones, respectively, in the two hemispheres, which are interactive zones between the westerly flow and the monsoon flow yearly and decadally. They are outer edge active zones of the GM.

Monsoons occur over the tropics of southern and eastern Asia and northern Australia, parts of western and central Africa, Mexico, the southwest USA, and parts of South America and South Africa (IPCC 2007). The climatic GM domain has been identified utilizing two criteria: difference of local maximum and

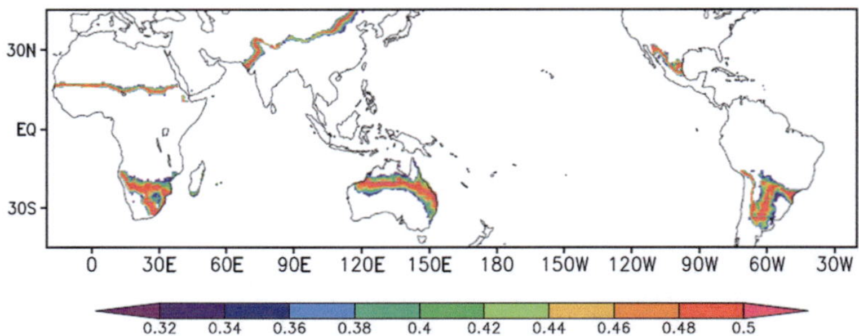

Fig. 4.20 Outer edge active zones of the GM calculated by CRU TS 3.1 precipitation with the spatial 0.5° × 0.5° latitude–longitude grid in the period of 1901–2008

minimum pentad mean precipitation rates exceeding 4 mm day^{-1}, and wind reversal of low-level cross-equatorial flow. Regional monsoon intensities show interannual and interdecadal variability. On the other hand, regional monsoon extents (or the GM domain) should also have interannual and interdecadal variability due to the interaction between the low-level cross-equatorial flow and westerlies near the out boundary of the GM domain. It means that in some years and decades the monsoon precipitation can reach higher latitude when the low-level cross-equatorial flow is stronger and the westerly is weaker.

Using the CRU TS 3.1 precipitation data with the spatial 0.5° × 0.5° latitude–longitude grid for 1901–2008, the GM swing zone is identified in Fig. 4.20. It is clear that the swing zone is an outer edge active zone of the GM. To determine the outer edge active zone, we check whether the precipitation of each grid is related to monsoon precipitation. For each year, we set the site (grid) as 1 if the monsoon precipitation arrives and otherwise we set the site as 0 if the monsoon precipitation cannot reach. Using the definition of Wang and Ding (2006), most years are signed as 1. In the wide area under the low-level westerly, the yearly precipitation rate is low, so many grids are signed as 0. Then, the active zone is determined by a large SD (>0.3) of monsoon precipitation, represented by 1 and 0. The width of active zone is about 10–20 latitudes in the southern continents over South Africa, northern Australia, and the subtropical South America. In the NH, a narrow zonally oriented zone is along the subtropical North Africa. In the Central America, a narrow southeast–northwest zone is located in eastern Mexico. In Asia, an outer edge active zone of summer monsoon is detected from northern India to Northwest China and North China.

We further examine the relationship between the outer edge active zones of the GM and the regions of semi-arid climates. Figure 4.21 shows the regions with semi-arid climates classified as "hot semi-arid" and "cold semi-arid" . To delineate "hot semi-arid climates" from "cold semi-arid climates" , there are three widely used isotherms: either an annual mean temperature of 18 °C, or a mean temperature of 0 °C or −3 °C in the coldest month.

Hot semi-arid climates tend to be located in the tropics and subtropics. These climates usually have hot, sometimes extremely hot, summers and mild to warm

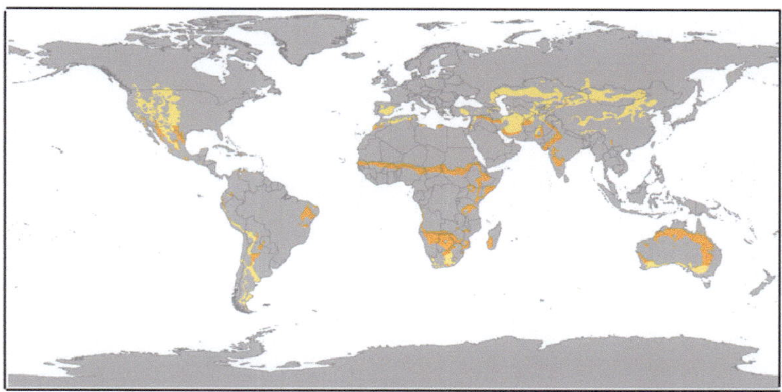

Fig. 4.21 Regions with semi-arid climates classified as "hot semi-arid" (*orange*) and "cold semi-arid" (*yellow*). Sourced from https://en.wikipedia.org/wiki/Semi-arid_climate

winters. Hot semi-arid climates are most commonly found around the fringes of subtropical deserts. The most common variant of a hot semi-arid climate, found in regions such as West Africa, India, parts of Mexico, bordering areas in Texas, parts of Southern California, and small parts of Pakistan, experiences the seasonal effects of monsoons. They have a short but well-defined wet season, but are not sufficiently wet overall to qualify as a tropical savanna climate. In Australia, a large portion of the outback surrounding the central deserts lies within the hot semi-arid climate regime. Hot semi-arid climates can also be found in sections of South America such as the sertão and on the poleward side of the arid deserts.

Cold semi-arid climates tend to be located in temperate zones. They are typically found in continental interiors far from large bodies of water. Cold semi-arid climates usually feature hot and dry summers. Unlike hot semi-arid climates, areas with cold semi-arid climates tend to have cold–dry winters with snowfall and hot-–wet summers. Areas featuring cold semi-arid climates tend to have higher elevations than areas with hot semi-arid climates, and are sometimes subject to major temperature swings between day and night. Cold semi-arid climates are most commonly found in Asia and North America. However, they can also be found in Northern Africa, South Africa, Europe (primarily in Spain), sections of South America, and sections of interior southern Australia.

In comparison between Figs. 4.20 and 4.21, it is noted that the above outer edge active zones of GM are basically consistent with the regions of semi-arid climates classified as "hot semi-arid" in Africa and Australia. The largest difference is in China, where the monsoon outer edge active zone is a cold semi-arid climate.

Annual precipitation anomalies over outer edge active zones from 1950 to 2008 in the NH and the SH are illustrated in Fig. 4.22. The semi-arid land is characterized by little annual precipitation, usually from 200 to 400 mm. The maximum anomaly of annual precipitation (Fig. 4.22a) is about 70 mm in the NH and about 190 mm in the SH (Fig. 4.22b). The interannual and interdecadal variability of annual precipitation is significant in the two hemispheres.

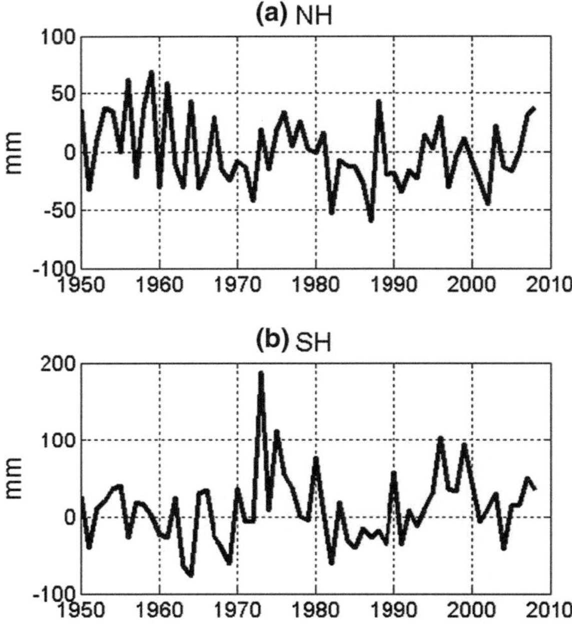

Fig. 4.22 Annual precipitation anomalies (mm) over outer edge active zones from 1950 to 2008 in **a** the NH and **b** the SH

4.8 East Asian Monsoon Outer Edge Active Zone

In East Asia, the monsoon outer edge active zone is mainly located in northwestern China (Fig. 4.23). Annual precipitation over the outer edge active zone of East Asian summer monsoon since 1960 is comparable from the CRU grid data and observed 560-station data in Fig. 4.24. Their average values of annual precipitation are about 200–400 mm. One-third of years do not reach the monsoon precipitation criterion along the monsoon outer edge active zone.

The East Asian monsoon, or the East Asian subtropical monsoon, is mainly located along the Yangtze River from China to Japan. The unique topography, particularly the Tibetan Plateau, has a strong influence on the intensity and location of the subtropical monsoon activity (Yu et al. 2004; Wu et al. 2007; Tang et al. 2007). Relevant to the Tibetan Plateau topography, there are two air flows: the westerly to its north and the southerly in eastern China. The two branches of air flows converge on the lee side of the Tibetan Plateau. That forms a subtropical frontal zone from East China to Japan, which moves gradually from South China to the Yangtze River, and to North China from May to late July. The features of the frontal rainfall in the subtropical monsoon region are different from those of the tropical monsoon rainfall. For the tropical monsoon, the two significant features are the monsoon onset date and the monsoon intensity, whereas the subtropical monsoon has an additional feature: change in location. This is especially important, as in some years the monsoon rainfall cannot reach North China due to weak monsoon

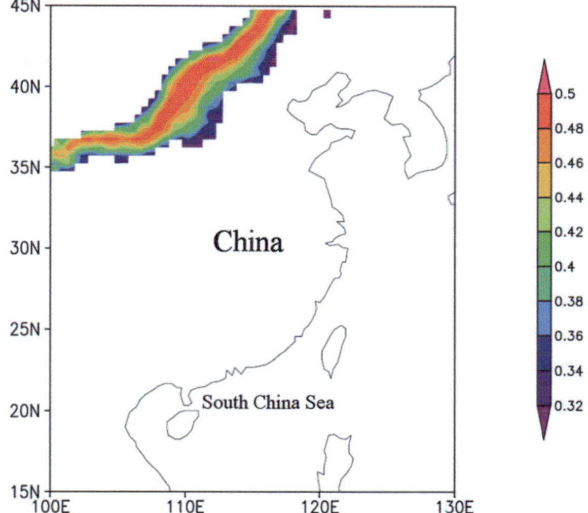

Fig. 4.23 Outer edge active zone of East Asian summer monsoon. *Shading* indicates the SD of monsoon precipitation larger than 0.3

flow. These location changes can lead to severe floods or droughts in some years and can even produce decadal climate events or interdecadal dry–wet regime alternations.

In North China, the northernmost location of the subtropical summer monsoon changes from year to year and from decade to decade. These changes form a summer monsoon northernmost outer edge or marginal active zone (Hu and Qian 2007; Qian et al. 2007), which are also the crop–pasture crossing zone, biological vulnerable zone, and climate sensitive zone (Fu 2003). Weather extremes and climate anomalies tend to occur along the active zone (Shi et al. 1994), so the study on its variability is the most important for the sustainable economic and life-supporting developments in North China.

The climatological northernmost location of the subtropical summer monsoon has been studied by many meteorological scientists in China. There are three types of definition to describe the climatological northernmost location of the summer monsoon in eastern China. One is based on the nature of the air mass, including both temperature and moisture. The other two are based on the monsoon rainfall and the lower tropospheric southwesterlies over East Asia. Tu and Huang (1944) constructed a combined variable, regarding the fact that the northernmost location of the summer monsoon in eastern China is a crossing section where a warm–humid air mass and a cold–dry air mass meet. Shi and Chao (1982) used the pentad mean pseudo-equivalent potential temperature at 850 hPa from June to September with the isotherm θ_{se} = 332 K as the critical value and found that the climatological northernmost latitude of the summer monsoon in North China is about 40°N. Wang et al. (1999) has proposed a definition of the summer monsoon including both the intensity of the southwesterly and the extent of both temperature and moisture.

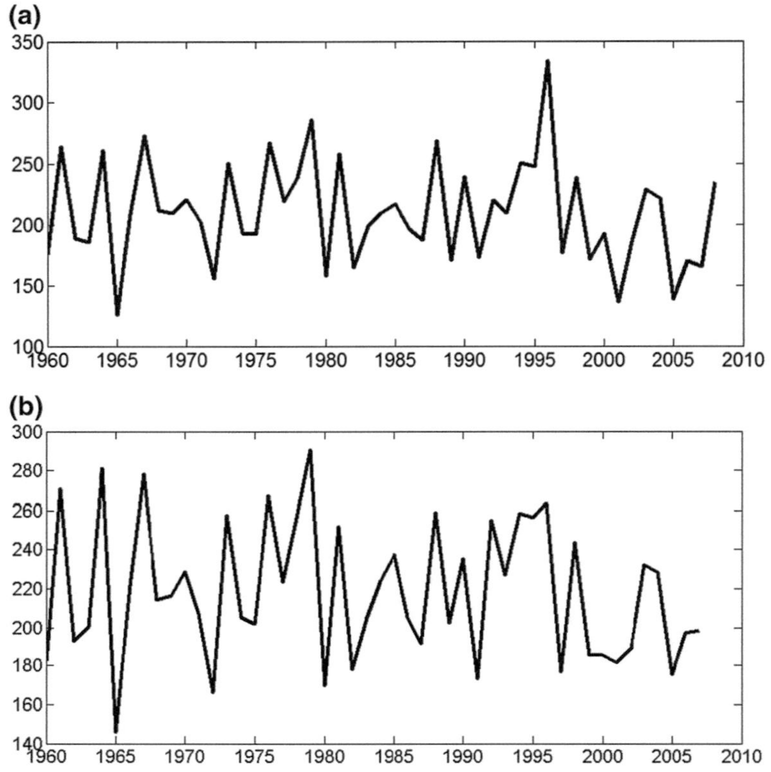

Fig. 4.24 Annual precipitation (mm) over the outer edge active zone of East Asian summer monsoon since 1960, calculated by **a** CRU data and **b** observed 560-station data in China

Based on the definition, Wu et al. (2005) found that there is interdecadal variation for the position of the northernmost location in eastern China during the last 50 years.

Monsoon precipitation is a joint result of circulation (wind) and moisture (vapor transportation) caused by temperature contrast. Gao (1962) identified that the climatological location of the monsoon northernmost edge is basically to the north of 40°N and its latitude gradually extends from 40°N in the west to 50°N at 120°E, regarding the largest pentad rainfall belt and the difference of the relative humidity between summer and winter.

In identification of the lengths of rainy season, we need to develop criteria to define the beginning and the end of the monsoon rainy season. Several precipitation criteria have been used in previous studies to define the start of the rainy season, including 4 mm day^{-1} (Qian and Lee 2000), 5 mm day^{-1} (Wang and Lin 2002), and 6 mm day^{-1} (Qian et al. 2002b). Figure 4.25 shows the evolutions of the climate isochrones of 4, 5, and 7 mm day^{-1} precipitation. Climatologically, precipitation of 4 or 5 mm day^{-1} initially starts from Southeast China in February and

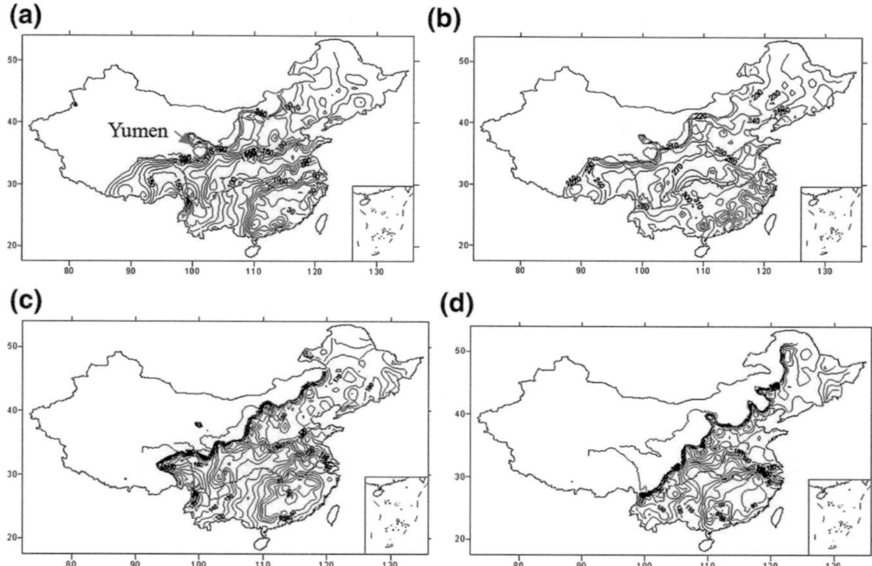

Fig. 4.25 Dates of the northward advance (or southward retreat) of the precipitation isochrones (as the calendar dates from the 1 January) averaged from 1960 to 2000 based on the daily precipitation data (7-day running mean is used for the precipitation) from 486 stations in China: **a** northward advance of 4 mm day^{-1} precipitation isochrones, **b** southward retreat of 4 mm day^{-1} precipitation isochrones, **c** northward advance of 5 mm day^{-1} precipitation isochrones, and **d** northward advance of 7 mm day^{-1} precipitation isochrones. Sourced from Qian et al. (2009)

then the isochrone moves northward in four steps along eastern China. From the calendar dates 30 to 60, the isochrones of 4 or 5 mm day^{-1} precipitation show spatial clustering basically along the lower Yangtze River for about one month (Fig. 4.25a, c). This exhibits relatively dry weather of less than 4 mm day^{-1} precipitation to the north of the isochrones and wet weather of more than 4 mm day^{-1} precipitation to the south. From the calendar dates 70 to 110 in March and April, the isochrones are spatially clustering along the Huaihe River (north of the lower Yangtze River). In May and early June from the calendar dates 120 to 160, the isochrones are clustering along the lower Yellow River. Finally, they arrive at the northernmost location from the central Tibetan Plateau, the upper Yellow River, and the north part of North China on the calendar dates 180–210.

During AD 688–742, Zhihuan Wang, a Chinese poet, wrote in his poetry "Liangzhou words": The Yellow River is far on the clouds with a lonely city surrounding by steep maintains. Shepherd's flute does not need to blame willows because the spring breeze never crosses the Yumen Pass. In his words, the spring breeze is the warm and moist southerly. It is clear in Fig. 4.25a that the Yumen Pass is located at the northwesternmost outer edge active zone of East Asian summer monsoon (Qian 2004).

The four steps of the northward advance of precipitation are only observed from the 4 and 5 mm day^{-1} precipitation isochrones, but only two steps are noted from the 7 mm day^{-1} precipitation isochrones. The climatological northernmost location of the summer monsoon can be identified from the last step of the precipitation advance for the 4 and 5 mm day^{-1} precipitation isochrones in July. This northernmost location of precipitation divides China into two parts: the summer monsoon region to its southeast and the non-monsoon region to its northwest.

The 4 and 5 mm day^{-1} precipitation isochrones remain along the northernmost location for about one month from the calendar dates 210 to 240 in August (Fig. 4.25a, b). The southward retreat of the 4 mm day^{-1} precipitation isochrones is faster than its northward advance particularly near the eastern coast. From the calendar dates 280 to 300 (in October), it retreats to the lower Yangtze River. The autumn precipitation can be noted in Southwest China. It is interesting that the zone between the last locations of the 4 and 7 mm day^{-1} precipitation covers the semi-arid region in northern and western China, in which the annual mean precipitation ranges from 200 to 600 mm (Qian and Lin 2005).

Depending on the strength of monsoon, the northernmost locations of various precipitation intensities vary from year to year, leading to large interannual dry–wet variations in the monsoon outer edge active region. As an example, the average precipitation from May to September over this region in 1964 was three times larger than that in 1965. The precipitation isoline of 7 mm day^{-1} reached the monsoon outer edge active region in 1964 but not in 1965 (Fig. 4.26). As a matter of fact, in northwestern China, 7 mm day^{-1} precipitation was observed at only a few stations. On the contrary, 7 mm day^{-1} precipitation arrives in southeastern China every year. Therefore, the monsoon outer edge active region is a transitional zone that separates the non-monsoon area to its northwest and the monsoon area to its southeast. Thus, it is considered as the northern fringe of East Asian summer monsoon. The 850-hPa flow is also shown in Fig. 4.26. It is noted that stronger southerly corresponds to larger precipitation in the monsoon outer edge region in Northwest China.

Stronger and weaker southerlies are critical to the annual harvest in China. In BD 23–22 centuries, Emperor Shun in ancient China had the "Southerly wind Song": The southerly in place brings my men's warming and the southerly in time brings my people's harvest. Strong and weak southerlies can be compared in 1964 and 1965, respectively (Fig. 4.26).

4.9 Millennium Monsoon Dry–Wet Modes

Monthly precipitation datasets of 160 stations in China, the monthly winds, and humidity data derived from the NCEP reanalysis 1 are used to construct the relationship between six summer dry–wet modes in eastern China and the northward advance of summer monsoon airflow in East Asia. A millennial series of the monsoon dry–wet mode index is reconstructed based on six summer dry–wet modes in eastern China since 950 AD. A high (low) index indicates that the strong

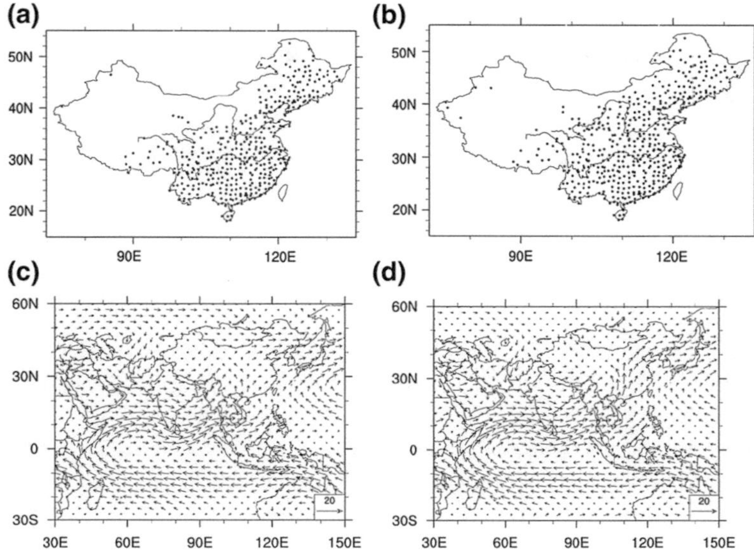

Fig. 4.26 Stations that recorded 7 mm day^{-1} precipitation from May to September in **a** 1965 and **b** 1964. *Each dot* represents one station. Wind at 850-hPa (m s^{-1}) from May to September in **c** 1965 and **d** 1964. Sourced from Qian et al. (2012)

(weak) East Asian summer monsoon airflow can reach North (South) China and cause above (below) normal precipitation. Interdecadal periodic variations, such as the approximate 70-year oscillation, can be found in the dry–wet index series. In the last millennium, the North China has experienced persistent decadal wet and dry periods. Currently, the dry–wet mode index is at a low level on the interdecadal timescale, so above-normal precipitation and below-normal precipitation are, respectively, observed in South China and North China.

The unique topography in China, particularly the Tibetan Plateau in the southwest, has a strong influence on the intensity and location of the subtropical monsoon. It is well known that the westerly dry–cool flow to the north of the Tibetan Plateau and the southerly warm–humid monsoon flow over eastern China converge in the leeward side of the Tibetan Plateau and form a subtropical frontal zone and rainy belt extending from eastern China to Japan. Climatologically, this frontal zone gradually moves from South China to the Yangtze River valley, and even to North China during May and late July. However, the advance of the rainy belt and regional precipitation differs from year to year. Thus, the precipitation departure (or dry–wet grades) can be used to compare the differences of interregional rainfall. Five dry–wet grades (Table 4.2) based on station's climatic precipitation and its SD from May to September in the period 1951–2008 are calculated using the monthly 160-station precipitation dataset in China. Details of the calculation have been described by Qian et al. (2003b). Lower values of the dry–wet grades indicate greater amounts of precipitation while higher values indicate less precipitation.

Table 4.2 Five dry–wet grades and precipitation departures

Dry–wet grades	Formula (range of R_i)
2 (flood)	$R_i > (R + 1.17\sigma)$
1 (wet)	$(R + 0.33\sigma) < R_i \leq (R + 1.17\sigma)$
0 (normal)	$(R - 0.33\sigma) < R_i \leq (R + 0.33\sigma)$
−1 (dry)	$(R - 1.17\sigma) < R_i \leq (R - 0.33\sigma)$
−2 (drought)	$R_i \leq (R - 1.17\sigma)$

R_i is the total summer (May–September) rainfall in the specific year, R and σ are the climatic average of the total summer rainfall and its standard deviation in the period of 1951–1990

Climatic departures of precipitation percentage in each summer (May–September) of 1951–2012 are calculated and the dry–wet distributions of 62 typical summers are grouped into the six dry–wet modes (Table 4.3). Typical summers with above-normal precipitation in eastern China occurred only in 1954 and 1998. The spatial distributions of six typical dry–wet modes are shown in Fig. 4.27. Mode 1 is a dry summer with below-normal precipitation in eastern China such as in 1972. Mode 2 shows the above-normal rainfall in the south of the Yangtze River and below-normal rainfall in its north, for example in 1999. Above-normal rainfall along the Yangtze River and below-normal rainfall in North and South China depict mode 3 in 1980. Above-normal rainfall in eastern China, as in 1973, is referred to as mode 4. Mode 4 and mode 1 show an opposite spatial distribution of summer precipitation. Mode 5, such as in 1963, shows a below-normal rainfall in the south of the Yangtze River and above-normal rainfall in its north. Mode 5 has an opposite distribution to the mode 2. Mode 6, such as in 1959, shows below-normal rainfall

Table 4.3 Six dry–wet modes (index) and corresponding years from 1951 to 2012

Index	Precipitation distribution	Year
1	Below-normal rainfall in Eastern China	1951, 1957, 1960, 1965, 1966, 1972, 1974, 1981, 1986, 1992, 2001, 2008, 2009
2	Above-normal rainfall in the south and below-normal rainfall in the north of the Yangtze River	1952, 1962, 1968, 1970, 1975, 1993, 1994, 1997, 1999, 2002, 2006, 2010
3	Above-normal rainfall along the Yangtze River but below-normal rainfall in North and South China	1969, 1979, 1980, 1982, 1983, 1987, 1989, 1991, 2000, 2005, 2007
4	Above-normal rainfall in Eastern China	1954, 1955, 1973, 1984, 1996, 1998
5	Below-normal rainfall in the south and above-normal rainfall in the north of the Yangtze River	1956, 1958, 1963, 1964, 1967, 1971, 1978, 1985, 1988, 1990, 2003, 2004, 2011
6	Below-normal rainfall along the Yangtze River but above-normal rainfall in North and South China	1953, 1959, 1961, 1976, 1977, 1995, 2012

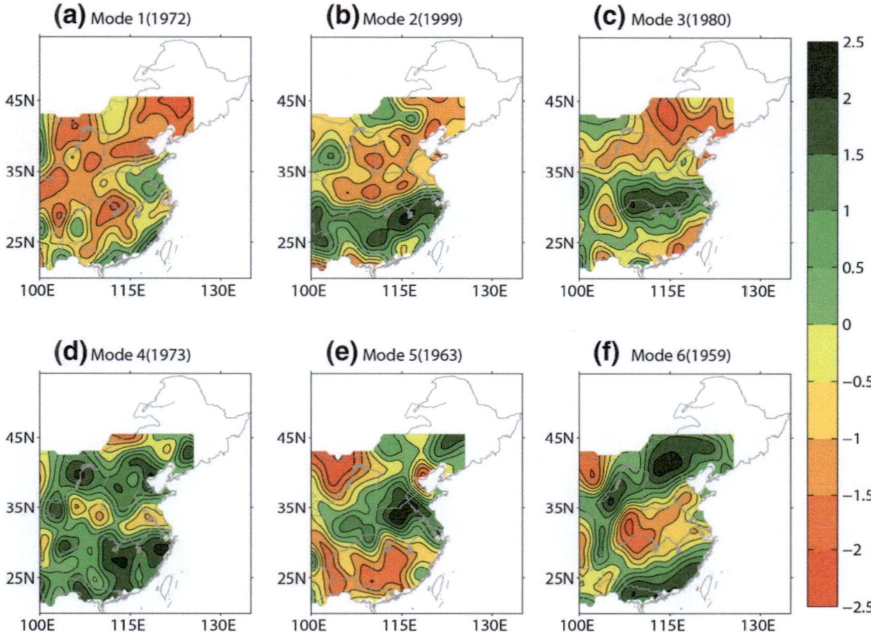

Fig. 4.27 Six dry–wet modes based on the summer (May–September) precipitation with five dry–wet grades in Table 4.2 and six typical dry–wet years in **a** 1972 for mode 1, **b** 1999 for mode 2, **c** 1980 for mode 3, **d** 1973 for mode 4, **e** 1963 for mode 5, and **f** 1959 for mode 6

along the Yangtze River but above-normal rainfall in North and South China and has an opposite pattern to mode 3. The six modes basically indicate the northward advance of above-normal precipitation.

Figure 4.28 shows the lower tropospheric moisture transportation. It illustrates that (1) the dry mode 1 in eastern China corresponds to below-normal moisture transportation over China; (2) more precipitation in the south of Yangtze River corresponds to more moisture transportation reaching the southern coast; (3) more precipitation over the Yangtze River corresponds to more moisture transportation reaching the river; (4) the wet mode 4 in eastern China corresponds to above-normal moisture transportation over the China land; (5) more precipitation over North China corresponds to more moisture transportation reaching North China; and (6) below-normal precipitation over the Yangtze River and above-normal precipitation in North China and South China corresponds to more moisture transportation reaching North China and South China than that over the Yangtze River. The relationships between the data presented in Figs. 4.27 and 4.28 become clear by analyzing these dry–wet modes and the lower tropospheric moisture transportation in eastern China. The moisture transportation in eastern China mainly comes from the BOB and the SCS. The monsoon forms by the thermal contrast between the nearby sea and land. Moisture transportation shown in Fig. 4.28 indicates that the thermal contrast is not only from the Pacific and Asian

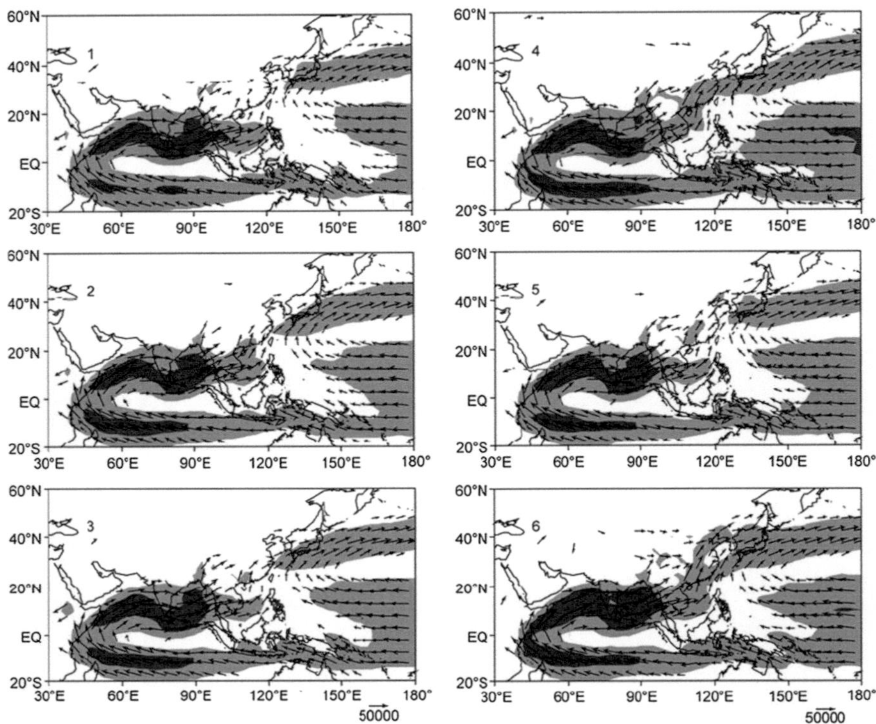

Fig. 4.28 Composites of lower tropospheric water vapor transportation integrated from the surface to 500 hPa based on six sets of typical dry–wet years. *Vectors* indicate the direction of moisture transports, *heavy and light shadings* indicate moisture transportation greater than 60,000 and 30,000 g m^{-1} s^{-1}, respectively. Sourced from Qian et al. (2011)

land, but also from the peninsular sea–land thermal contrasts, such as those in the Indian Peninsula and the BOB along with the ICP and the SCS.

After the six modes have been determined, the series of summer dry–wet mode indexes (DWM index) for 1951–2012 are shown in Fig. 4.29. The decadal variation of DWM index or monsoon intensity can be briefly separated into three periods including 1953–1967, 1979–1991, and 1993–2002, respectively, denoted by the three hollow arrows in Fig. 4.29. During 1953–1967, 1979–1991, and 1993–2002, the index mostly, respectively, reached 4–6, 3–4, and 1–2.

Figure 4.30 plots the spatial distributions of the summer (May–September) rainfall percentage (%) with respect to its climatic average from 1951 to 2012 in eastern China during three periods of 1953–1967, 1979–1991, and 1993–2002, respectively. It shows that a wet domain and a dry domain were, respectively, located in North and South China during 1953–1967. The wet domain then gradually shifted southward along the Yangtze River valley during 1979–1991, and then in South China during 1993–2002.

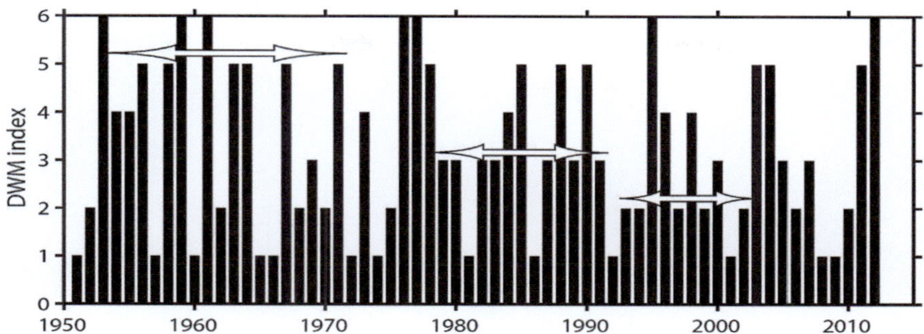

Fig. 4.29 Series of summer dry–wet mode index for 1951–2012. *Three hollow arrows* denote periods of 1953–1967, 1979–1991, and 1993–2002, respectively

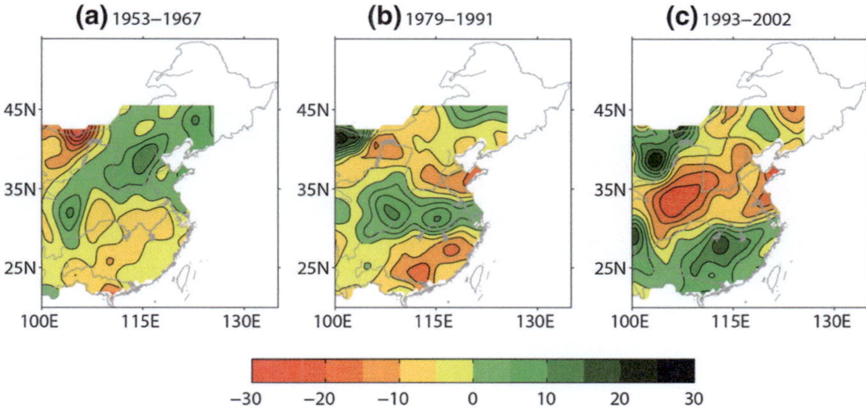

Fig. 4.30 Composites of summer (May–September) rainfall percentage (%) with respect to its climatic average from 1951 to 2012 in eastern China during three periods of **a** 1953–1967, **b** 1979–1991, and **c** 1993–2002, respectively

We further examine the connection between the decadal dry–wet distribution and summer water vapor transportation of the three periods. It is clear that the strongest water vapor transportation to North China occurred in the period of 1953–1967 (Fig. 4.31a). The northernmost location of water vapor transportation is indicated by the dashed line which separates the cross-equatorial monsoon flow from the westerly. From the period of 1953–1967 (Fig. 4.31a) to the periods of 1979–1991 (Fig. 4.31b) and 1993–2002 (Fig. 4.31c), the southwesterly weakened and moved southward gradually. The difference of summer water vapor transportation between the two periods of 1953–1967 and 1993–2002 is given in Fig. 4.32. There is an area of decreased water vapor transportation located from Southwest China to Northeast China.

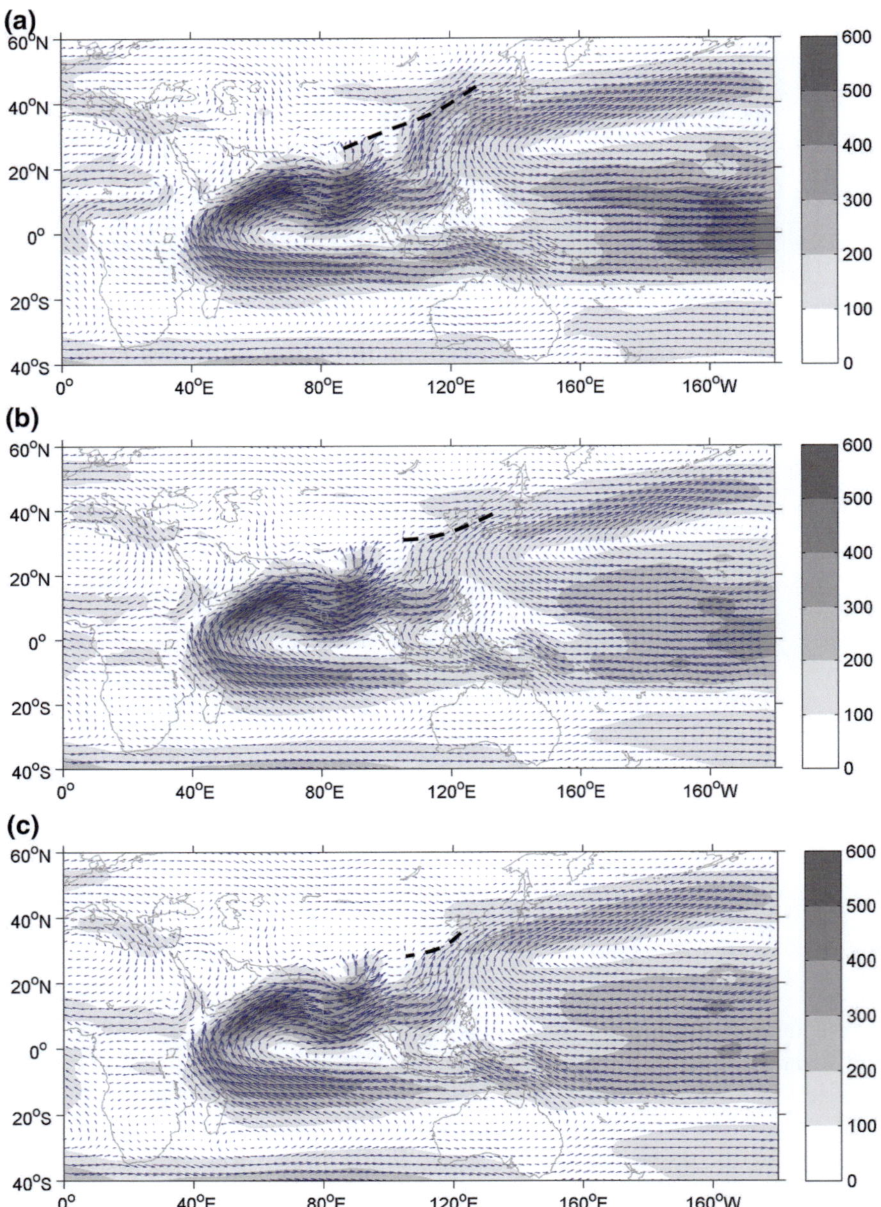

Fig. 4.31 Summer (JJA) water vapor transportation (*shading*, 100 kg m^{-1} s^{-1} interval) averaged from the surface to 300 hPa in the three periods of **a** 1953–1967, **b** 1979–1991, and **c** 1993–2002, respectively. The *dashed line* indicates the northernmost boundary of the cross-equatorial monsoon flow

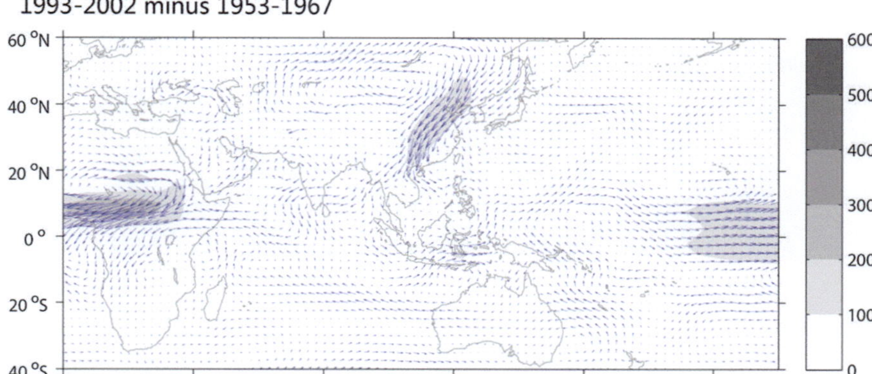

Fig. 4.32 Difference of summer (JJA) water vapor transportation (kg m^{-1} s^{-1}) between 1953–1967 and 1993–2002, namely by 1993–2002 minus 1953–1967

In a comparative analysis, Wang (1988) examined the empirical orthogonal function (EOF) derived from annual records of dryness/wetness in China for 500 years since 1470, and from instrumental precipitation collected during recent decades. His EOF results confirmed that the two datasets are nearly identical in spatial variations, which indicates that the conversion from the instrumentation data to the wet–dry intensity scale is reliable. While examining the data consistency, Wang et al. (1987) explored additional historical sources of dry–wet data and extended the historical rainfall records from 1470 back to 950 AD. Wang extended his series to the last 1000 years with six dry–wet anomalous modes that are similar to those shown in Fig. 4.27 in eastern China. Using Wang's annual dry–wet anomalous modes and the instrumental precipitation data, Qian et al. (2003b) updated these modes to 1999 and extracted five dry–wet grades in the three regions over North China, the Yangtze River, and South China. Quasi-periodic variations of dry–wet series for each individual region can be easily identified. However, it is difficult to determine the spatial structure of the dry–wet distribution and its evolution. Qian et al. (2011) updated Wang's dry–wet modes from 950 to 2008 and transformed each annual mode into one of six numbers based on Fig. 4.27. Figure 4.33a plots the East Asian summer DWM index series from 950 to 2008 with modes indicated by numbers 1–6 in Fig. 4.27. The standardized departures for the last millennium with 7-point and 21-point running averages are shown in Fig. 4.33b.

In the last millennium, the weakest summer monsoon was found around 1200 AD during an arid period and Song–Yuan dynastic alteration. In the late 1920s, severe dry climate events leading to locust swarms caused widespread crop failure and persistent famine in North China. In the last 500 years, summer monsoon flows that reached Northwest China had interannual and interdecadal variability, which resulted in drought and flood in the northwestern region over the upper Yellow River valley (Qian et al. 2003a). In the 1990s, the strength of the East Asian summer

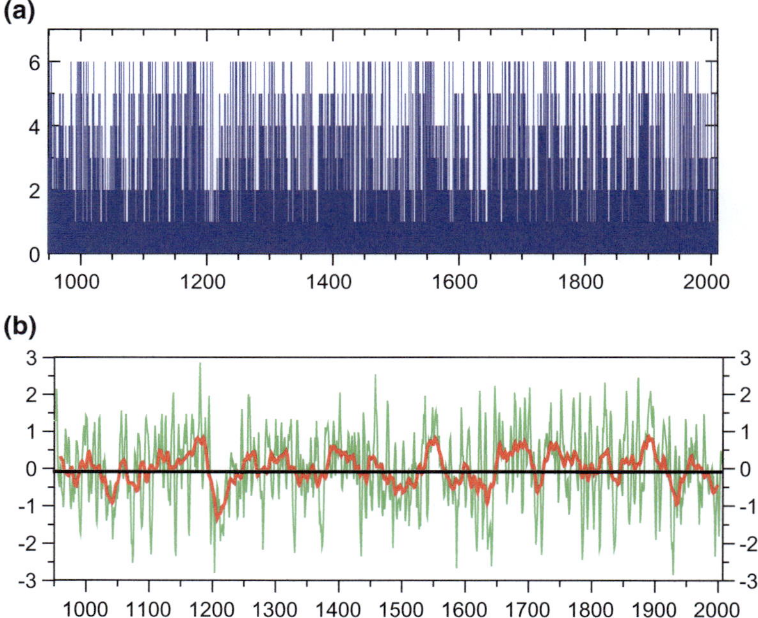

Fig. 4.33 Summer monsoon dry–wet index series in East Asia. **a** Original DWM index series from AD 950 to 2008. **b** The standardized departure of DWM index with 7-point and 21-point running averages. Redrawn from Qian et al. (2011)

monsoon decreased and the dry climate resulted in long-term low runoff in the Yellow River. During the last 200 years, the East Asian summer monsoon had been the strongest and could arrive at the northern location in the later period of the nineteenth century. Wavelet transform analysis showed that there were several periodic oscillations with timescales of about 23.3 and 67 years along a century time scale (figures not shown). An oscillation with a 10.1-year timescale reached the 0.05 significance level. The last weakening of summer monsoon was mainly influenced by the interdecadal (67 years) oscillation caused by the changes of large-scale sea–land thermal contrast. In the summer monsoon area of eastern China, regional dry–wet variation has a quasi-70-year oscillation (Qian et al. 2003a, b).

Questions

1. In the following three monsoon indices, the thermal contrast change is a thermodynamical index, the reversal of wind direction is a dynamical index, and the dry–wet alteration is a climatological index, which one of them is the result and which one is the cause?
2. Why should monsoon be defined as a region where its dry–wet alteration must be associated with prevailing wind reversal of low-level cross-equatorial flow?
3. In the mid-latitude of North Pacific, there is a dry–wet alteration region together with prevailing wind reversal. Why isn't it categorized as a monsoon region?

4. How many monsoon regions can be objectively divided from the GM domain? Are there any other methods to divide the GM domain?
5. What is the importance of studying the outer edge active zones of the GM?
6. A millennium series of global air temperature has been constructed to describe the global climate change. How can a millennium global precipitation series be constructed to describe the global climate change?

References

Adams DK, Comrie AC (1997) The North American monsoon. Bull Am Meteorol Soc 78(10):2197–2213

An Z, Wu G, Li J, Sun Y, Liu Y, Zhou W, Cai Y, Duan A, Li L, Mao J (2015) Global monsoon dynamics and climate change. Ann Rev Earth Planet Sci 43:29–77. doi:10.1146/annurev-earth-060313-054623

Balasubramaniyan R, Hüllermeier E, Weskamp N, Kmper J (2005) Clustering of gene expression data using a local shape-based similarity measure. Bioinform/Comput Appl Biosci 21: 1069–1077. doi:10.1093/bioinformatics/bti095

Barrett BS, Longoria MIE (2013) Variability of precipitation and temperature in Guanajuato, Mexico. Atmsfera 26(4):521–536

Benjamin S, Serge J (2003) The west African monsoon dynamics. Part II: the 'preonset' and 'onset' of the summer monsoon. J Clim 16:3407–3427

Chang CP, Ding YH, Lau GNC, et al. (2011) The Global Monsoon System: Research and Forecast (2nd). Singapore: World Scientific Publishing, pp 594

Chang CP, Wang Z, McBride J, Liu CH (2005) Annual cycle of Southeast Asia-Maritime continent rainfall and the asymmetric monsoon transition. J Clim 18(2):287–301. doi:10.1175/JCLI-3257.1

Chao WC, Chen B (2001) The origin of monsoons. J Atmos Sci 58(22):3497–3507

Chen L, Zhu Q, Luo H, He J, Dong M, Feng Z (1991) The East Asian Monsoon. Beijing Meteorological Press, Beijing (in Chinese)

Christiansen B (2007) Atmospheric circulation regimes: can cluster analysis provide the number? J Clim 20:2229–2250. doi:10.1175/JCLI4107.1

Draxler RR, Stunder B, Rolph G, Taylor A (1999) HYSPLIT4 users guide. NOAA Technical Memorandum ERL ARL 230:35

Fu CB (2003) Potential impacts of human-induced land cover change on East Asia monsoon. Global and Planetary Change, 37:219–229

Gao H, Xue F (2006) Seasonal variation of the cross-equatorial flows and their influences on the onset of South China Sea summer monsoon. Clim Environ Res 11(1):57–68

Gao YX (1962) Several Issues about East Asia Monsoon. Science Press, China

Gasse F, Van Campo E (1994) Abrupt post-glacial climate events in West Asia and North Africa monsoon domains. Earth Planet Sci Lett 126(4):435–456

Godfred-Spenning CR, Reason CJC (2002) Interannual variability of lower-tropospheric moisture transport during the Australian monsoon. Int J Climatol 22(5):509–532

Hartigan JA, Wong MA (1979) Algorithm AS 136: a *k*-means clustering algorithm. Appl Stat 28(1):100–108. doi:10.2307/2346830

Holland G (1995) Scale interaction in the western Pacific monsoon. Meteorol Atmos Phys 56(1–2):57–79

Hu H, Qian W (2007) Identifying the northernmost summer monsoon location in East Asia. Prog Nat Sci 17(7):812–820

IPCC (2007) Climate Change 2007. Working Group I Contribution to the Fourth Assessment Report of the IPCC Intergovernmental Panel on Climate Change. Cambridge University Press, New York

Jiang N, Qian WH, Leung JCH (2016) The global monsoon division combining the k-means clustering method and low-level cross-equatorial flow. Clim Dyn. doi:10.1007/s00382-015-2967-5

Johnson NC (2013) How many ENSO flavors can we distinguish? J Clim 26(13):4816–4827

Johnson NC, Feldstein SB (2010) The continuum of North Pacific sea level pressure patterns: intraseasonal, interannual, and interdecadal variability. J Clim 23:851–867. doi:10.1175/2009JCLI3099.1

Khromov SP (1957) Die geographische verbreitung der monsune. Petermanns Geogr Mitt 101:234–237

Kumar MRR, Shenoi SSC, Schluessel P (1999) On the role of the cross equatorial flow on summer monsoon rainfall over India using NCEP/NCAR Reanalysis Data. Meteorol Atmos Phys 70(3–4):201–213

Lau K, Yang S (1997) Climatology and interannual variability of the Southeast Asian summer monsoon. Adv Atmos Sci 14(2):141–162

Li C, Li S (2014) Interannual seesaw between the Somali and the Australian cross-equatorial flows and its connection to the East Asian summer monsoon. J Clim 27:3966–3981. doi:10.1175/JCLI-D-13-00288.1

Li J, Zeng Q (2002) A unified monsoon index. Geophys Res Lett 29(8):115-1–115-4

Li J, Zeng Q (2003) A new monsoon index and the geographical distribution of the global monsoons. Adv Atmos Sci 20(2):299–302

Li J, Zeng Q (2005) A new monsoon index, its interannual variability and relation with monsoon precipitation. Climate Environ Res 10(3):351–365

Lin R, Zhou T, Qian Y (2014) Evaluation of global monsoon precipitation changes based on five reanalysis datasets. J Clim 27(3):1271–1289

Lin YW, Chou C (2014) The role of the New Guinea cross-equatorial flow in the interannual variability of the western North Pacific summer monsoon. Environ Res Lett 9(4):044003. doi:10.1088/1748-9326/9/4/044003

Liu H, Ni Z (2009) Clustering method of time series based on EMD and *k*-means algorithm. Pattern Recognit Artif Intell 22(5):803–808

Mao J, Chan JCL (2005) Intraseasonal variability of the South China Sea Summer Monsoon. J Clim 18:2388–2402. doi:10.1175/ JCLI3395.1

McBride JL (1987) The Australian summer monsoon. In: Chang CP, Krishnamurty T (eds) Monsoon Meteorology. Oxford University Press, pp 203–231

Michelangeli PA, Vautard R, Legras B (1995) Weather regimes: recurrence and quasi stationarity. J Atmos Sci 52:1237–1256

Nguyen DQ, Renwick J, McGregor J (2014) Variations of monsoon rainfall: a simple unified index. Geophys Res Lett 41(2):575–581

Nicholson SE, Kim J (1997) The relationship of the El Niño-Southern Oscillation to African rainfall. Int J Climatol 17(2):117–135

Qian WH (2000) Dry/wet alternation and global monsoon. Geophys Res Lett 27(22):3679–3682

Qian WH (2004) Synoptic Meteorology. Peking University Press, pp 265

Qian WH, Chen DL, Zhu YF, Shen HY (2003a) Temporal and spatial variability of dryness/wetness in China during the last 530 years. Theor Appl Climatol 76:13–29

Qian WH, Deng Y, Zhu YF, Dong WJ (2002a) Demarcating the worldwide monsoon. Theor Appl Climatol 71(1–2):1–16

Qian WH, Hu Q, Zhu YF, Lee DK. (2003b) Centennial-scale dry-wet variations in East Asia. Clim Dyn 21:77–89

Qian WH, Jiang N (2015) The global monsoon definition using the difference of local minimum and maximum pentad precipitation rates associated with cross-equatorial flow reversal. Theor Appl Climatol 124(3):883–901. doi:10.1007/s00704-015-1448-4

Qian WH, Lee DK (2000) Seasonal march of Asian summer monsoon. Int J Climatol 20(11): 1371–1386

Qian WH, Tang SQ (2010) Identifying global monsoon troughs and global atmospheric centers of action on a pentad scale. Atmospheric and Oceanic Science Letters 3(1):1–6

Qian WH, Yang S (2000) Onset of the regional monsoon over Southeast Asia. Meteorol Atmos Phys 75(1–2):29–38

Qian WH, Ye Q, Zhu YF (1998a) Monsoonal oscillation revealed by the upper-troposphere water vapor band brightness temperature. Chinese Science Bulletin 43(17):1489–1494

Qian WH, Zhu YF, Xie A, Ye Q (1998b) Seasonal and interannual variations of upper tropospheric water vapor band brightness temperature over the global monsoon regions. Adv Atmos Sci 15:337–345

Qian WH, Kang HS, Lee DK (2002b) Distribution of seasonal rainfall in the East Asian monsoon region. Theor Appl Climatol 73(3–4):151–168

Qian WH, Lin X (2005) Regional trends in recent precipitation indices in China. Meteor Atmos Phys 90:193–207

Qian WH, Lin X, Zhu YF, Xu Y, Fu JL (2007) Climatic regime shift and decadal anomalous events in China. Clim Change 84(2):167–189

Qian WH, Ding T, Hu HR, Lin X, Qin AM (2009) An overview of dry-wet climate variability among monsoon-westerly regions and the monsoon northernmost marginal active zone in China. Adv Atmos Sci 26:630–641

Qian WH, Shan XL, Chen D, Zhu CW, Zhu YF (2012) Droughts near the northern fringe of the East Asian summer monsoon in China during 1470–2003. Clim Change 110:373–383

Qian WH, Zhu YF, Tang SQ (2011) Reconstructed index of summer monsoon dry-wet modes in East Asia for the last millennium. Chin Sci Bull 56:3019–3027

Ramage CS (1971) Monsoon meteorology (vol. 15 of international geophysics series). Academic Press.

Riddle E, Stoner M, Johnson N, L'Heureux M, Collins D, Feldstein S (2013) The impact of the MJO on clusters of wintertime circulation anomalies over the North American region. Clim Dyn 40(7–8):1749–1766

Salinger MJ, Mcgree S, Beucher F, Power SB, Delage F (2014) A new index for variations in the position of the South Pacific convergence zone 1910/11-2011/2012. Clim Dyn 43(3–4):1–12

Sankar-Rao M (1966) Equations for global monsoons and toroidal circulations in the σ-coordinate system. Pure and Applied Geophysics 65(1):196–215

Sankar-Rao M (1970) On global monsoons–further results. Tellus 22:648–654

Sankar-Rao M, Saltzman B (1969) On a steady state theory of global monsoons. Tellus 21(3): 308–330

Shi SW, Chao JM (1982) Summer monsoon over north of China. Conference Papers of Countrywide Academic Meeting about Tropical Summer Monsoon. Yunnan People Press, pp 72–85 (in Chinese)

Shi ZT, Zhang LY, Su GW (1994) Natural disasters and their formation causes on Chinese monsoon marginal belt. Journal of Catastrophology 9:59–64 (in Chinese)

Song C, Pei T (2012) Research progress in time series clustering methods based on characteristics. Prog Geogr 31(10):1307–1317

Tanaka M (1992) Intraseasonal oscillation and the onset and retreat dates of the summer monsoon over east, southeast Asia and the western Pacific region using GMS high cloud amount data. J Meteorol Soc Jpn 70(1B):613–629

Tang X, Sun GW, Qian WH (2007) Study on the northernmost marginal active zone of Asian summer monsoon. Meteorological Press, Beijing, pp 122

Tao S, Chen L (1987) A review of recent research on the East Asian summer monsoon in China. In: Monsoon Meteorology. Oxford University Press

Trenberth KE, Stepaniak DP, Caron JM (2000) The global monsoon as seen through the divergent atmospheric circulation. J Clim 13(22):3969–3993

Tu CW, Huang SS (1944) The advance and withdraw of East Asia summer monsoon. Acta Meteorologica Sinica, 18:1–20 (in Chinese)

Ueda H, Yasunari T, Kawamura R (1995) Abrupt seasonal change of large-scale convective activity over the western Pacific. J Meteorol Soc Jpn 73(4):795–809
Vincent DG (1994) The South Pacific Convergence Zone (SPCZ): a review. Mon Weather Rev 122(9):1949–1970. doi:10.1175/1520-0493
Voyeikov AI (1879) Regional monsoon climate over East Asia: Amur region, Baikal, Manchuria, Eastern Mongolia, China, Japan, etc. Proc Imp Russ Geogr Soc 15:321–410
Wang AY, Wu CS, Lin WS, Yang Y, Feng RQ, Gu ZM, Liang JJ (1999) The definition of the advance and retreat of the summer monsoon in China. Plateau Meteorology 18:400–408 (in Chinese)
Wang B (1994) Climatic regimes of tropical convection and rainfall. J Clim 7(7):1109–1118
Wang B, Ding Q (2006) Changes in global monsoon precipitation over the past 56 years. doi:10.1029/2005GL025347
Wang B, Ding Q (2008) Global monsoon: dominant mode of annual variation in the tropics. Dyn Atmos Oceans 44(3):165–183
Wang B, Fan Z (1999) Choice of South Asian summer monsoon indices. Bull Am Meteorol Soc 80(4):629–638
Wang B, Lin H (2002) Rainy season of the Asian-Pacific summer monsoon. J Clim 15(4):386–398
Wang B, Lin H, Zhang Y, Lu MM (2004) Definition of South China Sea monsoon onset and commencement of the East Asia summer monsoon. J Clim 17(4):699–710
Wang B, Liu J, Kim HJ, Webster PJ, Yim SY (2012) Recent change of the global monsoon precipitation (1979–2008). Clim Dyn 39(5):1123–1135
Wang H, Fu R (2002) Cross-equatorial flow and seasonal cycle of precipitation over South America. J Clim 15(13):1591–1608
Wang B, Wu Z, Li J, Liu J, Chang CP, Ding Y, Wu G (2008) How to measure the strength of the East Asian summer monsoon. J Clim 21(17):4449–4463
Wang H, Xue F (2003) The interannual variability of Somali jet and its influences on the inter-hemispheric water vapor transport and the East Asian summer rainfall. Chin J Geophys 46(1):11–20
Wang J, Li M (1982) Cross-equator flow from Australia and monsoon over China. Chin J Atmos Sci 6:1–10
Wang P, Wang B, Cheng H, Fasullo J, Guo Z, Kiefer T, Liu Z (2014) The global monsoon across time scales: is there coherent variability of regional monsoons? Clim Past Discuss 10: 2163–2291
Wang SW(1988) Wetness variation in China over the last 500 years. In: Zhang J, ed. The Reconstruction of Climate in China for Historical Times. Beijing: Science Press, 1988, pp 66–78
Wang SW, Zhao ZC, Chen ZH, et al (1987) Drought/flood variations for the last two thousand years in China and comparison with global climatic change. In: Ye D Z, Fu C B, Chao J P, et al., eds. The Climate of China and Global Climate. Springer Berlin Heidelberg New York: China Ocean Press, pp 20–29
Webster PJ (1981) Monsoons. Sci Am 245:108–118
Webster PJ (1987) The elementary monsoon. In: Fein JS, Stephens PL (eds). Monsoons John Wiley, pp 3–32
Webster PJ, Magana VO, Palmer T, Shukla J, Tomas R, Yanai Mu, Yasunari T (1998) Monsoons: Processes, predictability, and the prospects for prediction. J Geophys Res Oceans 103(C7): 14451–14510
Wiel KVD, Matthews AJ, Joshi MM, Stevens DP (2015) Why the South Pacific Convergence Zone is diagonal. Clim Dyn. doi:10.1007/s00382-015-2668-0
Wu CG, Liu HS, Xie A (2005) Interdecadal characteristics of the influence of northward shift and intensity of summer monsoon on precipitation over North China in summer. Plateau Meteorology 24:656–665 (in Chinese)
Wu GX and Coauthors (2007) The influence of mechanical and thermal forcing by the Tibetan Plateau on Asian climate. Journal of Hydrometeorology 8(4):770–789

Xie L (2013) Vertical structure of summer Somali cross-equatorial flow and its relationship with South Asian monsoon. J Meteorol Sci 33(1):37–42

Xue F, Zeng Q, Huang R, Li C, Lu R, Zhou T (2015) Recent advances in monsoon studies in China. Adv Atmos Sci 32(2):206–229

Yao F, Zhao X (2015) Changes in spatiotemporal pattern of precipitation over China during 1980–2012. Environ Earth Sci 73(4):1649–1662

Yu RC, Wang B, Zhou TJ (2004) Climate effects of the deep continental stratus clouds generated by the Tibetan Plateau. J Clim 17(13):2702–2713

Zeng Q, Li J (2002) Interactions between the Northern and Southern Hemispheric atmospheres and the essence of monsoon. Chin J Atmos Sci 26(4):433–448

Zeng X, Lu E (2004) Globally unified monsoon onset and retreat indexes. J Clim 17(11): 2241–2248

Zhang S, Wang B (2008) Global summer monsoon rainy seasons. Int J Climatol 28(12): 1563–1578

Zhao P, Jiang P, Zhou X, Zhu C (2009) Modeling impacts of East Asian ocean-land thermal contrast on spring southwesterly winds and rainfall in eastern China. Chin Sci Bull 54(24):4733–4741

Zhou J, Lau K (1998) Does a monsoon climate exist over South America? J Clim 11(5): 1020–1040

Zhou T, Zhang L, Li H (2008) Changes in global land monsoon area and total rainfall accumulation over the last half century. Geophys Res Lett 35(16):134–143

Zhou W, Chan JCL (2007) ENSO and the South China Sea summer monsoon onset. Int J Climatol 27(2):157–167

Zhu C, Zhou X, Zhao P, Chen L, He J (2011) Onset of East Asian subtropical summer monsoon and rainy season in China. Sci China Earth Sci 54(12):1845–1853

Chapter 5
Waves and Vortices

The atmosphere, as an important rotating fluid system in the Earth, is often described by two terms: wave and vortex. An example of decomposing atmospheric variables into climatic component and anomaly is demonstrated in this chapter. We will apply the decomposition method on geopotentional height, temperature, and wind obtained from the European Centre for Medium-Range Weather Forecasts (ECMWF) Interim Reanalysis (ERA-Interim) data at 0000 UTC 2 July 1998. We first depict the climatic component and anomaly of these variables in the lower stratosphere (or the tropopause), the upper troposphere, and the lower troposphere, respectively. Then, we show the vertical structures to indicate surface weather extremes. Traditional synoptic chart and anomalous synoptic chart are compared to show that the latter can better locate surface weather extremes vertically. The strongest signal that indicates surface weather extremes is the center of height anomalies at the upper troposphere.

5.1 Lower Stratosphere Decomposition

The tropopause is near the level of 100 hPa, below the lower stratosphere. Figure 5.1 plots the total wind vectors $\vec{V}_{\{d,y\}}(\lambda, \varphi, t)$ as well as the total height $H_{\{d,y\}}(\lambda, \varphi, t)$ and the total temperature $T_{\{d,y\}}(\lambda, \varphi, t)$ at 100 hPa on 2 July 1998. There are westerly long waves (dashed yellow line) in the middle and high latitudes along the zonal circle in each hemisphere. In boreal summer, the westerly long waves have larger amplitudes in the Northern Hemisphere (NH) than that in the Southern Hemisphere (SH). It shows that the westerly velocity in the SH is stronger than that in the NH. Several closed cyclones (L/l) and anticyclones (H/h) can be observed. The largest anticyclonic circulation (H1) is located in South Asia. Another anticyclonic circulation (H2) and a cyclonic circulation (l1) are located in the subtropical North America and the middle Pacific. An anticyclonic circulation (h) and a

© Springer Nature Singapore Pte Ltd. 2017
W. Qian, *Temporal Climatology and Anomalous Weather Analysis*,
Springer Atmospheric Sciences, DOI 10.1007/978-981-10-3641-5_5

cyclonic circulation (l) are respectively observed in the North Atlantic sector. The three letters (v1, v2, and v3) in the subtropical NW Pacific indicate three vortices. Our goal of decomposition is to discover the causes of these circulation systems in nature. The anticyclone (H1) is predominant indicated by the intense high center in South Asia, namely the South Asian High (SAH) in Fig. 5.1b. Other anticyclones and cyclones are not apparent in the total height field at 100 hPa. Among the three subtropical vortices, only the low center (v2) is visible. At the tropopause, temperature is lower in the lower latitudes and along the SH polar zone but higher in the mid-high latitudes of NH and the mid latitudes of SH.

The climatic wind at 100 hPa shows two anticyclones (H1 and H2) centered in South Asia and North America (Fig. 5.2a). Strong westerly is prevailing in the SH without closed circulation systems. In the NH middle latitudes, there is a climatic westerly waves with several troughs. These well-known westerly waves or troughs move from day-to-day with a seasonal cycle. The climatic temperature shows warm zones in high latitudes of the NH and middle latitudes of the SH but cold zones in lower latitudes and high latitudes of the SH (Fig. 5.2b). The climatic height and wind satisfy the geostrophic wind relation. In the central North Pacific, there is a well-known middle-oceanic trough connected with the cyclone L1. The SAH (H1) and the subtropical North American High (H2) climatologically exist at 100 hPa in boreal summer.

Using the total wind minus the climatic wind, we obtain wind anomalies in Fig. 5.3a. It is interesting to find that the anticyclone (h3), cyclone (l1), and the three subtropical vortices (v1, v2, and v3) can be comparable between total- and anomaly-based winds. It is noted that the two anticyclones (H1 and H2) existed in Figs. 5.1a and 5.2a cannot be found in Fig. 5.3a. Many anticyclones and cyclones in Fig. 5.3a are not clear in Fig. 5.1a. For example, the climatic anticyclone (H1) in Fig. 5.2a and the anomalous anticyclone (hx, Fig. 5.3a) over the Mediterranean Sea commonly exists within the SAH indicated by the total wind (Fig. 5.1a). In other words, those anomalous anticyclones and cyclones become clear after removing their climatic components from the total wind. The westerly in the SH is mostly straight (Fig. 5.2a). After removing this climatic flow, vortices become more apparent in the SH (Fig. 5.3a).

These anomalous circulation systems in wind anomalies are also observed in height and temperature anomalies (Fig. 5.3b). Usually, positive (negative) centers of height anomalies are accompanied with cold (warm) centers of temperature anomalies at 100 hPa. The two basic features are found from the above analyses. The westerly with several troughs is climatologically prevailing in the mid-high latitudes except there are two climatic subtropical anticyclone systems (H1 and H2) and cyclone (L1) caused by the continent-ocean contrast in the climatic synoptic chart (Fig. 5.2). Differently, the anomalous vortices including anomalous anticyclones and cyclones are detected in the anomalous synoptic chart (Fig. 5.3). Initially, any one of numerical models is expected to well predict the wind, temperature, and height as shown in Fig. 5.1. Actually, we only expect to have a correct prediction of wind, temperature, height anomalies in Fig. 5.3, rather than the climatic westerly with ridges or troughs of the climatic temperature and height (Fig. 5.2).

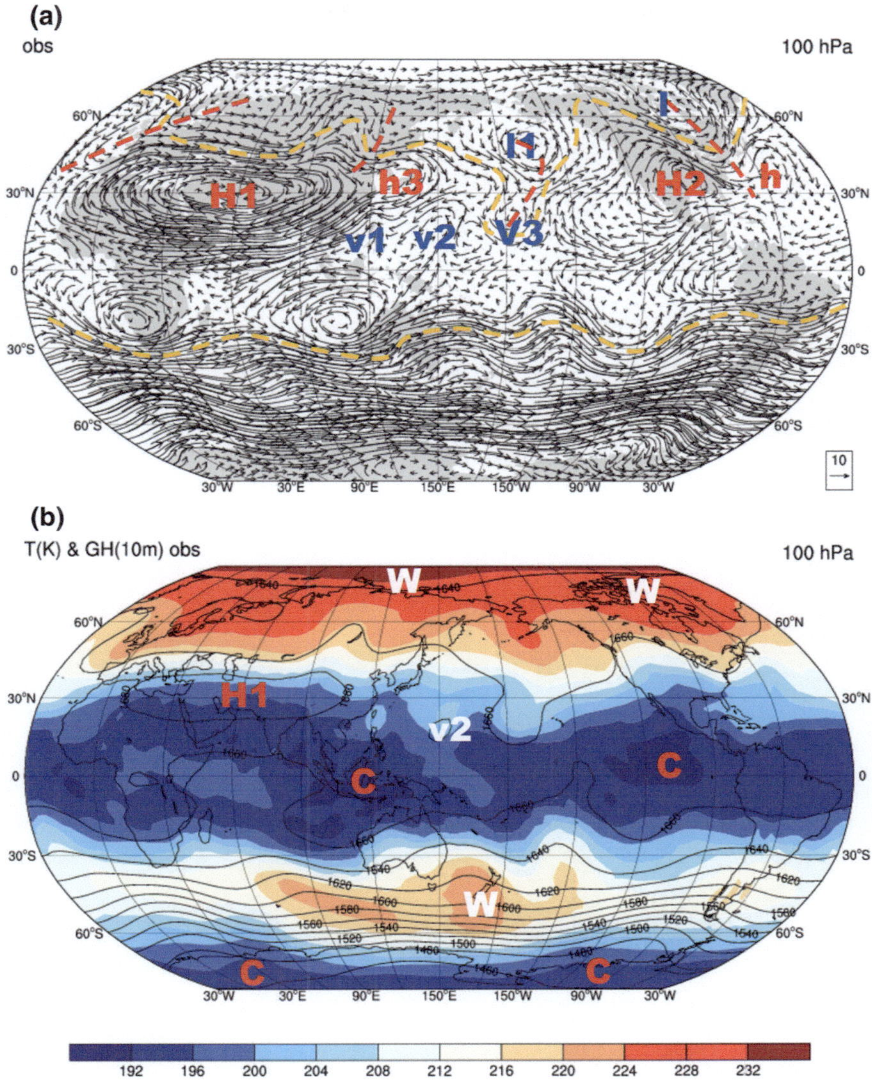

Fig. 5.1 **a** Total wind $\vec{V}_{\{d,y\}}(\lambda, \varphi, t)$ (m/s), **b** total height $H_{\{d,y\}}(\lambda, \varphi, t)$ (*solid contour*, 20×10 gpm interval), and total temperature $T_{\{d,y\}}(\lambda, \varphi, t)$ (*shading*, 4 K interval) at 100 hPa and 0000 UTC 2 July 1998 using the ERA-Interim data. The *heavy yellow-dashed lines* in **a** denote the westerly long waves. Letter "v" is the center of subtropical vortex. Letters "H/h" and "L/l" indicate the centers of anticyclone (*high*) and cyclone (*low*), while "W" and "C" are warm and cold centers of temperature. The *yellow-dashed lines* in **a** denote the westerly long waves with troughs (*red dashed line*)

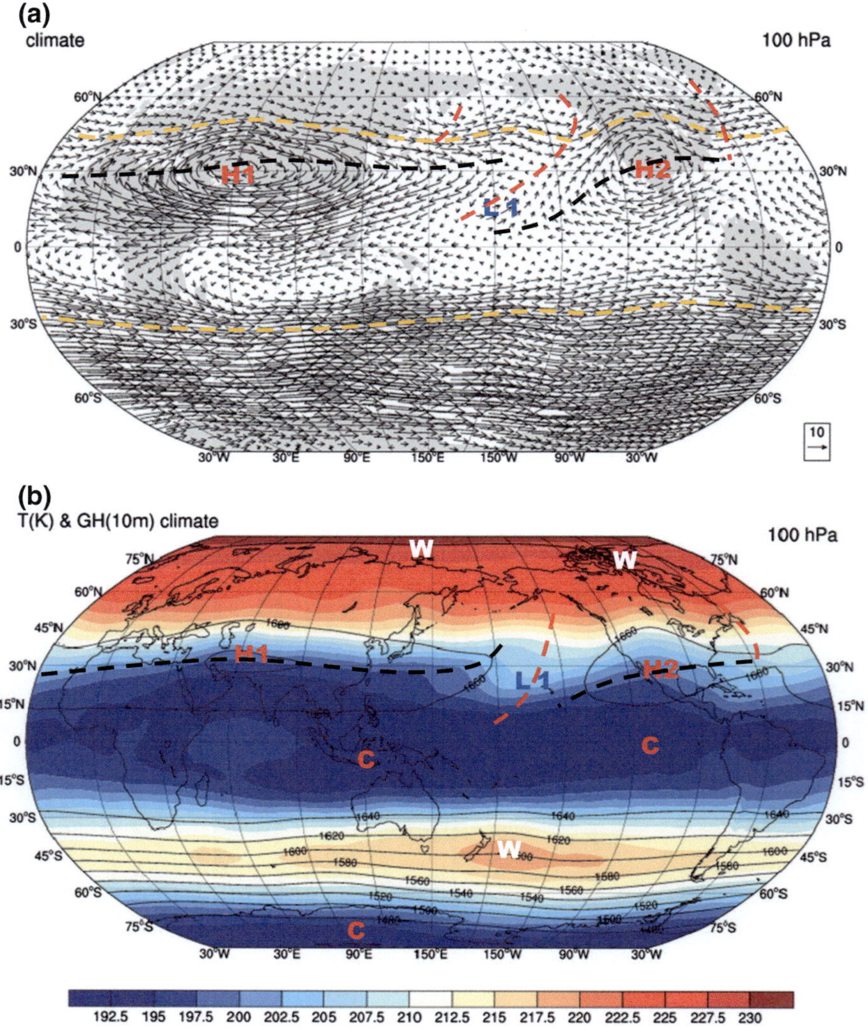

Fig. 5.2 Same as in Fig. 5.1 except of **a** the climatic wind $\overrightarrow{\tilde{V}}_{\{d,y\}}(\lambda, \varphi, t)$, **b** the climatic height $\tilde{H}_{\{d,y\}}(\lambda, \varphi, t)$ and the climatic temperature $\tilde{T}_{\{d,y\}}(\lambda, \varphi, t)$ at 100 hPa and 0000 UTC 2 July. The heavy *black–dashed lines* are ridges of subtropical highs

The global zonally averaged climatic wind $\left[\overrightarrow{\tilde{V}}_d(\varphi, t)\right]$, height $[\tilde{H}_d(\varphi, t)]$ and temperature $[\tilde{T}_d(\varphi, t)]$ at 100 hPa at 00000 UTC 2 July using the 30-year (1981–2010) ERA-Interim data are shown in Fig. 5.4. The strongest westerly zone is located between 20°S and 90°S (Fig. 5.4a). In the NH, there are a westerly zone in the middle latitudes and an easterly zone in the lower latitudes. These westerly and

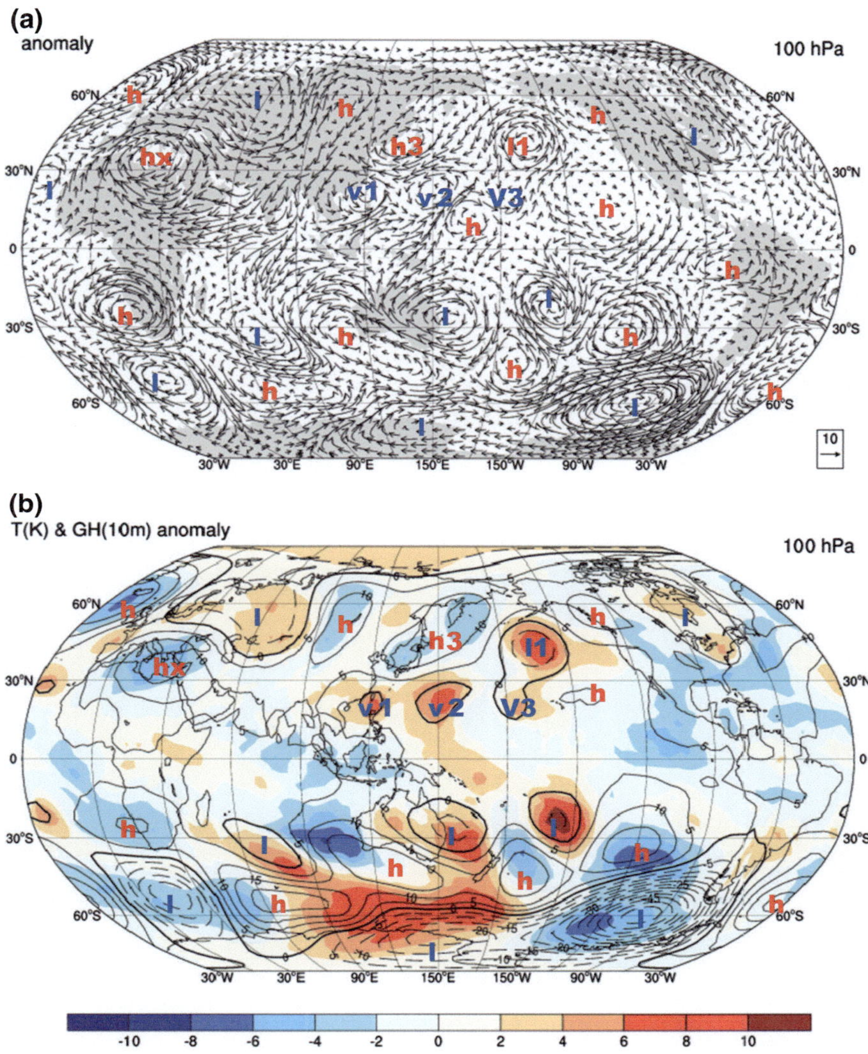

Fig. 5.3 Same as in Fig. 5.1 except of **a** wind anomaly $\overrightarrow{V}'_{\{d,y\}}(\lambda, \varphi, t)$ (m/s), **b** height anomaly $H'_{\{d,y\}}(\lambda, \varphi, t)$ (*contour*, 5 × 10gpm interval), and temperature anomaly $T'_{\{d,y\}}(\lambda, \varphi, t)$ (*red-yellow shading* for positive and *blue shading* for negative, 1 K interval) at 100 hPa and 0000 UTC 2 July 1998

easterly zones are responses to meridional height and temperature gradients (Fig. 5.4b). Indeed, all the meridional gradients and zonal distributions of these three variables are responses to the daily solar forcing.

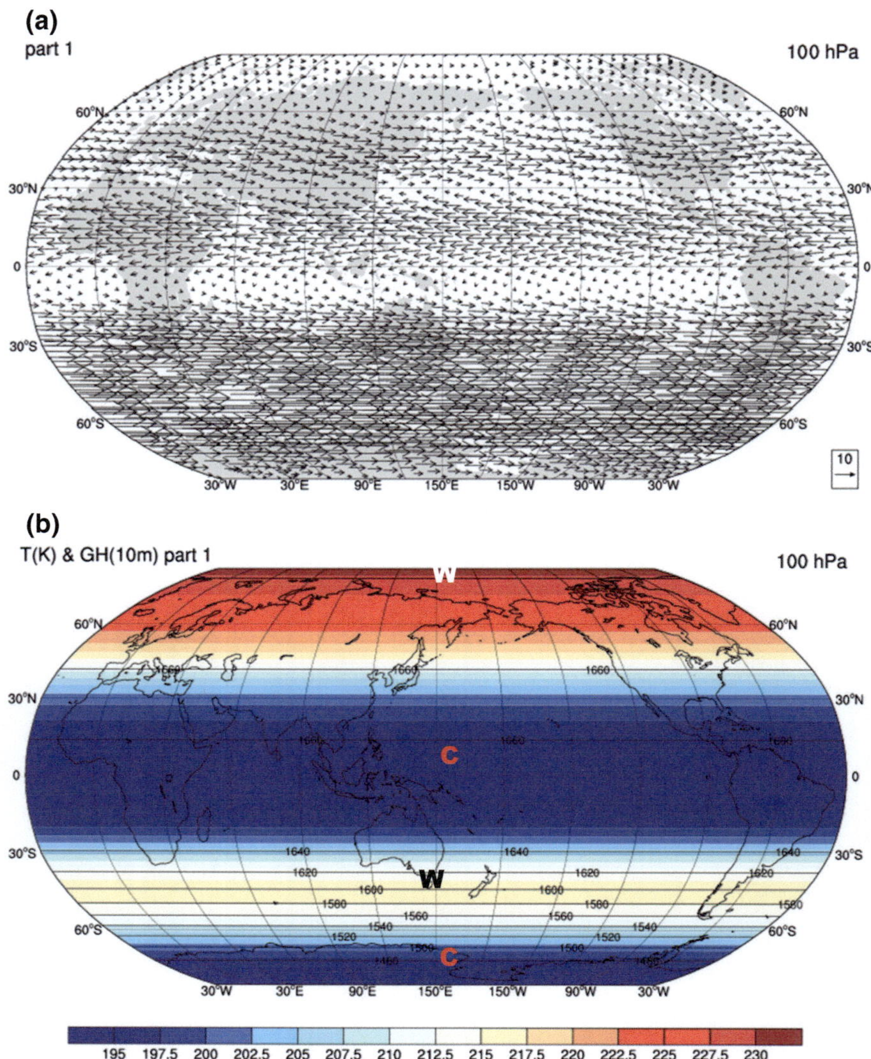

Fig. 5.4 Zonally averaged climatic **a** wind $\left[\overrightarrow{\tilde{V}}_d(\varphi, t)\right]$ (m/s), **b** height $\left[\tilde{H}_d(\varphi, t)\right]$ (*contour*, 20 × 10 gpm interval) and temperature $\left[\tilde{T}_d(\varphi, t)\right]$ (*shading*, 2.5 K interval) at 100 hPa and 0000 UTC 2 July based on the 30-year (1981–2010) ERA-Interim data

Climatic zonal wind departure or asymmetric wind at 100 hPa shows two anticyclones centered in South Asia and North America and two cyclones centered in the subtropical North Pacific and the subtropical North Atlantic (Fig. 5.5a). These circulations of climatic zonal wind departure are corresponding to the sea-land heating contrasts and height departures (Fig. 5.5b). Two pairs of

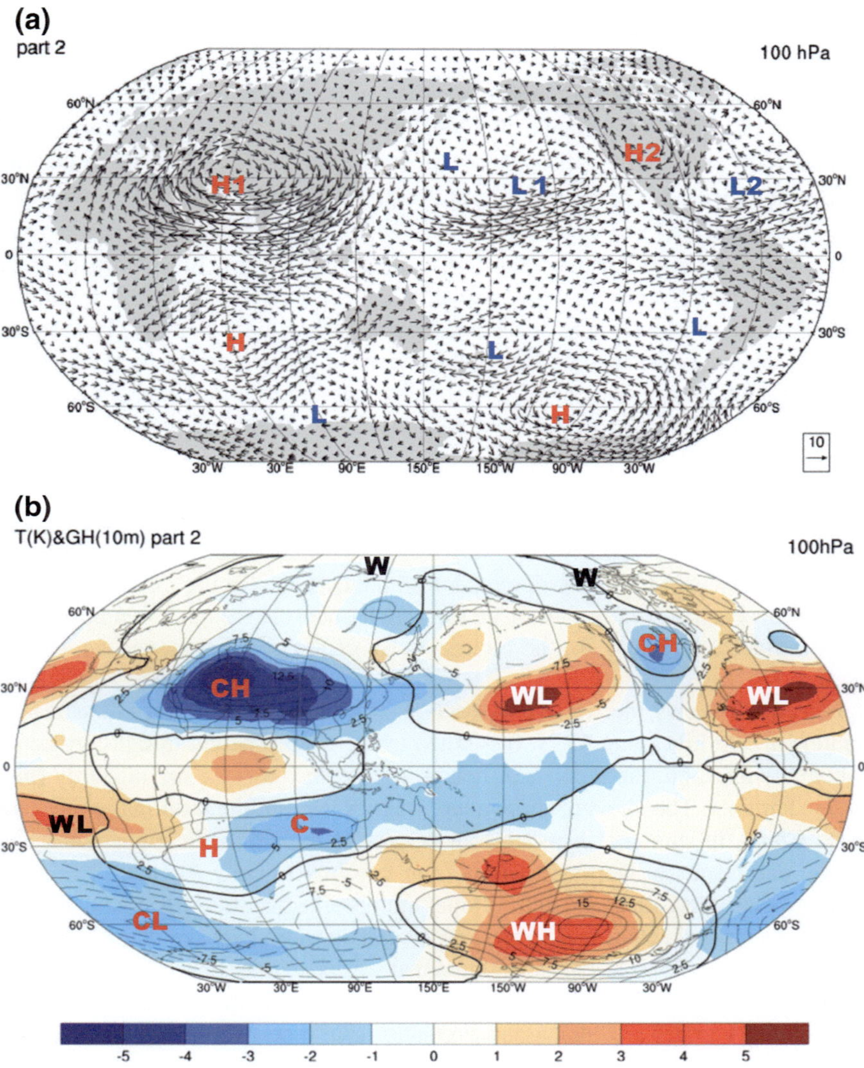

Fig. 5.5 Climatic zonal departures of **a** wind $\overrightarrow{V}^*_d(\lambda, \varphi, t)$ (m/s), **b** height $H^*_d(\lambda, \varphi, t)$(*solid* and *dashed contour*, 2.5 × 10 gpm interval), and temperature $T^*_d(\lambda, \varphi, t)$ (*shading*, 1 K interval) at 100 hPa and 0000 UTC 2 July based on the 30-year (1981–2010) ERA-Interim data. Letters "H/L" indicate anticyclonic/cyclonic centers in **a**. Letters "WH/CH" are warm-high/cold-high centers and "CL/WL" indicate cold-low/warm-low centers in **b**

oscillations with positive/negative height departures and cold/cold temperature departures are situated in the Asia-Pacific region and the North American-Atlantic region. In the SH, only a pair of warm-high and cold-low is separated in mid-high latitudes.

5.2 Upper Tropospheric Decomposition

From the traditional synoptic analysis, 300 hPa is regarded as the upper troposphere. In our synoptic decomposition analysis, 300 hPa is basically considered as the middle level from the troposphere to the stratosphere. Figure 5.6 shows the total wind vectors $\vec{V}_{\{d,y\}}(\lambda, \varphi, t)$ as well as the height $H_{\{d,y\}}(\lambda, \varphi, t)$ and the temperature $T_{\{d,y\}}(\lambda, \varphi, t)$ at 300 hPa on 2 July 1998. There are two bands of westerlies with long waves or troughs in the mid-high latitudes of the two hemispheres. These troughs are slowly moving eastward with the westerly flow. In the traditional synoptic analysis, forecasters concern the evolution of these troughs. The famous Rossby theory can well describe the motion of these troughs near 500 hPa. Cyclones and anticyclones as well as height and temperature centers are mainly accompanied and located within these westerly flows. These complex systems need to be decomposed.

Figure 5.7 shows the climatic wind, which is caused by meridional differences of solar insolation and sea-land heating contrast. The two westerly bands with waves or troughs are clearly observed in the two hemispheres. The six troughs (dashed red lines) in Fig. 5.6 belong to the climatic troughs in Fig. 5.7. One of them is over the Pacific, and is named as the middle-oceanic trough extending to the lower latitudes. In Fig. 5.7b, the highest warm center is located over the Tibetan Plateau.

Figure 5.8 shows the time-zonal section of climatic height and climatic temperature at 300 hPa along 45°N. It is noted that 2–3 deep height troughs are active in boreal winter, but the number of height troughs increases to six on 2 July (blue line) in summer. The climatic temperature reaches the minimum during winter over North China. In summer, there are four centers of high climatic temperature. The highest one occurs in Northeast China from late July to early August.

Many closed cyclones and anticyclones are observed at 300 hPa as shown in the anomalous synoptic chart (Fig. 5.9). These anomalous systems are mainly active in the total westerly flows in Fig. 5.6a, but zonal westerly is not observed in Fig. 5.9a. It is clear that the zonal westerly with waves or troughs is the climatic flow, while those closed vortices are anomalous circulations. In other words, the climatic flow and the anomalous circulations together form the observed total winds. The climatic flow with waves or troughs can be identified from Eq. 1.6.2, so forecasters only need to predict the anomalous circulations or those closed vortices.

In Fig. 5.9a, there are 12 anomalous anticyclones and 13 anomalous cyclones in the NH, while 9 anomalous anticyclones and 13 anomalous cyclones are found in the SH. Theses anomalous circulation systems observed at 300 hPa and 0000 UTC 2 July 1998 should be useful information for aviation at that level and at that moment globally. In Fig. 5.9b, there are 7 positive centers and 13 negative centers of height anomalies in the NH, while 7 positive centers and 11 negative centers of height anomalies are found in the SH. These centers of height anomalies coincide with warm and cold centers of temperature anomalies. It will be discussed in Chap. 7 that these warm and cold centers are directly linked with surface

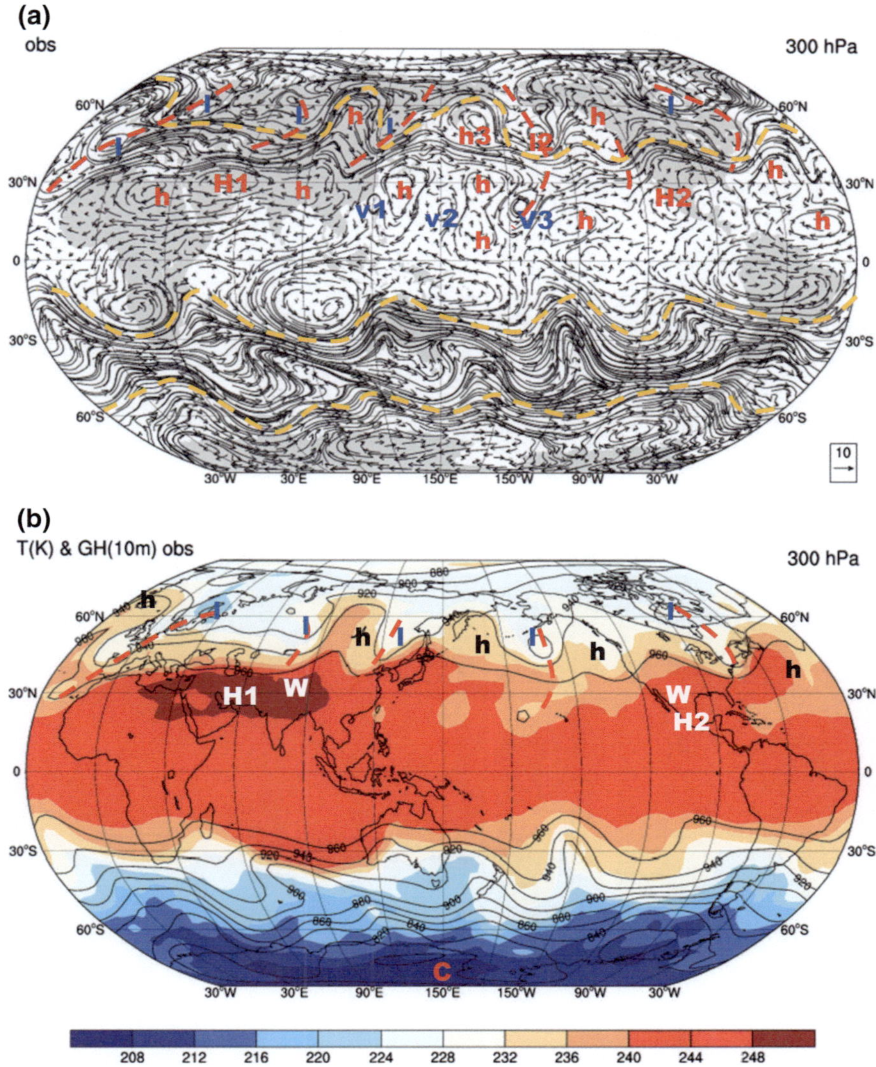

Fig. 5.6 Same as in Fig. 5.1 except for **a** total wind as well as **b** total height and total temperature at 300 hPa and 0000 UTC 2 July 1998

temperature extremes. Based on the anomalous synoptic chart, we can identify the number of weather extremes over the whole world.

The climatic zonal departures of 300 hPa wind show two global anticyclonic trains in the subtropical SH and NH, respectively (Fig. 5.10a). The two anticyclonic trains are consistent with two temperature and height trains in the two hemispheres (Fig. 5.10b). The strongest positive center of climatic zonal height and temperature departures is located in South Asia. Two strong negative centers of climatic zonal

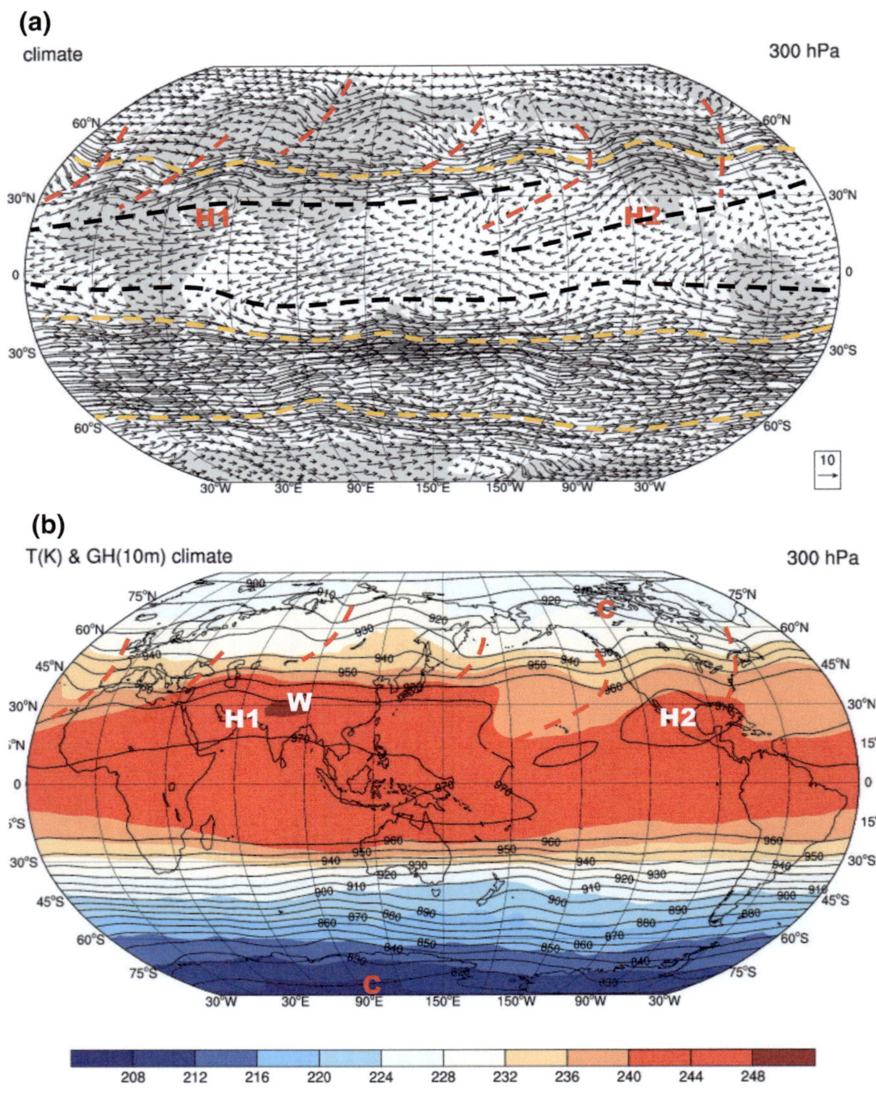

Fig. 5.7 Same as in Fig. 5.2 except of **a** the climatic wind $\vec{V}_d(\lambda, \varphi, t)$, **b** the climatic height $\tilde{H}_d(\lambda, \varphi, t)$ and the climatic temperature $\tilde{T}_d(\lambda, \varphi, t)$ at 300 hPa and 0000 UTC 2 July

height and temperature departures appear in the subtropical North Pacific and the subtropical North Atlantic. Clearly, theses climatic zonal departure centers of circulation, height, and temperature are caused by sea-land heating contrast. It is interesting to compare the signs of climatic zonal departures between 100 and 300 hPa. Warm and cold centers are respectively corresponding well with negative and positive departures of height in the NH at 100 hPa. But, the climatic zonal

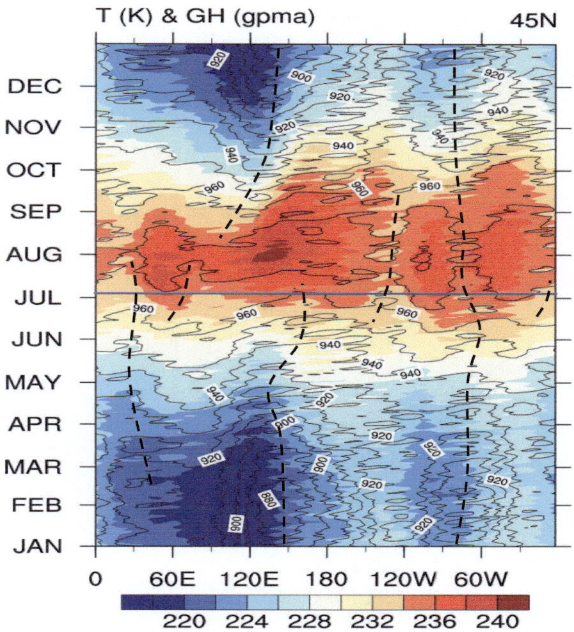

Fig. 5.8 Time-zonal section of climatic height (*contour*, 10 × 10 gpm interval) and climatic temperature (*shading*, 2 K interval) at 300 hPa along 45°N. The *dashed lines* indicate height troughs. At 0000 UTC 2 July (*blue line*), there are 6 weak toughs at the same time

temperature departure changes its sign at 300 hPa in the NH. For example, a CH center in South Asia and a WL center in the subtropical Pacific at 100 hPa (Fig. 5.5b) have changed their signs to a WH center in South Asia and a CL center in the subtropical Pacific at 300 hPa (Fig. 5.10b). It implies that the structure of circulation systems changes vertically.

5.3 Lower Tropospheric Decomposition

Usually, 700 hPa is considered as the lower troposphere. Figure 5.11 shows the total wind vectors $\vec{V}_{\{d,y\}}(\lambda, \varphi, t)$ as well as the total height $H_{\{d,y\}}(\lambda, \varphi, t)$ and the total temperature $T_{\{d,y\}}(\lambda, \varphi, t)$ at 700 hPa on July 2, 1998. This is the commonly used traditional synoptic chart. Cyclones and anticyclones as well as the height and temperature centers can be observed in the traditional synoptic charts. The westerly Rossby-wave flow is clearly observed in the middle latitudes of the SH. The structures of height, temperature, and circulation systems are complex from the traditional synoptic chart, so we need to decompose and compare them. The relationship among these systems at 100, 300, and 700 hPa is also not clear.

The climatic wind $\vec{\tilde{V}}_{\{d,y\}}(\lambda, \varphi, t)$, the climatic height $\tilde{H}_d(\lambda, \varphi, t)$ and the climatic temperature $\tilde{T}_d(\lambda, \varphi, t)$ at 700 hPa and 0000 UTC 2 July are shown in Fig. 5.12. These flows are caused by the meridional difference of solar radiation and the

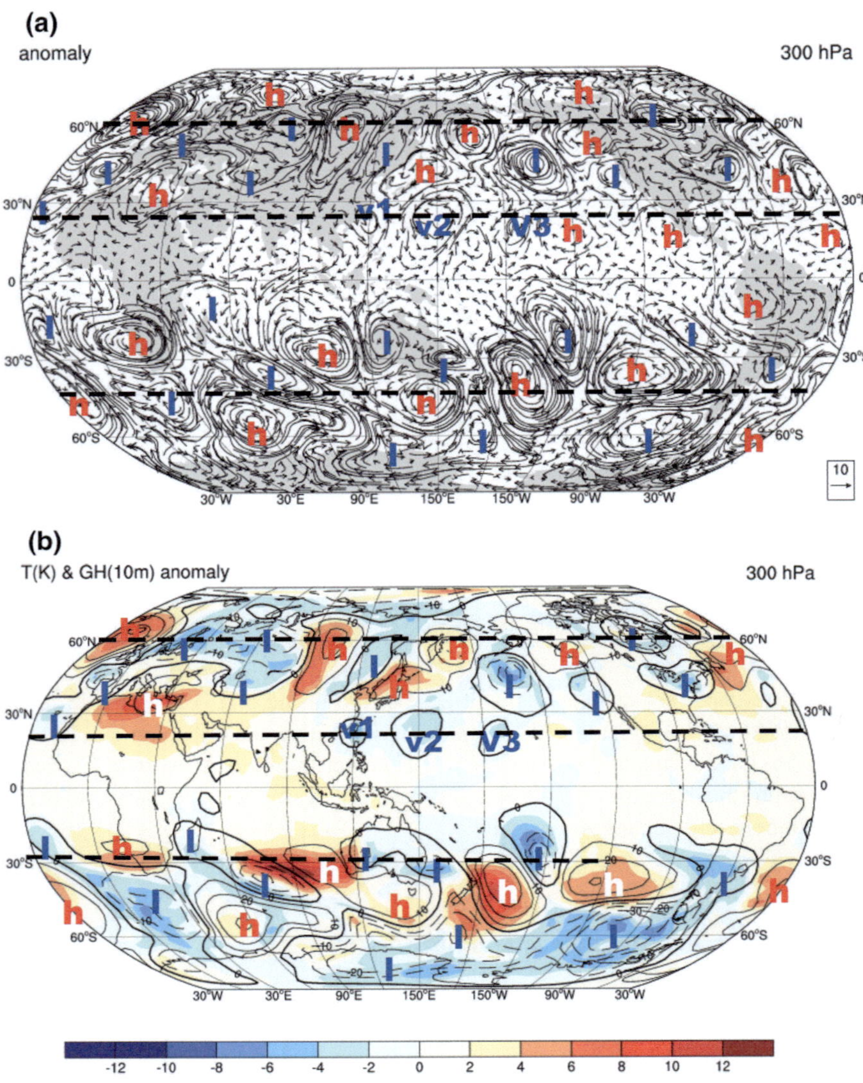

Fig. 5.9 Same as in Fig. 5.3 except of **a** the wind anomaly $\overrightarrow{V}'_{\{d,y\}}(\lambda, \varphi, t)$ (m/s) as well as **b** the height anomaly $H'_{\{d,y\}}(\lambda, \varphi, t)$ (*contour*, 10 × 10 gpm interval) and temperature anomaly $T'_{\{d,y\}}(\lambda, \varphi, t)$ (*red-yellow shading* for positive and *blue shading* for negative, 1 K interval) at 300 hPa and 0000 UTC 2 July 1998

sea-land heating contrast. The two heavy solid lines with arrow are the cross-equatorial flows. The zonal westerly with the Rossby waves (or troughs) is clearly observed in the middle latitudes. Three branches of southwesterly, southeasterly and westerly converge in the East Asian region (Fig. 5.12a). In Asia, three monsoon troughs are clearly active in the Arabian Sea, the South China Sea (SCS), and the subtropical East Asia from eastern China to the Korea Peninsula. All these

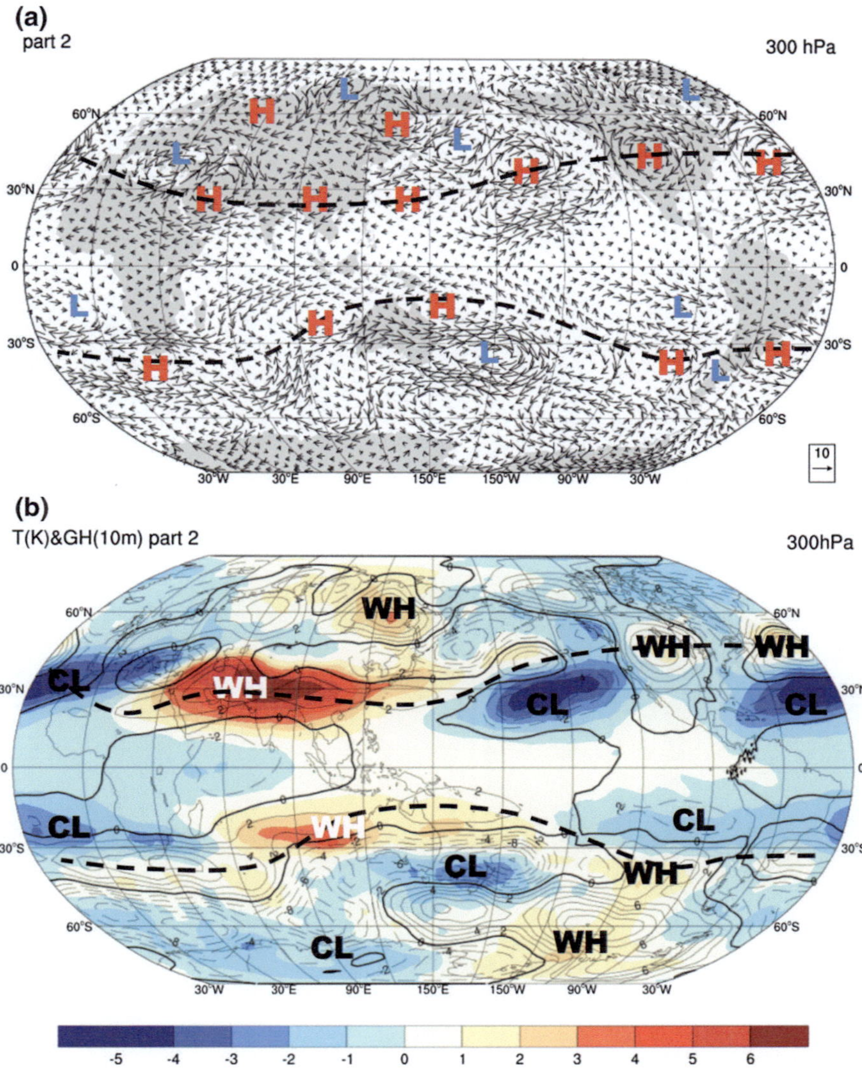

Fig. 5.10 Same as in Fig. 5.5 except of climatic zonal departures of **a** wind $\overrightarrow{V}^*_d(\lambda, \varphi, t)$ (m/s), **b** height $H^*_d(\lambda, \varphi, t)$ (*contour*, 2 × 10 gpm interval) and temperature $T^*_d(\lambda, \varphi, t)$ (*shading*, 1 K interval) at 300 hPa and 0000 UTC 2 July. Letters "H/L" denote anticyclonic/cyclonic centers in **a**, while letters "WH" and "CL" indicate warm-high centers and cold-low centers in **b**

climatic systems and the monsoon troughs migrate gradually. The comprehensive heating difference in solar radiation and sea-land heating can be observed in Fig. 5.12b, where three warm temperature centers are located in South Asia, subtropical North Africa, and subtropical North America. In the Antarctic area, there is a pair of climatic high and low.

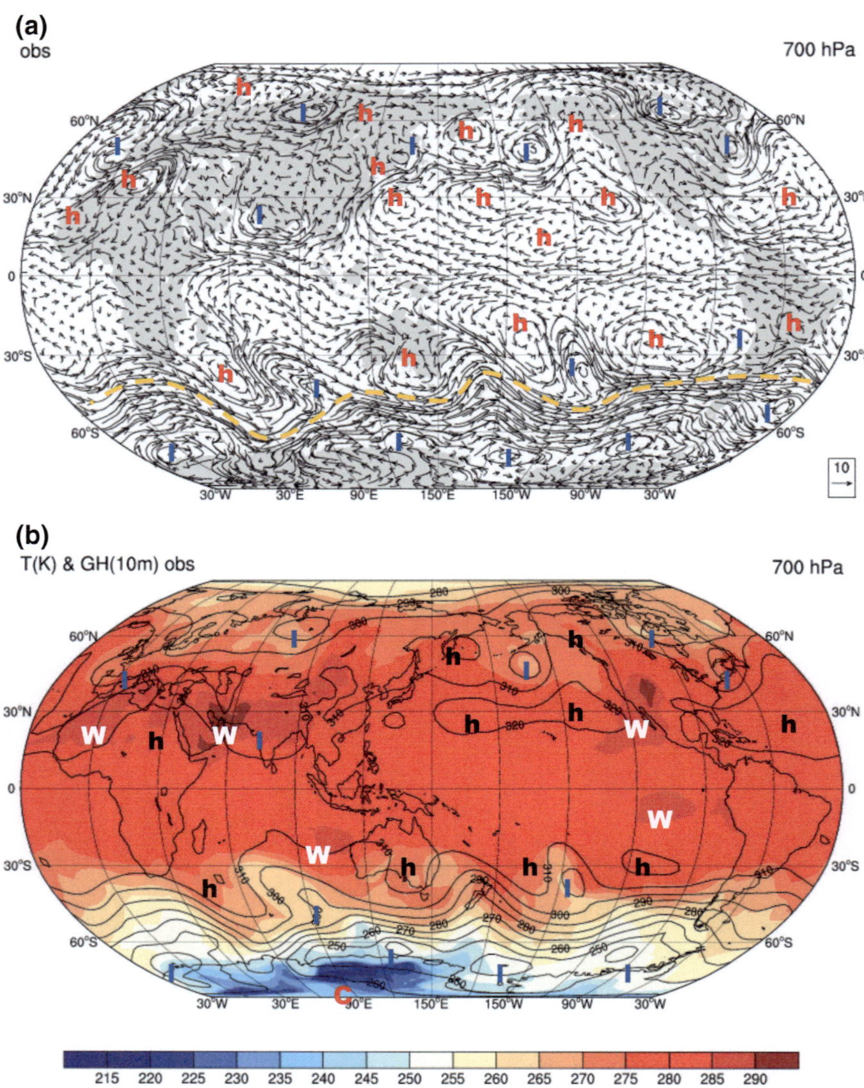

Fig. 5.11 Same as in Fig. 5.1 except of **a** total wind (m/s), **b** total height (*contour*, 10 × 10 gpm interval), and total temperature (*shading*, 5 K interval) at 700 hPa and 0000 UTC 2 July 1998

The anomalous circulation systems in Fig. 5.13 should be directly linked with the surface weather extremes such as regional rainfalls, heat waves, and cold surges. From South Asia to East Asia, there is an anomalous shear. Three subtropical cyclones (v1, v2, and v3) in the Northwest Pacific cannot be observed in the total wind field (Fig. 5.11a). However, the three subtropical vortices are clearly observed

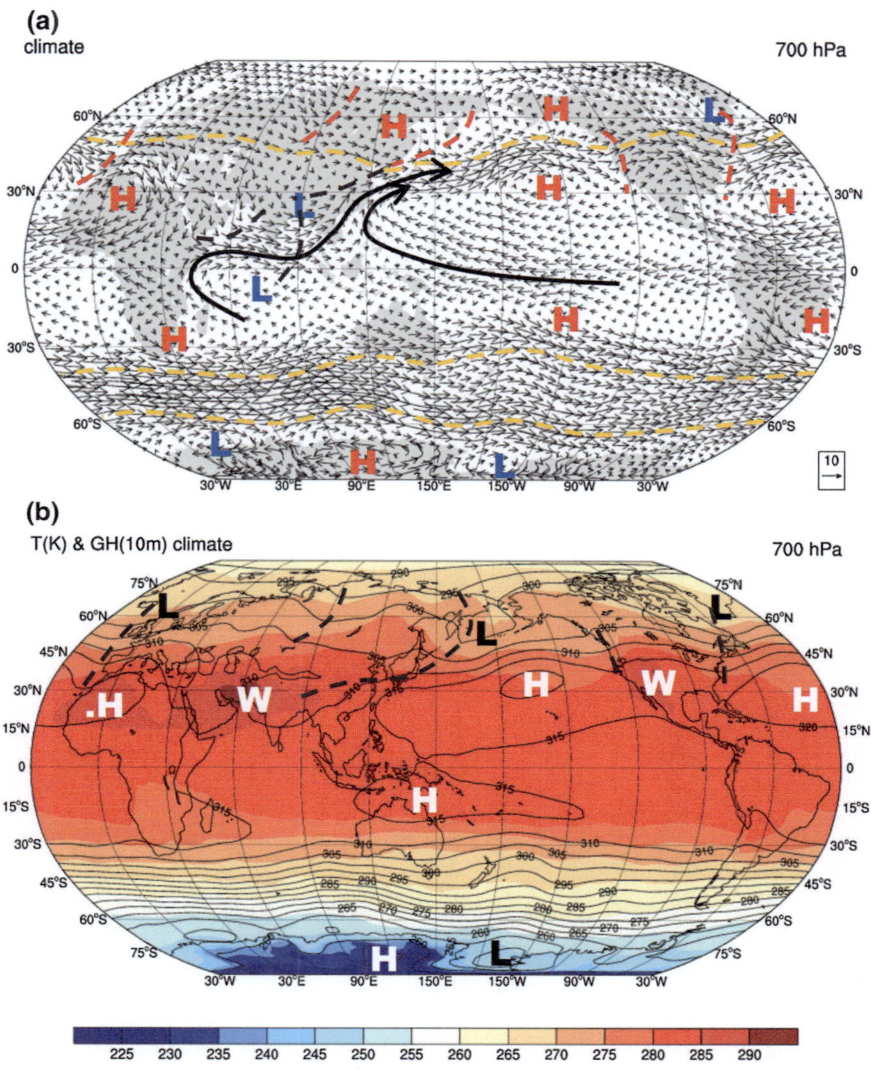

Fig. 5.12 Same as in Fig. 5.1 except of **a** the climatic wind $\vec{V}_d(\lambda, \varphi, t)$ (m/s), **b** the climatic height $\tilde{H}_d(\lambda, \varphi, t)$ (*contour*, 5×10 gpm interval) and the climatic temperature $\tilde{T}_d(\lambda, \varphi, t)$ (*shading*, 5 K interval) at 700 hPa and 0000 UTC 2 July. The *thick black arrow* indicates the cross-equatorial flow and the *heavy dashed lines* are monsoon troughs

in the anomalous wind field (Fig. 5.13a). The low (high) centers of height anomalies are corresponding with cold (warm) centers of temperature anomalies (Fig. 5.13b). The configuration between temperature and height anomalies in the lower troposphere (700 hPa) is opposite to that at 100 hPa (Fig. 5.3b).

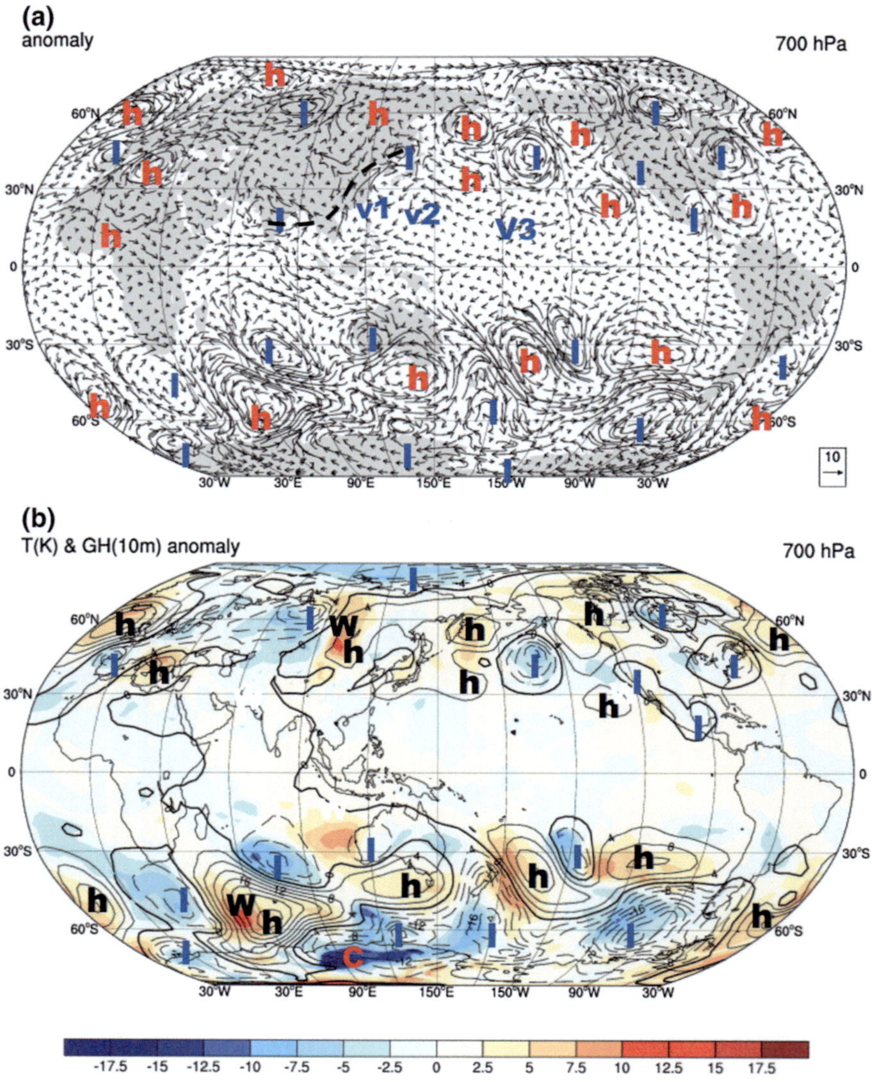

Fig. 5.13 Same as in Fig. 5.3 except of **a** the wind anomaly $\overrightarrow{V}'_{\{d,y\}}(\lambda, \varphi, t)$ (m/s) as well as **b** the height anomaly $H'_{\{d,y\}}(\lambda, \varphi, t)$ (*contour*, 4 × 10 gpm interval) and temperature anomaly $T'_{\{d,y\}}(\lambda, \varphi, t)$ (*red-yellow shading* for positive and *blue shading* for negative, 2.5 K interval) at 700 hPa and 0000 UTC 2 July 1998

5.4 Other Level and Vertical Decompositions

In the stratosphere, anomalous circulation systems are also strong. For example at 10 hPa, daily anomalous systems are strong (Fig. 5.14c) when comparing with the total wind field in Fig. 5.14a and the climatic wind field in Fig. 5.14b. Those anomalous systems can also be clearly observed at 5 hPa (Fig. 5.15c) and at 1 hPa (Fig. 5.16c). The climatic circulation systems are mainly anticyclones in the tropical zone.

To examine the connection between anomalous systems in the troposphere and surface weather anomalies, Fig. 5.17 shows the surface air temperature anomalies (SATAs) at the 2 m level and 0000 UTC 2 July 1998 and the precipitation percentage anomalies in pentad 37 (30 June– 4 July) of 1998. The wave-like distributions of SATAs and precipitation anomalies can be observed in the SH. Four warm and dry areas are indicated by H1–H4 in the SH along 30.75°S. In the NH, four warm (W1, W2, W3, and W4) and four dry slices (D1, D2, D3, and D4) are also marked along 60°N.

5.5 Vertical Sections of Total and Anomalous Variables

The surface weather extremes and variable anomalies from the troposphere to the stratosphere have been depicted in the above subsections. This subsection gives the vertical connection between the surface weather extremes and the atmospheric variable anomalies. Figure 5.18 shows the vertical sections of total height and total temperature as well as their anomalies along 60°N at 0000 UTC 2 July 1998. It is difficult to find any signals from the total field (Fig. 5.18a) to indicate the surface cold/warn and dry/wet anomalies in Fig. 5.17. It is clear from the anomalous field (Fig. 5.18b), where the four positive centers of height anomalies at the upper troposphere can well indicate the surface warm and dry anomalies shown in Fig. 5.17. An area of warm and dry anomalies (W1 and D1 in Fig. 5.17) at the surface result from a warm air column (red shading) below a positive center of height anomalies at the upper troposphere (Fig. 5.18b). In contrast, an area of cold and wet anomalies (C1 in Fig. 5.17) at the surface result from a cold air column (blue shading) below a negative center (L1 in Fig. 5.18b) of height anomalies at the upper troposphere. The height and temperature anomalies satisfy the hydrostatic equilibrium in the vertical direction.

Similarly, Fig. 5.19 shows the vertical sections of total height and total temperature as well as their anomalies along 30.75°S at 0000 UTC 2 July 1998. It is also difficult to indicate the surface cold/warn and dry/wet anomalies (Fig. 5.17) from the total field (Fig. 5.19a). However, it is clear from the anomalous field (Fig. 5.19b) that four centers of height anomalies at the upper troposphere can well indicate the surface cold/warn and dry/wet anomalies in Fig. 5.17. The height and temperature anomalies vertically satisfy the hydrostatic equilibrium.

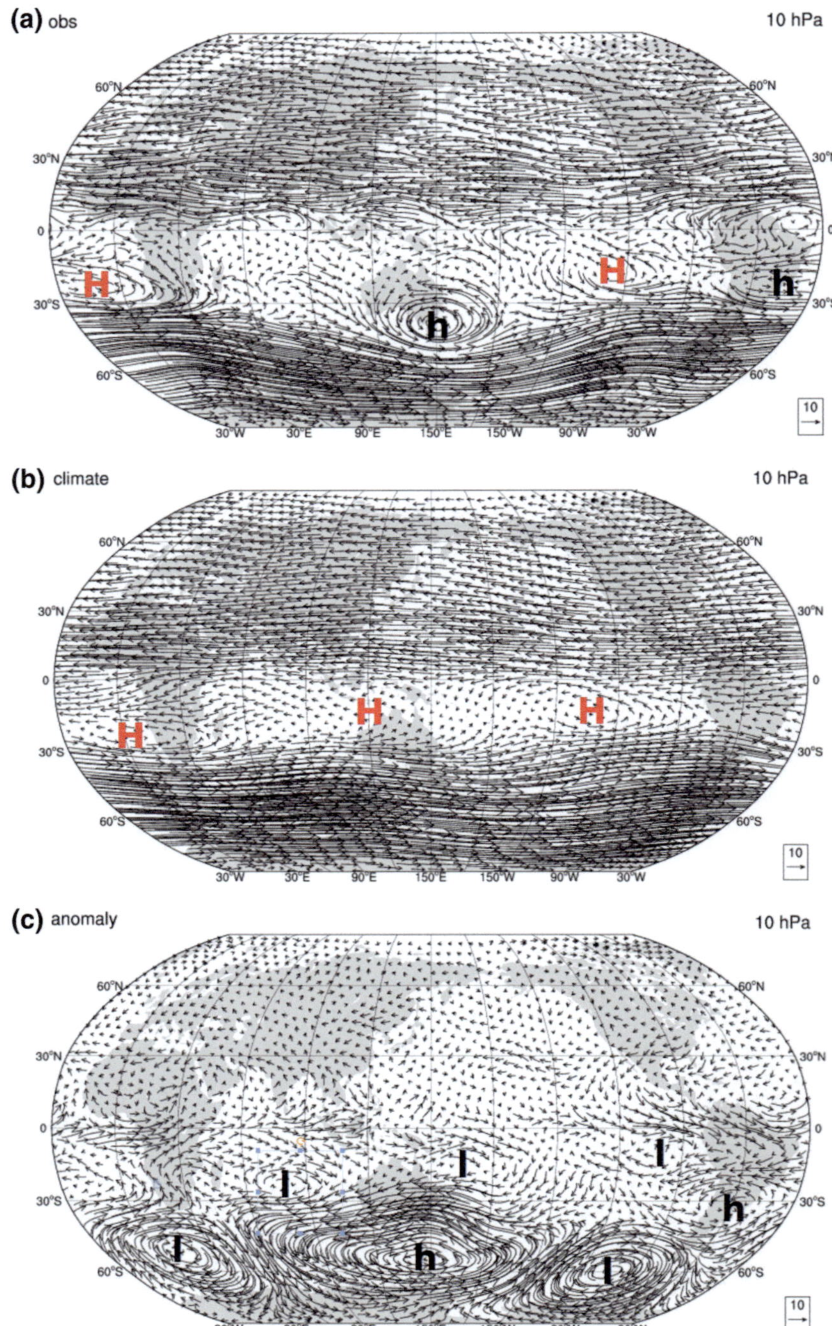

Fig. 5.14 **a** Total wind (m/s), **b** climatic wind (m/s) and **c** anomalous wind (m/s) at 10 hPa and 0000 UTC 2 July 1998

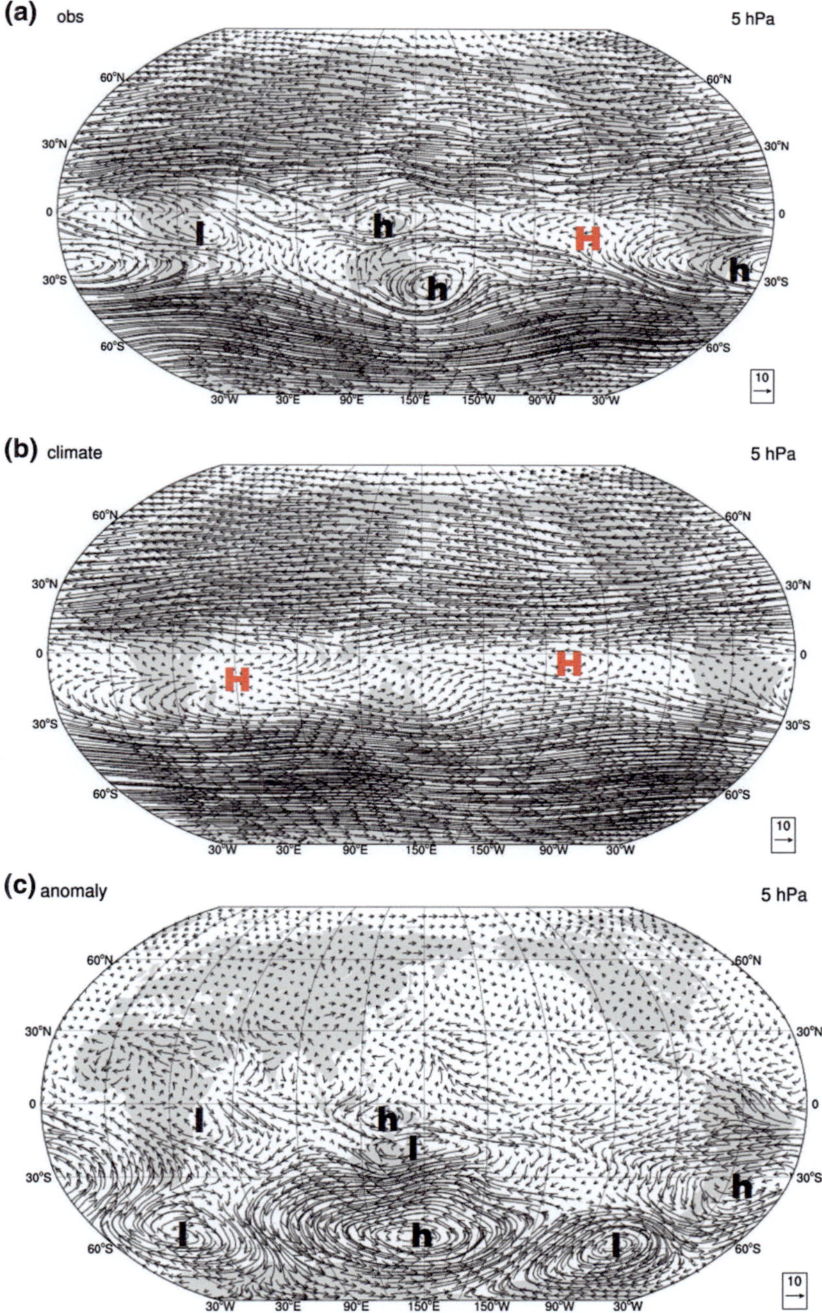

Fig. 5.15 **a** Total wind (m/s), **b** climatic wind (m/s) and **c** anomalous wind (m/s) at 5 hPa and 0000 UTC 2 July 1998

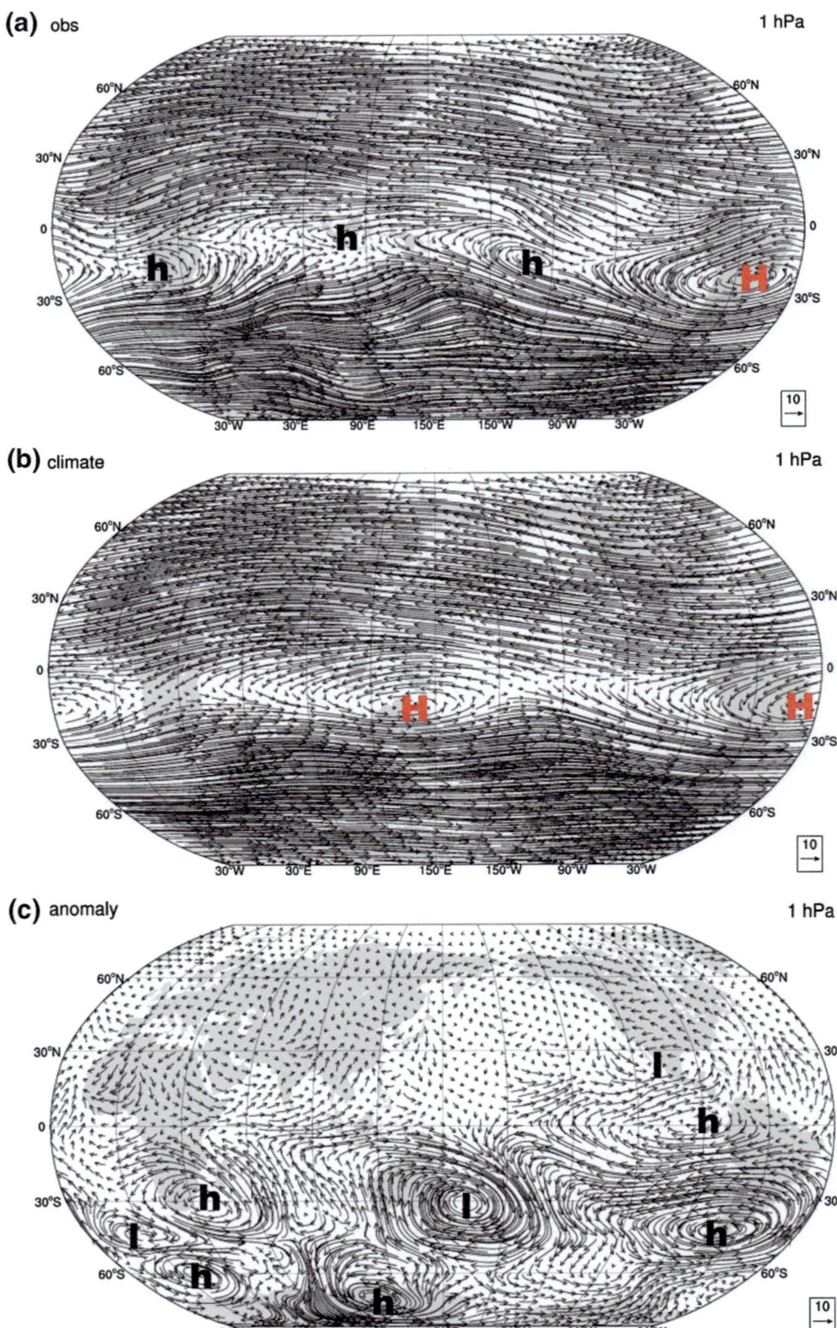

Fig. 5.16 **a** Total wind (m/s), **b** climatic wind (m/s) and **c** anomalous wind (m/s) at 1 hPa and 0000 UTC 2 July 1998

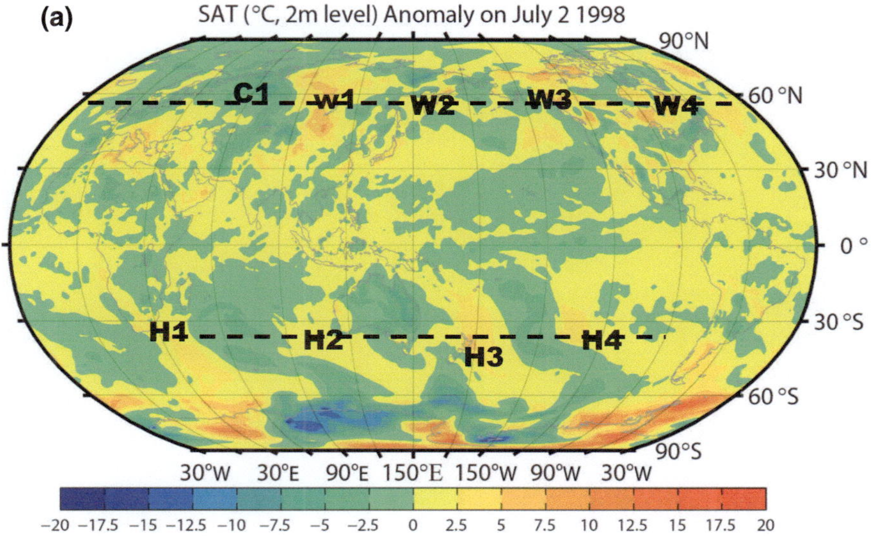

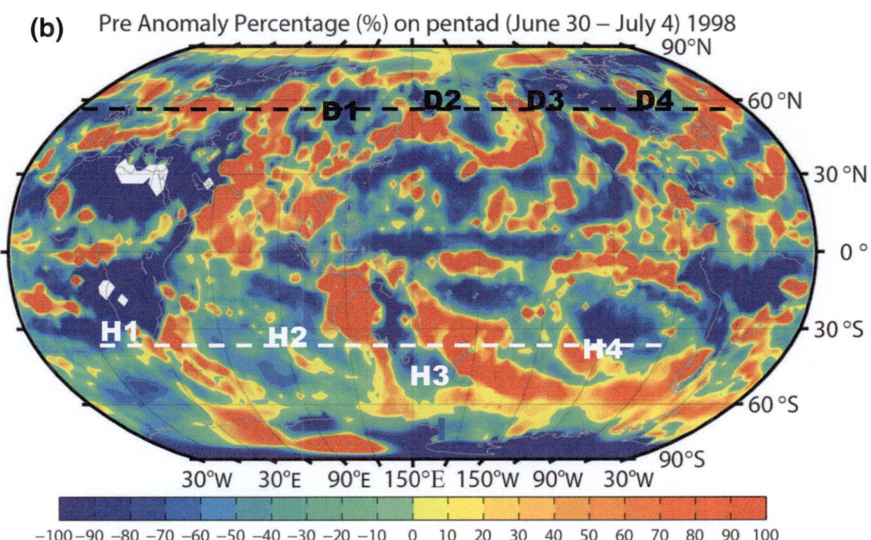

Fig. 5.17 **a** The SATAs (*shading*, 2.5 °C interval) at the 2 m level and 0000 UTC 2 July 1998. **b** The pentad-mean precipitation percentage anomalies (*shading*, 10% interval) in pentad 37 (30 June– 4 July) of 1998. The *heavy dashed line* in the SH is along 30.75°S with four warm and four dry slices (H1, H2, H3, and H4). The *heavy dashed line* in the NH is along 60°N with four warm (W1, W2, W3, and W4) and four dry slices (D1, D2, D3, and D4)

This vertical pattern of height and temperature anomalies satisfies the hydrostatic equilibrium relationship, i.e., the temperature anomalies T' can be derived from the vertical gradient of height anomalies H'

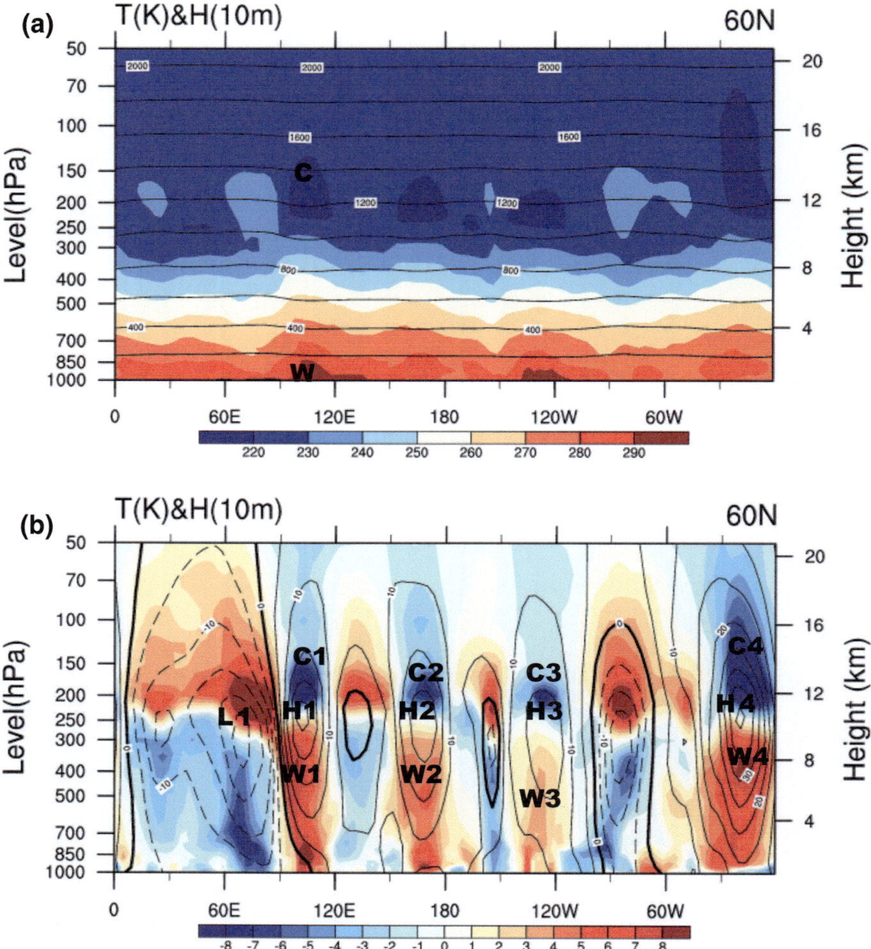

Fig. 5.18 **a** Vertical section of total height (*contour*, 200 × 10 gpm interval) and total temperature (*shading*, 10 K interval) along 60°N at 0000 UTC 2 July 1998. **b** Same as **a** except for height anomalies (*contour*, 4 × 10 gpm interval) and temperature anomalies (*shading*, 2 K interval). Symbols "H1" to "H4" indicate four positive centers of height anomalies at the *upper* troposphere. L1 is a negative center of height anomalies at the *upper* troposphere. Letters "W/C" are warm/cold centers of temperature anomalies

$$T' = -\frac{g}{R}\frac{\partial H'}{\partial \ln p} \tag{5.5.1}$$

where g is the gravitational acceleration, R is the ideal gas constant and p is the pressure level. For example, a positive height anomalous center (H1) in Fig. 5.18b is observed at 250 hPa in Siberia (105°E, 60°N). The height anomalies below 250 hPa increase (decrease) with altitude (pressure) and have the largest vertical

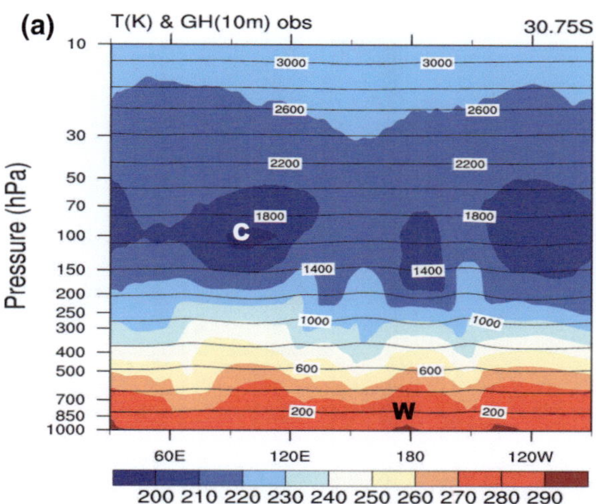

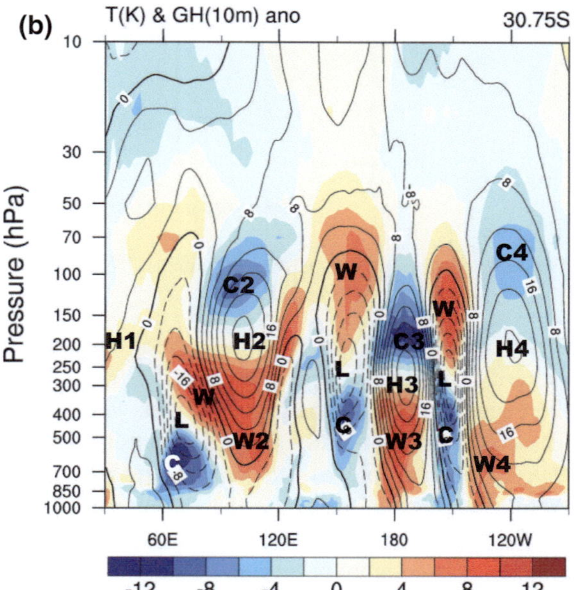

Fig. 5.19 Same as in Fig. 5.18 except along 30.75°S at 0000 UTC 2 July 1998

gradient $-\frac{g}{R}\frac{\partial H'}{\partial \ln p}$ near the surface (Fig. 5.18b). Therefore, a strong warm center "W1" is detected near the surface by Eq. 5.5.1, and it can explain the strongest surface warm (W1) and dry (D1) centers in Fig. 5.17.

In the previous anomaly-based synoptic charts (Fig. 5.13a), three vortices (v1, v2, and v3) in the subtropical NW Pacific are obvious. The three vortices are not observed from the total height and total temperature in Fig. 5.20a. However, they are clearly observed in the anomaly-based synoptic chart in Fig. 5.20b. Among

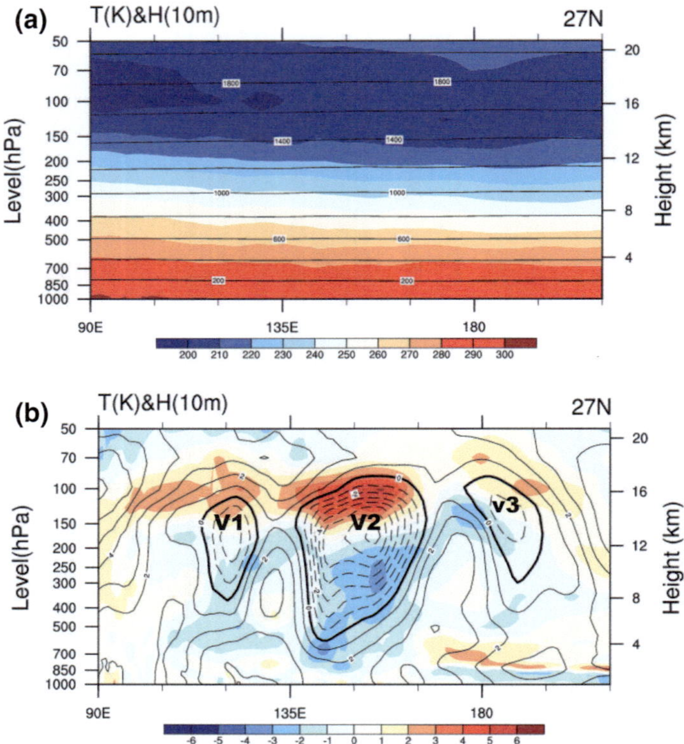

Fig. 5.20 Same as in Fig. 5.18 except along 27°N at 0000 UTC 2 July 1998

them, the strongest vortex is the central one (v2). At the altitude of three vortex centers, no temperature anomalies can be observed. A warm vortex is clear above the negative center of height anomalies while a cold vortex is below this center. These systems are usually referred to as cold vortices in traditional synoptic charts. Indeed, they are cold vortices only below the negative centers of height anomalies. Above the negative centers of height anomalies, they become warm vortices.

The vertical sections show that the maximum positive (negative) center of height anomalies is commonly located at the upper troposphere. The positive (negative) height anomalous center separates a warm (cold) air column below in the troposphere and a cold (warm) air column above in the stratosphere. This implies that we should not only focus on the signals in the troposphere to indicate the surface weather extremes. A surface weather extreme is directly associated with the anomalies over the whole atmosphere. The strongest signal of height anomalies is always observed at the upper troposphere.

5.6 Regional Total- and Anomaly-Based Synoptic Charts

In daily weather forecasting, traditional synoptic charts from both observations and model outputs are still applied in current practices (Fig. 5.21). Tropical and extra-tropical weather systems are clearly separated along 30°N at 500 hPa and 0000 UTC 2 July 1998 (Fig. 5.21a). In the westerly zone beyond 30°N, there are troughs and ridges along with blocking highs (anticyclones) with higher temperature and cutting lows (cyclones) with lower temperature. In the tropical zone, major weather systems are a warm cyclonic system centered at the Indian Peninsula, a trough extending southwestward in the Arabian Sea, and two warm subtropical anticyclones located in the NW Pacific, with an easterly wave between them. In

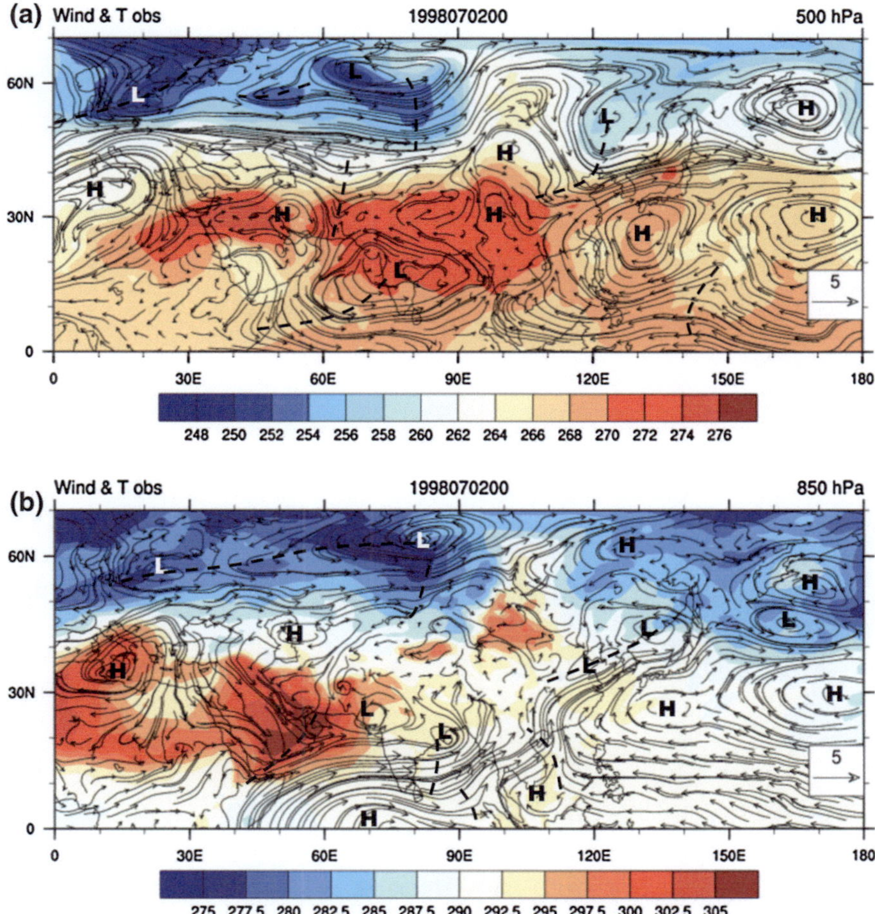

Fig. 5.21 Total wind (m/s) and total temperature (*shading*, 2.5 K interval) at **a** 500 hPa and **b** 850 hPa at 0000 UTC 2 July 1998 using the ERA-Interim data. The *dashed line* denotes the trough, shear, or *convergence line*. Letters "H" and "L" indicate high (anticyclone) and low (cyclone) systems

western China, there is a ridge. It is clear that there is interaction between tropical and extra-tropical systems in eastern China. At 850 hPa (Fig. 5.21b), two cold-low systems exist from Europe to western Russia and two cold-high systems cover from eastern Russia to the Bering Sea in high latitudes. There is a train of subtropical anticyclones at 850 hPa with their centers from the Arabian Sea, the Indi-China Peninsula to the NW Pacific, so the southwesterly cross-equatorial flow is very strong in the northern boundary of this anticyclone train. During this day, severe rains should easily occur in the Bay of Bengal and from the Yellow River to Northeast China due to the strong dynamical rainfall conditions from the three cyclones with strong southwesterly flow.

The climatic component and anomaly commonly exist in the total field. The climatic component is comfortably observed in Fig. 5.22. At 500 hPa (Fig. 5.22a) four troughs with temperature waves exist along with the high-latitude westerly,

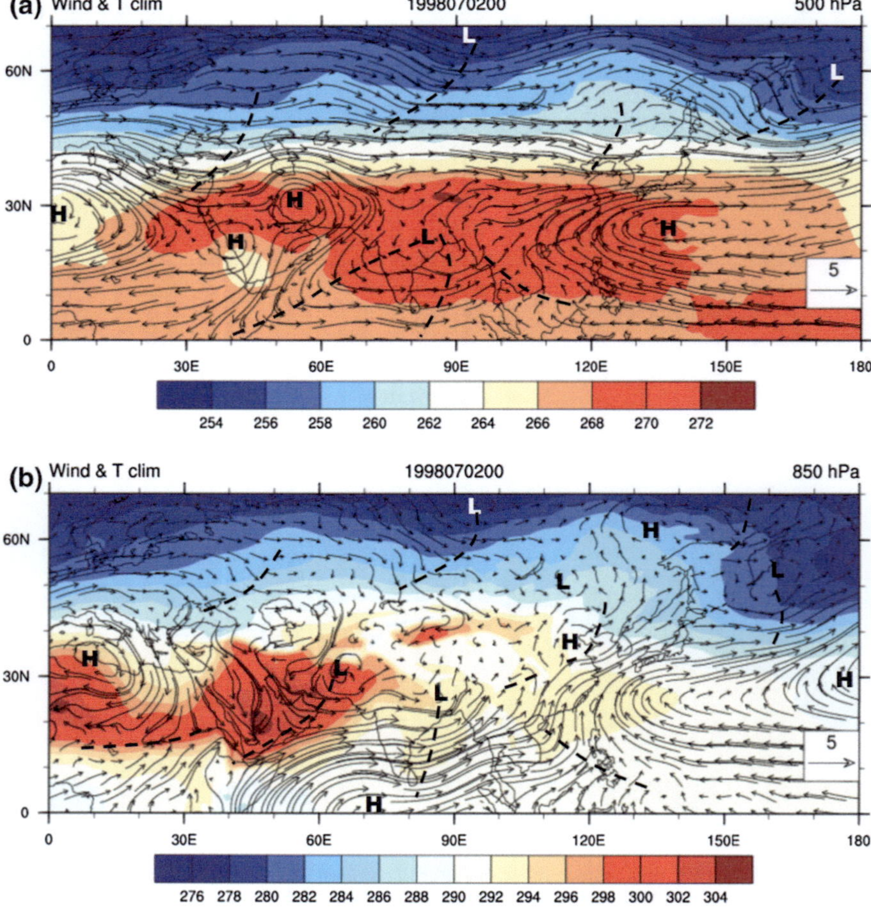

Fig. 5.22 Same as in Fig. 5.21 except of climatic wind (m/s) and climatic temperature (*shading*, 2 K interval) at **a** 500 hPa and **b** 850 hPa at 0000 UTC 2 July

while four anticyclones and one cyclone are situated in lower latitudes. The convergence between the westerly and subtropical flow is climatologically located along the Yellow River. At 850 hPa (Fig. 5.22b), four troughs with temperature waves are basically similar to those at 500 hPa, while four anticyclones and two cyclones are situated in lower latitudes. The convergence between the westerly and subtropical flow is climatologically located along the Yellow River. In lower latitudes, only two anticyclones at 850 hPa in North Africa and the Northwest Pacific are comparable with those at 500 hPa. In South Asia and East Asia, the climatic circulation is predominant for three troughs in the Arabian coast line, the Bay of Bengal and the SCS. In China, there are two shear lines from Southwest China to Northeast China along the Yellow River. It is interesting that an anticyclone is climatologically

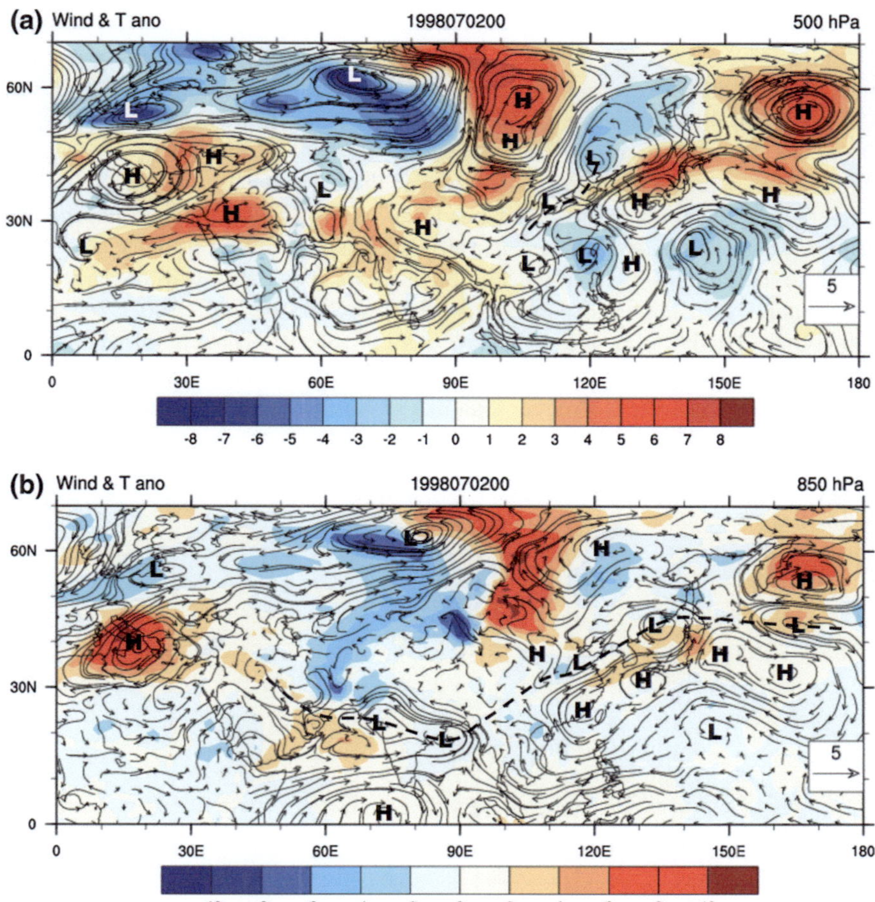

Fig. 5.23 Same as in Fig. 5.21 except of anomalous wind (m/s) and anomalous temperature (*shading*, 2 K interval) at **a** 500 hPa and **b** 850 hPa at 0000 UTC 2 July 1998

exhibited in North China. Therefore, the climatic circulation at lower troposphere and 0000 UTC 2 July is favorable to rains along this shear line in China.

The anomalous synoptic charts have the ability to indicate surface weather extremes. After removing the climatic component from the total fields, anomalous synoptic chart is shown in Fig. 5.23. Many vortices or anomalous systems, usually warm anticyclones and cold cyclones are observed at 500 hPa (Fig. 5.23a). Two anomalous cold cyclones and two anomalous warm anticyclones exist in high latitudes, while there is not any vortex in the tropics. Many vortices are active in subtropical and middle latitudes. From the SCS to the NW Pacific, three anomalous anticyclones and three anomalous cyclones in Fig. 5.23a are superposed on a climatic subtropical anticyclone in Fig. 5.22a. Two anomalous cyclones from the Yellow River to Northeast China are separating southwesterly in their southern side and northeasterly in their northern side. When comparing the anomalous systems at 850 hPa (Fig. 5.23b) and 500 hPa (Fig. 5.23a), it can be noted that the warm anticyclones and cold cyclones are deeper in high latitudes. Four anticyclones form a family from South China to Japan and the subtropical NW Pacific in Fig. 5.23b. Over the Taiwan Strait, an anticyclone is located in the lower troposphere (Fig. 5.23b), while a cyclone is at the upper troposphere (Fig. 5.23a). There is a convergence line along the five cyclones from South Asia to East Asia and the NW Pacific. These anomalous systems can be traced hourly and even daily.

As an example of application, Fig. 5.24 shows the relationship of daily mean precipitation with total and anomalous winds at 700 hPa and 0000 UTC 2 July 1998 (Qian 2013). It is clear that the large rainfall zone is located in the right side of the shear and the left side of the low-level jet (Fig. 5.24a). The southwesterly in the mid-low Yangtze River is stronger than that in North China. The situation changes

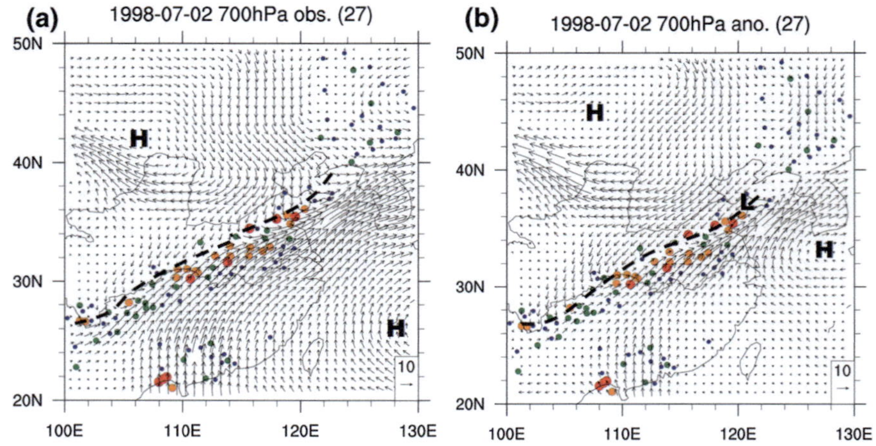

Fig. 5.24 **a** Total wind (m/s) and **b** anomalous wind (m/s) at 700 hPa and 0000 UTC 2 July 1998 using the ERA-Interim data. *Blue*, *green*, *yellow*, *red*, and *purple dots* indicate the amounts of rainfall with ranges: 10–24.9 mm, 25–49.9 mm, 50–99.9 mm, 100–249.9 mm, and exceeding 250 mm per 24 h at 0000 UTC 2 July 1998. The *dashed line* denotes the shear or convergence line. Letters "H" and "L" indicate high (anticyclone) and low (cyclone) circulation systems. Modified from Qian (2013) © China Meteorological Press, with permission

where the anomalous shear is just along the heavy rainfall zone (Fig. 5.24b) while the southwesterly and northeasterly are equal in its two sides. Therefore, the anomalous shear or the anomalous convergence line can well locate the heavy rainfall zone in practice.

We strongly suggest that vertical sections of synoptic chart should be used in daily weather analysis. Figure 5.25 shows the vertical-meridional section of total height and climatic height as well as total wind and climatic wind along 116.25°E at 0000 UTC 2 July 1998. Similar feature is observed from total height and climatic height as well as total temperature and climatic temperature (Fig. 5.25a, b) except that there is a higher temperature near the surface in the south of the rainfall (Fig. 5.25a). Therefore, there is no obvious signal to indicate the heavy rainfall in

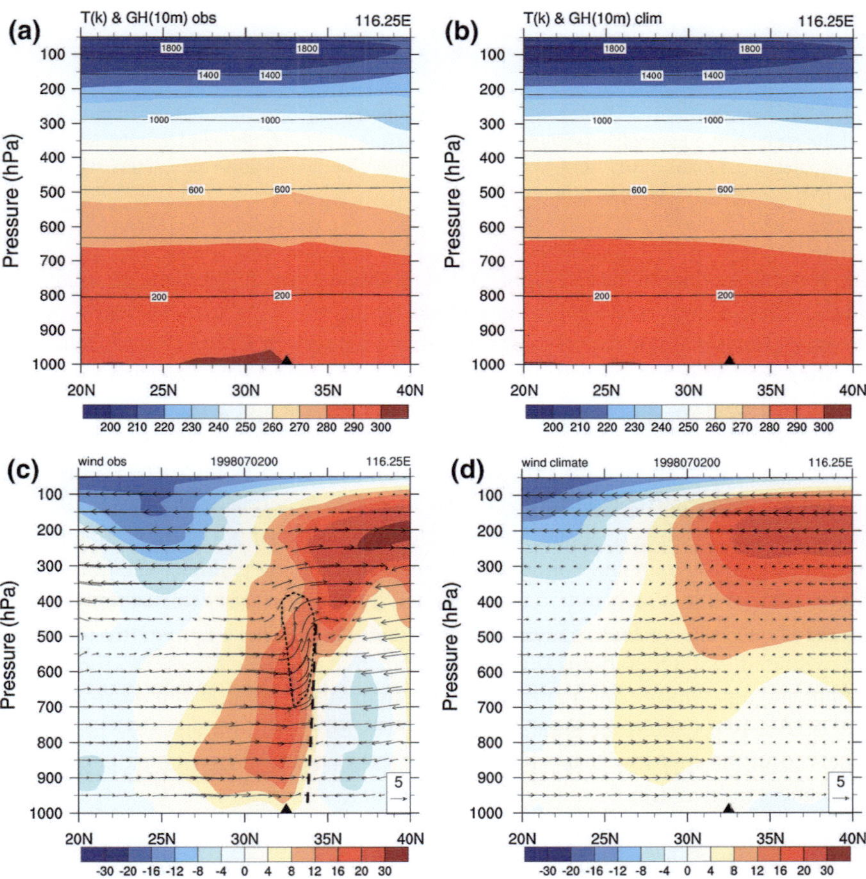

Fig. 5.25 Vertical-meridional section of **a** total height (*contour*, 200 × 10 gpm) and **b** climatic height (*contour*, 200 × 10 gpm) as well as **c** total wind (m/s) and **d** climatic wind (m/s) along 116.25°E at 0000 UTC 2 July 1998 using the ERA-Interim data. In **a** and **b**, *shading* denotes total and climatic temperatures (10 K interval). In **c** and **d**, *red* and *blue shadings* respectively indicate westerly and easterly velocities (4 m/s interval) while the *dotted line* covers the ascending velocity (0.5 Pa s^{-1}) by enlarging 10 times

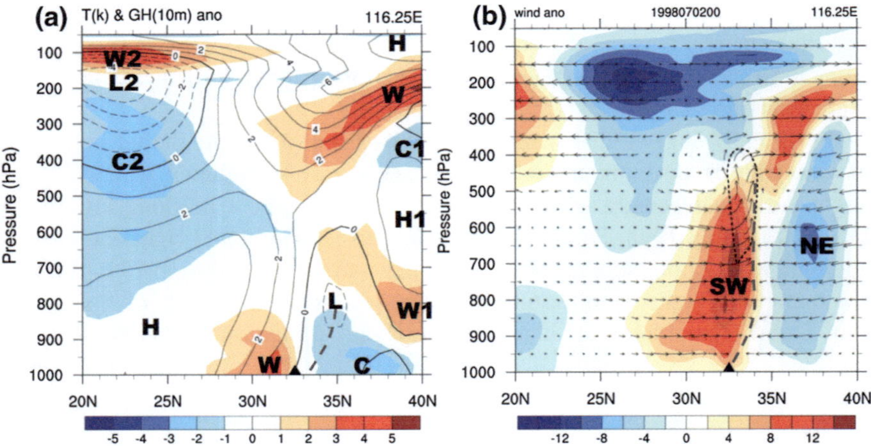

Fig. 5.26 Same as in Fig. 5.25a, c, but for **a** the height anomalies (*contour*, 1 × 10 gpm interval) and the temperature anomalies (*shading*, 1 K interval) as well as **b** the wind anomalies (*shading*, 2 m/s interval). Letters "H/L" and "W/C" indicate high/low and warm/cold centers while "SW" and "NE" are southwesterly and northeasterly

the vertical section of total height and temperature. In the section of total wind (Fig. 5.25c), a convergence line is located in the northern side of heavy rainfall because climatic southwesterly is predominant in eastern China (Fig. 5.25d).

After removing the climatic height, temperature, and wind, their anomalies are shown in Fig. 5.26. In Fig. 5.26a, a warm center (W1) and a cold center (C1) are located below and above the positive center (H1) of height anomalies at 600 hPa in North China. In South China, however, a warm center (W2) and a cold center (C2) are located above and below the negative center (L2) of height anomalies at 200 hPa. In the lower troposphere, there are warm center and positive height center to the south of the heavy rain and cold center and negative height center to the north of the heavy rain, while the heavy rain is located at the surface front of temperature anomalies and the surface trough of height anomalies. These symmetric structures are exhibited in the total fields (Fig. 5.25a) but are not visible directly to forecasters. Height and temperature anomalies satisfy the hydrostatic equilibrium relationship. In Fig. 5.26b, two centers of anomalous southwesterly and northeasterly are clearly observed in the mid-low troposphere, while their convergence line extends downward and southward to the place of heavy rainfall.

5.7 Examples of Model Prediction

Here, we give a series of anomaly-based synoptic charts, which are predicted by the Global Forecast System (GFS) of the National Centers for Environmental Prediction (NCEP). Figure 5.27 shows the daily prediction of height and

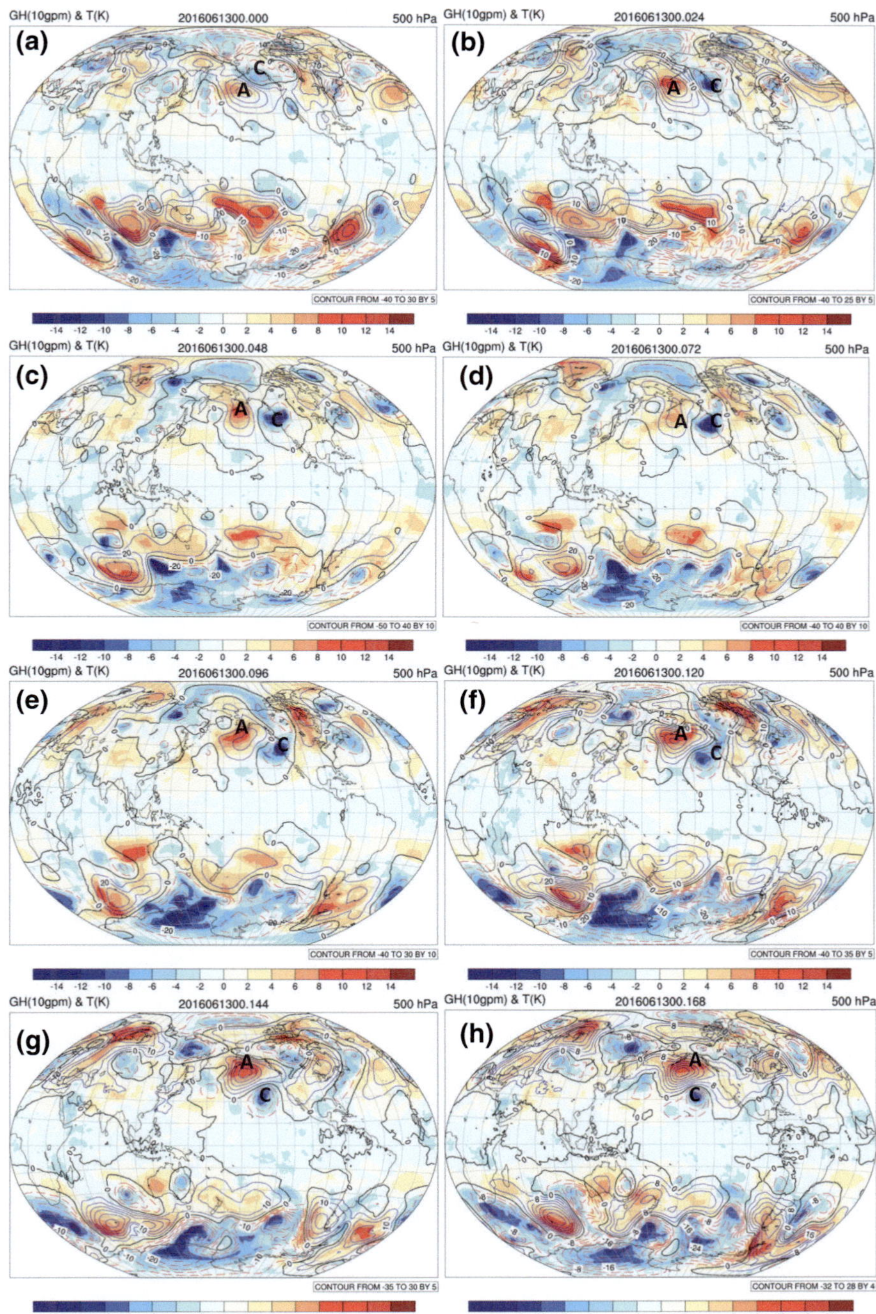

Fig. 5.27 Height anomalies (*contour*, 5 × 10 gpm interval) and temperature anomalies (*shading*, 2 K interval) predicted by GFS initiated from 0000 UTC 13 June 2016 at 500 hPa for following 360 h (15 days). Letter "A" indicates the center of warm and positive height anomalies, while the letter "C" is the center of cold and negative height anomalies

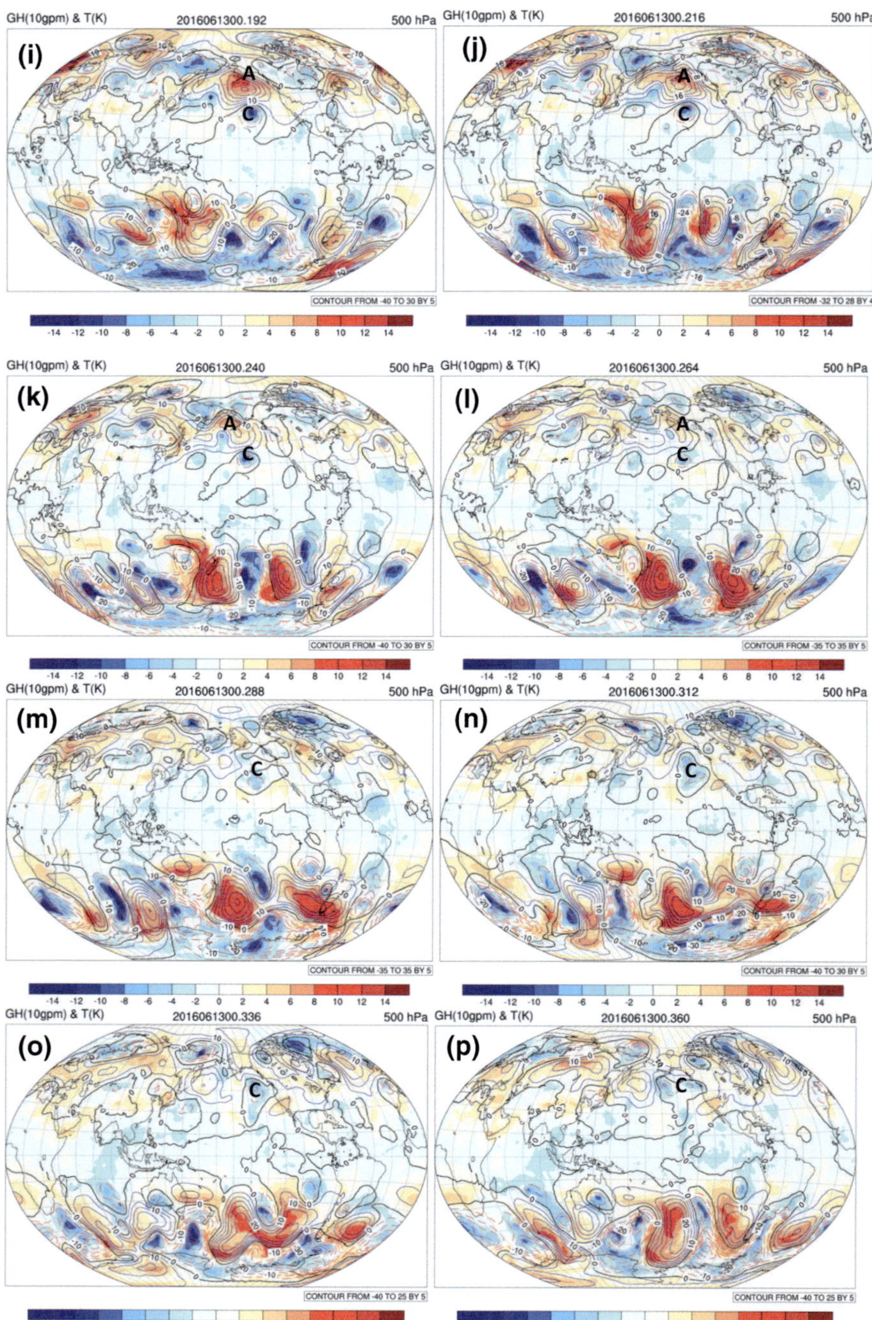

Fig. 5.27 (continued)

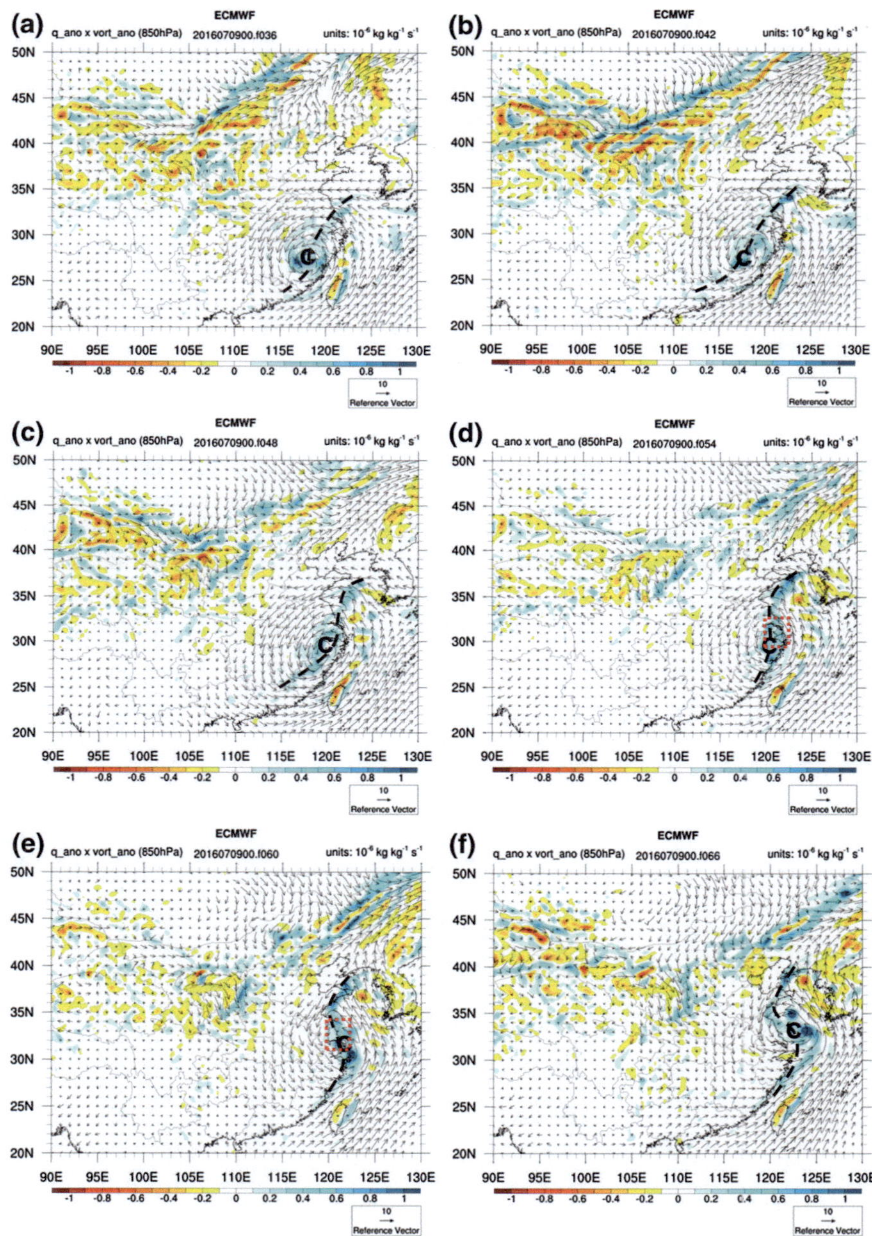

Fig. 5.28 Wind anomalies (m/s) and moist vorticity anomaly (*shading*, 10^{-6} kg kg^{-1} s^{-1}) at 850 hPa predicted by the ECMWF model initiated from 0000 UTC 9 July 2016 for the following **a** 36, **b** 42, **c** 48, **d** 54, **e** 60, and **f** 66 h. The *heavy dashed line* is the *convergence line* of anomalous flows in (**c**) and (**e**)

temperature anomalies at 500 hPa initiated from 0000 UTC 13 June 2016 for following 360 h (15 days). Anomalous systems are stronger in the mid-high latitudes in the SH than those in the NH in boreal summer. Along the tropics, anomalous systems are shallow. We follow a pair of positive center (A) and negative center (C). The former is an anticyclonic (high) warm center and the latter is a cyclonic (low) cold center. They are first observed over the Northeast Pacific and the northwestern coast of North America, respectively (Fig. 5.27a). The two centers experience a clockwise rotation up to 264 h (Fig. 5.27l). A warm center and a cold center at the surface can be traced by following the positive and negative centers at 500 hPa. Other positive and negative centers of height anomalies as well as warm and cold centers can also be traced in Fig. 5.27.

Figure 5.28 is another example of anomaly-based synoptic chart at 850 hPa predicted by the ECMWF model initiated from 0000 UTC 9 July 2016 for 36–66 h. The anomalous low system in Fig. 5.28a, b is a remnant of Typhoon Nepartak. It generated in the south of Guan at 0800 Beijing Time on 7 July 2016 and moved northwestward. It made landfall on Taiwan at 0550 Beijing Time on 8 July 2016 and then hit Fujian on the next day. It reduced to a tropical depression in the morning of 10 July 2016. Its hit Fujian and caused 69 deaths and 6 missing.

We calculate a combined index multiplying by vorticity anomaly and specific humidity anomaly, $D'Q' = \left(\frac{\partial v'}{\partial x} - \frac{\partial u'}{\partial y}\right)q'$, which is named as moist vorticity anomaly (MVA) by Qian et al. (2016). It is found that the ECMWF model predicted a positive MVA axis along the anomalous convergence line crossing the center of the tropical depression (Fig. 5.28). A regional heavy rainfall occurred from local afternoon to evening (the red-dotted box in Fig. 5.28d, e). The Nantong University station recorded 82.2 and 151.2 mm/h precipitation rates at 16:00 and 17:00 local time. This heavy rainfall was a case of successful predictions on 10 July 2016 based on Fig. 5.28.

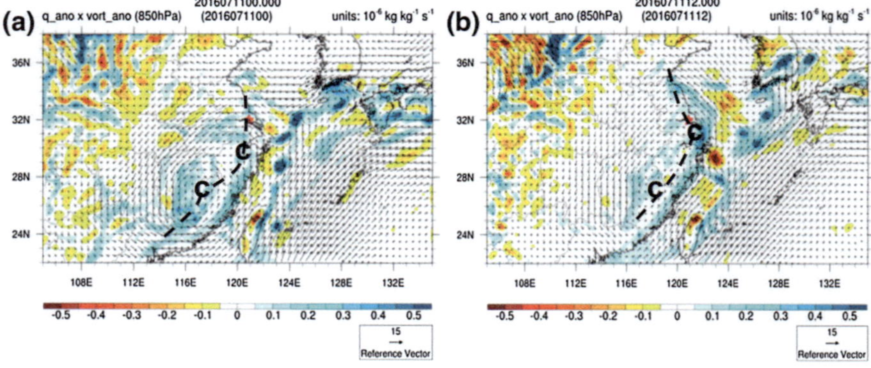

Fig. 5.29 Same as in Fig. 5.28c, e but from the ERA-Interim product

The analysis from the ERA-Interim product shows that the ECMWF model well predicted an induced low vortex in Zhejiang Province but the tropical depression dissipated in Fig. 5.28c when comparing with the product in Fig. 5.29a. The convergence line crossing the low center and the positive MVA can be comparable between Figs. 5.28e and 5.29b.

Questions

(1) Why do waves or troughs exist in daily climatic charts and vortices in anomaly-based synoptic charts? And why are waves, troughs, and vortices all observed in traditional synoptic charts?
(2) Why are anomalous cold-high (or warm-low) circulation systems observed at the upper troposphere and why do anomalous warm-high (or cold-low) systems exist in the mid-low troposphere?
(3) Why can forecasters find all the surface weather extremes from the positive and negative centers of height anomalies near 300 hPa at a moment?
(4) Can we locate the surface weather extremes two weeks in advance if the current numerical model can correctly predict the total height 15 days in advance?

References

Qian WH (2013) Atlas of anomalous circulations associated with regional heavy Rainfall in China. China Meteorological Press, Beijing, pp 227

Qian WH, Jiang N, Du J (2016) Anomaly-based weather analysis versus traditional total-field-based weather analysis for depicting regional heavy rain events. Wea Forcast 31:71–93

Chapter 6
Regional Convective Events

Regional persistent rainfalls, one of the high-impact weather extremes in China, can cause mud slides in maintain regions and floods in plains. Tornadoes, which occur frequently in the central United States and eastern China, are also weather extremes during the warm season. Traditional synoptic charts drawn by basic atmospheric variables and diagnostic parameters are usually used in short-term forecasting of weather extremes. The advantages of anomaly-based synoptic analyses over the traditional total ones in locating regional heavy rains (RHRs) and tornado-producing systems will be demonstrated in this chapter. The two different synoptic charts and diagnostic parameters are compared in a regional rain event on 1 July 1991 in China. Statistical results are made from 41 RHR cases from 1998 and other cases in eastern China. Analyses are also done on tornado-producing anomaly systems in central United States on 27 April 2011 and in eastern China on 23 June 2016. The basic anomalous variables and the diagnostic parameters can also be used in post-processing of model forecast outputs to potentially helping predicting weather extremes.

6.1 Site and Regional Rainfall Events in China[1]

Usually, the severities of heavy rain disasters caused by a single station and adjacent stations are rather different in many cases. A site heavy rain (SHR) day is defined as the daily 24-h (from 1200 UTC to the next 1200 UTC) precipitation amount exceeding 50 mm at a station as graded in Table 6.1. In the eastern part of China (Fig. 6.1), the averaged distance between two adjacent stations is 66 km and the largest distance is 163 km. Thus, the criterion for a RHR event is at least two adjacent SHR stations within 200 km. The regional coverage of a daily RHR event

[1]Modified article originally published in Shan et al. (2015). Published with kind permission of © Springer Science + Business Media Dordrecht 2015. All Rights Reserved.

© Springer Nature Singapore Pte Ltd. 2017

W. Qian, *Temporal Climatology and Anomalous Weather Analysis*,
Springer Atmospheric Sciences, DOI 10.1007/978-981-10-3641-5_6

can also be extended, if there are stations with the precipitation amount from 25 to 50 mm day^{-1} connecting with adjacent heavy rain stations. The detail about how to identify a daily RHR event is shown in the four regional cases below. In short, a RHR event is defined as at least two adjacent stations' precipitations (less than 200 km) exceeding 50 mm day^{-1} on the same day.

Shan et al. (2015) examined the number of SHR days in China in terms of the precipitation grade in Table 6.1 based on historical records. Figure 6.2 shows the annual mean frequency of SHR days averaged over 1960–2010. The annual mean frequency of SHR days in northwestern China is very low. A frequency zone with 1–3 SHR days per year is oriented in southwest–northeast direction from the upper Yangtze River valley (Southwest China) to North China (the low Yellow River) and the southern part of Northeast China. Two high-frequency centers of SHR days (5–10 days per year) are located in South China and the south of lower Yangtze River. The highest frequency of SHR days (above 10 days/year) is found along the coast of South China and over the Hainan Island.

In China, the number of SHR days shows significant interannual variability for all stations and for the selected 460 stations respectively from 1960 to 2010 (Fig. 6.3). The annual mean frequency of SHR days is about 190 days for all stations. Two maximal SHR 221 and 218 days occurred in 1983 while two minimal SHR 173 and 161 days were in 2004 in the records of all stations and 460 stations,

Table 6.1 Precipitation grades of 24-h precipitation (from 1200 UTC to the next 1200 UTC)

Grade	Precipitation amount (mm/24 h)
Small	0.1–9.9
Medium	10.0–24.9
Large	25.0–49.9
Heavy	50.0–99.9
Very heavy	100.0–249.9
Heavy downpour	≥250.0

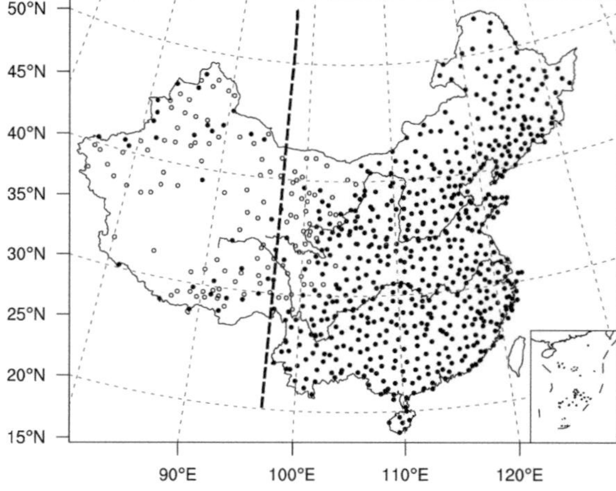

Fig. 6.1 The distribution of 754 stations in China. *Solid* and *open circles* indicate stations with and without heavy rainfall days from 1960 to 2010, respectively. The *heavy dashed line* denotes 97.5°E, which divides China into eastern and western parts

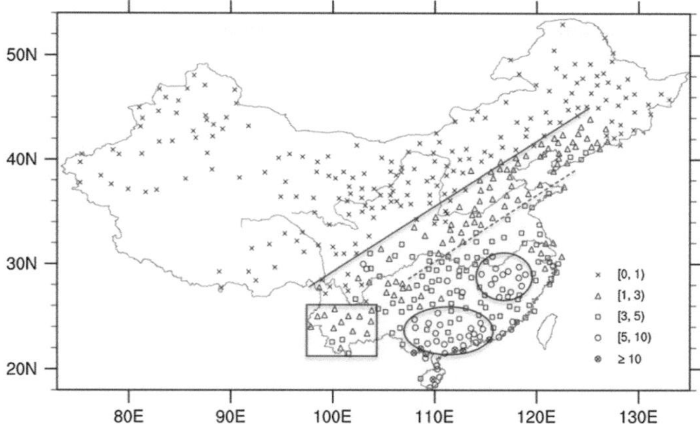

Fig. 6.2 Annual mean frequency (days) of SHR days (≥ 50 mm day^{-1}) at 460 stations averaged over 1960–2010. *Two solid* and *dashed straight lines* as well as a *box* in Southwest China and *two closed circles* in South China and the south of lower Yangtze River distinguish different frequency regions. Sourced from Shan et al. (2015) © Springer Science + Business Media Dordrecht 2015, with permission

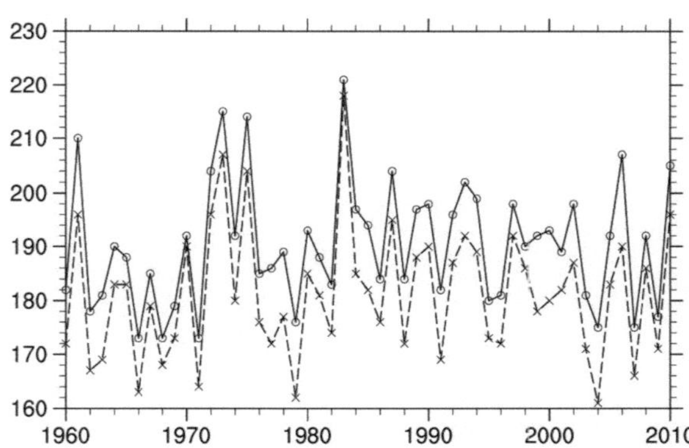

Fig. 6.3 Annul frequency of SHR days in China from 1960 to 2010. *Solid line* (days) and *dashed line* (days) denote all stations and 460 stations used in the statistical analysis, respectively. Sourced from Shan et al. (2015) © Springer Science + Business Media Dordrecht 2015, with permission

respectively. The lower frequency of SHR days is found during 1962–1971. There are significant interannual and interdecadal variability of SHR days in China. These SHR days should be mostly associated with synoptic systems such as tropical cyclone, subtropical cyclone, and low-level convergence (or shear) lines. In mainland China, most SHR days occur in the eastern summer monsoon region, i.e., east of 97.5°E. During the period of 1960–2010, total 93 SHR days occurred in the region of Xinjiang and Tibet, while 9686 SHR days occurred in the eastern part of China.

Using the definition of daily RHR event, Fig. 6.4 shows four daily cases occurred on 16 August 1994 and 12 July 2010. On 16 August 1994, two daily RHR events occurred along the coast of South China (13 heavy rain stations and 18 large

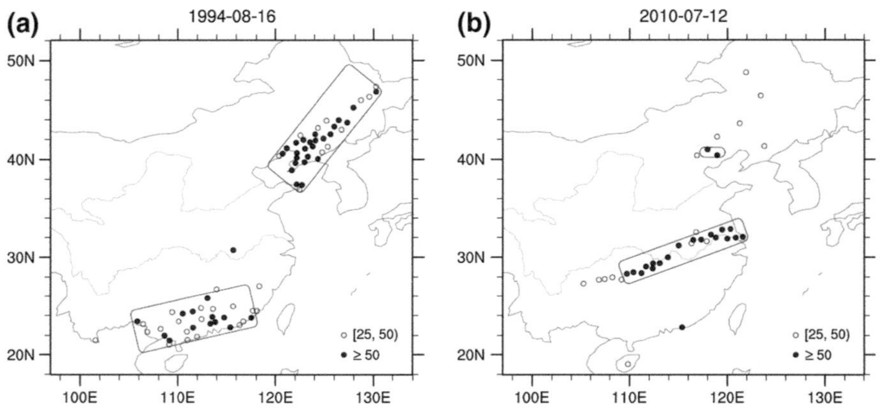

Fig. 6.4 Daily RHR events on **a** 16 August 1994 and **b** 12 July 2010. *Solid circles* and *open circles* indicate stations with precipitation more than 50 mm day^{-1} and from 25 to 50 mm day^{-1}, respectively. Sourced from Shan et al. (2015) © Springer Science + Business Media Dordrecht 2015, with permission

Table 6.2 Daily RHR events with adjacent stations from 2 to 10, ≥ 5 and ≥ 10 during 1960–2010

Adjacent stations	2	3	4	5	6	7	8	9	10	≥ 5	≥ 10
Regional daily rain events	3640	1891	1232	826	595	497	387	316	278	4280	1659
Annual mean events	71	37	24	16	11	9	7	6	5	83	32
Rate (%)	33.0	16.1	11.2	7.5	5.4	4.5	3.5	2.9	2.5	38.8	15.0

Sourced from Shan et al. (2015) © Springer Science + Business Media Dordrecht 2015, with permission

rain stations nearby) and Northeast China (25 heavy rain stations and 11 large rain stations nearby), respectively. The heavy rain at a single station in the lower Yangtze River is considered as a single site event. On 12 July 2010, a daily RHR band (18 heavy rain stations and 8 large rain stations nearby) was along the middle to lower Yangtze River and another small RHR event with only two heavy rain stations in North China. Total 8206 daily RHR events were identified in 460 stations during 1960–2010.

All daily RHR events during 1960–2010 are identified by the same procedure as that for the four events shown in Fig. 6.4. Daily RHR events with 2–10 adjacent stations and more than 10 adjacent stations are respectively listed in Table 6.2. Among all daily RHR events, 61.2% occurred with adjacent 2–4 stations, while 38.8 and 15% with at least 5 and 10 adjacent stations, respectively. Every year,

Table 6.3 Same as in Table 6.2 except of monthly mean daily RHR event frequency of adjacent stations ≥ 5 and ≥ 10 during 1960–2010

Adjacent stations	January	February	March	April	May	June	July	August	September	October	November	December
≥ 5	14	20	79	282	533	884	1010	812	385	173	67	21
≥ 10	7	5	28	85	212	428	403	283	137	50	15	6

Sourced from Shan et al. (2015) © Springer Science + Business Media Dordrecht 2015, with permission

there are, in average, 32 daily RHR events with at least 10 adjacent stations in China. The monthly mean daily RHR event frequencies of adjacent stations ≥ 5 and ≥ 10 during 1960–2010 are respectively listed in Table 6.3. Seasonally, maximal 1010 daily RHR events (adjacent stations exceeding 5) occur in July while minimal 14 daily RHR events in January.

The top 20 daily RHR events according to the number of adjacent stations are listed in Table 6.4. On 19 April 1996, the largest extent of daily RHR event covered 50 stations in the south of Yangtze River and South China. On 6 September 1962, a daily RHR event with 37 adjacent stations in the lower Yangtze River showed a regional average rainfall 110.8 mm day^{-1}. All the top daily 20 RHR events occurred from April to October. There were 1, 3, 1, 11, and 4 daily RHR events in the 1960s, 1970s, 1980s, 1990s, and 2000s, respectively. It is noted that most events of RHR occurred in the 1990s. The top 20 daily RHR events are categorized according to their triggering mechanisms in terms of anomaly-based synoptic systems. Among the top 20 daily RHR events, 4 cases are associated with convergence lines and 12 cases are linked with subtropical cyclones and tropical cyclones. Four cases are not categorized since they occurred before 1981, which is beyond the coverage of the ERA-Interim data used.

Annual frequencies of daily RHR events with 2–4 adjacent stations and ≥ 5 adjacent stations in eastern China are shown in Fig. 6.5. The annual frequency of daily RHR events with 2–4 adjacent stations has interannual variability. Differently, the annual frequency of daily RHR events (≥ 5 adjacent stations) shows a decadal variability with four major peaks in the last four decades and a long-term increasing trend particularly since the late 1970s. Three peaks reaching 110 times of daily RHR events are observed in 1983, 1998, and 2010. It is well known that many persistent RHR events lasted for consecutive days and severe disasters frequently occurred in the two summers of 1998 and 2010 (National Climate Center of China 1998; Qu et al. 2010; Wang et al. 2010; Shen et al. 2011; Min et al. 2011; Wang and Zhang 2011; Wikipedia 2014; Qian et al. 2013). Decadal anomalous events from dry–wet and warm–cold alterations are one of the multi-decadal climate oscillations in China (Qian et al. 2007), which should have close linkage with the RHR events.

Table 6.4 The dates, number of adjacent stations, average rainfall (mm day^{-1}), and the three types of triggering systems of the top 20 daily RHR events from 1981–2010

Range	Date(d/m/y)	Stations	Rain (mm)	Triggering systems
1	19/Apr/1996	50	87.2	Convergence line
2	17/Jun/1994	39	88.2	Subtropical cyclone
3	13/Jun/2008	39	87.7	Subtropical cyclone
4	6/Sep/1962	37	110.8	/
5	24/Jun/1998	37	92.9	Subtropical cyclone
6	19 /Aug/1997	36	104.6	Tropical cyclone
7	19/Jun/1998	36	80.8	Convergence line
8	2/Jun/1976	36	77.7	/
9	1/Jul/1990	35	75.7	Subtropical cyclone
10	5/Oct/1983	35	75.7	Convergence line
11	30/Jul/1975	34	105.0	/
12	10/Jun/1994	34	91.3	Subtropical cyclone
13	6/Jul/1992	34	83.7	Convergence line
14	26/May/1999	34	76.6	Subtropical cyclone
15	14/May/2002	34	73.4	Subtropical cyclone
16	20/Aug/1997	33	109.7	Tropical cyclone
17	19/Aug/2002	33	88.9	Tropical cyclone
18	31/Aug/1992	33	87.0	Tropical cyclone
19	24/Jun/1979	33	78.6	/
20	13/May/2002	33	76.0	Subtropical cyclone

Sourced from Shan et al. (2015) © Springer Science + Business Media Dordrecht 2015, with permission

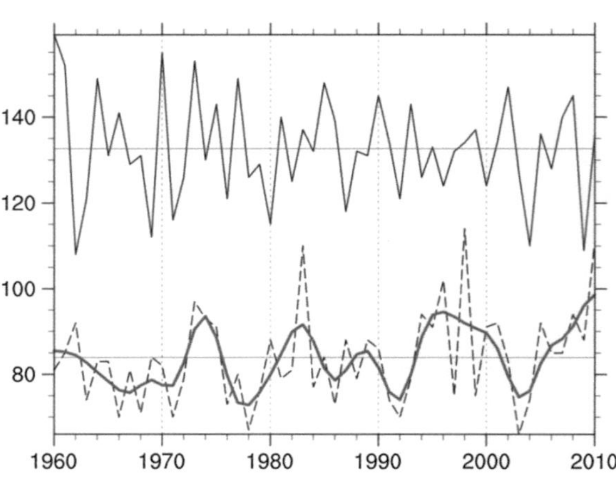

Fig. 6.5 Annual frequencies of daily RHR events with 2–4 adjacent stations (*thin solid line*) and ≥ 5 adjacent stations (*dashed line*) in eastern China (east of 97.5°E). The *heavy line* denotes 11-point weighted running mean. Sourced from Shan et al. (2015) © Springer Science + Business Media Dordrecht 2015, with permission

6.2 Comparison Between Total and Anomalous Fields[2]

Comparison between total and anomalous fields of basic atmospheric variables (height, temperature, wind, and humidity) has been examined by (Qian et al. 2016a). Total wind vectors $\vec{V}_{\{d,y\}}(\lambda, \varphi, t)$ are decomposed into climatic wind vectors $\overline{\vec{V}}_{\{d\}}(\lambda, \varphi, t)$ and anomalous wind vectors $\vec{V}'_{\{d,y\}}(\lambda, \varphi, t)$. Height, temperature, and humidity are also decomposed into their climatic components and anomalies. These four basic variables are used to indicate regional rainfalls. A case on 1 July 1991 and other cases in eastern China are respectively compared.

On 1 July 1991, 22 stations were recorded rainfall exceeding 50 mm day^{-1} with a maximum precipitation of 201.8 mm day^{-1} and an average precipitation of 96.5 mm day^{-1}. Figure 6.6 shows the total, anomalous, and climatic winds at 850 hPa and 0000 UTC 1 July 1991. In the total wind field (Fig. 6.6a), a heavy rain band was located on the left side of the low-level jet (LLJ) and near the horizontal low-level shear (LLS) line, where the southwesterly was much stronger on the southern side of the shear line. There were no cyclonic systems associated with this rain band. By contrast, two synoptic-scale cyclonic vortices, one over the upper Yangtze River Valley ("C1") and another over South Korea ("C2"), are clearly observed in the anomalous wind field (Fig. 6.6b), which are connected by an anomalous wind shear line along the Yangtze River. Two anticyclone centers ("A1" and "A2") are situated at each side of the anomalous wind convergence line over the lower Yellow River and the subtropical Northwest (NW) Pacific, respectively. The anomalous convergence line overlaps with the heavy rain band. Although the LLS lines in both total and anomalous fields indicate the rain band location equally well, the anomalous lows and highs can only be revealed in the anomalous wind field. The LLS lines in the climatic wind are apparently not a good indicator for the heavy rain band (Fig. 6.6c) as one would expect.

Figure 6.7 compares the vertical–latitude cross-sections of the total and anomalous height–temperature along 114.75°E (corresponding to the dashed line in Fig. 6.6) at 0000 UTC 1 July 1991. To have a comprehensive comparison, two different scales (regional and local) are provided. Both total and anomalous temperature fields have an equally visible signal connecting to the heavy rain event, but the former shows a large temperature gradient while the latter depicts a clash of anomalous cold and warm air masses in the lower troposphere. However, there is no obvious signal that allows for identification of the rain band from the total height distribution in regional and local scales (Fig. 6.7a1, a2). By contrast, the height anomalies show a clear signal indicating the rain band, where an axis of negative height anomalies extends downward and southward from the upper troposphere to the heavy rain area in two scales (Fig. 6.7b1, b2). These features indicate that vertical coupling between the atmospheric systems over lower and upper

[2]Modified article originally published in Qian et al. (2016a). Published with kind permission of © American Meteorological Society. All Rights Reserved.

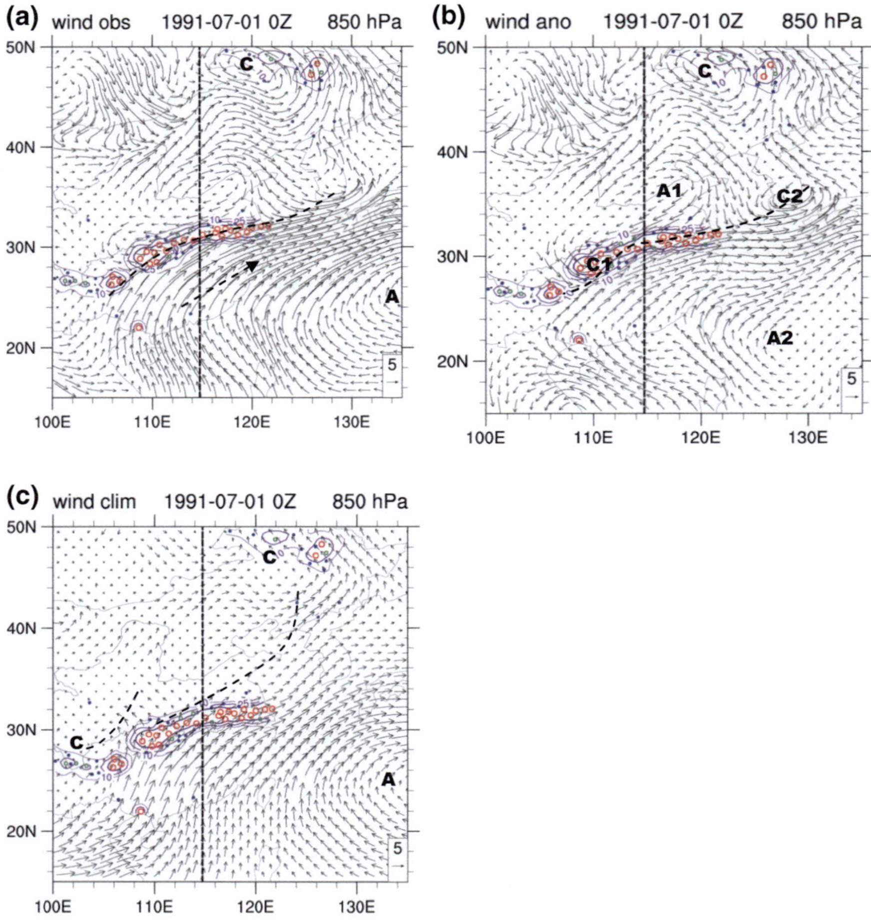

Fig. 6.6 **a** Total, **b** anomalous, and **c** climatic winds (m s^{-1}) at 850 hPa and 0000 UTC 1 July 1991. *Red*, *green*, and *blue open circles* (*dots*) indicate the stations with rainfall over 50, 25–50, and 10–25 mm day^{-1}, respectively. *Purple contours* are the isohyets of 10, 25, and 50 mm day^{-1} using the interpolated gridded precipitation (Chen et al. 2010). In **a** the *heavy dashed arrow* and *heavy dashed line* indicate the southwesterly LLJ stream and LLS line, respectively. In **b** the *heavy dashed line* indicates the anomalous convergence line. In **c** the *heavy dashed line* indicates the LLS line. Letters "A" and "C" denote the anticyclone and cyclone centers. The *straight dashed line* denotes 114.75°E longitude. Sourced from Qian et al. (2016a) © American Meteorological Society, with permission

troposphere might play a role in the formation of this heavy rain event. Apparently, this vertical coupling information cannot be easily seen from the total height field. This demonstrates that some small but important features are hard to be seen in the presence of climatological background in a total field, but it can be vividly revealed after the climatology is removed. This is an added value of anomalous approach to the total field based analysis. An area of scattered rains is located from North China

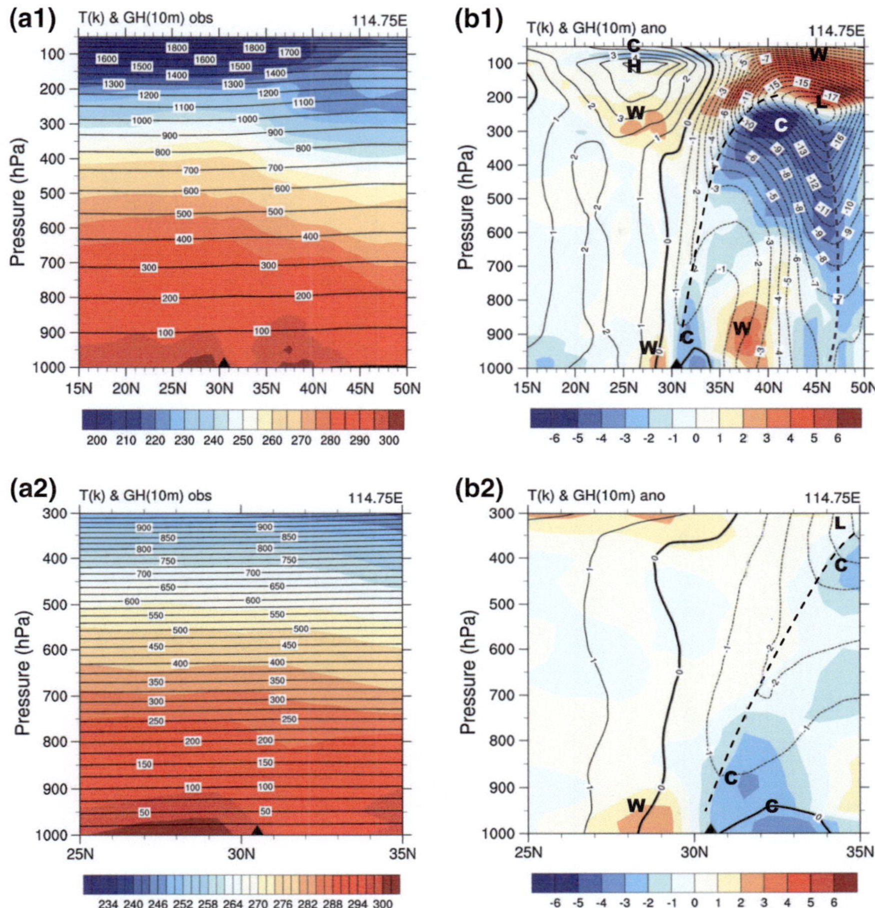

Fig. 6.7 Vertical–latitude cross-sections of **a1**–**a2** total and **b1**–**b2** anomalous height (*contour*) and temperature (*shading*) in regional and local scales, along 114.75°E at 0000 UTC 1 July 1991. The *heavy dashed line* denotes the axis of height anomalies in (**b1** and **b2**). The location of the surface rain band is indicated by the symbol *filled triangle*. Different *contour* intervals are used for different scales: In **a1** 100 × 10 gpm interval for height and 5 K interval for temperature; in **b1** 1 × 10 gpm interval for height anomaly and 1 K interval for temperature anomaly. In **a2** 25 × 10 gpm and 2 K are used; in **b2** 1 × 10 gpm and 1 K are used. Letters "H/L" and "W/C" indicate the centers of height and temperature anomalies, respectively. Sourced from Qian et al. (2016a) © American Meteorological Society, with permission

to Northeast China (Fig. 6.6). Its anomalous signal can also be found from another axis of height anomalies in Fig. 6.7b1.

Since both the height and temperature anomalies have clear signals in the lower troposphere as shown in Fig. 6.7b, the horizontal distributions of the total, anomalous, and climatic height–temperature distributions at 925 hPa and 0000 UTC 1 July 1991 are compared in Fig. 6.8. In Fig. 6.8a, there was a large

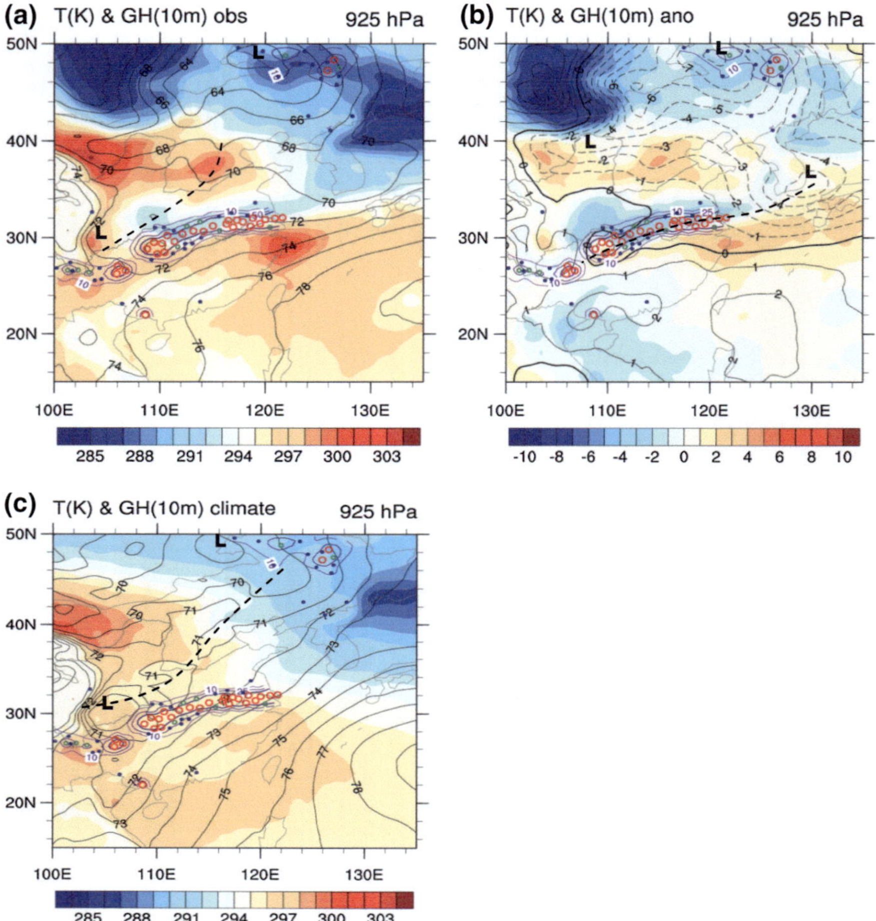

Fig. 6.8 **a** Total, **b** anomalous and **c** climatic height and temperature at 925 hPa and 0000 UTC 1 July 1991. The *red*, *green*, and *blue open circles* (*dots*) indicate the stations with rainfall exceeding 50, 25–50, and 10–25 mm day^{-1}, respectively. In **a** and **c** *contour* is for height (10 gpm interval) and color shading is for temperature (1 K interval). In **b** *contour* is for height anomalies (10 gpm interval) and shading is for temperature anomalies (1 K interval). The *heavy dashed line* denotes height trough. *Purple contours* are the isohyets of 0, 15, and 50 mm day^{-1} using the interpolated gridded precipitation. Sourced from (Qian et al. (2016a) © American Meteorological Society, with permission

temperature gradient zone lying along the rain band. But in Fig. 6.8b, the rain band can be seen as a clash result of anomalous warm and cold air masses. In the 925-hPa height field, the north boundary of the heavy rain band is better defined by the narrow anomalous trough (Fig. 6.8b), while it is located on the northern edge of a large total height gradient zone (Fig. 6.8a). Therefore, the rain band width, especially its northern boundary, is easier to be defined by the narrow anomalous

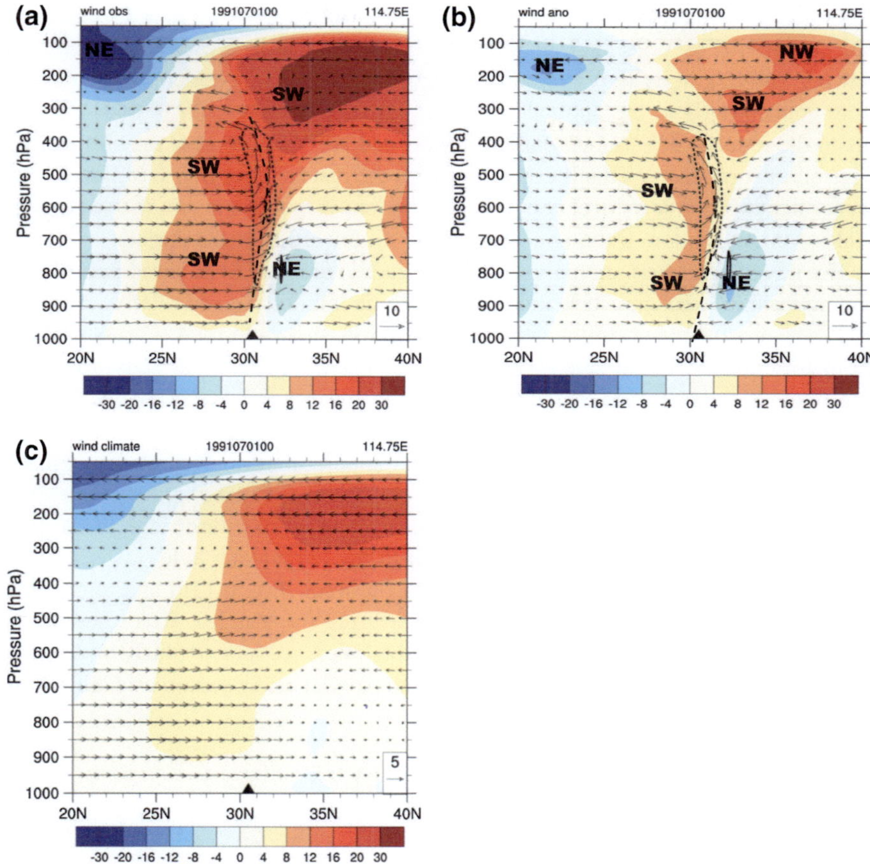

Fig. 6.9 Vertical–latitude cross-sections of **a** total, **b** anomalous, and **c** climatic winds along 114.75°E at 0000 UTC 1 July 1991. The heavy *rain area* is indicated by the symbol *filled triangle*. Westerly and easterly velocities are shaded in *red* and *blue* colors (m s^{-1}). In **a** and **b** the fine *dotted lines* denote the ascending area with vertical velocity exceeding 0.5 Pa s^{-1} (enlarged by 10 times), and the *heavy dashed line* denotes the convergence line. Symbols "SW", "NW", and "NE" denote the southwesterly, northwesterly and northeasterly velocities, respectively. Sourced from Qian et al. (2016a) © American Meteorological Society, with permission

trough than by the wide pressure uniform area in the total height field. In the climatic field (Fig. 6.8c), there is no well-defined signal indicating the rain band as expected. Figure 6.8b also shows an extension of the anomalous trough to the southern South Korea with an anomalous low center. Although the dataset utilized here does not contain rainfall records outside China, the South Korean observations confirm that a RHR event did occur in the southern South Korea in response to this anomalous vortex.

The vertical–latitude cross-sections of total and anomalous winds along 114.75°E at 0000 UTC 1 July 1991 are compared in Fig. 6.9. The convergence of winds and ascending motion are evident in the mid-low troposphere over the heavy rain area in both the total (Fig. 6.9a) and anomalous (Fig. 6.9b) winds. However, the contrast between southwesterly (SW) velocity and northeasterly (NE) velocity is clearer and more concentrated in the anomalous field, particularly between 950 and 400 hPa, due to the reduced SW velocity and the enhanced NE velocity after the climatology (Fig. 6.9c) is removed. This is consistent with the results from Fig. 6.6. Besides the deeper depth of the low-level convergence, the convergence zone in the anomalous field is vertically more tilted and shifts more southward near the surface.

The advantage of an anomaly-based approach over total field based approach is also noticeable in the moisture field. Both specific humidity q and relative humidity (RH) q/q_s are examined. In the vertical distribution of q, both total (Fig. 6.10a) and anomalous (Fig. 6.10b) fields show an axis of maximum value directly above the heavy rain area. However, the structure is more visible and well organized in the anomalous field: the axis of positive moisture anomaly is accompanied by a negative moisture anomaly in both northern and southern sides. Due to the climatologically high value (black contours in Fig. 6.10a), the distribution of q is almost opposite in the total and anomalous fields near 25°N: a moisture ridge in the total field and a moisture trough (negative value) in the anomalous field.

In the horizontal distribution of RH, both the total (Fig. 6.11a) and anomalous (Fig. 6.11b) fields do have a high RH belt covering the Yangtze River heavy rain band. However, the high RH belt in the total field covers a much wider area extending from Vietnam all the way to the Yangtze River via southwest China (area

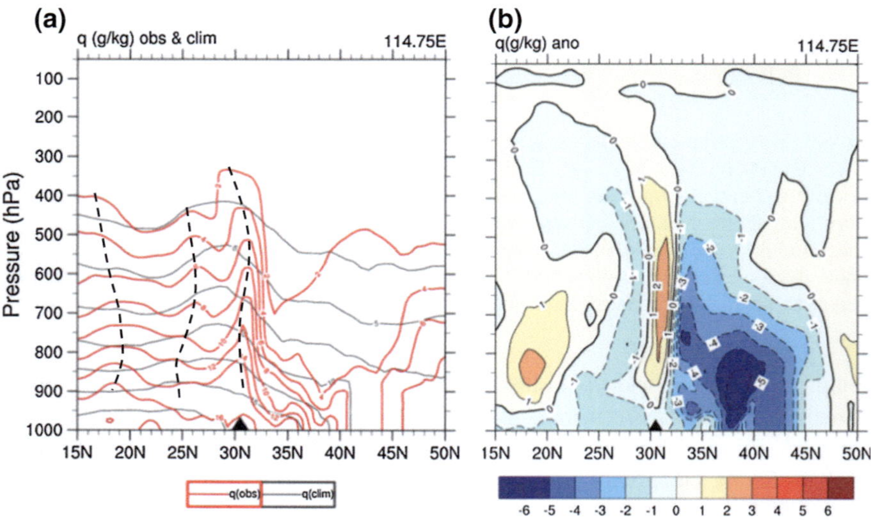

Fig. 6.10 Vertical–latitude cross-sections of specific humidity (q) along 114.75°E at 0000 UTC 1 July 1991. **a** Total specific humidity (q, *red contour* in 2 g/kg interval) and its climatic component (*black contour*, 2.5 g/kg interval). **b** Specific humidity anomaly (*shaded* with 1 g/kg *contour* interval). The *thick dashed lines* in **a** represent the axes of maximum total specific humidity. Sourced from Qian et al. (2016a) © American Meteorological Society, with permission

II + I in Fig. 6.11a). The RH within this area is almost uniform (with a maximum in the non-heavy rain area II) and indistinguishably covers both the heavy rain and non-heavy rain regions. In contrast, the anomalous RH shows a maximal zone which is concentrated mainly in the heavy rain area (area I in Fig. 6.11b). This certainly provides a more meaningful signal to forecasters. The reason why the false alarm area II is absent in the anomalous field is that Vietnam and southwest China are climatologically wetter than the Yangtze River valley (Fig. 6.11c). After the climate background is removed, an anomalous signal is revealed which has a closer relation with daily weather anomalies. This anomalous signal is even clearer if it is expressed by the ratio of the total RH to the climatic RH (Fig. 6.11d). The climatic component mainly reflects the large-scale wet southwest Asian monsoon and dry air mass in the northwest China desert region, which is slightly correlated with the heavy rain event. The horizontal distribution of specific humidity field is generally similar to that of relative humidity in both total (cf. Fig. 6.11e, a) and anomalous (cf. Fig. 6.11f, b) components. The heavy rain band is collocated with a maximum band of positive specific humidity anomalies. One advantage of using q instead of RH exists in climatologically dry region, i.e., the anomalous wet zone in terms of RH is now replaced by an anomalous dry zone in terms of q in the NW corner where no rain was observed. This positive RH anomaly is probably caused by the abnormally lower temperature (as indicated by Fig. 6.8b) rather than by excessive moisture content. Since dynamical factors are not considered, not all areas with anomalous high moisture are expected to have heavy rain events (the impact of combining moisture and dynamical factors together has been investigated in detail in Qian et al. 2015a).

We further give other cases, but only total winds and anomalous winds are compared. Figure 6.12 shows the total wind and anomalous wind at 850 hPa and 0000 UTC 3 July 2009. A regional daily heavy rain event satisfies the definition with more than 10 adjacent stations (less than 200 km apart), meeting the RHR criteria in southern China. However, this regional rainfall event is caused by two anomalous cyclonic systems in the lower troposphere. Two cyclonic centers are clearer in the anomaly-based synoptic chart (Fig. 6.12b) than those in total-based synoptic chart (Fig. 6.12a). Figure 6.12a shows a typical lower level flow indicating the location of regional rainfall, where the heavy rain is located in the right side of LLS and in the left side of LLJ in the traditional synoptic chart (Zhu 1975; Huang 1981). The spatial configuration is different from the anomaly-based synoptic chart in which the RHR is along the anomalous shear in Fig. 6.12b. The LLJ disappears in Fig. 6.12b, but the anomalous northeasterly is stronger in the anomaly-based synoptic chart. The two anomalous low vortices, which indicate the two centers of heavy rain, show that the anomaly-based synoptic chart is useful in practice to correctly locate the high-impact event.

Another case shows a strong cold flow outbreaking in Fig. 6.13a. The rain band is located in the right side of westerly trough and in the left side of LLJ in the traditional synoptic chart. In the anomalous field, there is no the LLJ and the anomalous flow is northerly, so the shear is very near the rain band in South China.

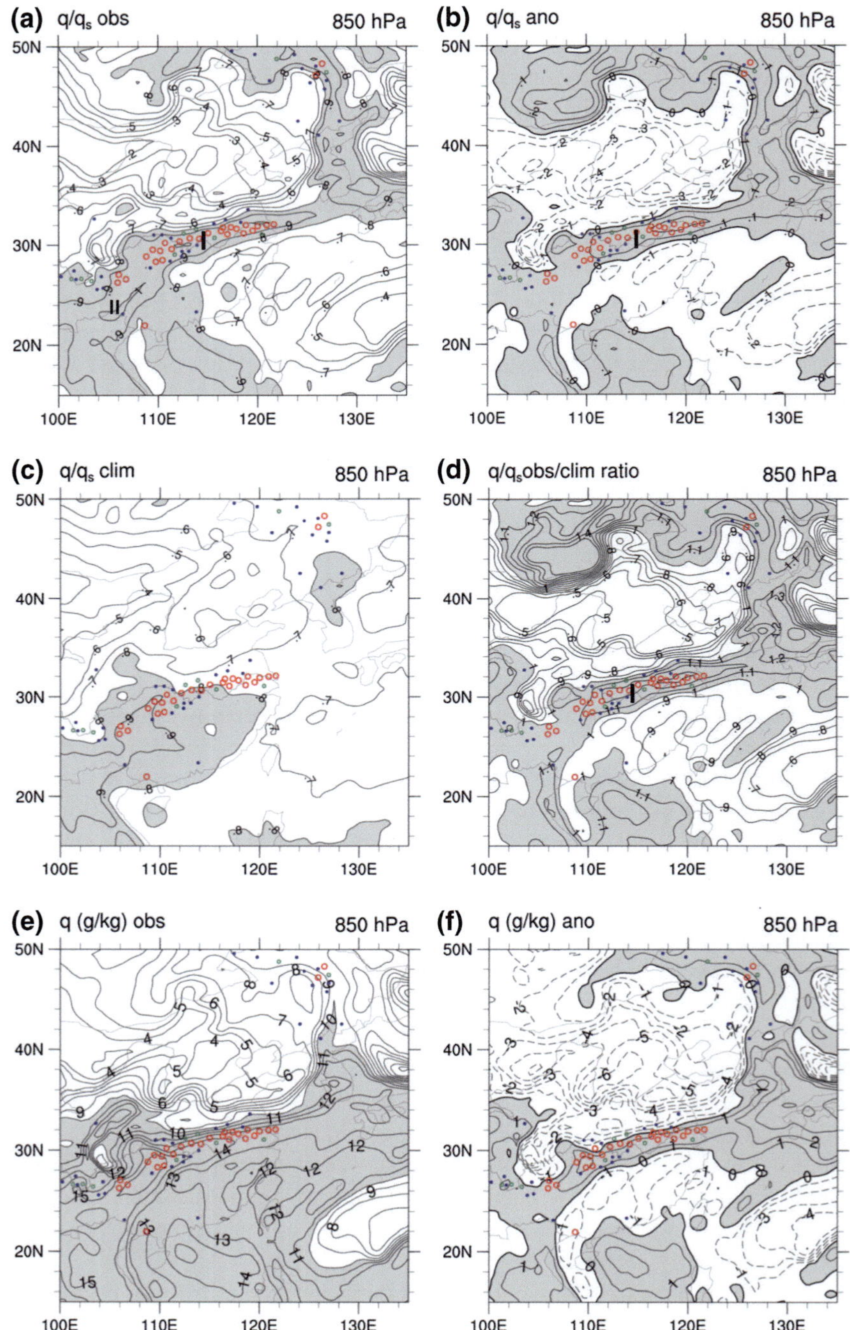
(a) q/qs obs 850 hPa
(b) q/qs ano 850 hPa
(c) q/qs clim 850 hPa
(d) q/qsobs/clim ratio 850 hPa
(e) q (g/kg) obs 850 hPa
(f) q (g/kg) ano 850 hPa
50N
40N
30N
20N
100E
110E
120E
130E

◀ **Fig. 6.11** Horizontal distributions of **a** total, **b** anomalous, and **c** climatic relative humidity RH (q/q_s) at 850 hPa at 0000 UTC 1 July 1991, in 0.1 or 10% *contour* interval. **d** The ratio of the total RH to the climatic RH, in 0.1 *contour* interval. Total and anomalous specific humidity q are respectively shown in **e** and **f**, in 1 g/kg contour interval. *Red*, *green*, and *blue open circles* (*dots*) indicate the stations with rainfall over 50, 25–50, and 10–25 mm day^{-1}, respectively. Sourced from Qian et al. (2016a) © American Meteorological Society, with permission

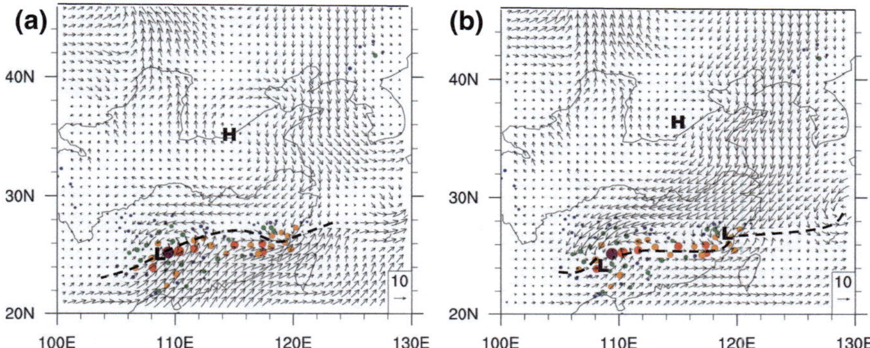

Fig. 6.12 **a** Total wind (m/s) and **b** anomalous wind (m/s) at 850 hPa at 0000 UTC 3 July 2009 using the ERA-Interim data. *Blue*, *green*, *yellow*, *red*, and *purple dots* are amounts of rainfall with ranges: 10–24.9, 25–49.9, 50–99.9, 100–249.9 mm, and exceeding 250 mm per 24 h at 0000 UTC 3 July 2009. The *dashed line* denotes the shear or convergence line. Letters "H" and "L" indicate high and low circulation systems. Modified from Qian (2013) © China Meteorological Press, with permission

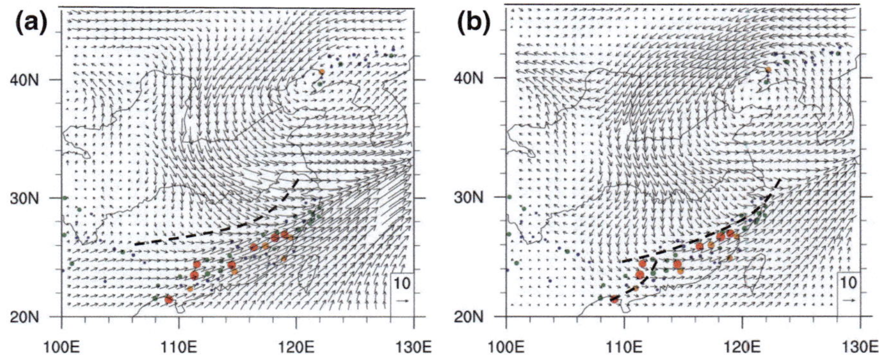

Fig. 6.13 Same as in Fig. 6.12 except at 700 hPa and 0000 UTC 21 June 1996. Panel **a** is the total-based synoptic chart and panel **b** is the anomaly-based synoptic chart. Modified from Qian (2013) © China Meteorological Press, with permission

We have plotted 603 daily RHR fall events in eastern China over the period of 1981–2010 (Qian 2013). Figure 6.14 depicts a 5-day episode of RHRs. The LLJ is observed from 19 to 22 June 2005 at 850 and 925 hPa. The rain band is located in the right side of LLS and in the left side of LLJ in the traditional synoptic chart for

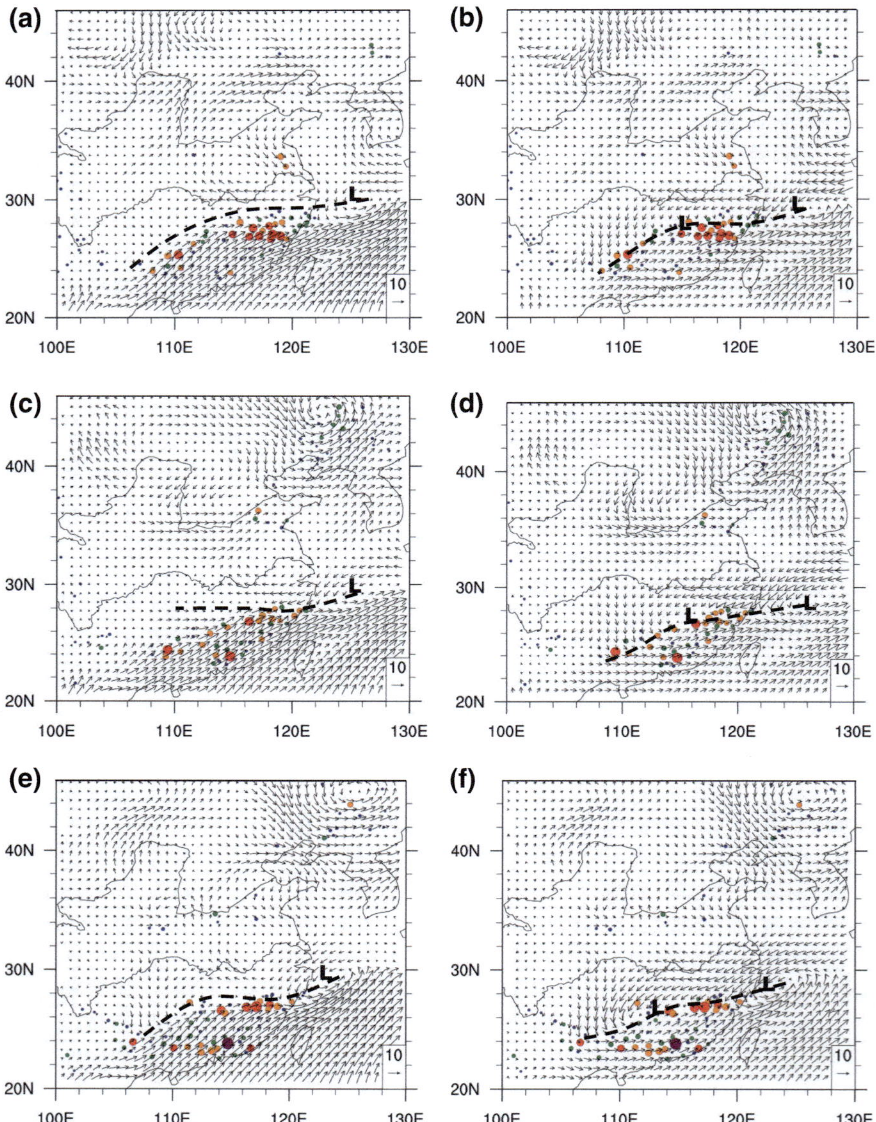

Fig. 6.14 Same as in Fig. 6.12 except at **a**, **b** 850 hPa at 0000 UTC 19, **c**, **d** 850 hPa at 0000 UTC 20, **e**, **f** 850 hPa at 0000 UTC 21, **g**, **h** 925 hPa at 0000 UTC 22, and **i**, **j** 925 hPa at 0000 UTC 23, June 2005. The *left panel* is the total-based synoptic chart and the *right panel* is the anomaly-based synoptic chart. Modified from Qian (2013) © China Meteorological Press, with permission

all the five days. A closed cyclone exists in the East China Sea during the five days, but no cyclone is observed along the LLS. Over all these days, heavy rains mainly occur in the strong southwesterly flow in total-based synoptic charts. However, RHRs in all the five days are coincided with the anomalous convergence line.

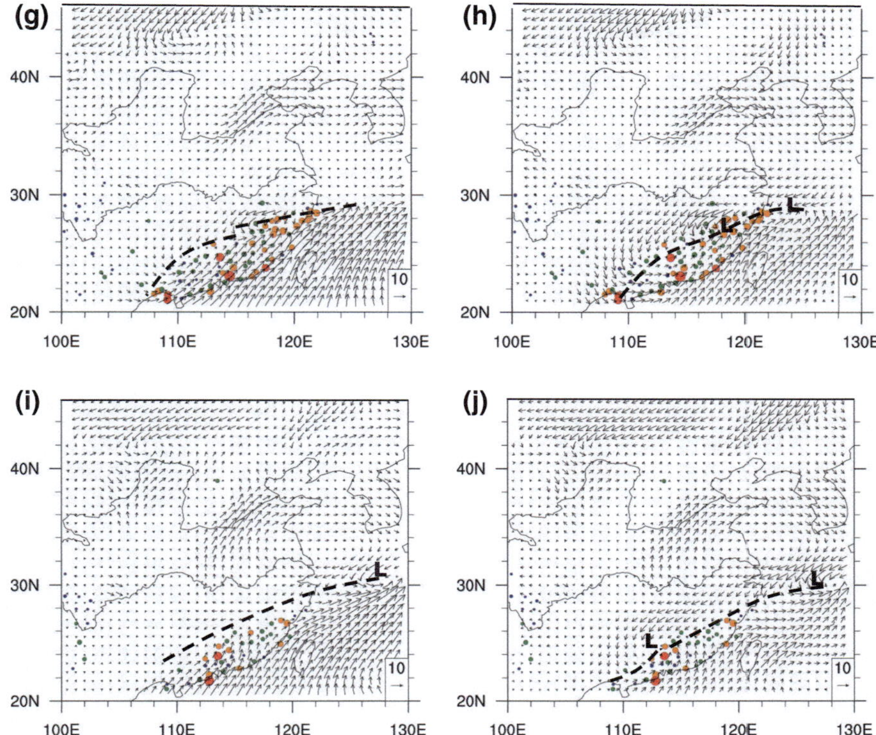

Fig. 6.14 (continued)

An anomalous low vortex propagates eastward along the anomalous convergence line. During the five days, the southwesterly total flow is significantly stronger than that of easterly flow near the LLS in the total-based synoptic chart, so it can give misleading information that the LLS and rain band should shift northward. A fact shows that the anomalous convergence line and rain band slowly shift southward in South China. We believe that these anomalous systems in the anomaly-based synoptic chart are dynamically attributed to surface weather extremes.

We examine another case of 3-day heavy rains in Fig. 6.15. The heavy rain band and the LLS originate along the Yangtze River and then shift southward to its western part. The rain band is always located in the southern side of the LLS during the three days and no cyclone can be observed along the LLS in the mainland China. In the anomaly-based synoptic chart, however, anomalous west-southwesterly and anomalous east-northeasterly form a contrast with similar intensity but opposite directions. An eastward-propagating low vortex is observed along the anomalous convergence line during the three days.

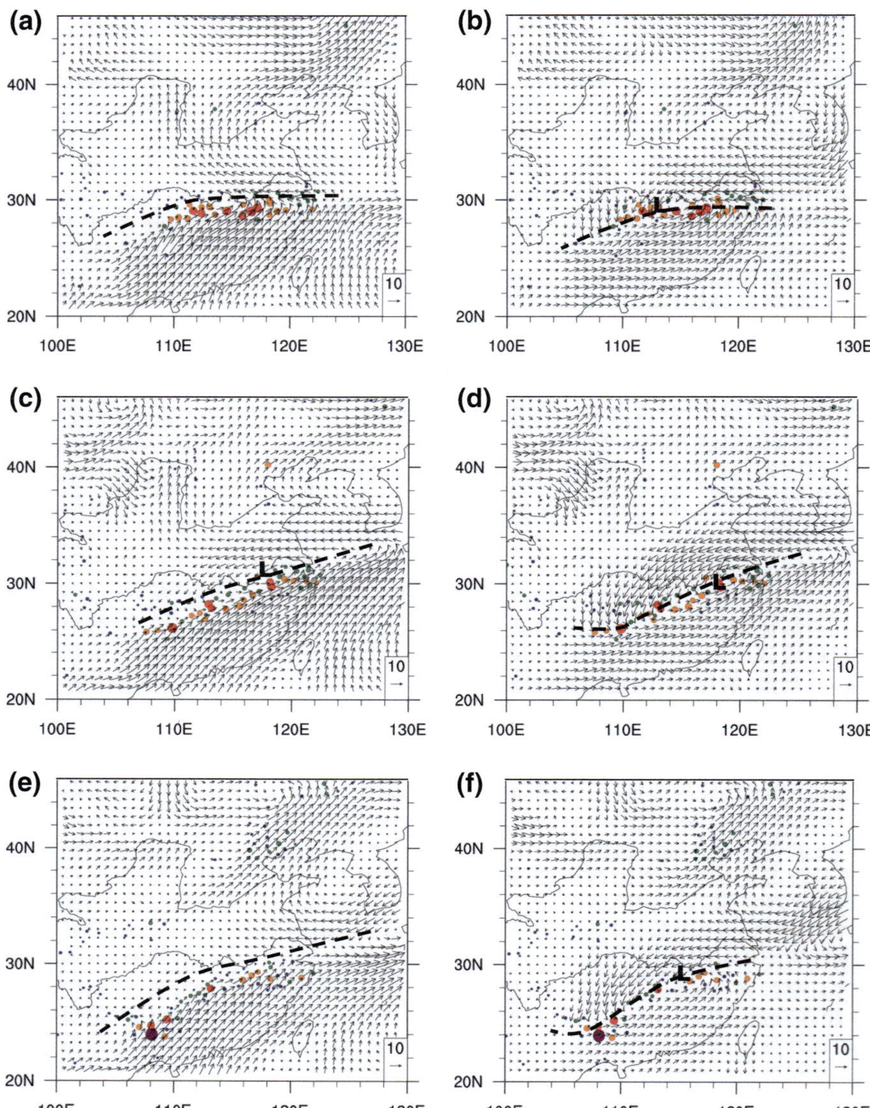

Fig. 6.15 Same as in Fig. 6.12 except at 850 hPa at **a**, **b** 0000 UTC 3 July, **c**, **d** 0000 UTC 4 July, and **e**, **f** 0000 UTC 5 July, 1993. The *left panel* is the total-based synoptic chart and the *right panel* is the anomaly-based synoptic chart. Modified from Qian (2013) © China Meteorological Press, with permission

6.3 Normalized Anomaly and Raw Anomaly[3]

Two different anomaly-based fields, normalized anomaly (NA) and raw anomaly (RA), are applied for different purposes, which have been studied in many previous works. Jiang et al. (2016) used the NA and RA in the analysis of an extreme rainfall event ("7.21" Beijing extreme rainfall), which hit Beijing and northern Hebei Province during the period of 1000 CST (China Standard Time, or 0200 UTC) 21 July to 0400 CST 22 July 2012. This event has a record amount of 460 mm in Fangshan. Several rain gauge stations receive hourly rainfall rates exceeding 85 mm, and leave 112 dead resulting from drowning and landslides. Which weather system causes this rain event attracts many investigators (Chen et al. 2012; Sun et al. 2012, 2013; Zhang et al. 2013; Meng et al. 2013; Du et al. 2014). They have given detailed analyses mainly from synoptic-scale system to mesoscale convective system based on various observations and model-predicted outputs. Comparisons based on NA and RA analyses are conducted on this case by Jiang et al. (2016). The CMORPH (CPC MORPHing technique) precipitation dataset at 3-hourly temporal and 0.25° spatial resolution, derived from satellites (Joyce et al. 2004) and the European Center for Medium-Range Weather Forecasts (ECMWF) Re-Analysis Interim data (ERA-Interim) (Dee et al. 2011) are used.

The "7.21" Beijing extreme rainfall case is reviewed in Fig. 6.16, which plots the surface wind at the 10 m level and temperature at the 2 m level as well as the precipitation rate four times a day. The rainfall process and the convergence system with six sectors are dominantly propagated eastward in 24 h. A spot of precipitation rate more than 5 mm/h was located in west side (40°N, 110°E) of Beijing on 0800 CST 21 July 2012 (Fig. 6.16a). The precipitation at this spot enhanced more than 8 mm/h and moved eastward to Beijing (39.75°N, 116.00°E) at 1400 CST 21 July (Fig. 6.16b). It continued to intensify and move eastward slowly in the next 12 h (Fig. 6.16c, d). The rain band with several spots oriented from southwest to northeast surrounding Beijing can be divided into several regions or sectors by six convergence or shear lines. The strongest rainfall spot passed Beijing from 1400 CST to 0200 CST 21 July. The eastward rain spot is well driven by the convergence system marked by the letter "C". Six sectors marked by numbers 1–6 are divided by shear lines in each panel of Fig. 6.16. The Beijing extreme rainfall is located ahead of the convergence "C" in sector 2 between two shears. Sector 2 is directly influenced by the easterly and southeasterly, which are separated by southerly in sector 3 and northeasterly in sector 6. Sector 4 is another major rain area located in the southwesterly. There are no rains in sectors 1 and 3 where northerly and southerly flows are observed, respectively. The northwesterly in sector 5 was small and finally combined with the southwesterly in sector 4 at 0200 CST 22 July.

Anomaly-based field analyses have been widely accepted and practiced for the climate study by the community for decades. Many climatic anomalous phenomena

[3]Modified article originally published in Jiang et al. (2016). Published with kind permission of © Springer Science + Business Media Dordrecht 2016. All Rights Reserved.

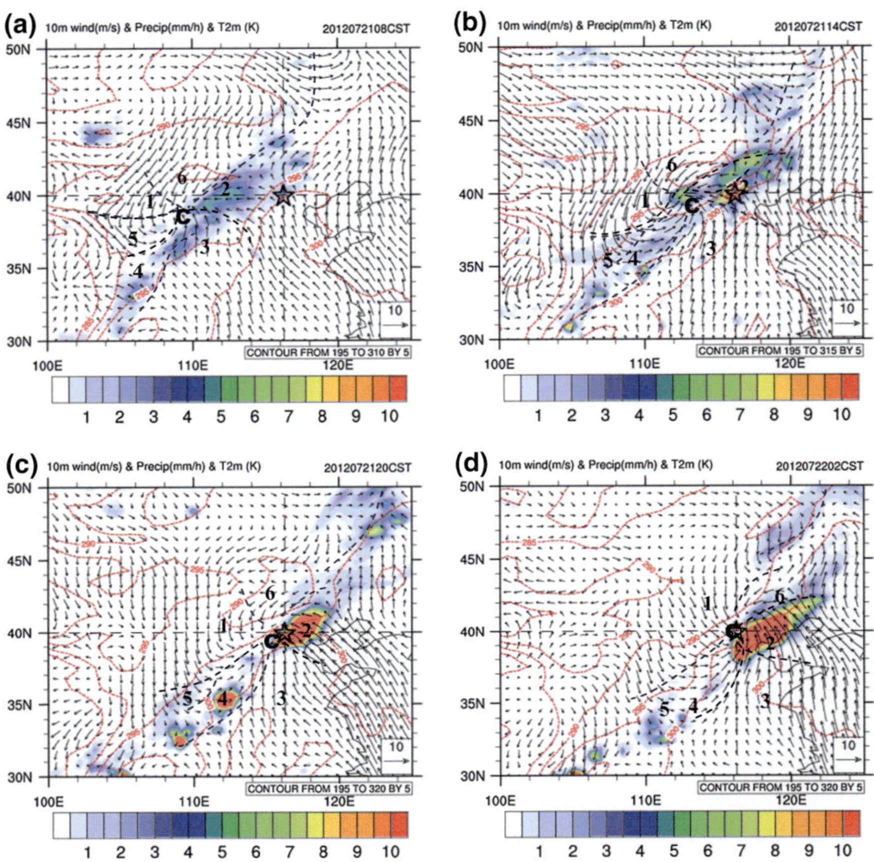

Fig. 6.16 Horizontal distributions of winds at the 10 m level, precipitation (0.5 mm/h interval, *shading*), and temperature (5 K interval, *red contour*) at the 2 m level at **a** 0800 CST 21, **b** 1400 CST 21, **c** 2000 CST 21, and **d** 0200 CST 22 July 2012. The star is the location of Beijing and the letter "C" is the convergence center while six numbers indicate six flow sectors separated by shear lines. Sourced from Jiang et al. (2016) © Springer Science + Business Media Dordrecht 2016, with permission

such as some oscillations and teleconnections have been revealed through the application of anomaly-based field analyses (Wallace et al. 1981; Rogers 1984; Moron et al. 1998; Goswami et al. 2001), which helps us to understand climatic anomalies better. In recent years, the anomaly-based field analyses are introduced to the synoptic study, which have been proved to be useful in weather analysis and forecasts. Particularly two different anomaly-based fields, NA and RA, are applied in different purposes, which have been studied in many works.

The use of NA fields or standardized anomalies (SAs) in identifying high-impact weather events has been demonstrated in many previous studies (Hart and Grumm 2001; Grumm and Hart 2001; Junker et al. 2008, 2009; Graham and Grumm 2010; Du et al. 2014). Grumm (2011a) showed the values of SAs in identifying the extreme and enduring Russian heat wave during July–August 2010. At the same time, the flow of an anticyclone and a deep cyclone beneath was associated with the extreme flooding in Pakistan in July 2010 (Galarneau et al. 2012; Grumm 2011b).

In recent years, a simple approach which directly applies RA fields has been proposed to use in short-, medium- and extended-range forecasts of extreme weather events (Chou 2012; Qian 2012), such as freezing rains (Qian and Zhang 2012), winter low temperature (Jiang et al. 2014), summer heat waves (Ding and Qian 2012), summer cold spells (Qian and Jiang 2014), spring–autumn low temperature (Qian et al. 2015b), RHRs (Qian et al. 2013, 2015a; Shan et al. 2015), unusual typhoon tracks in the South China Sea (SCS; Qian et al. 2014) and East China Sea (Huang et al. 2015), and even in the improvement of numerical weather prediction (Peng et al. 2013).

Obviously, both anomaly-based fields have been used for many cases from their own aspect. Few works focus on them together. So what are the consistency and difference between the NA and RA fields? Can we combine them together to improve the forecast skills of extreme weather events? To answer the above two questions, a detailed analysis has been performed on the Beijing extreme rainfall of 21 July 2012 as an example case (Jiang et al. 2016).

An atmospheric total field $F_{d,y}(\lambda, \varphi, p, t)$ such as geopotential height, temperature, specific humidity, and wind u–v components at diurnal time t (24 h a day) on calendar date d in year y at a spatial point of longitude λ, latitude φ and pressure level p, is decomposed into a climatic field $\widetilde{F}_d(\lambda, \varphi, p, t)$ and an anomalous field $F'_{d,y}(\lambda, \varphi, p, t)$, namely relative anomaly (RA) following Qian (2012) and Qian et al. (2014):

$$F_{d,y}(\lambda, \varphi, p, t) = \widetilde{F}_d(\lambda, \varphi, p, t) + F'_{d,y}(\lambda, \varphi, p, t) \quad (6.3.1)$$

the climatic field is estimated by averaging 30 years (1981–2010) based on the reanalysis data on calendar date d and at diurnal time t,

$$\widetilde{F}_d(\lambda, \varphi, p, t) = \sum_{y=1981}^{2010} F_{d,y}(\lambda, \varphi, p, t)/30 \quad (6.3.2)$$

It is assumed that the positive and negative anomalies of meteorological variables at a specific grid point and a given time cancel each other during the 30 year period to approximate the quasi-static climatic state. The climatic state (or climatology) defined by Eq. 6.3.2 varies from hour to hour and from day to day with diurnal and seasonal cycles.

The NAs were proposed by Grumm and Hart (2001). It can be written as

$$NA = (F - M)/\sigma, \quad (6.3.3)$$

where $F = F_{d,y}(\lambda, \varphi, p, t)$ is one of basic variables from the reanalysis data at each grid point, $M = \widetilde{F}_d(\lambda, \varphi, p, t)$ is the mean of the specified variable at each grid point and σ is the standard deviation (SD) of the variable at each grid point from 1981 to 2010. If we take $\sigma = 1$ in every grid, the NA field becomes the RA field.

Other two indices derived from the basic variables of NA are defined as:

$$mNA_{\mathrm{var}}(\lambda, \varphi, t) = \max(NA_{\mathrm{var}}(\lambda, \varphi, p, t)) \tag{6.3.4}$$

$$MNA(\lambda, \varphi, t) = \max(mNA_{\mathrm{var}}(\lambda, \varphi)) \tag{6.3.5}$$

Here var represents basic variables such as specific humidity (q), wind components (u, v), temperature (T), and geopotential height (z). The term $mNA_{\mathrm{var}}(\lambda, \varphi, t)$ indicates the maximum value of $NA_{\mathrm{var}}(\lambda, \varphi, p, t)$ for a specified variable (var) at a specified grid of all the vertical levels (p). The term $MNA(\lambda, \varphi, t)$ is the maximum value at a specified grid of all the variables of $mNA_{\mathrm{var}}(\lambda, \varphi, t)$. The most extreme events will be those that have large departures from climatology extending through the full depth of the troposphere for each of basic variables (Hart and Grumm 2001). So the term $MNA(\lambda, \varphi, t)$ acts as an index to indicate the maximum departure from climatology of all the basic variables through the whole vertical levels at a given grid.

From the previous results, there are usually anomalous signals in the high levels corresponding to the surface extreme weather events (Hart and Grumm 2001; Shan et al. 2015; Qian et al. 2016a). To examine the possible linkage between the Beijing rain and the aloft atmospheric variables, Fig. 6.17 compares the vertical–latitude cross-sections of the height–temperature along 39.75°N at 2000 CST (1200 UTC) 21 July 2012. Traditional and climatic synoptic charts of height and temperature are shown in Fig. 6.17a, b. No clear significant signals can be found to indicate the strong rainfall place from both the total (or climatic) height and temperature except that there were some total temperature fluctuations. The centers of SD in Fig. 6.17c reveal the locations where large amplitudes of height and temperature variability often occur. Climatologically at 2000 CST 21 July, large SD centers (M_H) of height are located in the upper troposphere at around 200 hPa while large SD centers (M_T) of temperature are respectively found around 100, 300, and 925 hPa (Fig. 6.17c).

The departures from climatology, which is closely related to the high-impact weather events, cannot be detected easily from the traditional synoptic charts in Fig. 6.17a, b. But utilizing anomaly-based fields, the signals can be clearly observed from Fig. 6.17d, e. The large departure centers are well corresponding to the locations of SD centers as high incidence areas of weather events. A positive center (H1) of maximum height anomaly (MHA) is located at around 200 hPa, while a negative center (C1) and positive center (W1) of temperature anomalies are respectively found above and below the positive MHA center right over the place of heavy rain. On the other hand, a positive center (W0) and a negative center (C0) of temperature anomalies are situated above and below a negative center (L0) of height anomalies near 700 hPa. Horizontally, a boundary between the anomalous cold air mass (C0) and the anomalous warm air mass (W) can be used to explain the rain band. It is confirmed that the rain band is a clash result of anomalous warm and cold air masses in the lower troposphere.

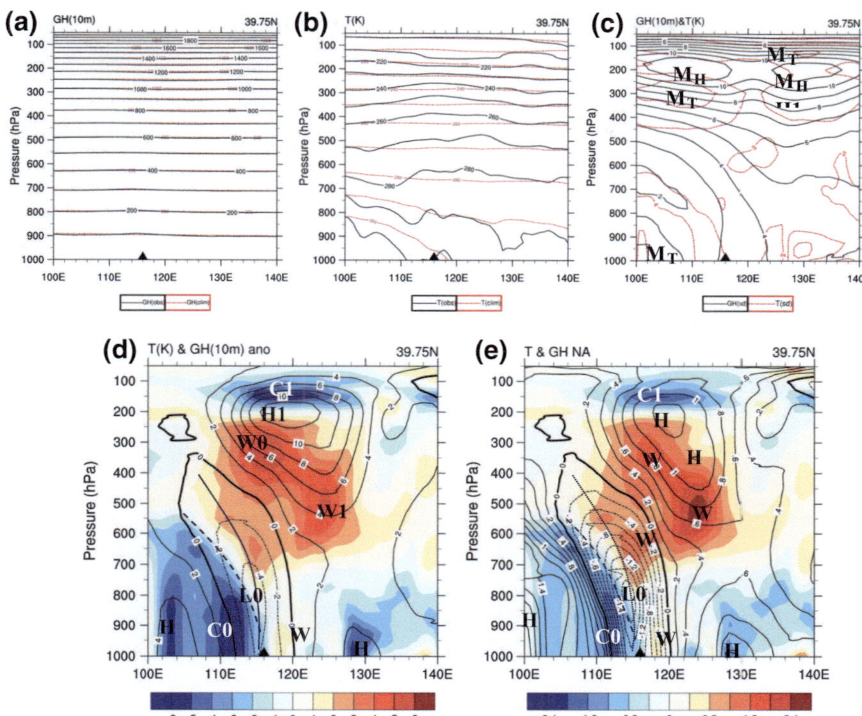

Fig. 6.17 Vertical sections of **a** total and climatic height (100×10 m interval), **b** total and climatic temperature (10 K interval), **c** height–temperature SD (10 m and 1 K intervals), **d** height (*contour*, 2×10 m interval) and temperature (*shading*, 1 K interval) RAs and **e** height (*contour*, 0.2 SD interval) and temperature (*shading*, 0.4 SD interval) NAs (*shading* with *solid* and *dashed lines*, 0.2 SD interval) along 39.75°N at 2000 CST 21 July 2012. Sourced from Jiang et al. (2016) © Springer Science + Business Media Dordrecht 2016, with permission

It is interesting to note that there are two pairs of positive and negative centers of temperature anomalies centered around two centers of different height anomalies right over the place of heavy rain. These height and temperature RAs vertically satisfy the hydrostatic equilibrium and can well indicate the surface high-impact weather events such as typhoon vortices (Qian et al. 2014; Huang et al. 2015) and heat waves and cold spells in different seasons (Qian and Jiang, 2014; Qian et al. 2015b, 2016b). It means that the temperature anomalies can be derived from height anomalies in the troposphere and the stratosphere. Figure 6.17d is referred to as the anomaly-based synoptic chart which can clearly express potential signals to indicate surface high-impact weather events than those of traditional synoptic chart. The maximum height anomalous center in Fig. 6.17d may act as an indicator to the event.

If the height–temperature RAs in Fig. 6.17d are divided by their SDs, the two set systems of height–temperature anomalies in Fig. 6.17e are still identified but their central locations are reformed. The positive and negative height NAs reach 1.2 SDs

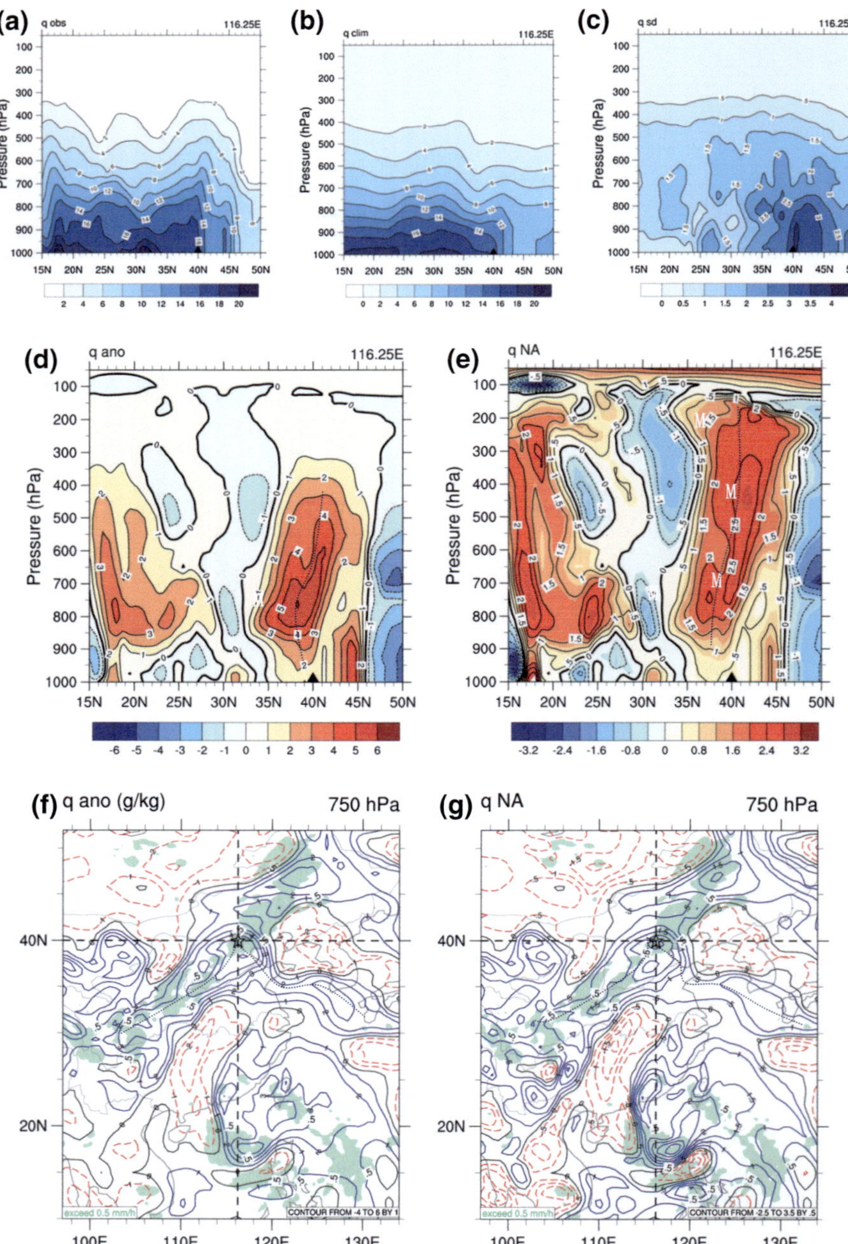

Fig. 6.18 Vertical sections of **a** total specific humidity (2 g/kg interval), **b** climatic specific humidity (2 g/kg interval), **c** specific humidity standard deviation (0.5 g/kg interval), **d** specific humidity RAs (1 g/kg interval) and **e** specific humidity NAs (0.4 SD interval) along 116.25°E at 2000 CST 21 July 2012. Letter "M" indicates the large center of specific humidity anomalies in **d** and **e**. Horizontal distributions of specific humidity at 750 hPa at 2000 CST 21 July 2012 for **f** specific humidity RAs (1 g/kg interval) and **g** NAs (0.5 interval). The *green shading* indicates precipitation rate. Sourced from Jiang et al. (2016) © Springer Science + Business Media Dordrecht 2016, with permission

while the positive and negative temperature NAs reach 1.6–2.0 SDs. The height and temperature NAs correspond to the heavy rain. In Fig. 6.17e, though height NAs and temperature NAs do not satisfy the hydrostatic relationship, the NA fields can be used to measure the abnormality of the event. The rainfall location can be well indicated by anomalous temperature contrast (i.e., the boundary of anomalous warm and cold air masses) only below 850 hPa, but several large centers of temperature NAs are observed in the mid-high troposphere, even in the stratosphere.

The high specific humidity plays an important role in this event. From the total field of specific humidity, no obvious signal can be found to indicate the rain band (Fig. 6.18a). Two centers of specific humidity in the low level are located far away from the rain band, which is nearly consistent with the climatic distributions (Fig. 6.18b). The SD chart shows a center right over the rain band (Fig. 6.18c), which reveals a large variation center of the specific humidity in the region of the rain band associated with high-frequency heavy rain events in this season (Zhang et al. 2013). A positive center of humidity RAs and NAs in Fig. 6.18d, e well corresponds with the rain band in Beijing. A positive axis of humidity RAs is centered from 400 to 900 hPa in the lower troposphere but a positive axis of humidity NAs extends to the 100 hPa in the upper troposphere. The variance of central NAs exceeds three times and can declare the intensity of rain event from the humidity condition. The specific humidity RAs show a single center at around 750 hPa just over the heavy rain while the specific humidity NAs show several centers in the troposphere over the heavy rain place.

Figure 6.18f, g show the horizontal distributions of the specific humidity at 750 hPa at 2000 CST 21 July 2012. The humidity RAs and NAs show that the rain band is well located within the positive humidity RAs and NAs. In the RA field, the primary source of the vapor is from the southeast of China, which supplies more vapor than the southwesterly transportation. However, from the humidity NA field, the abnormal degree of the transportation from the southeasterly and southwesterly is equal (Fig. 6.18g). Beijing is just located in the high moisture region where the two transportation trajectories converge (Sun et al. 2013).

From Fig. 6.17d, e, this heavy rain is not only well associated with a set synoptic-scale system of height and temperature RAs in the upper troposphere but also well associated with a set sub-synoptic-scale system of height and temperature RAs in the lower troposphere. Thus, the horizontal distributions of height and temperature are drawn at 850 hPa. In Fig. 6.19a, the total height and total temperature at 2000 CST 21 July 2012 show that three low centers are respectively located in the western side of Beijing (L1), southwest China (L2), and the SCS (L3). The low center L2 is stronger than the low center L1 while the strongest one L3 is a typhoon system. The Beijing heavy rain is located within warm sector (Zhang et al. 2013) while the cold air mass in northwestern side is not strong. After removing the climatic height and temperature (Fig. 6.19b), the height RAs show three negative centers and the weakest one is L2 (Fig. 6.19c). These temperature centers (Fig. 6.19a) are consistent with the climatic field (Fig. 6.19b), where two temperature centers (W1 and W2) are comparable. In contrast, the two types of height–temperature RA and NA fields can well indicate the heavy rain band

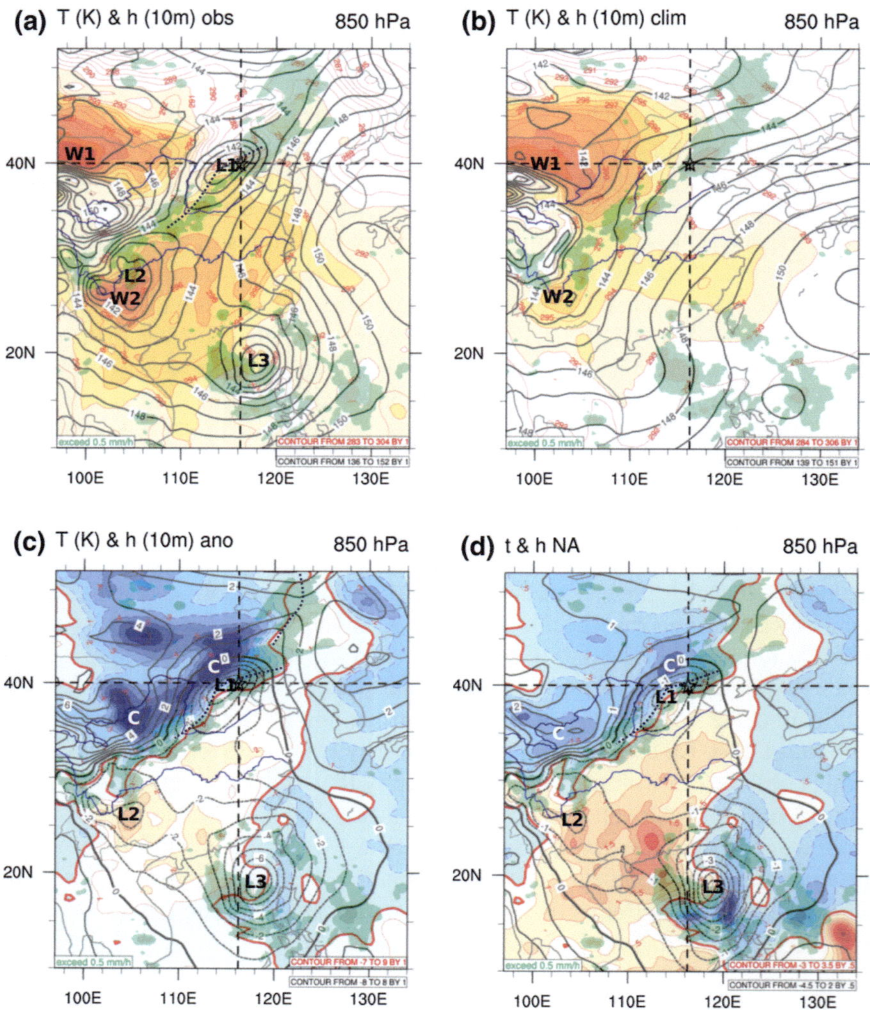

Fig. 6.19 Horizontal distributions of height–temperature at 850 hPa and 2000 CST 21 July 2012 for **a** total height (1×10 m interval, *contour*) and total temperature (1 K interval, *shading*), **b** climatic height (1 × 10 m interval, *contour*) and climatic temperature (1 K interval, *shading*), **c** height RAs (1×10 m interval, *contour*) and temperature RAs (1 K interval, *shading*), and **d** height NAs (0.5 interval) and temperature NAs (0.5 interval). The *green shading* indicates precipitation rate (0.5 mm/h interval) and the *dark green shading* indicates the centers of the precipitation rate. The *thick red* (*black*) *line* indicates the zero line of the temperature (height) field. Sourced from Jiang et al. (2016) © Springer Science + Business Media Dordrecht 2016, with permission

(Fig. 6.19c, d). Three negative height anomalous centers are associated with the rain band. The rain band is also located along the large contrast area between anomalous temperatures. The zero lines (or the boundary of anomalous cold–warm air masses) of the temperature RAs and NAs pass through the rain band (heavy red line) in North China. The biggest abnormal center of height RAs is the typhoon system.

In summary, both the two anomalous fields, NA and RA fields, have advantages over the traditional fields to indicate the "7.21" Beijing extreme rain event. The RA fields can be used to enlarge the event signals with a clear structure. The temperature RAs can be derived from the height RAs through the hydrostatic equilibrium. Meanwhile, a maximum height anomalous core can be found over the event location. The NA fields can be used to measure and compare the abnormalities of different variables for a certain case. The information obtained from the NA fields can be utilized to analyze the impact of the event comprehensively. Zero isolines of both temperature RA and NA fields can locate the rain belts well.

In order to improve the forecast skills, evolution features of the event process associated with height and temperature RAs and NAs are illustrated here. From Fig. 6.17d, e, it can be identified that this heavy rainfall is directly associated with a low-tropospheric vortex, so we plot the vertical-time section of height and temperature from 0800 CST 16 to 0200 CST 26 July 2012 with 6 h interval at the grid of 40°N and 116°E. As shown in Fig. 6.17a, b, there are not any signals to indicate the Beijing heavy rainfall from total height and total temperature fields, so it is also similar to the vertical-time section in Fig. 6.20a. Several fluctuations of temperature field can be found in the total field evolution. But the rain band (marked by the black triangle) cannot be detected. However, the height–temperature RAs and NAs have a strong signal process. In Fig. 6.20b, a positive center and a negative center of height RAs started from 0800 CST 21 and peaked at 0200 CST 22 July 2012. After this rainfall process, a positive MHA center extends downward without a negative center of RAs in the lower troposphere, so there is a warm air column below the MHA resulting in heat anomalous event in the surface. In Fig. 6.20c, the evolution and central intensity of height–temperature NAs are a bit different from that of height–temperature RAs. The duration of negative height NAs from 0800 CST 21 and peaking at 0200 CST 22 July 2012 are the same as height RAs but the first positive MHA center is relatively weak and the second positive MHA is not apparent.

From the features analyzed above, the height RAs at 200 and 850 hPa are selected to represent the opposite structure of height anomalies between the upper and lower troposphere. Due to the hydrostatic relationship mentioned above, height RAs at the two selected levels represent the vertical distribution of the temperature RAs. It is clearly noted that the rain event is featured by a positive value (blue line) of height RAs at 200 hPa and a negative value (red-dashed line) at 850 hPa (Fig. 6.20d). The Beijing extreme rainfall on 21 July 2012 is associated with this shading process. During the shading period, the *MNA* exceeding 2 SDs is indicated by the red and blue shadings. As formulated in Eqs. 6.3.4 and 6.3.5, a comprehensive abnormal degree is summarized from all five variables in the upper panel of Fig. 6.20d. Therefore, an extreme weather event can be indicated by height anomalies and its abnormal degree can be accessed by the maximum of NAs fields.

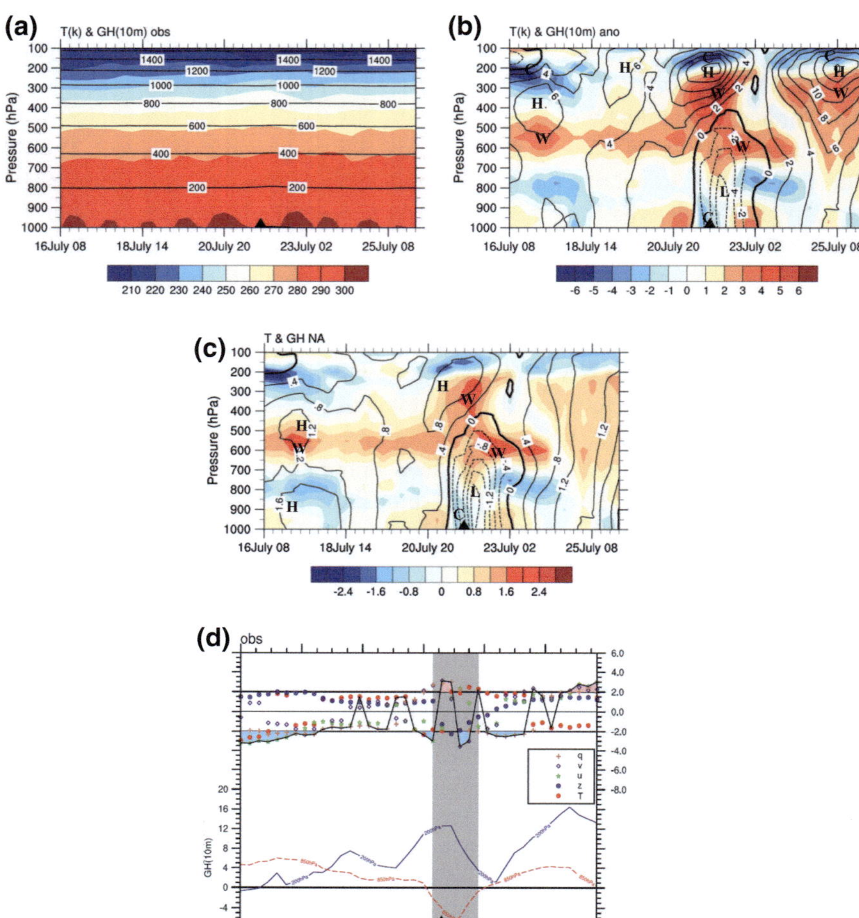

Fig. 6.20 Vertical (pressure)-time sections of **a** total height (*contour*, 200 × 10 interval) and total temperature (*shading*, 10 K interval), **b** height RAs (*contour*, 2×10 m interval) and temperature RAs (*shading*, 1 K interval), **c** height NAs (*contour*, 0.4 SD interval) and temperature NAs (*shading*, 0.4 SD interval). In the lower panel of **d**, height RAs series at 850 hPa and 200 hPa from 0800 CST 16 to 0200 UTC 26 July 2012 with 6 h interval at 40°N and 116°E. In the upper panel of **d**, different *dots* indicate the five variables of *mNA*s, the *MNA* is sketched in *black line*, and the *red* and *blue shadings* above 2.0 and below −2.0 indicate where *MNA* exceeds 2 SDs. In **d** the *gray shaded* column denotes the period satisfied by the negative height RAs at 850 hPa and positive height RAs at 200 hPa. Sourced from Jiang et al. (2016) © Springer Science + Business Media Dordrecht 2016, with permission

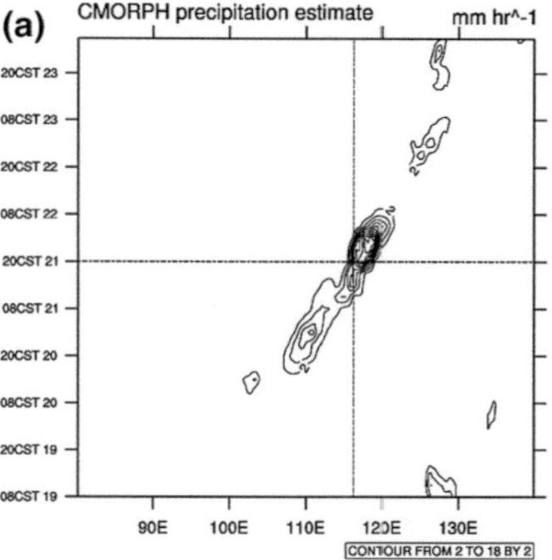

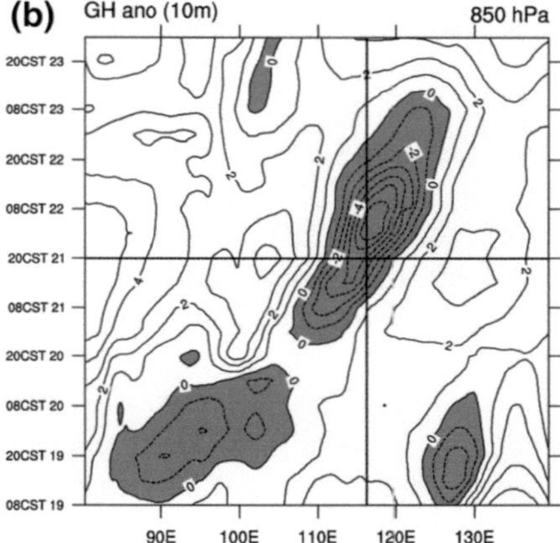

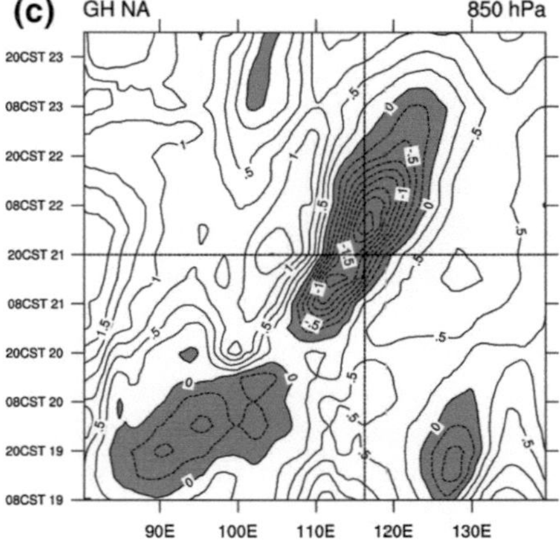

Fig. 6.21 Zonal-time sections of **a** precipitation rate (2 mm/h interval), **b** height RA (10 m interval) at 850 hPa, **c** height NA (0.25 SD interval) at 850 hPa from 0000 UTC 19 to 1800 UTC 23 July 2012, average from 39 to 41°N. The negative values of **b** and **c** are shaded. Sourced from Jiang et al. (2016) © Springer Science + Business Media Dordrecht 2016, with permission

Figure 6.21 shows zonal-time sections of precipitation rate and height RA and NA fields. The rain belt and the negative height RA and NA propagate eastward along Beijing's latitude. The height RA field is consistent with the height NA field for the same positive (negative) sign. The strongest rain occurred before the negative center of height RAs and NAs arrived in Beijing for about 12 h. Here the evolution features summarized in Fig. 6.20d are addressed and suggested to be used as a tool to improve the forecast skills. Based on the analysis of the RA fields, the two levels are selected to represent the whole level structure. The multiple variable NAs can reveal the abnormalities of the event comprehensively.

6.4 Heavy Rains and Anomalous Circulation Patterns[4]

Regional heavy rainfall is an extreme event which occurs in all seasons in eastern China. Statistical result shows that there are totally 3640 RHR events that satisfy 2 adjacent stations more than 50 mm/d during 1960–2010, which is equivalent to 71 events in average per year (Shan et al. 2015). During this period, there are 1659 heavy rainfall events covering more than 10 adjacent stations, equivalent to 32 events per year in average. RHR events can be measured in terms of duration (24 h), coverage (10 stations) and intensity (50 mm). The definition of heavy rainfall event is widely concerned by public and government. How to predict and locate a regional heavy rainfall event is a critical duty for forecasters in daily practice. A basic tool is the traditional synoptic chart, which can be drawn from station observations and numerical model predictions. By observing circulation patterns such as low pressure systems, cyclones, shears, reversed troughs and fronts drawn in synoptic charts, forecasters subjectively deduce where is possibly to experience rainfall in following several hours and days. The rain location usually occurs in the right side of a LLS and the left side of a LLJ (Zhu 1975; Huang 1981). Modern numerical models make quantitative rain prediction even from short-term to medium-range period. However, their predicted locations are sometimes far from observations (Qian et al. 2013). A total precipitation is the sum of climatic precipitation and anomalous precipitation. The large portion of a RHR results from anomalous precipitation, so only anomaly-based atmospheric variables can be used to locate it. Many cases show that the anomaly-based synoptic analysis is better than that the traditional total-based synoptic analysis in locating the heavy rain (Qian et al. 2013, 2015a, 2016a; Shan et al. 2015). We here introduce seven anomalous synoptic patterns describing how to apply the anomaly-based synoptic chart to locate RHRs in eastern China.

Two datasets are used to construct this division of patterns. The first one is the observed precipitation from 754 stations in mainland China. The stations are relatively uniformly spaced in eastern China and sparsely distributed in western China.

[4]Modified article originally published in Qian et al. (2016c). Published with kind permission of © Meteorological Monthly. All Rights Reserved.

This observed precipitation data is interpolated to a 0.5° latitude–longitude grid using the ordinary Kriging method by Chen et al. (2010). The second dataset is the ERA-Interim with a 0.75° (~ 80 km) latitude–longitude grid and standard pressure levels from 1000 to 50 hPa (Dee et al. 2011). In order to locate a heavy rain, the climatology, which includes a diurnal cycle and a seasonal cycle, is removed from the observed total field.

Relying on the traditional synoptic chart, circulation patterns such as shear line and cyclones are widely used by forecasters to deduce possible locations of regional heavy rainfall. The book, Atlas of anomalous circulations associated with RHRs in China, has collected all regional heavy rainfall cases from 1980 to 2000 in eastern China (Qian 2013). Qian et al. (2016c) focused on 41 daily RHR cases occurred in 1998 to describe seven different circulation patterns in eastern China. Seven patterns of the anomalous circulation systems at 850 hPa linked with the RHRs are identified and divided in four geographical areas: South China, the Yangtze River valley, North China, and Northeast China. Three patterns are identified in South China: the anomalous shear line, the anomalous reversed trough and the anomalous cyclone. Two patterns are identified in the Yangtze River valley: anomalous cyclone along with anomalous shear line and trough. In North China and Northeast China, there is only one patter: anomalous cyclone. Composite analysis is made by averaging the identified RHR events. Thus, the magnitude of the composite precipitations will be weaker than that of a single RHR event.

6.4.1 Three Anomalous Circulation Patterns in South China

The first anomalous circulation pattern in South China is the anomalous shear line. In 1998, there were four daily RHR events, respectively on 9 June, 14 June, 21 June, and 29 July 1998 (Table 6.5). Qian et al. (2015a) found that the anomalous

Table 6.5 Synoptic systems of 41 RHR days in 1998

Synoptic systems	Dates	Total days
Shear line in South China	9 Jun, 14 Jun, 21 Jun, 29 Jul	4
Vortex in South China	14 Jan, 9 Mar, 14 May, 24 Jul	4
Reversed trough in South China	16 Feb, 17 Feb, 8 Mar, 22 May, 24 Jun	5
Shear and trough along the Yangtze River	2 May, 13 Jun, 19 Jun, 23 Jun, 25 Jun, 26 Jun, 3 Jul, 22 Jul, 27 Aug	9
Vortex along the Yangtze River	11 Apr, 23 May, 12 Jun, 30 Jun, 23 Jul, 30 Jul	6
Vortex in North China	1 Jul, 2 Jul, 4 Aug, 7 Aug, 15 Aug, 22 Aug	6
Vortex in Northeast China	6 Jul, 7 Jul, 8 Jul, 13 Jul, 14 Jul, 15 Jul, 5 Aug	7

Sourced from Qian et al. (2016c) © Meteorological monthly, with permission

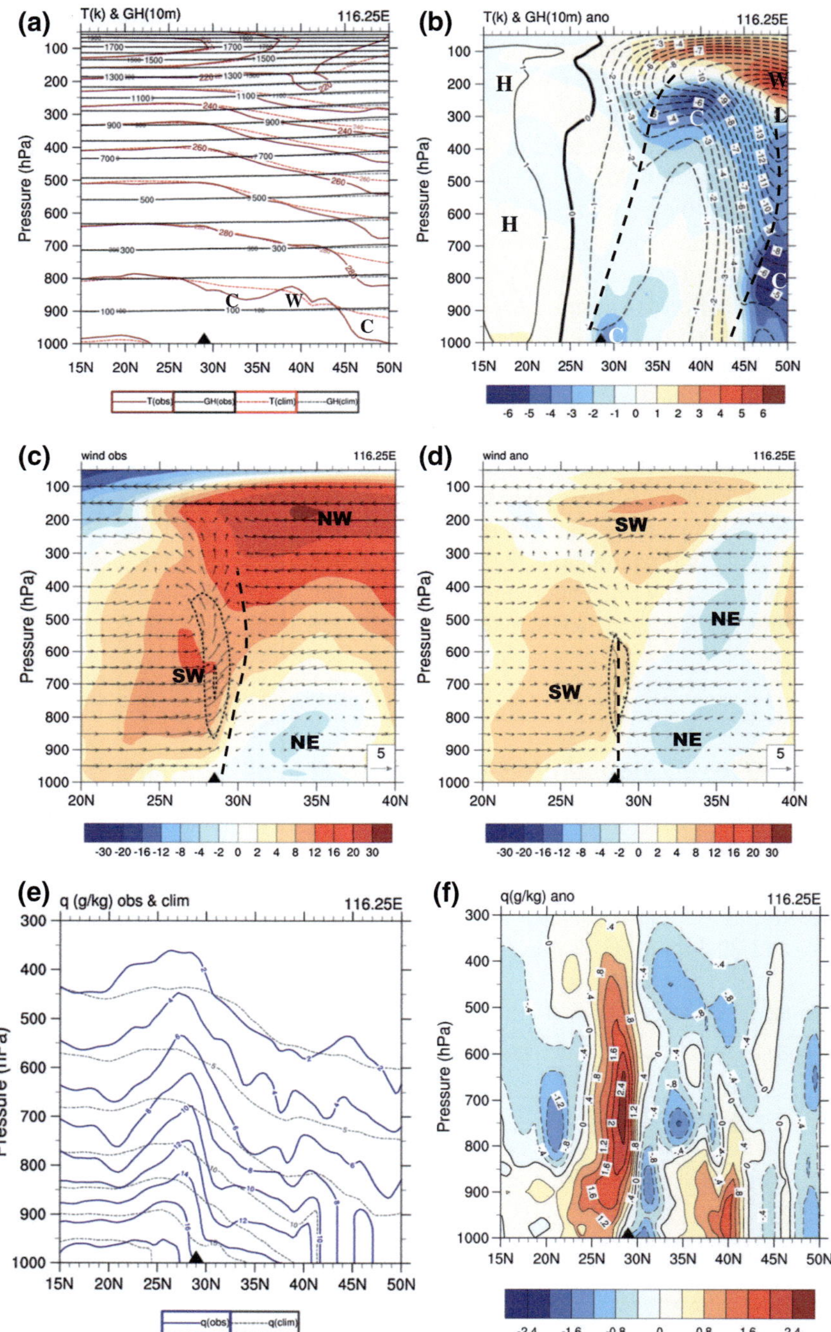
(a)
T(k) & GH(10m)
116.25E
Pressure (hPa)
C
W
C
T(obs)
GH(obs)
T(clim)
GH(clim)
(b)
T(k) & GH(10m) ano
116.25E
H
H
W
C
L
C
C
(c)
wind obs
116.25E
NW
SW
NE
(d)
wind ano
116.25E
SW
NE
SW
NE
(e)
q (g/kg) obs & clim
116.25E
q(obs)
q(clim)
(f)
q(g/kg) ano
116.25E

◀ **Fig. 6.22** Vertical–latitude cross-sections of four basic variables along 116.25°E for the composite of shear in South China in 1998 at 0000 UTC. **a** Total height (*solid-black contour*, 100 × 10 gpm interval) and climatic height (*dashed-black contour*, 100 × 10 gpm interval) as well as total temperature (*solid-red contour*, 10 K interval) and climatic temperature (*dashed-red contour*, 10 K interval). **b** Height anomalies (*contour*, 10 gpm interval) and temperature anomalies (*shading*, 1 K interval). **c** Total winds (m s^{-1}) and **d** wind anomalies (m s^{-1}). **e** Total specific humidity (*q*, *blue contour*, 2 g/kg interval) as well as its climatic component (*black dashed contour*, 2.5 g/kg interval). **f** Specific humidity anomalies (*shading*, 1 g/kg interval). The heavy *rain area* is indicated by the symbol *filled triangle*. In **c** and **d** the *thin dotted line* denotes the ascending velocity (0.25 Pa s^{-1} interval, enlarged 10 times). The *heavy dashed line* is the axis of anomalous height trough in **b** and wind convergence line in **c** and **d**. Letters "C/W" and "L/H" indicate temperature colder/warmer centers and height lower/higher centers. Letters "SW", "NW", and "NE" indicate the southwesterly, northwesterly and northeasterly, respectively. Sourced from Qian et al. (2016c) © Meteorological Monthly, with permission, with permission

circulation system at 0000 UTC can present a central time point to indicate the daily 24-h precipitation for a regional heavy rainfall event. Thus, a composite at 0000 UTC of the above 4 days is conducted in the following analyses.

Figure 6.22 shows the vertical–latitude cross-sections of four basic variables along 116.25°E for the composite of shear line in South China in 1998. In Fig. 6.22a, some deviations indicated by letters "C" and "W" from the total temperature and climatic temperature curves can be noted, but the deviation between the total height and climatic height is not visible because they contain the climatic component which overwhelms the anomalies superimposed on it. If forecasters can locate the rainfall using this traditional synoptic chart (Fig. 6.22a), it should strongly rely on their experiences accumulated for a long-term period. However, these experiences cannot be explained in words. Differently, Fig. 6.22b is an anomalous synoptic chart which is the difference between total height/temperature and climatic height/temperature. It is clear that the surface rain is a downward extension of anomalous height axis (heavy dashed line) from the upper troposphere and along the boundary of anomalous warm–cold air masses in the low troposphere. Another pair of anomalous temperature boundary and anomalous height axis is observed near 45°N, which indicates a rain area in Northeast China.

The total wind shows a vertically asymmetrical distribution surrounding the surface heavy rain with a strong southwesterly in the middle troposphere and a strong northwesterly at the upper troposphere (Fig. 6.22c). Differently, a symmetrical distribution or convergence of wind anomalies depicts in Fig. 6.22d, where the anomalous southwesterly and anomalous northeasterly are respectively located in the two sides of the heavy rain. Figure 6.22e is the total and climatic specific humidity distributions in the vertical direction. Relatively large humidity is noted from the climatic and total field near the rain area. A deep humidity column located over the rain area in South China and another shallow humidity column in Northeast China are clearly shown by the humidity anomalies (Fig. 6.22f).

In Fig. 6.22, Qian et al. (2016c) plotted these anomalous fields at 850 hPa. The goal is to show how theses anomalous systems and variable anomalies are better to

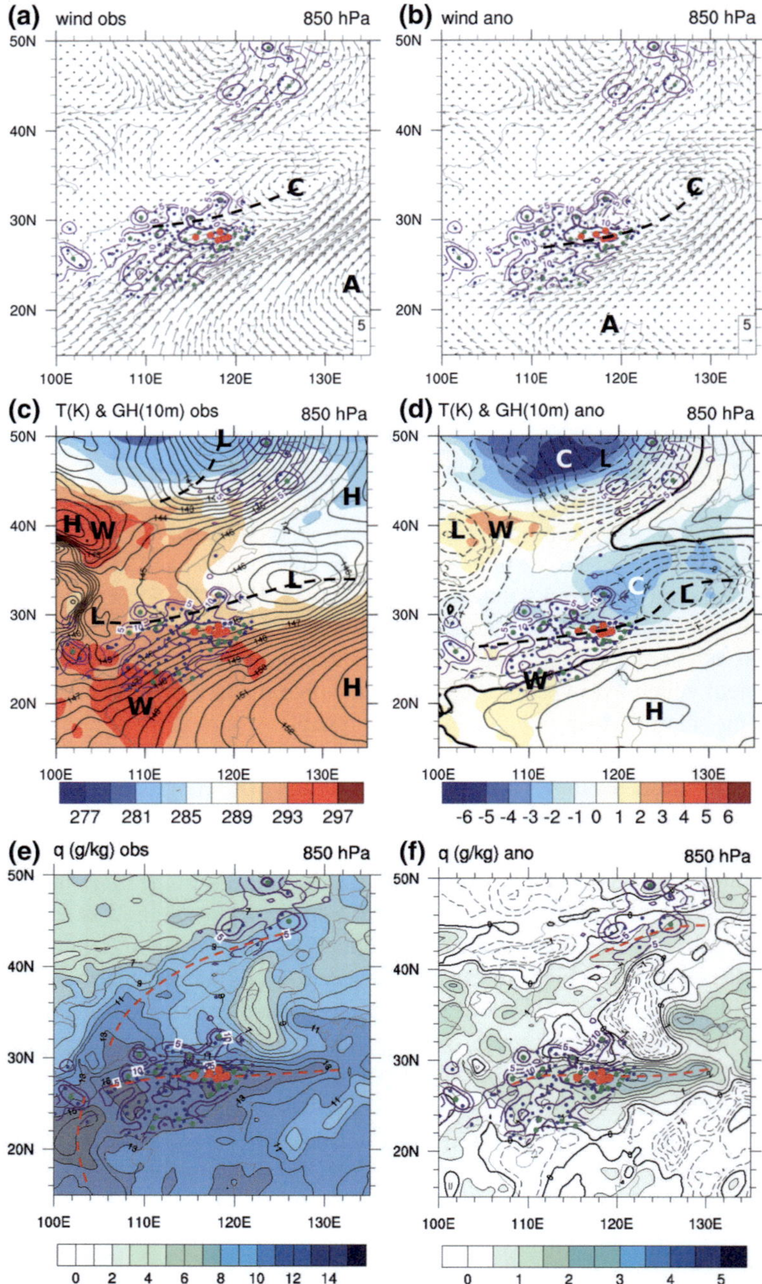
(a) wind obs 850 hPa
(b) wind ano 850 hPa
(c) T(K) & GH(10m) obs 850 hPa
(d) T(K) & GH(10m) ano 850 hPa
(e) q (g/kg) obs 850 hPa
(f) q (g/kg) ano 850 hPa
50N
40N
30N
20N
100E
110E
120E
130E
277 281 285 289 293 297
-6 -5 -4 -3 -2 -1 0 1 2 3 4 5 6
0 2 4 6 8 10 12 14
0 1 2 3 4 5

◀ **Fig. 6.23** Composites of four basic variables at 850 hPa of the four shear cases in South China at 0000 UTC. **a** Total winds (m s^{-1}) and **b** wind anomalies (m s^{-1}) with letters C and A indicating the centers of convergence and divergence. **c** Total height (*contour*, 0.5 × 10 gpm interval) and total temperature (*color shading*, 1 K interval), **d** height anomalies (*contour*, 0.5 × 10 gpm interval) and temperature anomalies (*color shading*, 1 K interval) with letters C and W indicating the cold and warm centers. **e** Total specific humidity (1 g/kg interval) and **f** anomalies (1 g/kg interval). *Purple lines* are the daily mean precipitation amount with 5, 10, and 25 mm indicated by grid dataset. *Blue*, *green*, and *red dots* are the daily mean precipitation of 10–25 mm, 25–50 mm, and exceeding 50 mm, respectively. In **e** and **f**, the *red-dashed line* indicates the axis of high specific humidity. Sourced from Qian et al. (2016c) © Meteorological Monthly, with permission

locate heavy rain than other analyses. Figure 6.23a is a composite of heavy rains associated with the 850 hPa total wind. The rain center is located in the right side of a wind convergence line and the left side of a LLJ. However, the LLJ disappears in the anomalous winds as shown in Fig. 6.23b, where the heavy rain band is along the convergence line of anomalous winds. In the two sides of the convergence line of anomalous winds, the wind directions are opposite and the wind velocities are basically equal.

At the 850-hPa total height field, the rain band is located in the south side of a height trough, which connects two lows respectively centered in Southwest China and the Yellow Sea, while the total temperature has no obvious signal to locate the rain band. An anomalous synoptic chart and a total synoptic chart at 850 hPa can be compared in Fig. 6.23c, d. An anomalous height trough and a boundary of anomalous warm–cold air masses extend exactly along the rain band in South China. In Fig. 6.23d, the low in Southwest China disappears, implying that the low is a climatic system rather than an anomalous system. At the 850 hPa total humidity (Fig. 6.23e), there are three large specific humidity bands respectively in Southwest China, South China, and Northwest China, but only a large humidity band corresponds with the South China rain. In Fig. 6.23f, the anomalous humidity field shows that the two large humidity bands can better indicate the two rain bands respectively in South China and Northeast China.

The second anomalous circulation pattern in South China is the anomalous vortex. Four daily RHR events occurred respectively on 14 January, 9 March, 14 May, and 24 July 1998 (in Table 6.5). Figure 6.24 shows the vertical–latitude cross-sections of two basic variables along 116.25°E for the composite vortex in South China at 0000 UTC. The vortex cannot be clearly indicated by the total height and the total temperature in Fig. 6.24a except that there are several warm "W" and cold "C" deviations between the total temperature and the climatic temperature. The height anomalies show that a deep anomalous vortex is centered over the rain area in South China. The heavy rain is not only a downward axis of anomalous height trough but also along the boundary of anomalous warm–cold air masses in the lower troposphere.

Similar to the height and temperature anomalies in Fig. 6.24b, a comparison of total and anomalous fields at 850 hPa is shown in Fig. 6.25. The total wind in Fig. 6.25a shows that the heavy rains fall in the right side of shear line and the left side of LLJ. Differently, the anomalous wind in Fig. 6.25b shows that the heavy

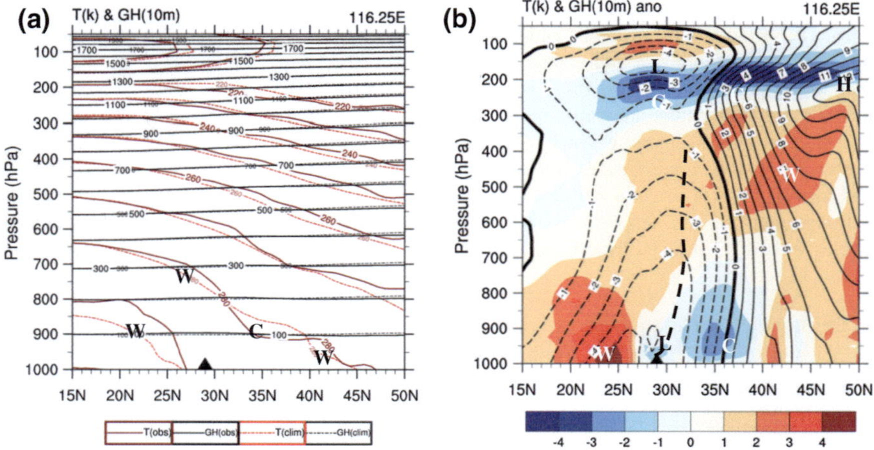

Fig. 6.24 Same as in Fig. 6.22a, b but for the heavy rain of South China vortices along 116.25°E. Sourced from Qian et al. (2016c) © Meteorological Monthly, with permission

rains fall in a cyclonic center under a calm environment but surrounding the anomalous southwesterly wind in the south side, while the anomalous southeasterly and northeasterly winds in northeastern and northwestern sides, respectively. An anomalous shear line indicates the heavy rain points well. The cyclone indicated by the total wind and the total height is similar in Fig. 6.25a, c in which the heavy rains fall in the southern side of height trough. However, the heavy rains are not only centered in the anomalous cyclone but also associated with the anomalous height trough and the anomalous temperature boundary in Fig. 6.25d. There are two large specific humidity bands in South China and Southwest China (Fig. 6.25e), but only the band in South China corresponds well with the heavy rains (Fig. 6.25f).

In Southwest China, a strong southwesterly warm advection centered at 850 hPa forms a surface reversed trough. In Fig. 6.26a, c, the total wind, the total temperature, and the total height fields at 850 hPa show that the southwesterly warm advection is clear located in the warm section between the two shear lines (troughs). Stronger southwesterly winds are observed in the total wind field than those in the anomalous wind field (Fig. 6.26a, b). The rains mainly fall in the eastern side of the southwest vortex in Fig. 6.26c, but the rains fall in the area of a low in Fig. 6.26d. In the 850-hPa temperature anomalies, the anomalous cold air does not invade to the anomalous low. In the total humidity field (Fig. 6.26e), there are two large humidity areas respectively located in South China and western China, but only one large humidity area exists in South China. The latter well corresponds to the rains in South China (Fig. 6.26f).

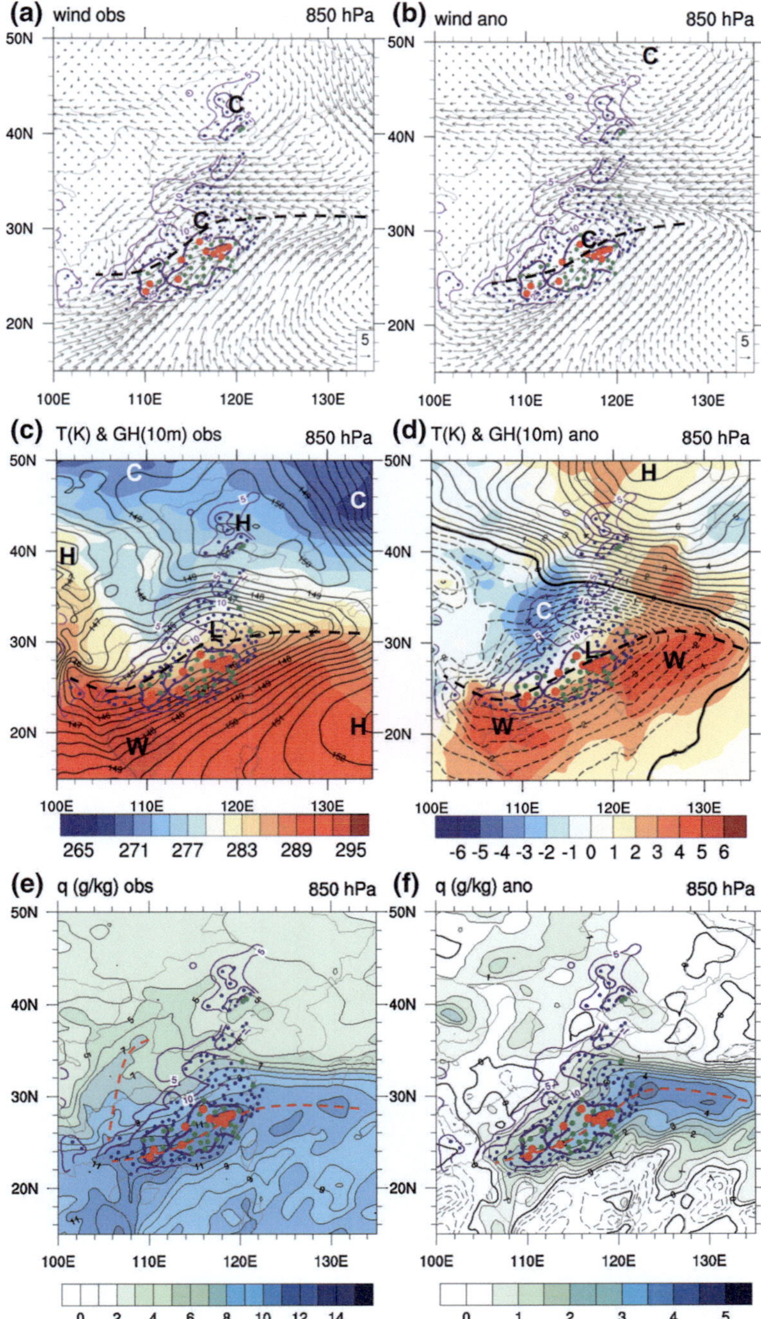

Fig. 6.25 Same as in Fig. 6.23 but for the heavy rain of South China vortices. Sourced from Qian et al. (2016c) © Meteorological Monthly, with permission

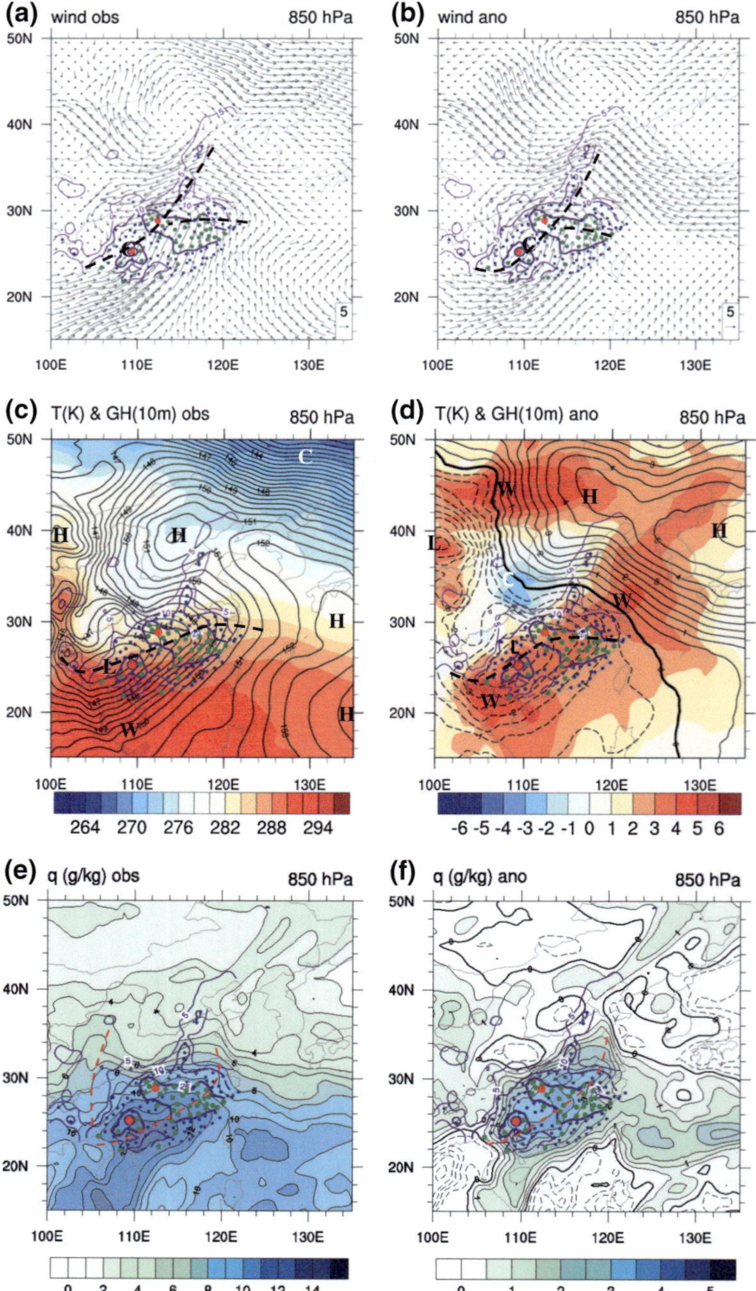

Fig. 6.26 Same as in Fig. 6.23 but for the heavy rain of reversed trough in South China. Sourced from Qian et al. (2016c) © Meteorological Monthly, with permission

6.4.2 Two Anomalous Circulation Patterns along the Yangtze River Valley

The first anomalous circulation pattern along the Yangtze River valley is the anomalous shear line and trough constructed. Nine daily RHR events occurred from May to July 1998 (Table 6.5). The composite circulation shows that a shear line is observed along the upper to middle Yangtze River and a trough from the Yellow Sea to the lower Yangtze River (Fig. 6.27a). The southwesterly flow is stronger in their south side while the northeasterly flow is weaker in their north side, so the concentrated rain band is located in their south side. Differently, the southwesterly flow and the northerly flow show opposite directions but equivalent intensities surrounding the rains along with the anomalous shear line and trough (Fig. 6.27b).

In the 850-hPa total height field (Fig. 6.27c), the southwest vortex is clear with concentrated rains in its east side along the height trough. Another trough is an extension of low located in Northeast China. In the anomalous height and temperature fields (Fig. 6.27d), the southwest vortex disappears instead of an anomalous vortex in the middle Yangtze River while an anomalous low is centered in the middle Yellow Sea. The rain band is well indicated by the two anomalous troughs along the Yangtze River. In the total humidity field (Fig. 6.27e) there are two large humidity areas respectively located in South China and from Southwest China to North China, but the rain band is located along the Yangtze River. After removing the climatic humidity, a large zone of anomalous humidity well corresponds with the rain band along the Yangtze River (Fig. 6.27f).

The second anomalous circulation pattern along the Yangtze River valley is the Jianghuai cyclone. Six daily RHR events occurred from April to July 1998 (Table 6.5). There are several differences between Fig. 6.28a, b. The cyclonic center in the total flow is located in the middle Yangtze River while it is in the lower river in the anomalous flow. The southwesterly flow is stronger in the total wind while the southwesterly and northeasterly flows show opposite directions but equivalent intensities surrounding the rains along the anomalous shear line.

A southwestern vortex is observed from the total height field in the upper Yangtze River with rains in the south side of the total height trough (Fig. 6.28c). The relationship between trough and rain band changes in the height anomalous field (Fig. 6.28d). The rains are located in the anomalous cyclone along the anomalous trough or the anomalous contrast of warm–cold air masses in the mid-low Yangtze River valley. In the total humidity field (Fig. 6.28e), there are three large humidity zones respectively located in Southwest China, along the Yangtze River and North China, but the major rains are located along the Yangtze River. After removing the climatic humidity, the large zone of anomalous humidity corresponds with the rains along the Yangtze River (Fig. 6.28f).

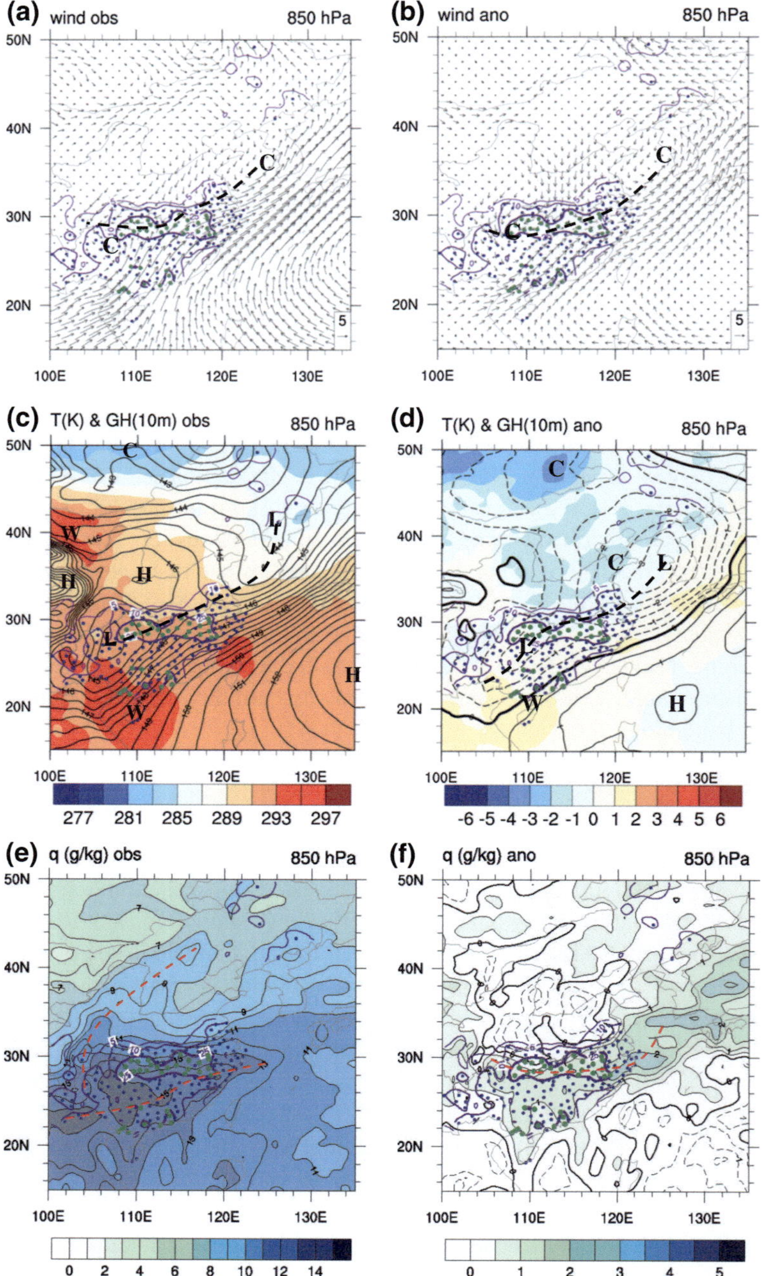

Fig. 6.27 Same as in Fig. 6.23 but for the heavy rain of shear and trough along the Yangtze River. Sourced from Qian et al. (2016c) © Meteorological Monthly, with permission

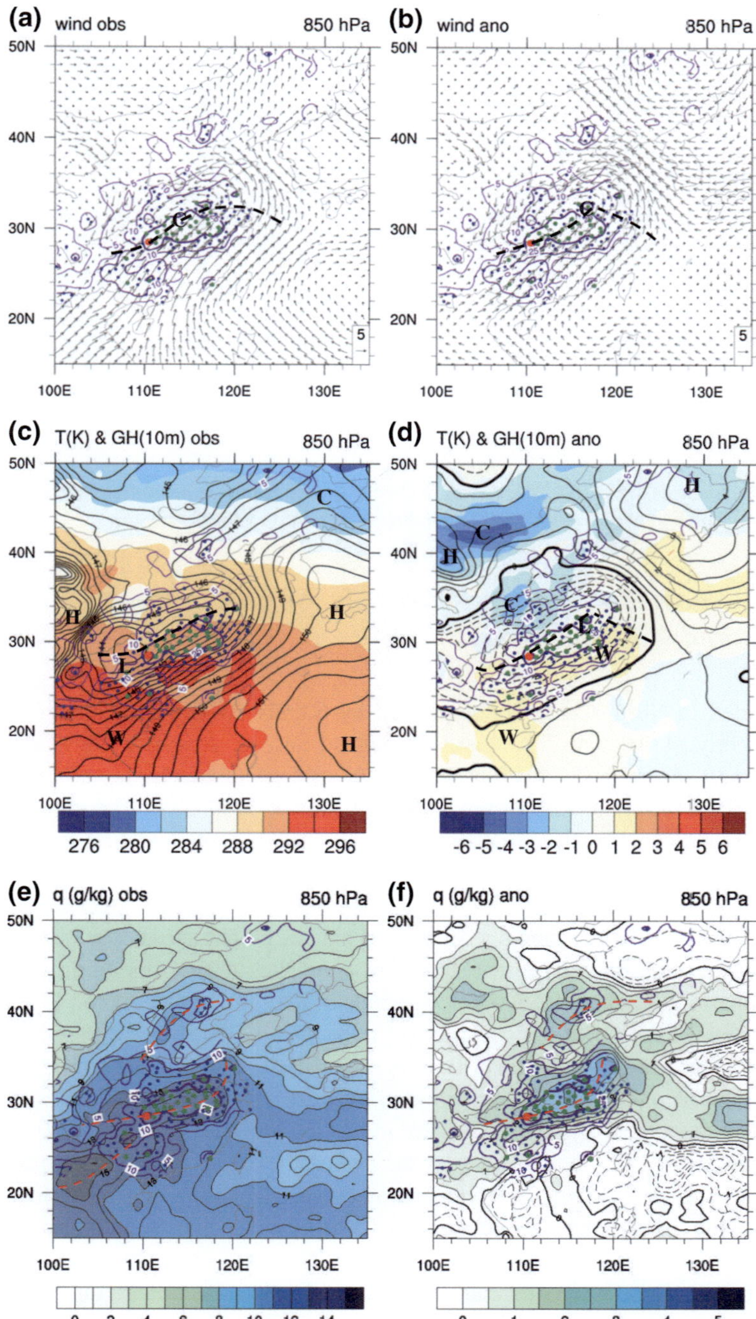

Fig. 6.28 Same as in Fig. 6.23 but for the heavy rain of vortices along the Yangtze River. Sourced from Qian et al. (2016c) © Meteorological Monthly, with permission

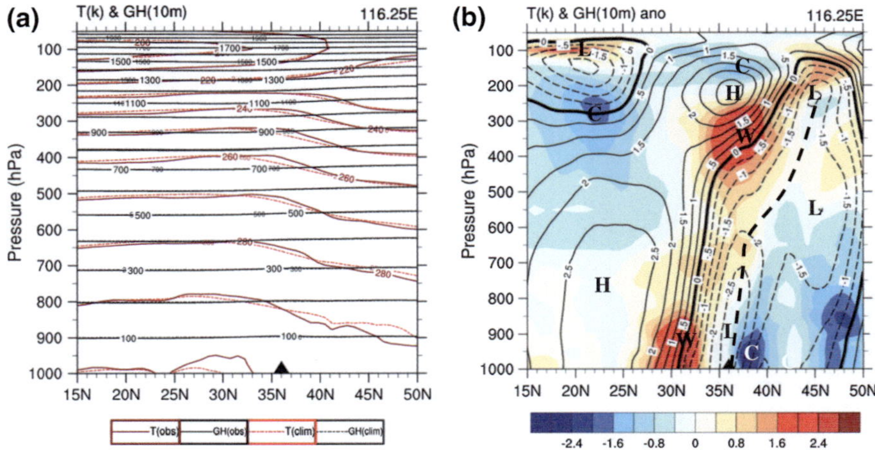

Fig. 6.29 Same as in Fig. 6.22a, b but for the heavy rain of North China vortices along 116.25°E. Sourced from Qian et al. (2016c) © Meteorological Monthly, with permission

6.4.3 *Anomalous Vortex Patterns in North China and Northeast China*

The vortex in North China is composited by 6 daily RHR events occurring on 1 July, 2 July, 4 August, 7 August, 15 August, and 22 August 1998 (Table 6.5). Figure 6.29 shows the vertical–latitude cross-sections of two basic variables along 116.25°E for the composite vortex in North China at 0000 UTC. The vortex cannot be clearly indicated by the total/climatic heights and the total/climatic temperatures in Fig. 6.29a. The height anomalies show that a deep anomalous vortex is centered over the rain area in North China and extends upward to the upper troposphere (Fig. 6.29b). The heavy rain is a downward axis of height anomalous trough and a vertical extension boundary of temperature anomalies which tilts northward.

Comparing the total and anomalous winds at 850 hPa, the deviation is not large at the central location of the cyclone and rains (Fig. 6.30a, b). A difference is that the southwesterly (northeasterly) flow is stronger (weaker) in the total wind than that in anomalous wind. The climatic southwest vortex always exists and contributes to the total height field as shown in Fig. 6.30c. The rains are scattered in the warm section and in the south side of the cyclonic center. In the anomalous height field, the rains are centered in the anomalous vortex and along the anomalous shear line. On the other hand, the boundary of temperature anomalies (warm–cold air masses) also indicates the rain points. In Fig. 6.30e, the eastern China is covered by the high humidity. Differently, anomalous humidity is concentrated in North China and corresponds with the rains (Fig. 6.30f).

The total and anomalous fields of seven heavy rain events are compared in Fig. 6.31. The southwest vortex is clearly observed in the total flow except that there is a vortex in Northeast China (Fig. 6.31a). In the anomalous field

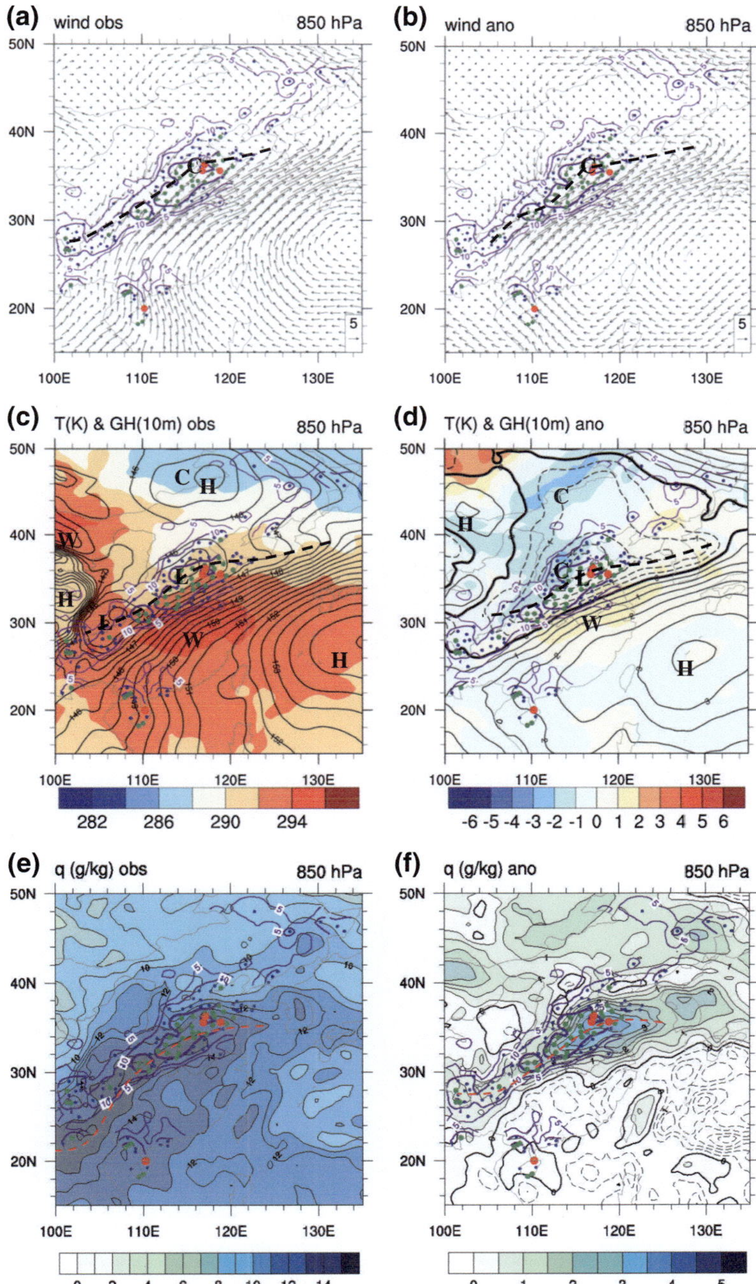

Fig. 6.30 Same as in Fig. 6.23 but for the heavy rain of North China vortices. Sourced from Qian et al. (2016c) © Meteorological Monthly, with permission

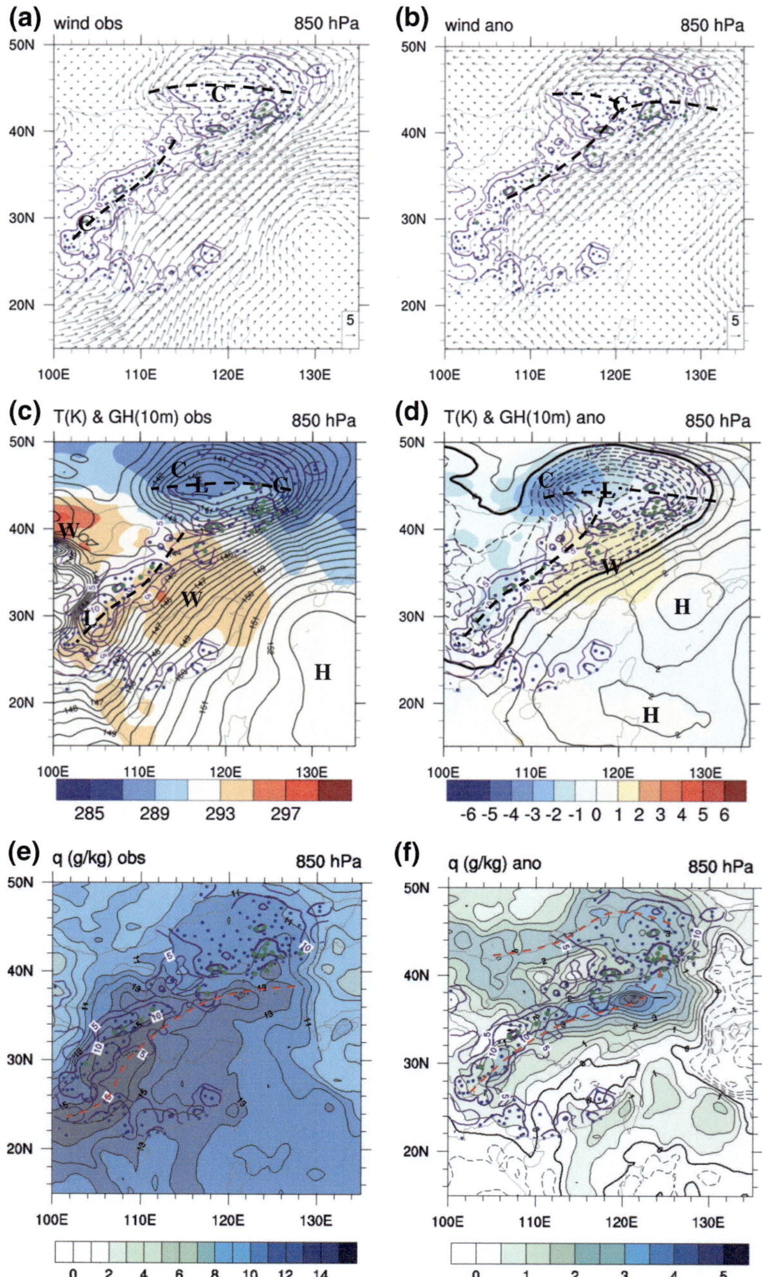

Fig. 6.31 Same as in Fig. 6.23 but for the heavy rain of Northeast China vortices. Sourced from Qian et al. (2016c) © Meteorological Monthly, with permission

(Fig. 6.31b), the southwest vortex disappears so the vortex in the Northeast China becomes clear, with symmetrical circulation to indicate rains. In the total flow, the total height also shows that there are a southwestern vortex and a cyclone in Northeast China (Fig. 6.31c). After removing the climatic height, the southwestern vortex disappears in Southwest China and the anomalous vortex in Northeast China becomes deeper (Fig. 6.31d). The coldest center is located in western side of the anomalous vortex and appears like an occluded cyclone. All scattered rains can be generally indicated by the boundary of temperature anomalies. The vortex of Northeast China in summer can form persist low temperature event (Qian and Jiang 2014). The vortex in Northeast China mostly exists in July and August, so the total specific humidity is larger and covers the whole China as shown in Fig. 6.31e. After removing the climatic humidity, the anomalous humidity with its axis well indicates those scattered rain points from Southwest China to Northeast China (Fig. 6.31f).

As shown in Table 6.5, RHRs are mainly caused by the shear (or convergence) lines and cyclones. During 1981–2010, 266 daily RHR events (with more than 10 adjacent stations) associated with anomalous convergence lines are identified in total. Their central positions, monthly histogram, and annual frequencies are shown in Fig. 6.32. The central location of a daily RHR event is calculated by averaging the longitude and latitude of all heavy rain stations in the RHR event. Figure 6.32a shows that the centers are dense in Jiangxi, Hunan, Anhui, and Zhejiang provinces along the middle to lower Yangtze River. There are several centers of daily RHR events in North China and the southern part of Northeast China. Seasonally, most daily RHR events occur in June (104 events) and July (77 events). In May and August, 33 and 18 daily RHR events are observed, respectively. Thus, the season of daily RHR events mainly occurs from May to August (Fig. 6.32b). An increasing trend (2.1 times per decade at a significance level 95%) of annual RHR events associated with anomalous convergence lines is observed from 1981 to 2010 (Fig. 6.32c). The average annual RHR events are about 9 with the minimum 3 RHR events in 1985 and the maximum 17 in 1995.

Among all the 181 RHR events in June and July, there are 25 events centered along the Yangtze River (30°N), which are listed in Table 6.6. There were 12 events in 1990s and only 5 events occurred in 1980s. It is noted that there were three daily events at the same calendar date on 3 July in 1990s (3 July 1992, 3 July 1993, and 3 July 1998).

Usually, anomalous synoptic systems are masked by climatic fields. For some extreme weather cases, anomalous synoptic systems are difficult to be identified by forecasters directly from the total-based synoptic chart. Most comparisons and analyses in recent years confirm that the anomaly-based synoptic analysis is better than the total-filed synoptic analysis in indicating the surface weather extremes such as RHR events. In an operational sense, seven circulation patterns based on the 41 heavy rain events in 1998 in eastern China are constructed by Qian et al. (2016c). They believe that these patterns can help forecasters to improve the ability of locate RHRs in China.

In the traditional synoptic chart of wind field, a RHR is usually located in the right side of the LLS line and the left side of the LLJ. In the anomalous synoptic

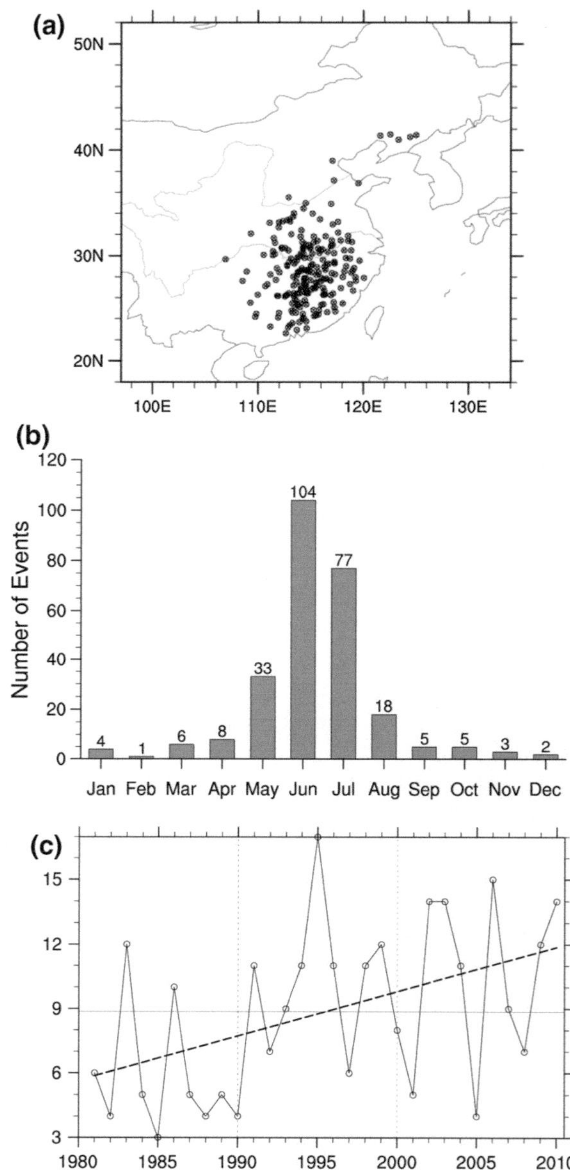

Fig. 6.32 The central locations **a**, monthly RHR events **b**, and annual RHR events **c** of the 266 daily RHR events (above adjacent 10 stations) associated with anomalous convergence lines from 1981 to 2010. Sourced from Shan et al. (2015) © Springer Science + Business Media Dordrecht 2015, with permission

chart, however, a RHR is just located along the anomalous shear line, which is surrounded by flows of opposite directions and equivalent intensities. In the traditional synoptic chart of height field, it contains the climatic southwestern vortex in Southwest China and the climatic subtropical high in the NW Pacific. These two climatic systems can largely influence the locations of trough, shear line, even cyclonic center in the total height field. After removing the climatic height, the

Table 6.6 Twenty-five RHR events centered in the Yangtze River (30°N) in June and July from 1981 to 2010

1981-07-11	1983-07-05	1985-06-04	1986-07-16	1987-07-02
1991-07-05	1992-07-03	1993-07-03	1995-07-07	1996-07-02
1996-07-04	1998-07-03	1999-06-26	1999-06-27	1999-06-28
1999-06-29	1999-06-30	2003-07-05	2003-07-06	2003-07-09
2003-07-10	2005-07-07	2007-07-09	2010-07-09	2010-07-12

Sourced from Shan et al. (2015) © Springer Science + Business Media Dordrecht 2015, with permission

anomalous height trough, shear line, and cyclonic center can well locate the rains. In the traditional synoptic chart of temperature field, there is no apparent signal to indicate a heavy rain location. After removing the climatic temperature, the boundary of temperature anomalies can indicate the scattered rain points. In summer, the climatic moisture is higher in eastern China as a part of East Asian summer monsoon. Thus, total specific humidity cannot be used to directly indicate the heavy rains. After removing the climatic humidity, the axis of anomalous humidity can display the scattered rain points.

Anomalous synoptic systems, which can be applied to indicate the surface weather extremes, are largely masked by the climatic field, so the traditional synoptic chart is not convenient for forecasters in practice. In eastern China, cyclones are often observed in South and North China. The anomalous height shows that the depth of anomalous vortex in South China is below at 400 hPa, while in its northern side over North China there is an anomalous high extending from the troposphere to the stratosphere. Differently, the depth of anomalous vortex in North China can extend upward to the upper troposphere, while in its southern side there is a deep anomalous high. For any time and any case, the height and temperature anomalies satisfy the hydrostatic relationship, which means that the temperature anomalies can be derived from the height anomalies from the troposphere to the stratosphere (Qian et al. 2015b, 2016a). It is evidently noted that a positive (negative) center of temperature anomalies is robustly located below (above) a positive (negative) center of height anomalies at the upper troposphere. Therefore, it is not optimal to choose height, temperature, wind, and humidity anomalies at 850 hPa. In the South China coastal area, the height and temperature anomalies are better at 925 hPa while the center of specific humidity anomalies is concentrated at 700 hPa. In Southwest China, variables chosen at 850 hPa may be worse because there is a climatic southwestern vortex and topography.

6.5 Moist Divergence and Moist Vorticity[5]

Properly introducing moisture effects into a dynamical parameter can significantly increase the parameter's ability to diagnose heavy rain locations. The "relative humidity based weighting" approach to extend the moist potential vorticity (MPV) to the generalized moist potential vorticity (GMPV) is analyzed and demonstrated. Following the same approach, two new diagnostic parameters, moist vorticity (MV) and moist divergence (MD), have been proposed by Qian et al. (2015a), incorporating moisture effects into the traditional vorticity and divergence. A RHR event that occurred along the Yangtze River on 1 July 1991 is analyzed as a case study, and a systematic evaluation is made on 41 RHR events during the notorious flooding year of 1998 in eastern China. Results show that after the moisture effects are properly incorporated, the ability of all three parameters to capture a heavy rain area is significantly improved: the GMPV is improved over the MPV by 194%, the MD over the divergence by 60%, and the MV over the vorticity by 34% in terms of the threat score (TS). The average TS is 0.270 for the MD, 0.262 for the MV, and 0.188 for the GMPV. The results from their study suggest that MV and MD can be used to post-process a model forecast to potentially locate heavy rain.

Heavy rain is an outcome of combined favorable dynamic and moisture conditions. It has been shown that combining dynamic and moisture factors together in a forecast method can greatly improve heavy rain prediction (e.g., Doswell et al. 1996), where precipitation efficiency and duration are another two important elements considered in their method. The extension of potential vorticity (PV) to moist potential vorticity (MPV) is an example of combining dynamic and moisture factors together (Bennetts and Hoskins 1979). Another example is the generalization of the MPV, i.e., generalized moist potential vorticity (GMPV, Gao et al. 2004a). How to properly include moisture effects into a dynamic factor is not obvious. Qian et al. (2015a) demonstrated a feasible way to include moisture effects within a dynamical parameter and how it improves the parameter's performance through a case study and systematic evaluation. The goal is not to suggest that the application of this parameter can replace a conceptual-model-based multi-scale forecast process, but simply to demonstrate how the incorporation of moisture effects in the parameter can improve the correlation between areal coverage of the parameter and observed rainfall location. For this goal, a RHR event that occurred on 1 July 1991 is used for a detailed analysis and method test. An independent dataset of 41 RHR cases from the notorious flooding year of 1998 in eastern China is used for a systematic evaluation to confirm the robustness of this approach.

The term "heavy rain" used refers to a RHR event. This observed precipitation data is interpolated to a 0.5° latitude–longitude grid using the ordinary Kriging

[5]Modified article originally published in Qian et al. (2015a). Published with kind permission of © American Meteorological Society. All Rights Reserved.

Table 6.7 Description of the 41 RHR events occurred in eastern China in 1998

Date	Num of sites that rainfall ≥25/50/100 mm/day	Regions	Synoptic-scale system	Date	Num of sites that rainfall ≥25/50/100 mm/day	Regions	Synoptic-scale system
Jan 14	68/23/0	S. Yangtze R	Cyclone	Jul 1	18/10/2	N. China	Cyclone
Feb 16	24/10/0	S. Yangtze R	Reversed trough	Jul 2	43/23/5	N. China	Cyclone
Feb 17	31/10/0	S. China	Reversed trough	Jul 3	26/12/0	Yangtze R	Shear
Mar 8	45/13/0	S. Yangtze R	Reversed trough	Jul 6	29/12/1	NE. China	Cyclone
Mar 9	43/16/0	S. China	Cyclone	Jul 7	15/11/0	NW. China	Cyclone
Apr 11	52/11/2	Yangtze R	Cyclone	Jul 7	24/10/2	NE. China	Cyclone
May 2	54/14/1	Yangtze R	Shear	Jul 8	21/10/2	NE. China	Cyclone
May 14	36/15/5	S. Yangtze R	Cyclone	Jul 13	31/14/3	NE. China	Cyclone
May 22	24/10/1	SW. China	Reversed trough	Jul 14	32/14/1	NE. China	Cyclone
May 23	45/14/4	Yangtze R	Cyclone	Jul 15	17/11/0	NE. China	Cyclone
Jun 9	34/11/2	S. Coast	Shear	Jul 22	16/11/4	Yangtze R	Shear
Jun 12	29/11/1	Yangtze R	Cyclone	Jul 23	42/22/5	Yangtze R	Cyclone
Jun 13	24/16/9	Yangtze R	Shear	Jul 24	39/15/3	S. Yangtze R	Cyclone
Jun 14	30/17/7	S. Yangtze R	Shear	Jul 29	26/14/3	S. Yangtze R	Cyclone
Jun 19	70/36/6	Yangtze R	Shear	Jul 30	28/12/3	Yangtze R	Cyclone
Jun 21	45/17/5	S. Yangtze R	Shear	Aug 4	34/18/7	N. China	Cyclone
Jun 23	62/22/3	Yangtze R	Shear	Aug 5	42/22/2	NE. China	Cyclone
Jun 24	70/37/10	S. Yangtze R	Reversed trough	Aug 7	24/11/1	N. China	Cyclone
Jun 25	57/25/5	Yangtze R	Shear	Aug 15	28/13/0	N. China	Cyclone
Jun 26	35/21/3	Yangtze R	Shear	Aug 22	23/12/2	N. China	Cyclone
Jun 30	16/10/2	Yangtze R	Cyclone	Aug 27	22/11/1	Yangtze R	Shear

Sourced from Shan et al. (2015) © Springer Science + Business Media Dordrecht 2015, with permission method by Chen et al. (2010), used for the threat score calculation. Given the ERA-Interim data is about 80 km in spatial resolution, only long-lasting (24 h) large-scale and widespread RHR events (more than 10 adjacent stations with observed rainfall exceeding 50 mm day^{-1}) caused by obvious synoptic-scale circulation systems (Qian 2013) are analyzed (see Table 6.7), while those isolated local heavy rain events probably caused by individual convective systems are excluded. Table 6.7 is more detailed than Table 6.5.

From 30 June to 12 July 1991, a prolonged heavy rain event occurred over the lower Yangtze River basin, in response to a series of shortwaves moving eastward across the area (Ding 1993). On 1 July 1991, 22 stations exceeded 50 mm day^{-1} rainfall with a maximum precipitation of 201.8 mm day^{-1} and an average precipitation of 96.5 mm day^{-1}. The event on 1 July (1200 UTC 30 June–1200 UTC 1 July) 1991 is used as a showcase. In the summer of 1998, heavy rain events frequently occurred over the Yangtze River basin, resulting in the most severe flooding on record. There were 41 RHR episodes during that summer based on the criterion of "at least 10 adjacent stations exceeding 50 mm day^{-1} (a 24 h period centered at 0000 UTC of a day) rainfall at each station". These 41 cases are listed in Table 6.7 with a brief description including the date and number of stations with precipitation exceeding 25 or 50 or 100 mm per day as well as associated synoptic systems. On 7 July 1998, there were two RHR events over Northwest and Northeast China. Among these 41 cases, most of them occurred along or south of the Yangtze River. They were mainly associated with synoptic frontal boundary (horizontal wind shear, convergence zone, and reversed trough) and extratropical cyclones (Table 6.7). These 41 cases are used for systematic evaluation.

The following two basic dynamical parameters are widely used in weather diagnostics:

$$\zeta = \left(\frac{\partial v}{\partial x} - \frac{\partial u}{\partial y}\right), \tag{6.5.1}$$

$$D = \left(\frac{\partial u}{\partial x} + \frac{\partial v}{\partial y}\right). \tag{6.5.2}$$

They are the vertical relative vorticity and horizontal divergence, respectively, expressed in the form of planar coordinates. Since heavy precipitation is closely associated with convergent and cyclonic flows in a moist environment, either dynamic or moisture factor alone should not be able to accurately depict the area of a heavy rain event. Figures 6.33 and 6.34 show the relative vorticity (Eq. 6.5.1) and divergence (Eq. 6.5.2) at 850 hPa every 6 h during the entire period (1200 UTC June 30–1200 UTC 1 July) of the heavy rain event on 1 July 1991. Although they are generally similar to each other at different times during the event, the midpoint 0000 UTC (Figs. 6.33c and 6.34c) seems to match the heavy rain area the best (at least as good as others). Therefore, the midpoint is used as a representative time in

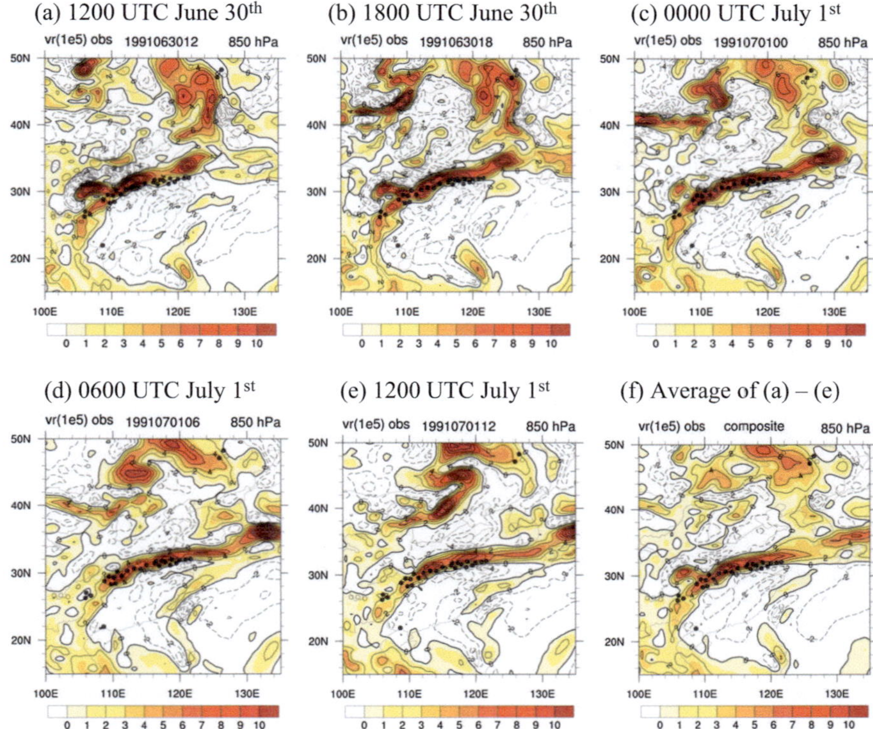

Fig. 6.33 Horizontal distribution of 850 hPa vorticity at **a** 1200, **b** 1800 UTC 30 June, **c** 0000, **d** 0600, and **e** 1200 UTC 1 July 1991 as well as **f** the average of **a**–**e**. The *contour* interval is $2 \times 10^{-5}\ s^{-1}$. *Solid* (*open*) *circles* indicate the stations with rainfall exceeding 50 (25–50) mm during the 24 h from 1200 UTC 30 June to 1200 UTC 1 July 1991. Sourced from Qian et al. (2015a) © American Meteorological Society, with permission

their study. In Figs. 6.33 and 6.34, there are two main areas of maximal "positive" vorticity (cyclonic flow) and "negative" divergence (convergence): one in South China along the Yangtze River and another in Northwest China around 40°N. The former coincides well with the heavy rain band, while the latter does not correspond to any precipitation band (due to the lack of favorable moisture condition). Figure 6.35 shows the relative humidity at 850 hPa at 0000 UTC 1 July 1991. The heavy rain band along the Yangtze River is well coincident with a maximum relative humidity area, while the maximum relative humidity area in the Southwest China does not correspond to any precipitation band (due to the lack of favorable dynamic condition). Figures 6.33 and 6.35 demonstrate that although a dynamic or moisture parameter alone can reflect heavy rain regions, it also exhibits "false alarm" regions. Physically, it implies that heavy rain requires both forcing for ascent and moisture to be coincident across a same area. Therefore, parameters combining both dynamic and moisture factors need to hopefully depict heavy rain areas more exclusively.

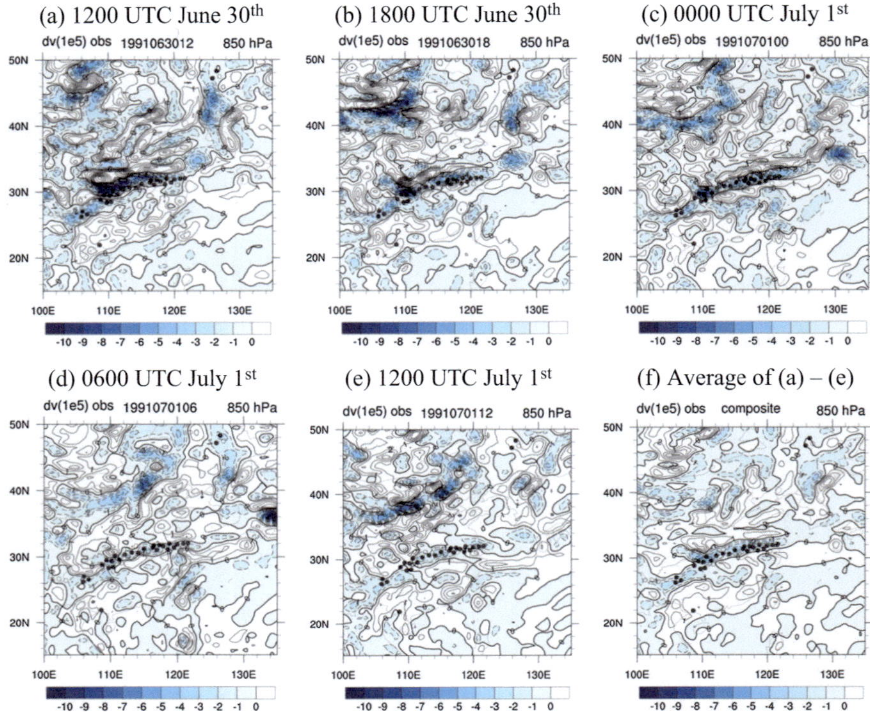

Fig. 6.34 Same as Fig. 6.33 but for the divergence at 850 hPa with $1 \times 10^{-5}\ s^{-1}$ *contour* interval. Sourced from Qian et al. (2015a) © American Meteorological Society, with permission

Fig. 6.35 Horizontal distribution of 850 hPa relative humidity (q/q_s, 0.1 intervals) at 0000 UTC 1 July 1991. *Solid* (*open*) *circles* indicate the stations with rainfall exceeding 50 (25–50) mm during the 24 h from 1200 UTC 30 June to 1200 UTC 1 July 1991. Sourced from Qian et al. (2015a) © American Meteorological Society, with permission

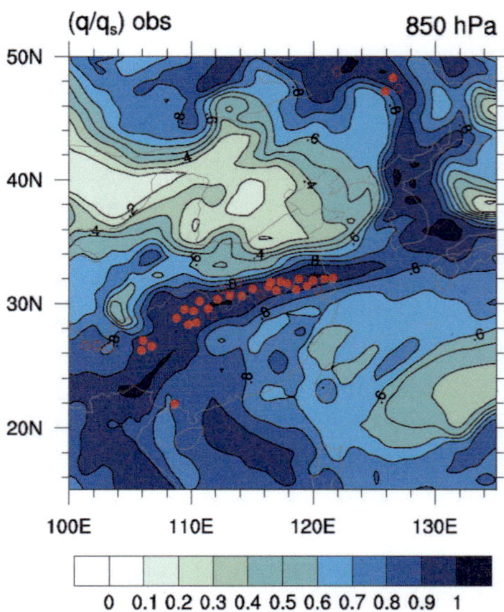

How to properly include moisture effects in a dynamical parameter? Potential vorticity (PV) is used as an example in this subsection. PV is a dynamical diagnostic parameter and can be derived from the combination of the first law of thermodynamics and the law of momentum conservation, which implies that PV can only be changed by diabetic heating or frictional processes. Equation 6.5.3 is the PV expression commonly used in meteorology:

$$\mathrm{PV} = \rho^{-1}\varsigma_a \cdot \nabla\theta \tag{6.5.3}$$

$$\varsigma_a = f + \varsigma = 2\Omega\,\sin\,\varphi + \left(\frac{\partial v}{\partial x} - \frac{\partial u}{\partial y}\right), \tag{6.5.4}$$

where ρ is the fluid (air) density, ς_a the absolute vorticity, and θ the potential temperature. PV is often used to diagnose observed atmospheric behavior and understand synoptic-scale dynamics and atmospheric numerical simulation results (Cao and Cho 1995; Shou and Li 1999; Schubert et al. 2001; Gao et al. 2004a, b; Deng and Gao 2009). Since PV is not conserved when latent heat release is taken into account in a saturated moist atmosphere, Bennetts and Hoskins (1979) first generalized PV into moist potential vorticity (MPV, Eq. 6.5.5) by replacing θ with the equivalent potential temperature θ_e in the PV equation, i.e.,

$$\mathrm{MPV} = \rho^{-1}\varsigma_a \cdot \nabla\theta_e \tag{6.5.5}$$

Schubert et al. (2001) proved an annihilation of the solenoidal term in the MPV equation, thus leading to a conservation of MPV in moist adiabatic and frictionless processes. The MPV at a pressure coordinate consists of two terms, MPV1 and MPV2:

$$\mathrm{MPV1} = -g(f + \varsigma)\frac{\partial\theta_e}{\partial p} = -g\left[\left(\frac{\partial v}{\partial x} - \frac{\partial u}{\partial y}\right) + f\right]\frac{\partial\theta_e}{\partial p} \tag{6.5.5a}$$

$$\mathrm{MPV2} = g\frac{\partial v}{\partial p} \cdot \frac{\partial\theta_e}{\partial x} - \frac{\partial u}{\partial p} \cdot \frac{\partial\theta_e}{\partial y} \tag{6.5.5b}$$

MPV1 is a product of absolute vorticity (related to ascent) and vertical gradient of equivalent potential temperature (related to atmospheric instability), representing the impact of convective instability on vorticity growth; and MPV2 is the product of vertical wind shear and horizontal gradient of equivalent potential temperature, representing the impact of vertical wind shear and baroclinicity on vorticity development (Wu et al. 1995).

The MPV has been extensively used in studies of conditional symmetric instability (Emanuel 1983, 1988; Bennetts and Sharp 1982; Schultz and Schumacher 1999) and the generation of MPV in extratropical cyclones (Cao and Cho 1995; Martínez-Alvarado et al. 2010). The MPV has also been used in the analysis of heavy rainfall (Shou and Li 1999; Clark et al. 2002; Gao et al. 2004a, b;

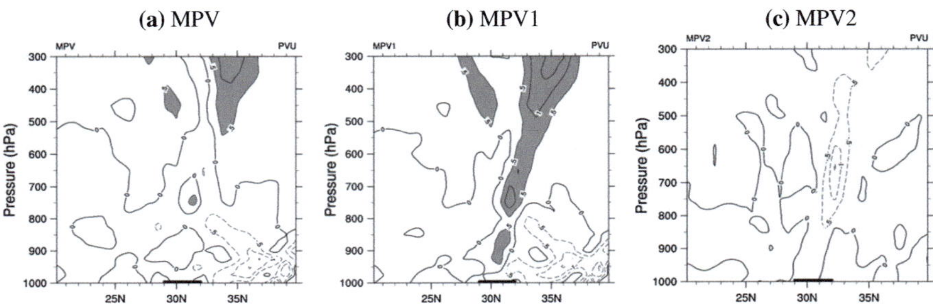

Fig. 6.36 Vertical–latitude cross-sections of **a** MPV, **b** MPV1, and **c** MPV2 (0.5 PVU interval) along 114.75°E at 0000 UTC 1 July 1991. The *thick horizontal line* at the surface represents the area of heavy rain. The *shading* indicates positive PV with value larger than 0.5 PVU. Sourced from Qian et al. (2015a) © American Meteorological Society, with permission

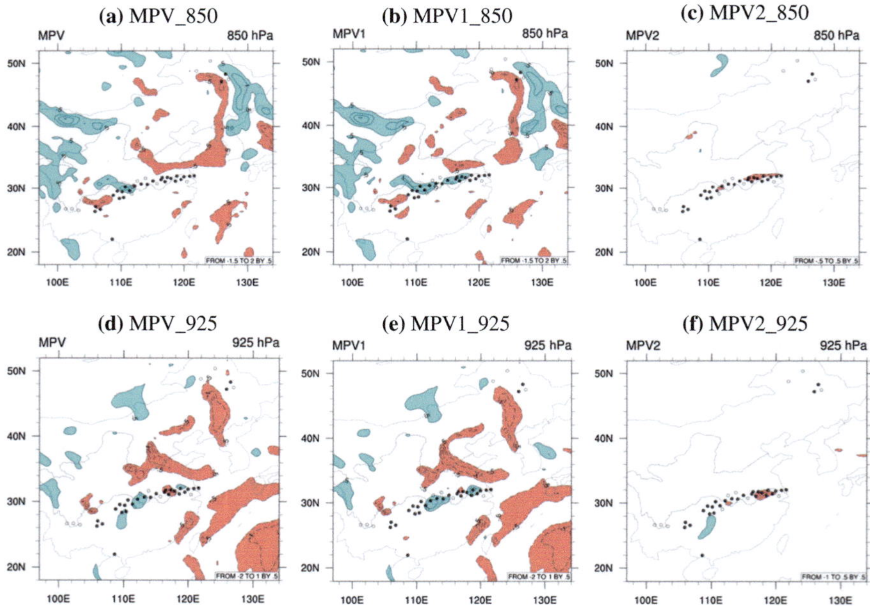

Fig. 6.37 **a** MPV, **b** MPV1, and **c** MPV2 (0.5 PVU interval) at 850 hPa at 0000 UTC 1 July 1991. Panels **d–f** are the same as **a–c** but at 925 hPa. *Solid* (*open*) *circles* indicate the stations with rainfall exceeding 50 (25–50) mm day^{-1}. Positive (negative) PV is in *solid* (*dashed*) *lines*, where cyan (*brown*) shading covers the value larger than 0.5 PVU (less than −0.5 PVU). Sourced from Qian et al. (2015a) © American Meteorological Society, with permission

Deng and Gao 2009; Novak et al. 2009; Zhou et al. 2010). Gao et al. (2004b) argued that the maximum MPV and the maximum surface rainfall are nearly collocated due to the impact of heat and mass forcing on the development of MPV. Therefore, the MPV can be used to track the propagation of rain system. Qian et al.

(2015a) systematically and quantitatively examine the performance of MPV in diagnosing heavy rain location. Figure 6.36 shows the vertical cross-section of the MPV along with its two components MPV1 and MPV2 for the RHR case on 1 July 1991. The MPV has virtually no ability to depict a heavy rain area (Fig. 6.36a) due to the opposite signals coming from the MPV1 (Fig. 6.36b) and MPV2 (Fig. 6.36c) over the heavy rain region. The two components can depict the heavy rain location better than the MPV itself, e.g., the maximal axes of positive MPV1 and negative MPV2 extend to the lower level, right over the heavy rain band. To examine their horizontal distribution, Fig. 6.37 shows the MPV and the two components at 850 and 925 hPa. Consistent with Fig. 6.36, the MPV (Fig. 6.37a, d) does not match the heavy rain area due to the opposite signs of the two components. Although the MPV1 (Fig. 6.37b, e) does cover most or parts of the heavy rain area, it is too widespread spatially, resulting in a high "false alarm" rate. The MPV2 (Fig. 6.37c, f) is the best among the three with much reduced false alarms but it misses many heavy rain spots such as the western end of the Yangtze River heavy rain band and the heavy rain area in Northeast of China. For the MPV, MPV1, and MPV2, their performance is better at 850 hPa than that at 925 hPa. Therefore, they will be evaluated at 850 hPa, hereafter.

To improve the capability of the MPV in depicting heavy rain locations, Gao et al. (2004a) proposed a generalized MPV (GMPV) by tweaking the moisture effects in the MPV. The rationale behind their modification is that an assumed generally "saturated (if defined as 100% in relative humidity)" air mass might not be uniformly saturated in the real atmosphere, e.g., relative humidity of lighter fog could be less than 100% and different from that of denser fog (Cao and Gao 2008). To reflect this nonuniformity of moisture condition, Gao et al. (2004a) defined a generalized equivalent potential temperature θ^* by multiplying a relative humidity (q/q_s) based weight (called "condensation probability function") in the formula of the equivalent potential temperature θ_e under saturated situation as follows

$$\theta^* = \theta \cdot \exp\left(\frac{Lq_s}{c_pT}\left(\frac{q}{q_s}\right)^k\right), \tag{6.5.6}$$

where k is an empirical constant varying from 1 to 20, representing an exponential power of relative humidity. Theoretically and statistically, the changes in spatial distribution of $(q/q_s)^k$ become slower when $k \geq 9$ (Wang et al. 1999). If k = 1 or 0, it becomes the regular equivalent potential temperature of unsaturated or saturated air. This new definition distinguishes the relative contribution of saturated water vapor based on its contained moisture content. Furthermore, by replacing equivalent potential temperature θ_e with the generalized equivalent potential temperature θ^* in Eq. 6.5.5, the MPV can thus be extended to GMPV:

$$\text{GMPV} = \rho^{-1}\varsigma_a \cdot \nabla\theta^*. \tag{6.5.7}$$

If $k = 1$, it becomes the original MPV. A similar expression and application can also be found in Korty and Schneider (2007). Following Eq. 6.5.5, the two components of the GMPV, GMPV1, and GMPV2, can be expressed as

$$\mathrm{GMPV1} = -g(\varsigma + f)\frac{\partial \theta^*}{\partial p} = -g\left[\left(\frac{\partial v}{\partial x} - \frac{\partial u}{\partial y}\right) + f\right]\frac{\partial \theta^*}{\partial p} \tag{6.5.7a}$$

$$\mathrm{GMPV2} = g\frac{\partial v}{\partial p}\cdot\frac{\partial \theta^*}{\partial x} - \frac{\partial u}{\partial p}\cdot\frac{\partial \theta^*}{\partial y} \tag{6.5.7b}$$

Similar to MPV1 and MPV2, GMPV1 is related to the impact of convective instability and GMPV2 is related to vertical wind shear and horizontal distribution of heat and moisture ("moist" baroclinicity). It was reported that the GMPV and their components have been applied to diagnose heavy rain in several previous studies (Gao et al. 2002, 2004b; Deng and Gao 2009; Zhou et al. 2010). They used the maximum GMPV to trace heavy rain locations. But no systematic quantitative evaluation has been done in these studies. The k value controls the moisture impact area. When $k = 1$, moisture impact spreads toward the largest area; and when k increases, moisture impact will be concentrated over a smaller area given $q/q_s \le 1.0$. Figure 6.38 shows the impact of varying the k value on the relationship between relative humidity (q/q_s) and heavy rainfall in horizontal view (Fig. 6.38a–c) and in vertical cross-section (Fig. 6.38d–e) for the case on 1 July 1991. It is noted that the impact area is too wide when $k = 1$, it misses some heavy rain spots when $k = 20$ (e.g., Fig. 6.38c), and it covers all the heavy rain spots when $k = 10$. Therefore, $k = 10$ is used throughout their study. To examine if the GMPV is superior to MPV in diagnosing heavy rain locations, Fig. 6.39 shows the vertical–latitude cross-section of the GMPV along with its two components for the case on 1 July 1991. Comparing Fig. 6.39 with Fig. 6.36, the improvement of the GMPV (Fig. 6.39a) over the MPV (Fig. 6.36a) is obvious. Not only the two components (Fig. 6.39b, c) but the GMPV itself (Fig. 6.39a) now also has a strong signal pointing to the heavy rain location, because the two components have the same sign in the GMPV, while there is no clear signal in the MPV (Fig. 6.36a). Unlike the MPV, the strongest signals from the GMPV concentrate in the lower troposphere, so the heavy rain location can be easily traced around 925 hPa. Figure 6.40 is their horizontal distributions at 850 and 925 hPa. Apparently, the GMPV (Fig. 6.40a, d) is influenced mostly by the GMPV1 (Fig. 6.40b, e). Both of them cover the heavy rain regions well although their areal extents are a bit too wide, particularly in Southwest and Northeast China. Similar to the MPV2, the GMPV2 (Fig. 6.40c, f) has the smallest coverage among the three with much reduced false alarms, but resulting in many "misses". For the GMPV, GMPV1, and GMPV2, their performance is better at 925 hPa than 850 hPa (more "misses" at 850 hPa by positive PV area). At the same time, all of them cover the heavy rain areas better than the MPV, MPV1, and MPV2 at both levels (cf. Fig. 6.37). Given the better performance of GMPV, GMPV1, and GMPV 2 at 925 hPa, they will be evaluated at 925 hPa hereafter.

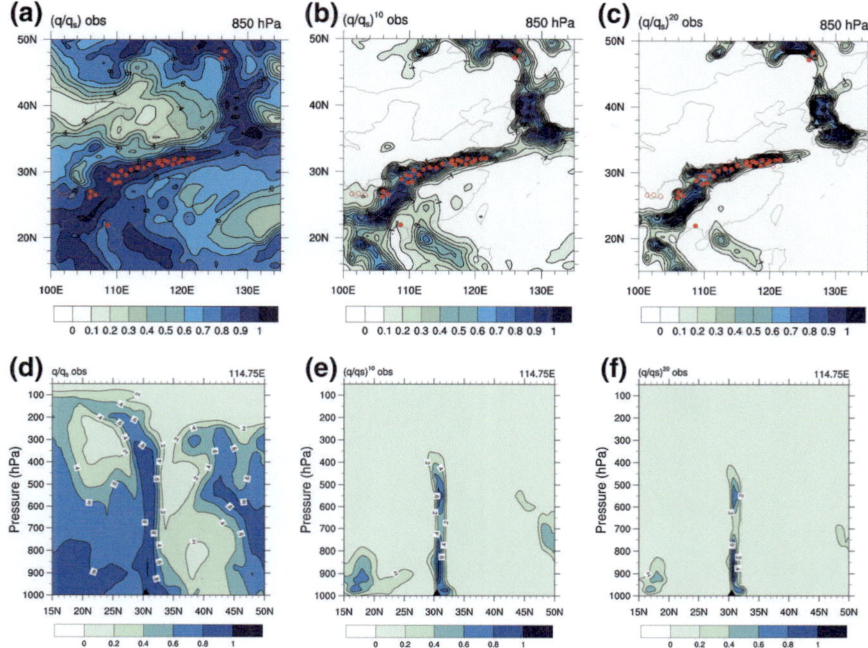

Fig. 6.38 Impact of different k values on the relative humidity based weight at 0000 UTC 1 July 1991. **a** $k = 1$, **b** $k = 10$, and **c** $k = 20$ for horizontal distribution (0.1 interval); and **d** $k = 1$, **e** $k = 10$, and **f** $k = 20$ for vertical distribution (0.2 interval). *Solid* (*open*) *circles* indicate the stations with rainfall exceeding 50 (25–50) mm day^{-1} in **a–c**; and the heavy *rain area* is indicated by the symbol *filled triangle* in **d–f**. Sourced from Qian et al. (2015a) © American Meteorological Society, with permission

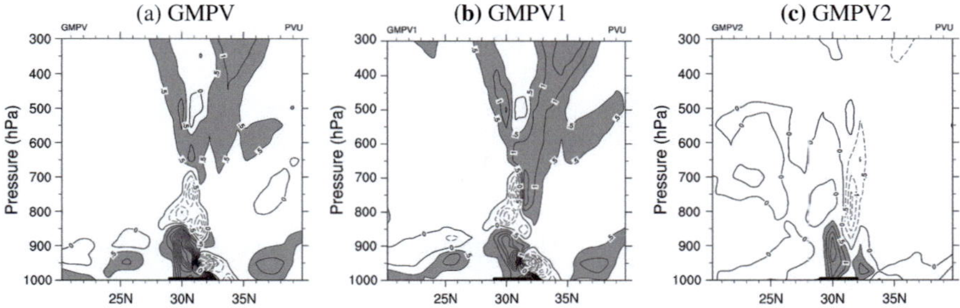

Fig. 6.39 Same as Fig. 6.36 but for **a** GMPV, **b** GMPV1, and **c** GMPV2 ($k = 10$). Sourced from Qian et al. (2015a) © American Meteorological Society, with permission

To find a robust result, GMPV over the MPV are systematically compared based on the independent 41 RHR cases from 1998 (Table 6.7). TS (Palmer and Allen 1949) in Eq. 6.5.8 is calculated to quantitatively measure how well a parameter overlaps with a heavy precipitation area ≥ 25 mm day^{-1} based on all grid points

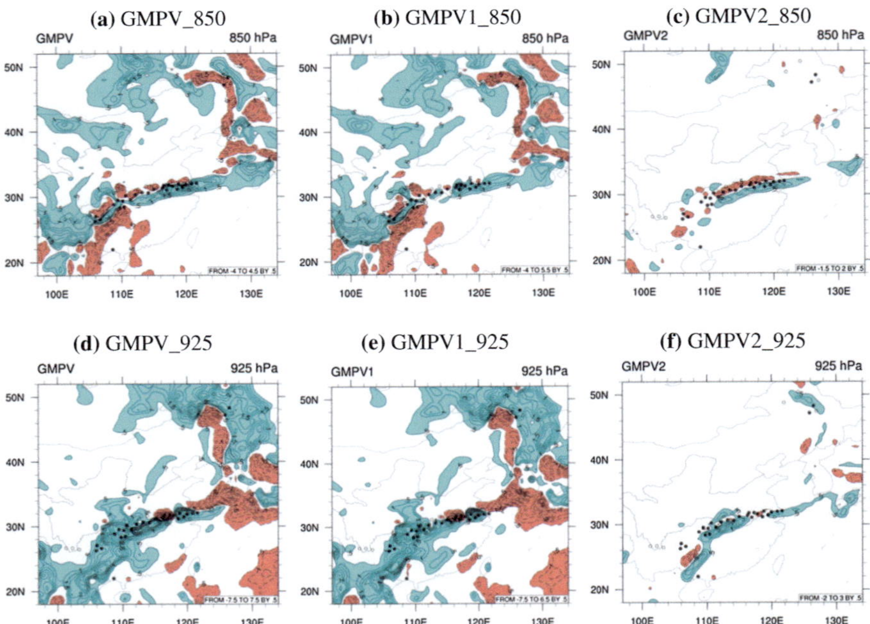

Fig. 6.40 Same as Fig. 6.37 but for GMPV, GMPV1, and GMPV2 (k = 10). Sourced from Qian et al. (2015a) © American Meteorological Society, with permission

within China on a 0.5° latitude–longitude grid. From Eq. 6.5.8, TS takes missing (M) and false alarm (FA) values into account besides counting overlapping or hitting areas (H):

$$TS = \frac{H}{FA + H + M}. \tag{6.5.8}$$

To be fair to all parameters, an optimal threshold (maximized the TS for a parameter, Fig. 6.41) is first determined for each parameter based on the 41-case average, then the optimal threshold is used to define the area of each corresponding parameter like MPV and GMPV, etc. The TS result averaged over the 41 cases is shown in Fig. 6.42 and Table 6.8. The improvement of the GMPV over the MPV is significant: about a 194% increase in TS for the GMPV, 16% for the GMPV1, and 18% for the GPMV2. The improvement of the GMPV over the MPV is statistically significant at a 99% confidence level based on the student t-test, while the improvement of the GMPV1 (GMPV2) over the MPV1 (MPV2) is statistically significant at a 65% (80%) confidence level. The less significant improvement for the two components than for the total term is consistent with the result of the case on 1 July 1991 (cf. Figs. 6.36 and 6.39 or Figs. 6.37 and 6.40).

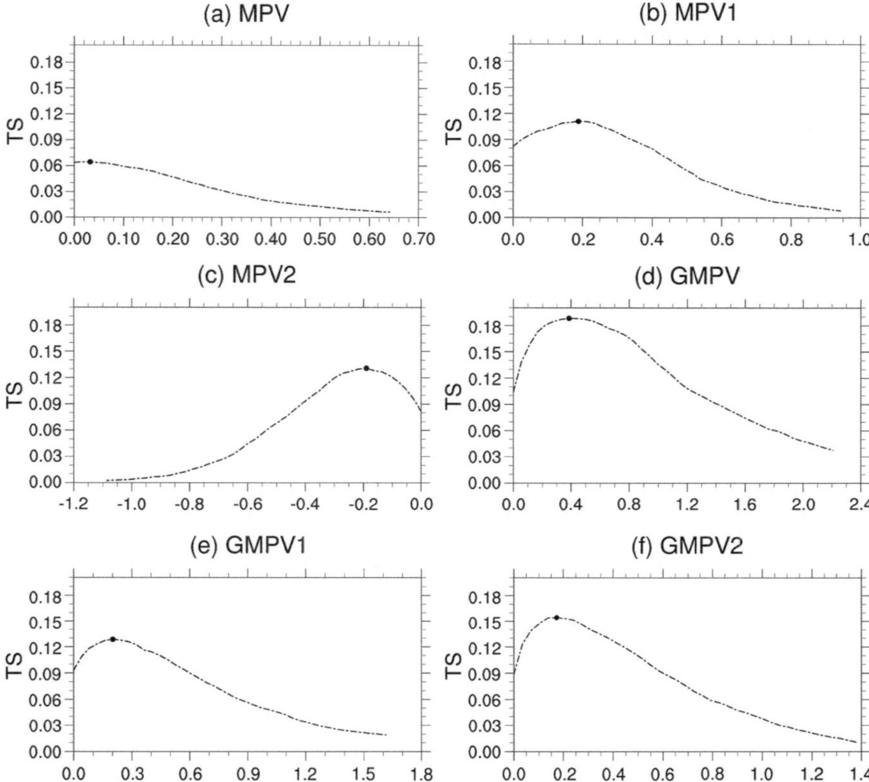

Fig. 6.41 Variation of TS with the threshold of parameters, averaged over the 41 heavy rain cases in eastern China during 1998. The *black dots* correspond to the optimal thresholds respectively for **a** MPV, **b** MPV1, **c** MPV2, **d** GMPV, **e** GMPV1, and **f** GMPV2. Sourced from Qian et al. (2015a) © American Meteorological Society, with permission

Fig. 6.42 Average TSs of various parameters from the MPV and GMPV groups in depicting heavy precipitation area (≥ 25 mm day^{-1}) based on 41 heavy rain cases in eastern China during 1998. The TS difference between the MPV and the GMPV is statistically significant at the 99% confidence level based on the Student *t*-test. Sourced from Qian et al. (2015a) © American Meteorological Society, with permission

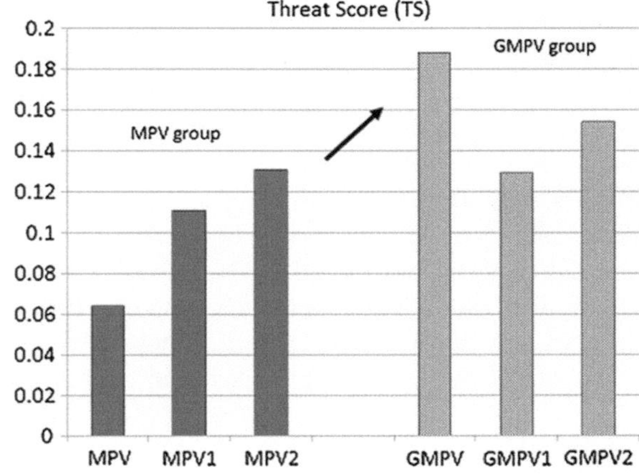

Table 6.8 TSs and its SD of various parameters in depicting heavy rain locations ($\geq$25 mm day^{-1}) based on 41 cases in eastern China in 1998

Parameter		Threshold	TS_{mean}	TS_{best} (date)	TS_{worst} (date)	SD	Statistical significance (*t*-test)
Divergence (850 hPa)	Div	−2.0e−5	0.169	0.320 (Mar 9)	0.000 (Jul 29)	0.0804	_
	MD (div× (q/qs)^10)	−0.5e−5	0.270	0.548 (Jun 26)	0.073 (Jul 29)	0.0868	99% (over Div)
Vorticity (850 hPa)	Vor	4.3e−5	0.196	0.384 (Jul 2)	0.010 (Mar 8)	0.0819	_
	MV (vort × (q/qs)^10)	1.4e−5	0.262	0.498 (Jun 26)	0.121 (Feb 16)	0.0895	99% (over Vor)
MPV (850 hPa)	MPV	0.03	0.064	0.171 (Jul 6)	0.001 (Mar 8)	0.0370	_
	MPV1	0.2	0.111	0.309 (Jan 14)	0.000 (Jul 24)	0.0827	_
	MPV2	−0.2	0.131	0.369 (Aug 22)	0.000 (Jul 22)	0.0955	_
GMPV (925 hPa, k = 10)	GMPV	0.4	0.188	0.382 (May 23)	0.067 (Aug 15)	0.0636	99% (over MPV)
	GMPV1	0.2	0.129	0.317 (Jan 14)	0.019 (Feb 17)	0.0713	65% (over MPV1)
	GMPV2	0.2	0.154	0.315 (Jun 9)	0.022 (Mar 8)	0.0637	80% (over MPV2)

TS_{mean}, TS_{best} and TS_{worst} are the mean, best and worst scores for a parameter. Statistical significance of an improvement is based on a *t*-test. Optimal threshold used for each parameter is also listed based on the average of 41 cases. Sourced from Qian et al. (2015a) © American Meteorological Society, with permission

The reason why the GMPV is superior to the MPV is the more realistic treatment of moisture in the heavy rain area from replacing the equivalent potential temperature by the generalized equivalent potential temperature. To demonstrate this, Figs. 6.43 and 6.44 compare the horizontal and vertical distributions and gradients of θ_e and θ^* of the case on 1 July 1990. Horizontally θ^* is more concentrated around the heavy rain areas than θ_e (Figs. 6.43 and 6.44a, b), while the major signal from $\partial\theta^*/\partial p$ is vertically more concentrated near the lower level than that from $\partial\theta_e/\partial p$ (Figs. 6.44c, d). That explains why the GMPV has stronger signals near the heavy rain area and the lower level than the MPV, as we observe from Figs. 6.39 and 6.40 (cf. Figs. 6.36 and 6.37). For the total terms (MPV and GMPV), the reason that GMPV is superior is that both the GMPV1 and GMPV2 have the same sign, so they add together to form a more obvious signal (Figs. 6.39 and 6.40), while MPV1 and MPV2 are opposite in sign in most areas smearing the signal

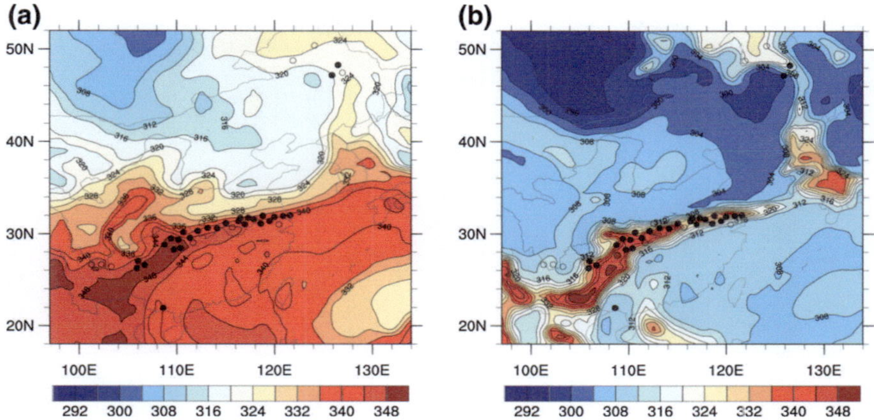

Fig. 6.43 **a** The equivalent potential temperature θ_e and **b** the generalized equivalent potential temperature θ^* ($k = 10$) at 850 hPa and 0000 UTC 1 July 1991. Interval is 4 K. Sourced from Qian et al. (2015a) © American Meteorological Society, with permission

(Figs. 6.36 and 6.37). This difference originates also from the difference between θ_e and θ^*. For MPV1 (Fig. 6.37a) and GMPV1 (Fig. 6.39a), they remain the same sign because both $\partial\theta_e/\partial p$ and $\partial\theta^*/\partial p$ are negative in the lower level over the heavy rain area (Fig. 6.44c, d) reflecting convective instability. For the MPV2 (Fig. 6.37b) and GMPV2 (Fig. 6.39b), they have opposite sign because $\partial\theta_e/\partial y$ and $\partial\theta^*/\partial y$ have opposite sign (Fig. 6.44e, f). The $\partial\theta_e/\partial y(\partial\theta^*/\partial y)$ relates more on heat content (moisture content) and is negative (positive) over the heavy rain area. The impact from the zonal gradients $\partial\theta_e/\partial x$ and $\partial\theta^*/\partial x$ should be small due to the zonal orientation of both θ_e and θ^*(Fig. 6.43).

Following the approach of the GMPV, Qian et al. (2015a) extended the vorticity (Eq. 6.5.1) and divergence (Eq. 6.5.2) to the MV (Eq. 6.5.9) and MD (Eq. 6.5.10) by multiplying the same relative humidity based weight:

$$\mathrm{MV} = \left(\frac{\partial v}{\partial x} - \frac{\partial u}{\partial y}\right) \cdot \left(\frac{q}{q_s}\right)^k, \tag{6.5.9}$$

$$\mathrm{MD} = \left(\frac{\partial u}{\partial x} + \frac{\partial v}{\partial y}\right) \cdot \left(\frac{q}{q_s}\right)^k, \tag{6.5.10}$$

where $k = 0, 1, 2, \ldots$, indicates the relative contribution of moisture. When $k = 0$, moisture has no impact and the MV and MD become the original dry vorticity and divergence. Similar to the GMPV and our case analysis, $k = 10$ is used in the MV and MD. To examine if the MV and MD improve over the original relative vorticity and divergence, Fig. 6.45 shows the latitude–vertical cross-section through the heavy rain band along 114.75°E at 0000 UTC 1 July 1991, which reveals that the signals from both the MV (Fig. 6.45b) and MD (Fig. 6.45d) are weaker (less

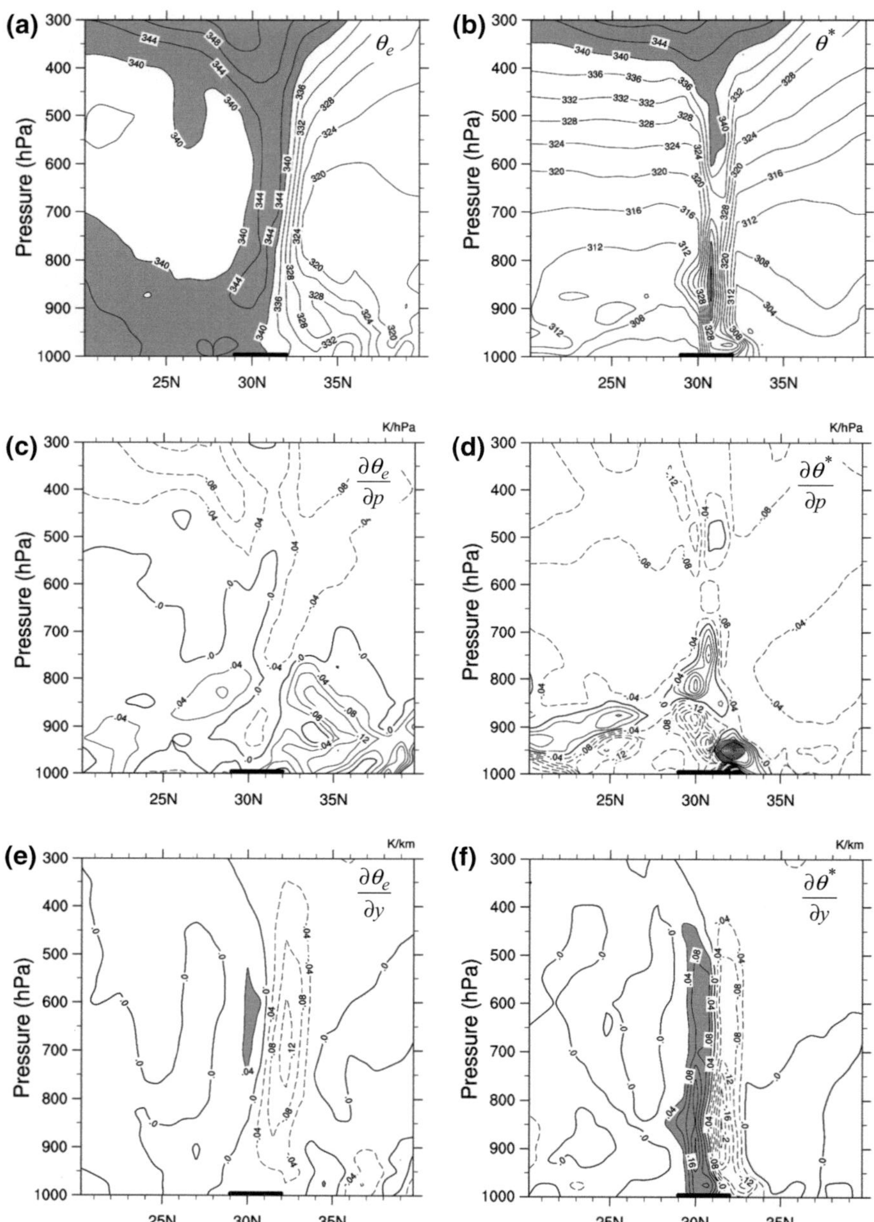

Fig. 6.44 Vertical–latitude cross-sections of **a** the equivalent potential temperature θ_e, **b** the generalized equivalent potential temperature θ^*(4 K interval), **c** the vertical gradient of θ_e, **d** the vertical gradient of θ^*(0.04 K hPa^{-1} interval), **e** the horizontal gradient (from south to north direction) of θ_e, and **f** the horizontal gradient of θ^*(0.04 K/km interval), along 114.75°E at 0000 UTC 1 July 1991. The heavy rain band is highlighted by the heavy *solid line*. Sourced from Qian et al. (2015a) © American Meteorological Society, with permission

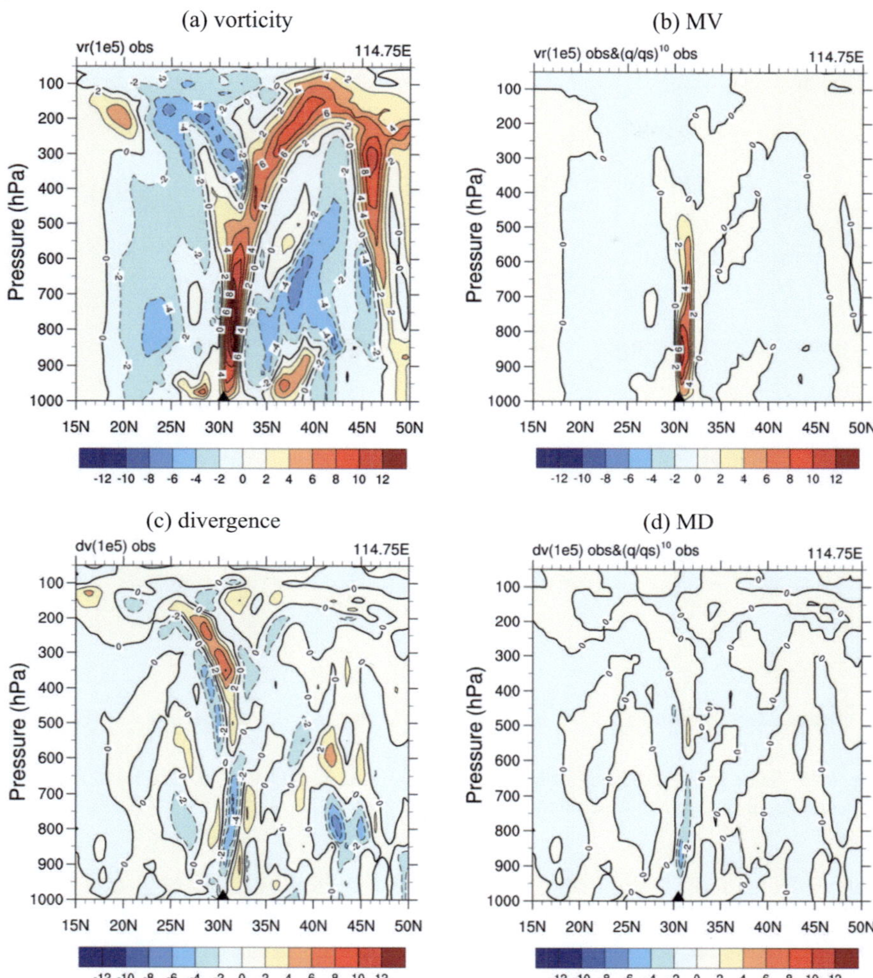

Fig. 6.45 Vertical–latitude cross-sections of **a** vorticity, **b** MV, **c** divergence, and **d** MD along 114.75°E at 0000 UTC 1 July 1991. The *contour* interval is 2×10^{-5} s^{-1}. The heavy rain area is indicated by the symbol *filled triangle*. Sourced from Qian et al. (2015a) © American Meteorological Society, with permission

sharper) but more concentrated around the heavy rain area compared to the vorticity (Fig. 6.45a) and divergence (Fig. 6.45c), due to the concentrated distribution of the relative humidity based weight $(q/q_s)^{10}$ (Fig. 6.38e). For example, the high vorticity at upper levels and at multiple lower level centers (near 28°, 37°, and 49°N where no heavy rain occurred, Fig. 6.45a) is much reduced or disappeared in the MV. Similar improvement is found for the MD over the divergence. This concentration of signal can be more clearly seen in the horizontal view of the fields (Figs. 6.46 and 6.47). Since the maxima of the MV and MD are located around 850 hPa,

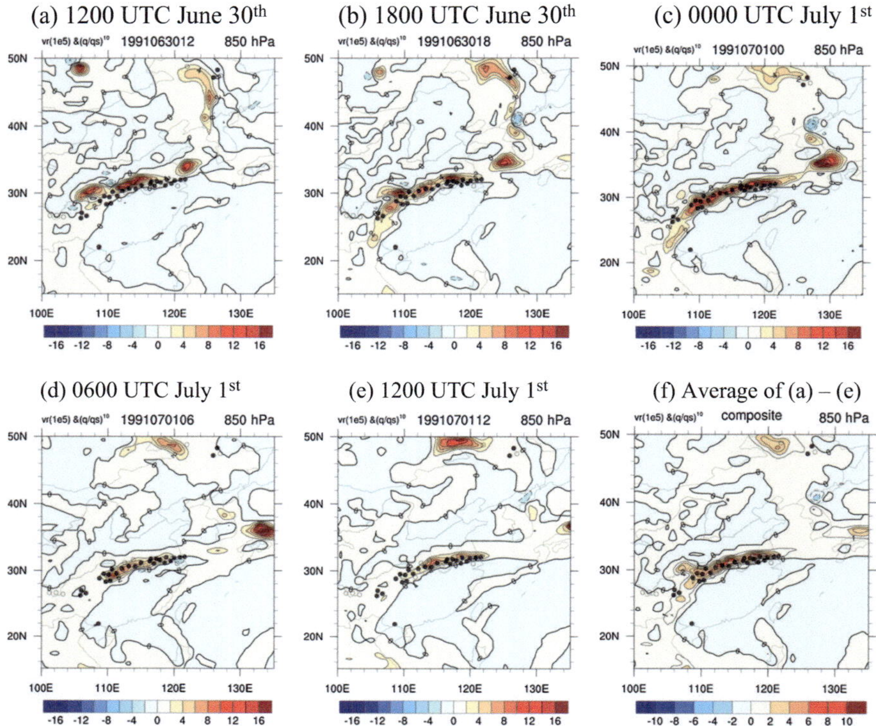

Fig. 6.46 Same as Fig. 6.33 but for MV (k = 10). Sourced from Qian et al. (2015a) © American Meteorological Society, with permission

Figs. 6.46 and 6.47 show the horizontal distributions of the MV and MD at 850 hPa every 6 h during the entire period of the heavy rain event on 1 July 1991. The improvement of a dynamic moisture combined parameter (MV and MD) is obvious over either a dynamic-only parameter (relative vorticity and divergence) or a moisture-only parameter (relative humidity). For example, the spurious areal coverage in Northwest China indicated by the original relative vorticity (Fig. 6.33) and divergence (Fig. 6.34) has now disappeared and the spurious areal coverage in Southwest China indicated by the original humidity field (Fig. 6.35) has also been eliminated. The heavy rain band along the Yangtze River is well indicated by both the MV and MD. This is true for all the five time points during the event. Other global observed precipitation datasets (figures not shown) indicate that the maximum center located in the south tip of the Korean Peninsula in both the MV and MD does overlap with an area of heavy rain events, although the data is missing in the China observation record.

For a quantitative evaluation, TS is again calculated based on the 41 RHR cases to measure how well a vorticity- or a divergence-related parameter overlaps with a heavy precipitation area (≥ 25 mm day^{-1}) within China. Similar to the situation of

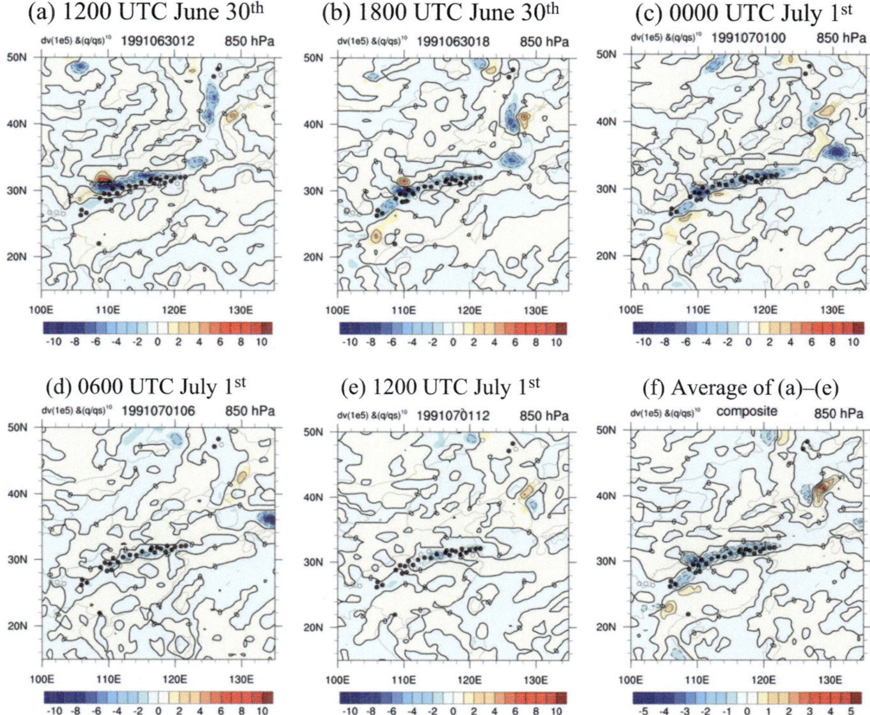

Fig. 6.47 Same as Fig. 6.34 but for MD (k = 10). Sourced from Qian et al. (2015a) © American Meteorological Society, with permission

vorticity and divergence in Figs. 6.33 and 6.34, although the MV and MD are similar to each other in general at different times during the event, the midpoint 0000 UTC (Figs. 6.46c and 6.47c) seems to match the heavy rain area the best (at least as good as others). Therefore, the midpoint is again used as a representative time for a RHR event in the following statistics. The optimal threshold for each parameter is shown in Fig. 6.48. Based on these optimal thresholds, Fig. 6.49 shows the average TS results for the three groups. It is noted that the MPV and GMPV group is the same as Fig. 6.42 and is listed for comparison purposes. After incorporating the effects of moisture into the relative vorticity and divergence, the improvements of MV or MD over the vorticity or divergence are significant: about a 60% increase in TS for MD (from 0.169 to 0.27) and 24% for MV (0.196–0.262). This result is also statistically significant at the 99% confidence level. The average TS as well as the best, worst scores, and their variance (SD with respect to mean) for each parameter out of the 41 cases is also listed in Table 6.8 along with the statistical confidence levels.

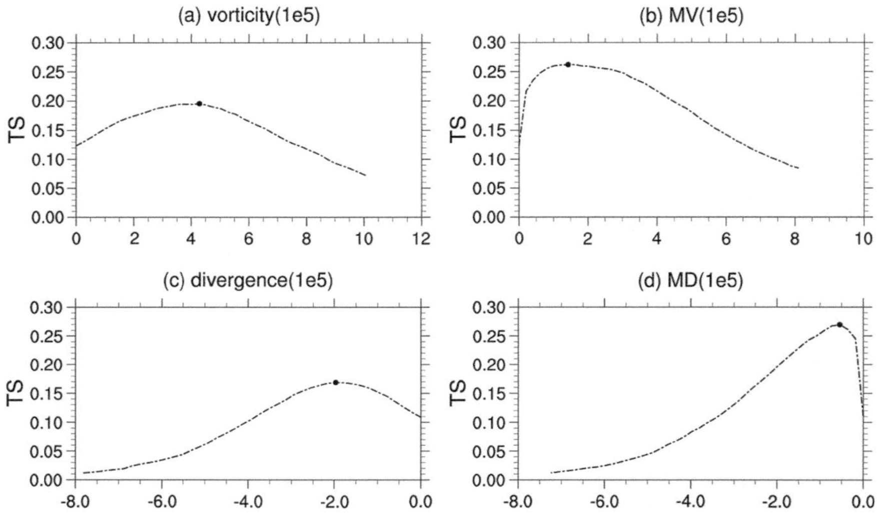

Fig. 6.48 Same as Fig. 6.41 but for **a** vorticity, **b** MV, **c** divergence, and **d** MD. Sourced from Qian et al. (2015a) © American Meteorological Society, with permission

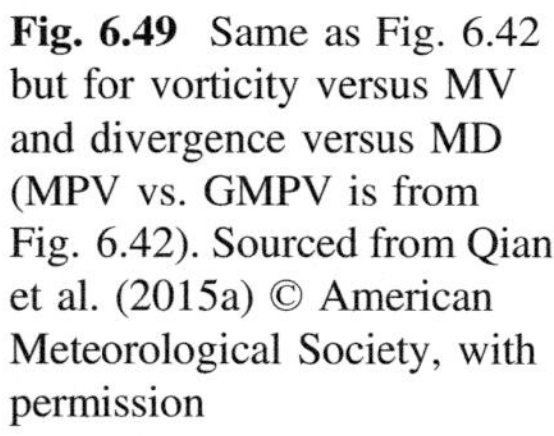

Fig. 6.49 Same as Fig. 6.42 but for vorticity versus MV and divergence versus MD (MPV vs. GMPV is from Fig. 6.42). Sourced from Qian et al. (2015a) © American Meteorological Society, with permission

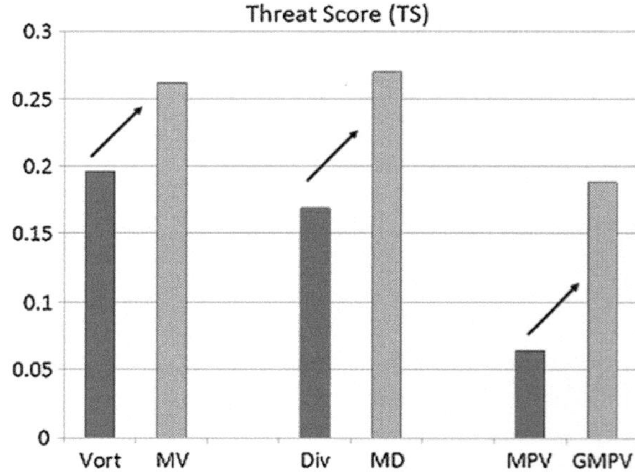

6.6 Nonconservative Thermodynamic Parameters[6]

In this subsection, a few nonconservative thermodynamic parameters including relative vorticity (Eq. 6.5.1), divergence (Eq. 6.5.2), MV (Eq. 6.5.9), MD (Eq. 6.5.10) , specific humidity q, and the two products of divergence and vorticity

[6]Modified article originally published in Qian et al. (2016a). Published with kind permission of © American Meteorological Society. All Rights Reserved.

directly multiplying by q (DQ in 6.6.1 and VQ in 6.6.2) will be examined. These are given by (Qian et al. 2016a):

$$\mathrm{DQ} = \left(\frac{\partial u}{\partial x} + \frac{\partial v}{\partial y}\right) \cdot q, \tag{6.6.1}$$

$$\mathrm{VQ} = \left(\frac{\partial v}{\partial x} - \frac{\partial u}{\partial y}\right) \cdot q, \tag{6.6.2}$$

where $k = 10$ is used in the MD and MV following Qian et al. (2015a). The divergence and vorticity can be expressed in total, climatic, and anomalous components, i.e., $D = \left(\frac{\partial u}{\partial x} + \frac{\partial v}{\partial y}\right)$, $\widetilde{D} = \left(\frac{\partial \widetilde{u}}{\partial x} + \frac{\partial \widetilde{v}}{\partial y}\right)$ and $D' = \left(\frac{\partial u'}{\partial x} + \frac{\partial v'}{\partial y}\right)$ for divergence; and $\zeta = \left(\frac{\partial v}{\partial x} - \frac{\partial u}{\partial y}\right)$, $\widetilde{\zeta} = \left(\frac{\partial \widetilde{v}}{\partial x} - \frac{\partial \widetilde{u}}{\partial y}\right)$ and $\zeta' = \left(\frac{\partial v'}{\partial x} - \frac{\partial u'}{\partial y}\right)$ for vorticity.

Figures 6.50a, b compare the total-based and anomaly-based divergences at 850 hPa at 0000 UTC 1 July 1991. Generally, they are quite similar to each other. In the total-based divergence, there are three bands of convergence located, respectively, over North China (I), the Yangtze River Valley (II), and South China (III), where only the one along the Yangtze River (II) is associated with heavy rain (Fig. 6.50a). The convergence band in South China (III), however, is largely reduced in the anomalous divergence (Fig. 6.50b) because it is caused by the climatic wind (figures not shown). Therefore, the anomalous divergence has a slightly better indication of heavy rain than the total divergence. As for the strong convergence band in North China (I) revealed in the anomalous divergence field, it does not produce heavy rain due to the lack of atmospheric moisture. Figure 6.50c, d compare the total-based and anomaly-based vorticity at 850 hPa at 0000 UTC 1 July 1991. There are totally four bands of positive vorticity located over North China (I), Northeast China (II), the Yangtze River Valley (III), and Southwest China (IV). As with divergence, anomalous vorticity is generally similar to the total vorticity in the large picture with only two differences: one is the elimination of the southwest extension of band (IV), and the other is a slightly southward shift of the same band (III) in the anomalous vorticity due to the removal of the climatic component. The elimination of the southwest extension (IV) can be deemed as an improvement, while the southward shift is too small to have a meaningful impact on this case. The fact that the positive vorticity band in North China (I) does not correspond to any heavy rain is again due to the lack of atmospheric moisture over that region (Fig. 6.11).

The MD and MV are evaluated in four forms: purely total-based (using total wind and total moisture, denoted as *_TT), purely anomaly-based (anomalous wind and anomalous moisture, *_AA) and two hybrid versions of mixed total- and anomaly-based (total wind and anomalous moisture, *_TA; anomalous wind and total moisture, *_AT), where * can be replaced by either MD or MV. Figure 6.51 shows the four forms of MD at 850 hPa, where the heavy rain band is generally collocated with the strongest convergence zone in all four forms. Although the four

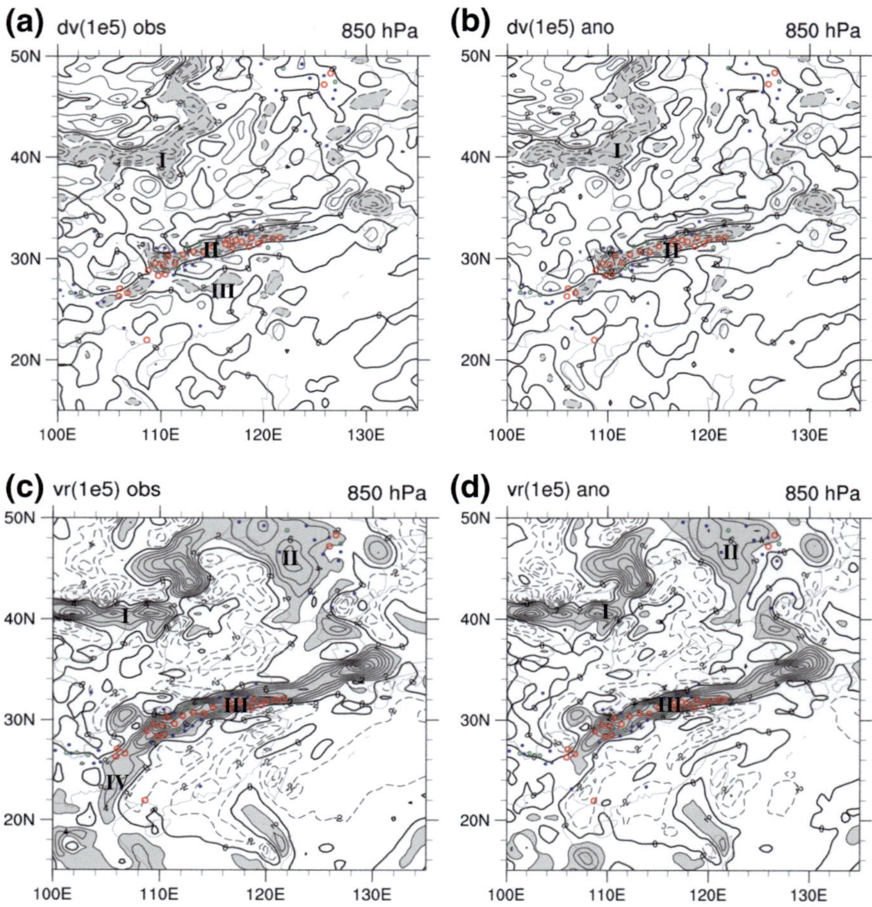

Fig. 6.50 **a** Total and **b** anomalous divergence; **c** total and **d** anomalous relative vorticity at 850 hPa at 0000 UTC 1 July 1991. *contour* interval is 2×10^{-5} s^{-1}. *I*, *II*, *III*, and *IV* denote the convergence bands or positive vorticity bands. The areas of divergence (vorticity) lower than -2×10^{-5} s^{-1} (greater than 2×10^{-5} s^{-1}) are shaded. *Red*, *green*, and *blue open circles* (*dots*) indicate the stations with rainfall over 50, 25–50, and 10–25 mm day^{-1}, respectively. Sourced from Qian et al. (2016a) © American Meteorological Society, with permission

versions are similar in general, the two with total moisture (MD_TT and MD_AT) have a stronger convergence zone and are better than the two with anomalous moisture (MD_TA and MD_AA). For example, MD_TT and MD_AT cover the southwest tail of the heavy rain band better than MD_TA and MD_AA. A very similar result is found for the MV at 850 hPa (Fig. 6.52). The heavy rain is generally collocated with the maximum positive vorticity zone in all four forms. MV_TT (Fig. 6.52a) and MV_AT (Fig. 6.52d) are similar to each other and have a stronger signal. MV_TA (Fig. 6.52c) and MV_AA (Fig. 6.52b) are similar to each other with a weaker signal. Although MV_AT and MV_TT are generally similar,

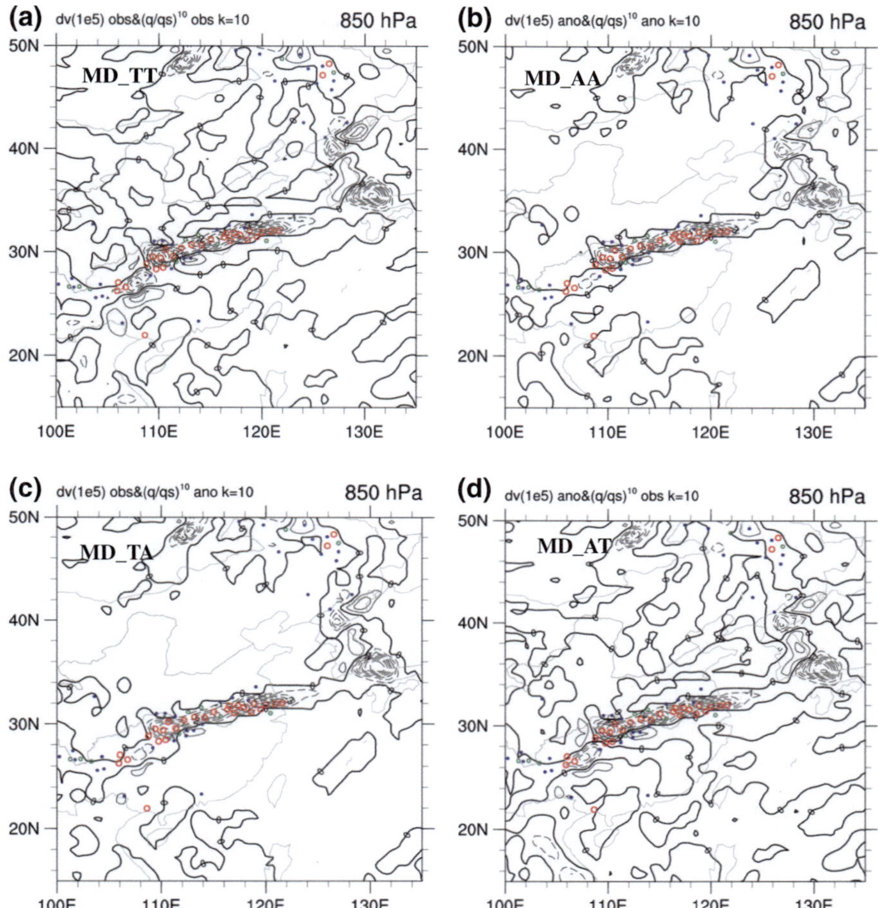

Fig. 6.51 MD at 850 hPa at 0000 UTC 1 July 1991: **a** Total-based divergence by total-based $(q/q_s)^k$, MD_TT; **b** anomaly-based divergence by anomaly-based $(q/q_s)^k$, MD_AA; **c** total-based divergence by anomaly-based $(q/q_s)^k$, MD_TA; and **d** anomaly-based divergence by total-based $(q/q_s)^k$, MD_AT. The *contour* interval is 1×10^{-5} s^{-1} and $k = 10$ is used in the calculations. *Red*, *green*, and *blue open circles* (*dots*) indicate the stations with rainfall over 50, 25–50, and 10–25 mm day^{-1}, respectively. Sourced from Qian et al. (2016a) © American Meteorological Society, with permission

the former is slightly better than the latter, because the climatologically induced "false alarm" tail in Southwest China is removed in MV_AT (cf. Fig. 6.52a, d). This implies that under some circumstances, the mixing of total and anomalous fields might have a better coverage with heavy rain area than either using the purely total field or purely anomaly field based methods for a compound parameter.

As for the MD and MV, the DQ and VQ are also examined in the four forms (*_TT, *_AA, *_TA and *_AT, where * can be replaced by either DQ or VQ). In the two forms (*_AA and *_TA) which involve anomalous q, only positive q' is

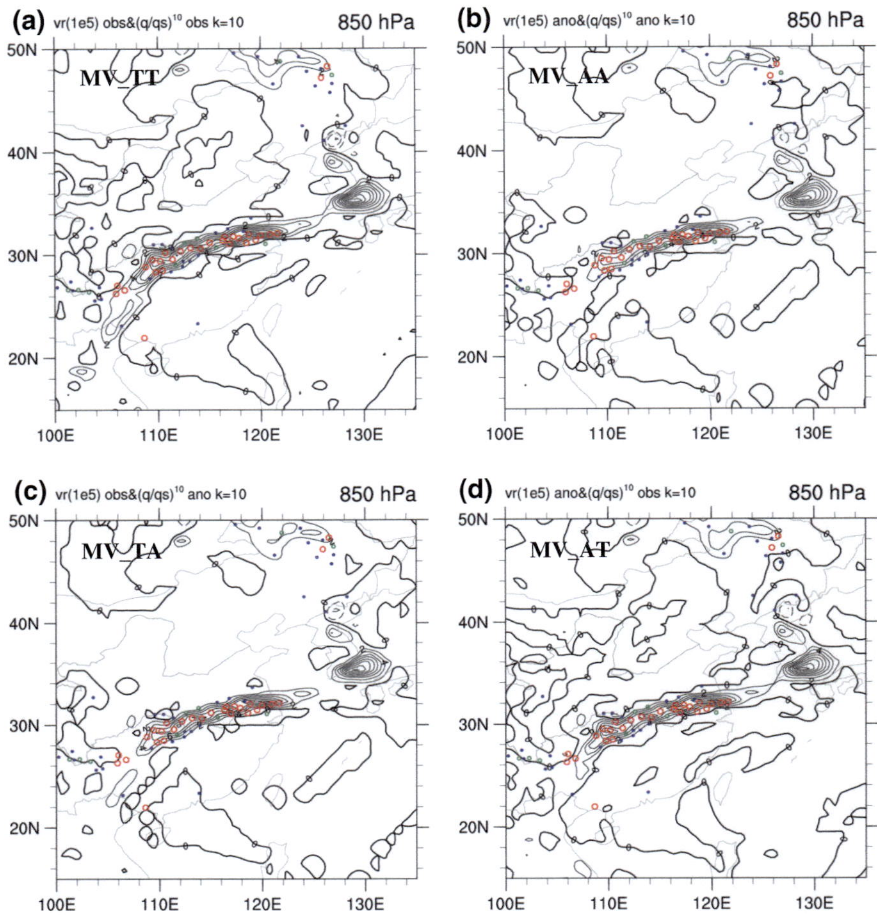

Fig. 6.52 Same as Fig. 6.51 but for MV with k = 10. The *contour* interval is 2×10^{-5} s^{-1}. Sourced from Qian et al. (2016a) © American Meteorological Society, with permission

applied in the calculation (i.e., negative q' is treated as zero) since heavy rain is our concern. The results of DQ and VQ are shown in Figs. 6.53 and 6.54, respectively. They show that all the four forms have a band of high value collocated with the heavy rain band. The two forms using anomalous q (*_AA and *_TA in Figs. 6.53b, c, and 6.54b, c) have a more concentrated signal singling out the heavy rain band because only positive q' is used in the calculation.

To have a robust and quantitative result, Qian et al. (2016a) have systematically examined other 41 heavy rain events that occurred in eastern China during 1998. TS (Palmer and Allen 1949) is used to quantitatively measure how well an area defined by a parameter overlaps with a heavy precipitation area of exceeding 25 mm day^{-1} over China. To be fair for each parameter, an optimal threshold value is estimated first for each parameter, which gives the highest TS on average of the 41 cases

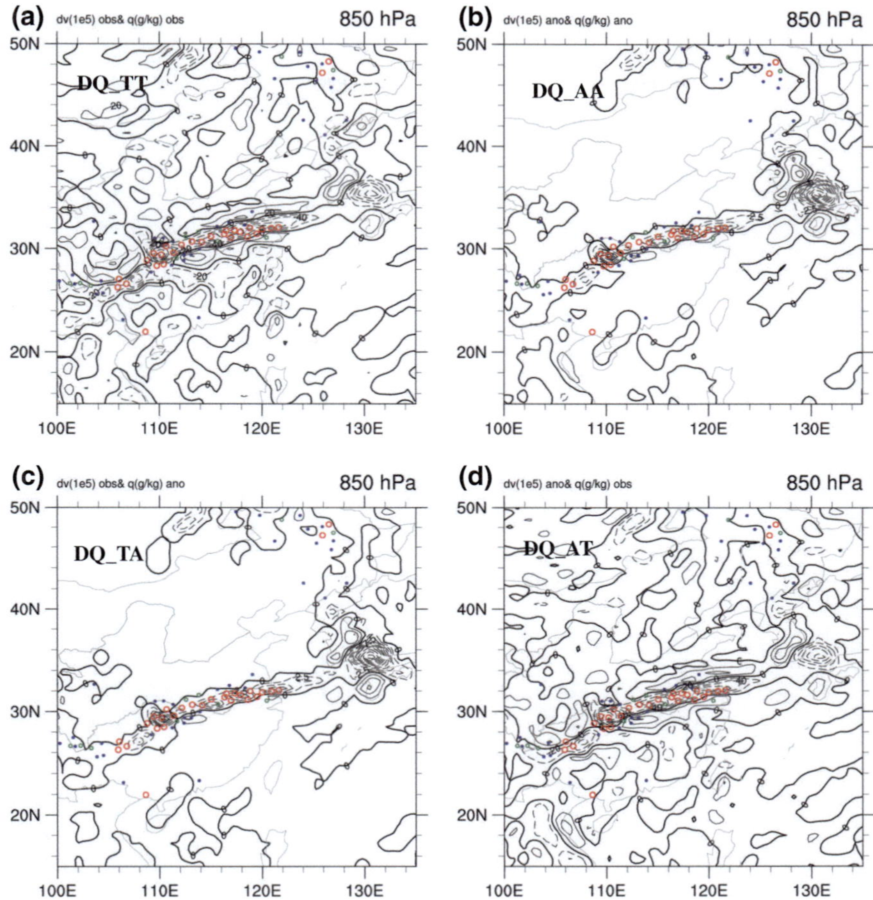

Fig. 6.53 Same as Fig. 6.51 but for DQ. The *contour* intervals are 20×10^{-5} s^{-1} g/kg in **a** and **d**, and 2.5×10^{-5} s^{-1} g/kg in **b** and **c**. Sourced from Qian et al. (2016a) © American Meteorological Society, with permission

(Table 6.9). The TS results are summarized by Fig. 6.55a, which reiterates the above case findings. The TSs suggest that the performances are generally similar between the total and anomalous versions for these parameters. The hybrid version of mixed anomalous wind and total moisture based MV_AT and VQ_AT covers the heavy rain area slightly better than either the purely total-based MV_TT and VQ_TT or purely anomaly-based MV_AA and VQ_AA (but the differences are not statistically significant).

To better understand the performance or TS result, the probability of detection (POD, Fig. 6.55b), areal bias (Fig. 6.55c), and false alarm ratio (FAR, Fig. 6.55d) are calculated using the same optimal threshold in TS. From Fig. 6.55, when moisture (q) is used alone to depict heavy rain area, it is often too wide in area (high

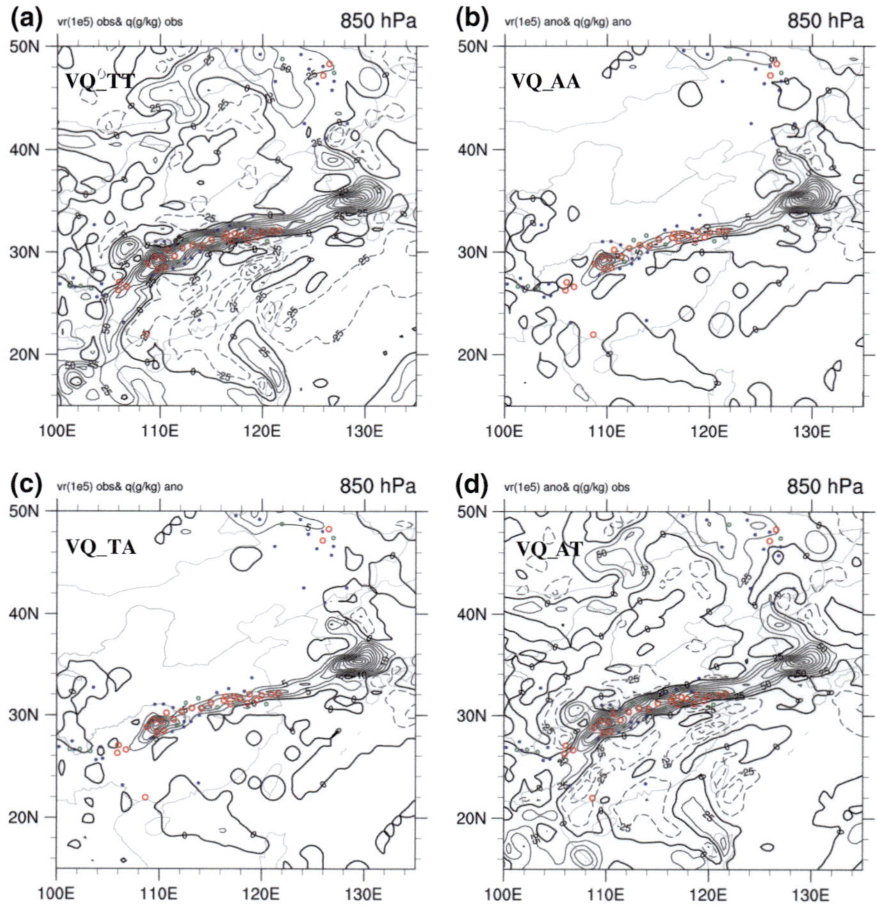

Fig. 6.54 Same as Fig. 6.51 but for VQ. The *contour* intervals are 25×10^{-5} s^{-1} g/kg in **a** and **d**, and 5×10^{-5} s^{-1} g/kg in **b** and **c**. Sourced from Qian et al. (2016a) © American Meteorological Society, with permission

bias) and leads to high FAR although its POD is high. The FAR results in a worse performance or a lower TS. When dynamic factor (divergence and vorticity) is used alone, it has higher FAR and lower detection rate, which implies that its location does not match the heavy rain area well and it leads to a lower TS. When moisture and dynamics factors are considered together (MV, MD, VQ, and DQ), the FAR is reduced and POD is increased, which results in better performance with a higher TS. The MD and MV are even more superior to the DQ and VQ. Therefore, Qian et al. (2015a) recommended to combine dynamical and moisture factors together to obtain a hybrid compound parameter to increase the ability in depicting heavy rain area.

Table 6.9 Average TS, POD, Bias, and FAR of MPV and specific humidity q at 850 hPa, GMPV at 925 hPa, as well as various forms of divergence and vorticity at 850 hPa in depicting heavy rain locations (≥ 25 mm day^{-1}) based on 41 cases in eastern China during 1998

Parameter		POD	Bias	FAR	TS	Optimal threshold
Divergence (850 hPa)	Div	0.295	1.12	0.707	0.169	−2.0e−5 s^{-1}
	Div′	0.355	1.60	0.757	0.167	−1.4e−5
Moist divergence (850 hPa, $k = 10$)	MD_TT (div × q/q_s)	0.488	1.37	0.612	0.270	−0.5e−5
	MD_TA (div × q'/qs')	0.449	1.22	0.592	0.263	−0.5e−5
	MD_AT (div′ × q/q_s)	0.476	1.33	0.610	0.268	−0.5e−5
	MD_AA (div′ × q'/qs')	0.441	1.23	0.603	0.257	−0.4e−5
Vorticity (850 hPa)	Vort	0.357	1.27	0.694	0.196	4.3e−5
	Vort′	0.403	1.51	0.711	0.201	3.4e−5
Moist Vorticity (850 hPa, $k = 10$)	MV_TT (vort × q/q_s)	0.538	1.749	0.656	0.262	1.4e−5
	MV_TA (vort × q'/q_s')	0.561	1.846	0.658	0.266	0.9e−5
	MV_AT (vort′ × q/q_s)	0.511	1.507	0.619	0.274	1.5e−5
	MV_AA (vort′ × q'/q_s')	0.534	1.700	0.646	0.267	0.9e−5
MPV (850 hPa)	MPV	0.330	4.992	0.926	0.064	0.03 PVU
	MPV1	0.239	1.467	0.824	0.111	0.2
	MPV2	0.280	1.316	0.777	0.131	−0.2
	MPV′	0.322	4.286	0.915	0.007	0.05
	MPV1′	0.244	1.572	0.834	0.109	0.2
	MPV2′	0.249	1.311	0.814	0.116	−0.1
GMPV (925 hPa, $k = 10$)	GMPV	0.420	1.748	0.739	0.188	0.4
	GMPV1	0.315	1.935	0.822	0.129	0.2
	GMPV2	0.343	1.720	0.774	0.154	0.2
	GMPV′	0.464	2.043	0.756	0.189	0.3
	GMPV1′	0.326	2.080	0.827	0.127	0.2
	GMPV2′	0.316	1.827	0.811	0.132	0.1
Specific humidity (850 hPa)	q	0.674	3.902	0.734	0.179	12.9 g/kg
	q'	0.582	2.915	0.758	0.192	2.3
DQ (850 hPa)	DQ_TT (div × q)	0.349	0.962	0.631	0.219	−21.8e −5 s^{-1}g/kg
	DQ_TA (div × q')	0.419	1.300	0.637	0.228	−2.9e−5
	DQ_AT (div′ × q)	0.369	1.088	0.648	0.219	−17.3e−5
	DQ_AA (div′ × q')	0.441	1.500	0.671	0.222	−2.1e−5

(continued)

Table 6.9 (continued)

Parameter		POD	Bias	FAR	TS	Optimal threshold
VQ (850 hPa)	VQ_TT (vort × q)	0.434	1.333	0.659	0.231	45.2e−5
	VQ_TA (vort × q')	0.469	1.562	0.656	0.240	6.79e−5
	VQ_AT (vort′ × q)	0.458	1.302	0.636	0.250	38.4e−5
	VQ_AA (vort′ × q')	0.528	1.863	0.679	0.244	4.7e−5

The optimal thresholds used for each parameter are listed. Sourced from Qian et al. (2016a) © American Meteorological Society, with permission

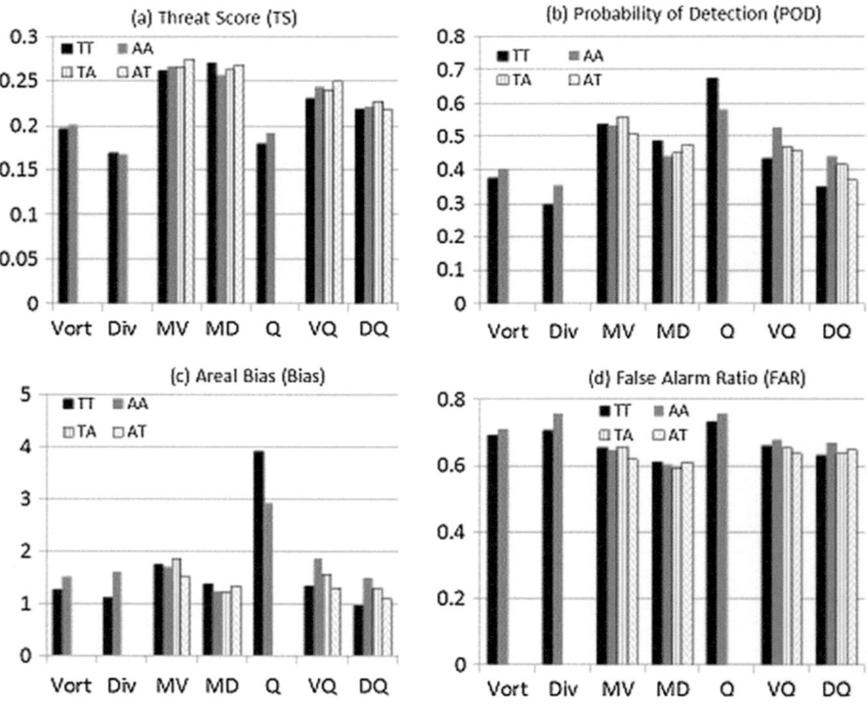

Fig. 6.55 **a** Average TSs, **b** POD, **c** areal bias (Bias), and **d** FAR of various vorticity and divergence forms as well as specific humidity (total field in *black bar*; anomalous versions in *gray* and other lighter color) in depicting heavy precipitation area (≥ 25 mm day^{-1}) based on a total of 41 heavy rain cases in eastern China during 1998. Sourced from Qian et al. (2016a) © American Meteorological Society, with permission

Figure 6.56a shows the average TSs based on the 41 cases under each parameter's optimal threshold (see also Table 6.9), which include conservative and nonconservative thermodynamic parameters. For GMPV, the anomaly-based GMPV′ has similar performance of depicting heavy rain area to the total-based GMPV, the GMPV1′ and GMPV2′ are slightly worse than the GMPV1 and GMPV2. For MPV, the MPV′, MPV1′, and MPV2′ are found to be slightly worse than the MPV, MPV1, and MPV2 overall. Since the scales represented in the anomalous field are much more detailed than those in the total field, TS which penalizes more for misses than false alarms probably favors more on total field than anomalous field. TS for anomalous field is slightly worse than total field, indicating that they have similar performance. POD, areal bias, and FAR are also calculated for MPV and GMPV. These scores are generally similar between the total- and anomaly-based fields (Fig. 6.56b–d). When one interprets the results between figures and scores, it should keep in mind that the same contour interval was normally used for both total and anomalous fields in figures but the TS is calculated based on optimal threshold of each individual parameter.

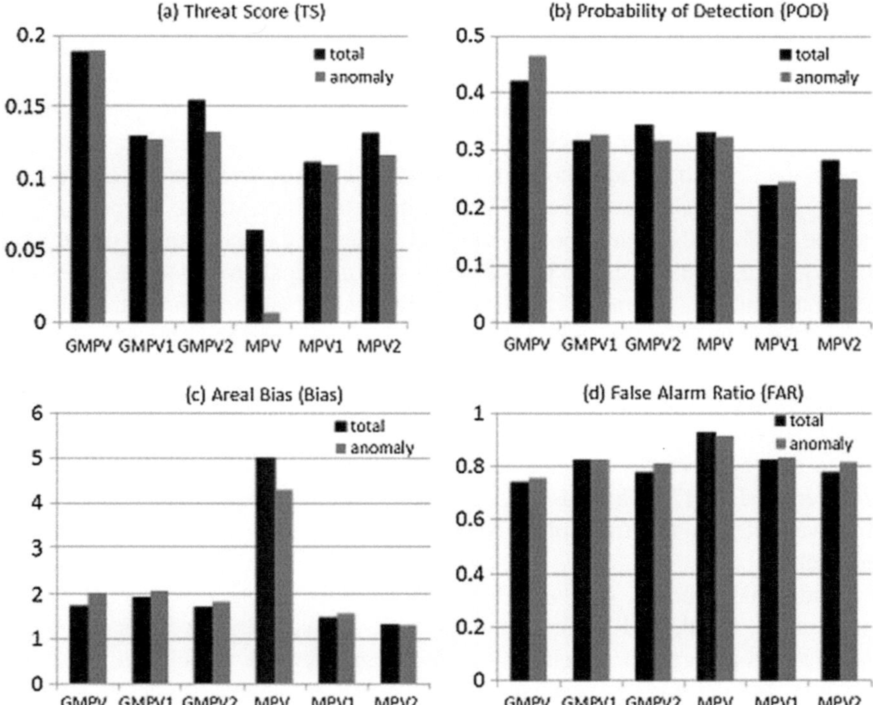

Fig. 6.56 **a** Average TSs, **b** POD, **c** Bias, and **d** FAR of various MPV and GMPV forms (total field in *black bar*; anomalous field in *gray bar*) in depicting heavy precipitation area ($\geq$25 mm day^{-1}) based on 41 heavy rain cases in eastern China during 1998. Sourced from Qian et al. (2016a) © American Meteorological Society, with permission

6.7 Tornado-Producing Anomalous Systems in the Central US

Tornadoes occur frequently in the central United States between the Rocky Mountains and Appalachian Mountains during the warm seasons. The southeastern U.S. region is particularly prone to violent and long-track tornadoes. Most tornadoes were recorded during the spring in 2011, particularly a record number (around 758) in April (Simmons and Sutter 2012). A few tornado outbreaks accounted for the majority of the most damaging and fatal tornadoes, including extended outbreaks (Knupp et al. 2014). During 14–16 April 2011, there were about 170 tornadoes, mostly EF-0 to EF-2 on the enhanced Fujita (EF) scale over primarily three regions: Oklahoma (OK)–Arkansas (AR), southern Mississippi (MS)/Alabama (AL), and the mid-Atlantic. During 25–28 April 2011, 350 tornadoes occurred from Texas (TX) to eastern Virginia (VA), causing 321 fatalities.

Although a tornado warning is usually limited within 10–20 min, many studies always adhere to pursuing their goal to find notable synoptic precursors (Fujita et al. 1970; Agee et al. 1975; Ferguson et al. 1987; Lee et al. 2006; Metz and Bosart 2010; Corfidi et al. 2010). Environmental conditions for the forecast implications of tornado outbreaks have also been paid wide attention (Thompson and Edwards 2000; Doswell et al. 2006; Bunkers et al. 2006; Thompson et al. 2012; Weisman et al. 2013). All these studies analyze their cases of tornado outbreaks by using the traditional synoptic charts and dynamical (or thermodynamical) diagnostic methods. As summarized by Mercer et al. (2012), these studies characterize synoptic conditions prior to the tornado outbreaks with three key features: (1) upper level trough axis west of the outbreak location that contributes to the deepening of a surface cyclone through synoptic-scale processes (e.g., using quasi-geostrophic theory), (2) upper level stream jet collocated with the rapid drop of surface pressure, which intensifies a synoptic cyclone and increases vertical shear in the troposphere, and (3) subsequent formation of a deep surface cyclone that incorporates a plume of enhanced moisture to develop large buoyant instability in the warm sector. Shortly, 500-hPa trough, deepening surface low, and enhancing moisture are the three major features during the developing process of a tornado outbreak. However, operational forecasting applications of these studies have not been fully developed recently (Mercer et al. 2012). The reason is that these synoptic-scale systems revealed by total variables in the traditional synoptic chart are not able to locate tornado precisely.

Traditional synoptic charts of basic variables including geopotential height, temperature, wind, and mixing ratio (specific humidity) as well as diagnostic parameters of divergence, relative vorticity, and vertical velocity at 300, 500, 850 hPa, and the surface level have been analyzed in previous studies (Thompson and Edwards 2000; Mercer et al. 2012; Weisman et al. 2013; Knupp et al. 2014). However, since tornado outbreaks belong to weather extremes, an anomaly-based approach could give a better picture on the connections between surface weather extremes and tornado-producing anomalous systems and conditions aloft.

A new approach with a set of basic anomalous variables and dynamical anomalous indices which improve the accuracy of locating surface weather extremes, such as tornado outbreaks, is introduced and examined in this section. In this approach, an atmospheric variable is decomposed into a temporal climatic component and an anomaly (Qian et al. 2014). Two datasets are used in here. The first is the CMORPH precipitation dataset with 3-hourly temporal and 0.25° horizontal resolution, derived from satellite observations (Joyce et al. 2004). The second dataset is the ERA-Interim data (Dee et al. 2011).

The Glossary of Meteorology (Glickman and Todd 2000) describes a tornado outbreak as "multiple tornado occurrences within a single synoptic-scale system." Thus, it is critical to identify such synoptic-scale systems to help predict tornadoes in practice as a favorable background condition. Here, we illustrate how to forecast and alert where severe tornado-producing systems and conditions could possibly occur from both reanalysis products and model predictions. During 25–28 April 2011, most tornadoes occurred between 32 and 38°N in the southern US. On 25 April, around 64 tornadoes (only four with EF-2 or greater intensity) were spawned between northeast TX, western Tennessee (TN), and Kentucky (KY). On 26 April, approximately 50 tornadoes were documented from eastern TX to the western portions of TN and KY. Figure 6.57 illustrates the vertical time–pressure sections of total height and temperature as well as height and temperature anomalies averaged over 32–38°N at 95°W. Three processes of tornado outbreak are indicated by triangle symbols during three days from 25 to 27, April 2011. No apparent signal can indicate the three processes from the total height except three total temperature fluctuations in the lower level in Fig. 6.57a. After removing the climatic height and temperature, three processes can be indicated by the transition of warm to cold anomalies and negative height anomalies near the surface (Fig. 6.57b).

Tornado outbreaks and heavy rainfalls are the result of strong convective storms. As shown in Fig. 6.58a, the central location of tornado outbreak and heavy rainfall is occurrence near a strong horizontal gradient of total temperature but no obvious signal from the total height can indicate the convective system. The horizontal gradients of total temperature and total height can also be noted from Fig. 6.58b over the place of tornado outbreak and heavy rainfall. After removing the temporal climatic height and temperature, Fig. 6.58c depicts the anomalous height and temperature fields at 925 hPa. It is clear that both tornadoes and rainfalls are occurrence near the boundary of anomalous warm–cold air masses (C0 and W0). The boundary is also along the trough of height anomalies at 925 hPa.

Figure 6.59 illustrates the vertical distributions of height anomalies and temperature anomalies along 93°W on 25 and 26 April 2011. In Fig. 6.59a the tornado-producing low (L) forms near the boundary between the anomalous warm and cold air masses in the lower troposphere. From Fig. 6.58a, b, strong convective weathers (tornadoes and rainfalls) form in the region of large horizontal gradients of total temperature and total height. However, Figs. 6.58c and 6.59a show that the strong convective weathers are the clash result of anomalous warm and cold air masses (C0 and W0) in the lower troposphere. The vertical pattern of height and temperature anomalies associated with the convective weathers in Fig. 6.59a is

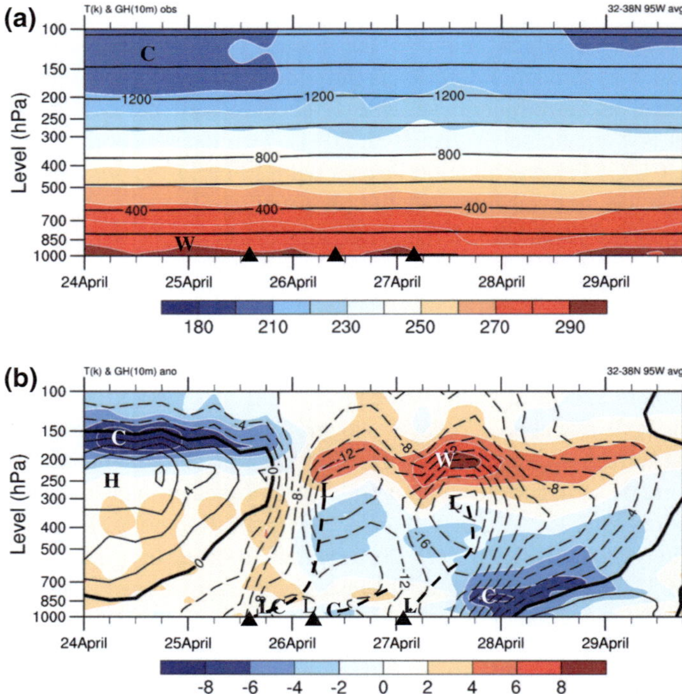

Fig. 6.57 Vertical time–pressure sections of **a** total height (*contour*, 200 × 10 gpm interval) and temperature (*shading*, 10 K interval) as well as **b** height anomalies (*contours*, 2 × 10 gpm interval) and temperature anomalies (*shading*, 2 K interval) averaged over 32–38°N along 95°W. Symbol "*filled triangle*" indicates the central position of tornado outbreak. The *heavy dashed line* is the trough axis of height anomalies and links two low centers. Letters "H/L" and "W/C" indicate the centers of height and temperature anomalies, respectively

similar with that of "7.21" Beijing extreme rainfall in Fig. 6.17d. The low center (L) is located below a positive center (H) of height anomalies at the upper troposphere (Fig. 6.59a). This vertical configuration is also observed in Fig. 6.59b–d except a strengthened low (L) and a weakened positive center (H) of height anomalies. The clash of anomalous warm and cold air masses is clear from 25 to 26 April 2011.

The number of tornadoes over a 24-h period (midnight–midnight central daylight time, CDT) of 27 April 2011 reached 199 and was the third process in Fig. 6.57b. Tornado fatalities were 316 [the most in a 24-h period since 21 March 1932 reported by NOAA (2011)], and injuries exceeded 2700. As reported by Simmons and Sutter (2012), tornadoes on 27 April 2011 across the southern United States set a record for the most tornadoes occurred over a 24-h period and were the second deadliest in U.S. history. Three distinct episodes of tornadoes over this 24-h period (0500 UTC 27 April to 0500 UTC 28 April) showed: (1) a mesoscale convective system (MCS) that evolved to an intense quasi-linear convective system

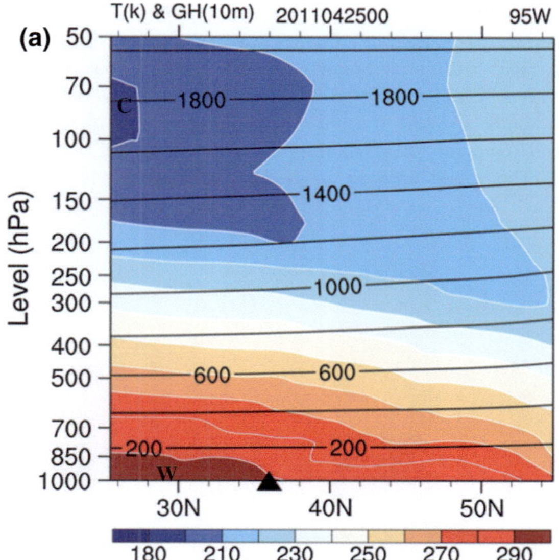
(a)
T(k) & GH(10m)
2011042500
95W
Level (hPa)
50
70
100
150
200
250
300
400
500
700
850
1000
1800
1400
1000
600
200
C
W
30N
40N
50N
180
210
230
250
270
290

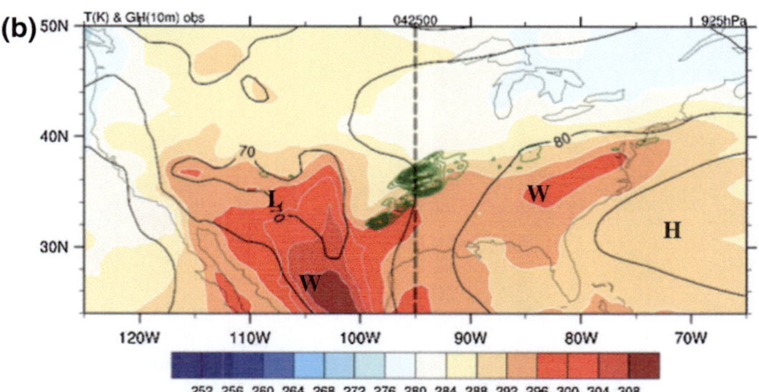
(b)
T(K) & GH(10m) obs
042500
925hPa
50N
40N
30N
70
80
L
W
H
120W
110W
100W
90W
80W
70W
252
256
260
264
268
272
276
280
284
288
292
296
300
304
308

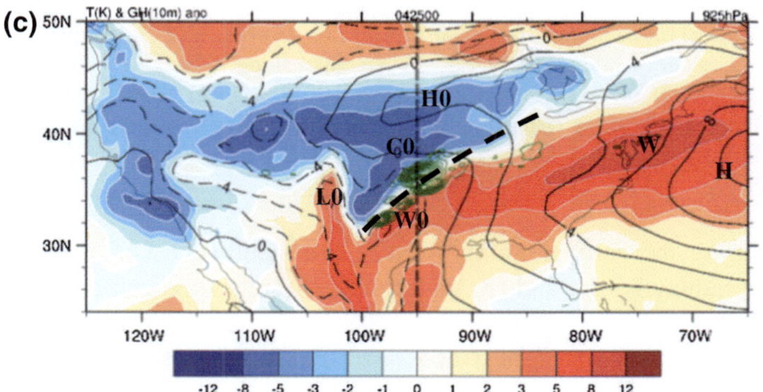
(c)
T(K) & GH(10m) ano
042500
925hPa
50N
40N
30N
H0
C0
L0
W0
W
H
0
4
-4
8
120W
110W
100W
90W
80W
70W
-12
-8
-5
-3
-2
-1
0
1
2
3
5
8
12

◀ **Fig. 6.58** **a** Traditional vertical synoptic chart with total height (*contour*, 200 × 10 gpm interval) and temperature (*shading*, 10 K interval) along 95°W at 0000 UTC 25 April 2011. **b** Traditional horizontal synoptic chart with total height (*contour*, 5 × 10 gpm interval) and temperature (*shading*, 4 K interval) at 925 hPa at 0000 UTC 25 April 2011. **c** The same as **b** except anomalous synoptic chart with height anomaly (*contour*, 2 × 10 gpm interval) and temperature anomaly (*shading*, 1 K interval). Symbol "*filled triangle*" indicates the central position of tornado outbreak and heavy rainfall. The *green contour* in **b** and **c** indicates the hourly precipitation rate (2.5 mm h^{-1})

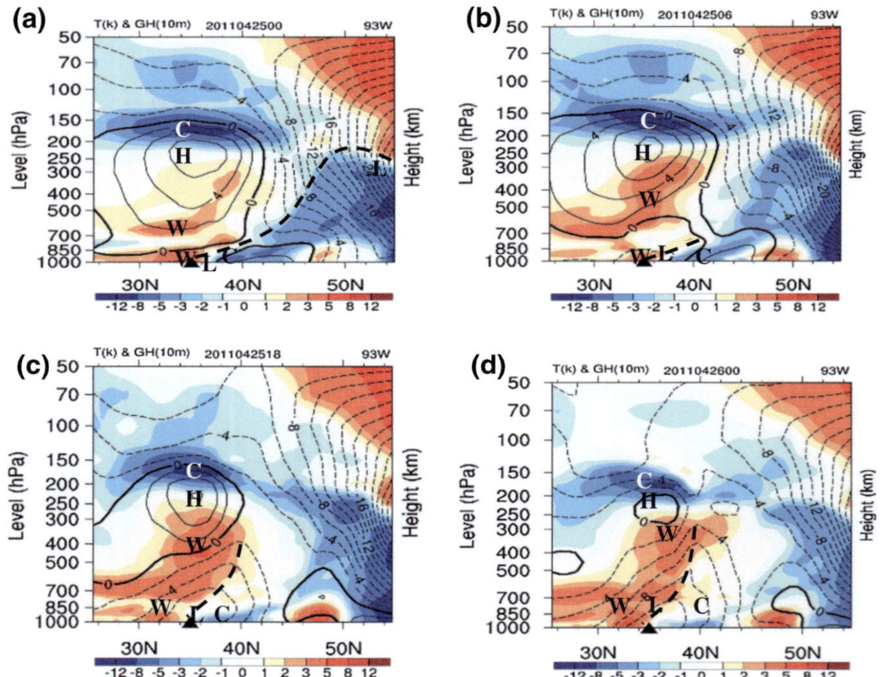

Fig. 6.59 Vertical distributions of height anomalies (*contours*, 2 × 10 gpm interval) and temperature anomalies (*shading*, 1, 2, and 3 K intervals) along 93°W at **a** 0000 UTC 25, **b** 0600 UTC 25, **c** 1800 UTC 25, and **d** 0000 UTC 26, April 2011. Symbol "*filled triangle*" indicates the central position of tornadoes and rainfalls while *thick dashed lines* indicate the trough axis of height anomalies

(QLCS) during the early morning hours, (2) a smaller QLCS over northern AL during the midday hours, and (3) widespread discrete supercell storms during the afternoon–evening period (Knupp et al. 2014).

On 27 April, most strong tornadoes were generated in the night. According to the statistics, 76 tornadoes occurred mainly in the morning from 5:03 to 13:55 UTC, 7 tornadoes in the afternoon from 16:15 to 17:05 UTC, and 116 tornadoes from 18:36 UTC on 27 April to 4:54 UTC on 28 April 2011 during that night. In the afternoon, only 7 tornadoes occurred with weakened precipitation rate, while most 116 tornadoes accompanied with restrengthened precipitation rate during the night. Therefore, three episodes were clearly indicated by the intensity of rainfalls and tornadoes from 27 to 28 April 2011.

Most strong tornados were occurrence from 1800 UTC 27 to 0000 UTC 28 April and covered over 32–38°N and 84–88°W. Figure 6.60 illustrates the vertical time–pressure sections of total height and temperature as well as height and temperature anomalies averaged over 32–38°N along 88°W (Tennessee and Kentucky). Several temperature fluctuations could be observed from Fig. 6.60a in the lower troposphere, but there was no any total height signal to indicate the occurrence of tornadoes in the midnight of 27 April. Differently, Fig. 6.60b indicates that there was a process of height and temperature anomalies near the surface with a transition at 1800 UTC 27 April 2011. The surface anomalous low (Lt) seemed to link with the tornado outbreak of the strongest episode. On 24 April, a positive center (H0) of height anomalies is found at the upper troposphere with a warm center (W0) below and a cold center (C0) above. This is a vertical pattern of height and temperature anomalies. Under this vertical pattern, Tennessee and Kentucky experienced a process of surface positive temperature anomalies. The four processes with the positive temperature anomalies in the mid-low troposphere and the positive center of height anomalies at the upper troposphere occurred consecutively from 24 to 27 April 2011. Starting from early 25 April, a negative signal of height anomalies appeared first near the surface and then deepened with time. On 27 April, a pair of positive (H1) and negative (Lt) centers of height anomalies appeared vertically in the upper troposphere and the surface. The positive center (H1) of height anomalies separated a warm center (W1) below and a cold center (C1) above from the mid-upper troposphere to the lower stratosphere (Fig. 6.60b). In the lower troposphere, a warm center (Wt) was located above the surface low (Lt). It implies that this vertical pattern and its evolution of height and temperature anomalies with their height centers and warm–cold centers are directly associated with the tornado outbreak. It is clear that the tornado outbreak occurred at the moment when the surface anomalous low reaches the deepest from a transition of warm to cold anomalies. The tornado outbreak activities ended after the surface anomalous low (Lt) reduced and the cold air column deepened below the negative center (L2) of height anomalies. The surface anomalous low (Lt), the upper anomalous low (L2), and all warm and cold anomalous centers moved eastward. This process could be found from the surface anomalous low (Lt) located at 88°W at 1800 UTC 27 April (Fig. 6.60b).

Figures 6.57 and 6.60 show that tornado outbreaks and heavy rainfalls are directly associated with the evolution of anomalous synoptic systems. To examine the surface anomalous low (Lt) in Fig. 6.60b, height and temperature anomalies at 925 hPa are shown in Fig. 6.61. Rainfall and tornados occurred near the boundary of anomalous warm–cold air masses and the trough of height anomalies at 925 hPa with an anomalous low center (Lt) moving northeastward from 1200 UTC 27 to 1800 UTC 27, 0000 UTC 28, and 0600 UTC 28 April 2011 (Fig. 6.61a–d). Figure 6.61a–D clearly show that rainfall and tornados were clearly associated with

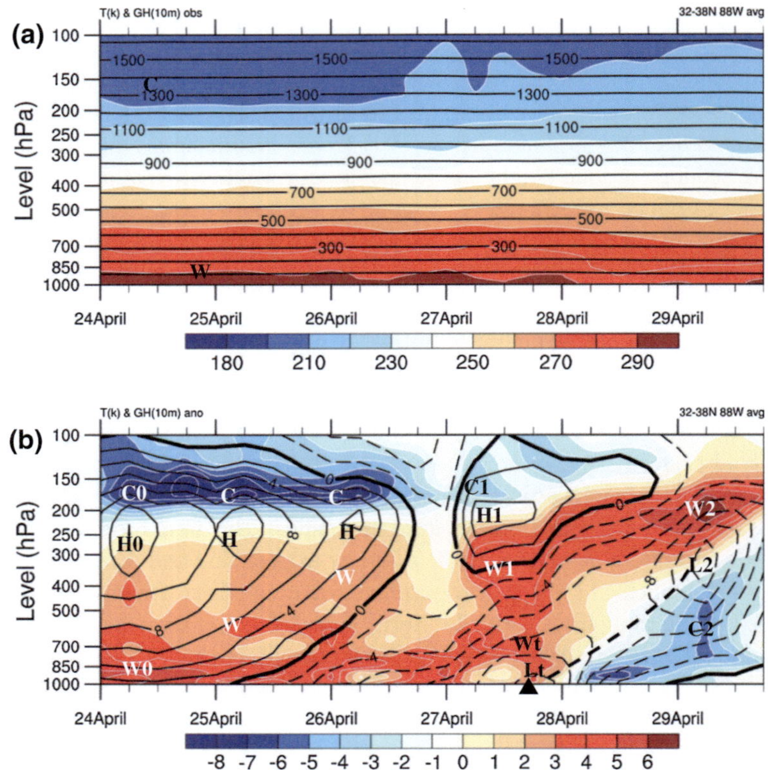

Fig. 6.60 Same as in Fig. 6.57 except along 88°W

an anomalous ascending flow (green dashed line) at the troposphere. The surface anomalous low (Lt) and the upper anomalous low (L2) were located in the western side of heavy rainfalls and tornadoes. The upper anomalous low (L2) was stronger than that of the surface anomaly low (Lt). A trough axis (heavy dashed line) vertically linked the two anomalous low centers. An anomalous cold air mass in the northwest part and an anomalous warm air mass in the southeast part are spatially observed in the mid-low troposphere. Reversely, two anomalous warm and cold air masses are also spatially observed from the upper troposphere to the stratosphere. Clearly, a negative center (C2) and a positive center (W2) of temperature anomalies were located below and above a negative center (L2) of height anomalies at the upper troposphere, respectively. A positive center (W3) and a negative center (C3) of temperature anomalies were located below and above a positive center (H3) of height anomalies at the upper troposphere, respectively. The spatial distribution of temperature anomalies can be derived by the height anomalies because they satisfy the hydrostatic equilibrium relationship.

The vertical pattern of height and temperature anomalies with a pair of centers of height anomalies at the upper troposphere and an anomalous low near the surface

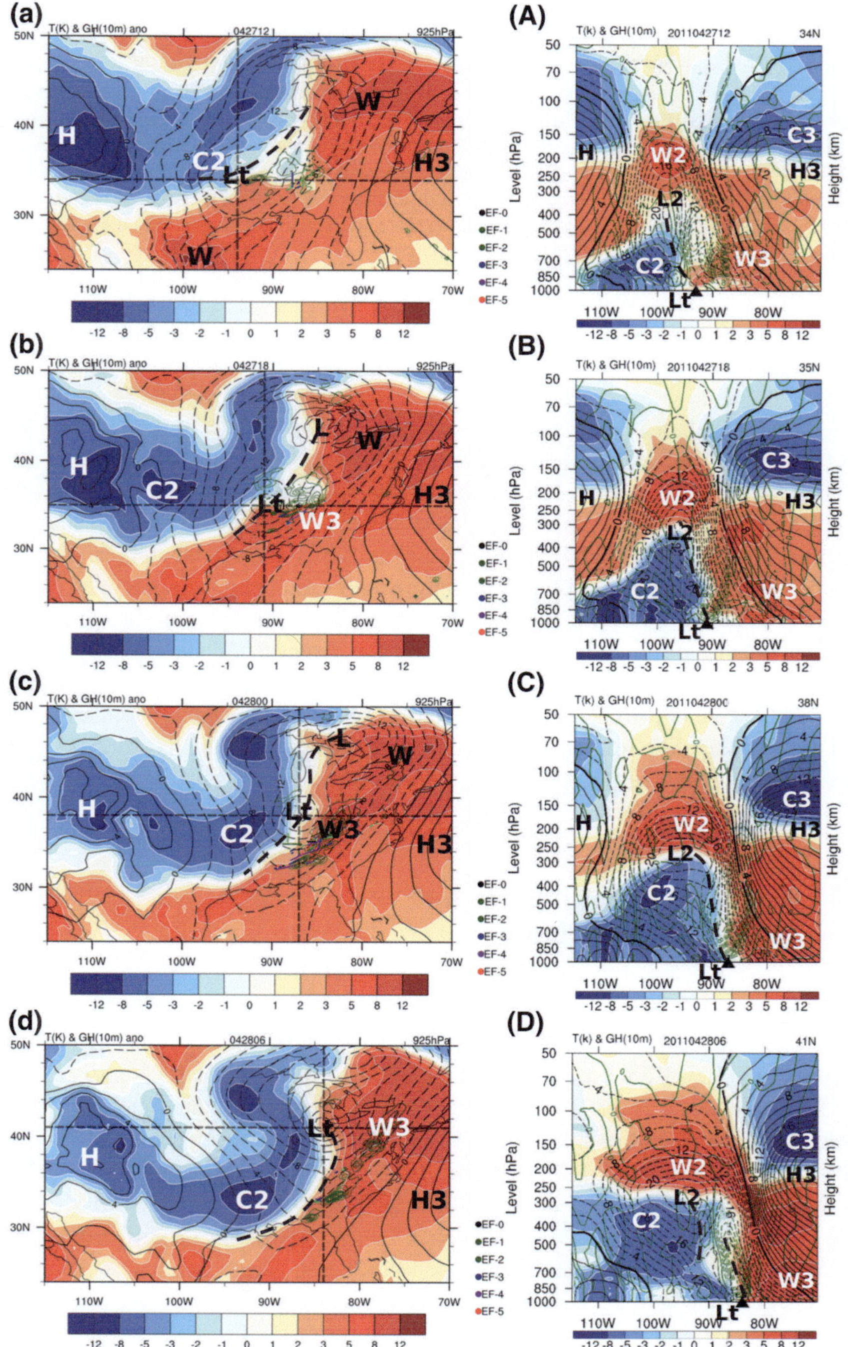
(a)
T(K) & GH(10m) ano
042712
925hPa
H
C2
Lt
W
W
H3
(A)
T(k) & GH(10m) 2011042712
34N
H
W2
C3
H3
L2
C2
W3
Lt
(b)
042718
L
W
H
C2
Lt
W3
H3
(B)
2011042718
35N
(c)
042800
(C)
2011042800
38N
(d)
042806
(D)
2011042806
41N
Level (hPa)
Height (km)
EF-0
EF-1
EF-2
EF-3
EF-4
EF-5
110W
100W
90W
80W
70W
50N
40N
30N
-12 -8 -5 -3 -2 -1 0 1 2 3 5 8 12

◀ **Fig. 6.61** Horizontal height anomalies (*contours*, 2 × 10 gpm interval) and temperature anomalies (*shading*, 1 K interval) at 925 hPa at **a** 1200 UTC 27, **b** 1800 UTC 27, **c** 0000 UTC 28, and **d** 0600 UTC 28 April 2011, with vertical distributions along **A** 34°N, **B** 35°N, **C** 38°N, and **D** 41°N. From **a** to **d**, *green contour* indicates the hourly precipitation rate (2.5 mm h^{-1}) and the *thick dashed line* denotes the trough of height anomalies along with the boundary of temperature anomalies. *Different colored lines* indicate different intensities and tracks of tornadoes within 6 h. From **A** to **D**, symbol "*filled triangle*" indicates the central position of the surface anomalous low (Lt) while *thick dashed lines* indicate the trough axis of height anomalies, the *green solid* and *dashed contours* respectively indicate anomalous sinking and rising pressure velocities (0.2 Pa s^{-1} interval)

persisted from 1200 UTC to 1800 UTC 27 April and 0000 UTC 28 April, except that there was a strengthening process before 0000 UTC 28 April. The strengthening of rising flow was also clear in the process from Fig. 6.61a–c and Fig. 6.61A–D. Thus, we believe that most strong tornadoes occurred from 1800 UTC 27 to 0000 UTC 28 April in the midnight was a result of the strengthening of rising flow, which was a release of anomalous potential energy in the mid-low troposphere. The intensities of tornadoes, rainfall, ascending flow, and cold–warm contrast started to weaken since 0600 UTC 28 April 2011 (Fig. 6.61d and D). It is clear that this vertical–horizontal pattern of height–temperature anomalies can help to explain severe weather events, such as tornado outbreaks and heavy rainfalls. The severe weather events are directly associated with the intensity of anomalous cold and warm air masses. It means that the tornado outbreaks and heavy rainfalls on 27 April 2011 were occurrence near the boundary of anomalous cold–warm air masses and are also the clash result of them in the mid-low troposphere. Tornadoes usually occur in the anomalous warm side.

6.8 Tornado-Producing Anomalous Systems in the Eastern China

Tornadoes often occur in warm season in the eastern China but their frequency is relatively lower than that in the central and southern US. We here give a case occurred at 14:30 CST (0600 UTC) 23 June 2016 in the Funing County (33°26′–33°59′N, 119°27′–119°58′E) of Jiangsu Province in eastern China. The tornado was scaled EF-4 and fatalities were 99. Figure 6.62 shows the surface wind observed at 14:00 Beijing Time (0600 UTC) 23 June 2016. The tornado appeared from 14:20 to 14:50 for about 30 min and moved 30 km as indicated by the yellow-dashed line (Fig. 6.62). There were three convergence lines respectively in northern, central, and eastern Jiangsu Province. The tornado-producing wind system was associated with the central convergence line and its path ended in the southeasterly flow in the eastern shore. The northern convergence line accompanied with cold air and heavy rainfall. In past 50 years, 11 tornado events occurred mainly along the central and eastern convergence lines.

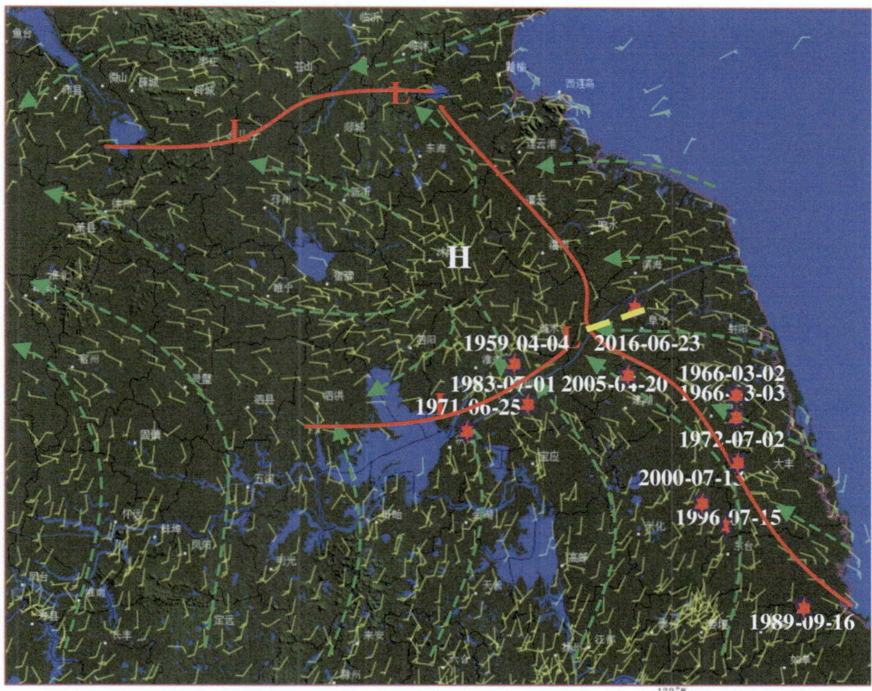

Fig. 6.62 Wind field observed by surface automatic weather stations at 14:00 Beijing Time 23 June 2016 near Jiangsu Province. *Green dashed arrows* denote streamlines and *red solid lines* denote convergence line. "L" and "H" indicate the cyclonic and anticyclonic centers. The dates and the *red symbol* denote the 11 tornado events occurred in the past 50 years. The *yellow-dashed line* indicates the path of tornado occurred on 23 June 2016. Modified from Qian et al. (2017)

Figure 6.63 shows the vertical–latitude cross sections of total height–temperature and height–temperature anomalies along 120°E. In the upper panel, total height and total temperature from the surface to the upper troposphere had no obvious signals to indicate the tornado except that there was a horizontal temperature gradient in the mid-low troposphere (Figs. 6.63a–c). On the contrary, there were obvious evolution signals of height and temperature anomalies and vertical velocity anomaly in the lower panel. First, there was a strengthening center of height anomalies at the upper troposphere and a warming center of temperature anomalies in the lower troposphere, which satisfied the hydrostatic relationship. It means that the surface warming was directly caused by the strengthening height anomaly at the upper troposphere. At the moment that the tornado occurred, a surface low of height anomalies and ascending flow were observed at the location of tornado (Fig. 6.63e). The tornado occurred near the boundary between anomalous warm and cold air masses in the lower troposphere. It can be compared that the clash of anomalous cold–warm air masses was weaker in China for this case than that in the central US. The strengthening rising flow in the warm side from 0600 UTC to 1200 UTC was

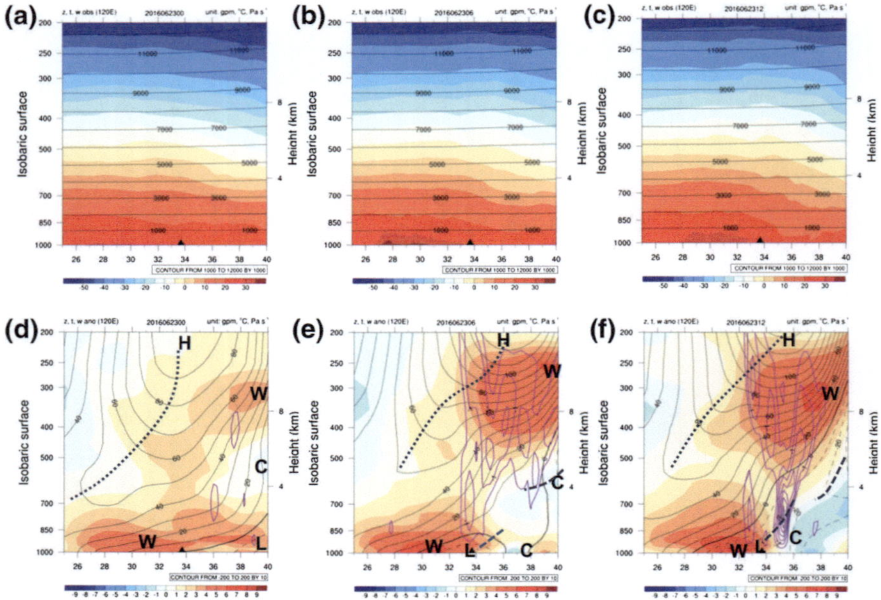

Fig. 6.63 Vertical–latitude sections of total height (*contour*, 1000 gpm interval) and total temperature (*shading*, 5 °C interval) along 120 °E at **a** 0000 UTC, **b** 0600 UTC and **c** 1200 UTC June 2016. **d–f** are the same as **a–c** except of height anomalies (*black contour*, 10 gpm interval) and temperature anomalies (*shading*, 1 °C interval). In **d–f**, *heavy dotted* and *dashed lines* are axes of positive and negative height anomalies while the *purple contour* indicates the ascending velocity (0.5 Pa s^{-1}). Letters "H/L" and "W/C" are centers of height and temperature anomalies, respectively. Symbol "*filled triangle*" indicates the location of tornado. The data used are from the GFS. Modified from Qian et al. (2017)

also associated with the anomalous cold–warm contrast in the lower troposphere. At 1200 UTC 23 June 2016 (Fig. 6.63f), the northern convergence line arrived at Funing County and caused the heavy rainfall while the central convergence line disappeared.

A vertical time–pressure section of total height and temperature over the location of tornado is given in Fig. 6.64a. There was not any signal to indicate the tornado at 0600 UTC 23 June since 18 June 2016 except there was warming process. After removing their climatic height and climatic temperature, several important systems of height and temperature anomalies could be clearly identified since 20 June 2016 (Fig. 6.64b). Three axes of maximum height anomalies are observed at around 0000 UTC 20, 0000 UTC 21, and 1800 UTC 22 June before three positive centers of height anomalies at the upper troposphere, three warm centers of temperature anomalies at the upper and lower troposphere, as well as three low centers of height anomalies near the surface. The tornado and rainfall corresponded well with the third low center before the cold air mass invaded from the lower troposphere in late 23 June 2016.

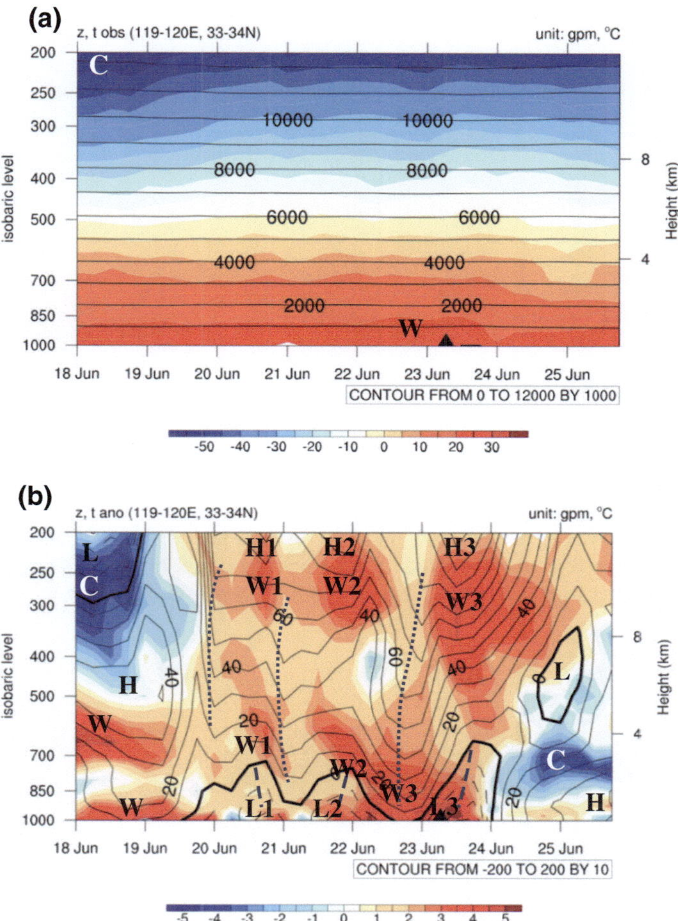

Fig. 6.64 Vertical time sections of **a** total height (*contour*, 1000 gpm interval) and temperature (*shading*, 5 °C interval) as well as **b** height anomalies (*contour*, 10 gpm interval) and temperature anomalies (*shading*, 0.5 °C interval) averaged over 33–34°N and 119–120°E (where the tornado occurred). Symbol "*filled triangle*" indicates the occurring time of tornado. *Heavy dotted* and *dashed lines* are axes of positive and negative height anomalies. The data are derived from the GFS. Modified from Qian et al. (2017)

In the traditional synoptic chart of 850-hPa height and temperature (Fig. 6.65a–c), we can find that a low center moved eastward along 34–35°N and a large gradient zone of temperature moved southeastward, but the height trough and the large temperature gradient zone did not arrive at the tornado place. After removing the climatic temperature and height, a trough of height anomalies at 925 hPa and the axis of MV anomaly at 850 hPa intersected with the surface central wind convergence line at the location of tornado (Fig. 6.65e). This feature cannot be found from the traditional synoptic chart. Figure 6.63d–f and 6.65d–f illustrate a process of tornado and rainfall

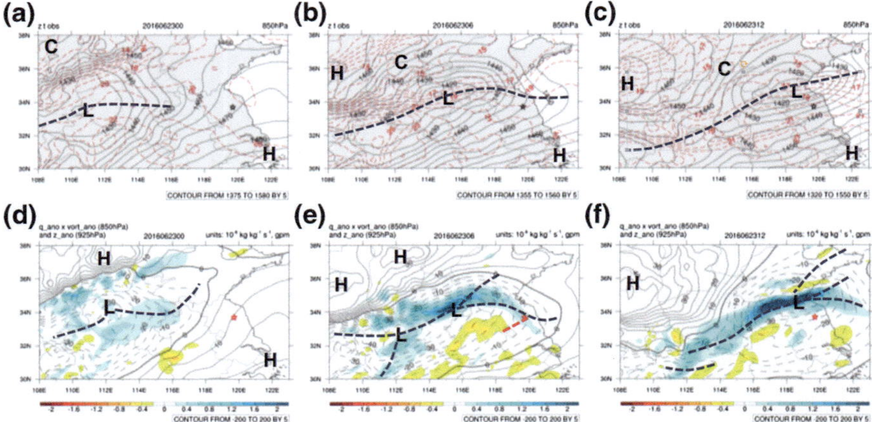

Fig. 6.65 Total height (*black solid contour*, 5 gpm interval) and total temperature (*red-dashed contour*, 1 °C) at 850 hPa at **a** 0000 UTC, **b** 0600 UTC, and **c** 1200 UTC 23 June 2016. **d–f** are the height anomaly (*contour*, 5 gpm interval) at 925 hPa and MV anomaly (*shading*, 10^{-6} kg kg^{-1} s^{-1}) at 850 hPa. *Thick dashed lines* indicate height troughs. The *red-dashed line* is the surface central wind convergence line in Fig. 6.62. Modified from Qian et al. (2017)

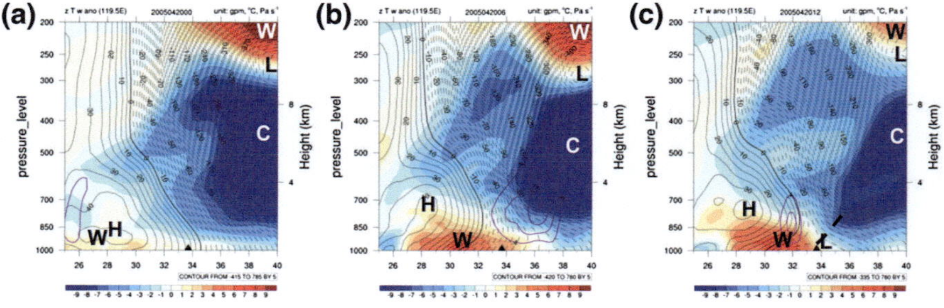

Fig. 6.66 Same as in Fig. 6.63d–f but at **a** 0000UTC, **b** 0600UTC and **c** 1200 UTC 20 April 2005. Modified from Qian et al. (2017)

from the morning to the afternoon and the evening with different anomalous systems and indices.

As marked in Fig. 6.62, another case of tornado occurred in the afternoon (after 1600 CST) of 20 April 2005. Figure 6.66 depicts the process of height and temperature anomalies vertically on 20 April 2005. A strengthening contrast of anomalous warm–cold air masses is clearly observed from the morning to the afternoon. The tornado was associated with the anomalous trough axis of height anomalies and the clash of anomalous warm–cold air masses in the afternoon.

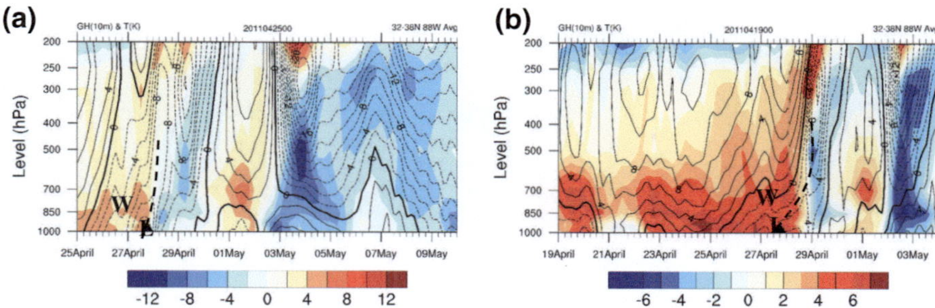

Fig. 6.67 Same as in Fig. 6.60b along 88°W, except of model prediction initiated at **a** 0000 UTC 25 April 2011 for leading 2 days and **b** 0000 UTC 19 April 2011 for leading 8 days

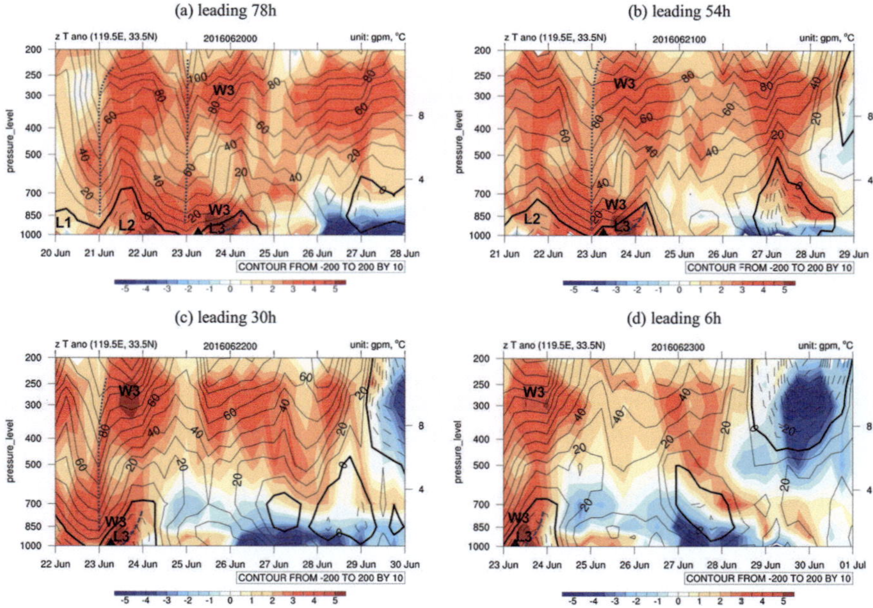

Fig. 6.68 Same as in Fig. 6.64b except of the ECMWF model prediction for leading **a** 78 h, **b** 54 h, **c** 30 h, and **d** 6 h in advance

6.9 Model Prediction Ability

We hope to examine the model prediction ability from two aspects: (1) basic variable and (2) diagnostic parameters or indices. The model-predicted data are obtained from the ECMWF. The 6-hourly forecasts, including 60 steps (+1 to +15 days ahead), which are generated only from 0000 UTC on each day before the tornado occurred are examined. The variables selected are similar to those of the reanalysis

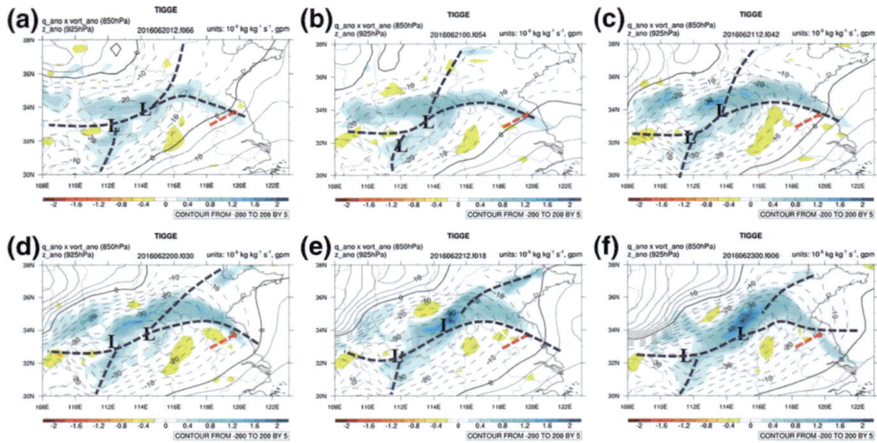

Fig. 6.69 Same as in Fig. 6.65e, except of the ECMWF model prediction for leading **a** 66 h, **b** 54 h, **c** 42 h, **d** 30 h, **e** 18 h, and **f** 6 h in advance. Modified from Qian et al. (2017)

datasets and have the same spatial resolution. For further details of the forecast models and data, the reader can refer to http://tigge.ecmwf.int.

The process studied in the previous sections shows an advantage of the anomaly-based analyses in diagnosing locations of tornado-producing system. In this section, we will apply the same method to the products from model prediction and show that the location of tornado-producing systems can be detected from model products. The ECMWF model predictions are employed in this section and compared with the ERA-Interim products. Figure 6.67 shows the time–vertical evolution of height and temperature anomalies along 88°W to compare with Fig. 6.60b. On 27 April 2011 (Fig. 6.67a, b), a trough axis of height anomalies accompanied with a transition of temperature anomalies from warm to cold air masses in the mid-low troposphere was exactly predicted by the model initiated at 0000 UTC 25 and 0000 UTC 19 April for leading 2 and 8 days, respectively.

We further examine the predictions of the ECMWF model to compare with the process of tornado-producing system in Fig. 6.64b. For leading 78 h (Fig. 6.68a), a low (L3) and warm center (W3) were well predicted on 23 June 2016. From leading 54 to 6 h (Fig. 6.68b–d), the low (L3) and the warm center (W3) were stably predicted on 23 June 2016.

Figure 6.69 depicts the results of ECMWF model prediction for leading 6–66 h to compare with Fig. 6.65e. The axis (blue shading) of MV anomaly and the trough (dashed line) of height anomalies intersected at the tornado location with the central surface convergence line (red-dashed line). The two signals indicated to Funing County were stable since the leading 66 h to the tornado occurrence at 0600 UTC 23 June 2016. From Figs. 6.68 and 6.69, the ECMWF model well predicted the trough axis feature for leading 78 and 66 h respectively in advance. This anomalous system and indices can be used to indicate a potential time–space

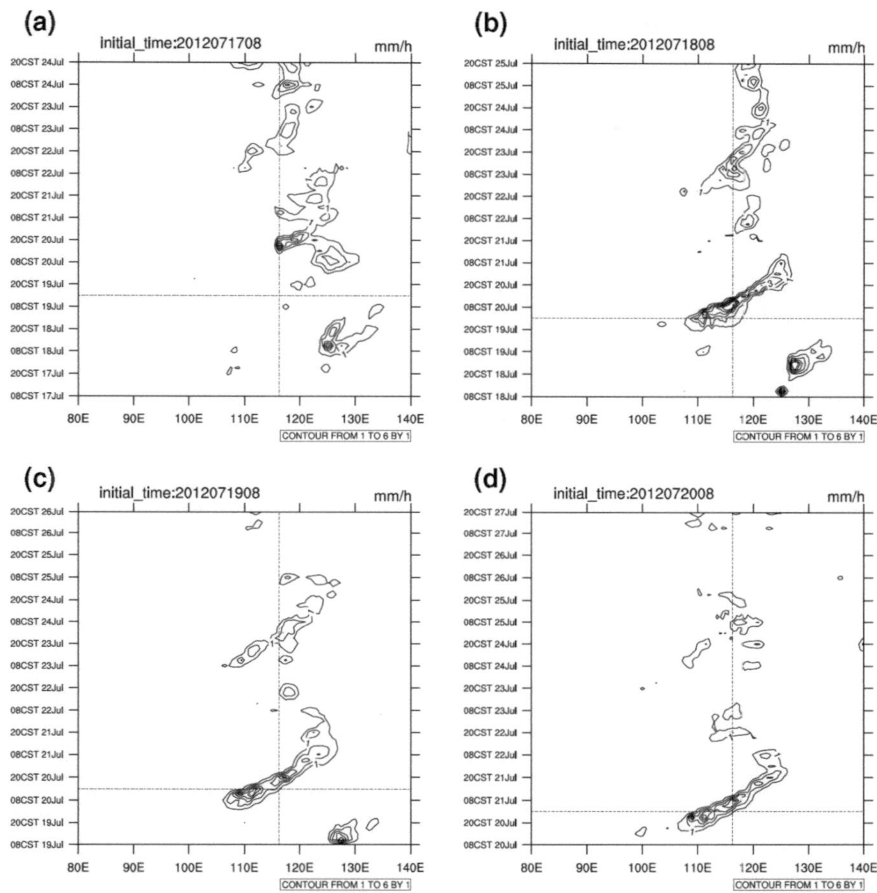

Fig. 6.70 The same as Fig. 6.21a except for the model rain predictions initiated from **a** 0800 CST 17, **b** 0800 CST 18, **c** 0800 CST 19, and **d** 0800 CST 20 July 2012. Sourced from Jiang et al. (2016) © Springer Science + Business Media Dordrecht 2016

domain of tornadic occurrence, but they cannot predict the exact location and duration of tornadic occurrence.

Another case of model prediction ability is compared with Fig. 6.21, a zonal-time sections of precipitation rate and height RA at 850 hPa from 0000 UTC 19 to 1800 UTC 23 July 2012, averaged over 39–41°N for the Beijing heavy rain. No precipitation process can be found in Fig. 6.70a for leading 4 days when the model prediction was initiated from 0800 CST 17 July 2012. Since 0800 CST 18 July 2012 (refer to the 2000 CST 21 July 2012, +84 h lead), the model predicted a precipitation process propagating through Beijing, which is comparable with observation (Fig. 6.21a vs. 6.70b). Therefore, the model rain prediction was limited within 84 h (3.5 days). Similar to Fig. 6.21b, Jiang et al. (2016) examined the ability of model in predicting height RAs. A better result can be obtained from the

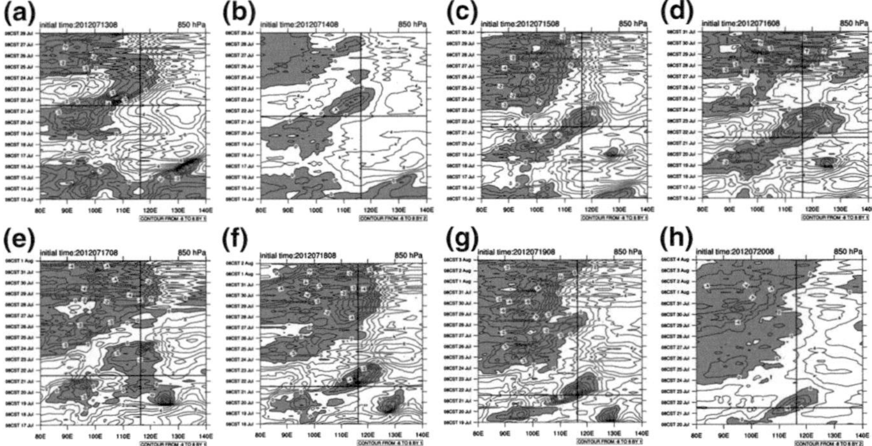

Fig. 6.71 The same as Fig. 6.21b except for the height RA model predictions initiated from **a** 0800 CST 13, **b** 0800 CST 14, **c** 0800 CST 15, **d** 0800 CST 16, **e** 0800 CST 17, **f** 0800 CST 18, **g** 0800 CST 19, and **h** 0800 CST 20 July 2012. Sourced from Jiang et al. (2016) © Springer Science + Business Media Dordrecht 2016

forecast height RA field (Fig. 6.71) than the model-predicted precipitation. The propagation of the height RA field was predicted from the initial time 0800 CST 14 July (7.5 days in advance), but the location of negative height RA peak was predicted to the west of observation (Fig. 6.71b). Until from 0800 CST 15 July (leading 6.5 days), the predicted result of time and location can be comparable with reanalysis (Fig. 6.71c vs. 6.21b). But the negative RAs predicted from initial time 0800 CST 17 July were wider in the time dimension (Fig. 6.71e). From 0800 CST 18 to 0800 CST 19 and 0800 CST 20 July 2012, model prediction was well comparable with reanalysis.

Questions

1. What are the advantages of the anomaly-based synoptic analyses over the traditional ones to locate the RHRs and tornado-producing systems?
2. What is the difference between determining rainfall domain and the producing rain system?
3. Vertically, why does a heavy rainfall or a tornado (tornado outbreak) storm occur when the strengthening height anomaly is at the upper troposphere and the deepening low is near the surface?
4. How many dynamical parameters and combined indices can be used to indicate regional convective events?
5. Tornado is the strongest convective weather event. How can we find the tornado-producing system in advance based on the current forecast model outputs?

References

Agee E, Church C, Morris C, Snow J (1975) Some synoptic aspects and dynamic features of vortices associated with the tornado outbreak of 3 April 1974. Mon Wea Rev 103:318–333

Bennetts DA, Hoskins BJ (1979) Conditional symmetric instability—A possible explanation for frontal rainbands. Q J Roy Meteorol Soc 105:945–962

Bennetts DA, Sharp JC (1982) The relevance of conditional symmetric instability to the prediction of mesoscale frontal rainbands. Q J Roy Meteorol Soc 108:595–602

Bunkers MJ, Johnson JS, Czepyha LJ, Grzywacz JM, Klimowski BA, Hjelmfelt MR (2006) An observational examination of long-lived supercells. Part II: Environmental conditions and forecasting. Wea Forecast 21:689–714

Cao ZH, Cho HR (1995) Generation of moist potential vorticity in extratropical cyclones. J Atmos Sci 52:3263–3281

Cao J, Gao ST (2008) Generalized potential temperature in non-uniformly saturated atmosphere. Chin J Geophys 51(6):1651–1656 (in Chinese)

Chen DL, Ou TH, Gong LB, Xu CY, Li WJ, Ho CH, Qian WH (2010) Spatial interpolation of daily precipitation in China: 1951–2005. Adv Atmos Sci 27:1221–1232

Chen Y, Sun J, Xu J, Yang SN, Zong ZP, Chen T, Fang C, Sheng J (2012) Analysis and thinking on the extremes of the 21 July 2012 torrential rain in Beijing Part I: Observation and Thinking. Meteorol Mon 38(10):1255–1266 (in Chinese)

Chou JF (2012) An innovation study on the physical decomposition of numerical model atmospheric variables and their application in weather extreme events. Chin J Geophys 55(5):1433–1438 (in Chinese)

Clark JHE, James RP, Grumm RH (2002) A reexamination of the mechanisms responsible for banded precipitation. Mon Wea Rev 130:3074–3086

Clark AJ, Gao JD, Marsh PT, Smith T, Kain JS (2013) Tornado pathlength forecasts from 2010 to 2011 using ensemble updraft helicity. Wea Forecast 28(2):387–407

Corfidi S, Weiss S, Kain J, Corfidi S, Rabin R, Levit J (2010) Revisiting the 3–4 April 1974 Super Outbreak of tornadoes. Wea Forecast 25:465–510

Dee DP, at al (2011) The ERA-interim reanalysis: configuration and performance of the data assimilation system. Q J Roy Meteorol Soc 137(656):553–597

Deng G, Gao ST (2009) The theory of moist potential vorticity and its application in the diagnosis of typhoon rainfall and intensity. J Trop Meteorol 15:204–209

Ding T, Qian WH (2012) Statistical characteristics of heat wave precursors in China and model prediction. Chin J Geophys 55(5):1472–1486

Ding YH (1993) Research on the 1991 persistent, severe flood over Yangtze–Huai river valley. Chinese Meteorological Press, 255 pp (in Chinese)

Doswell CA, Brooks HE, Maddox RA (1996) Flash flood forecasting: An ingredients-based methodology. Wea Forecast 11:560–581

Doswell CA, Edwards R, Thompson RL, Hart JA, Crosbie KC (2006) A simple and flexible method for ranking severe weather events. Wea Forecast 21:939–951

Du J, Grumm RH, Deng G (2014) Ensemble anomaly forecasting approach to predicting extreme weather demonstrated by extremely heavy rain event in Beijing. Chin J Atmos Sci 38(4): 685–699 (in Chinese)

Emanuel KA (1983) The Lagrangian parcel dynamics of moist symmetric instability. J Atmos Sci 40:2368–2376

Emanuel KA (1988) Observational evidence of slantwise convective adjustment. Mon Wea Rev 116:1805–1816

Ferguson EW, Ostby F, Leftwich P (1987) The tornado season of 1985. Mon Wea Rev 115: 1437–1445

Fujita T, Bradbury DL, Thullenar CFV (1970) Palm Sunday tornadoes of April 11, 1965. Mon Wea Rev 98:29–69

Galarneau TJJ, Hamill TM, Dole RM, Perlwitz J (2012) A multiscale analysis of the extreme weather events over western Russia and northern Pakistan during July 2010. Mon Wea Rev 140(5):1639–1664

Gao ST, Lei T, Zhou YS (2002) Diagnostic analysis of moist potential vorticity anomaly in torrential rain systems. Quart J Appl Meteor 13:662–670 (in Chinese)

Gao ST, Wang XR, Zhou YS (2004a) Generation of generalized moist potential vorticity in a frictionless and moist adiabatic flow. Geophys Res Lett 31:L12113

Gao ST, Zhou YS, Cui XP, Dai GP (2004b) Impacts of cloud-induced mass forcing on the development of moist potential vorticity anomaly during torrential rains. Adv Atmos Sci 21:923–927

Glickman TS, Todd S (2000) Glossary of meteorology. American Meteorological Society, 855 pp

Goswami BN, Mohan RSA (2001) Intraseasonal oscillations and interannual variability of the Indian summer monsoon. J Clim 14(6):1180–1198

Graham RA, Grumm RH (2010) Utilizing normalized anomalies to assess synoptic-scale weather events in the Western United States. Wea Forecast 25(2):428–445

Grumm RH (2011a) The central European and Russian heat event of July-August 2010. Bull Amer Meteorol Soc 92(10):1285–1296

Grumm RH (2011b) New England record maker rain event of 29–30 March 2010. Electronic J Operational Meteor 12(4):1–31

Grumm RH, Hart RE (2001) Standardized anomalies applied to significant cold season weather events: Preliminary findings. Wea Forecast 16(6):736–754

Hart RE, Grumm RH (2001) Using normalized climatological anomalies to rank synoptic-scale events objectively. Mon Wea Rev 129(9):2426–2442

Huang J, Du J, Qian WH (2015) A comparison between a generalized beta-advection model and a classical beta-advection model in predicting and understanding unusual typhoon tracks in eastern China seas. Wea Forecast 30(3):771–792

Huang SS (1981) A diagnostic analysis of the formation and variation of the low-level jet during heavy-rain processes. Chin J Atmos Sci 5(2):123–135 (in Chinese)

Jiang N, Qian WH, Du J, Grumm RH, Fu JL (2016) A comprehensive approach from the raw and normalized anomalies to the analysis and prediction of the Beijing extreme rainfall on 21 July 2012. Nat Hazards submitted

Jiang M, Yu TT, Qian WH (2014) Analysis on atmospheric anomalous signals of winter low temperature and snowice storms in southern China. Chin J Atmos Sci 38(4):813–824 (in Chinese)

Joyce RJ, Janowiak JE, Arkin PA, Xie PP (2004) CMORPH: A method that produces global precipitation estimates from passive microwave and infrared data at high spatial and temporal resolution. J Hydrometeorology 5(3):487–503

Junker NW, Grumm RH, Hart R, Bosart LF, Bell KM, Pereira FJ (2008) Use of normalized anomaly fields to anticipate extreme rainfall in the mountains of northern California. Wea Forecast 23(3):336–356

Junker NW, Brennan MJ, Pereira F, Bodner MJ, Grumm RH (2009) Assessing the potential for rare precipitation events with standardized anomalies and ensemble guidance at the hydrometeorological prediction center. Bull Amer Meteorol Soc 90(4):445–453

Knupp KR, Murphy TA, Coleman TA, … Klockow K (2014) Meteorological overview of the devastating 27 April 2011 tornado outbreak. Bull Amer Meteorol Soc 95:1041–1062

Korty RL, Schneider T (2007) A climatology of the tropospheric thermal stratification using saturation potential vorticity. J Clim 20:5977–5991

Lee BD, Jewett BF, Wilhelmson RB (2006) The 19 April 1996 Illinois Tornado outbreak. Part II: Cell mergers and associated tornado incidence. Wea Forecast 21:449–464

Martínez-Alvarado O, Weidle F, Gray SL (2010) Sting jets in simulations of a real cyclone by two mesoscale models. Mon Wea Rev 138:4054–4075

Meng XF, Sun YG, Sarina, Yuan HM, Hasi (2013) Correlation between eastward developing of hetao cyclone and the severe rainstorm in Beijing on 21 July 2012. Meteorol Mon 39(12):1542–1549 (in Chinese)

Mercer AE, Shafer CE, Doswell CA, Leslie LM, Richman MB (2012) Synoptic composites of tornadic and nontornadic outbreaks. Mon Wea Rev 140:2590–2608

Metz N, Bosart L (2010) Derecho and MCS development, evolution, and multiscale interactions during 3–5 July 2003. Mon Wea Rev 138:3048–3070

Min AR, Liao YS, Yang JA (2011) Important heavy rain processes in China from April to October in 2010. Torrential Rain Disasters 30(1):90–96 (in Chinese)

Moron V, Vautard R, Ghil M (1998) Trends, interdecadal and interannual oscillations in global sea-surface temperatures. Clim Dyn 14(7–8):545–569

National Climate Center of China (1998) The catastrophic flood in China in 1998 and climate abnormality. China Meteorological Press, Beijing (in Chinese)

NOAA (2011) Service assessment: The historic tornadoes of April 2011. US Department of Commerce/NOAA/NWS Rep, 76 pp

Novak DR, Brian AC, McTaggart-Cowan R (2009) The role of moist processes in the formation and evolution of mesoscale snowbands within the comma head of Northeast U.S. cyclones. Mon Wea Rev 137:2662–2686

Palmer WC, Allen RA (1949) Note on the accuracy of forecasts concerning the rain problem. US Weather Bureau, 4 pp

Peng XD, Che YZ, Chang J (2013) A novel approach to improve numerical weather prediction skills by using anomaly integration and historical data. J Geophys Res Atmos 118(16):8814–8826

Qian WH (2012) Principles of Medium to Extended Range Weather Forecasts. Science Press, Beijing, 410 pp

Qian WH (2013) Atlas of anomalous circulations associated with regional heavy Rainfall in China. China Meteorological Press, Beijing, 227 pp

Qian WH, Jiang M (2014) Early signals of synoptic-scale atmospheric anomalies associated with the summer low temperature events in northeast China. Meteorol Atmos Phys 124(1):33–46

Qian WH, Zhang ZJ (2012) Precursors to predict low-temperature freezing-rain events in southern China. Chin J Geophys 55(5):1501–1512 (in Chinese)

Qian WH, Lin X, Zhu YF, Xu Y, Fu JL (2007) Climatic regime shift and decadal anomalous events in China. Clim Change 84(2):167–189

Qian WH, Li J, Shan XL (2013) Application of synoptic-scale anomalous winds predicted by medium-range weather forecast models on the regional heavy rainfall in China in 2010. Sci China Earth Sci 56(6):1059–1070

Qian WH, Shan XL, Liang HY, Huang J, Leung CH (2014) A generalized beta-advection model to improve unusual typhoon track prediction by decomposing total flow into climatic and anomalous flows. J Geophys Res Atmos 119(3):1097–1117

Qian WH, Du J, Shan XL, Jiang N (2015a) Incorporating the effects of moisture into a dynamical parameter: moist vorticity and moist divergence. Wea Forcast 30(6):1411–1428

Qian WH, Chen Y, Jiang M, Hu Q (2015b) An anomaly-based method for identifying signals of spring and autumn low temperature events in the Yangtze River Valley, China. J Appl Meteorol Climatol 54:1216–1233

Qian WH, Jiang N, Du J (2016a) Anomaly-based weather analysis versus traditional total-field-based weather analysis for depicting regional heavy rain events. Wea Forcast 31:71–93

Qian WH, Yu TT, Du J (2016b) A unified approach to trace surface heat and cold events by using height anomaly. Climate Dynamics 46(5):1647–1664

Qian WH, Jiang N, Du J (2016c) Seven anomalous synoptic patterns of regional heavy rain in eastern China. Meteorol Mon 42(6):656–667 (in Chinese)

Qian WH, Leung JCH, Jin RH, Fu JL, Wu FF, Kuang ZJ, Du J (2017) Application of anomalous variables on strong convective system analyses and model evaluation—a case study on tornado-producing anomalous systems near Lixia River, Jiangsu Province. Meteorol Mon 43(2):129–143 (in Chinese)

Qu XB, Zhang T, Liu XH, Zhou QL, Zhang XW (2010) Analysis on the meteorological causes for the 8 August 2010 massive mudslide in Zhouqu, Gansu Province. Meteorol Mon 36(10):102–105 (in Chinese)

Rogers JC (1984) The association between the North Atlantic Oscillation and the Southern Oscillation in the northern hemisphere. Mon Wea Rev 112(112):1999–2015

Schubert HW, Hausman SA, Garcia M, Ooyama KV, Kuo H (2001) Potential vorticity in a moist atmosphere. J Atmos Sci 58:3148–3157

Schultz DM, Schumacher PN (1999) The use and misuse of conditional symmetric instability. Mon Wea Rev 127:2709–2732

Shan XL, Jiang N, Qian WH (2015) Regional heavy rain locations associated with anomalous convergence lines in eastern China. Nat Hazards 77:1731–1750

Shen HY, Kuang YY, Zi L (2011) Genesis of 2010 storm–flood in Yangtze River Basin and its comparison with 1998 flood. Yangtze River 42(6):11–14 (in Chinese)

Shou SW, Li YH (1999) Study on moist potential vorticity and symmetric instability during a heavy rain event occurred in the Jiang-Huai Valley. Adv Atmos Sci 16:314–321

Simmons KM, Sutter D (2012) The 2011 tornadoes and the future of tornado research. Bull Amer Meteorol Soc 93, 959–961

Sun J, Chen Y, Yang SN, Dai K, Chen T, Yao R, Xu J (2012) Analysis and thinking on the extremes of the 21 July 2012 torrential rain in Beijing Part II: Preliminary causation analysis and thinking. Meteorol Mon 38(10):1267–1277 (in Chinese)

Sun JH, Zhao SX, Fu SM, Wang HJ, Zheng LL (2013) Multi-scale characteristics of record heavy rainfall over Beijing area on July 21, 2012. Chin J Atmos Sci 37(3):705–718

Thompson RL, Edwards R (2000) An overview of the environmental conditions and forecast implications of the 3 May 1999 tornado outbreak. Wea Forecast 15:682–699

Thompson RL, Smith BT, Grams JS, Dean AR, Broyles C (2012) Convective modes for significant severe thunderstorms in the contiguous United States. Part II: Supercell and QLCS tornado environments. Wea Forecast 27:1136–1154

Wallace JM, Gutzler DS (1981) Teleconnections in the geopotential height field during the northern hemisphere winter. Mon Wea Rev 109(4):784–812

Wang BS, Zhang Y (2011) Analysis of a large-scale severe precipitation event in Jilin Province from 27 to 29 July 2010. Torrential Rain Disasters 30(1):36–43 (in Chinese)

Wang XR, Wu KJ, Shi CE (1999) The introduction of condensation probability function and the dynamic equations on non-uniform saturated moist air. J Trop Meteorol 15:64–70

Wang XF, Xu M, Min AR, Zhan DY, Yang JA, Cui CG, Liao YS (2010) Analysis of precipitation and affecting systems features on persistent heavy rain in South China in May 2010. Torrential Rain Disasters 29(2):193–199 (in Chinese)

Weisman MA, Evans C, Bosart L (2013) The 8 May 2009 superderecho: Analysis of a real-time explicit forecast. Wea Forecast 28:863–892

Wikipedia (2014) 2010 Gansu mudslide—Wikipedia, the free encyclopedia. http://en.wikipedia.org/w/index.php?title=2010_Gansu_mudslide&oldid=609803997. Accessed 26 Sept 2014

Wu GX, Cai YP, Tang XJ (1995) Moist potential vorticity and slantwise vorticity development. Acta Meteor Sin 53:387–405 (in Chinese)

Zhang DL, Lin YH, Zhao P, Yu XD, Wang SQ, Kang HW, Ding YH (2013) The Beijing extreme rainfall of 21 July 2012: "Right results" but for wrong reasons. Geophys Res Lett 40(7): 1426–1431

Zhou GB, Cui CG, Gao ST (2010) Application of generalized moist potential vorticity to the analysis of a torrential rain case. Acta Meteor Sin 24:732–739

Zhu QG (1975) Low level jet and heavy rainfall. Inf Meteorol Sci Technol 21(8):12–18 (in Chinese)

Chapter 7
Heat Waves and Cold Events

A heat wave is a prolonged period of excessively hot weather, which may be accompanied by high humidity, especially in oceanic climate countries. A cold event (surge) is a weather phenomenon that is distinguished by a cooling of the air. Heat wave and cold surge are two types of surface temperature anomalous events, so we first need to define them and then to find their early signals from the atmosphere to indicate their coverage, intensity, and duration. This chapter shows that a positive (negative) center of height anomalies at the upper troposphere can indicate the surface heat wave (cold surge) because an abnormal warm (cold) air column extends downward from the height anomalous center to the surface. The temperature and height anomalies satisfy the hydrostatic equilibrium vertically. Standing-type heat wave events are observed in Western Europe while moving-type heat wave events often occur in China. Aloft height anomalous signals that indicate the onset of heat and cold events in China can be traced 2–3 weeks in advance, and the current model can predict this height signal for about 12 days in advance in case analysis. This suggests that the daily maximum height anomaly (MHA) at the upper troposphere can be used to indicate surface heat and cold events globally.

7.1 Definitions of Heat Waves and Cold Events[1]

Heat wave is a prolonged period of hot weather with surface air temperature (SAT) exceeding a threshold such as 35 °C, normally occurring in summer and autumn. Severe heat waves can cause economic and health problems, such as catastrophic crop failures, thousands of human and animal deaths from hyperthermia, and widespread power outages. The heat wave occurred in Europe in July

[1]Modified article originally published in *Qian WH, Yu TT, Du J (2016b) A unified approach to trace surface heat and cold events by using height anomaly. Climate Dyn 46: 1647–1664.* Published with kind permission © Springer-Verlag Berlin Heidelberg 2016. All Rights Reserved.

© Springer Nature Singapore Pte Ltd. 2017

W. Qian, *Temporal Climatology and Anomalous Weather Analysis*,
Springer Atmospheric Sciences, DOI 10.1007/978-981-10-3641-5_7

2013, for example, caused 700–800 loss of human life in the United Kingdom and Ireland (Silverman 2013). A nearly simultaneous but less devastating heat wave also persisted in eastern China (Sun et al. 2014). In contrast, a cold surge is a prolonged period of cold weather with SAT dropping below normal, normally occurred in winter and spring. Extreme cold temperature can cause injury or death to livestock and wildlife. Wintery precipitation can severely interrupt transportation and damage electrical power lines. In China, prolonged cold surges may occur even in summer (Wang and Wu 1997; Cui et al. 2007).

A decade ago, a heat wave was defined as a relative measure to a typical weather in a region and average temperature for a season (Meehl and Tebaldi 2004). The same consideration can be applied to cold surges, e.g., temperature is below a predefined threshold for a certain number of consecutive days. By this definition, temperatures that people from a hotter climate region consider normal can be termed as a heat wave in a cooler climate region (Robinson 2001). Therefore, a local climatology of temporal temperature evolution needs to be established in order to define heat waves and cold events. To accurately define these events, daily climatology is needed since low temporal resolution will wipe out detailed information.

For a particular location or a region, a heat wave event can be defined based on its historical daily maximum temperature observations. In China, SAT exceeding 35 °C for more than 2 days at a same location is considered as a heat wave for that location (Liu et al. 2008). A regional heat wave event is defined if at least five adjacent locations (stations) meet this criterion during the same time period (Ding and Qian 2011).

A case is shown how to define heat and cold events as well as surface general temperature anomalies. Figure 7.1 shows site surface maximum temperature observations and their anomalies on 10 September 1999 in the mainland China. The anomalies are derived after removing daily climatology from observation. A regional heat wave event with the highest surface maximum temperature exceeding 36 °C occurred on 10 September 1999 in Central China (Fig. 7.1a). A positive center (>10 °C) of surface maximum temperature anomalies is located in the box of 30–35°N, 107.5–112.5°E in Central China while a negative center (<−2 °C, red-dashed contour) is situated in Northwest China (Fig. 7.1b).

Within the box area as indicated in Fig. 7.1, the average surface maximum temperature series, anomalous series, and climatic mean series are shown in Fig. 7.2. The climatic mean series has a decreasing trend from about 32 °C on 1 August to 20 °C on 31 October. The anomalous series has many positive and negative processes with different lengths and intensities. The longest duration of positive anomalies is the strongest heat wave centered on 10 September 1999 when the surface maximum temperature reached 36.27 °C. This high temperature is harmful for human activity. The second longest duration of a positive surface maximum temperature anomaly centered on 25 September 1999 and its maximum temperature (30.79 °C) did not reach the 35 °C criterion of heat wave, but this was

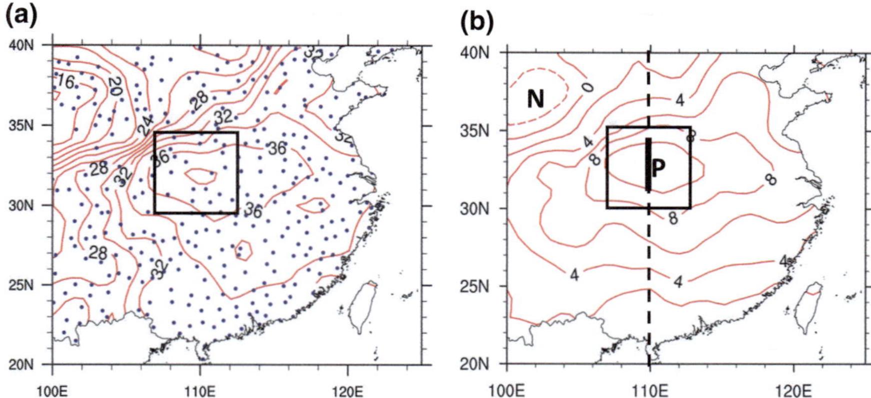

Fig. 7.1 **a** Site surface maximum temperature (*red contour*, 2 °C interval) and **b** maximum temperature anomalies (*red contour*, 2 °C interval) on 10 September 1999. In **a**: 333 stations are marked by *blue dots* and 20 stations are located in the box (30–35°N, 107.5–112.5°E). In **b**: *Heavy solid* and *dashed lines* cross the center of high temperature anomalies along 110°E longitude, letters "P" and "N" denote the centers of positive and negative temperature anomalies, respectively. Sourced from Qian et al. (2016b) © Springer-Verlag Berlin Heidelberg 2016, with permission

also an abnormal weather process for forecasting. In Fig. 7.2, a rapid drop of temperature was also observed during 2–5 October 1999. These processes are important in routine forecasting practice.

The above processes can also be identified from the time series of observed surface minimum, maximum, and daily mean temperature anomalies (figures not shown). In the following analysis, observed daily mean SAT anomaly (SATA) is

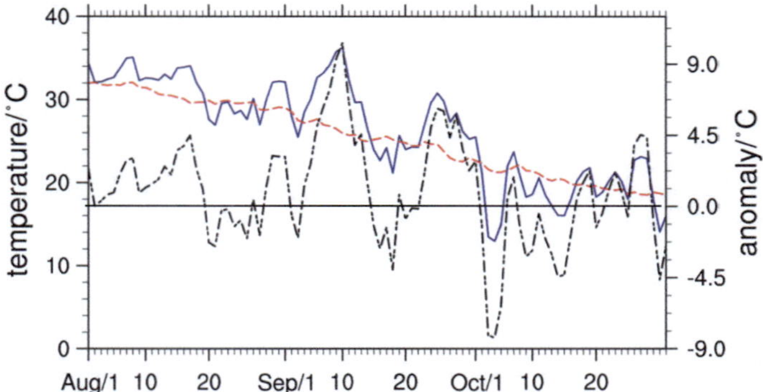

Fig. 7.2 Daily area-averaged 20-station observational series (*blue solid line*, °C), climatic series (*red-dashed line*, °C), and anomalous series (*black dot-dashed line*, °C) calculated by the surface maximum temperature in the box of 30–35°N, 107.5–112.5°E from 1 August to 31 October 1999. Sourced from Qian et al. (2016b) © Springer-Verlag Berlin Heidelberg 2016, with permission

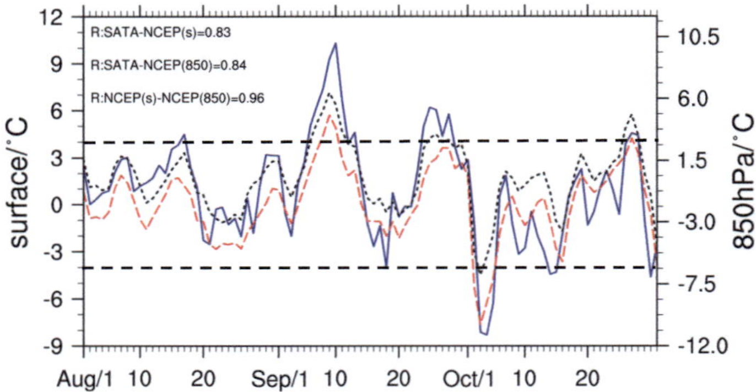

Fig. 7.3 Daily mean SATA series (*red solid line*, °C), NCEP surface temperature anomalous series (*red-dashed line*, °C), and NCEP 850-hPa temperature anomalous series (*black dotted line*, °C) averaged over the box of 30–35°N, 107.5–112.5°E. Correlation values of three series are 0.83, 0.84, and 0.96. Two heavy-dashed lines (+4 and −4 °C) denote the criterion of heat and cold events, respectively. Sourced from Qian et al. (2016b) © Springer-Verlag Berlin Heidelberg 2016, with permission

commonly used in the description of warm and cold processes. Figure 7.3 shows a comparison among other three series of observed daily mean SATA, NCEP surface temperature anomalies, and NCEP 850-hPa temperature anomalies averaged over 30–35°N, 107.5–112.5°E. The warm and cold processes are captured by the three series. Although the amplitude of daily mean SATA series from observational stations is larger than that of NCEP surface temperature anomalous series, the three series have the same phase of fluctuations with high correlations ranging from 0.83 to 0.96.

Using the daily climatology instead of seasonal mean, Qian et al. (2016b) defined a general criterion of regional warm (cold) anomaly and a special criterion of regional heat (cold) event as distinguished in Fig. 7.3. For this region, the daily mean SATAs exceeding (below) climatology 4 °C (−4 °C) and lasting 3 or more days are defined as a heat (cold) event and otherwise lasting 3 or more days as warm (cold) anomalies. Based on the criteria (heavy-dashed line), two heat events centered, respectively, on 10 and 25 September and one cold event from 2 to 5 October 1999 can be determined. Other seven warm anomalies and five cold anomalies are also identified.

The above definition of heat waves and cold events is widely used in China. Qian et al. (2016b) gave another case of definition in western Russia. When analyzing the Russian summer heat waves in 2010, some studies have used observational SATs at the Moscow station (e.g., Grumm 2011; Rahmstorf and Coumou 2011; Dole et al. 2011), and others have used SATs at the 2 m level from the ERA-Interim product (Schubert et al. 2014). They compared the SATs at the 0.995-sigma level from the NCEP reanalysis 1 (R1) and the SATs at the 2 m level from the ERA-Interim product. Figure 7.4 compares an absolute threshold and three relative thresholds used to describe the heat wave coverage, with SATs at the

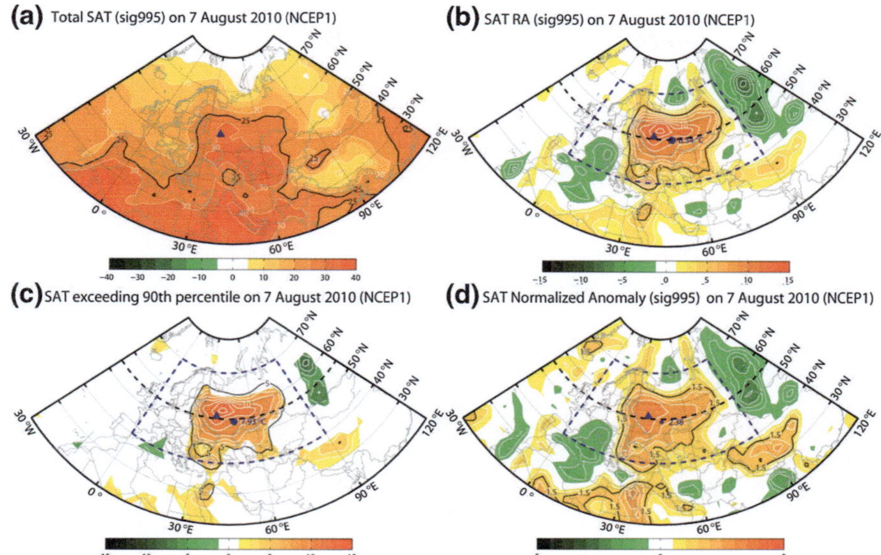

Fig. 7.4 Spatial distributions of **a** total SAT (*shading*, 5 °C interval), **b** SAT RAs (*shading*, 1.25 °C interval), **c** SAT RAs exceeding the 90th percentile and below the 10th percentile (*shading*, positive for above the 90th percentile, negative for below the 10th percentile), and **d** SAT NA (*shading*, 0.5 SD) on 7 August 2010 from NCEP R1 data. The *black solid line* indicates the 25 °C isotherm of total SAT in **a**, the 5 °C isotherm of SAT RAs in **b** and **c**, and the 1.5 SDs isotherm of SAT NAs in **d**, respectively. The *blue dot* is the daily weighted central location of the heat wave event. The value near the dot is the intensity averaged within the 5 °C isotherm in **b**, the 90th percentile in **c** and 1.5 SDs in **d**. The Moscow location is indicated by the blue triangle. The *blue dashed box* delineates western Eurasia (40–70°N, 0–90°E). A *black-dashed line* crossing the daily heat wave center denotes 55°N for vertical-zonal section analysis. Sourced from Qian et al. (2016a) © 2016 John Wiley & Sons, Inc., with permission

0.995-sigma level on 7 August 2010. Figure 7.4a highlights the areas where SATs exceed 25 °C (absolute threshold). The maximum SAT center, with values exceeding 40 °C, is located in the Middle East. Moscow is located in the northwestern part of the second SAT center. The differences between the coverage resulting from the three relative thresholds are shown in Fig. 7.4b–d. A large region with SAT relative anomalies (RAs) exceeding 5 °C at a spatial scale over 2000 km is centered near Moscow (Fig. 7.4b). Another small area with SAT RAs exceeding 5 °C is centered in the Middle East. The distribution of the SATs above the 90th percentile in Fig. 7.4c is similar to that in Fig. 7.4b. In Fig. 7.4d, normalized anomalies (NAs) of three positive areas exceed 1.5 standard deviations (SDs). The largest one is similar to the large heat wave areas in western Russia in Fig. 7.4b, c. However, there are also two areas with SDs greater than 1.5 in western China and the Middle East. The differences in coverage between the absolute threshold and the three relative thresholds are clearly illustrated in Fig. 7.4. A wave-like pattern as a result of the three relative thresholds is zonally-oriented in Eurasia, with a positive center over western Russia among two negative centers in Siberia and southern Europe. Thus, the RA and the 90th percentile thresholds consistently indicate the areas with heat waves in western Russia. In Siberia and southern Europe, two cold events occurred at the same time.

As an example, Qian et al. (2016b) calculated the daily central location of a heat wave in Fig. 7.4b by weighting all longitude or latitude grid points in the coverage (C) with positive SAT RAs exceeding 5 °C using the following formulas:

$$\text{lat} = \sum\nolimits_{i=1}^{C} (\text{RA}_i \times \text{lat}_i) / \sum\nolimits_{i=1}^{C} \text{RA}_i, \tag{7.1.1}$$

$$\text{lon} = \sum\nolimits_{j=1}^{C} (\text{RA}_j \times \text{lon}_j) / \sum\nolimits_{j=1}^{C} \text{RA}_j. \tag{7.1.2}$$

The lat_i and lon_j are the values of latitude i and longitude j of any grid points within the daily heat wave coverage (C) of SAT RAs exceeding 5 °C. The heavy blue dot near Moscow was the weighted central location of the heat wave on 7 August 2010, which had a regional-mean SAT RA of 8.54 °C in the coverage (Fig. 7.4b). The weighted central locations of the RA, the percentile, and the NA are all located to the east of Moscow. The regional-mean SAT RA is 7.93 °C in Fig. 7.4c and 2.38 SDs in Fig. 7.4d.

7.2 Summer Moving-Type Heat Waves[2]

Climatology with diurnal cycle and seasonal cycle has been the basic state used in the analysis of extreme rainfall and the track prediction of unusual tropical cyclones in our previous studies. For the events of daily SATA, daily mean atmospheric variables are used to find possible early signals. Since anomaly-based weather analysis has advantages over total-based weather analysis (Qian et al. 2013, 2014), the anomaly-based approach is used to study heat wave and cold events. An daily mean atmospheric variable $F_{d,y}(\lambda, \varphi, p, t)$ on calendar date t in a year y at a spatial point of longitude λ, latitude φ, and pressure level p can be decomposed into its daily mean climatic field $\tilde{F}_d(\lambda, \varphi, p)$ and temporal deviation $\tilde{F}'_{d,y}(\lambda, \varphi, p, t)$

$$F_{d,y}(\lambda, \varphi, p, t) = \tilde{F}_d(\lambda, \varphi, p) + F'_{d,y}(\lambda, \varphi, p, t) \tag{7.2.1}$$

[2]Modified article originally published in *Chen Y, Hu Q, Yang YM, Qian WH (2016) Anomaly based analysis of extreme heat waves in Eastern China during 1981–2013. Inter J Climatol. doi:* 10.1002/joc.4724. Published with kind permission © 2016 John Wiley & Sons, Inc. All Rights Reserved.

where the subscript d indicates daily mean. The daily mean climatic state is obtained by a 30-year (1981–2010) average of any observed or reanalysis variable on the calendar date t,

$$\tilde{F}_d(\lambda,\varphi,p) = \sum_{y=1981}^{2010} F_{d,y}(\lambda,\varphi,p,t)/30. \tag{7.2.2}$$

It is assumed that positive and negative anomalies of a variable at a specific grid point and at a calendar date cancel each other during the 30-year period to approximate the quasi-static climatic state, reflecting only the final equilibrium state between solar radiation and surface conditions. The climatic state defined by Eq. 7.2.2 varies from day to day and is different from monthly mean (or seasonal-mean) climatology. Monthly mean and seasonal-mean anomalies have been applied in short-term climate predictions by modeling (Xing and Chao 1982; Zebiak and Cane 1987) and empirical approach (Van den Dool 2007) as well as in climatological analysis of extreme events (Meehl and Tebaldi 2004). This is because climate is often defined using long-term means over a certain time interval, such as monthly, seasonal, and annual averages (Peixoto and Oort 1992).

Usually, the NCEP/NCAR R1 with a 2.5° × 2.5° latitude–longitude resolution (Kalnay et al. 1996) and the ERA-Interim with a 0.75° latitude–longitude grid (Dee et al. 2011) are mostly used in our analyses. Other data are the observed daily mean/maximum/minimum SATs at 549 stations in mainland China (Li and Yan 2009). The intensity of a heat wave is measured by the average of daily mean SAT (DMSAT) anomalies over the region. The severity of a heat wave event (a societal aspect) would depend on its intensity, duration, coverage, and location (populated urban setting versus remote and sparsely populated areas). Chen et al. (2016) focused on the regional heat wave events that have a spatial coverage larger than 300 × 500 km (1.5×10^5 km^2) and persist for at least 5 consecutive days in eastern China.

As an example, the most recent extremely hot summer in eastern China was in 2013 with several excessive heat wave events in June, July, and August (Tang et al. 2014). This example was also studied by Chen et al. (2016). On 18 June 2013, the DMSAT exceeded 35 °C along the Yangtze River valley with two centers in the upper and the middle valley, marked by "A" and "B," respectively (Fig. 7.5a). The DMSAT at those centers reached 40.0 °C, becoming one of the highest DMSATs for the upper Yangtze River valley since 1961.

For this event, Chen et al. (2016) first calculated the DMSAT climatology in the heat wave period using daily mean climatic state. After removing the daily mean climatic state, they obtained the DMSAT anomalies and show their spatial distribution in Fig. 7.5b. The DMSAT anomalies also show two centers in the upper and middle Yangtze River valley. The DMSAT anomalies exceed 5 °C in most of the area between 25°N and 35°N. Following the convention, they used the DMSAT and its anomaly as the "absolute" and "relative" thresholds to characterize the heat waves. The absolute threshold measures the prevailing local condition for heat waves, and

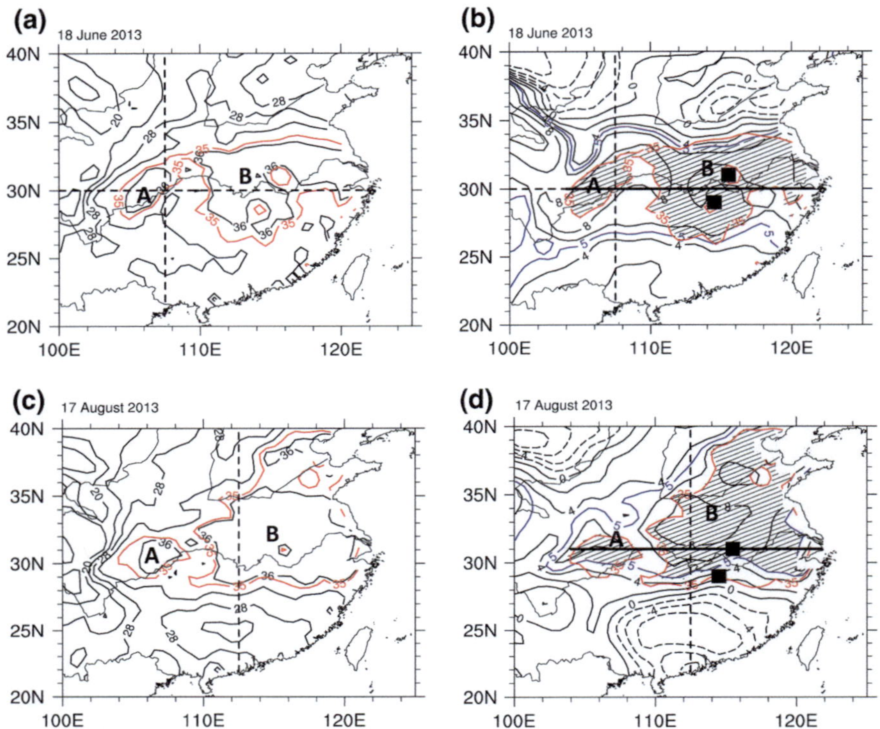

Fig. 7.5 Distribution of **a** DMSAT (*black-solid contour*, 4 °C interval) and **b** DMSAT anomalies (*black solid* and *dashed contour*, 2 °C interval) on 18 June 2013. The *vertical* and *horizontal-dashed lines,* respectively, mark 107.5°E and 30°N crossing the center A in **b**. The 35 °C *contour line* is highlighted in *red*. In **b**, the blue line denotes 5 °C anomaly and the *thick black line* along 30°N indicates the spatial length of heat wave location with DMSAT exceeding 35 °C and anomalies above 5 °C. Panels **c** and **d** are same as **a** and **b** except on 17 August 2013 with a heavy-dashed line along 112.5°E. The *shading area* in **b** and **d** indicates the heat wave region. Letters "A" and "B" represent two centers of DMSTA anomalies in **b** and **d**. *Black squares* mark the local mountains which are not included in the heat wave region. Sourced from Chen et al. (2016) © 2016 John Wiley & Sons, Inc., with permission

the relative threshold ensures that the event meets the intensity requirement. In that study, they focused on the regions where both the absolute threshold (DMSAT $\geq$ 35 °C) and the relative threshold (DMSAT anomaly $\geq$ 5 °C) are satisfied. Figure 7.5b shows such a region. It covers 8.2×10^5 km^2 with a regional average DMSAT anomaly of 7.29 °C.

A similar anomalous condition occurred on 17 August 2013 when high surface temperatures affected large portions of central and eastern China (Fig. 7.5c). There are two regions with DMSAT exceeding 35 °C (centered at "A" and "B" in Fig. 7.5c). The shaded regions in Fig. 7.5d are areas undergoing a heat wave. They are 1.2×10^5 and 5.7×10^5 km^2 in size and with average DMSAT anomalies of 6.07 and 7.08 °C, respectively.

Figure 7.6 shows the time series of the coverage of DMSAT anomalies ≥ 5.0 °C and the average intensity from 1 June to 31 August 2013. The intensity series (red line) stops when the coverage area is zero. The red line shows a sequence of six extreme heat processes. The first process ended on 4 June, and the last one was from 22 to 29 August. The longest process was from 23 July to 20 August. During these processes, three periods, the ones peaked on 19 June (marked by "A" in Fig. 7.6), on 11 July (marked by "B"), and on 17 August (marked by "C") had their spatial coverage, intensity, and duration. The DMSAT anomaly on 19 June was 7.81 °C, and it was 6.91 °C on 17 August 2013.

Two sets of vertical cross sections in the heat wave region on 18 June, one along 30°N from 45° to 150°E and the other along 107.5°E from 10° to 60°N (both across the heat wave region A in Fig. 7.5b), are shown in Fig. 7.7. Figure 7.7a displays the cross section of observed geopotential height and temperature (without decomposition) along 30°N. In Fig. 7.7a, observed geopotential heights at different altitudes are similar to their climatic values (black solid contour vs. black dashed contour) on that day. Observed temperature only deviates from its climatic value in the stratosphere and the lower troposphere over the surface heat wave region (red solid contour vs. red-dashed contour). In contrast, the anomalies in both geopotential and temperature in Fig. 7.7b reveal outstanding features characterizing the heat wave. There are two warm centers, marked as "W_1" and "W_2", of temperature anomalies in the lower troposphere right over the surface heat wave region. Along with these temperature anomalies, there are two centers (H_1 and H_2) of positive geopotential height anomaly in the upper troposphere. The center of the geopotential anomalies extends from the upper troposphere downward to the centers of positive temperature anomalies in the lower troposphere over the surface heat wave regions (marked by "A" and "B" in Fig. 7.5a, b). Clearly, the two centers of high temperature at surface are directly connected to the two centers of positive temperature anomalies in the lower troposphere and the two centers of positive geopotential height anomalies in the upper troposphere. Similar relationship of temperature and geopotential anomalies also occurred during the event on 17 August. Because the two centers (H_1 and H_2 in Fig. 7.7b) of positive geopotential anomalies in the upper troposphere are in the same synoptic system, the two shaded

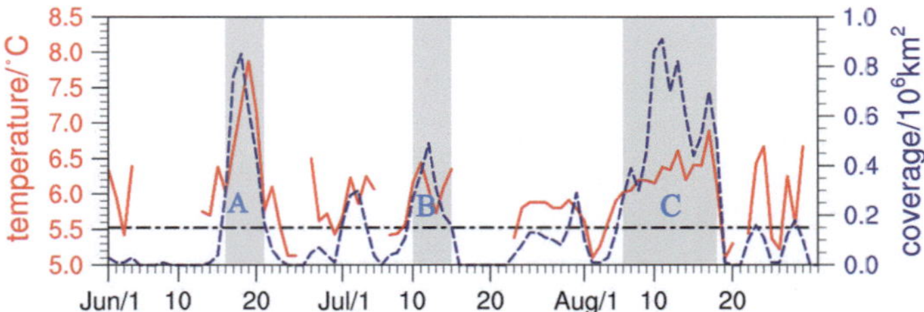

Fig. 7.6 Daily time series of the coverage (*blue-dashed line* with its unit 10^6 km^2) and the intensity (*red line* with its unit °C) from 1 June to 31 August 2013. The *grey bars* denote time periods of the three heat wave events indicated by letters "A," "B," and "C." Sourced from Chen et al. (2016) © 2016 John Wiley & Sons, Inc., with permission

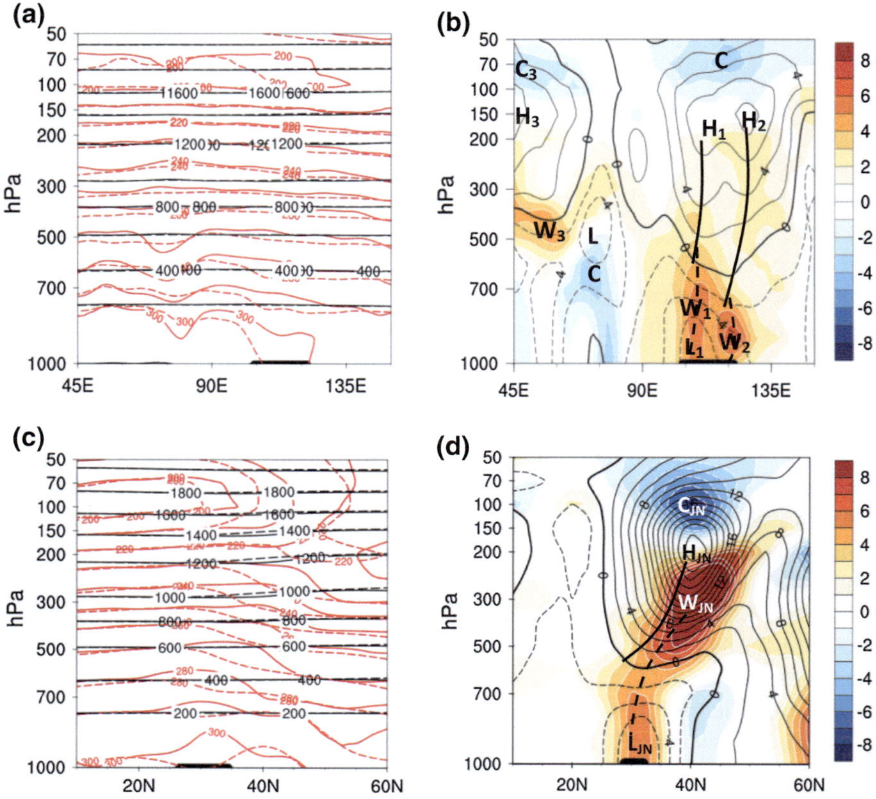

Fig. 7.7 **a** Vertical section of total and climatic geopotential heights (*black-solid* and *black-dashed contour* respectively, 200 × 10 gpm interval) as well as total and climatic temperatures (*red-solid* and *red-dashed contours*\ respectively, 10 K interval) along 30°N on 18 June 2013 using the NCEP R1. **b** Same as **a** except for height anomalies (*contour*, 2 × 10 gpm interval) and temperature anomalies (*shading*, 1 K interval). **c** and **d** are the same as **a** and **b** except along 107.5°E. Letters "H," "L," "W," and "C" denote centers of height and temperature anomalies. *Heavy solid lines* and *dashed lines* are axes of height anomalies and temperature anomalies, respectively. The *heavy bar* denotes the width of heat wave at the surface. Sourced from Chen et al. (2016) © 2016 John Wiley & Sons, Inc., with permission

regions (Fig. 7.5b) at the surface are simply showing the spatial variations of the system that produced the heat wave.

Figures 7.7c, d show the same contrast between the observed geopotential height/temperature and their anomalies along the vertical section crossing the heat wave center A in Fig. 7.5b. The striking differences between the anomalous fields in Fig. 7.7d and the un-decomposed fields in Fig. 7.7c again show the advantage of the anomalous fields in depicting the signature of the heat wave. The surface heat wave region is a downward extension of a positive temperature anomalous center (W_{JN}) below a strong positive center (H_{JN}) of geopotential height anomalies in the upper troposphere. The reason for the lack of capability to capture these features of heat wave by the un-decomposed values (in Fig. 7.7a, c) is that they contain the climatic component which overwhelms the anomalies superimposed on it. These differences in Fig. 7.7 show that those anomalies can describe the heat wave with less ambiguity.

The distinctive patterns in temperature and geopotential anomalies in Fig. 7.7d indicate that the surface heat wave is a downward extension of the positive temperature anomalies associated with strong positive height anomalies in the upper troposphere, around 200 hPa in this case. Meanwhile, there is a negative temperature anomalous center (C_{JN}) in the stratosphere over the surface heat wave region and is exactly above the positive height anomaly at 200 hPa, as required by the hydrostatic equilibrium. This vertical structure along 107.5°E in Figs. 7.5b and 7.7d is also found along 112.5°E in Fig. 7.5d crossing the other heat wave center B. They also noticed that the center of the geopotential height anomalies at 200 hPa is about 10° latitudes to the north of the surface heat wave center (Fig. 7.7d). This northward tilt of the vertical axis of the height anomalies suggests that the upper level center of height anomalies in North China is responsible for the surface heat wave in southeastern China. Thus, it is possible to identify and trace positive anomalies of geopotential height at 200 hPa and use them to provide early warnings for potential heat waves in China.

To illustrate this notion for heat wave prediction, they traced the geopotential height anomalous center (H_{JN}) of the heat wave event in Fig. 7.7d. The disturbance leading to this anomalous center was first identified on 5 June, marked by the green dot in Fig. 7.8a, 11 days before the onset of that heat wave. From 5 to 11 June, this disturbance moved eastward from Northeast China to the Japan Sea (Fig. 7.8b, c). On 15 June (Fig. 7.8d), it split into two anticyclonic systems: one in the west moving southwestward to North China and one in the east stable over the Japan Sea. On the following days, the latter moved to the east of Japan. The former stalled in North China (Fig. 7.8e, f), resulting in the development of the surface heat wave on 16 June.

Similar processes can be traced using the anomaly-based method for the heat wave occurred in mid-August 2013. On 11 August, a region of positive geopotential height anomalies appeared over Mongolia. The height anomalous center is marked by a green dot and letter "H_{A3}" in Fig. 7.9a. On the following days, the positive anomalous region expanded and intensified while moving southeastward slowly to northeastern China. It stalled in eastern China on 15 and 16 August, resulting in development of the heat wave on 17 August.

Before the establishment of "H_{A3}" in the August heat wave, there was a sequence of positive geopotential height anomalies that had been already influencing eastern China and caused hot days. Tracks of all those disturbances or centers of height anomalies are shown in Fig. 7.10a, b. This heat wave was an outcome of the sequence of those anomalies, which explained the multiple temperature peaks in Fig. 7.6 during that heat wave event. On 22 July (Fig. 7.10a), 15 days before the onset of the heat wave, a high pressure anomalous center (H_{A1}) appeared in western China. This center moved east and then south in the following days till 4 August. After that, it reversed northward and westerward and arrived at

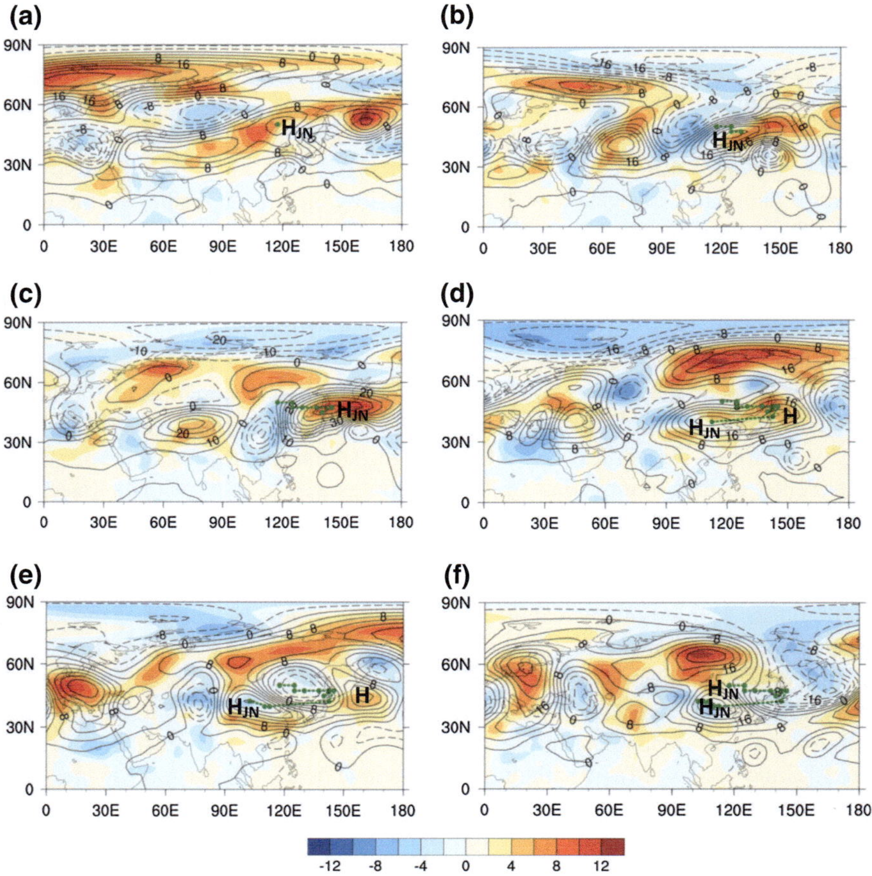

Fig. 7.8 Anomaly-based synoptic charts showing the height anomalies at 200 hPa (positive in *solid contour* and negative in *dashed contour*, 4 × 10 gpm interval), and the temperature anomalies at 850 hPa (*red shading* for positive and *blue shading* for negative, 2 K interval) on **a** 5, **b** 8, **c** 11, **d** 15, **e** 18, and **f** 21 June 2013 using the NCEP R1 product. Letter "H" indicates the positive centers of height anomalies at 200 hPa. The *green line* is the track of height anomalous center H_{JN}. Sourced from Chen et al. (2016) © 2016 John Wiley & Sons, Inc., with permission

North China on 6 August when the heat wave began. The height anomaly (H_{A1}) that hit China in August followed a relatively complex X-shaped track, which increased difficulty in its predition. After its influence ended on 9 August, the second (H_{A2}) and the third (H_{A3}) high anomalous centers moved into the region from Kazakhstan to Xinjinag (far western China) on 10 and 15 August, respectively, and prolonged that heat wave event (Fig. 7.10b).

A similar issue also presented itself in the heat wave in June 2013. The disturbance of that heat wave followed a zigzag track before arriving in North China (H_{JN} in Fig. 7.10c). The key point in this track was on 15 June after the disturbance split into two centers, and the west one moved southwestward to China. After

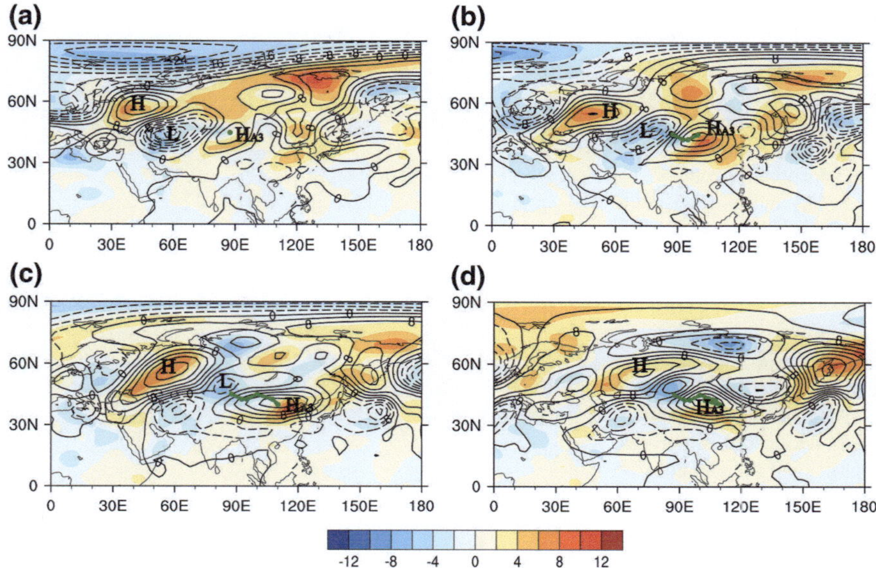

Fig. 7.9 Same as Fig. 7.8, but on **a** 11, **b** 14, **c** 16, and **d** 18 August 2013, respectively. Sourced from Chen et al. (2016) © 2016 John Wiley & Sons, Inc., with permission

arriving in the North China, it again stalled for about a week to cause the heat wave event. For the heat wave event in July (Fig. 7.10d), the signal (H_{JL}) occurred on 9 July, one day before the onset date, and stalled in the North China for 7 days. To predict these processes, it is needed to apply this anomaly-based method to medium-range model products which contain over 2-week long predictions of geopotential height field. After extracting the heights anomalies from model predictions, one can detect and trace the disturbance centers in the predicted weather maps (as that in Fig. 7.9) and provide early warnings for potential heat waves.

Using the definition of heat wave, Chen et al. (2016) identified 15 heat wave events in eastern China in the 33 years from 1981 to 2013. The daily intensity and coverage series during these summers are shown in Fig. 7.11 (except for the summer in 2013 which is shown in Fig. 7.6). In the 1980s, prolonged heat waves occurred in 1981 and 1988. In the 1990s, one event was in 1994 and one in 1997. In contrast to the four heat wave events in the 1980s and 1990s, there was a sharp increase in the number of heat wave events in the twenty-first century when eleven events occurred in 8 of the 13 summers, two in 2006 and three in 2013. All heat wave events in eastern China during the 33 years are summarized in Table 7.1. The average duration of the heat waves is 6.8 days, with the longest event lasting for 13 days.

The more frequent occurrences of heat waves in the 13 years of the twenty-first century suggest average higher temperatures in those summers. Indeed, according to our definition, there were 12 heat wave days in 2006 and 25 days in 2013, and the summers of those years were among the hottest in the record of eastern China.

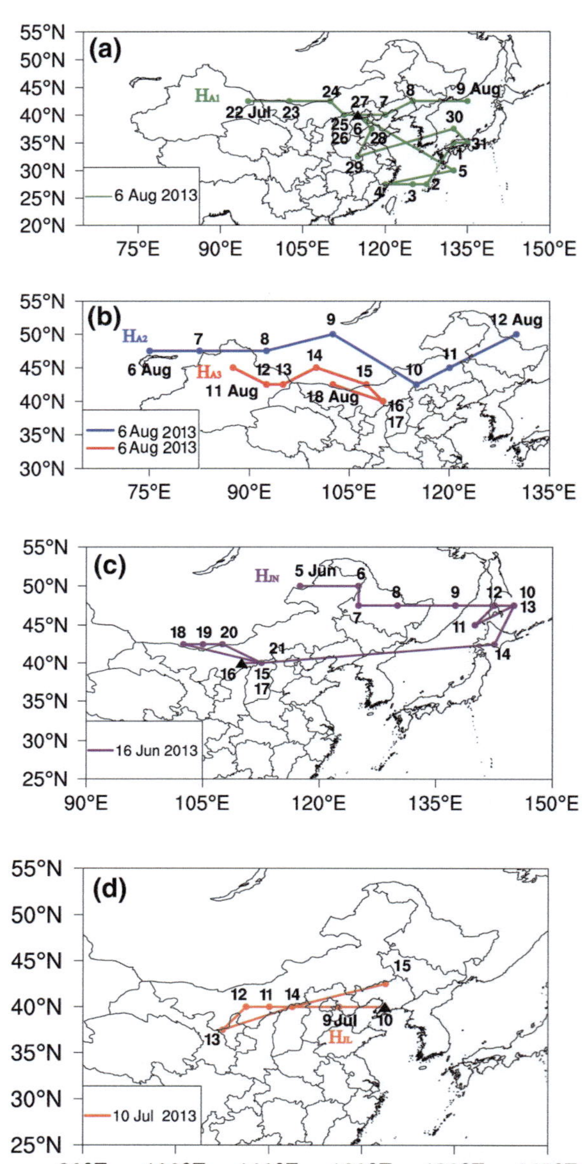

Fig. 7.10 Tracks of height anomalous centers for three heat wave events in 2013 using the NCEP R1 product. Tracks in **a** and **b,** respectively, correspond to three disturbance centers (H_{A1}, H_{A2} and H_{A3}) related to the heat wave in August 2013. Tracks in **c** and **d** depict the movement of two disturbance centers (H_{JN} and H_{JL}) that are, respectively, related to the heat waves in June and July 2013. For each track, the start date denotes the day when the height anomalous center first appears, and the end date indicates the time when the center weakens or an event ends. The onset days of heat wave events are annotated in the *left corner*. *Black triangle* marks the position of anomalous center on the onset day of a heat wave event. Sourced from Chen et al. (2016) © 2016 John Wiley & Sons, Inc., with permission

While the frequency has increased, the duration of heat wave events is similar in the recent decades, except for 2013.

For each heat wave in Table 7.1, Chen et al. (2016) examined the vertical and horizontal distributions of geopotential height and temperature anomalies. The results show that every heat wave event is a direct consequence of strong positive temperature anomaly in the air column between the surface and a positive center of geopotential height anomalies in the upper troposphere. Since these centers of

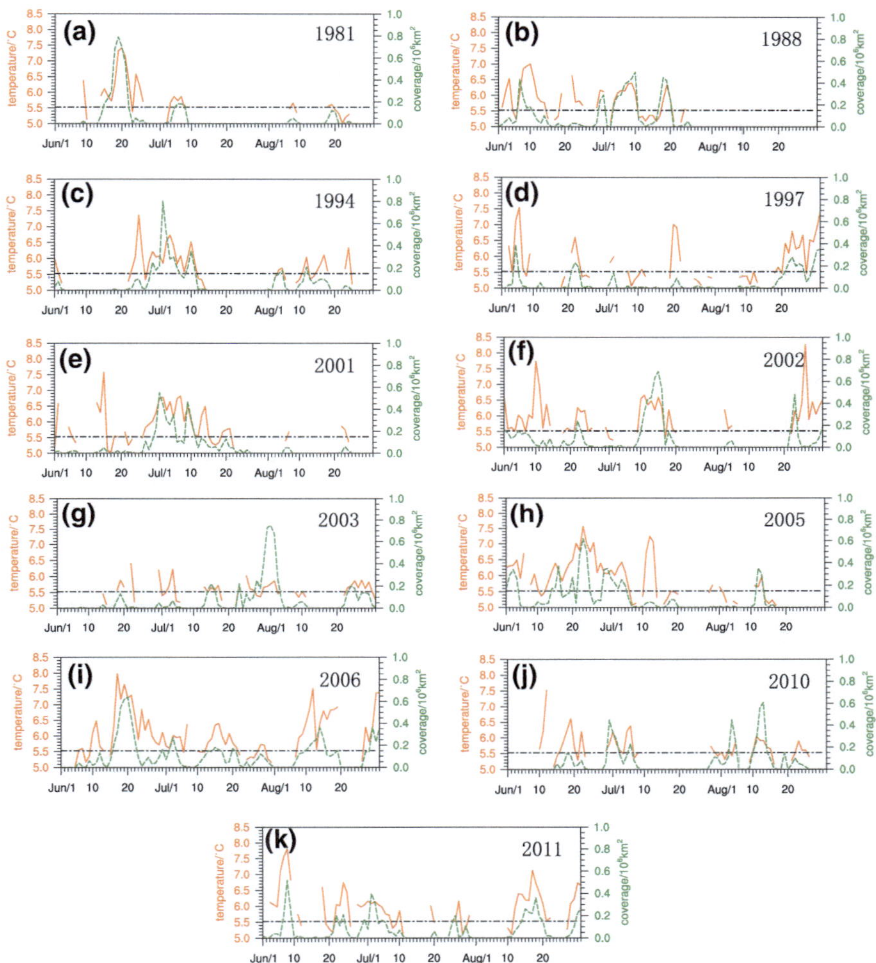

Fig. 7.11 Same as in Fig. 7.6, but for **a** 1981, **b** 1988, **c** 1994, **d** 1997, **e** 2001, **f** 2002, **g** 2003, **h** 2005, **i** 2006, **j** 2010, and **k** 2011. Sourced from Chen et al. (2016) © 2016 John Wiley & Sons, Inc., with permission

height and temperature anomalies represent concentrated heat by synoptic-scale eddy processes, the increasing frequency of heat waves can imply an increase in such processes since the twenty-first century.

The disturbance tracks of the 15 heat wave events listed in Table 7.1 are shown in Figs. 7.10 and 7.12. Figure 7.10 shows the three events in 2013 and Fig. 7.12 shows the tracks of the other 12 events. By examining these tracks, they found three focused areas for initial appearances of height and temperature anomalies leading to those heat waves: Eastern Europe, Central Asia (mainly in Uzbekistan and Kazakhstan) and Eastern Asia (around Mongolia). For the 12 events in years other than 2013, Chen et al. (2016) divided them into three groups according to the initial

Table 7.1 Heatwave events in eastern China over 1981–2013

Year	Start date	Peak Date	Duration	Year	Start date	Peak date	Duration
1981	15 Jun	20 Jun	7	2006	16 Jun	17 Jun	7
1988	4 Jul	9 Jul	7	2006	11 Aug	12 Aug	5
1994	29 Jun	4 Jul	8	2010	10 Aug	11 Aug	5
1997	22 Aug	23 Aug	5	2011	14 Aug	17 Aug	7
2001	30 Jun	2 Jul	6	2013	16 Jun	19 Jun	6
2002	11 Jul	11 Jul	6	2013	10 Jul	11 Jul	6
2003	28 Jul	2 Aug	7	2013	6 Aug	17 Aug	13
2005	29 Jun	4 Jul	7	Average	15 Jul	18 Jul	6.8

Sourced from Chen et al. (2016) © 2016 John Wiley & Sons, Inc., with permission

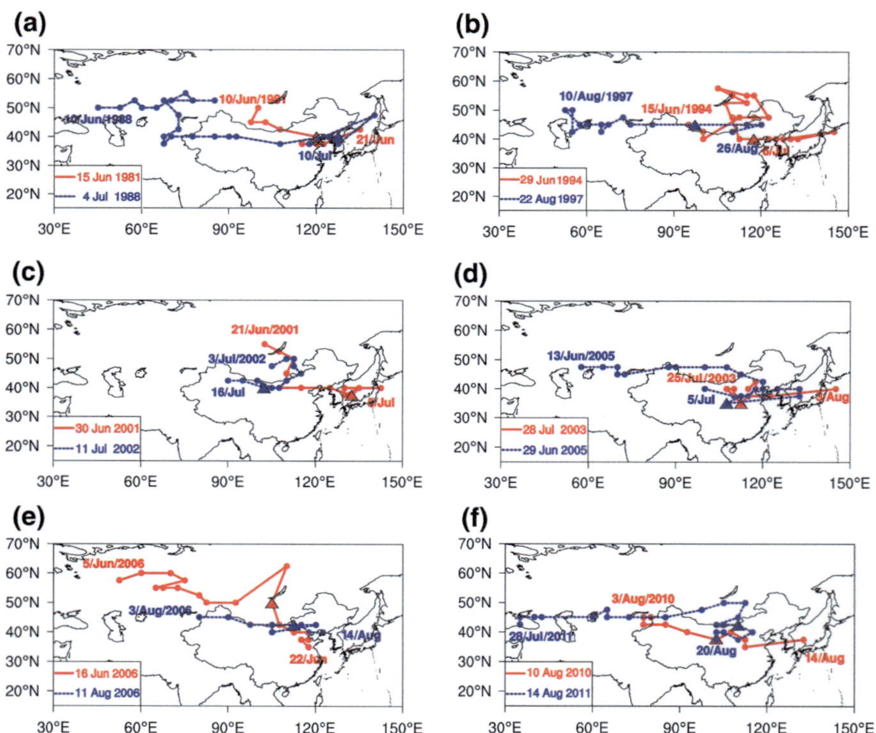

Fig. 7.12 Same as in Fig. 7.10 but for height anomalous center tracks of other 12 heat wave events. Sourced from Chen et al. (2016) © 2016 John Wiley & Sons, Inc., with permission

locations of their geopotential anomalies, and analyze each group for their development in detail.

The first group includes the events in July 1988 (Case 1), June 2006 (Case 2), and August 2011 (Case 3), which developed from disturbances originating in

Europe. The track of Case 1 is shown by a blue line in Fig. 7.12a. The initial disturbance was in the Volga River valley on 10 June 1988, 24 days before the onset of the heat wave event in eastern China. It propagated eastward, turned southward for 9 days, and then to the east to the Japan Sea. On 3 July 1988, the system was forced westward, apparently going through a process similar to that in the event in June 2013 (Fig. 7.10b). It stalled in eastern China for a week and caused a local heat wave.

The red line in Fig. 7.12e shows an overall northwest–southeast track for Case 2. In this case, the heat wave started in eastern China on 16 June when the disturbance was forced southward from Mongolia. After that, this disturbance continued to move southward while intensifying and spreading the heat wave.

Case 3 went through a course rather similar to that in Case 2. The initial disturbance developed on 28 July 2011 in Eastern Europe. It propagated eastward along the blue line track in Fig. 7.12f to central Mongolia on 10 August. Then it turned southward and moved to North China on 14 August when the heat wave started. After that, this disturbance continued spreading southward into North Central China. Similar patterns in vertical cross sections are shown and the axes of height and temperature anomalies tilt to North China in the vertical direction.

The second group of the heat waves consists of the events in August 1997 (Case 4), June 2005 (Case 5), August 2006 (Case 6) and August 2010 (Case 7). Their initial disturbances can be traced to central Asia and, in general, propagated from west to east. The two tracks of Case 4 and Case 5 are shown by the blue lines in Fig. 7.12b, d, respectively, starting over Uzbekistan and propagating eastward into Mongolia over a week. Both cases show a forced reverse in propagation direction (to the west), shortly before or after their corresponding heat waves set off. Case 6 (blue line in Fig. 7.12e) and Case 7 (red line in Fig. 7.12f) have their tracks over Kazakhstan about a week before the onset of their corresponding heat wave events. These disturbances propagated eastward and amplified, and later stalled over North China causing week-long heat wave events.

The third group has the remaining five events occurred in June 1981 (Case 8), June 1994 (Case 9), June 2001 (Case 10), July 2002 (Case 11), and July 2003 (Case 12). They are characterized by a track starting from eastern Asia. In Case 8 (red line in Fig. 7.12a), the track started in Mongolia 5 days before the onset date of the heat wave. It is interesting that this case also showed a reversed propagation a few days after the heat wave started. A similar process occurred in Case 10 (red line in Fig. 7.12c). It is noted that the centers of height anomalies at 200 hPa in Cases 8 and 10 were located to the northeast of the surface heat wave, largely owing to northeast upward tilt of the height and temperature anomalies.

For the rest of the cases in this group, the tracks of Case 9 (red line in Fig. 7.12b) and Case 11 (blue line in Fig. 7.12c) started over Mongolia. They had rather complex tracks with varying propagation speed and direction before arriving in North China. In Case 12 (red line in Fig. 7.12d), its track initiated in North China and developed locally for 3 days before the heat wave event started. In the wake of that event, the anomalies in height and temperature faded while propagating eastward. In Cases 9, 11, and 12, the locations of upper level anomalous centers in height and temperature were also found northwest of the surface heat wave.

To summarize, it has been shown that it is possible to find and trace the disturbances that develop into heat waves in China. In traditional synoptic charts, heat waves are under strong ridge (blocking high) which has a strong center on the north of the dome of warm air. Considering the advantage of the anomaly-based weather analyses, a composite of height anomalies at 200 hPa and temperature anomalies at 850 hPa on the day of peak intensity for the 15 heat wave events in Table 7.1 is shown in Fig. 7.13. At 200 hPa, positive height anomalies are in the eastern and Central China. Negative height anomalies are in the northeast and south of the

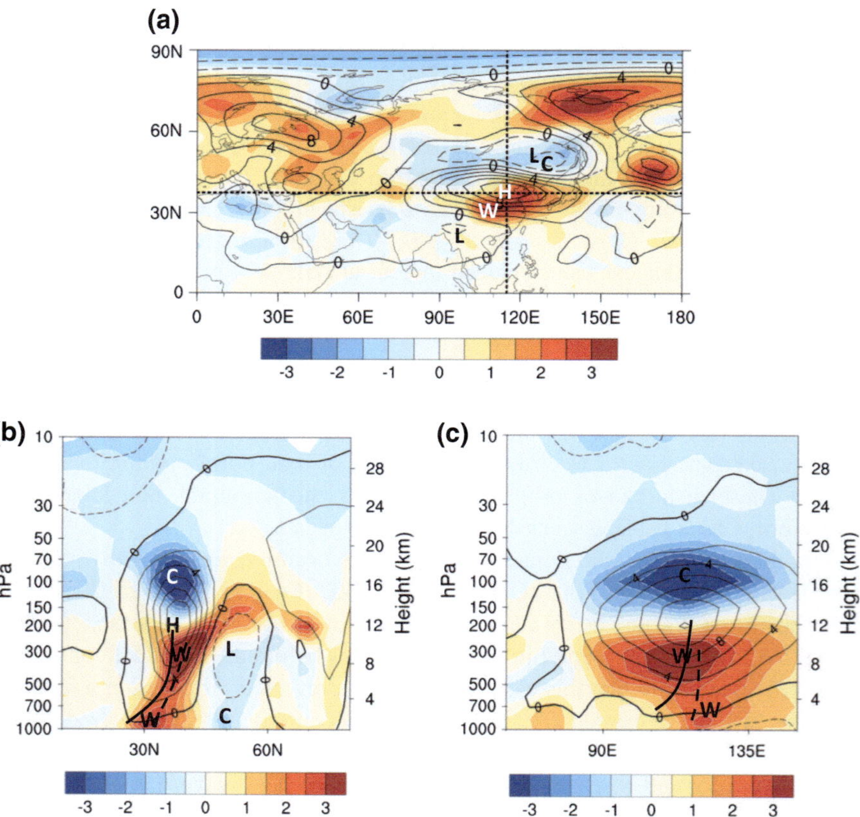

Fig. 7.13 **a** Composite height anomalies (*contour*, 2 × 10 gpm interval) at 200 hPa and temperature anomalies (*shading*, 0.5 K interval) at 850 hPa for 15 heat wave events on their peak dates using the NCEP R1 product. In **a**, the thick-dashed lines denote 115°E longitude and 37.5°N latitude respectively. Vertical cross sections of composite height (*contour*, 2 × 10 gpm interval) and temperature (*shading*, 0.5 K interval) anomalies along **b** 115°E longitude and **c** 37.5°N latitude. In **b** and **c**, the *heavy solid* and *dashed lines* indicate the axes of height and temperature anomalies respectively. Sourced from Chen et al. (2016) © 2016 John Wiley & Sons, Inc., with permission

positive anomalous region, with the one in the northeast being relatively stronger. Meanwhile, at 850 hPa, a positive center of temperature anomalies develops under the 200-hPa positive height anomaly (slightly off to its southwest), and negative anomaly in temperature is found under the negative height anomaly at 200 hPa. These overlapped positions of the height and temperature anomalies at the peak intensity of the heat waves are further shown in the vertical cross sections along 115°E (Fig. 7.13b) and 37.5°N (Fig. 7.13c). They both show a northeastward tilted vertical axis of height/temperature anomalies over the surface heat wave region. This composite structure of the heat waves shows an advantage of the anomaly-based analyses in diagnosing and predicting heat waves in China.

7.3 Heat Waves in Model Predictions[3]

Current forecasts of heat waves, as well as other extreme weather conditions, rely on predictions made by numerical weather prediction (NWP) models at major operational centers, such as the European Centre for Medium-Range Weather Forecasts (ECMWF) and the U.S. National Centers for Environmental Prediction (NCEP) Global Forecast Systems (GFS). These models provide 1–2 week predictions of daily atmospheric temperatures, geopotential height, moisture, winds, and precipitation. These variables, depicting conditions in the atmosphere and at the surface, have to be examined and integrated with conditions special for regions to make predictions of heat waves. In China, such post-processing in making heat wave predictions remains in development.

The case studies in the previous sections show that initial height and temperature anomalies leading to the heat wave events in eastern China can be traced back 1–2 weeks before the onset of those events using our anomaly-based analysis method. Chen et al. (2016) applied this method to model predictions and showed that this method can aid us in locating heat wave using model predictions. The model predictions are from the NCEP and ECMWF, which are employed in comparison with the NCEP reanalysis and the ERA-Interim products. Chen et al. (2016) used the heat wave from 16 to 21 June 2013 in eastern China as an example to illustrate the utility of the anomaly-based method in interpreting the model predictions.

As shown in Fig. 7.6, the strongest heat wave event occurred in June 2013, which has been analyzed in detail using the reanalysis products. Recall that for the case in June 2013, the anomaly-based method identified the initial height anomalies in the reanalysis data 11 days before the onset of the heat wave (Figs. 7.8 and 7.10c). Applying the same method to the NCEP predictions (data), Chen et al.

---

[3]Modified article originally published in *Chen Y, Hu Q, Yang YM, Qian WH (2016) Anomaly based analysis of extreme heat waves in Eastern China during 1981–2013. Inter J Climatol. doi:* 10.1002/joc.4724. Published with kind permission © 2016 John Wiley & Sons, Inc. All Rights Reserved.

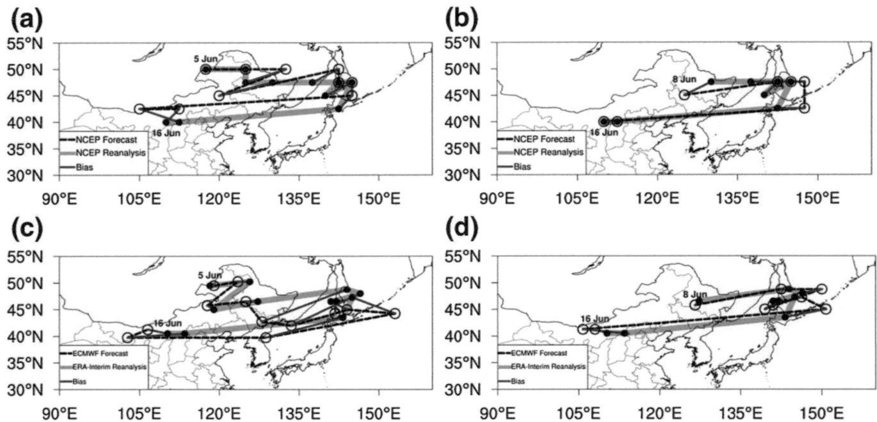

Fig. 7.14 Comparisons between the reanalysis tracks (*solid grey lines*) and the predicted tracks (*dashed black lines*) of height anomalous centers. In **a** and **b**, the predicted tracks are made by NCEP, respectively, initiated at 1200 UTC on 4 June and 8 June 2013, compared with the NCEP R1 product. Panels **c** and **d** are the same as **a** and **b**, but made by ECMWF, compared with the ERA-Interim product. The *blue lines* indicate the prediction deviation. Sourced from Chen et al. (2016) © 2016 John Wiley & Sons, Inc., with permission

(2016) found that the NCEP is able to predict the center location of geopotential height anomaly at 200 hPa on each day from 5 to 16 June 2013. Figure 7.14a, b compare the tracks of anomalous centers extracted from the NCEP reanalysis product and NCEP prediction. The model prediction was made at 1200 UTC on 4 June 2013 for the result shown in Fig. 7.14a and on 8 June for the result shown in Fig. 7.14b. The zigzag track from 5 to 16 June was well predicted 12 days before the onset of the heat wave, although there were deviations in exact location on those days. Similar analysis is made for the same event using predictions from the ECMWF and compared with tracks derived from ERA-Interim product. The results are shown in Fig. 7.14c, d. Although ECMWF also predicted a zigzag track on 4 June (12 days before the onset of the heat wave), the predicted center location of the anomalies (Fig. 7.14c) had larger deviations than those made by NCEP (Fig. 7.14a).

Chen et al. (2016) further examined the deviation between the predictions and the reanalysis products in horizontal distributions of height anomalies on the (16 June) of this heat wave. For the central location of height anomalies on 16 June, the deviation between the NCEP prediction and the NCEP R1 was smaller than that between ECMWF prediction and ERA-Interim product (blue lines in Fig. 7.14b, d). It is impressive that these predictions are made 8 days ahead of the event and the deviation is less than 2.5° in latitude and longitude. Moreover, both models predicted negative height anomalies at 200 hPa to the west of the positive anomalous center (Fig. 7.15a, b). In the vertical direction, northward tilted profiles of height and temperature anomalies predicted by the models (Fig. 7.15c, d) are comparable to the reanalysis data (Fig. 7.7d).

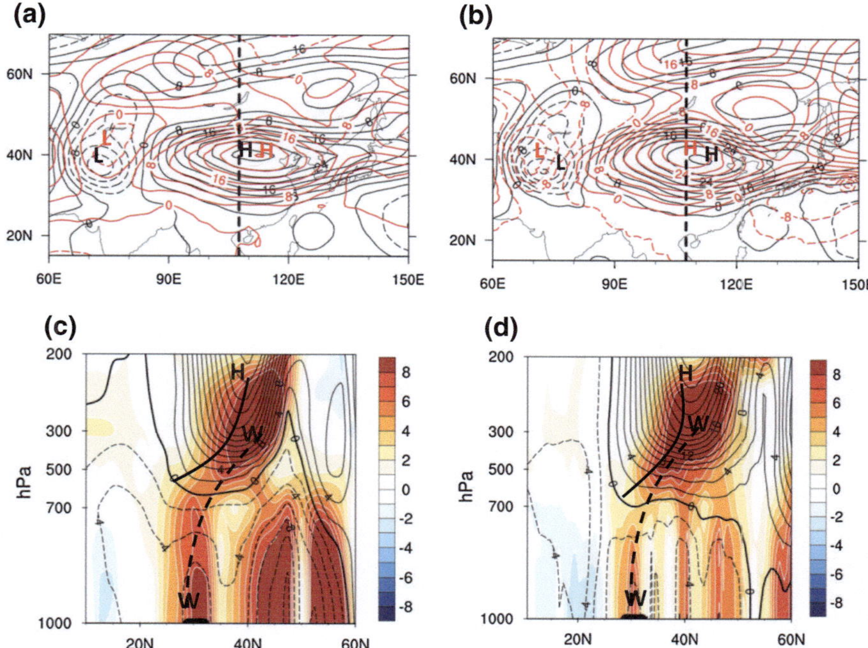

Fig. 7.15 **a** Height anomalies at 200 hPa on 16 June 2013 derived from NCEP R1 product (*black contour*, 4 × 10 gpm interval) and the NCEP prediction (*red contour*, 4 × 10 gpm interval) initiated at 1200 UTC 8 June 2013. **b** Same as **a**, but from the ERA-Interim dataset and the ECMWF prediction initiated at 1200 UTC 8 June 2013. Panels **c** and **d** are the same as Fig. 7.7d, but at 1200 UTC 16 June and derived from the NCEP and ECMWF predictions initiated at 1200 UTC 8 June 2013 and related to their reanalysis products. Sourced from Chen et al. (2016) © 2016 John Wiley & Sons, Inc., with permission

Figures 7.16 and 7.17 give the accuracy in the predictions for central locations of geopotential height anomalies at different initial time in NCEP and ECMWF predictions, respectively. If the predicted location of the NCEP (ECMWF) is $\leq 2.5°$ in latitude and longitude from the location derived from the NCEP R1 (ERA-Interim) product, or in other words, it falls within 3 × 3 (7 × 7 for the ECMWF prediction) grid array centered over the location derived from the reanalysis data, Chen et al. (2016) considered it as a "correct" prediction. Figure 7.16a shows that the central location of height anomalies on 5 June was correctly predicted by the NCEP at 1200 UTC on 3 June. The central location on 6 June was also correctly predicted at 1200 UTC on 3 June. From 7 to 10 June, the center of height anomalies was predicted but its location is off from the location derived from the reanalysis by up to 5° in latitude or longitude. As for the onset date on 16 June, the central location of height anomalies was correctly predicted on 4 June (12 days in advance). From Figs. 7.16 and 7.17, they found that both models can predict the occurrence of the height anomalous center 2 days in advance (i.e., on 3 June), and its propagation from 5 to 16 June was predicted since 3 June. But neither of the two models can always accurately capture

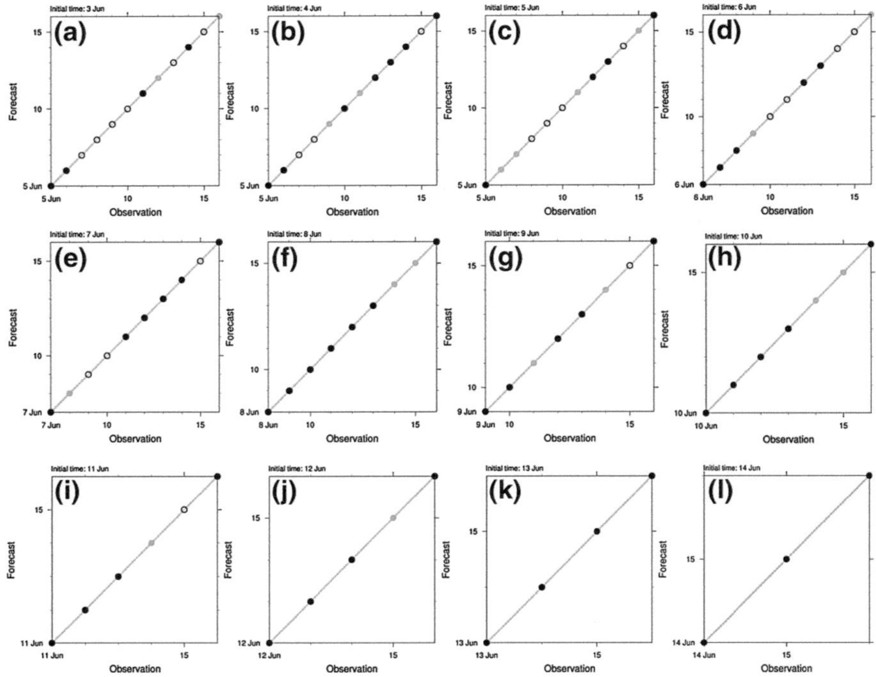

Fig. 7.16 Accuracy of the central locations of height anomalies in NCEP prediction, respectively, initiated from 1200 UTC 3–14 June 2013. *Black dots*, *circles*, and *gray dots*, respectively, indicate that the deviation between the predicted and the NCEP R1 product derived central locations is $\leq 2.5°$ (within 9-point matrix), >5°, and between them in latitude and longitude. Sourced from Chen et al. (2016) © 2016 John Wiley & Sons, Inc., with permission

the location of this center and deviations in predictions exist. For the NCEP prediction results, the largest deviation appeared in the prediction for 7 June at initial time 1200 UTC 3 June (Fig. 7.16a). The predicted central location was 15° to the east of the center derived from the NCEP R1 product. For ECMWF prediction, the largest deviation occurred at initial time 1200 UTC on 5 June (Fig. 7.17d), when the predicted center on 14 June was 18° longitude to the east of the one derived from the ERA-Interim product. However, the predicted horizontal distributions of geopotential height anomalies are comparable to those from the reanalysis even though deviations between the model predicted and reanalysis data derived central locations can be large.

These evaluations of the two models for the development of the height anomalies leading to the heat waves suggest a potential utility of the anomaly-based method. It can help improve interpretation of model products to identify and trace early signals of heat waves in eastern China. This utility certainly relies on the model ability to accurately predict the location of the early disturbance center. When the model fails to predict the location of the disturbance, this analysis method is helpless.

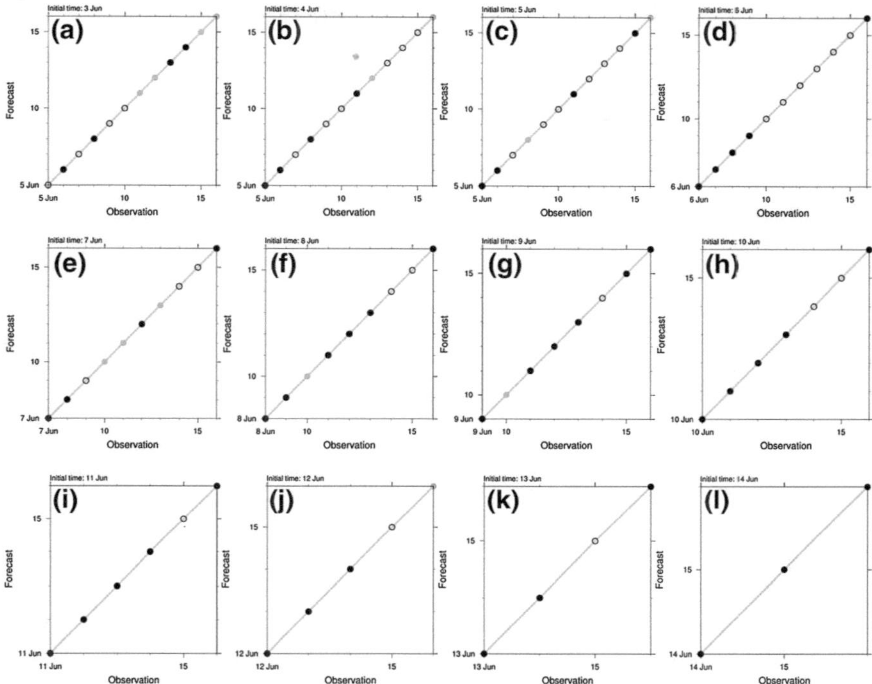

Fig. 7.17 Same as Fig. 7.16, but the ERA-Interim dataset and the ECMWF prediction. Sourced from Chen et al. (2016) © 2016 John Wiley & Sons, Inc., with permission

7.4 Spring and Autumn Low Temperature Events[4]

One of the high-impact weather events adversely affecting agriculture in the mid- to lower Yangtze River valley in China is anomalous low temperature (LT) persisting in the valley for 3 or more days during spring or autumn. Spring and autumn low temperature events in the mid- to lower Yangtze River valley in China were studied by Qian et al. (2015). The climate in southern China allows for double or even triple cropping systems for rice each year. Early April is when the spring rice crop is planted. The crop grows to maturity and is harvested in July. Following the harvest, the autumn rice crop is planted in early August, and harvested in middle October. Persistent LT events for prolonged periods in April and May can severely damage sprouting in rice growth, leading to replanting or loss of crops. Similarly, prolonged LT events in late September and early October will stop the flowering and kernel filling and ruin the autumn rice production. In spring, such events of prolonged

---

[4]Modified article originally published in *Qian WH, Chen Y, Jiang M, Hu Q (2015) An anomaly-based method for identifying signals of spring and autumn low temperature events in the Yangtze River Valley, China. J Appl Meteorol Climatol 54: 1216–1233*. Published with kind permission © American Meteorological Society. All Rights Reserved.

period of unseasonably cold temperature are associated with cold surges that interrupt the climatic (or normal) spring warming process. They are often referred to as "returning-cold". In autumn, similar abnormal events, occurring when cold air dips south with strong and cold winds, are called "cold dew wind", because of the accompanied strong cold wind during the events. These meteorological events have been recognized, defined, and documented at the China Meteorological Administration (State Standardization Committee 2012). They have been examined and studied over years to provide warnings to farmers who may apply mitigation strategies to reduce the impacts of such events on rice production (Du 2005; Wang et al. 2009; Chen et al. 2013).

Because such events are caused by extreme cold SAT anomalies of the spring or fall seasonal average, the strategy is to find a connection between the cold SAT anomalies and anomalies of other atmospheric variables, while the latter can be derived from observations and NWP model outputs. Although a standardized-anomaly analysis, which is suitable for all climate zones, has gradually attracted the attentions of the forecasters and researchers alike in the last decade (Easterling et al. 2000; Yan and Yang 2000; Grumm and Hart 2001; Hart and Grumm 2001), it has not been applied in forecasting the extreme cold events in crop growing seasons in China. The study area of Qian et al. (2015) for the LT events focuses on the regions of the Chongqing District, Hunan, Jiangxi, Hubei, and Zhejiang provinces, southern parts of Anhui and Jiangsu provinces, and the northern part of Fujian province (the boxed area in Fig. 7.18). Within this area, there are 88 stations among the 549 stations in mainland China.

April is the vegetative growth season after the spring rice crop is planted in the mid- to lower Yangtze River valley. During this time, the spike differentiation in rice growth will be inhibited if the daily mean SAT is colder than 12 °C according to the study of Chen et al. (2013). For autumn rice crop in the valley, its flowering and filling processes occur in September. In this period, if the daily mean SAT is

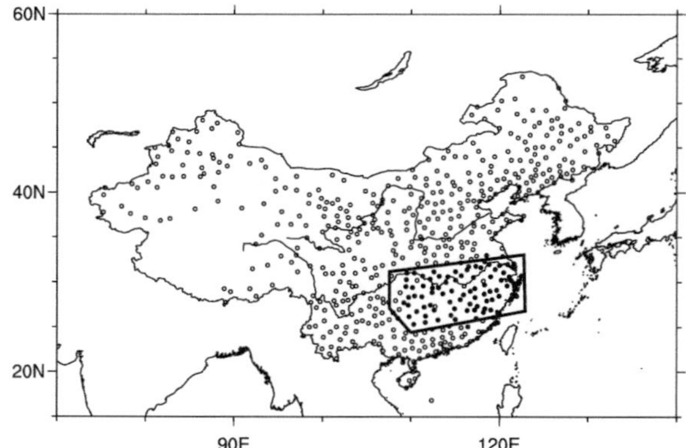

Fig. 7.18 The 549 surface observational stations in mainland China (*open* and *solid circles*). The *solid circles* mark the 88 stations inside the *boxed* study domain. Several *open circles* inside the *box mark* the stations in complex terrains near the Yangtze River and are excluded from this study. Sourced from Qian et al. (2015) © American Meteorological Society, with permission

colder than 20 °C, such processes may be terminated and resulting in empty and shriveled grains, and even loss of the harvest. Because of this critical role of the low temperature in this period, the temperature threshold from 1 September to 10 October of each year has been documented by China Meteorological Administration for agricultural use (Feng et al. 1985).

To provide a background or reference for the LT events in the two seasons, Fig. 7.19 shows the climatology (1979–2008) of daily mean SAT averaged from the 88 stations along the valley in spring and autumn. In spring (Fig. 7.19a), the daily mean SAT has exceeded 12 °C since 26 March. Climatologically, the temperature will smoothly increase from that day onward, and reach 13.8 °C on 1 April. Thus, when the daily mean temperature at one station is colder than 12 °C in April or May, the area represented by the station has a cold surface with damagingly low temperature for spring rice crop. In autumn (Fig. 7.19b), the daily mean SAT gradually decreases from above to below 20 °C on 8 October. When the daily mean temperature at a station is colder than 20 °C in late September or early October, the area represented by that station has a cold surface with damagingly low temperature for autumn rice crop.

In a previous study (Chen et al. 2013) and government document (Feng et al. 1985), daily mean temperature colder than 12 °C in spring and colder than 20 °C in autumn were the criteria of LT events. Three criteria in time, space, and intensity were used to determine an LT event by Zhang and Qian (2011). In this study, a LT event is defined as a period of three or more consecutive days (time) over an area with five or more adjacent stations (space) of daily mean temperature colder than

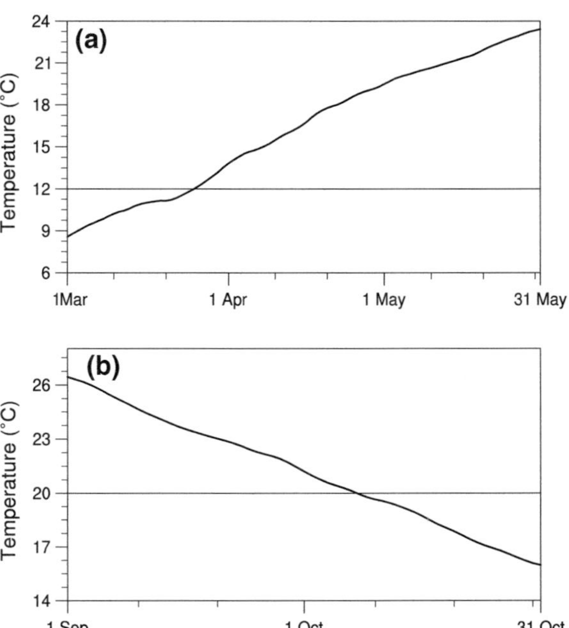

Fig. 7.19 Climatic daily mean SAT (°C) averaged from all 88 stations along the valley in **a** spring (1 March to 31 May) and **b** autumn (1 September to 31 October). Sourced from Qian et al. (2015) © American Meteorological Society, with permission

12 °C in spring and colder than 20 °C in autumn having a negative deviation (intensity) to the climatology in the past 30 years (1979–2008).

Using these data and the definition, the historical autumn LT events have been identified. Furthermore, these LT events are categorized into cold dry events and cold wet events according to the absence or presence of precipitation during the LT events (State Standardization Committee 2012). The difference between the cold wet and cold dry events is that the former occur when cold air invades the valley from the north and forms a stationary front along the valley with a warmer air mass in its south. Along the stationary or quasi-stationary front, it is rainy under the low surface temperatures. The latter occurs in the absence of such stationary front. The weather processes in the cold dry events are primarily strong cold air mass moving from the north to the valley and even to the south China coast. The entire region including the river valley and southern China is under a cold-high pressure system. The cold-high pressure system brings, in addition to the cold temperature, clear skies and strong northerly surface winds. The clear skies during nocturnal hours enhance surface cooling and create rather low surface temperatures that are particularly harmful to rice crops. Cold dry events often cause severer cold injury to rice plants than cold wet events. That is another reason for the state standardization in China to separately categorize the cold dry events and the cold wet events.

The onset date of a regional LT event in autumn in the valley is defined as the day when more than one half (≥ 44) of the stations in the valley (Fig. 7.18)

Table 7.2 17 Regional dry LT events in autumn

Start date (day mon year)	Duration (days)	Coverage (stations)	Intensity (anomaly, °C)	*R*
2 Oct 1960	8	52	−2.42	0.15
30 Sep 1967	10	82	−6.09	0.27
30 Sep 1969	11	76	−3.88	0.09
29 Sep 1970	10	84	−5.92	0.29
3 Oct 1972	4	69	−2.79	0.25
6 Oct 1973	5	67	−2.48	0.3
4 Oct 1979	7	75	−2.33	0.01
2 Oct 1985	5	60	−4.06	0.17
30 Sep 1986	4	70	−3.67	0.14
26 Sep 1988	3	51	−3.36	0.05
5 Oct 1988	5	67	−1.84	0.3
7 Oct 1989	4	61	−1.79	0.08
5 Oct 1991	6	63	−1.45	0.16
30 Sep 1993	5	58	−2.63	0.23
6 Oct 1993	3	53	−0.83	0.03
6 Oct 2002	5	71	−2	0.11
1 Oct 2004	***8***	***86***	***−4.93***	***0.08***

The case highlighted in bold and italic font is used in vertical and tracing analyses

Sourced from Qian et al. (2015) © American Meteorological Society, with permission reaching their site LT event threshold. A regional cold wet event is distinguished from a regional cold dry event by the ratio of the number of rainy sites (N_r) to the total number of sites (N) in the LT event,

$$R = \frac{N_r}{N}. \tag{7.4.1}$$

In the study of Zhang et al. (1995), three types of LT events were used: cold dry events with $R \leq 0.3$, cold wet events with $R \geq 0.7$, and hybrid cold events with $0.3 < R < 0.7$. Qian et al. (2015) categorized the cold events into two groups based on the R values: cold dry events with $R \leq 0.3$, and cold wet event for $R > 0.3$. Using these criteria, they identified 17 regional cold dry events and 27 regional cold

Table 7.3 Same as Table 7.2, but for 27 regional wet LT events in autumn

Start date (day mon year)	Duration (days)	Coverage (stations)	Intensity (anomaly, °C)	R
4 Oct 1963	7	68	−2.46	0.47
6 Oct 1964	5	79	−2.31	0.6
6 Oct 1965	5	73	−2.7	0.8
28 Sep 1968	4	65	−3.89	0.31
1 Oct 1971	10	78	−3.47	0.47
23 Sep 1972	3	73	−4.95	0.61
19 Sep 1974	3	63	−5.47	0.76
4 Oct 1974	3	53	−2.66	0.92
8 Oct 1974	3	52	−1.77	0.58
6 Oct 1977	5	78	−3.8	0.39
4 Oct 1978	3	51	−2.22	0.55
22 Sep 1980	4	74	−4.29	0.71
12 Sep 1981	3	61	−5.36	0.93
1 Oct 1981	10	68	−7.14	0.77
27 Sep 1982	3	45	−6.57	0.52
6 Oct 1983	4	63	−1.87	0.55
3 Oct 1984	***8***	***86***	***−5.43***	***0.73***
4 Oct 1992	7	83	−5.15	0.4
3 Oct 1994	5	64	−2.27	0.64
3 Oct 1995	8	79	−3.61	0.61
6 Oct 1996	5	84	−6	0.79
20 Sep 1997	3	55	−5.09	0.41
25 Sep 1997	5	73	−4.86	0.46
4 Oct 1997	4	70	−1.95	0.54
3 Oct 1999	4	53	−4.74	0.35
7 Oct 2001	3	53	−1.26	0.7
3 Oct 2003	5	64	−4.45	0.51

Sourced from Qian et al. (2015) © American Meteorological Society, with permission wet events from 1 September to 10 October in 1960–2008 (Tables 7.2 and 7.3). In Table 7.2, the smallest $R = 0.01$ indicates the driest event that started on 4 October 1979 and lasted for 7 days. On the opposite extreme, in Table 7.3, the largest $R = 0.93$ indicates the wettest event that started on 12 September 1981 and lasted for 3 days. To summarize, there were 7, 11, 11, and 11 (dry and wet) LT events between 1 September and 10 October in the 1960s, 1970s, 1980s, and 1990s, respectively. From 2000 to 2008, there were a total of 4 regional LT events.

Among 48 LT events in spring (Table 7.4), 47 of them were in April and only 1 occurred in May. Because spring is the season when stationary fronts occur frequently, bringing precipitation to the valley (Zhang and Wang 2008; Zhu et al. 2011), among the 48 LT events, 46 were cold wet events ($R > 0.3$). The longest cold wet event ($R = 0.63$) in Table 7.4 started on 4 April 1993 and lasted for 11 days. There were 15, 11, 12, and 7 spring LT events in the 1960s, 1970s, 1980s, and 1990s, respectively. These numbers suggest a decreasing trend of spring LT events in the valley over the recent decades. Three special LT events, including the two widest cases on 1 October 2004 and on 3 October 1984 and the longest cold wet event starting on 4 April 1993, listed in the three tables are used to examine their spatial structures and early signals.

In the 17 autumn cold dry events, the one starting on 1 October 2004 had the largest spatial coverage (86 stations were affected) from 3 to 4 October 2004 in the mid- to lower Yangtze River valley. Qian et al. (2015) has chosen this event as an example, and used the anomaly-based method to examine the weather conditions from 1 to 8 October 2004 and detected early synoptic disturbances for this cold dry event. The number of stations where the SAT fell below 20 °C and their average daily mean SAT anomaly (SATA) series from 1 to 8 October 2004 are shown in Fig. 7.20a. The daily mean SATA at a station is defined by the difference between the observed daily mean SAT and the climatic daily mean SAT from 1979 to 2008. In this cold dry event, the lowest regional (with 86 stations) daily mean SATA was −4.93 °C, recorded on 3 October 2004. The distribution of the SATA on 3 October 2004 (peak intensity of this event) was shown in Fig. 7.20b. There were three surface LT centers (LTC) in the upper Yellow River valley, the upper-Yangtze River, and the mid- to lower Yangtze River valley, respectively. The processes focus on affecting the valley inside the study region (Fig. 7.18).

By analyzing the conventional synoptic charts, they found that on 23 September 2004 there was a height trough along with cold temperature at 500 hPa over Siberia (60°N, 90°E). By 0000 UTC 1 October 2004, the height trough leading a temperature trough had propagated eastward to North China at 500 hPa (Fig. 7.21a). At 850 hPa (Fig. 7.21b), the height trough reached the eastern coast of China. The associated temperature trough was located in the west of height trough, forming a flow pattern favorable for strong cold air advection in the lower troposphere. At 0000 UTC 3 October 2004 (Fig. 7.21c), the cold air advection at 500 hPa started to weaken when a tropical cyclone moved northward in the East China Sea (Fig. 7.21d). The SATA series in Fig. 7.20a was sensitive to the strengthening and weakening of cold air advection with a peak on 3 October 2004.

Table 7.4 Same as Tables 7.2 and 7.3, but for 48 regional LT events in late spring

Start date	Duration	Coverage	Intensity	Start date	Duration	Coverage	Intensity
11 Apr 1960	6	44	−2.66	**20 Apr 1978**	**3**	**27**	**−3.07**
6 May 1960	3	12	−2.45	1 Apr 1979	5	35	−2.4
2 Apr 1962	**4**	**74**	**−4.57**	**13 Apr 1980**	**3**	**27**	**−4.9**
10 Apr 1962	3	15	−1.43	24 Apr 1980	3	52	−2.79
6 Apr 1963	**4**	**77**	**−5.49**	2 Apr 1981	4	44	−2.5
18 Apr 1963	5	33	−1.42	**3 Apr 1982**	**4**	**64**	**−3.09**
6 Apr 1964	**5**	**70**	**−4.21**	8 Apr 1982	3	16	−2.44
3 Apr 1965	5	61	−2.97	15 Apr 1983	3	24	−2.34
10 Apr 1965	3	14	−2.09	5 Apr 1984	8	28	−2.44
4 Apr 1966	9	39	−2.53	4 Apr 1985	3	12	−1.88
4 Apr 1967	3	11	−1.68	**1 Apr 1987**	**3**	**14**	**−4.15**
10 Apr 1967	9	37	−1.64	**11 Apr 1987**	**6**	**86**	**−3.51**
9 Apr 1968	**3**	**19**	**−4.26**	**6 Apr 1988**	**3**	**33**	**−4.33**
3 Apr 1969	**4**	**76**	**−5.48**	11 Apr 1989	4	30	−1.54
16 Apr 1969	**3**	**16**	**−3.62**	**3 Apr 1990**	**3**	**15**	**−3.15**
10 Apr 1970	5	42	−1.95	***4 Apr 1993***	***11***	***66***	***−3.66***
16 Apr 1971	3	15	−1.78	9 Apr 1994	3	39	−2.82
1 Apr 1972	**4**	**30**	**−5.93**	1Apr 1995	4	25	−2.48
7 Apr 1972	4	46	−2.84	**1 Apr 1996**	**5**	**30**	**−3.89**
1 Apr 1973	4	19	−2.9	8 Apr 1996	8	64	−2.25
1 Apr 1974	**3**	**15**	**−3.35**	**3 Apr 1997**	**3**	**15**	**−3.4**
1 Apr 1975	3	10	−2.37	25 Apr 2002	4	30	−1.68
2 Apr 1976	**4**	**41**	**−3.52**	9 Apr 2005	3	16	−1.39
7 Apr 1976	5	35	−1.74	**12 Apr 2006**	**4**	**48**	**−3.72**

Bold−faced 20 events are used in individual and composite analyses. The event of the longest duration is highlighted in bold and italic font and used as an example in the analysis

Sourced from Qian et al. (2015) © American Meteorological Society, with permission

To demonstrate the utility of the anomaly-based method in capturing this LT event and its development, the vertical cross sections of daily mean height and temperature along 115°E on 3 October 2004 are shown in Fig. 7.22. Figure 7.22a is the vertical cross section of the total height and temperature (black solid and red-dashed contours, respectively) routinely used in daily weather analysis. In Fig. 7.22a, the climatological values were derived from 1979 to 2008 of height and temperature on 3 October along the same cross section (heavy solid and dashed contours, respectively). Compared to the climatic temperature, the total temperature showed a cold anomaly in the mid- to lower-troposphere over the valley, while the magnitude and the spatial structure of this cold anomaly were unclear. The climatic height and the actual height have rather large basic values nearly equal to each other, so their differences are too small to indicate the cold event clearly. A slight difference between the climatological height (heavy solid contour) and the total

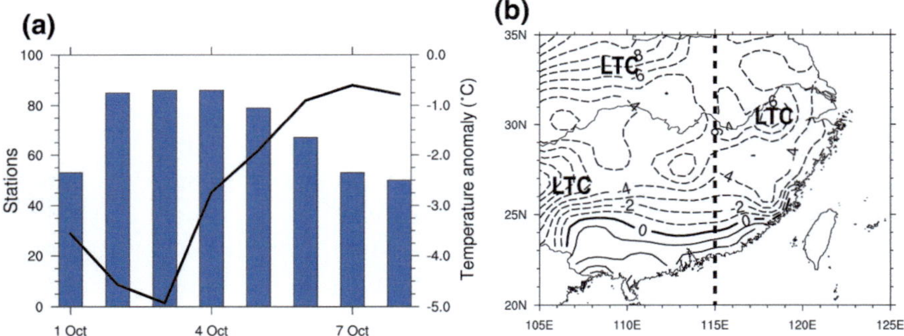

Fig. 7.20 A regional cold dry event from 1 to 8 October 2004 in the mid- and lower-Yangtze River valley. **a** Number of stations (*bar*) with daily mean temperature below 20 °C, and their average daily mean SATA (*solid line*, °C). **b** Distribution of SATA (*contour*, 1 °C interval) on 3 October 2004. "*LTC*" indicates the low-temperature centers. The *heavy-dashed line* marks the 115°E along which vertical cross sections. Sourced from Qian et al. (2015) © American Meteorological Society, with permission

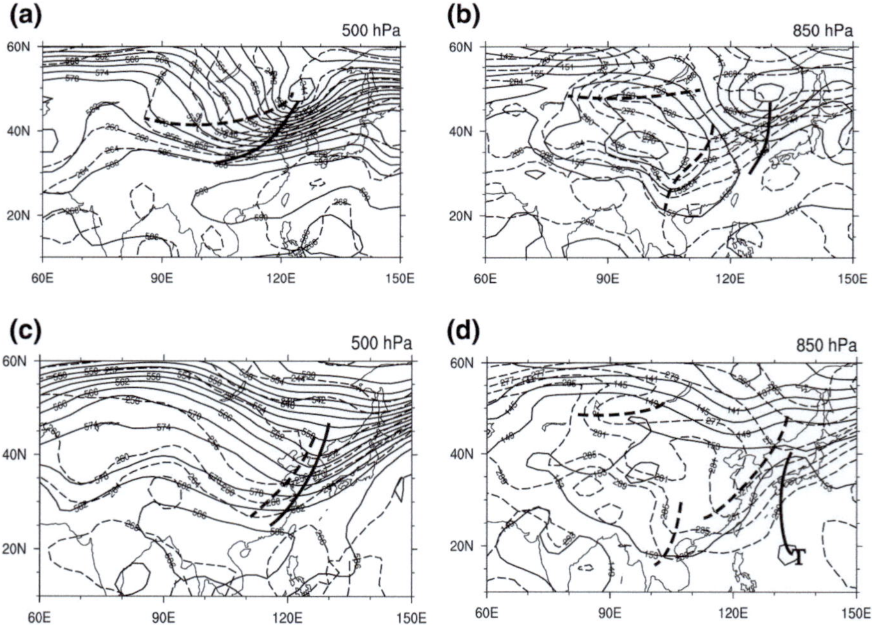

Fig. 7.21 Conventional synoptic charts. Geopotential height (*solid contour*, 40 gpm interval) and temperature (*dashed contour*, 4 K interval) at **a** 500 hPa and **b** 850 hPa at 0000 UTC 1 October 2004. **c** and **d** are the same but for 0000 UTC 3 October 2004. The *heavy solid line* indicates geopotential height trough and the heavy-dashed line marks the temperature trough. The capital letter "T" in **d** is the location of tropical cyclone. Sourced from Qian et al. (2015) © American Meteorological Society, with permission

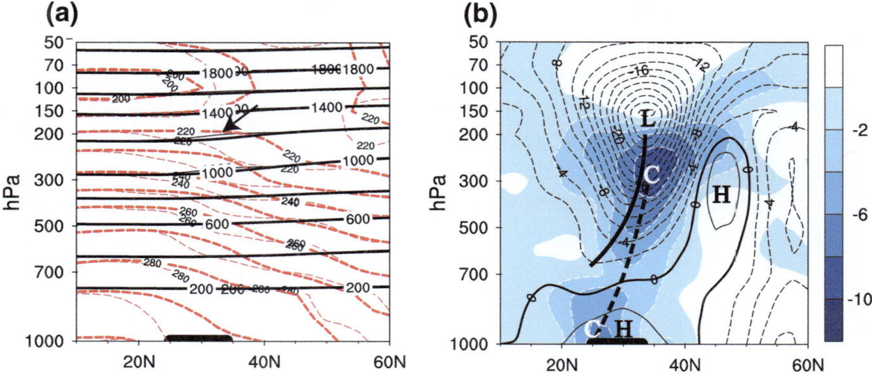

Fig. 7.22 Vertical cross section of geopotential height and temperature on 3 October 2004. **a** Total and climatic height (*black* and *heavy solid contours*, 2000 gpm interval), and total and climatic temperature (*red* and *heavy dashed contours*, 10 K interval). **b** Anomalies of height (*contour*, 20 gpm interval) and temperature (*shading*, 2 K interval) along 115°E on 3 October 2004. The mid- to lower-Yangtze River valley is highlighted by the *heavy solid line* along the abscissa. In **a**, the *arrow* indicates the difference between the total and climatic heights or between the *black solid* and *dashed contours*. In **b**, the *thick solid line* indicates the trough in height anomalies, and the *thick-dashed line* is the axis of cold temperature anomalies. Letters "H" and "L" mark the positive and negative centers in height anomalies, respectively, and "C" is the center of negative temperature anomalies. In **b**, *white areas* are positive temperature anomalies. Sourced from Qian et al. (2015) © American Meteorological Society, with permission

height (black solid contour) marked by an arrow in Fig. 7.22a can be found near 200 hPa over the valley.

In contrast, such deviations were clearly by the anomalies of height and temperature on 3 October 2004 (Fig. 7.22b). These anomalies showed unequivocally a strong cold air pool over the valley on 3 October 2004. This capability of Fig. 7.22b demonstrated its utility in revealing this LT event. As shown, this particular event was part of such an anomalous synoptic system which had a negative center of height anomalies in the upper troposphere, a negative (cold) center of temperature anomalies in the middle troposphere, and another cold center in the lower troposphere. It is noted that the anomalous surface cold air and high pressure were coupled with a negative center of height anomalies in the upper troposphere. The strong negative height anomalies in the upper troposphere were related to the strong cold air mass (negative temperature anomaly) below it, as required by the hydrostatic equilibrium. The vertical structure of a negative center of height anomalies at the upper troposphere, a positive center of height anomalies near the cold temperature surface, and a cold air column between these two centers was a typical vertical pattern, namely the LT vertical pattern, for the LT events in the valley. In the north of the LT vertical pattern, there was a positive center of height anomalies which is useful to distinguish between cold dry and cold wet events. This anomalous structure of height and temperature in Fig. 7.22b was also observable in the conventional synoptic chart in Fig. 7.22a but it can hardly be directly distinguished.

These striking differences from the analysis of the vertical profile of the anomalies in height and temperature suggest that it is possible to trace, by this method, the early development and propagation of the set of negative temperature anomalies at middle-lower troposphere and a negative center of height anomalies at upper troposphere (Qian and Zhang 2012; Zhang and Qian 2012). From daily height and temperature anomalies, their centers of SD are, respectively, located between 200 and 300 hPa and around 850 hPa in middle and high latitudes. In this method, such anomalous centers in the troposphere are the target signals, or

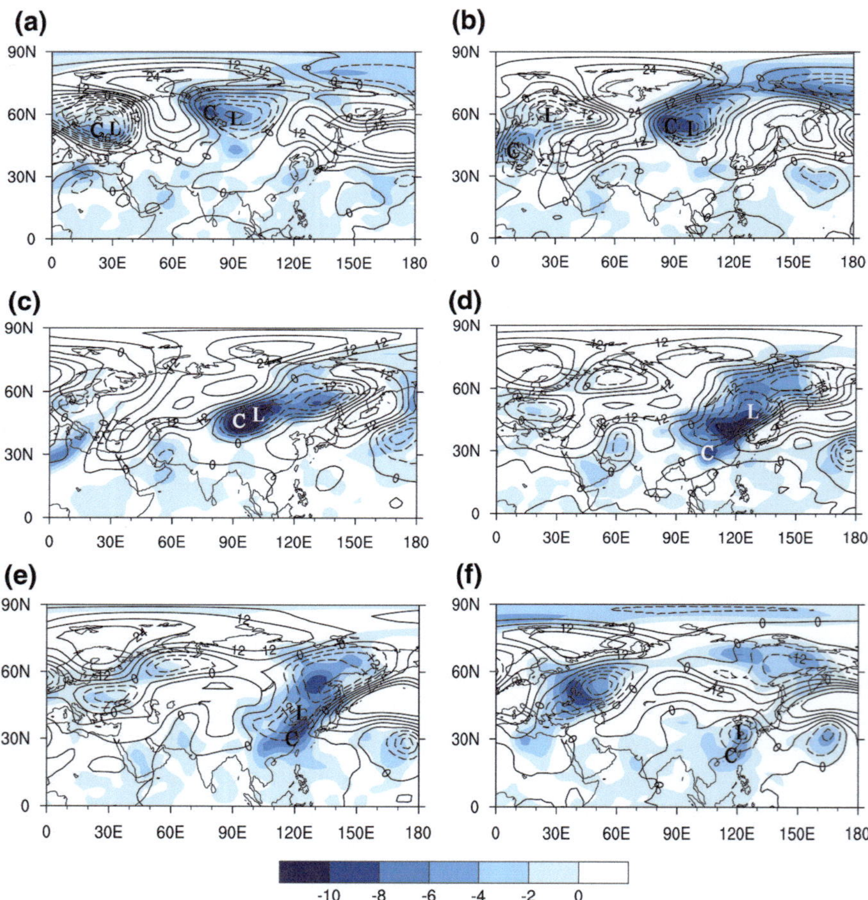

Fig. 7.23 Anomaly-based synoptic charts showing the anomalies of height (*contour*, 60 gpm interval) at 300 hPa, and the anomalies of temperature (*shading*, 2 K interval) at 850 hPa on **a** 23 September, **b** 25 September, **c** 29 September, **d** 1 October (starting date for this LT event in the valley), **e** 2 October, and **f** 4 October 2004, respectively. Letters "L" and "C" in the panels indicate the negative centers of height anomalies at 300 hPa and the negative centers of temperature anomalies at 850 hPa, respectively. *White areas* represent the positive temperature anomalies. Sourced from Qian et al. (2015) © American Meteorological Society, with permission

sources, for autumn LT events in the valley. For this event on 3 October 2004, a negative center of height anomalies at 300 hPa and a negative center of temperature anomalies at 850 hPa could be traced back day by day to Siberia (90°E, 60°N) on 23 September 2004. As depicted in Fig. 7.23a, a pair of negative centers of temperature anomalies at 850 hPa and height anomalies at 300 hPa had established on 23 September 2004. They propagated southeastward into eastern China on 1 October 2004 (Fig. 7.23d), causing the particular LT event in the valley. In Fig. 7.23d, e, the center of height anomalies at 300 hPa was located in North China, while a cold center at 850 hPa extended to the mid- and lower Yangtze River. The location of the negative temperature anomalous center at the lower troposphere located in the south of the height anomalous center at the upper troposphere can be clearly observed in Fig. 7.22b. This example showed that the negative centers of temperature anomalies at 850 hPa and height anomalies at 300 hPa from the region in eastern China can be traced back about 8 days to Siberia. However, another pair of negative centers of temperature anomalies at 850 hPa and height anomalies at 300 hPa in Western Europe gradually weakened from 23 September (Fig. 7.23a) to 25 September (Fig. 7.23b) and finally dissipated on 29 September (Fig. 7.23c). If these signals are presented in a forecast model, they may provide forecasters early warning information for a potential LT event.

Additional analyses on the vertical structure and horizontal distribution of height and temperature anomalies as done on the above case were conducted for other 16 autumn cold dry events in the valley. Among total 17 dry events, 20 early signals of initial disturbances were traced with pairs of negative centers of height and temperature anomalies at 300 and 850 hPa, respectively. Their initial locations were shown by the open dots in Fig. 7.24a. 17 of them further developed and amplified

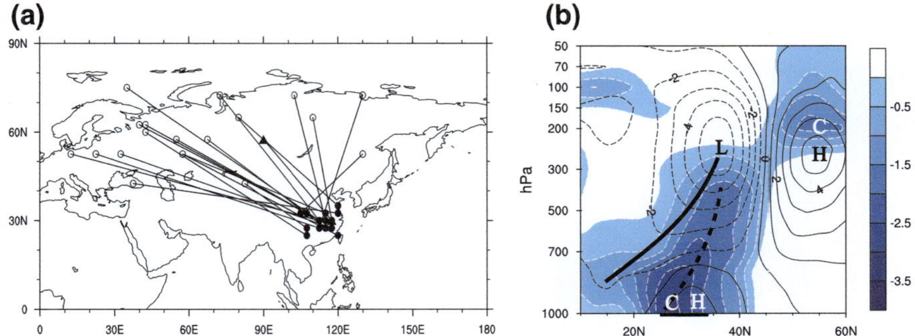

Fig. 7.24 **a** Locations of initial disturbances in height and temperature (*20 open dots*) and the corresponding positions of 17 autumn cold dry events (*solid dots*), based on the signal tracing from anomalies of height and temperature. The *black triangle* in **a** indicates the case shown in Figs. 7.22 and 7.23. **b** Composite vertical cross section of height anomalies (*contour*, 10 gpm interval) and temperature anomalies (*shading*, 0.5 K interval) along 115°E of the 17 cold dry events. The *thick solid line* indicates the trough of height anomalies and the *thick-dashed line* denotes an axis of negative temperature anomalies. In **b**, *white areas* are positive temperature anomalies. Sourced from Qian et al. (2015) © American Meteorological Society, with permission

while propagating to eastern China, resulting in 17 cold dry events. Among them, three initial disturbances merged with others during their propagation processes. Many remote disturbances in the focused domain did not propagate to eastern China. The evolution of the initial disturbance marked by a black triangle in Fig. 7.24a was the case illustrated in Fig. 7.23, which caused the largest coverage dry event in October 2004. The average length from the disturbance initiation to the occurrence of cold dry events in the valley was 12.5 days, a remarkable lead time to indicate those events.

A composite of vertical cross sections along 115°E is shown in Fig. 7.24b for all the 17 cold dry events when their initial disturbances amplify and arrive at the solid dot locations in Fig. 7.24a. The vertical structure in Fig. 7.24b has all the key features as described in Fig. 7.22b, indicating a strong cold temperature anomaly centered near 850 hPa, underneath a negative center of height anomalies between 200 and 300 hPa. A center of cold-high pressure is shown near the surface in the valley. The LT vertical pattern is also clearly observed from this dry composite of 17 cases with a positive center of height anomalies to its north in the upper troposphere.

As a special case, the wet event with the largest coverage occurred from 3 to 10 October 1984 and influenced a broad area (86 stations) in the mid- to lower Yangtze River valley. It is also examined using this anomaly-based synoptic analysis method. In this wet LT event, the lowest regional daily mean SATA was −5.43 °C, with $R = 0.73$, peaked on 5 October 1984 (Table 7.3 and Fig. 7.25a). The regional distribution of the SATA on 5 October 1984 is shown in Fig. 7.25b. A major LT center (indicated by LTC) was located in the south of the Yangtze River.

For this cold wet event, its cold advection at 500 hPa (Fig. 7.26a) is weaker than that in the cold dry event shown in Fig. 7.21a. This weak cold advection from northwest, and the warm air mass in the south, allowed the formation of a lower tropospheric stationary front. Several isothermal lines were zonally-oriented along the Yangtze River from 3 to 5 October 1984 (Fig. 7.26b, d). The stationary front brought cold rain and anomalously low temperature along the mid- to lower

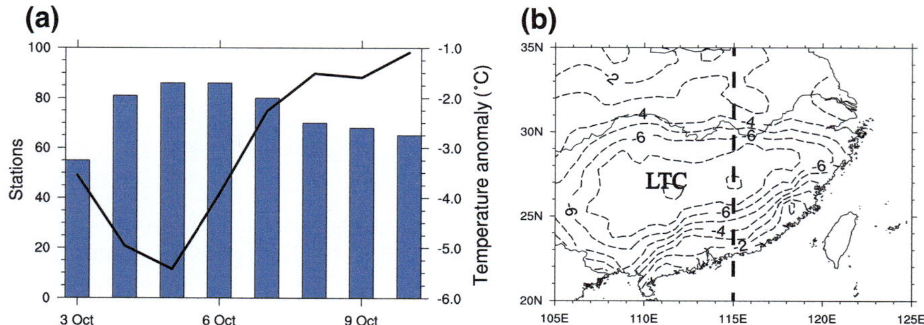

Fig. 7.25 Same as in Fig. 7.20, but for the cold wet event occurring from 3 to 10 October 1984 in south of the Yangtze River. Sourced from Qian et al. (2015) © American Meteorological Society, with permission

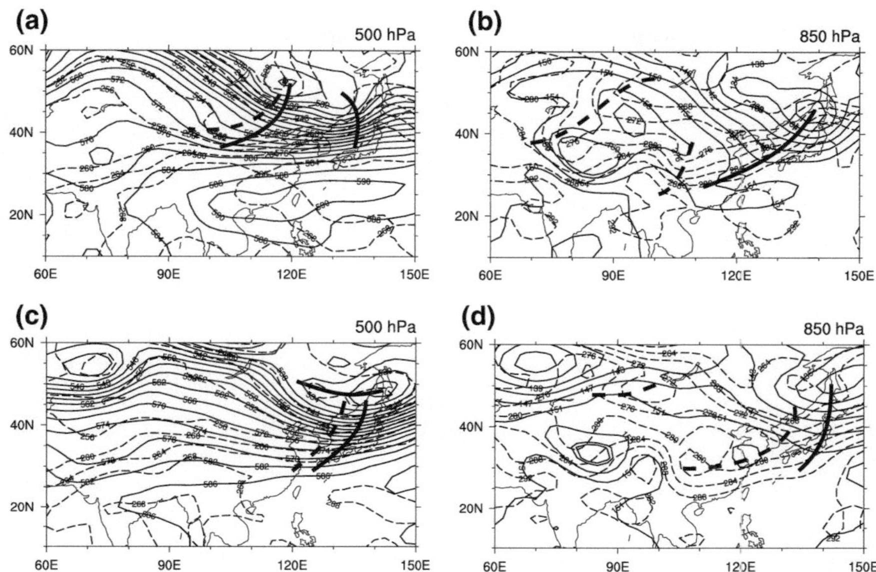

Fig. 7.26 Same as in Fig. 7.21, but at **a** 500 hPa and **b** 850 hPa at 0000 UTC 3 October 1984, as well as at **c** 500 hPa and **d** 850 hPa at 0000 UTC 5 October 1984. Sourced from Qian et al. (2015) © American Meteorological Society, with permission

Yangtze River valley. Similarly, two vertical cross sections are shown along 115°E for this cold wet event in Fig. 7.27. Figure 7.27a shows the cross section constructed from the traditional way and Fig. 7.27b shows the cross section constructed from the anomaly-based method. Again, Fig. 7.27b reveals the axis of cold temperature anomalies extending from a negative center of height anomalies at 300 hPa to a positive center of height anomalies near the surface. The LT vertical pattern is also noted over the valley, but there is a negative center of height anomalies at the upper troposphere over North China and a weak warm temperature anomaly in its south. This feature of wet event is different from that of dry event in Fig. 7.22b.

The centers of height anomalies at 300 hPa and temperature anomalies at 850 hPa in Fig. 7.27b could also be traced back day by day to a disturbance featured with a pool of anomalously cold air in northern Europe (55°E, 70°N) on 24 September 1984 (Fig. 7.28). On 24 September 1984, there were five large centers of disturbances over the Eurasian continent (Fig. 7.28a). Among them, a pair of negative centers arrived Northeast China and its cold air influenced North China. However, only the one from northern Europe amplified and propagated southeastward to North China on 2 October 1984 (Fig. 7.28d). From 2 to 4 October, this air mass of strong cold temperature anomalies split into two centers, and one of them moved south into eastern China and caused the cold wet event in the valley. The results in Fig. 7.28 also confirm that the pair of negative centers of temperature anomalies at 850 hPa and height anomalies at 300 hPa serves as the early indication

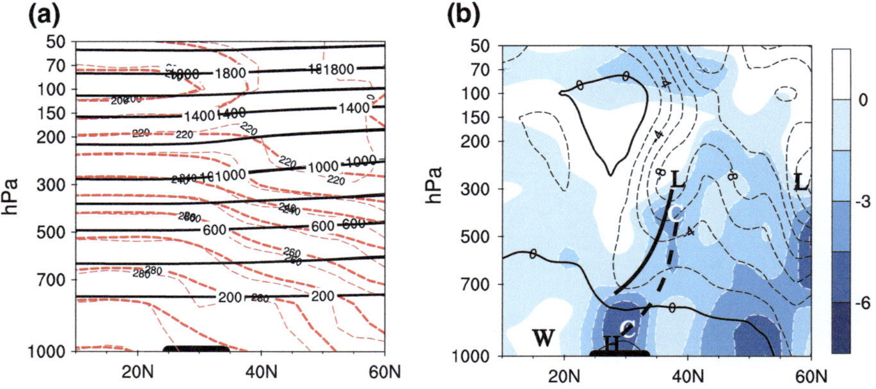

Fig. 7.27 Same as in Fig. 7.22, but for the cold wet event on 5 October 1984. In **b**, the letter "W" denotes a weak warm center. Sourced from Qian et al. (2015) © American Meteorological Society, with permission

of disturbances that can develop while propagating into China, and cause cold wet event in the valley leading about 9 days.

For all the cold wet LT events listed in Table 7.3, 30 pairs of initial disturbances of negative height anomalies at 300 hPa and negative temperature anomalies at 850 hPa were identified and traced as done in Fig. 7.28. Their geographical locations are shown by the open circles in Fig. 7.29a. Of those initial disturbances, 27 amplified while propagating into the eastern China and caused cold wet events in the valley (marked by solid dots in Fig. 7.29a). Three disturbances among them merged with others during their propagations. The average time length between the appearance of the initial disturbances and the occurrence of the cold wet events in the valley was 8.8 days. The composite of the vertical cross sections for these autumn cold wet events is shown in Fig. 7.29b, delineating similar features to those in Fig. 7.24 for autumn cold dry events. The LT vertical pattern with a negative center of height anomalies at the upper troposphere, a positive center of height anomalies near the cold surface, and cold air column between these two centers is also typical in the mid- to lower Yangtze River valley. There are major differences between the cold dry and cold wet events in autumn. Comparisons between Figs. 7.24b and 7.29b reveal that: (a) the cold dry events have a much larger body of cold air in the valley than that for the cold wet events, (b) in the cold dry events, there is a positive center of height anomalies located on the northern side of the cold air column, and (c) a region with weaker positive height anomaly is found on the southern side of the cold air column in the cold wet events. In the former, the positive height anomaly in the north pressures the cold air mass deepening into southern China (Fig. 7.24b). In the latter, the contrast of pressure in the valley forms a stationary front and favors precipitation (Fig. 7.29b).

There were 46 cold wet events ($R > 0.3$) among total 48 LT events in spring during the study period of 1960–2008 (Table 7.4). Most weather systems in spring in southern China are characterized by interactions between air masses of

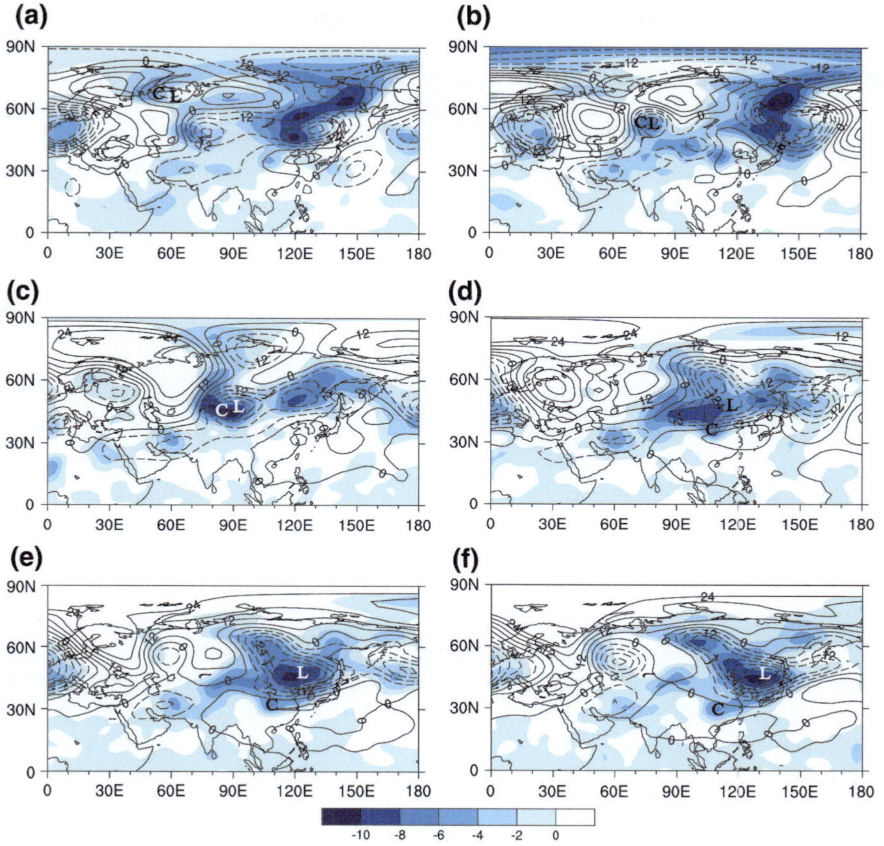

Fig. 7.28 Same as in Fig. 7.23, but on **a** 24 September, **b** 27 September, **c** 30 September, **d** 2 October, **e** 3 October (starting date of the event), and **f** 4 October 1984, respectively. Sourced from Qian et al. (2015) © American Meteorological Society, with permission

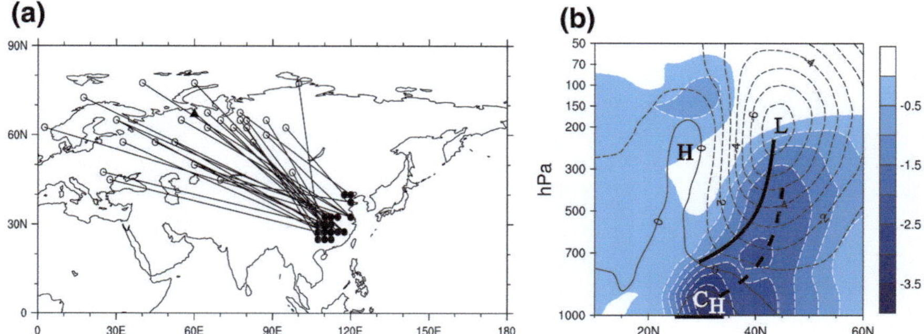

Fig. 7.29 Same as in Fig. 7.24, but for 27 cold wet events in autumn. The *black triangle* in **a** indicates the case shown in Figs. 7.27 and 7.28. Sourced from Qian et al. (2015) © American Meteorological Society, with permission

contrasting temperatures and moistures, and associated with precipitation. Thus, for the spring LT events, we only focus on the cold wet events. Among the 46 late spring cold wet events, the longest duration was the case from 4 to 14 April 1993 for about 11 days. It happened with a stationary front established in the Yangtze River valley ($R = 0.63$). In Fig. 7.30, a similar set of figures is presented as in Fig. 7.20 to illustrate the process and influence of this particular cold wet event. Figure 7.30a shows the number of stations affected by this wet event and the time series of their average daily SAT anomaly. On 4 April (the starting date), there were 56 stations affected by the low temperature. The lowest temperature was reported on 5 April among 62 stations in the valley with the daily SATA −3.66 °C. The spatial distribution of the SATA on 5 April 1993 is shown in Fig. 7.30b. There were two low temperature centers, one in the mid- to lower Yangtze River valley and the other in the upper Yellow River valley. The temperature anomaly at the low temperature centers in the Yangtze River valley reached −6 °C on 5 April 1993. Although the affected region diminished after 6 April, other two cooling processes peaked on 8 and 12 April 1993, extending the LT event's duration. Clearly, the persistence of LT events such as this case was caused by three cooling processes. A similar case was the LT freezing-rain event occurred in southern China from 10 January to 2 February 2008, undergoing four cooling processes (Qian and Fu 2010).

Figure 7.31 shows a similar comparison between the cross sections of temperature and geopotential height derived from the conventional analysis method (Fig. 7.31a) and that derived from the anomaly-based method (Fig. 7.31b). The difference between these cross sections is outstanding. Again, results from the anomaly-based method clearly depict the cold wet event whereas the results from the conventional method give a rather fuzzy description of the event. In Fig. 7.31b, the LT vertical pattern with a set of a negative height anomaly centered at around 300 hPa and a negative temperature anomaly centered at 850 hPa clearly indicates unequivocally the cold wet event. Interestingly, this cold wet event has its vertical profile of anomalies (Fig. 7.31b) nearly identical to the composite profile for cold

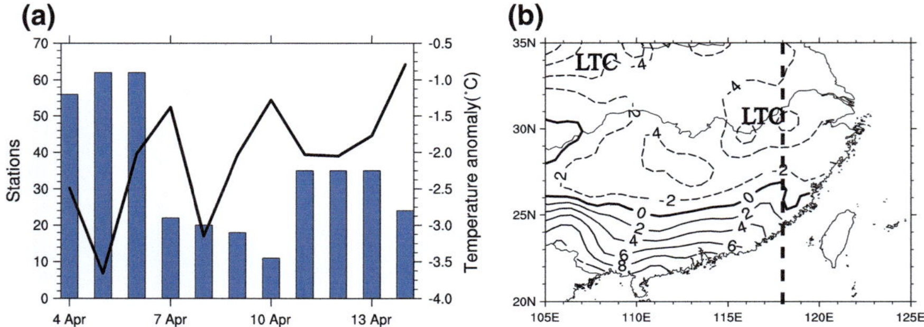

Fig. 7.30 Same as in Fig. 7.20, but for **a** the spring cold wet event from 4 to 14 April 1993 and **b** the distribution of SATA (°C) on 5 April 1993. "LTC" indicates a low-temperature center. Sourced from Qian et al. (2015) © American Meteorological Society, with permission

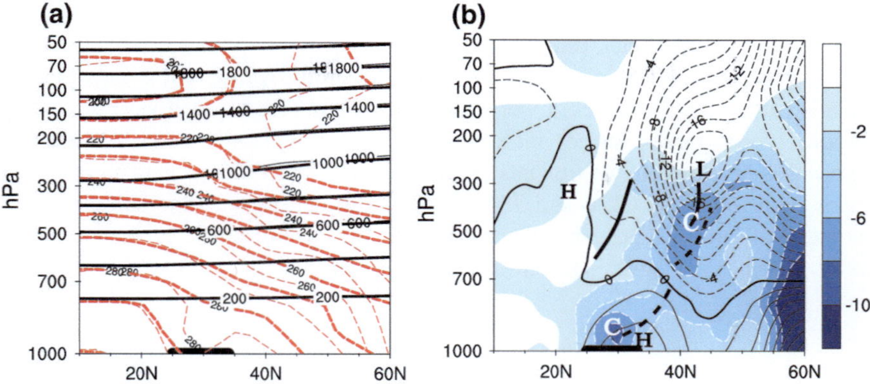

Fig. 7.31 Same as in Fig. 7.22, but for the spring cold wet event on 5 April 1993 and along 117.5°E. Sourced from Qian et al. (2015) © American Meteorological Society, with permission

wet events in autumn (Fig. 7.29b). The similarity that a weaker positive height anomaly is on the southern side of the cold air column suggests the consistency of these cold wet events in the mid- to lower Yangtze River valley in spring and autumn.

Again, using the anomaly-based method, one can also trace the early disturbances of this spring cold wet event in the valley. Among the 46 cold wet events in late spring (Table 7.4), there were 20 cases with average daily SATA colder than −3.0 °C. Qian et al. (2015) examined these events using the anomaly-based method to trace the initial disturbances that developed and caused those events. Figure 7.32a shows the geographical locations of initial disturbances in the height anomalies at 300 hPa and the temperature anomalies at 850 hPa that caused those 20 spring LT events. The average leading time length from the initial disturbances to the occurrence of 20 spring LT events in the valley was 6.9 days. The earliest disturbance was identified 13 days ahead of the spring LT event. The initial

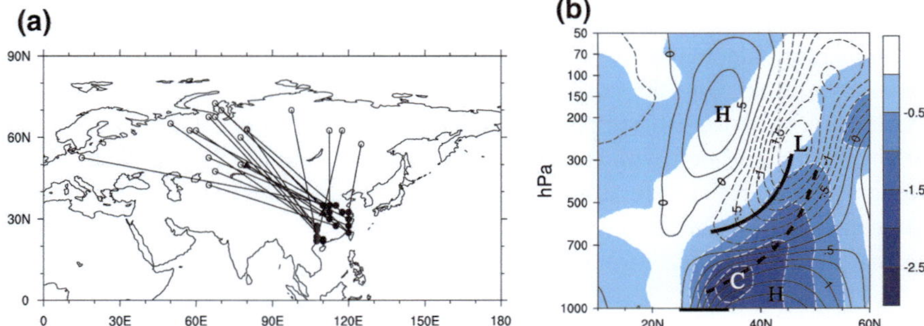

Fig. 7.32 Same as in Fig. 7.24, but for 20 strong cold wet events in spring. In **b**, the *contour* interval of geopotential height anomalies is 2.5 gpm. Sourced from Qian et al. (2015) © American Meteorological Society, with permission

location of the disturbance resulting in the case from 4 to 6 April 1993 was marked in Fig. 7.32a by the black triangle.

The composite LT vertical pattern of the 20 strong spring wet events is comparable to the one for the cold wet events in autumn (Figs. 7.32b vs. Fig. 7.29b). Detailed comparisons further indicate that the spring cold wet events are less vigorous than the cold wet events in autumn (see the difference in temperature scales between Figs. 7.32b and 7.29b). Their differences near the surface height anomalies further suggest that the cold air in the spring events is shallower and concentrates in the lower portion of the troposphere.

7.5 Positive Height Anomalies and Warm Events[5]

In the analyses of all the above warm and cold events, centers of height anomalies at the upper troposphere are critically associated with the surface weather extremes. Therefore, Qian et al. (2016b) examined the relationship between them in details. A comparison between the total field, climatic field and anomaly-based weather charts are shown in Fig. 7.33. In the total field, there is an obvious deviation between total and climatic temperatures (red- and black-dashed contours indicated by an arrow at 850 hPa) but the deviation between total and climatic heights (red and black solid contours indicated by arrows near 150 and 300 hPa) is unclear (Fig. 7.33a). However, these deviations are clearly shown in the two anomaly-based weather charts (Fig. 7.33b, c), where there are different positive/negative centers of height/temperature anomalies in the vertical structure. A positive center of height anomalies at the upper troposphere is located right over the surface heat wave event indicated by the heavy bar. In the North China, the negative SATA is also indicated by a negative center of height anomalies at the upper troposphere (Fig. 7.33b).

It is clearly noted that the surface heat wave appeared along the downward and southward extension of positive height anomalies (Fig. 7.33b, heavy dash-dotted line) from the center of MHA at the upper troposphere. For the heat wave event, a negative center of temperature anomalies near the 100 hPa and two positive centers of temperature anomalies at 300 and 850 hPa in the mid-low troposphere are a dynamical response to the positive MHA centered around 200 hPa. The opposite dynamical response of temperature anomalies from troposphere to mid-low stratosphere associated with the negative MHA is located in North China, forming a surface low temperature anomaly. The surface high (low) temperature anomalies are coupled with a positive (negative) center of height anomalies in the upper troposphere. The strong positive (negative) height anomaly in the upper

[5]Modified article originally published in *Qian WH, Yu TT, Du J (2016b) A unified approach to trace surface heat and cold events by using height anomaly. Climate Dyn 46: 1647–1664.* Published with kind permission © Springer-Verlag Berlin Heidelberg 2016. All Rights Reserved.

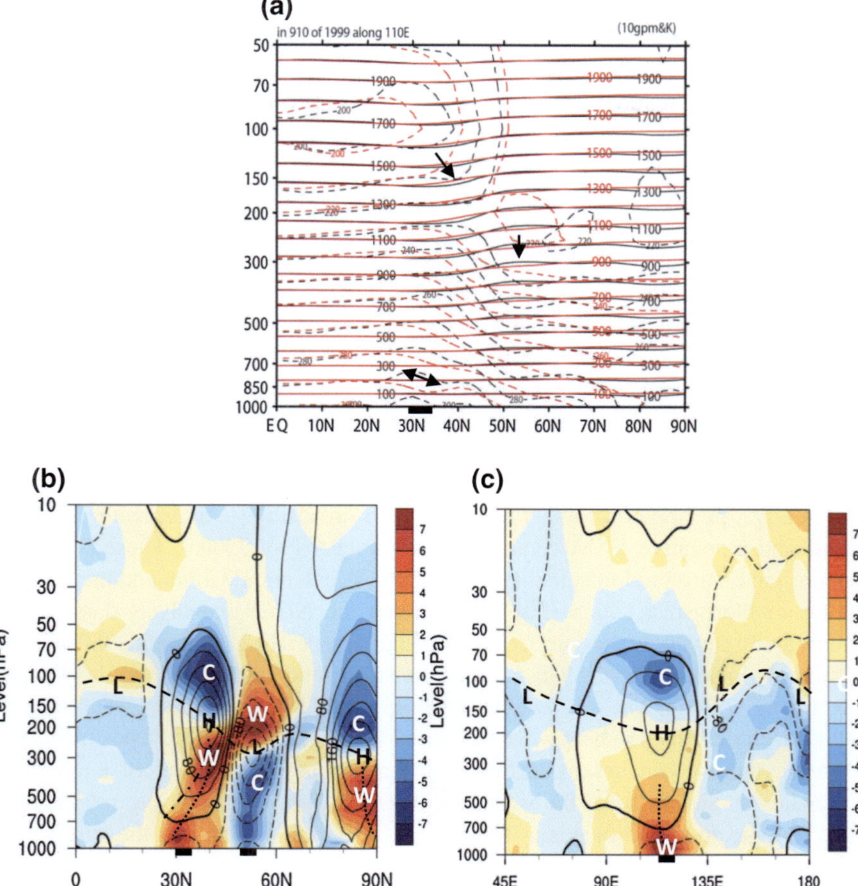

Fig. 7.33 Vertical cross sections of **a** total (*black-solid contour*) and climatic (*red-solid contour*) height (100 × 10 gpm interval) as well as total (*black-dashed contour*) and climatic (*red-dashed contour*) temperature (10 K interval), height anomaly (*contour*, 40 gpm interval) and temperature anomaly (*shading*, 1 K interval) along **b** 110°E and **c** 32.5°N on 10 September, 1999. The *heavy-dashed line* indicates the maximum height anomaly (MHA). The *dash-dotted line* denotes anomalous height axis and the *dotted line* is anomalous temperature axis. In **a**, *arrows* indicate the deviation between total and climatic heights (temperatures). The *heavy bar* indicates the positive SATA (30–35°N) in Central China and another bar indicates the negative SATA (47.5–52.5°N) in North China. In **b** and **c**, letters "H" and "L" as well as "W" and "C" indicate the positive and negative centers of height anomalies as well as the positive and negative centers of temperature anomalies, respectively. Sourced from Qian et al. (2016b) © Springer-Verlag Berlin Heidelberg 2016, with permission

troposphere is related to the strong positive (negative) temperature anomaly below it, as required by the hydrostatic relation. The vertical structure of a positive (negative) center of height anomalies at the upper troposphere and a warm (cold) air column below the MHA center is a typical vertical pattern, namely the vertical warm (cold) event pattern.

In Fig. 7.33b, there are two positive and two negative MHA centers but their altitudes vary with locations. It is observed that a center of positive (negative) MHAs at the upper troposphere corresponds to a positive (negative) center of temperature anomalies at around 850 hPa and a positive (negative) SATA at the surface. In other words, a positive (negative) SATA at the surface is associated with a positive (negative) center of MHAs at the upper troposphere. These standalone MHAs are connected by the heavy-dashed line in Fig. 7.33b, c. Using gridded NCEP reanalysis data, we have established a daily MHA database for the past 31 years. Therefore, all signals of surface heat and cold events as well as warm and cold anomalies can be searched from the MHA database.

A comparison between total-based and anomaly-based weather charts is shown in Fig. 7.34. Figure 7.34a is the total height at 200 hPa on 10 September 1999, which shows that the surface heat wave event is associated with a broad region over 12,500 gpm but the region is too large to indicate the exact coverage of the event. Differently, the height anomaly clearly shows a positive MHA center in North China and a negative MHA center in its northwest (Fig. 7.34b). These two MHA centers can be used to indicate the positive SATA in Central China and the negative SATA in Northwest China (Fig. 7.1b). From the NCEP reanalysis data, a high temperature center at the surface and a high temperature center at 850 hPa are commonly located over the Tibetan Plateau and extend eastward along the Yangtze River (Fig. 7.34c). The highest temperature from the total-based weather chart cannot well describe the heat wave in Central China. In Fig. 7.34d, a positive center of temperature anomalies in eastern China and a negative center of temperature anomalies at Siberia from the surface to 850 hPa, however, can be, respectively, used to indicate the surface heat wave event in Central China and the surface low temperature anomaly in Northwest China. The four panels in Fig. 7.34 show that the centers of MHA and 850-hPa temperature anomalies are better to indicate the surface temperature anomalies.

Traditionally, troughs indicated by dashed lines in the mid-high latitudes exist in the daily synoptic charts (Fig. 7.34a). After removing daily mean climatic fields, the remaining anomalous fields are not wave-like systems but standalone eddies (Fig. 7.34b). For example, height anomalies show multi-scale positive and negative centers with the maximum value alone located in upper troposphere at 100–400 hPa (Fig. 7.33b, c). The structure of height and temperature anomalies satisfies the hydrostatic equilibrium, which means that the temperature anomalies can be derived from the height anomalies in troposphere to stratosphere (Fig. 7.33b, c). The Rossby wave-like systems can be still observed in the climatic field (Fig. 7.34e) but their propagations have a seasonal cycle. Therefore, as a closed synoptic-scale system, the movement of any MHA center is influenced by daily large-scale climatic flow (Fig. 7.34e) and interacts with other nearby MHA centers (Fig. 7.34b). This is similar to a typhoon's movement, which is influenced by daily large-scale climatic steering flow and interacts with other nearby anomalous vortices (Qian et al. 2014).

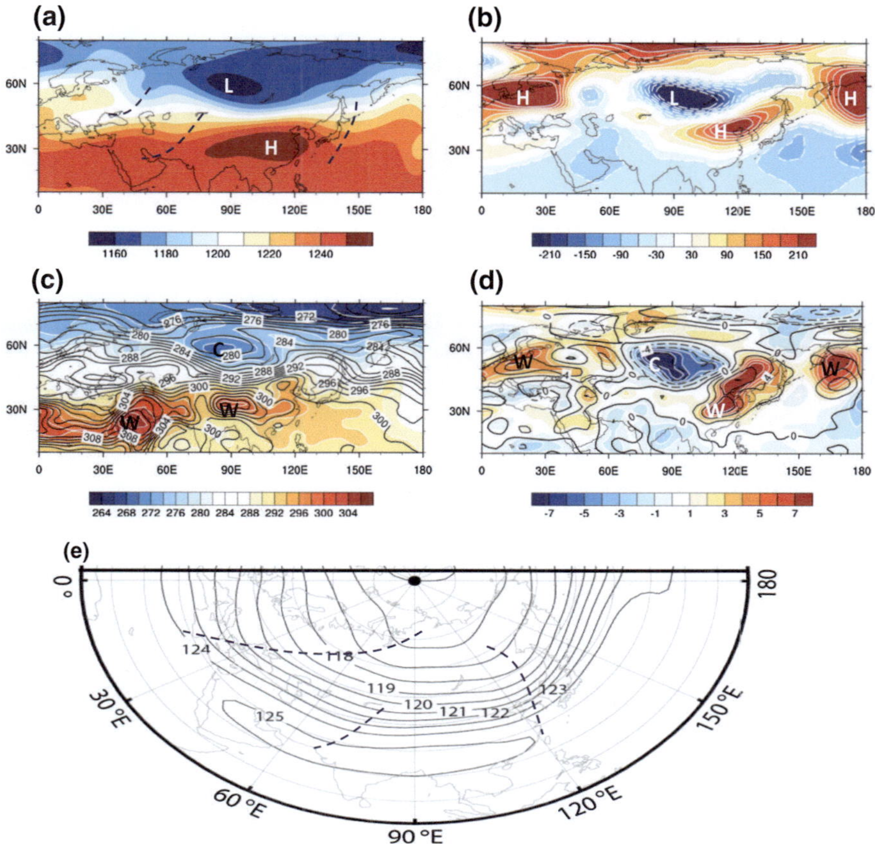

Fig. 7.34 **a** Daily geopotential height (*shading*, 100 gpm interval) at 200 hPa, **b** daily mean MHA (*shading*, 30 gpm interval), **c** 850 hPa temperature (*shading*, 2 K interval) and SAT (*contour*, 2 K interval), **d** 850 hPa temperature anomaly (*shading*, 1 °C interval) and daily mean SATA (*contour*, 2 °C interval), **e** climatic height (100 gpm interval) and climatic troughs (*dashed lines*) at 200 hPa on 10 September 1999. Letters "H/L" and "W/C" indicate the positive/negative centers of height anomalies and positive/negative centers of temperature anomalies, respectively. Sourced from Qian et al. (2016b) © Springer-Verlag Berlin Heidelberg 2016, with permission

To examine which MHA center is directly associated with the heat wave described above, Fig. 7.35 shows the daily MHA centers during August and September in 1999. In Fig. 7.35, only three tracks of MHA centers were connected from day to day. The yellow track of MHA center was first observed in North Europe, moving eastward with a complex path and finally arriving to the Middle East after a 21-day travel. The signal lasted longer than 21 days because its real initial position started from the Western Hemisphere. That is why there were more points at 2.5°E. The purple track of MHA center was first originated from the

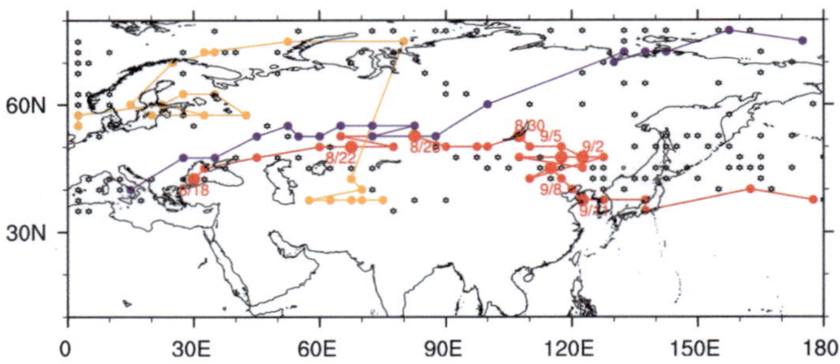

Fig. 7.35 Daily points of MHA centers during August and September in 1999. *Three color lines* show the tracks of three MHA centers. Along the *red line*, "8/18" and "8/22" indicate dates of 8 August and 22 August 1999. Sourced from Qian et al. (2016b) © Springer-Verlag Berlin Heidelberg 2016, with permission

Mediterranean Sea and moved northeastward with different speed for 19 days. For the red track of MHA center, it was also originated in the Mediterranean Sea and moved eastward for about 8 days before arriving at Northwest China on 28 August, 1999. From 30 August to 12 September, this center stayed in North China for 14 days. As seen from Fig. 7.3, it formed a heat wave event from 3 to 13 September 1999 in Central China.

To spatially and temporally examine the relations of nearby MHA systems associated with the temperature anomalies at 850 hPa and the surface temperature anomalies, eight dates marked by big-red dots are shown in Fig. 7.35. Four big-red dots indicated the MHA center in North China and the other four experienced eastward propagation from 18 to 30 August 1999. It is noted that the positive MHA and positive temperature anomalies in the mid-low troposphere are vertically consistent from day to day. A pair of MHA and temperature anomalies, marked by capital letters "H" and "W," is used to describe the signal propagation. Other centers of positive/negative MHA and temperature anomalies exist in the vicinity near the signals of "H" and "W." From 22 to 30 August 1999, a negative MHA center and a negative center of temperature anomalies were located in eastern China and linked with a weak cold event as shown in Fig. 7.2. As a negative MHA center and a negative temperature anomaly were located in Northwest China from 8 to 11 September, the SAT in Northwest China was lower than normal as shown in Fig. 7.1b. On these days, the strongest positive MHA and heat wave event were observed in eastern China (Fig. 7.36f, g).

Spatial frequency distributions of daily positive MHA centers from August to September during 1980–2010 are shown in Fig. 7.37. Most positive MHA centers with the value larger than 120 gpm are located between 35°N and 70°N in the Northern Hemisphere. There are several areas of high-frequency positive MHA centers. The highest frequency of positive MHA center exceeding 20 days during

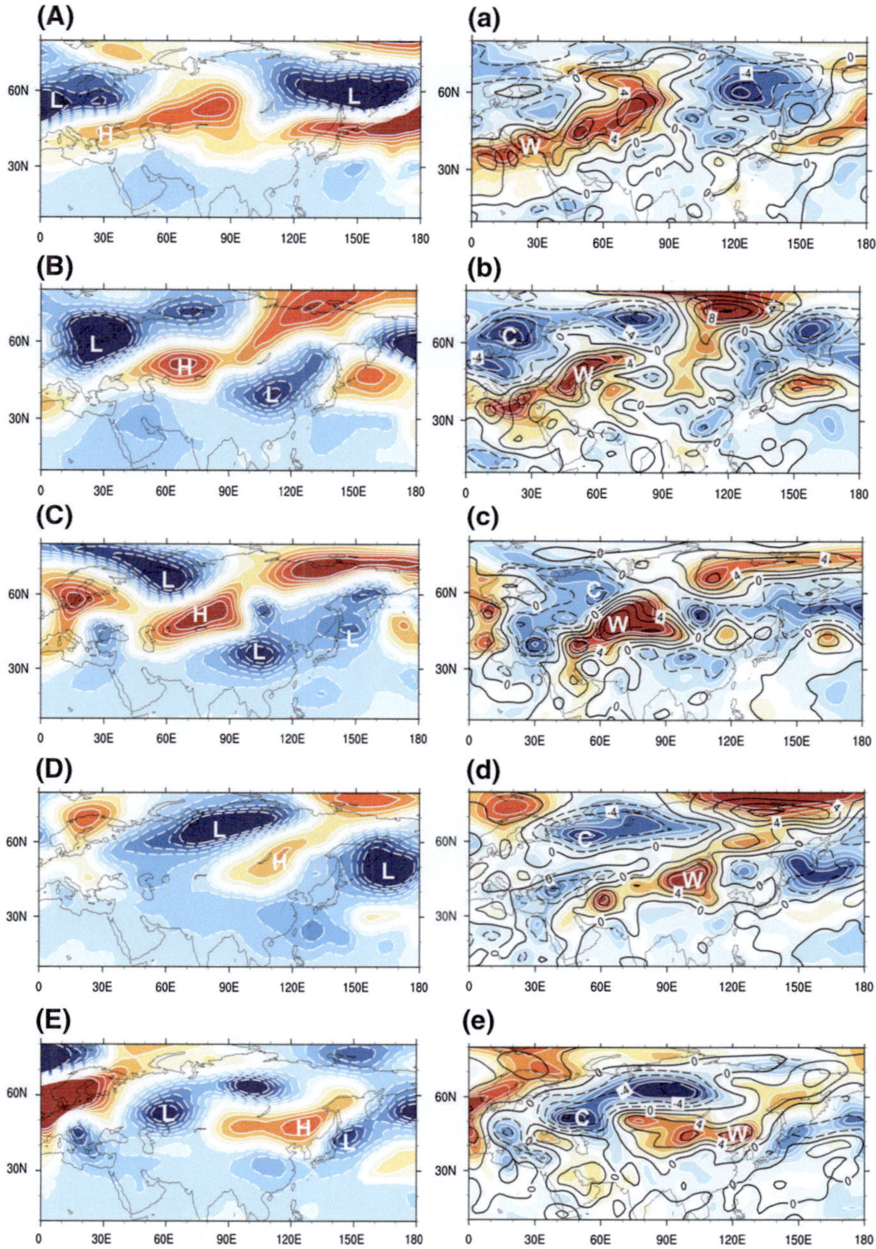
(A)
(a)
(B)
(b)
(C)
(c)
(D)
(d)
(E)
(e)
60N
30N
0
30E
60E
90E
120E
150E
180
L
H
W
C

◀ **Fig. 7.36** Spatial distributions of MHA (*left panel*, 30 gpm interval) as well as temperature anomalies (*shading*, 1 K interval) at 850 hPa (*right panel*) and the surface (*right panel*, *contour*, 2 °C interval) on **A**, **a** 18 August, **B**, **b** 22 August, **C**, **c** 26 August, **D**, **d** 30 August, **E**, **e** 2 September, **F**, **f** 5 September, **G**, **g** 8 September, and **H**, **h** 11 September 1999, respectively. Letters "H" and "L" as well as "W" and "C" indicate the positive and negative centers of height anomalies as well as positive and negative centers of temperature anomalies, respectively. Sourced from Qian et al. (2016b) © Springer-Verlag Berlin Heidelberg 2016, with permission

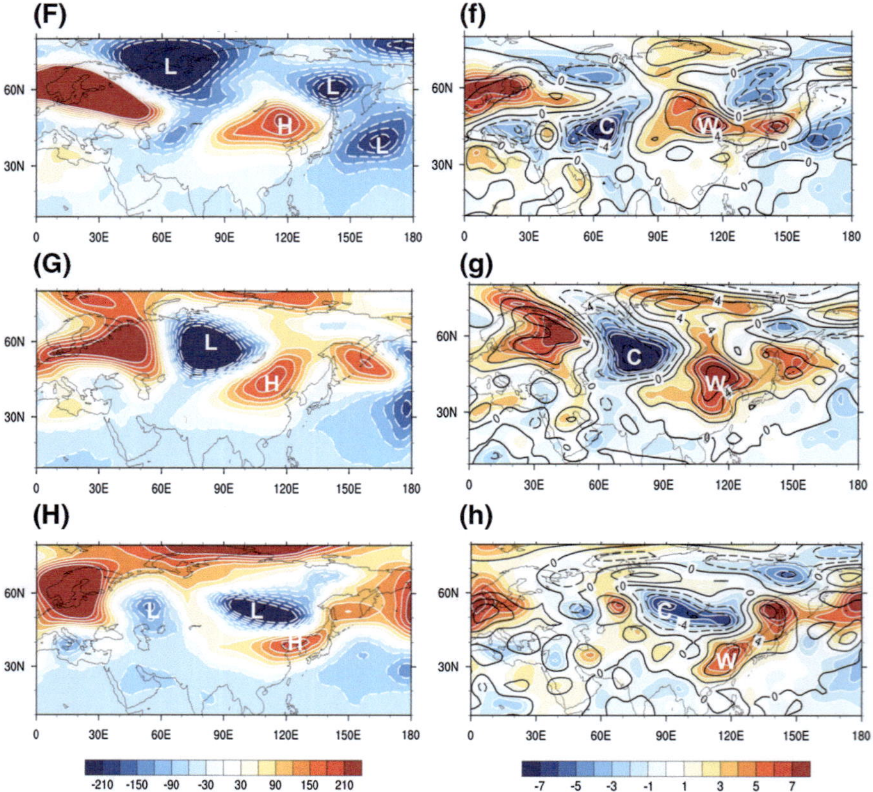

Fig. 7.36 (continued)

the three decades is observed in North China. Other two high frequency centers of positive MHA days are located in the East Sea of Japan and the Middle East. A high-frequency center in the area of 2.5–5.0°E and 55–60°N may be unreal because no other MHAs are used in its western areas. The highest frequency center of MHA from Europe to Asia is also observed in North China, with the central value of 227 days for all seasons and years during 1980–2010.

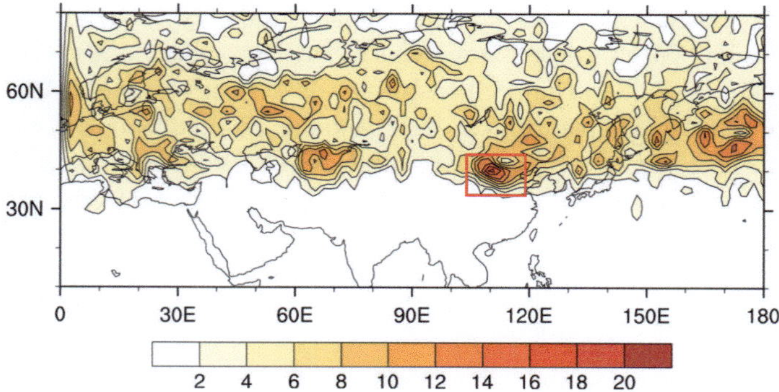

Fig. 7.37 Frequency distributions (days/grid) of daily positive MHA centers (≥ 120 gpm) from August to September during 1980–2010. *Red-line box* denotes the area (35–45°N, 105–120°E) in North China. Sourced from Qian et al. (2016b) © Springer-Verlag Berlin Heidelberg 2016, with permission

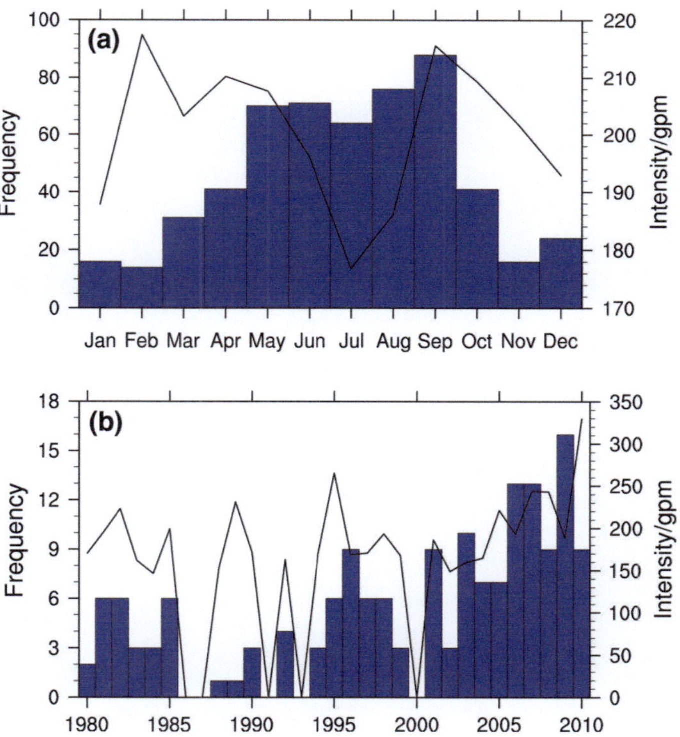

Fig. 7.38 **a** Monthly frequency (*bar*) and monthly mean intensity (*solid line*, gpm) of daily positive MHA centers from January to December during 1980–2010 in North China, **b** annual frequency (*bar*) and intensity (*solid line*, gpm) of daily positive MHA centers from August to September during 1980–2010 in North China. Sourced from Qian et al. (2016b) © Springer-Verlag Berlin Heidelberg 2016, with permission

For this high-frequency area in North China, its monthly frequency and intensity of daily positive MHA centers from January to December during 1980–2010 is illustrated in Fig. 7.38a. The high-frequency days of positive MHA is noted in August (76 days) and September (88 days) and the low-frequency days in February (14 days). There are two periods with high MHA intensity from September to November and from February to May while a month of low MHA intensity in July. Increasing trends of frequency and intensity for the daily positive MHA centers in two months (August and September) are noted from 1980 to 2010, particularly in the last decade (Fig. 7.38b). The highest frequency was observed in 2009 and the strongest intensity occurred in 2010.

It is interesting to examine the relationship between positive MHA centers and surface temperature anomalies regionally. Figure 7.39 shows the positive MHA centers with the value larger than 120 gpm in North China (35–45°N, 105–120°E) and regional-averaged daily mean SATA in Central China (25–35°N, 105–120°E). Generally, higher central values of positive MHA in North China correspond to higher SATAs in Central China. Their statistical relationship between the SATA and MHA can be written as a regressed model

$$\text{SATA} = a \times \text{MHA} + b, \tag{7.5.1}$$

where $a = 0.01345403$ and $b = -0.7464675$. But, there are 21 points (days) showing a negative relation with SATA below the zero line in Fig. 7.39. These differences will be examined in more details. The total 164 days in North China contain 52 positive MHA cases, among which three cases have been traced in Fig. 7.35 as examples.

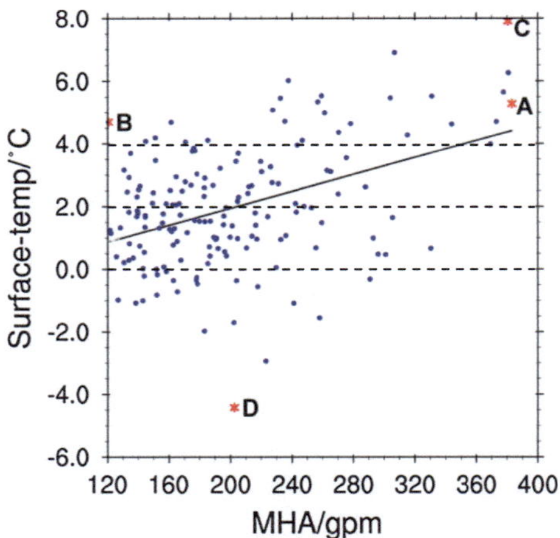

Fig. 7.39 Positive MHA (gpm) centers over the region (35–45°N, 105–120°E) of North China and regional-averaged daily mean SATAs (°C) in Central China (25–35°N, 105–120°E). The *solid line* is a linear regression to 164 points. Four groups of SATA intervals separated by *dashed lines* which are below 0, 0–2, 2–4 °C, and above 4 °C, respectively. Capital letters "A," "B," "C," and "D" denote four special days for MHA and SATA. Sourced from Qian et al. (2016b) © Springer-Verlag Berlin Heidelberg 2016, with permission

Four groups of SATA are categorized in Fig. 7.39 with their temperature anomalies below 0, 0–2, 2–4 °C, and above 4 °C, respectively. For each group, the composite of height and temperature anomalies along 112.5°E is shown in Fig. 7.40. Their vertical structures of height and temperature anomalies are similar to those shown in Fig. 7.33b. A positive center of temperature anomalies is located around 400 hPa and extends its one or two warm axes to the surface. Positive temperature anomalies are dynamically located below the positive MHA center at around 200 hPa. The axes of the downward warm anomalies vary in different groups and cases. Figure 7.40a shows the group composite of largest SATA in Central China by 26 points with an average SATA of 5.1 °C. For this group, the positive MHA is also the largest with an average value of 264.1 gpm. This implies that heat wave event (regional SATA > 4 °C) should be associated with a larger central value of MHA (>264.1 gpm for average). The positive MHA center at 200 hPa and temperature anomaly center at around 400 hPa along with their axes

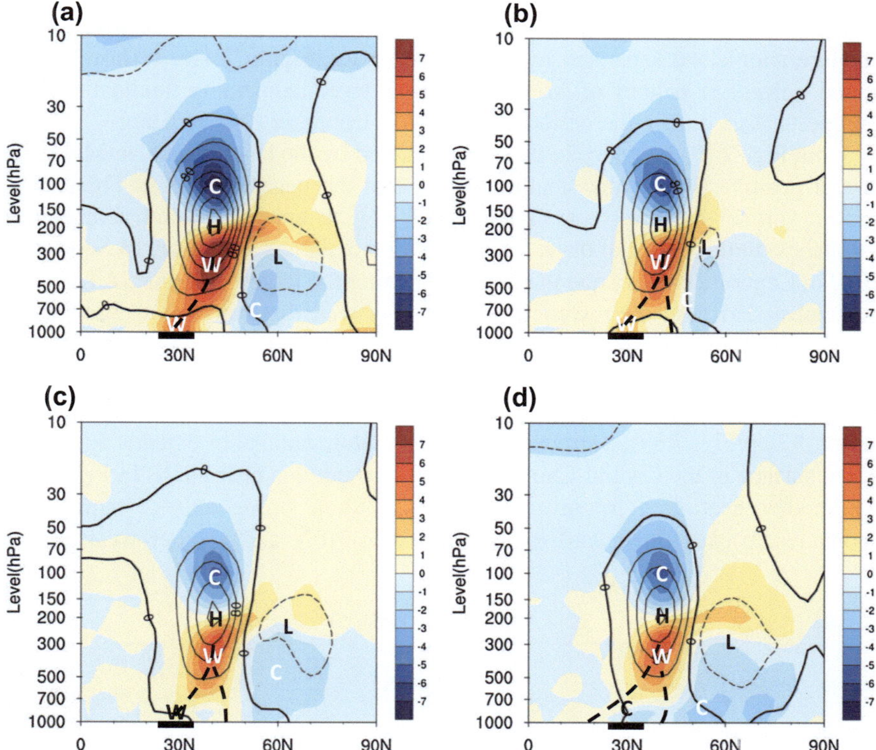

Fig. 7.40 Composites of height (*contour*, 40 gpm interval) and temperature (*shading*, 1 K interval) anomalies along 112.5°E over **a** 26 days, **b** 49 days, **c** 68 days, and **d** 21 days with their temperature anomalies above 4, 2–4, 0–2 °C and below 0 °C, respectively, as separated in Fig. 7.39. The *heavy bar* indicates the place (25°N–35°N) in Central China. The *heavy-dashed line* indicates the axis of positive temperature anomalies. Capital letters "H/L" and "W/C" indicate the positive/negative centers of height anomalies and positive/negative centers of temperature anomalies, respectively. Sourced from Qian et al. (2016b) © Springer-Verlag Berlin Heidelberg 2016, with permission

can be used to indicate the positive SATA in the Central China. Figure 7.40b, c show the two groups with SATA intervals of 0–2 and 2–4 °C in Central China. This shows the relation between surface moderate warm anomaly and MHA. For the two groups, the vertical structure is mostly the same as in Fig. 7.40a except with relatively weak MHAs and temperature anomalies. The last group has a negative relation between the positive MHA and negative SATA in the Central China (Fig. 7.40d). For this group (21 days or 15 cases), the average MHA is also weaker than that of other three groups, while the two warm axes in the mid-low troposphere extend downward to South China and North China rather than to Central China. It means that the location of surface warm temperature anomalies depends upon the vertical structure of tropospheric height–temperature anomalies for weak MHAs.

The vertical pattern in Central China is also confirmed by the composite of 26 points above 4 °C in Fig. 7.40a. The positive center of temperature anomalies around 400 hPa extends down-southward to the surface. Horizontally, a positive center of MHA is located in North China (Fig. 7.41a) and linked to a warm anomaly in Central China from low troposphere (850 hPa) to the surface (Fig. 7.41b). From Europe to Asia, there are two positive centers of MHA (temperature anomalies) separated by a negative center. Figure 7.41a, b show a horizontal consistent pattern of MHA and lower level temperature anomalies from Europe to Asia during the period of heat wave events in Central China.

From Fig. 7.39, four special days can be noted by the largest positive MHA, the largest positive SATA, the smallest MHA, and the negative SATA. Figure 7.42 shows vertical–latitude cross sections of height anomalies and temperature anomalies on three special days. The day with the highest SATA on 18 September 2010 in Central China was on the second day after the largest positive SATA on 17 September 2010, so their vertical distributions were basically similar to that in Fig. 7.33b. Along the meridional direction, they all exhibited two axes of temperature anomalies corresponding to two axes of height anomalies, but their intensities and locations varied from case to case. In Fig. 7.42a for a heat event (point “C” in Fig. 7.39), a strong axis of temperature anomalies extends downward to the surface in the Central China and another weak one near 80°N. In Fig. 7.42b for another heat event (point “B” in Fig. 7.39), a strong axis of temperature anomalies extends downward to the surface in 70°N and another weak one in

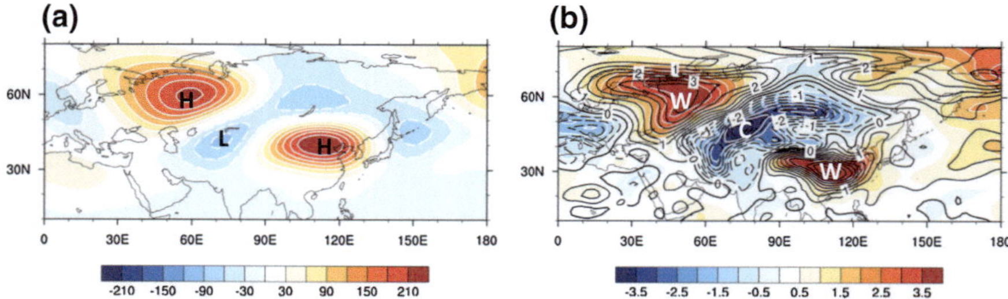

Fig. 7.41 Two horizontal distributions of **a** MHA (30 gpm interval) and **b** temperature anomalies at 850 hPa (*shading*, 0.5 K interval) and the surface (*contour*, 0.5 °C interval) based on the composite of 26 days with temperature anomalies above 4 °C. Sourced from Qian et al. (2016b) © Springer-Verlag Berlin Heidelberg 2016, with permission

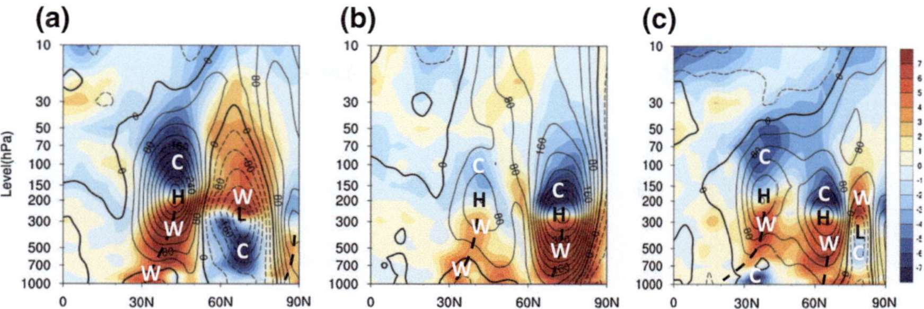

Fig. 7.42 Vertical–latitude cross sections of height anomalies (*contour*, 40 gpm interval) and temperature anomalies (*red* and *blue shadings*, 1 K interval) along **a** 115°E on 17 September 2010, **b** 107.5°E on 15 September 2003, and **c** 112.5°E on 27 September 2008. The *heavy-dashed line* indicates the axis of positive temperature anomalies. Sourced from Qian et al. (2016b) © Springer-Verlag Berlin Heidelberg 2016, with permission

Central China. In Fig. 7.42c, a strong axis of temperature anomalies extends downward to the surface in 65°N and a weak one in South China instead of a weak negative temperature anomaly in Central China and the lower troposphere, so a cold center of SATAs forms in Central China (point "D" in Fig. 7.39).

Most positive MHA centers in North China can be used to indicate possible positive SATA in Central or South China and can also be traced 8 days ahead in average from their initial signals in Europe. But, sometimes, positive MHA centers in North China extend an axis of height and temperature anomalies down-southward to South China and another axis to the surface of North China instead of a negative temperature in Central China. Those large tilted relations have a rate 21/164 (days/total days) or 15/52 (cases/total cases) associated with relatively weak MHA centers in North China. It is certain that a positive MHA arriving at North China will induce a positive SATA in either Central or South China. The actual locations, however, depend upon the vertical structure of height–temperature anomalies in the troposphere. This provides anticipation for predicting the heat event in Central China using the MHA signal. Reversely, during the period from 1980 to 2010 in August and September, 15 cases of heat wave events with the daily maximum temperature exceeding 35 °C over 23 adjacent stations occurred in Central China (25–35°N, 105–120°E). These cases as dry heat waves were listed in the atlas of extreme climate events (Qian 2011). For these events, their early signals from MHAs can be traced with average 9 days in advance.

The positive MHAs arriving in North China (35–45°N, 105–120°E) of leading 10 days and forming heat waves in August and September are listed in Table 7.5. It is noted that positive MHA signal appears 10 day prior to arriving in North China for 16 out of 52 cases. The longest leading days of positive MHA signal were 20 when it originated from the Mediterranean Sea (42.5°N, 30°E) and arrived in North China on 7 September 1999 and lasted for 4 days with an intensity of 232.3 gpm. Only one case occurred in the 1980s and five cases in the 1990s, but the number increased to ten in the 2000s. In Table 7.5, some cases with positive MHAs stayed

Table 7.5 The positive MHA centers arriving in North China (35–45°N, 105–120°E) of leading at least 10 days in August and September

Arriving date (y/m/d)	Days	Intensity (gpm)	First position	Lead. Days
80/S/13	2	203.8	40°N, 37.5°E	11
96/A/7	3	155.6	47.5°N, 65°E	10
97/A/25	2	204.6	45°N, 67.5°E	10
98/A/23	2	174.1	40°N, 60°E	11
98/S/11	4	258	45°N, 47.5°E	12
99/S/7	4	232.3	42.5°N, 30°E	20
01/S/13	9	270	57.5°N, 70°E	12
02/A/1	1	185.3	50°N, 77.5°E	14
02/S/1	2	133.7	42.5°N, 62.5°E	10
03/S/15	2	151.1	52.5°N, 60°E	15
05/S/19	2	237.5	42.5°N, 67.5°E	16
07/S/20	8	305.4	55°N, 52.5°E	18
07/S/25	5	369.2	50°N, 42.5°E	18
08/A/20	1	151	42.5°N, 35°E	17
10/A/2	2	301	62.5°N, 35°E	19
10/S/15	6	383	55°N, 60°E	12

Sourced from Qian et al. (2016b) © Springer-Verlag Berlin Heidelberg 2016, with permission

in North China for only 1–2 days but heat events were longer than 3 days. This is because a heat event sometimes starts before a MHA center arrives and persists after it leaves.

7.6 Negative Height Anomalies and Cold Events[6]

Spatial frequency distributions of daily negative MHA centers in two periods of January–February and October–December during the 1980–2010 are shown in Fig. 7.43 by Qian et al. (2016b). Most negative MHA centers with their values lower than −120 gpm are located between 20°N and 70°N in the Northern Hemisphere. There are several areas of high-frequency of negative MHA centers. Two areas are chosen in North China and South China. Other high-frequency centers of negative MHA are located in the Mediterranean Sea, Southwest Asia, and from Japan to Northwest Pacific. Another high-frequency center along 2.5°E is probably unreal due to the limitation of grid data.

[6]Modified article originally published in *Qian WH, Yu TT, Du J (2016b) A unified approach to trace surface heat and cold events by using height anomaly. Climate Dyn 46: 1647–1664.* Published with kind permission © Springer-Verlag Berlin Heidelberg 2016. All Rights Reserved.

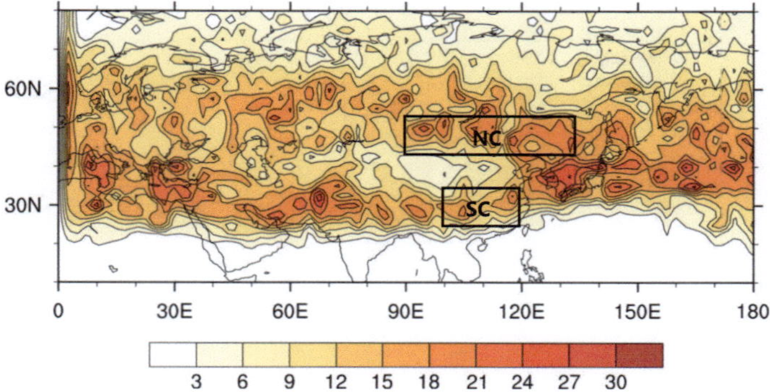

Fig. 7.43 Same as Fig. 7.37 except frequency distributions (days/grid) of daily negative MHA centers (≤ -120 gpm) in January–February and October–December during 1980–2010. *Two boxes* of *NC* (45–55°N, 90–135°E) and *SC* (25–35°N, 100–120°E) denote two areas of North China and South China. Sourced from Qian et al. (2016b) © Springer-Verlag Berlin Heidelberg 2016, with permission

For the two high-frequency areas, monthly frequency and intensity of daily negative MHA centers as well as annual frequency and intensity of daily negative MHA centers in North China from October to November and in South China from January to February during 1980–2010 are illustrated in Fig. 7.44. The monthly frequency difference of MHA is small in North China with two relatively high frequencies in October (190 days) and November (188 days) while the strongest monthly intensity is in February (−270 gpm). The monthly frequency difference of MHA is large in South China with two high frequencies in January (71 days) and February (58 days) while the intensity difference is comparatively small with an average of about −174 gpm. For the series of annual frequency and intensity in North China (Oct–Nov) and South China (Jan–Feb), inter-annual variability is rather large (Fig. 7.44b). In North China, higher frequency reaching 20 days is, respectively, observed in 1980, 1981, 2000, and 2002. In South China, higher frequency reaching 10 days is, respectively, noted in 1983, 1985, and 1992.

It is also interesting to examine the relationship between negative MHA centers and SATAs regionally. Figure 7.45a shows the negative MHA centers and regional-averaged SATAs in North China (45–55°N, 90–135°E). Generally, larger central values of negative MHAs in North China correspond to larger negative SATAs. Statistical relationships between the SATA and MHA in North China can be written as a similar regressed model as Eq. 7.5.1, but with two different coefficients $a = 0.02501119$ and $b = 0.4243307$. 34 points (days) show opposite relations, with positive SATAs in North China. Figure 7.45b shows the negative MHA centers in South China (25–35°N, 100–120°E) and regional-averaged SATAs in the South China coast (17.5–27.5°N, 100–120°E). Again, their basic relationship is that larger central values of negative MHAs in South China correspond to larger negative SATAs in the South China coast. Their statistical relationship in South China has two different coefficients of $a = 0.01900313$ and $b = 1.695035$ in the linear

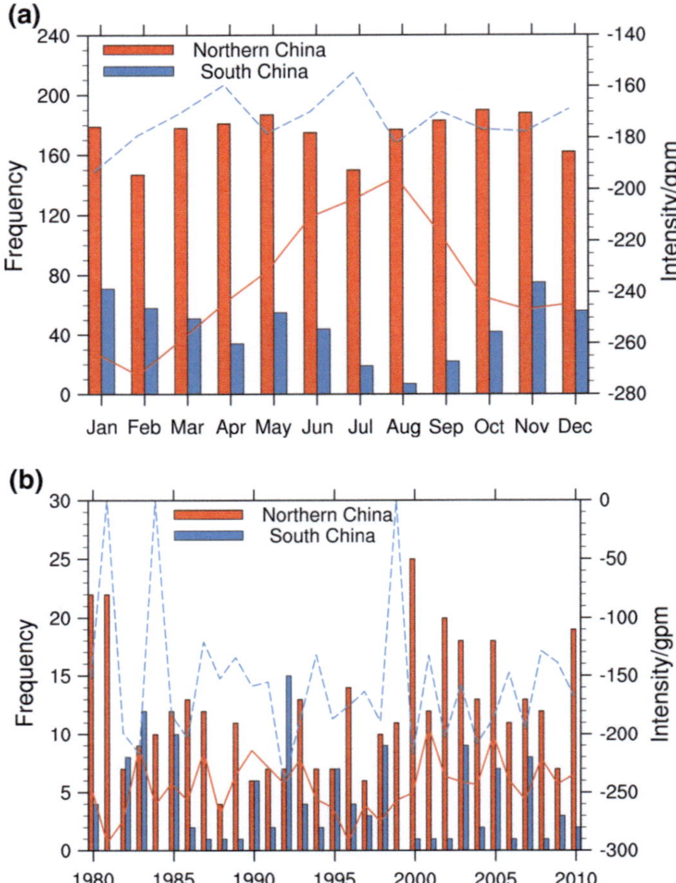

Fig. 7.44 Same as Fig. 7.38 except **a** monthly frequency (*bar*) and intensity (*solid line*, gpm) of daily negative MHA centers and **b** annual frequency (*bar*) and intensity (*solid line*, gpm) of daily negative MHA centers during 1980–2010 in North China for October and November and in South China for January and February. Modified from Qian et al. (2016b) © Springer-Verlag Berlin Heidelberg 2016, with permission

regressed model (Eq. 7.5.1). However, 31 points (days) show opposite relations with positive SATAs in the South China coast.

Four groups of SATA are also categorized in Fig. 7.45a with their temperature anomalies below −10 °C, −10 to −4 °C, −4 to 0 °C, and above 0 °C, respectively. For each group, the composite of height and temperature anomalies along 112.5°E is illustrated in Fig. 7.46. Figure 7.46a shows the group of lowest SATA in North China (55 points) with an average SATA of −13.2 °C. For this group, the negative MHA is the lowest with an average value of −302.4 gpm. The negative MHA center is located at 300 hPa and the negative center of temperature anomalies is at around 700 hPa, resulting in a cold event below them. Figure 7.46b, c examine the two groups of cold events and cold anomalies with the SATA intervals of −8 to −4

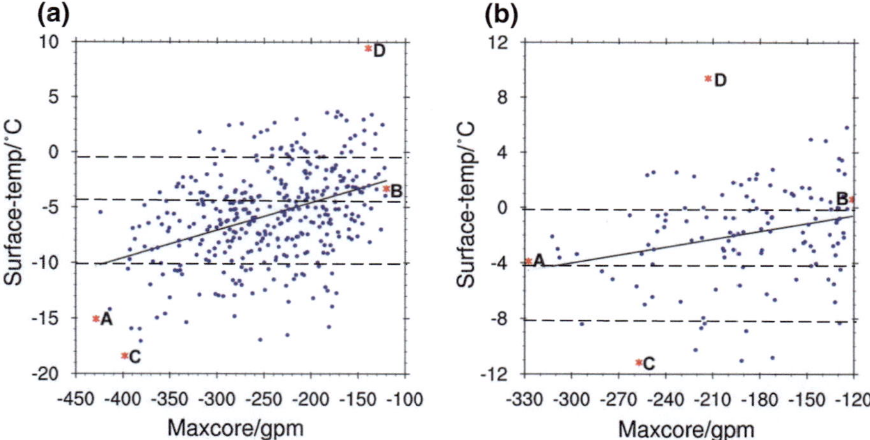

Fig. 7.45 Same as Fig. 7.39 except **a** negative MHA (gpm) centers and regional-averaged SATAs (°C) at North China (45–55°N, 90–135°E) as well as **b** negative MHA (gpm) centers at South China (25–35°N, 100–120°E) and regional-averaged SATAs (°C) in the South China coast (17.5–27.5°N, 100–120°E). In **a**, four groups of SATA intervals above 0 °C, 0 to −4 °C, −4 to −10 °C, and below −10 °C are separated by dashed lines, respectively. In **b**, four groups of SATA intervals above 0 °C, 0 to −4 °C, −4 to −8 °C, and below −8 °C are separated by dashed lines, respectively. Sourced from Qian et al. (2016b) © Springer-Verlag Berlin Heidelberg 2016, with permission

and −4 to 0 °C, respectively, in North China. The vertical structure is the same as in Fig. 7.46a except with relatively weaker MHAs and the SATAs. The above three groups describe the general relation between the cold events (anomalies) and MHAs in North China. The last group has an opposite relation: the negative MHAs are associated with positive SATAs in North China (Fig. 7.46d). For this group (34 points or 13 cases), the average MHA is weaker than that in other three groups, and the cold axis of negative temperature anomalies in the mid-low troposphere extends downward to Central China instead of North China.

For all 251 days (or points) with temperature anomalies below −4 °C in Fig. 7.45a, composite vertical cross sections of height and temperature anomalies along 50°N and 112.5°E in North China are, respectively, illustrated in Fig. 7.47a, b. The negative center of temperature anomalies around 700 hPa is located below the negative center of MHA around 300 hPa and extends down-westward to the surface. A negative MHA center, horizontally, corresponds well with the cold anomaly in North China from low troposphere (850 hPa) to the surface (Fig. 7.47c, d). From western Russia to East Asia, there are two positive centers of MHA and temperature anomaly separated by a negative center in North China. The vertical and horizontal patterns in North China are confirmed by the composite of the 251 points below −4 °C in Fig. 7.47.

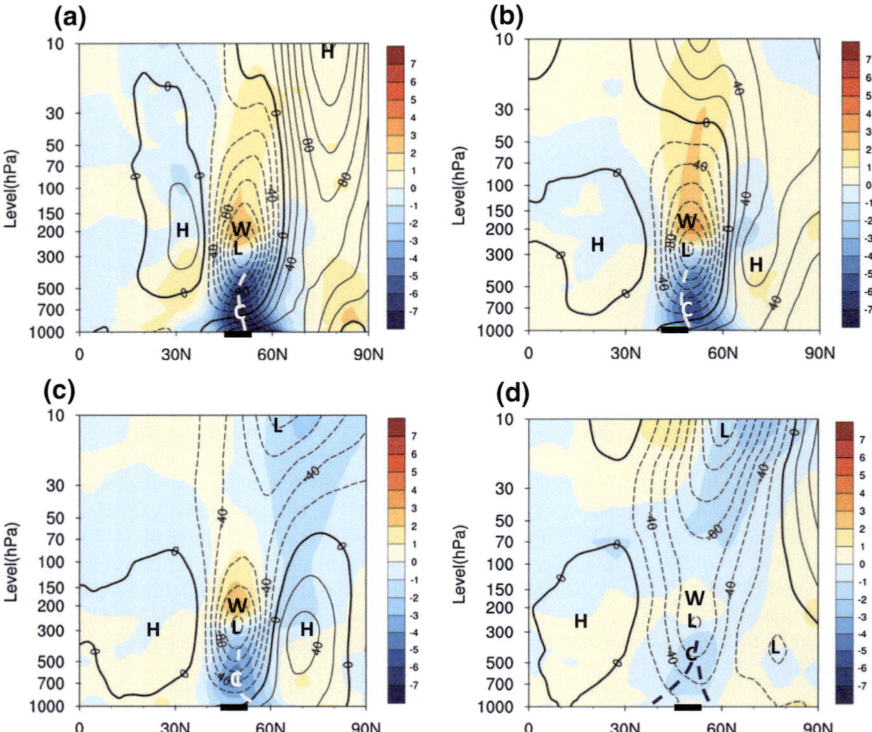

Fig. 7.46 Same as Fig. 7.40 except composites of height and temperature anomalies crossing North China along 112.5°E over **a** 55 days, **b** 196 days, **c** 93 days, and **d** 34 days with their temperature anomalies below −10 °C, −10 to −4 °C, −4 to 0 °C, and above 0 °C as separated in Fig. 7.45a. The *heavy bar* indicates the North China (45–55°N). The *dashed line* indicates the axis of negative temperature anomalies. Sourced from Qian et al. (2016b) © Springer-Verlag Berlin Heidelberg 2016, with permission

All cases from October and November, with a traceable negative MHA signal at least 10 days prior to its arrival in the North China (45–55°N, 90–135°E) and formation of the cold events, are listed in Table 7.6. In 16 out of 97 cases, a negative MHA signal appears 10 days prior to its arrival in the North China. The longest leading negative MHA signal was more than 19 days, which was first detected from Europe (55°N, 2.5°E) and arrived in North China on 24 October 1996 and lasted for 2 days with an intensity of −210.7 gpm. There were eight cases in the 1980s and other eight cases in the 1990s and 2000s.

Other four groups of SATA are categorized in Fig. 7.45b with their temperature anomalies below −8, −8 to −4, −4 to 0 °C, and above 0 °C, respectively. For each group, the composite of height and temperature anomalies along 110°E is shown in

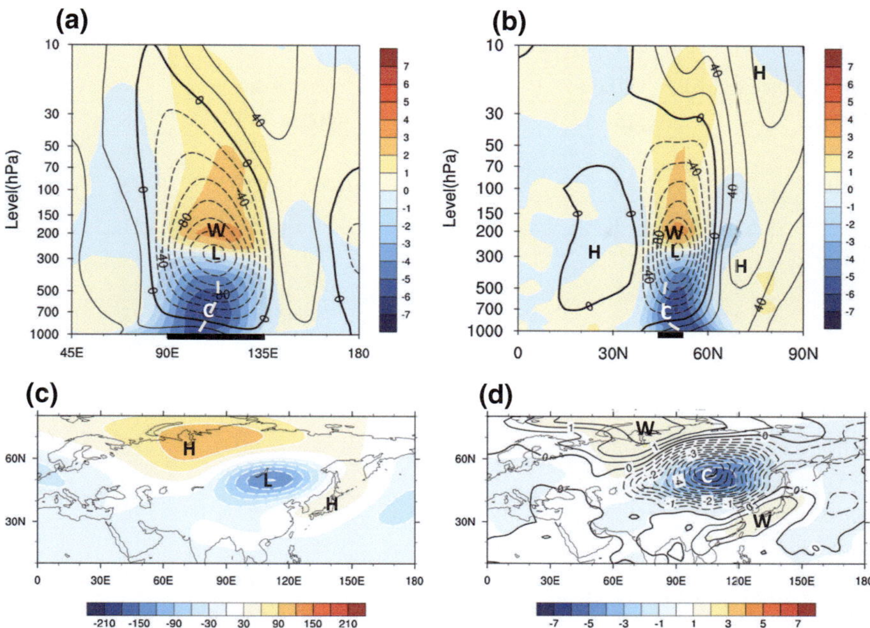

Fig. 7.47 All 251-day composites of height anomalies (*contour*, 20 gpm interval) and temperature anomalies (*shading*, 1 K interval) with temperature anomalies below −4 °C for *two vertical cross sections* along **a** 50°N and **b** 112.5°E in North China as well as *two horizontal distributions* of **c** MHA (*shading*, 30 gpm interval) and **d** temperature anomalies at 850 hPa (*shading*, 1 K interval) and the surface (*contour*, 0.5 °C interval). The *heavy bar* in **a** and **b** indicates the place (90–135°E, 45–55°N) in North China. The heavy-dashed line indicates the axis of negative temperature anomalies. Qian et al. (2016b) © Springer-Verlag Berlin Heidelberg 2016, with permission

Fig. 7.48. A negative center of temperature anomalies is located around 300–400 hPa and extends downward to the surface, forming a negative SATA in the South China coast. This negative temperature anomaly is also located below the negative MHA center at around 200 hPa in South China. The places of the downward-extended cold anomaly vary in different groups, cases, and intensities. Figure 7.48a shows the group of lowest SATA in the South China coast of 10 points with an average SATA of −5.8 °C. For the lowest SATA group, the negative MHA is also the lowest with an average value of −207.3 gpm. The negative center of MHAs at 200 hPa and the negative center of temperature anomalies at around 400 hPa result in cold events in the South China coast. Figure 7.48b shows the second group of cold events with SATA intervals −8 to −4 °C in the South China coast. The vertical structure is the same as that in Fig. 7.48a, except with relatively weaker MHAs and temperature anomalies. The third group in Fig. 7.48c illustrates

Table 7.6 The negative MHA arriving in the North China (45–55°N, 90–135°E) of leading at least 10 days in October and November

Arriving date (y/m/d)	Days	Intensity (gpm)	First position	Lead. Days
81/N/22	5	−246.9	60°N, 20°E	18
82/O/31	2	−210.7	55°N, 2.5°E	19
84/O/16	3	−328.3	55°N, 62.5°E	13
84/N/10	3	−321.6	60°N, 72.5°E	11
85/N/22	6	−328.4	57.5°N, 15°E	12
87/O/16	4	−235.5	60°N, 67.5°E	12
87/N/23	3	−205.9	70°N, 37.5°E	18
89/N/5	8	−301.5	57.5°N, 52.5°E	10
93/N/14	11	−281.4	67.5°N, 62.5°E	12
96/O/24	4	−382.6	70°N, 42.5°E	11
98/N/29	2	−283.1	52.5°N, 7.5°E	15
99/O/14	3	−394.7	75°N, 20°E	10
00/N/2	11	−398.2	60°N, 5°E	10
02/N/11	5	−336.3	55°N, 20°E	15
09/O/18	1	−244.1	67.5°N, 17.5°E	11
10/N/12	3	−314.4	62.5°N, 2.5°E	10

Sourced from Qian et al. (2016b) © Springer-Verlag Berlin Heidelberg 2016, with permission

the relation between weak cold anomalies (−4 to 0 °C) and the negative MHAs in South China. An axis of negative temperature anomalies also clearly extends downward from the negative MHA center to the surface, forming weak cold anomalies. The last group has a negative relation: a negative MHA in South China is associated with a positive SATA in the South China coast (Fig. 7.48d). For this group (31 points or 18 cases), the average MHA is also weaker than that of other three groups, and a cold axis in the mid-low troposphere extends downward to the South China Sea instead of the South China coast and even without reaching the surface.

For all 31 days (or points) with temperature anomalies below −4 °C in Fig. 7.45b, composite vertical cross sections of height and temperature anomalies along 30°N and 110°E crossing the South China are illustrated in Fig. 7.49a, b. The negative center of temperature anomalies around 300–400 hPa is located below the negative MHA center around 200 hPa and extends down-southward to the surface. Horizontally, a negative (positive) MHA center corresponds well with the cold (warm) anomaly in the South China coast (Siberia) from lower troposphere (850 hPa) to the surface (Fig. 7.49c, d). The vertical and horizontal patterns in South China are also confirmed by the composite of 31 points below −4 °C in Fig. 7.49.

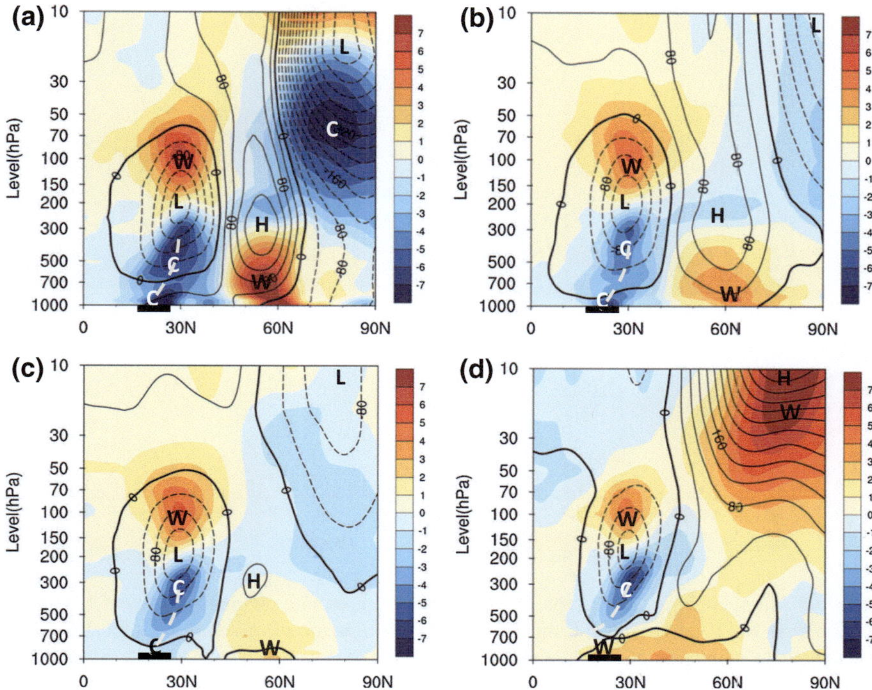

Fig. 7.48 Same as Fig. 7.40 except composites of height and temperature anomalies crossing South China along 110°E over **a** 10 days, **b** 21 days, **c** 65 days, and **d** 31 days with their temperature anomalies below −8 °C, −8 to −4 °C, −4 to 0 °C, and above 0 °C as separated in Fig. 7.45b. The *heavy bar* indicates the South China coast (17.5–27.5°N). The *heavy-dashed line* denotes the axis of negative temperature anomalies. Sourced from Qian et al. (2016b) © Springer-Verlag Berlin Heidelberg 2016, with permission

Since the leading days of negative MHA arriving in South China (25–35°N, 100–120°E) are quite long, Table 7.7 lists only those cases which have caused the cold events with more than 15 leading days in January and February. Out of 54 negative MHA cases, 9 of them arrived in South China leading longer than 15 days. The earliest signal of negative MHA was leading 26 days. It originated from the Mediterranean Sea (40°N, 10°E) and arrived in South China on 17 January 2004 and lasted for 2 days with an intensity of −240.2 gpm. There were 3, 5, and 1 such cases occurred during the 1980s, 1990s, and 2000s, respectively. It implies that the number of negative MHA events arriving in South China possibly has decreased in the last three decades.

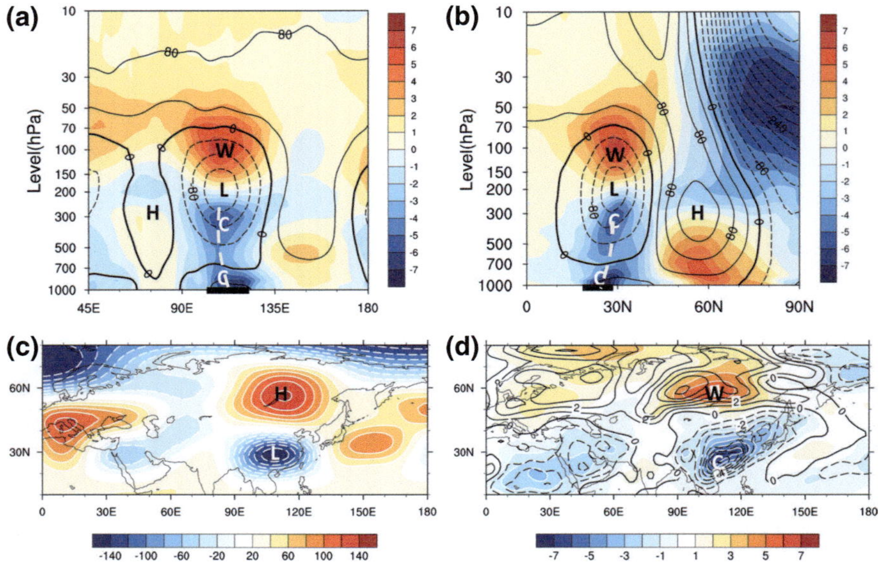

Fig. 7.49 Same as Fig. 7.47 except all 31-day with temperature anomalies below −4 °C along **a** 30°N and **b** 110°E in South China as well as *two horizontal distributions* of **c** MHA and **d** temperature anomalies at 850 hPa and the surface. The *heavy bar* indicates the South China coast (17.5–27.5°N, 100–120°E). The *heavy-dashed line* in **a** and **b** denotes the axis of negative temperature anomalies. Sourced from Qian et al. (2016b) © Springer-Verlag Berlin Heidelberg 2016, with permission

Table 7.7 The negative MHA arriving in the South China (25–35°N, 100–120°E) leading at least 15 days in January and February

Arriving date (y/m/d)	Days	Intensity (gpm)	First position	Lead. Days
80/F/26	1	−138.3	55°N, 42.5°E	17
82/F/10	6	−280.7	37.5°N, 2.5°E	23
83/J/19	5	−327.7	75°N, 2.5°E	19
90/F/25	2	−215.1	67.5°N, 2.5°E	17
91/J/26	2	−182.7	62.5°N, 20°E	15
92/F/6	7	−255.5	65°N, 30°E	22
97/J/19	1	−204.1	17.5°N, 50°E	20
98/J/27	1	−153.3	57.5°N, 20°E	20
04/J/17	2	−240.2	40°N, 10°E	26

Sourced from Qian et al. (2016b) © Springer-Verlag Berlin Heidelberg 2016, with permission

7.7 Summer Standing-Type Heat Events[7]

Standing-type heat events are characterized with stable central locations at a limited area such as in western Russia. In the summer of 2010, western Russia experienced extreme heat, which was noted for its exceptional spatial spread, long duration, high intensity and impacts. Qian et al. (2016a) used the anomaly-based approach to decompose atmospheric variables into daily climatic components and daily anomalies from two reanalysis datasets. They used this approach to analyze all summer regional heat wave events with a spatial scale wider than 2000 km and the duration longer than 5 days over western Eurasia from 1980 to 2014. Results demonstrate that the rapid increase of regional heat wave events over western Eurasia since 2010 is a direct response to the increasing frequency of large-scale, quasi-stationary positive centers of MHAs in the upper troposphere.

Western Russia experienced a record hot summer with a series of heat wave events in 2010, resulting in over 15,000 deaths and many wildfires, and inflicting large economic losses in Russia (WMO 2011; Matsueda 2011; Lau and Kim 2011). Many studies have examined this summer (Grumm 2011; Schubert et al. 2011; Matsueda, 2011; Dole et al. 2011; Schneidereit et al. 2012; Ruti et al. 2014). From monthly to decadal time scales, Schubert et al. (2014) used recent reanalysis products and model simulations to study the relationship among northern Eurasian heat waves, droughts, and long lasting blocking highs. In an attribution study, Dole et al. (2011) reported that the 2010 Russian heat waves were largely natural in origin, whereas Rahmstorf and Coumou (2011) concluded that they were anthropogenic. Addressing this issue first requires a historical examination of the frequency, coverage, and intensity of extreme events in a region over the past 30 years at least.

A strategy for analyzing extreme surface weather events, including heat waves, is to find a connection between SATAs and dynamic indicators of other atmospheric variables aloft. The relationship between this persistent heat wave and temperature–pressure fields, as well as their vertical structure across the troposphere and stratosphere are still not clear. Many previous studies of atmospheric variables were limited to certain levels, for example examining temperature at 850 hPa, geopotential height at 500 hPa, and winds at 200 or 300 hPa (Dole et al. 2011; Lau and Kim 2011; Screen and Simmonds 2013; Watanabe et al. 2013; Katsafados et al. 2014). Therefore, from a spatial view, Qian et al. (2016a) focused on the three-dimensional structures of relevant atmospheric variables to study this heat wave.

The daily NA of atmospheric variable fields has been proven to be useful for diagnosing and predicting surface weather extremes. For example, the effectiveness of NAs, also called standardized anomalies (SAs), in identifying high-impact

---

[7]Modified article originally published in *Qian WH, Wu KJ, Leung JCH (2016a) Three-dimensional structure and long-term trend of heat wave events in western Eurasia revealed with an anomaly-based approach. Inter J Climatol doi:* 10.1002/joc.4634. Published with kind permission © 2016 John Wiley & Sons, Inc. All Rights Reserved.

weather events and ranking their intensity has been demonstrated in previous studies by Grumm and others (Hart and Grumm 2001; Grumm and Hart 2001; Junker et al. 2008, 2009; Graham and Grumm 2010; Du et al. 2014). In particular, Grumm (2011) showed that SAs could help identify the extreme and enduring Russian heat waves in July and August of 2010. However, in recent years, a simpler approach that directly uses the raw anomaly (RA) of atmospheric variables has been introduced in the analysis of anomalously hot and cold events in China. These include winter low temperatures (Jiang et al. 2014; Qian et al. 2015) and summer cold spells (Qian and Jiang 2014). The NA is a ratio in which the RA is divided by its standard deviation (SD) at every spatial grid point. If the SD is defined as 1 at every spatial grid point, the NA becomes the RA. Generally, the RA is better in describing the spatial structure related to a surface anomalous weather event, while the NA is better in indicating the number of abnormality of an event relative to climatology.

Qian et al. (2016a) compared three different relative thresholds for all regional heat wave events in western Eurasia. A heat wave event is characterized by its coverage, intensity, and duration. A heat wave is defined as an event that covers an area larger than 4×10^6 km^2 (i.e., a spatial scale larger than 2000 km) with (1) SAT RAs reaching 5 °C with respect to the local climatology, (2) SATs exceeding the 90th percentile, or (3) SAT NAs larger than 1.5 SDs of daily distribution of 30 years at every grid point. When the daily coverage exceeds 4×10^6 km^2 for 5 consecutive days or 6 (7 or 8) consecutive days with 1 (2 or 3) days having coverage lower than 4×10^6 km^2 at most, a large-scale persistent heat event is identified. The 90th percentile criterion is a running mean of SATs centered on a calendar date for 5 consecutive days (Fang et al. 2008). Qian et al. (2016a) obtained similar results from a running mean of SATs centered on a calendar date for nine or eleven consecutive days. The daily intensity of a heat wave event is defined as the average of SAT RAs within the daily coverage. The duration of a heat wave event is defined by the number of consecutive days with daily coverage greater than 4×10^6 km^2. The daily moving speed of the center for a heat wave event is estimated as the distance between two weighted central locations between 2 consecutive days. The distance is measured in kilometers rather than latitude and longitude.

Similarly, Qian et al. (2016a) defined a large-scale persistent episode of geopotential height anomalies. That is an episode that must have a coverage greater than 4×10^6 km^2 for at least 5 consecutive days with threshold of (1) the height RA above 150 gpm, (2) the height above 90th percentile, or (3) the height NA over 1.5 SDs. They only analyzed SAT events and height anomalous episodes with daily weighted centers located in western Eurasia (40–70°N, 0–90°E).

Based on the definition in Sect. 7.1, Qian et al. (2016a) have chosen the RA approach to analyze all heat wave events during 1980–2014 in western Eurasia. The heat wave area on 7 August 2010 in western Russia was located in the central part of western Eurasia, so they expanded the study domain to western Eurasia (40–70°N, 0–90°E). Using the threshold of SAT RAs exceeding 5 °C with respect to their local climatology, Fig. 7.50a shows the time series of coverage, intensity, and

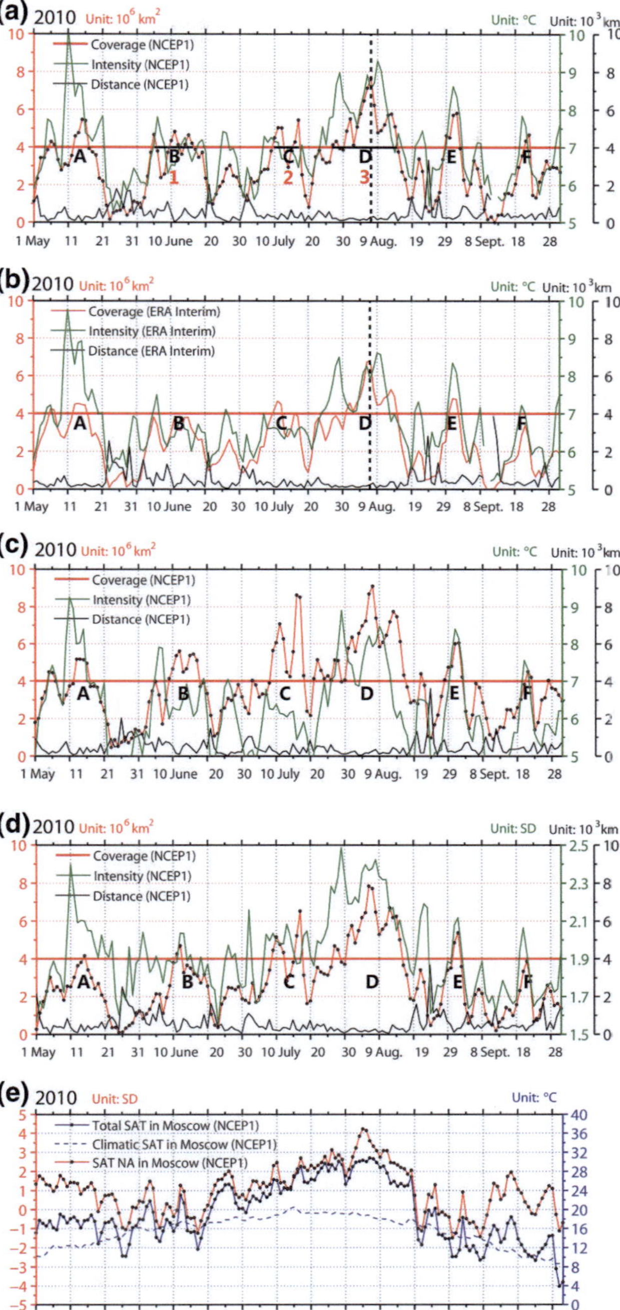
(a) 2010 Unit: 10^6 km^2
Unit: °C
Unit: 10^3 km
Coverage (NCEP1)
Intensity (NCEP1)
Distance (NCEP1)
A B C D E F
1 2 3
1 May 11 21 31 10 June 20 30 10 July 20 30 9 Aug. 19 29 8 Sept. 18 28
(b) 2010 Unit: 10^6 km^2
Unit: °C
Unit: 10^3 km
Coverage (ERA Interim)
Intensity (ERA Interim)
Distance (ERA Interim)
A B C D E F
1 May 11 21 31 10 June 20 30 10 July 20 30 9 Aug. 19 29 8 Sept. 18 28
(c) 2010 Unit: 10^6 km^2
Unit: °C
Unit: 10^3 km
Coverage (NCEP1)
Intensity (NCEP1)
Distance (NCEP1)
A B C D E F
1 May 11 21 31 10 June 20 30 10 July 20 30 9 Aug. 19 29 8 Sept. 18 28
(d) 2010 Unit: 10^6 km^2
Unit: SD
Unit: 10^3 km
Coverage (NCEP1)
Intensity (NCEP1)
Distance (NCEP1)
A B C D E F
1 May 11 21 31 10 June 20 30 10 July 20 30 9 Aug. 19 29 8 Sept. 18 28
(e) 2010 Unit: SD
Unit: °C
Total SAT in Moscow (NCEP1)
Climatic SAT in Moscow (NCEP1)
SAT NA in Moscow (NCEP1)
1 May 11 21 31 10 June 20 30 10 July 20 30 9 Aug. 19 29 8 Sept. 18 28

◂ **Fig. 7.50** Daily time series of the coverage (*red curve*, 10^6 km^2; each *black dot* represents 1 day) with SAT RAs exceeding 5 °C, the average intensity in the coverage area (*green line*, °C) and the consecutive 2-day moving distance (*black line*, 10^3 km) of the SAT RA center derived from **a** NCEP R1 product, and **b** ERA-Interim product. **c** and **d** are the same as **a** except using the 90th percentile threshold and the 1.5 SD threshold of SAT NAs, respectively. **e** shows the SAT (*blue line with black dots*), climatic SAT (*blue-dashed line*), and SAT NA series (*red line with black dots*) near Moscow at the grid point (37.5°E, 55°N). The *thick black-dashed line denotes* 7 August 2010 in **a** and **b**, and the *thick black solid lines* in **a** indicate three persistent large-scale heat wave events marked by *red numbers 1*, *2*, and *3*. The *capital letters* denote six positive SATA processes. The *thick-red straight line* denotes the areal coverage of 4 × 10^6 km^2. Sourced from Qian et al. (2016a) © 2016 John Wiley & Sons, Inc., with permission

duration in western Eurasia from 1 May to 30 September 2010. In total, 41 days have daily coverage exceeding 4 × 10^6 km^2. The heat wave with the longest duration (19 days) and an average intensity of +8.24 °C, occurred from 27 July to 14 August 2010 (the thick black solid line labeled 3). The strongest daily intensity (+10.12 °C) occurred on 11 May 2010 (green peak) but its coverage was smaller than 4 × 10^6 km^2.

Using the same approach but different reanalysis products, six heat processes (labeled with letters "A" to "F") can be compared between Fig. 7.50a, b. Figure 7.50c, d, respectively, show the results using the other two relative thresholds (the 90th percentile and the 1.5 SD). Three large-scale persistent heat wave events can be identified from the series of coverage and intensity in the first four panels. The first event, marked by number "1" (or letter "B"), occurred during June 10–16 in Fig. 7.50a and June 10–19 in Fig. 7.50c. The second and third events occurred during 10–17 July and 27 July–14 August in Fig. 7.50a. However, Fig. 7.50c shows that they combined as a single longer event from 9 July to 16 August. In other words, this long event can be separated into two smaller ones, from 9 to 18 July and 21 July–16 August (Fig. 7.50c). The fourth persistent large-scale event in Fig. 7.50c occurred from 28 August to 2 September (6 days), but from 29 August to 1 September (4 days) in Fig. 7.50a. Six heat processes (letters "A" to "F") were also apparent in Fig. 7.50d, with a major heat wave event persisting from late July to the middle of August. The correlation coefficients between the intensity series of Fig. 7.50a & b, a & c, and a & d are 0.93, 0.94, and 0.82, respectively, while the correlation coefficients between the coverage series are 0.96, 0.92, and 0.92. All correlations reach the 99% confidence level. This result shows that the frequency and duration of heat wave events derived from the three relative thresholds are roughly comparable.

The strongest heat wave event (letter "D" in Fig. 7.50) occurred in western Russia from late July to middle August 2010. In Moscow, the heat wave occurred from 15 July to 18 August, when the daily mean SAT exceeded 25 °C and the SAT NA exceeded 1.0 SD (Fig. 7.50e). The duration of the heat wave event was consistent with the hottest days in Moscow as described by Dole et al. (2011) and Grumm (2011). Usually, the coverage of a heat wave is larger and the intensity is stronger when the center of an event is quasi-stationary, moving at a slow speed or over a short distance.

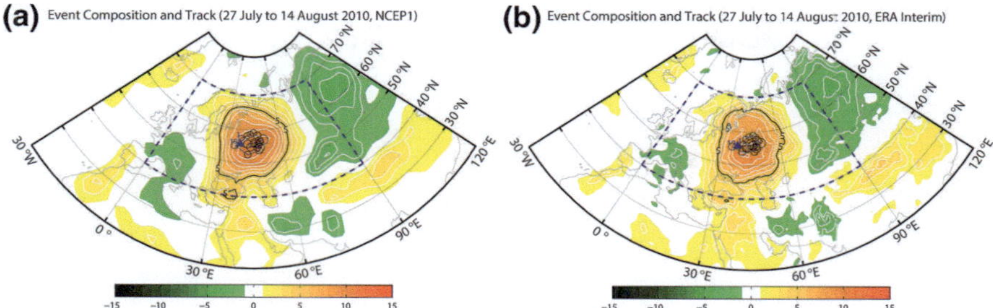

Fig. 7.51 Composites of SAT RAs using **a** NCEP R1 product at the 0.995 sigma level and **b** ERA-Interim product at the 2 m level over the period from 27 July to 14 August 2010. Western Eurasia (0–90°E, 40–70°N) is indicated by the *blue-dashed line*. The *black solid lines* in **a** and **b** surround the domain of SAT anomalies higher than 5 °C, and the *hollow circles* show the weighted central locations of daily SAT RAs exceeding 5 °C. A *larger hollow circle* denotes a stronger intensity. Sourced from Qian et al. (2016a) © 2016 John Wiley & Sons, Inc., with permission

Using the two reanalysis datasets (NCEP R1 and ERA-Interim), Fig. 7.51a, b show the SAT RA composites of the strongest event, from 27 July to 14 August 2010. Moscow (blue triangle) is surrounded by 19 daily weighted central points in the corresponding coverage with SAT RAs exceeded 5 °C. Another area of high SAT RAs is located in North China (30–40°N and 90–120°E). In addition to the areas with positive SAT RAs, two areas of negative SAT RAs are located in Siberia and southern Europe. A similar positive-negative pattern is also shown by the results of the percentile and the NA thresholds (figures not shown).

Since anomaly-based weather analysis has an advantage over total-based weather analysis (Qian and Jiang 2014; Qian et al. 2015), the anomaly-based approach is also used to study heat wave events in western Eurasia. The above analysis demonstrated that the largest daily coverage was observed in western Russia on 7 August 2010 (Fig. 7.50a–d). Figure 7.52 showed vertical cross sections of the observed (or reanalysis) versus climatic height and temperature as well as their RAs along 55°N crossing the daily heat wave center on that day. In the vertical section of the total field, there are obvious deviations of the total temperature from the climatic temperature (red and black dashed contours in Fig. 7.52a), but the deviations of total height from climatic height (red and black solid contours in Fig. 7.52a) are too small to be visible due to the large contour interval. Thus, these deviations in the traditional synoptic chart are not good indicators of the SAT RAs. However, detailed positive and negative centers of height and temperature RAs in the vertical structure from the troposphere to the stratosphere are clearly revealed in the anomaly-based RA synoptic chart (Fig. 7.52b).

The vertical structure of temperature and height RAs from the troposphere to the stratosphere shows that the surface heat wave in western Russia is associated with a strengthened warm air column below a positive center of height RAs. In contrast, the surface cold anomaly in Siberia is associated with a strengthened cold air column below a negative center of height RAs (Fig. 7.52b). The maximum positive

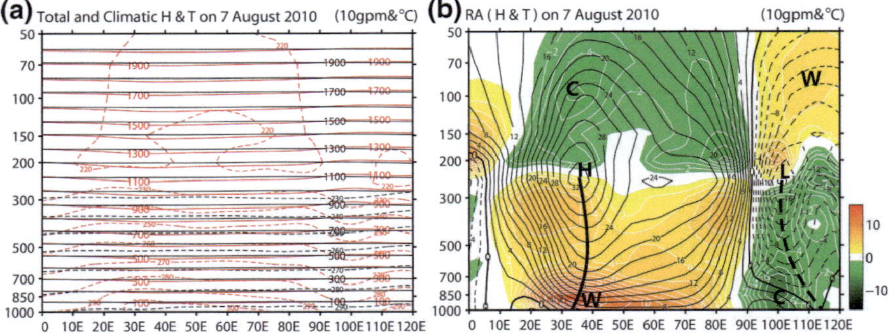

Fig. 7.52 Vertical cross sections of **a** total and climatic heights (*red-* and *black-solid contours*, respectively; 100 × 10 gpm interval) as well as total and climatic temperatures (*red-* and *black-dashed contours*, respectively; 10 °C interval) along 55°N on 7 August 2010 using NCEP R1 data. **b** Height RAs (*contour*, 2 × 10 gpm interval) and temperature RAs (*shading*, 1 °C interval). In **b**, the *thick solid line* and the *thick-dashed line* show the axes of height RAs extending downward from positive (*H*) and negative (*L*) centers of height anomalies, and the letters "W" and "C" indicate centers of positive and negative temperature RAs, respectively. Sourced from Qian et al. (2016a) © 2016 John Wiley & Sons, Inc., with permission

and negative centers of temperature RAs are located at approximately 925 hPa, which correspond to the surface heat and cold anomalous regions in western Russia and Siberia. Geopotential height anomalies at 700 and 500 hPa are not enough to describe surface extreme weather events. Figure 7.52b shows that a positive (or negative) center of the MHA in the upper troposphere is sufficient to indicate a surface weather extreme. For a surface heat (cold) anomalous event, a negative (positive) center of temperature anomalies near 100 hPa and a positive (negative) center of temperature anomalies at 925 hPa are a dynamical response to a positive (negative) MHA centered at approximately 250 hPa. A positive (negative) MHA center in the upper troposphere usually corresponds to a positive (negative) temperature anomaly below, satisfying the hydrostatic relationship. The vertical structure of a warm (cold) air column below the positive (negative) MHA center is a typical vertical pattern, namely the vertical warm (cold) anomalous pattern. This vertical pattern was also found in all cold and warm daily cases in eastern China (Qian et al. 2015; Qian et al. 2016b).

As illustrated in Fig. 7.52b, the maximum anomalous signals are approximately located at 300 hPa for height RAs and at 925 hPa for temperature RAs. Therefore, height RAs at 300 hPa and temperature RAs at 925 hPa are drawn to indicate SAT anomalies (Fig. 7.53a). On 7 August 2010, a wide area of positive height RAs at 300 hPa corresponded well to the surface heat wave in western Russia with two centers of negative height anomalies in Siberia and southern Europe (Fig. 7.53a). Using the 90th percentile threshold, a large area of height anomalies is also observed in western Russia at the same level (Fig. 7.53b). The two different thresholds show similar spatial coverage of height and temperature anomalies, indicating the surface heat wave.

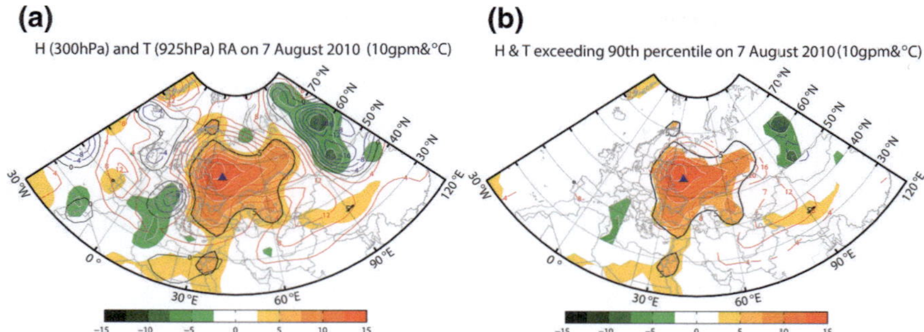

Fig. 7.53 **a** Height RAs (*contour*, 4 × 10 gpm interval) at 300 hPa and temperature RAs (*shading*, 2.5 °C interval) at 925 hPa. **b** Same as **a** but for height (*contour*) and temperature (*shading*) RAs exceeding the 90th percentile and below the 10th percentile (positive for above the 90th percentile, negative for below the 10th percentile) on 7 August 2010. The *thick black line* in **a** and **b** indicates the 5 °C isotherm of SAT RAs. Sourced from Qian et al. (2016a) © 2016 John Wiley & Sons, Inc., with permission

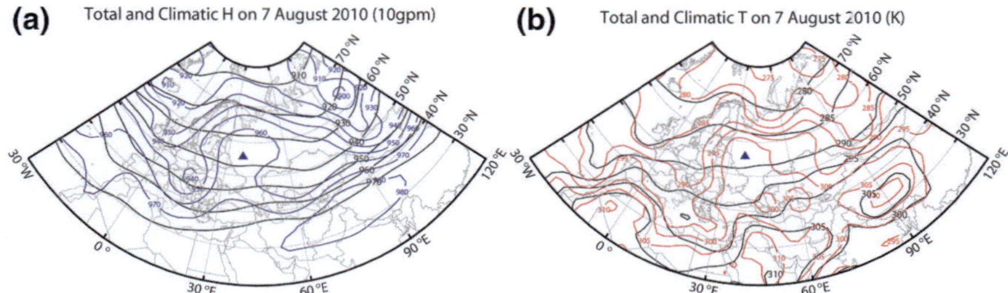

Fig. 7.54 Same as in Fig. 7.53 except for **a** total and climatic heights at 300 hPa (*blue* and *black contours*, respectively; 10 × 10 gpm interval) as well as **b** total and climatic temperatures at 925 hPa (*red* and *black contours*, respectively; 5 °C interval). Sourced from Qian et al. (2016a) © 2016 John Wiley & Sons, Inc., with permission

Traditional daily total-field synoptic charts of height at 300 hPa and temperature at 925 hPa on 7 August 2010 were compared with their climatic means in Fig. 7.54a, b. A height ridge in western Russia separates two height troughs located in Europe and Siberia at 300 hPa (Fig. 7.54a). Similarly, a warm ridge in western Russia separates two cold troughs in Europe and Siberia at 925 hPa (Fig. 7.54b). This is a blocking high in temperature–pressure fields illustrated by the traditional synoptic chart. Forecasters generally pay more attention to the synoptic system of blocking highs, which indicate anomalous and persistent weather events. However, an event's coverage and intensity cannot be objectively or quantitatively evaluated based on the distribution of the total height field or the total temperature field, as shown in Fig. 7.54. This means that it is also difficult to quantitatively determine the duration of a heat wave event using the traditional synoptic charts and absolute thresholds.

Figure 7.52b suggests that a surface heat wave event is directly connected to a standalone positive MHA center in the upper troposphere. In the middle latitudes, height RAs reach their maximum at approximately 300 hPa (Qian and Jiang 2014; Qian et al. 2015, 2016b). Thus, Qian et al. (2016a) attempted to define height RA episodes and relate them to the surface heat wave events. At 300 hPa, the daily coverage of a height RA episode is where the height RAs exceed the criterion of 150 gpm. They also calculated the daily average intensity and weighted central location within the coverage, which can obtain the distance that the height RA center moved over 2 consecutive days in western Eurasia from 1 May to 30 September 2010. The duration of a persistent height RA episode is defined by the number of consecutive days when the daily coverage is greater than 4×10^6 km^2, as defined for the surface heat wave event. Figure 7.55a shows daily time series of the coverage, intensity averaged over the coverage, and consecutive 2-day moving distance of the height anomalous center in western Eurasia. A large-scale persistent height RA episode can be identified by an area exceeding 4×10^6 km^2 for several consecutive

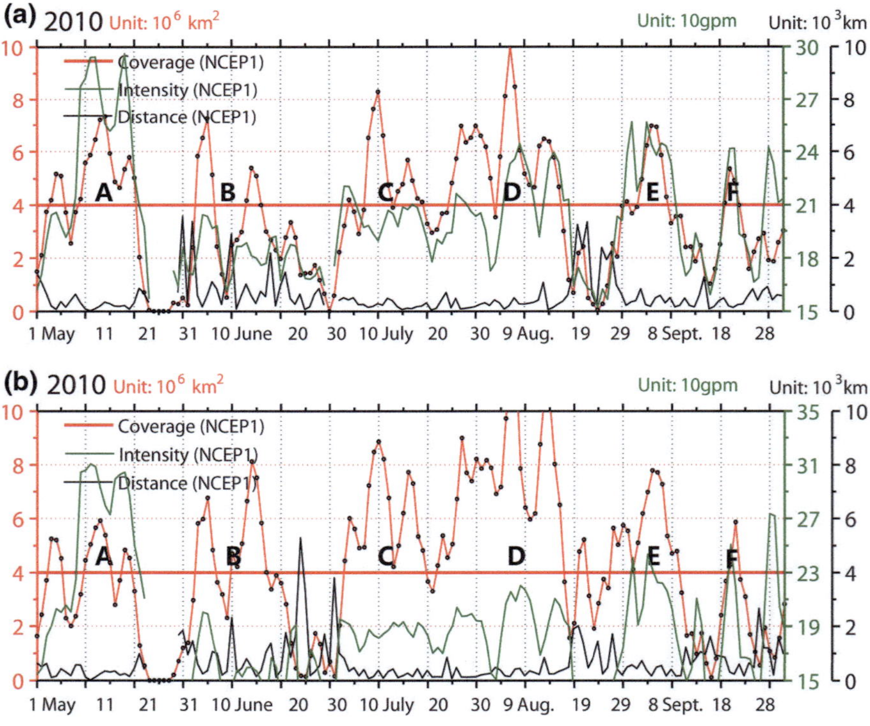

Fig. 7.55 **a** Daily time series of the coverage (*red line*, 10^6 km^2) with height RAs exceeding 150 gpm at 300 hPa, the intensity averaged over the coverage (*green line*, 10 gpm) and the consecutive 2-day moving distance (*black line*, 10^3 km) of height anomalous centers derived from NCEP R1 data in western Eurasia. **b** is the same as **a** except with positive height anomalies exceeding the 90th percentile. The *thick-red straight line* denotes the coverage of 4×10^6 km^2. Sourced from Qian et al. (2016a) © 2016 John Wiley & Sons, Inc., with permission

days. The six processes of major height RAs (indicated by letters "A" to "F") are comparable with those in Fig. 7.50a, b. It is noted that the surface heat wave events are generally associated with height RA episodes in the same region. A large and persistent positive coverage of height RAs aloft corresponds well with a large and persistent coverage of surface heat wave event at the surface. The correlation coefficients between the two coverage series and the two intensity series in Figs. 7.50a and 7.55a are 0.71 and 0.53, respectively, both reaching the 99% confidence level. This result is consistent with the conclusions of Chen and Newman (1998) and Schubert et al. (2011), who demonstrated that quasi-stationary Rossby waves precede the occurrence of severe heat waves in North America. These quasi-stationary Rossby waves or blocking highs are often seen in the middle-high latitudes (Matsueda 2011; Schneidereit et al. 2012; Ruti et al. 2014). In contrast, most positive and negative height RAs have been identified as moving systems in eastern China (Qian et al. 2015; Qian et al. 2016b), as described in Sects. 7.2, 7.5 and 7.6.

Three time series of coverage, intensity, and moving distance derived from positive height anomalies exceeding the 90th percentile are shown in Fig. 7.55b. The correlations between the two coverage series and two intensity series from Fig. 7.55a, b are 0.87 and 0.92, respectively, both reaching the 99% confidence level. This demonstrates that both different thresholds can be used to identify the same height anomaly or the same heat wave event.

Many studies have addressed the intensity and frequency of heat wave events from a regional to global scale (Rahmstorf and Coumou 2011; Peterson et al. 2012, 2013; Screen and Simmonds 2014). As in Fig. 7.50a, Qian et al. (2016a) only investigated the first 24 large-scale persistent heat wave processes with durations longer than 5 days for the last 35 years. Figure 7.56 and Table 7.8 show daily time series of the coverage with SAT RAs exceeding 5 °C and height RAs exceeding 150 gpm in western Eurasia, derived from the NCEP R1 product. In the 1980s, only one heat wave event occurred, a 7-day event in May of 1987. In the 1990s, four heat wave events, including a total of 24 days, reached the SAT RA threshold in 1991 and 1998. Nine heat wave events, including a total of 90 days over 6 years, occurred during the 2000s. During the first 5 years of the 2010s, 10 heat wave events (85 days) reaching the SAT RA threshold have already occurred in 2010, 2011, 2012, and 2014. The longest heat wave event persisted from 27 July to 14 August 2010 and its strength (defined as the sum of daily coverage multiplied by daily mean intensity) reached 830×10^6 km^2 °C. The hottest summer was in 2010, when three heat wave events occurred over 34 days and with a total strength of 1272×10^6 km^2 °C (Table 7.8). The second longest heat wave event occurred over 12–27 May 2014. A well-known heat wave event in 2003, which has been widely studied over the last decade (e.g., Beniston 2004; Garcia-Herrera et al. 2010), persisted from 28 August to 8 September and was the sixth strongest event. The daily intensity and coverage in 2003, measured by the SAT RAs and the height RAs, were weaker than those of the events in 2010 and 2014. These cases demonstrate that the intensity and frequency of large-scale persistent heat wave events in western Eurasia have increased over the last 35 years.

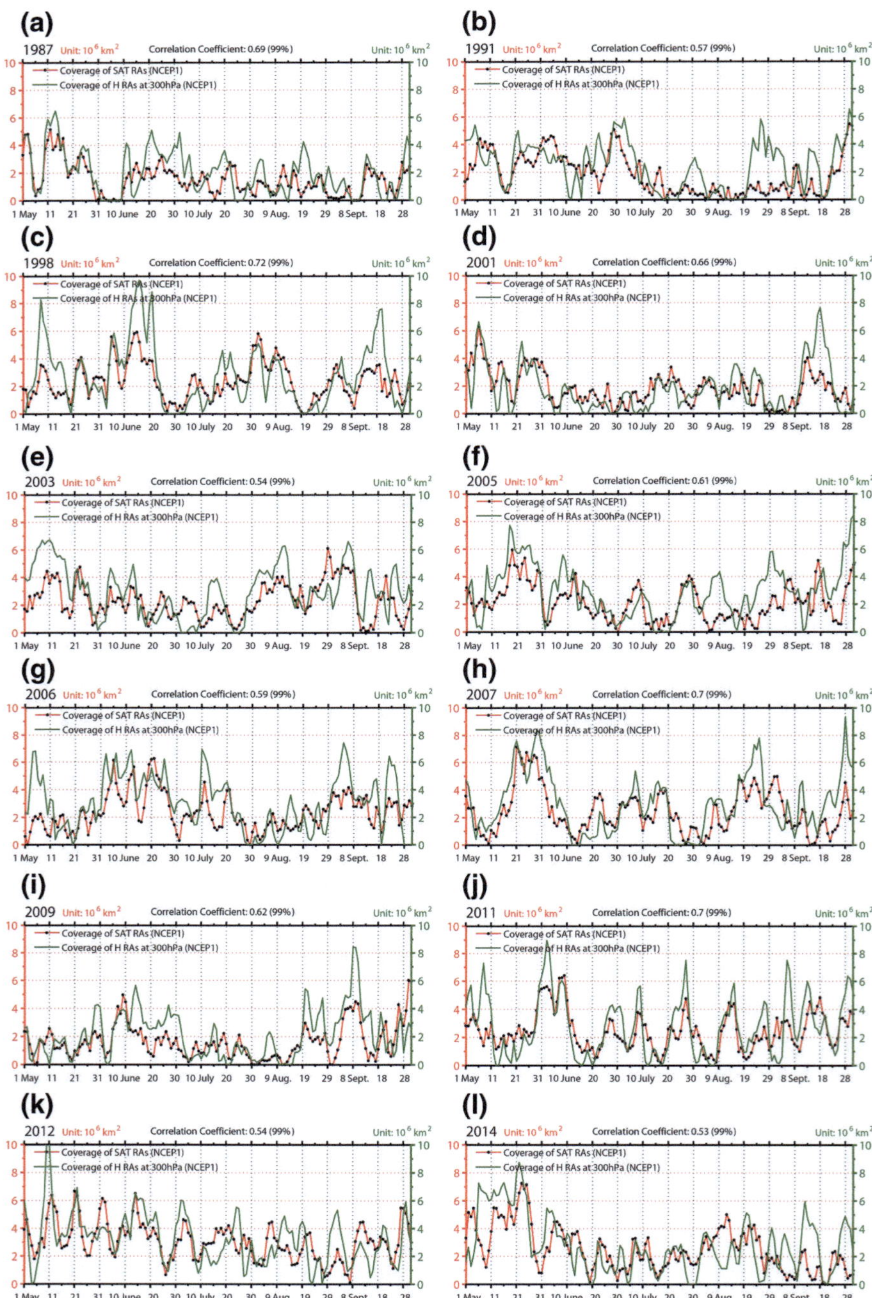

Fig. 7.56 Daily time series of the coverage (10^6 km^2) of SAT RAs (*solid line*) exceeding 5 °C and height RAs (*dashed line*) exceeding 150 gpm in western Eurasia for the other 12 summers from 1 May to 30 September in **a** 1987, **b** 1991, **c** 1998, **d** 2001, **e** 2003, **f** 2005, **g** 2006, **h** 2007, **i** 2009, **j** 2011, **k** 2012, and **l** 2014. Sourced from Qian et al. (2016a) © 2016 John Wiley & Sons, Inc., with permission

Table 7.8 Persistent (exceeding 5 days) regional (4×10^6 km^2) heat wave events organized by their start and end dates as well as durations (days) over western Eurasia (40–70°N, 0–90°E) during the summer (MJJAS) from 1980 to 2014

Year	Start–end dates	Duration/strength	Order	Year	Start–end dates	Duration/strength	Order
1987	11 May–17 May	7/210	15	2009	6 Sep–10 Sep	5/147	22
1991	6 May–12 May	7/226	12	2009	26 Sep–1 Oct	6/180	18
1991	1 Jun–5 Jun	5/217	14	2010	10 Jun–16 Jun	7/204	16
1998	12 Jun–18 Jun	7/264	8	2010	10 Jul–17 Jul	8/238	11
1998	31 Jul–4 Aug	5/163	20	2010	27 Jul–14 Aug	19/830	1
2001	3 May–7 May	5/175	19	2011	30 May–10 Jun	12/516	4
2003	10 May–14 May	5/136	24	2011	14 Sep–18 Sep	5/147	21
2003	28 Aug–8 Sep	12/416	6	2012	10 May–14 May	5/193	17
2005	18 May–24 May	7/261	9	2012	30 May–3 Jun	5/225	13
2006	18 Jun–25 Jun	8/286	7	2012	9 Jun–21 Jun	13/433	5
2007	20 May–1 Jun	13/545	3	2014	12 May–27 May	16/692	2
2007	17 Aug–24 Aug	8/250	10	2014	9 Aug–13 Aug	5/140	23

Event strength (10^6 km^2 °C) is calculated by the sum of daily coverage (10^6 km^2) multiplied by daily mean intensity (°C) for each event's duration

Sourced from Qian et al. (2016a) © 2016 John Wiley & Sons, Inc., with permission

Figures 7.50a and 7.55 show that surface heat wave processes correspond well with positive height RAs episodes aloft. Thus, the height RA episodes of other 12 summers are given in Fig. 7.56. The correlation coefficients between the two coverage series of SAT RAs and height RAs over western Eurasia in the 12 summers range from 0.53 to 0.72, all reaching the 99% confidence level. Similarly, correlation coefficients between the two intensity series of SAT RAs and height RAs over western Eurasia in the 12 summers range from 0.22 to 0.58, all reaching the 99% confidence level (Fig. 7.57). The correlation in 2010 is higher than that in the other 12 summers, indicating that stronger surface heat wave events are more highly correlated with the height RA signal in the upper troposphere. This simple relationship may provide evidence in support of previous studies, which suggested that more frequent and intense large-scale persistent heat waves and blocking high systems would occur as the planet warms (Meehl and Tebaldi 2004; Coumou and Robinson 2013; Christidis et al. 2005). The increasing frequency of height RA episodes in the upper troposphere can be explained by the recent Arctic temperature amplification (Serreze et al. 2009; Screen and Simmonds 2014; Cohen et al. 2014). Amplified warming in the Arctic leads to a decreased meridional temperature gradient, which should weaken upper-level zonal winds and cause amplified and slower-moving planetary waves such as height RA episodes at the middle-high latitudes.

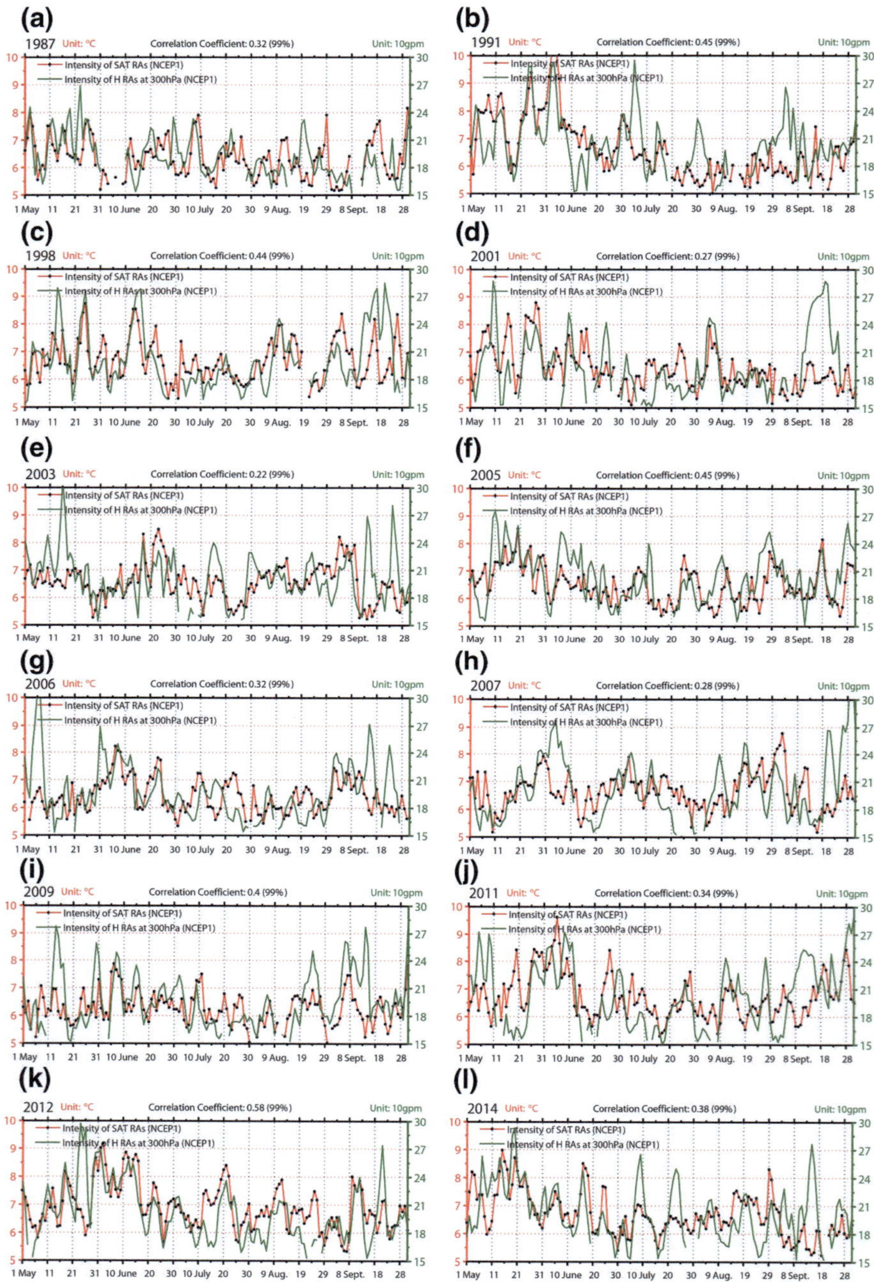

Fig. 7.57 Daily time series of the intensity of SAT RAs (*solid line*, °C) exceeding 5 °C and height RAs (*dashed line*, 10 gpm) exceeding 150 gpm in western Eurasia for the other 12 summers from May 1 to September 30 in **a** 1987, **b** 1991, **c** 1998, **d** 2001, **e** 2003, **f** 2005, **g** 2006, **h** 2007, **i** 2009, **j** 2011, **k** 2012, and **l** 2014. Sourced from Qian et al. (2016a) © 2016 John Wiley & Sons, Inc., with permission

In summary, two reanalysis products and three relative thresholds have been used to examine heat wave events in terms of three quantitative indices: coverage, intensity, and duration. Qian et al. (2016a) used the heat wave event in western Russia in 2010 as an example and then characterized all similar events in western Eurasia during 1980–2014. All heat wave events were identified with the NCEP reanalysis and the ERA-Interim product. The coverage of a regional heat wave event with a spatial scale (length and width) larger than 2000 km can be identified using three RA thresholds, where the SAT RAs exceed the 5 °C, the SAT exceeds the 90th percentile, and the SAT NAs exceed the 1.5 SDs, respectively. The first threshold is the anomaly after removing the daily climatic component from a total variable. The three relative thresholds can also be used for basic variables such as height and temperature aloft.

Surface heat waves are a type of anomalous or extreme weather events. It aims to find a corresponding anomalous signal from an aloft atmospheric variable to indicate the event. This signal exists in the traditional synoptic chart from two basic variables of height and temperature, but forecasters generally cannot quantitatively detect its intensity and coverage, because the climatic component constitutes a large component of the total variable. After removing the daily climatic component, height or temperature anomalies can be used to identify the signal clearly and quantitatively calculate its location and intensity.

Anomalous synoptic charts show a standalone center of height RAs located in the upper troposphere over a surface heat event. Furthermore, the hydrostatic equilibrium relationship between temperature and height RAs means that a surface heat wave (cold surge) event is accompanied by a positive (negative) MHA center aloft. The vertical structure of a warm (cold) air column below the MHA center is a typical vertical heat wave (cold surge) pattern.

The RA approach to examine historical heat wave events during 1980–2014 in western Eurasia shows that the frequency and intensity of summer heat wave events with a daily spatial scale greater than 2000 km for at least 5 consecutive days have increased, particularly since 2010. This rapid increase in large-scale persistent heat events over western Eurasia since 2010 is consistent with the increasing frequency of large-scale persistent positive height RA episodes in the upper troposphere.

Questions

1. How to define an event of regional heat wave (or clod surge) for disaster assessments and weather forecasts respectively?
2. Why are summer standing-type heat waves observed in high latitudes while moving-type heat events often occur in China?
3. How to find the early signals of heat wave and cold surge events?
4. What is the difference between early signals based on anomalous synoptic chart and prediction signals based on dynamic model of a heat wave or cold event?

References

Beniston M (2004) The 2003 heat wave in Europe: a shape of things to come? An analysis based on Swiss climatological data and model simulations. Geoph Res Lett 31(2): L02202

Chen F, Yang SB, Shen SH, Zhang M, Zhao YX (2013) Spatial and temporal distribution of spring cold damage in double cropping rice areas of the middle and lower reaches of the Yangtze River. Jiangsu J Agric. Sci 29: 540–547. (in Chinese)

Chen Y, Hu Q, Yang YM, Qian WH (2016) Anomaly based analysis of extreme heat waves in Eastern China during 1981–2013. Inter J Climatol. doi:10.1002/joc.4724

Chen P, Newman M (1998) Rossby wave propagation and the rapid development of upper-level anomalous anticyclones during the 1988 US drought. J Climate 11: 2491–2504

Chen F, Yang SB, Shen SH, Zhang M, Zhao YX (2013) Spatial and temporal distribution of spring cold damage in double cropping rice areas of the middle and lower reaches of the Yangtze River. Jiangsu J Agricul Sci 29: 540–547 (in Chinese)

Christidis N, Stott P, Brown S, Hegerl G, Caesar J (2005) Detection of changes in temperature extremes during the second half of the 20th century. Geoph Res Lett 32: L20716

Cohen J, Screen JA, Furtado JC, Barlow M, Whittleston D, Coumou D, Francis J, Dethloff K, Entekhabi D, Overland J, Jones J (2014) Recent Arctic amplification and extreme mid-latitude weather. Nature Geoscience 7: 627–637

Coumou D, Robinson A (2013) Historic and future increase in the global land area affected by monthly heat extremes. Enviro Res Lett 8(3): 034018

Cui J, Li J, Zhang AZ, Yan Q (2007) Advances in studies of summer low temperature in Northeast China. Meteorological Monthly, 33(4): 3–9 (in Chinese)

Dee DP, et al. (2011) The ERA-Interim reanalysis: configuration and performance of the data assimilation system. Q J R Meteorol Soc 137: 553–597

Ding T, Qian WH (2011) Geographical patterns and temporal variations of regional dry and wet heatwave events in China during 1960–2008. Adv Atmos Sci, 28(2): 322–337

Dole R, et al. (2011) Was there a basis for anticipating the 2010 Russian heat wave? Geophy Res Lett, 38: L06702

Du J, Grumm RH, Deng G (2014) Ensemble anomaly forecasting approach to predicting extreme weather demonstrated by extremely heavy rain event in Beijing. Chin J Atmos Sci 38(4): 685–699 (in Chinese)

Du YD (2005) Frost and high temperature injury in China. Natural Disasters and Extreme Events in Agriculture. Sivakumar MVK Ed, Springer 145–157

Easterling DR, Evans JL, Groisman PY, Karl TR, Kunkel KE, Ambenje P (2000) Observed variability and trends in extreme climate events: A brief review. Bull AmerMeteorol Soc 81: 417–425

Fang XQ, Wang AY, Fong SK, Lin WS, Liu J (2008) Changes of reanalysis-derived Northern Hemisphere summer warm extreme indices during 1948–2006 and links with climate variability. Global and Planetary Change 63: 67–78

Feng P Z, et al. (1985) China Major Meteorological Disasters Analysis. Meteorology Press, 271 pp

Garcia-Herrera R, Diaz J, Trigo RM, Luterbacher J, Fischer EM (2010) A review of the European summer heat wave of 2003. Critical Rev Environ Sci Technol 40(4): 267–306

Graham RA, Grumm RH (2010) Utilizing normalized anomalies to assess synoptic-scale weather events in the western United States. Wea Forecasting 25: 428–445

Grumm RH (2011) The central European and Russian heat event of July-August 2010. Bull Amer Meteorol Soc 92: 1285–1296

Grumm RH, Hart R (2001) Standardized anomalies applied to significant cold season weather events: Preliminary findings. Wea Forecasting 16: 736–754

Hart RE, Grumm RH (2001) Using normalized climatological anomalies to rank synoptic-scale events objectively. Mon Wea Rev 129: 2426–2442

Jiang M, Yu TT, Qian WH (2014) Analysis on atmospheric anomalous signals of winter low temperature and snow-ice storms in southern China. Chin J Atmos Sci 38(4): 813–824 (in Chinese)

Junker NW, Grumm RH, Hart R, Bosart LF, Bell KM, Pereira FJ (2008) Use of normalized anomaly fields to anticipate extreme rainfall in the mountains of northern California. Wea Forecasting 23(3): 336–356

Junker NW, Brennan MJ, Pereira F, Bodner MJ, Grumm RH (2009) Assessing the potential for rare precipitation events with standardized anomalies and ensemble guidance at the Hydrometeorological Prediction Center. Bull Amer Meteorol Soc 90(4): 445–453

Kalnay E, et al. (1996) The NCEP/NCAR 40-year reanalysis project. Bull Amer Meteorol Soc 77 (3):437–471

Katsafados P, Papadopoulos A, Varlas G, Papadopoulou E, Mavromatidis E (2014) Seasonal predictability of the 2010 Russian heat wave. Natural Hazards and Earth System Sciences 14: 1531–1542

Lau KM, Kim KM (2011) The 2010 Pakistan flood and Russian heat wave: teleconnection of hydrometeorological extremes. J Hydrometeorol 13: 392–403

Li Z, Yan ZW (2009) Homogenized daily mean/maximum/minimum temperature series for China from 1960 to 2008. Atmos Ocean Sci Lett 2(4): 237–243

Liu LL, Sun LH, Liao YM, Zhu YF, Zou XK, Wang YM, Yan JH (2008) Development and application of national prediction system for extreme high temperature. Meteorological Monthly, 34: 102–107 (in Chinese)

Matsueda M (2011) Predictability of Euro-Russian blocking in summer of 2010. Geophy Res Lett 38: L06801

Meehl GA, Tebaldi C (2004) More intense, more frequent, and longer lasting heat waves in the 21st Century. Science, 305(305): 994–997

Peixoto JP, Oort AH (1992) Physics of climate. American Institute of Physics Press, New York.

Peterson TC, Stott PA, Herring S (2012) Explaining extreme events of 2011 from a climate perspective. Bull Amer Meteorol Soc 93: 1041–1067

Peterson TC, Hoerling MP, Stott PA, Herring SC (2013) Explaining extreme events of 2012 from a climate perspective. Bull Amer Meteorol Soc 94: S1–S74

Qian WH (2011) Atlas of Climate Change and China Extreme Climate Events. Meteorological Press, Beijing

Qian WH, Fu JL (2010) Frontal genesis of moisture atmosphere during the early 2008 persistent freezing-rain event in southern China. Sci China-Earth Sci 53: 454–464

Qian WH, Jiang M (2014) Early signals of synoptic-scale atmospheric anomalies associated with the summer low temperature events in Northeast China. Meteorol Atmos Phys 124(1–2): 33–46

Qian WH, Zhang ZJ (2012) Precursors to predict low-temperature freezing-rain events in southern China. Chin J Geophy 55: 1501–1512 (in Chinese)

Qian WH, Li J, Shan XL (2013) Application of synoptic-scale anomalous winds predicted by medium-range weather forecast models on the regional heavy rainfall in China in 2010. Sci China-Earth Sci 56(6): 1059–1070

Qian WH, Shan XL, Liang HY, Huang J, Leung CH (2014) A generalized beta-advection model to improve unusual typhoon track prediction by decomposing total flow into climatic and anomalous flows. J Geophy Res Atmos 119(3): 1097–1117

Qian WH, Chen Y, Jiang M, Hu Q (2015) An anomaly-based method for identifying signals of spring and autumn low temperature events in the Yangtze River Valley, China. J Appl Meteorol Climatol 54: 1216–1233

Qian WH, Wu KJ, Leung JCH (2016a) Three-dimensional structure and long-term trend of heat wave events in western Eurasia revealed with an anomaly-based approach. Inter J Climatol doi:10.1002/joc.4634

Qian WH, Yu TT, Du J (2016b) A unified approach to trace surface heat and cold events by using height anomaly. Climate Dyn 46: 1647–1664

Rahmstorf S, Coumou D (2011) Increase of extreme events in a warming world. Proceedings of the National Academy of Sciences, 108(44):17905–17909

Robinson PJ (2001) On the definition of a heat wave. J Appl Meteorol, 40(4): 762–775

Ruti PM, Dell'Aquila A, Giorgi F (2014) Understanding and attributing the Euro-Russian summer blocking signatures. Atmos Sci Lett 15(3): 204–210

Schneidereit A, Schubert S, Vargin P, Lunkeit F, Zhu XH, Peters DHW, Fraedrich K (2012) Large-scale flow and the long-lasting blocking high over Russia: summer 2010. Mon Wea Rev 140(9): 2967–2981

Schubert SD, Wang H, Suarez MJ (2011) Warm season subseasonal variability and climate extremes in the Northern Hemisphere: the role of stationary Rossby waves. J Climate, 24(18): 4773–4792

Schubert SD, Wang HL, Koster RD, Suarez MJ, Groisman PY (2014) Northern Eurasian heat waves and droughts. J Climate 27(9): 3169–3207

Screen JA, Simmonds I (2013) Caution needed when linking weather extremes to amplified planetary waves. Proceedings of the National Academy of Sciences, 110(26), E2327

Screen JA, Simmonds I (2014) Amplified mid-latitude planetary waves favour particular regional weather extremes. Nature Climate Change, 4: 704–709

Serreze MC, Barrett AP, Stroeve JC, Kindig DM, Holland MM (2009) The emergence of surface-based Arctic amplification. Cryosphere 3: 11–19

Silverman, R. (2013) Heatwave deaths: 760 lives claimed by hot weather as high temperatures continue. The Telegraph http://www.telegraph.co.uk/news/weather/10187140/Heatwave-deaths-760-lives-claimedby-hot-weather-as-high-temperatures-continue.html (accessed 4 August 2015)

State Standardization Committee (2012) Low temperature disaster of southern rice, rapeseed and orange. Rep. GB/T 27959-2011, State Standardization Committee of China, 4 pp

Sun Y, Zhang XB, Zwiers, FW, Song LC, Wan H, Hu T, Yin H, Ren GY (2014) Rapid increase in the risk of extreme summer heat in Eastern China. Nature Climate Change, 4: 1082–1085

Tang T, Jin RH, Peng XY, Niu RY (2014) Analysis on extremely high temperature over southern China in summer 2013. Meteorological Monthly 40: 1207–1215 (in Chinese)

Van den Dool (2007) Empirical methods in short-term climate prediction. Oxford University Press: Oxford, UK

Wang JF, Wu GX (1997) Evolution and characteristics of the persistent cold summer in Northeast China. Scientia Atmospherica Sinica, 21: 523–532 (in Chinese)

Wang SW, Ma SQ, Chen L, Wang Q, Huang JB (2009) Low temperature and cold disasters. China Meteorological Press, 134 pp

Watanabe M, Shiogama H, Imada Y, Mori M, Ishii M, Kimoto M (2013) Event attribution of the August 2010 Russian heat wave. Scientific Online Lett Atmos Sola, 9: 65–68

WMO (2011) Weather extremes in a changing climate: hindsight on foresight. WMO No. 1075, World Meteorological Organization, Geneva 17 pp

Xing RN, Chao JP (1982) Seasonal forecasting experiments with a three-layer filtered model. Chin Sci Bull 12: 738–740 (in Chinese)

Yan ZW, Yang C (2000) Geographic patterns of climate extreme changes in China during 1951–1997. Clim Environ Res 5: 267–272 (in Chinese)

Zebiak SE, Cane MA (1987) A model El Niño-Southern Oscillation. Mon Wea Rev 115: 2262–2278

Zhang S, Wang B (2008) Global summer monsoon rainy seasons. Inter J Climatol 28: 1563–1578

Zhang ZJ, Qian WH (2011) Identifying regional prolonged low temperature events in China. Adv Atmos Sci 28: 338–351

Zhang ZJ, Qian WH (2012) Precursors of regional prolonged low temperature events in China during winter-half year. Chin J Atmos Sci 36: 1269–1279 (in Chinese)

Zhang SY, Liu XN, Sun AJ (1995) Climatic features of autumn low temperature damages. Meteorological Monthly 21: 21–24 (in Chinese)

Zhu CW, Zhou XJ, Zhao P, Chen LX, He JH (2011) Onset of East Asian subtropical summer monsoon and rainy season in China. Sci China-Earth Sci 54: 1845–1853

Chapter 8
Unusual Tropical Cyclones

Intensity and track of tropical cyclones (TCs) are two basic elements in operational forecasting. Using a novel approach of decomposing hourly total flow into temporal climatic and anomalous flows (and other anomalous variables), this chapter describes the intensity evolution of TCs and the development of a generalized beta-advection model (GBAM) to improve unusual typhoon track predictions within 2–3 days. The GBAM is written at an optimal level of the maximum vorticity anomaly (max-VA) and minimum divergence anomaly (min-DA). Typhoon Megi (2010) in the Northwest (NW) Pacific and Hurricane Sandy (2012) in the North Atlantic are used as two special cases to introduce the application of this model. The decomposition approach and the model are also applied in the analyses and post-predictions of TCs in the South China Sea (SCS) and East China Sea (ECS). The model can well predict the rapid right-turning TCs in the SCS and the rapid left-turning TCs in the ECS. The binary interaction of two TCs occurred in previous studies are reexamined and the theoretical binary interaction (Fujwhara effect) of two TCs is also simulated by the model.

8.1 Separating the Typhoon Vortex from Its Surrounding Flow

Previous studies have shown that a tropical cyclone (TC) vortex is considered to be embedded in and steered by a large-scale environmental flow (Kasahara 1957; Miller and Moore 1960; Holland 1983, 1984; Chen et al. 2002). The steering flow is the main factor affecting the movement of TCs and this concept has been generally accepted and applied by forecasters in the track prediction of typhoons and hurricanes in various basins (Chen and Ding 1979; Chan and Gray 1982; Deng et al. 2010; Roy and Kovordanyi 2012). In all synoptic analyses and model experimental predictions, one needs to know a TC vortex's size and intensity in

© Springer Nature Singapore Pte Ltd. 2017

W. Qian, *Temporal Climatology and Anomalous Weather Analysis*,
Springer Atmospheric Sciences, DOI 10.1007/978-981-10-3641-5_8

order to define the steering flow. In the traditional view, steering flow is obtained by removing a TC vortex from the total circulation or total flow on its spatial scale of a thousand kilometers. However, the actual circulation is much more complex because it contains the TC vortex (internal forcing) and surrounding flow (external forcing) as well as interactions with other vortices in the vicinity (Wang et al. 1998). Due to these factors, it is difficult to define a TC vortex for the purpose of removing the TC vortex to find the steering flow.

The empirical relations between TC movement and surrounding large-scale environmental flows usually vary with the definition of the environmental flows. Previous work attempted to extract the steering flow from observational total flow by spatially removing TC vortex as well as other small-scale disturbances (Deng et al. 2010; Singh et al. 2012; Galarneau and Davis 2013). George and Gray (1976) suggested that the flow near 700 hPa, or the average layer between 700 and 500 hPa is the best to represent the environmental wind that drives TC motion in the tropics, while Holland (1984) suggested that an optimum choice may be the vertical mean steering flow between 850 and 300 hPa. Some investigators then suggested that the deep-layer mean, from 1000 to 150 hPa or 100 hPa, steering flow typically approximates the actual TC motion (Sanders et al. 1980; Dong and Neumann 1986; Franklin 1990; Velden and Leslie 1991). Recently, Galarneau and Davis (2013) defined an optimal environmental steering flow with variable vertical extent of steering layer after TC vorticity and divergence are removed by a determined TC's radius. As demonstrated by numerical models, the optimum steering level increases with TC intensity (Velden and Leslie 1991) because a stronger TC normally has a deeper convection that must be steered by a higher layer of environmental flow. In many cases, however, the motion of strong TCs can depart from the deep-layer mean steering flow, which leads to a large error of path prediction. In all previous studies, the environmental flow is a large-scale time-mean flow after removing a residual small-scale flow associated with the TC vortex. To compute the steering flow for a TC, some previous studies first computed the environmental wind from 850 to 200 hPa with an increment of 50 hPa, and then determine the environmental wind with the TC vortex removed (e.g., Davis et al. 2008; Galarneau and Davis 2013). For each disturbance, the vortex is removed at all vertical levels between 850 and 200 hPa within a given radius using the Poisson equation for stream function and velocity potential with homogeneous boundary conditions applied at the edge of a computational domain. However, Trinh and Krishnamurti (1992) indicated that the prediction error of TC track is considerably sensitive to its initial size and intensity. Uncertainties in the radius of TC vortex and computational domain create unbalance and incompleteness when extracting steering flow from total flow. Therefore, it is difficult to obtain the realistic steering flow and TC vortex.

In order to find a useful definition separating the steering flow and TC vortex from a total flow, we first review the climatology as a reference. In previous studies, the monthly mean and seasonal mean climatology have been widely used in analysis as the reference for variations of atmospheric variables at monthly and seasonal timescales (Peixoto and Oort 1992; Van den Dool 2007; Zebiak and Cane

1987). In ancient China, a relatively complete reference is to separate the climatic seasonal cycle into as 24 solar terms in a half-month interval. To accurately define extreme weather events, daily climatology is needed since low temporal resolution will wipe out detailed information. For example, the average of three consecutive abnormal monthly mean temperatures can be normal for a seasonal mean temperature. Similarly, the average of 30 consecutive abnormal daily mean temperatures can give normal monthly mean temperature. These ideas can also be used to decompose any meteorological field into climatology and anomaly with an hourly interval on each of 365 solar (or calendar) days. The temporal climatology is quasi-stationary and reflects an equilibrium state between solar radiation and local geographical features, while the anomaly is a departure or deviation from the equilibrium state varying at any moment. Thus, the climatic seasonal cycle a year and diurnal cycle a day can be divided into $365 \times 24 = 8760$ terms. On the other hand, the average flow over a certain time interval, such as five or ten days, even months, refers to large-scale time-mean environmental flow, but not the climatic flow, because synoptic- and monthly scale circulation and temperature anomalies might exist.

As a matter of fact, the method of decomposing a total field into temporal climatic and anomalous components was proposed and used in anomaly forecasting in early 2000 (Grumm and Hart 2001; Hart and Grumm 2001) and gained popularity for high-impact weather forecasts in recent years (Junker et al. 2008, 2009; Graham and Grumm 2010; Grumm 2011a, b; Randy et al. 2013). Du et al. (2014) further expanded this concept into "ensemble anomaly forecasting" by combining anomaly forecasting with ensemble forecasts not only to identify abnormal events but also to quantify the confidence of an anomaly forecast itself. Similar to ensemble anomaly forecasting, an "extreme forecast index" is used at the European Centre for Medium-Range Weather Forecasts (ECMWF) (Lalaurette 2003; Zsoter 2006). To be suitable for all climate zones, the standardized anomaly (SA) is used in an anomaly forecast. In recent years, Qian and his colleagues found that a non-normalized or raw anomaly (RA) is also useful in short-, medium-, and extended-range forecasts of many extreme weather events, such as freezing rain (Qian and Zhang 2012) and heat waves (Ding and Qian 2012). This technique has also been used in the interpretation and application of numerical model outputs based on the ECMWF model and the NCEP Global Forecast System (GFS) (Qian et al. 2013). Using the concept of temporal variable anomaly, Peng et al. (2013) developed a novel approach which is capable to improve numerical weather prediction skills up to 15 days.

Super typhoon Megi (2010) is used to demonstrate the application of the novel approach of decomposing a total variable into a temporal climatic component and an anomaly. Megi was first identified by the Joint Typhoon Warning Center (JTWC) on 12 October 2010 as a tropical disturbance located southeast of Guam (Fig. 8.1). Before 16 October 2010, Megi gradually strengthened and moved northwestward along the easterly flows to the south of the NW Pacific subtropical high. On 17 October 2010, it strengthened and developed into a super typhoon. It later peaked with its wind velocity of 160 knots (296 km/h, 184 mph) and lowest

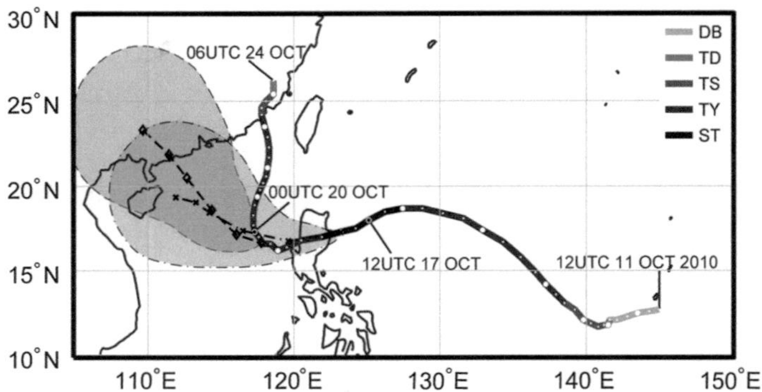

Fig. 8.1 The track of Typhoon Megi (2010_No.15) and two forecasts issued by China NMC at 0000 UTC 18 and 1200 UCT 19 October 2010. The *white spots* show the locations of Megi at every 6 h from 1200 UTC 11 October to 0600 UTC 24 October 2010. Two thin-*dashed lines* with *shadings* indicate the track forecasts. Megi's intensity is indicated by disturbance (DB), tropical depression (TD), tropical storm (TS), typhoon (TY) and super typhoon (ST). Sourced from Qian et al. (2014a) © 2014 World Scientific Publishing Co. Pte. Ltd., Singapore, with permission

surface pressure of 903 hPa. It made landfall on the northern Philippines on 18 October 2010. On 19 October, it reduced its speed and turned northwards abruptly. On 20 October, it strengthened with its maximum wind velocity reaching 115 knots (212 km/h, 132 mph) and surface pressure reducing to 937 hPa. On October 22, it weakened to a severe tropical storm approaching the Taiwan Strait. At 1255 Beijing Time on October 23, it made landfall over Zhangpu County in Fujian Province, Southeast China.

During Megi's lifetime, its track experienced three processes, moving west-southwestwards after formation, turning left on 14 to 17 October before its intensity peak in the northeast sea of the Philippines, and turning right after it moved into the SCS on 19 to 22 October 2010. Two forecasts from China National Meteorological Center (NMC) are plotted to compare with the track of Megi (Fig. 8.1). At 0000 UTC 18 October, operational forecast from China NMC showed that Megi would pass the northern Philippines, then enter the SCS and eventually approach the Hainan Island in three days. At 1200 UCT 19 October, NMC announced that Megi would move northwestwards and make landfall in the western coast of South China (Guangdong Province) on 23 October 2010. Meanwhile, ECMWF and JTWC made the same prediction (Xu 2011; Qian et al. 2012). However, none of them correctly predicted the rapid turning at that time.

To well explain the sudden right turning, we first compare the total height and temperature with their corresponding decomposed anomalies for the typhoon. Figure 8.2 shows the vertical cross sections of the total height and temperature as well as height and temperature anomalies along 117.00°E at 1200 UTC 20 October 2010 after Megi turned right. Clearly, both the height and temperature change with altitude at the location of Megi. But the distributions of the two total fields are unable to tell where the warm core and the strongest vortex are (Fig. 8.2a, b).

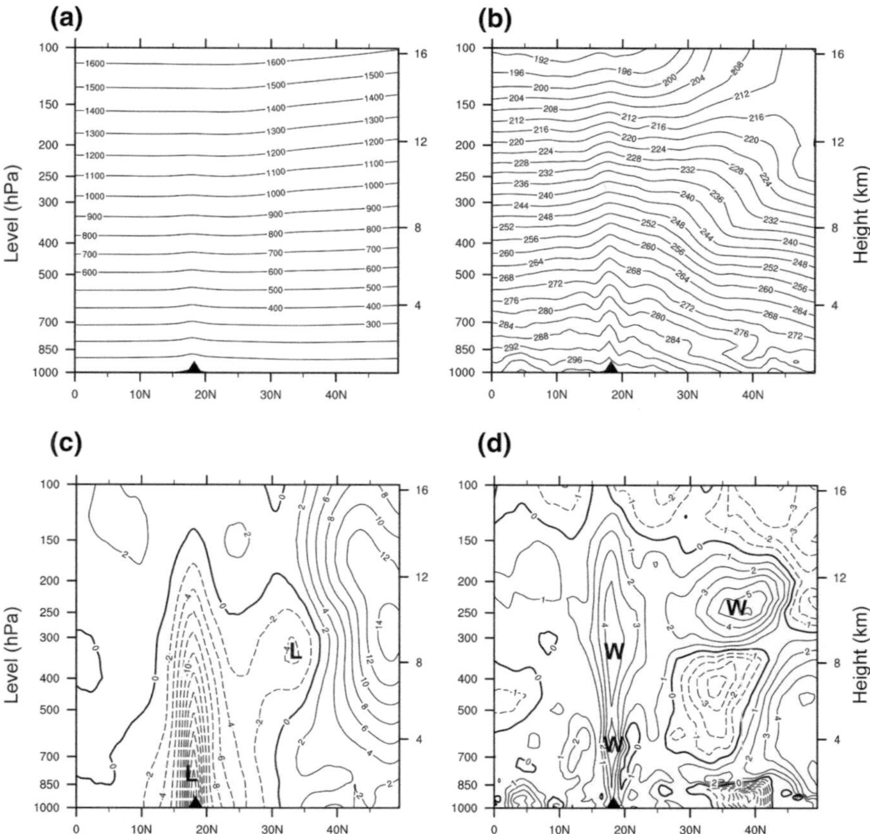

Fig. 8.2 Vertical cross sections of **a** total height (100 × 10 gpm interval) and **b** total temperature (4 K interval) as well as **c** height anomaly (2 × 10 gpm interval) and **d** temperature anomaly (1 K interval) along 117.00°E at 1200 UTC 20 October 2010. The location of Megi is indicated by the symbol *triangle*. The low centers of height anomalies and the warm centers of temperature anomalies are denoted by symbols "L" and "W", respectively. The data used are obtained from the ERA-Interim. Sourced from Qian et al. (2014b) © 2014 John Wiley & Sons, Inc., with permission

Therefore, the total height and temperature are decomposed into its climatic component and anomaly, respectively (Fig. 8.2c, d). The Megi vortex indicated by the height anomaly is very deep in the troposphere with its center located at around 850 hPa. At the same time, there is another "anomalous vortex" located in the middle latitude (to the north of Megi) at around 300 hPa (Fig. 8.2c). The temperature anomalies are quite different between the Megi vortex and this "anomalous vortex" (Fig. 8.2d). For the Megi vortex, the warm temperature anomaly extends to the whole troposphere with two warm centers at 300 and 700 hPa, respectively. For the "anomalous vortex", it displays a baroclinic structure with a warm center at 200–250 hPa and a cold center at 400–500 hPa (Fig. 8.2d). In the mid-high latitudes, usually, maximum negative (or positive) centers of height anomalies vertically appear near 300 hPa while a positive (or negative) center of temperature anomaly at around 200–250 hPa and negative (or positive) ones at 400 hPa or 850 hPa. The temperature anomalies can be dynamically derived or reconstructed from height anomalies by the hydrostatic equation. As done by Qian et al. (2014a) in their atlas, 322 typhoons occurred during 1979–2011 were decomposed using this approach at 850 and 300 hPa. Here, the total and climatic flows at 300 and 850 hPa are also chosen to illustrate the interaction between the Megi vortex and other anomalous vortices in the vicinity.

From 11 to 13 October 2010, a tropical disturbance (Megi) formed and moved west-southwestward basically along the easterly steering flow in the south of the subtropical high at the traditional synoptic charts (Fig. 8.3). In the first three days, no other tropical disturbances (cyclones) interacted with Megi, except an

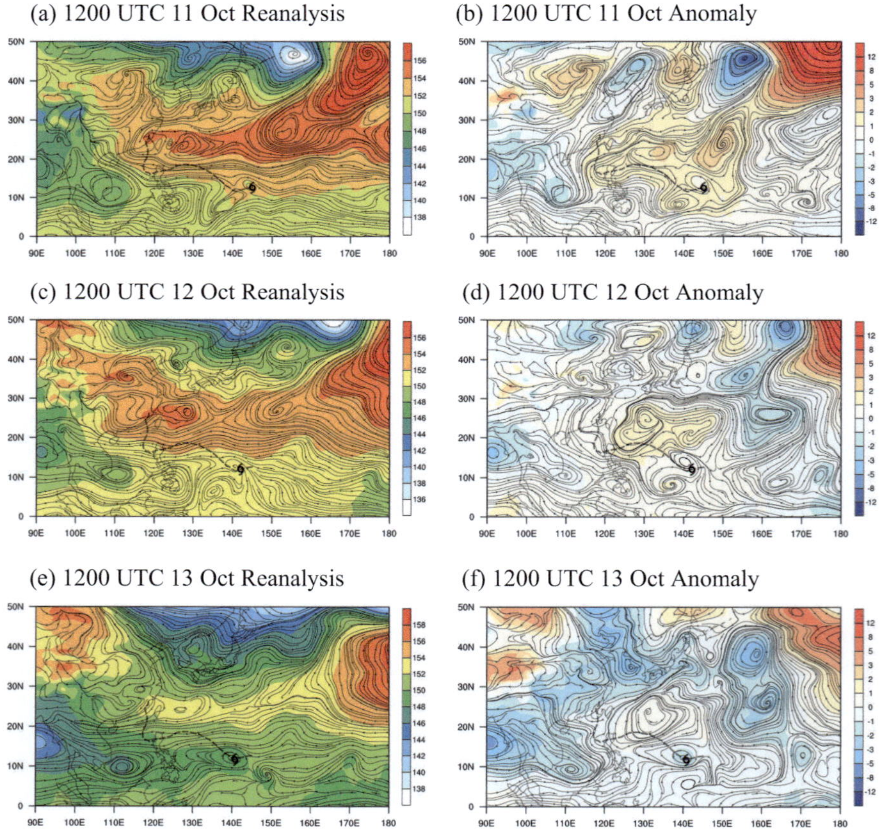

Fig. 8.3 Total flow and height (*left panel*, *shading*, 2 × 10 gpm interval) and anomalous flow and height (*right panel*, *shading*, 1 × 10, 2 × 10, 3 × 10 and 4 × 10 gpm intervals) at 850 hPa at **a** and **b** 1200 UTC 11 October, **c** and **d** at 1200 UTC 11 October, and **e** and **f** at 1200 UTC 11 October 2010. Symbol "ꝏ" indicates the position of Typhoon Megi. Sourced from Qian et al. (2014a) © 2014 World Scientific Publishing Co. Pte. Ltd., Singapore, with permission

anomalous anticyclone in its northern side moved westward and another anticyclone in its southern side moved eastward and weakened, while the Megi vortex moved west-southwestward. Indeed, the climatic steering flow (figures not shown) was easterly, so its motion in west-southwest direction may be influenced by the easterly steering flow and an anticyclonic center in its southern side.

Megi was turning left and slowly moving northwestward on 14 October to southwestward on 17 October in 2010. Its intensity was weak before 16 October and the negative height anomaly was clear in the low troposphere over the typhoon place as shown in Fig. 8.4. The Megi vortex was mainly located below 400 hPa while a low center existed in its southwestern side in the lower levels and another one in the upper troposphere. The two vortices in the lower troposphere circled each other in a counterclockwise manner as shown in Fig. 8.4. The two vortices merged at 1200 UTC 17 October 2010, so the Megi vortex reached its peak intensity (Figs. 8.5 and 8.6).

To confirm whether there are other vortices that influence the motion of Megi and to study the difference between the total flow and the climatic flow, we examine the 300-hPa circulations (winds) at 0000 UTC 19 October 2010. As shown in Fig. 8.7b, the ridge line (dotted line) of the subtropical high extends to the north of the typhoon and separates large-scale westerlies in the middle latitude and easterlies in the low latitude. Since the Megi vortex is located in the northeastern SCS, it should be influenced by the environmental easterlies. Figure 8.7a shows the temporal climatic wind, in which the climatic ridge of the NW Pacific subtropical high is approximately along the 20°N, i.e., the location of Megi vortex is in the climatic easterly. The climatic flow derived here is also physically different from any environmental steering flows extracted by time-mean flow or by spatially removing TC vortex as well as other small-scale disturbances. Figure 8.7c, e show a straightforward comparison between a 5-day-mean wind and a monthly mean

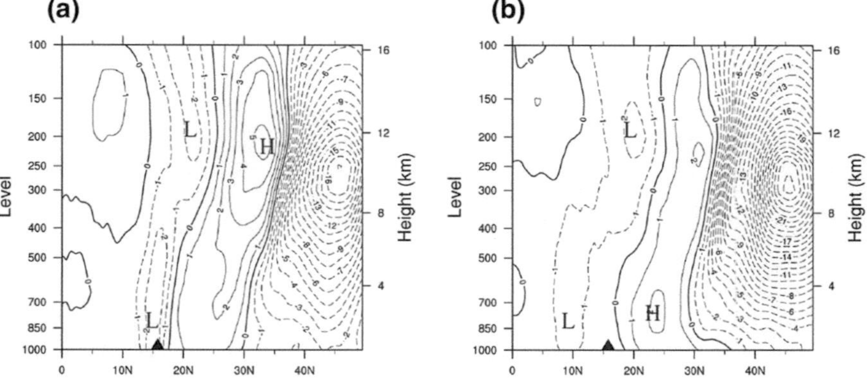

Fig. 8.4 Vertical meridional sections of geopotential height anomalies (1 × 10 gpm interval) along **a** 135.75°E and **b** 130.50°E at 1200 UTC 15 October 2010. The low and high centers of height anomalies are denoted by letters "L" and "H", respectively. Sourced from Qian et al. (2014a) © 2014 World Scientific Publishing Co. Pte. Ltd., Singapore, with permission

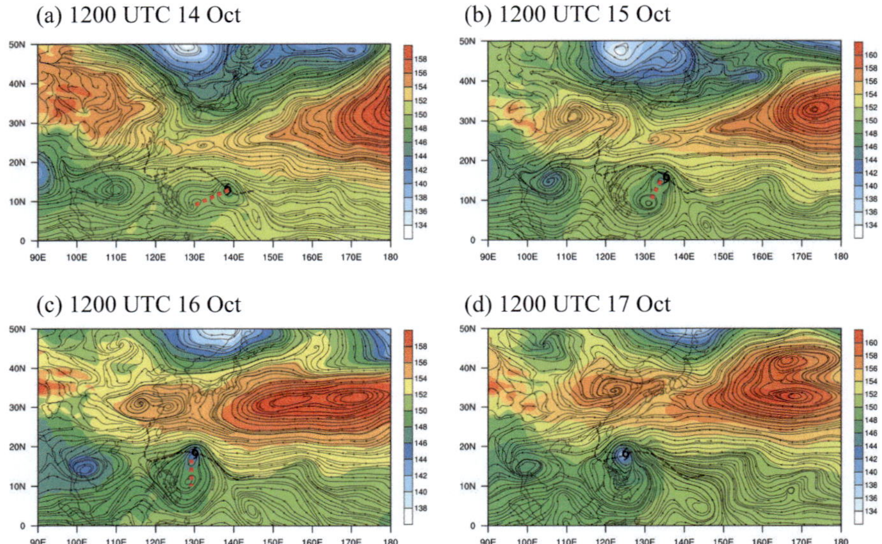

Fig. 8.5 Total flow and height (*shading*, 2 × 10 gpm interval) at **a** 1200 UTC 14, **b** 1200 UTC 15, **c** 1200 UTC 16, and **d** 1200 UTC 17 October 2010 at 850 hPa. The *red dotted line* connects two vortices. Sourced from Qian et al. (2014a) © 2014 World Scientific Publishing Co. Pte. Ltd., Singapore, with permission

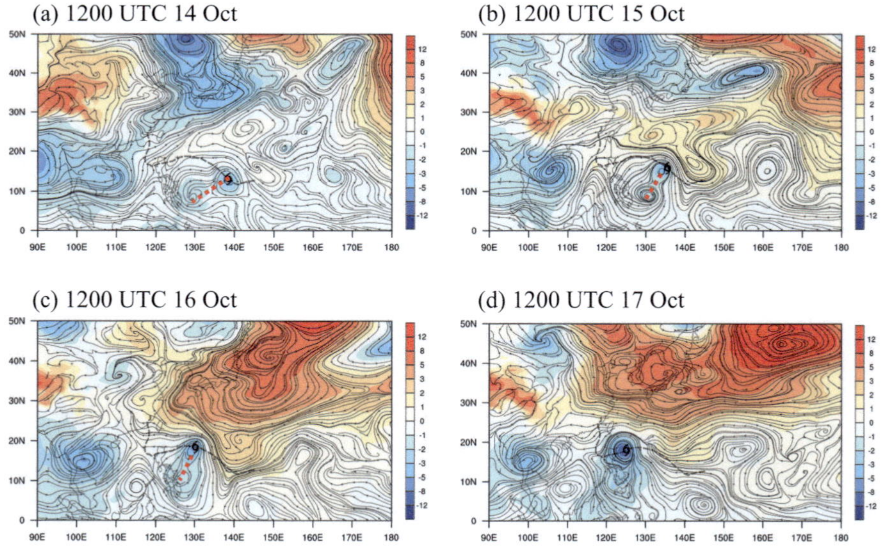

Fig. 8.6 Same as in Fig. 8.5 except of anomalous flow and height at 850 hPa. Sourced from Qian et al. (2014a) © 2014 World Scientific Publishing Co. Pte. Ltd., Singapore, with permission

(31-day) wind centered at 0000 UTC 19 October 2010 at 300 hPa. Locations of trough and ridge as well as the intensities of westerly are different as shown in Fig. 8.7a, c, and e. The intensities and extents of the residual Megi vortex and other residual vortices indicated by the residual height and flow are also different (Fig. 8.7d, f). A relatively weak residual Megi vortex (Fig. 8.7d) is located in a relatively strong large-scale cyclonic-like easterly steering flow (Fig. 8.7c). A relatively strong residual Megi vortex (Fig. 8.7f) is located in a relatively weak large-scale easterly steering flow (Fig. 8.7e). Therefore, the steering effects of the 5-day-mean wind and the monthly mean wind on the Megi motion are different. Figure 8.7g, h show another comparison between a low-pass spatially filtered environmental wind obtained by a spherical harmonic expansion with the wave number less than 15 and its residual height and wind at 300 hPa at 0000 UTC 19 October 2010. The low-pass spatially filtered wind (Fig. 8.7g) as well as its residual height and flow (Fig. 8.7h) are also different from the temporal climatic winds and the time-mean winds as well as other residual heights and flows. A relatively weak residual Megi vortex (Fig. 8.7h) is located in a relatively strong large-scale easterly steering flow (Fig. 8.7h).

The right turning of Megi occurred on 19 October 2010. Two troughs at 300 hPa can be traced in the large-scale total westerly over the mid-high latitudes from October 19 to 22, 2010 (Fig. 8.8). At 1200 UTC 19 October, the two troughs were located in Mongolia and Northwest China, respectively. From 20 to 22 October, the northern trough moved eastwards to North China and the Japan Sea, while the southern trough moved eastwards to Central and South China. In these four days, the ridge line of the NW Pacific subtropical high stayed to the north of the typhoon. Since the trough, the ridge, and the typhoon are three systems that interact with each other, it is difficult to reason whether the typhoon would continue moving westwards or turn northwards based on the total flow.

By the decomposition, the anomalous flow extracted from the total flow might be able to indicate the right turning of Megi in the northern SCS, as shown in Fig. 8.9. Two troughs in Fig. 8.8 became two vortices in Fig. 8.9. At 0000 UTC 19 October (Fig. 8.9a), two anomalous vortices marked by I and II respectively existed in Mongolia and western China. The northern anomalous vortex II moved eastwards in North China at 1200 UTC 19 October (Fig. 8.9b) and disappeared in Northeast China after 0000 UTC 20 October 2010 (Fig. 8.9c). The anomalous vortex I moved southeastwards and strengthened gradually from October 19 to 21, 2010. On the other hand, the Megi vortex turned northwards suddenly on 19 October 2010. Although no other TCs existed around Megi, there was an anomalous vortex in Southwest China. The distance between the two vortices had become shorter since 20 October 2010. They circled around each other counterclockwise and became closer to each other. As they got closer on 22 October 2010 (Fig. 8.9g, h), the two vortices finally merged at 1200 UTC 22 October 2010. Actually, there was another anomalous vortex III in east of the Philippines from 19 to 21 October 2010 and it also merged with the Megi vortex on 22 October 2010. Thus, multiple interactions were observed from these anomalous vortices in the vicinity of Typhoon Megi. During these days, however, the spatially filtered

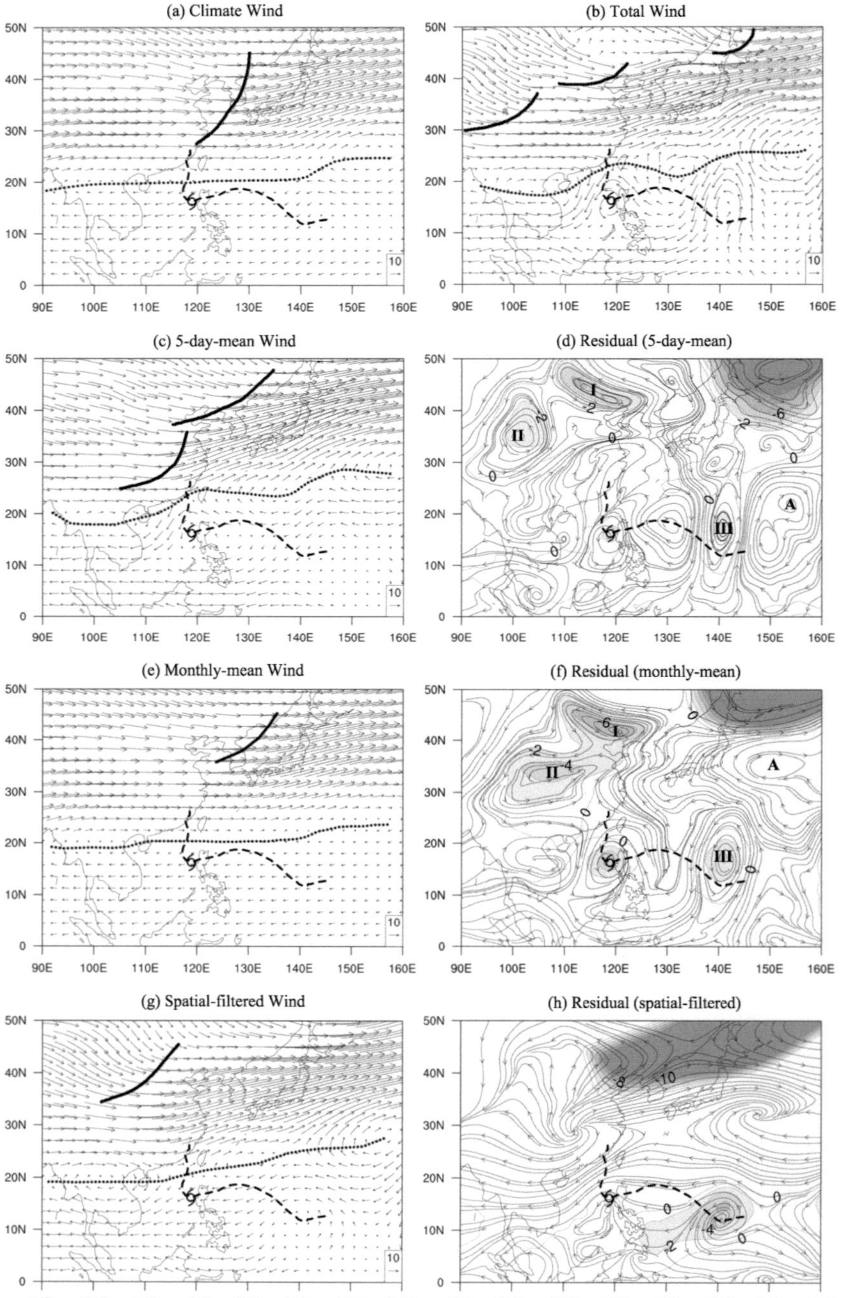
(a) Climate Wind
(b) Total Wind
(c) 5-day-mean Wind
(d) Residual (5-day-mean)
(e) Monthly-mean Wind
(f) Residual (monthly-mean)
(g) Spatial-filtered Wind
(h) Residual (spatial-filtered)
50N
40N
30N
20N
10N
0
90E
100E
110E
120E
130E
140E
150E
160E
10
I
II
III
A

◀ **Fig. 8.7** **a** Temporal climatic wind (ms^{-1}), **b** total wind (ms^{-1}), **c** 5-day-mean wind (ms^{-1}), **d** residual flow and residual height (*shading*, 2 × 10 gpm interval) related to the 5-day mean flow and height, **e** monthly-mean (31-day) wind (ms^{-1}), **f** residual flow and residual height (*shading*, 2 × 10 gpm interval) related to the monthly-mean flow and height, **g** spatial-filtered environmental wind, and **h** residual flow and residual height (*shading*, 2 × 10 gpm interval) related to the spatial-filtered environmental wind and height centered at 0000 UTC 19 October 2010 at 300 hPa. Only negative residual height is shaded with the interval of 20 gpm. The current location and the track of Megi are denoted by the symbol 6 and a heavy *dashed line*, respectively. The troughs in the westerly and the ridge line of subtropical high are indicated by *heavy-curved lines* and *dotted lines*. Three residual vortices and one residual anticyclonic center are indicated by I, II, III and A, respectively. Sourced from Qian et al. (2014b) © 2014 John Wiley & Sons, Inc., with permission

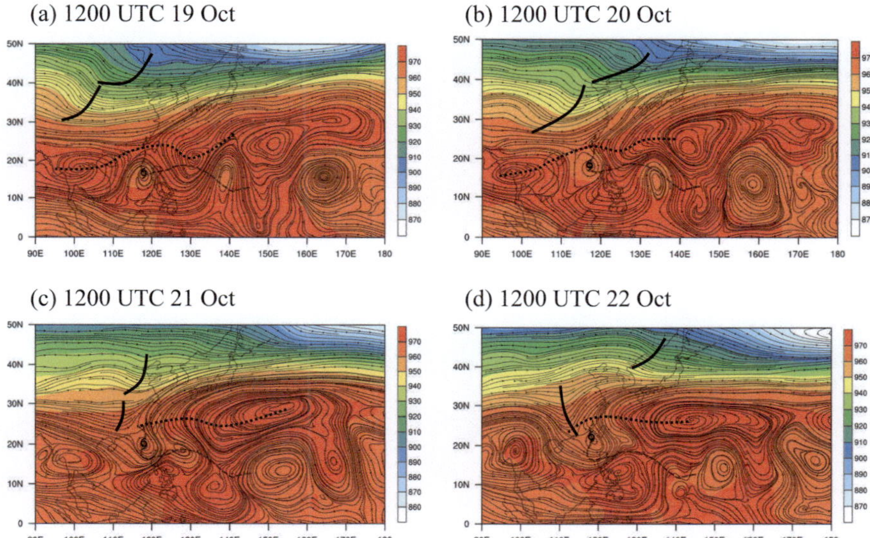

Fig. 8.8 Total flows and height (*shading*, 10 × 10 gpm interval) at 300 hPa at 1200 UTC on **a** 19, **b** 20, **c** 21, and **d** 22 October 2010. The troughs in the westerly and the ridge line of subtropical high are indicated by *heavy-curved lines* and *dotted lines*. The current location and the track of Megi are denoted by the symbol 6 and a *heavy-dashed line*, respectively. Sourced from Qian et al. (2014a) © 2014 World Scientific Publishing Co. Pte. Ltd., Singapore, with permission

residual heights and flows did not show any closed vortex in Mongolia and western China, such as shown in Fig. 8.7h.

Comparing Figs. 8.8 with 8.9, influences of the mid-latitude trough and the anomalous vortex on the motion of typhoon Megi can be revealed by different flows. The influence of the mid-latitude trough can be seen from the total flow (Fig. 8.8), where large-scale circulation systems (i.e., steering flows and the ridge of the subtropical high) and small-scale circulation systems coexist. The rotating and merging interactions among the Megi vortex and other anomalous vortices are clearly observed in the anomalous flow (Fig. 8.9) after the temporal climatic flow is filtered out from the total flow.

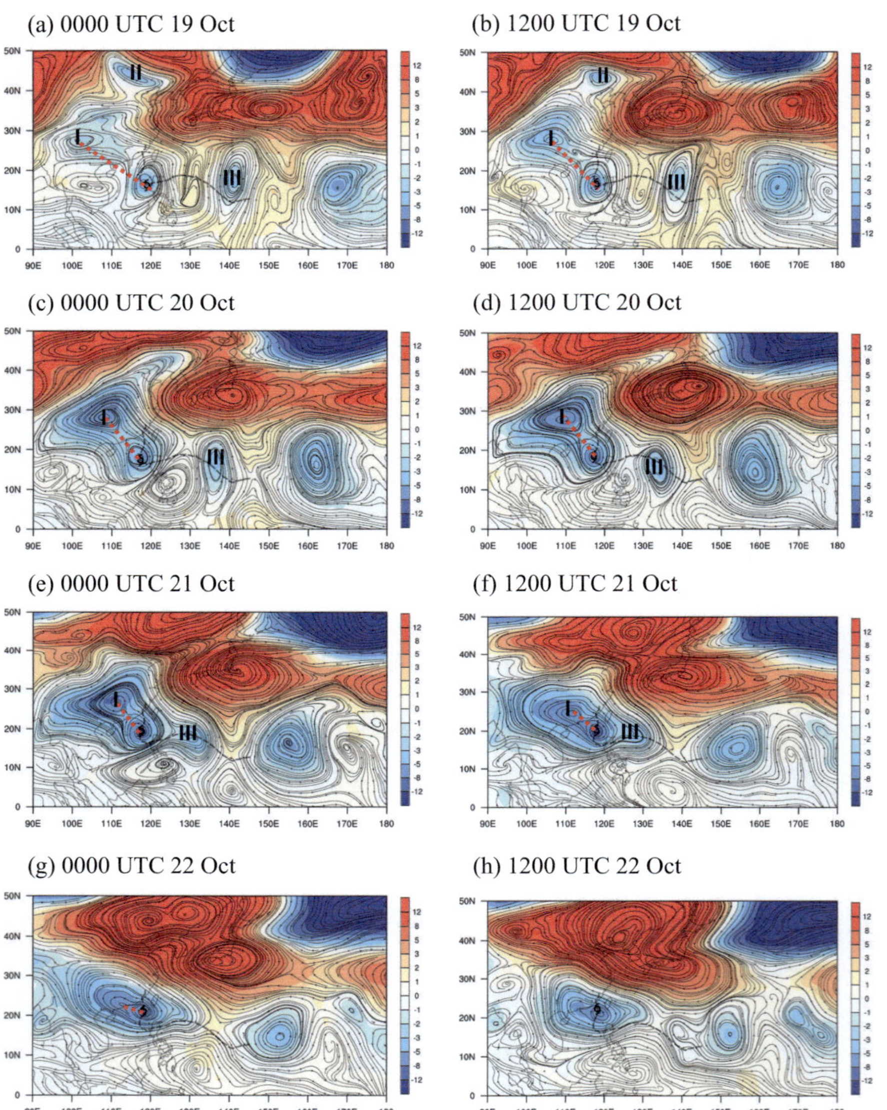

Fig. 8.9 Anomalous flow and height (*shading*, 1 × 10, 2 × 10, 3 × 10 and 4 × 10 gpm intervals) at 300 hPa at **a** 0000 UTC 19, **b** 1200 UTC 19, **c** 0000 UTC 20, **d** 1200 UTC 20, **e** 0000 UTC 21, **f** 1200 UTC 21, **g** 0000 UTC 22, and **h** 1200 UTC 22 October 2010. The *red dotted line* connects two vortices. Letters I, II and III indicate three vortices. Sourced from Qian et al. (2014a) © 2014 World Scientific Publishing Co. Pte. Ltd., Singapore., with permission

8.2 Evaluating Typhoon Intensification[1]

Intensity and location are two basic outputs of typhoon prediction. Qian et al. (2015) used an anomaly based approach to identify the intensity of TC. The super typhoon Megi (2010) was used as an example to describe the inconsistency of TC intensity estimated from four meteorological centers in China, Japan, and the United States. A hydrostatic equilibrium relationship for both height and temperature anomalies is described from the ECMWF Reanalysis Interim (ERA-Interim). Results show that the anomalous variables can be used in indicating the evolution of typhoon anomalous intensity. This work provides a useful reference for applying the output products of medium-range weather forecast models on the prediction of typhoon intensity.

Typhoon intensity is defined as the maximum wind speed in terms of either a 1-min or a 10-min average at the standard reference height of 10 meters. Another definition is based on its central sea surface pressure. The choice of averaging period, as well as the naming convention for classifying storms, differs across forecast centers and ocean basins. Two reanalysis datasets and one analysis dataset are used to identify the intensity evolution of Typhoon Megi (2010). The first dataset is the National Centers for Environmental Prediction Reanalysis 1 (NCEP R1) (Kalnay et al. 1996). It has a spatial resolution of $2.5 \times 2.5°$ in the horizontal longitude–latitude grids and 17 vertical levels. The second dataset is the ERA-Interim product (Dee et al. 2011). The ERA-Interim product has a spatial resolution of $0.75 \times 0.75°$ in the horizontal latitude–longitude grids and 37 vertical levels. The third is the NCEP FNL (Final Operational Global Analysis) with a spatial resolution of $1 \times 1°$ for 6 h interval.

The intensity of Megi was estimated from 11 to 22 October 2010 by four meteorological centers, i.e., the JTWC, China Meteorological Administration (CMA), Hong Kong Observatory (HKO) and Regional Specialized Meteorological Centre (RSMC) of Japan Meteorological Agency at Tokyo (TOKYO). The central pressure time series of Typhoon Megi (2010) based on JTWC, CMA, TOKYO, and HKO databases are given in Fig. 8.10. The deepest pressure was given by the TOKYO from 1800 UTC 17 to 0000 UTC 18 October 2010. The shallowest pressure was given by the JTWC at 1200 UTC 17 October 2010. There was the second peak of low pressure but their times are different determined by different centers from 1800 UTC 19 to 1200 UTC 21 October 2010.

For a comparison, Fig. 8.11 gives the time series of surface pressure, 1000-hPa height and 700–850 hPa average height, following the center of Typhoon Megi (2010) based on the FNL database. Three peaks can be identified respectively at 1200 UTC 17, 0600 UTC 18 and 0000–0600 UTC 20 October. The deepest

[1]Modified article originally published in *Qian W.H., Zhang G.W., Huang J. (2015) Intensity evolution of Typhoon Megi (2010) revealed from anomaly-based atmospheric variables. Mon Meteorol 41(7): 818–827.* Published with kind permission © Meteorological Monthly. All Rights Reserved.

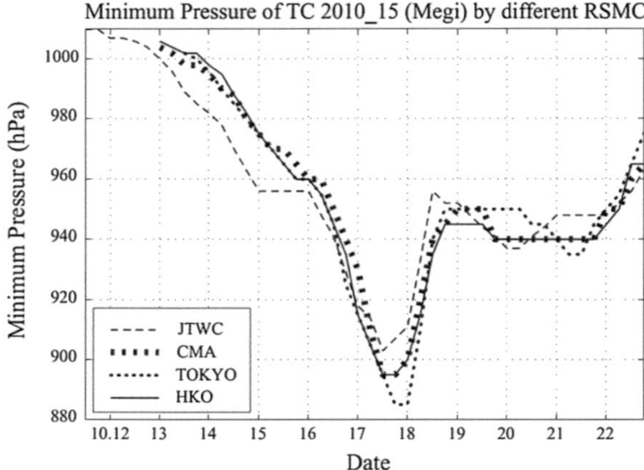

Fig. 8.10 Central pressure (unit: hPa) time series (6-h interval) of Typhoon Megi (2010) based on JTWC, CMA, TOKYO, and HKO database. Sourced from Qian et al. (2015) © Meteorological Monthly, with permission

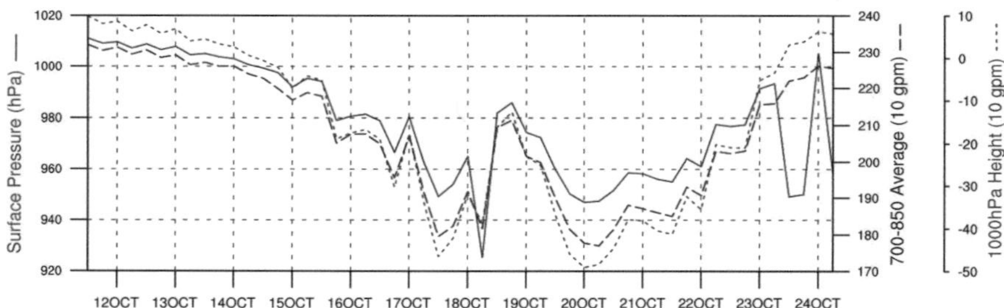

Fig. 8.11 The time series (6-h interval) of surface pressure (unit: hPa), 1000-hPa height (unit: 10 gpm), and 700–850 hPa average height (unit: 10 gpm) over the center of Typhoon Megi (2010) based on FNL database. Sourced from Qian et al. (2015) © Meteorological Monthly, with permission

pressure was at 0600 UTC 18 October. The lowest 1000-hPa height was at 0000 UTC 20 while the lowest 700–850-hPa height was at 0600 UTC 20 October. The three series showed that no one indicated the strongest at 1200 UTC 17 October from the FNL database.

Time series of total, climatic, and anomalous heights at 1000 hPa following the Megi center are given in Fig. 8.12a based on the ERA-Interim data. The total height series (solid line) had the lowest value at 0600 UTC 21 October while there was a diurnal cycle of climatic height with the lowest value at 0600 UTC 19 October. After removing the climatic height, the anomalous height at 1000 hPa displayed the lowest value at 0600 UTC 21 October. A similar evolution feature was also observed from the series of 700–850 hPa height, with the lowest value of total and anomalous heights at 0600 UTC 21 October (Fig. 8.12b). Figure 8.12c showed the time series

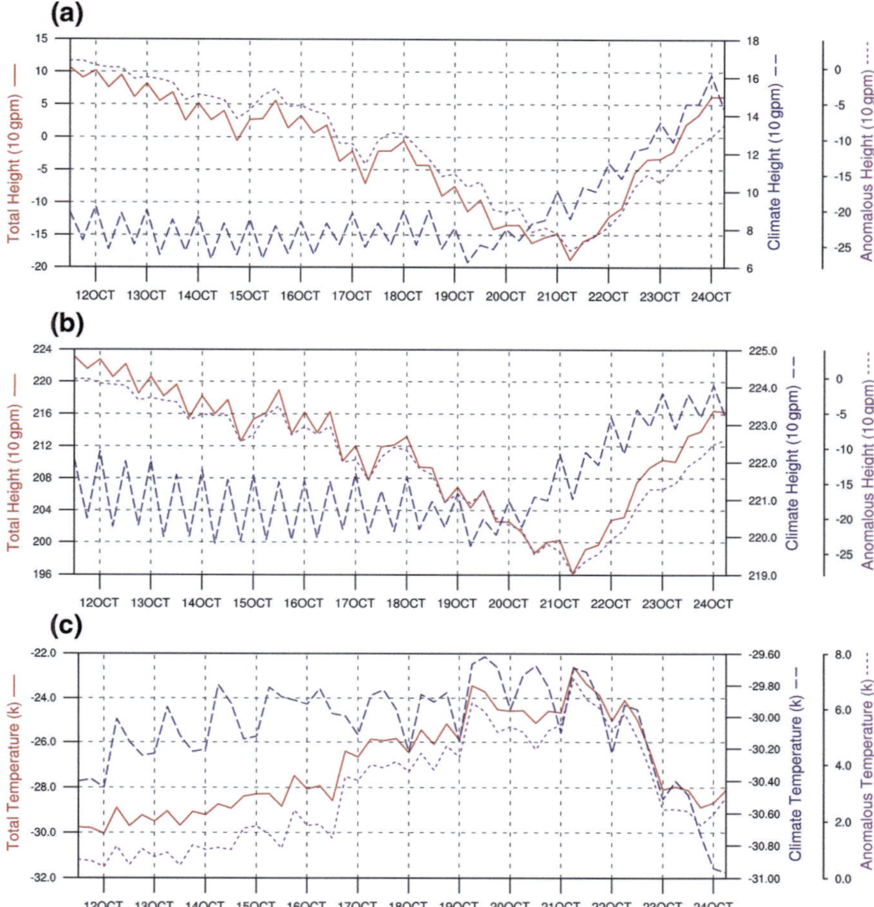

Fig. 8.12 Time series of variables over the center of Typhoon Megi (2010) based on the ERA-Interim product. Total, climatic, and anomalous heights (unit: 10 gpm) at **a** 1000 hPa, **b** 700–850 hPa. And **c** total, climatic and anomalous temperatures at 300 hPa (unit: K). Sourced from Qian et al. (2015) © Meteorological Monthly, with permission

of total, climatic, and anomalous temperatures at 300 hPa. The diurnal cycle can be identified from the climatic and total temperature series. The positive peak of anomalous temperature series was found at 0600 UTC 21 October (Fig. 8.12c).

Although the resolution of the NCEP R1 database is lower than the others, the lowest peak of height anomalies at 1000 hPa and 700–850 hPa was detected at 0000 UTC 21 October after removing its climatic height (Fig. 8.13a, b). The time difference of the two peak height anomalies is only 6 h from the two reanalysis products. The positive peak of anomalous temperature series from the NCEP R1 was found at 0600 UTC 20 October (Fig. 8.13c). The time difference of the two peak temperature anomalies is 1 day from the two reanalysis products. These series

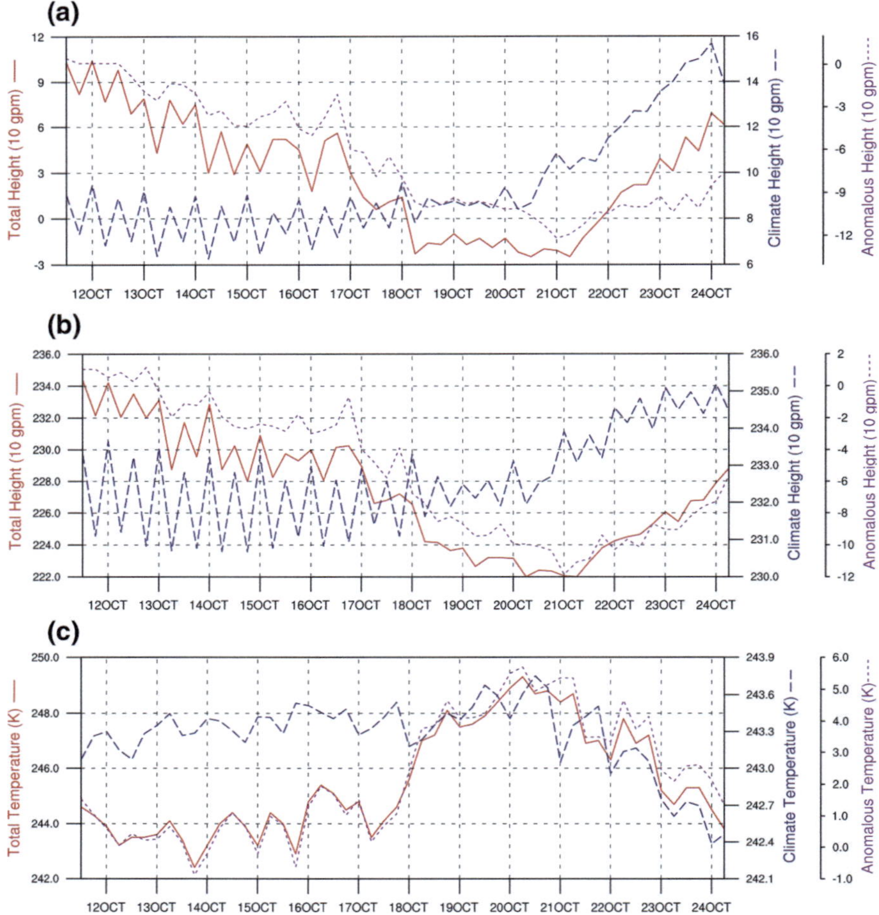

Fig. 8.13 Same as Fig. 8.12 except of time series of variables over the center of Typhoon Megi (2010) based on the NCEP R1 database. Sourced from Qian et al. (2015) © Meteorological Monthly, with permission

derived from the two reanalysis products and one analysis product indicated that the strongest intensity of Megi did not appear at 1200 UTC 17 October 2010. The differences of climatic and total variables can be compared from the two reanalysis products.

The above analysis showed that the date of strongest Megi's intensity estimated by the four meteorological centers was on 17 October, but it was on 21 October from the two reanalysis products. This estimation from the four meteorological centers was mainly based on the Dvorak (1975) technique over ocean. The Dvorak technique is a widely used system to estimate TC intensity, solely based on visible and infrared satellite images. As the technique matured through the 1970s and 1980s, measurement of cloud features became dominant in defining TC intensity and central pressure of the TC's low-pressure area. There are several visual patterns that a cyclone may take on, which define the upper and lower bounds on its intensity. However, each reanalysis product is developed to achieve specific goals with distinct parameterization schemes and resolutions, and nearly all of the reanalysis products include a variety of surface, upper air, and satellite observations (when available) using either a 3-Dimensional Variational (3DVar) or a 4-Dimensional Variational (4DVar) assimilation technique.

To estimate the intensity evolution of Megi, Qian et al. (2015) first gave the structure analysis at 1200 UTC 17 and 0600 UTC 21 October 2010. Figure 8.14 was the vertical–latitude section crossing the Megi center of total and climatic heights and climatic temperature using the ERA-Interim product along 125.25°E at 12 UTC 17 October 2010. The traditional synoptic chart (Fig. 8.14a) was similar to the vertical climatic chart (Fig. 8.14b). There were not any obvious features to indicate the Megi, except that the traditional synoptic chart had some small temperature fluctuations while the climatic chart was smooth.

After removing the climatic height and temperature from the total traditional weather chart, the height and temperature anomalies can clearly indicate the location and intensity of Megi. In Fig. 8.15a, the lowest center of height anomalies can locate the Megi. On the other hand, the temperature anomalies satisfy the hydrostatic equilibrium relationship with the height anomalies, so both of them can be used to indicate the Megi vortex. The hydrostatic relationship can be described by Eq. 8.2.1.

$$T' = -\frac{g}{R}\frac{\partial H'}{\partial \ln p} \approx -\frac{g}{R}\frac{(H'_{p=500} - H'_{p=250})}{(\ln 500 - \ln 250)} > 0 \tag{8.2.1}$$

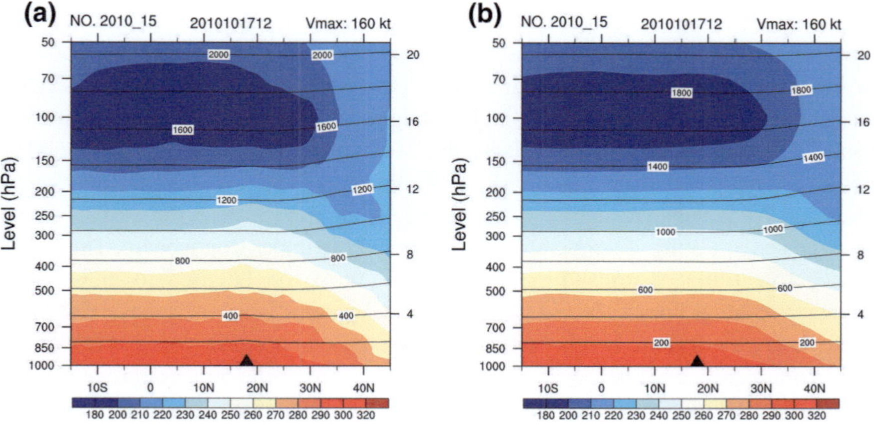

Fig. 8.14 The vertical–latitude section crossing the Megi center of **a** total and **b** climatic heights (*contour*, 200 × 10 gpm interval) and climatic temperature (*shading*, 10 and 20 K intervals) using the ERA-Interim product along 125.25°E at 1200 UTC 17 October 2010. *Triangle* indicates the location of Megi center. Sourced from Qian et al. (2015) © Meteorological Monthly, with permission

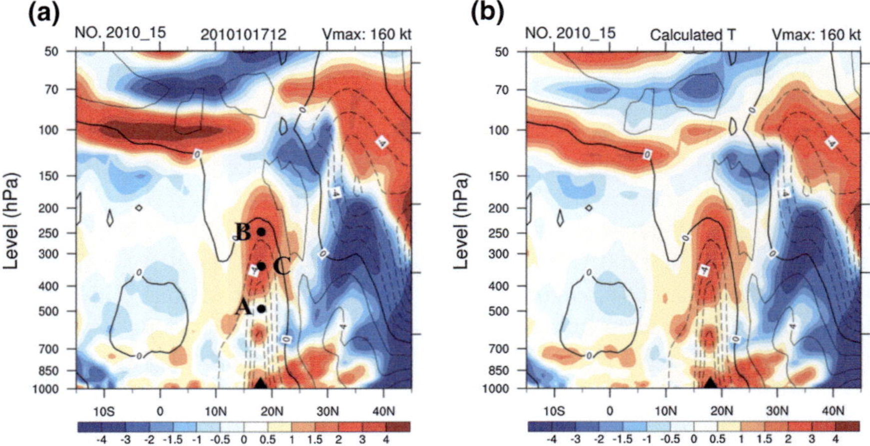

Fig. 8.15 Same as Fig. 8.14 except of **a** height anomalies (*contour*, 2 × 10 gpm interval) and temperature anomalies (*shading*, 0.5 and 1 K intervals) along with **b** height anomalies and calculated temperature anomalies by the relationship of hydrostatic balance at 1200 UTC 17 October 2010. Sourced from Qian et al. (2015) © Meteorological Monthly, with permission

It means that the temperature anomaly T' at the central point C can be calculated by Eq. 8.2.1, based on the lower point "A" at 500 hPa and the upper point "B" at 250 hPa. Figure 8.15b was the vertical distribution of temperature anomalies calculated by the hydrostatic equilibrium at 1200 UTC 17 October 2010. The vertical distributions of temperature anomalies from Fig. 8.15a, b were comparable.

Similar to that in Figs. 8.14 and 8.15, Qian et al. (2015) gave another set of total, climatic, and anomalous vertical cross sections of heights and temperatures in Fig. 8.16. In comparison between Figs. 8.16a and 8.14a, the total height does not have any signal to indicate the Megi vortex, except that a warm center is observed in the troposphere over the Megi center. However, the height and temperature anomalies, after removing the climatic height and temperature, can illustrate a useful structure or pattern to indicate the Megi vortex (Fig. 8.16c, d). The Megi vortex strengthened from 1200 UTC 17 October to 0600 UTC 21 October 2010, indicated by a negative center (Lm) of height anomalies near the surface and a positive center (Wm) of temperature anomalies at the upper troposphere. The values of negative center (Lm) and positive center (Wm) can be used to indicate the intensity of Megi vortex. In Fig. 8.16, a vertical pattern is robustly observed that a positive center (H) of height anomalies separates a positive center (W) of the temperature anomalies below and a negative enter (C) of the temperature anomalies above. Reversely, a negative center (L) of height anomalies separates a negative center (C) of the temperature anomalies below and a positive enter (W) of the temperature anomalies above.

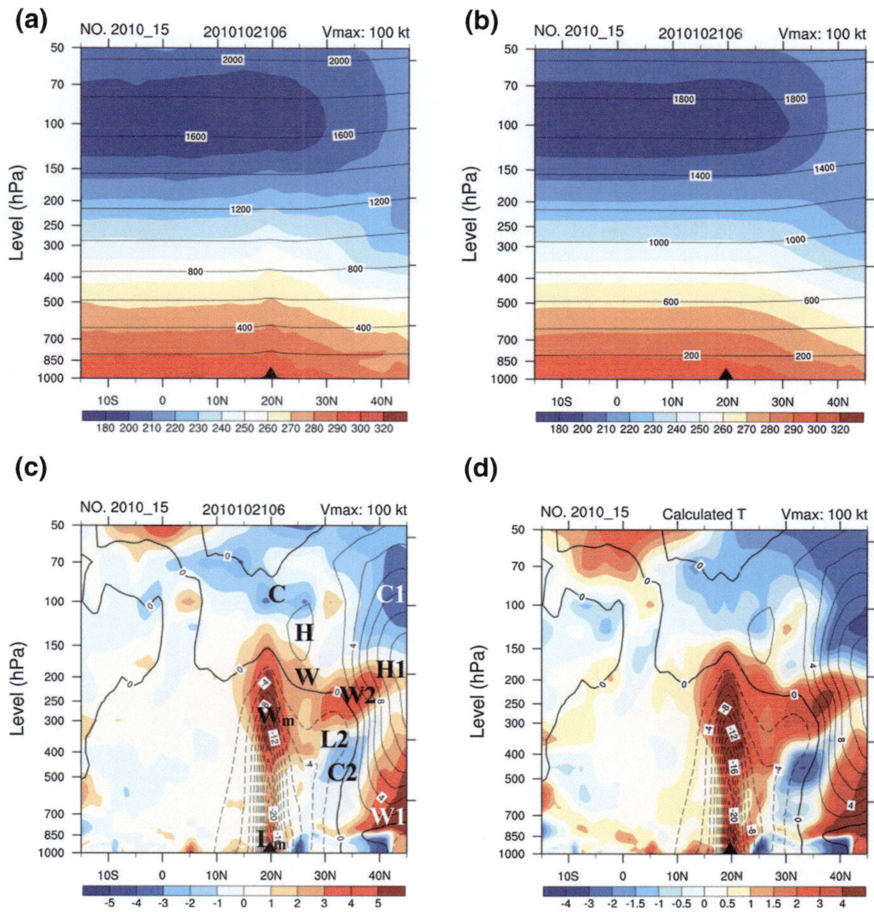

Fig. 8.16 **a**, **b** Same as in Fig. 8.14, **c** and **d** same as in Fig. 8.15, except along 117.75°E at 0600 UTC 21 October 2010. Sourced from Qian et al. (2015) © Meteorological Monthly, with permission

The two other datasets, the FNL analysis and NCEP R1, are used to compare with the ERA-Interim product. The FNL analysis (Fig. 8.17a) shows a stronger central intensity of Megi than that of the ERA-Interim product (Fig. 8.15a), while the weakest intensity is obtained from the NCEP R1 (Fig. 8.17b). The vertical structure or pattern of height and temperature anomalies is still observed from different datasets. The FNL analysis and NCEP R1 product also showed that the Megi vortex was strengthened from 1200UTC 17 to 0600 UTC 21 October 2010 (Fig. 8.18).

To examine the intensity evolution of Megi, Qian et al. (2015) plotted a time-vertical section of height and temperature anomalies following the center of Megi through its lifetime using the ERA-Interim product. Figure 8.19 showed that

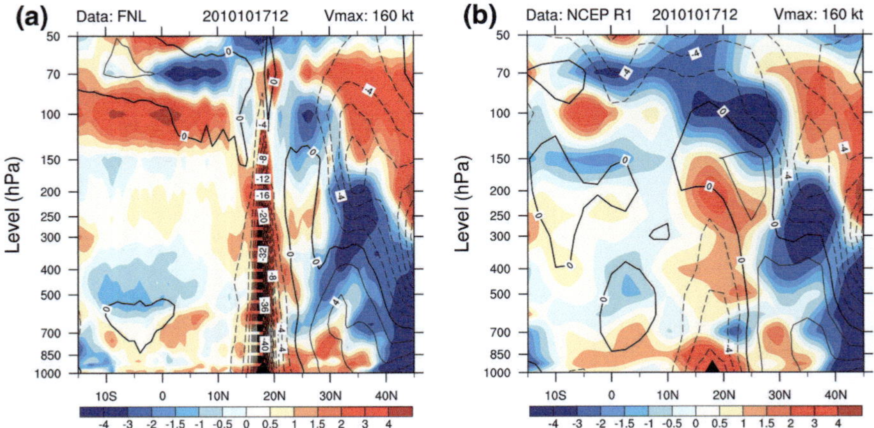

Fig. 8.17 The vertical–latitude sections crossing the Megi center of height anomalies (*contour*, 2 × 10 gpm interval) and temperature anomalies (*shading*, 0.5 and 1 K intervals) along 125.25°E at 1200 UTC 17 October 2010 based on the **a** FNL analysis and **b** NCEP R1 product. *Triangle* indicates the location of Megi center. Sourced from Qian et al. (2015) © Meteorological Monthly, with permission

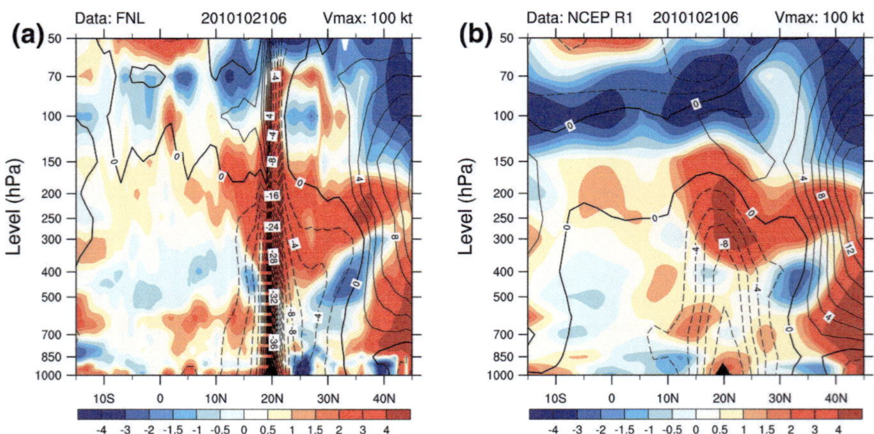

Fig. 8.18 Same as Fig. 8.17, except crossing section along 117.75°E at 0600 UTC 21 October 2010. Sourced from Qian et al. (2015) © Meteorological Monthly, with permission

the strongest intensity of Megi can be indicted by the 1000-hPa height at 0600 UTC 21 October 2010. At this time, the temperature anomaly also reached the highest (8 K) at the upper troposphere. This result shows that raw anomalies of height and temperature can be used to indicate the evolution of a TC during its lifetime.

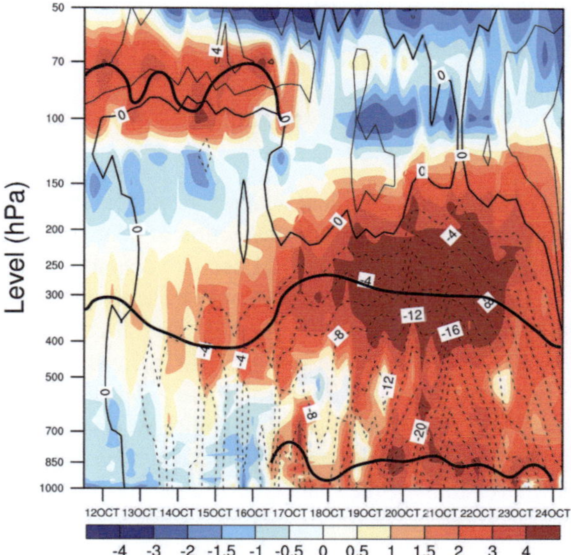

Fig. 8.19 Time-(6-h interval) vertical (hPa) cross-section of height anomalies (*contour*, 2 × 10 gpm interval) and temperature anomalies (*shading*, 0.5 and 1 K intervals) over the center of Megi through its lifetime using the ERA-Interim product. The *heavy solid line* indicates the axis of positive temperature anomalies. Sourced from Qian et al. (2015) © Meteorological Monthly, with permission

8.3 The Generalized Beta-Advection Model (GBAM)[2]

As described in the above section, a TC's motion is influenced by temporal climatic flows and interacted with other anomalous vortices. Thus, a simple model to experimentally predict and diagnose the TC tracks based on the above decomposition and the vorticity equation has been developed by Qian et al. (2014b) and Huang et al. (2015).

The barotropic vorticity equation approximately governs the horizontal flow under the consideration and is written in a planar coordinate system as follows:

$$\frac{\partial \varsigma}{\partial t} = -u\frac{\partial \varsigma}{\partial x} - v\frac{\partial \varsigma}{\partial y} - \beta v + F \tag{8.3.1}$$

where $\beta = (\mathrm{d}f/\mathrm{d}y) = 2\Omega\cos\varphi/a$, $f = 2\Omega\sin\varphi$, and $\Omega = 7.292 \times 10^{-5}\,\mathrm{rad/s}$ representing the angular speed of Earth's rotation. a and φ are the mean radius of the Earth and geographical latitude, respectively. F is a generic forcing term which includes forces at all scales and dissipation such as divergence, friction, and diabatic

[2]Modified article originally published in *Huang J., Du J., Qian W.H. (*2015*) A comparison between Generalized Beta-Advection Model and classical Beta-Advection Model in predicting and understanding unusual typhoon tracks in eastern China seas. Wea Forecasting 30:771–792.* Published with kind permission) © American Meteorological Society. All Rights Reserved.

processes. For non-divergent, frictionless, and adiabatic atmospheric motion, Eq. 8.3.1 can be simplified as

$$\frac{\partial\varsigma}{\partial t} = -u\frac{\partial\varsigma}{\partial x} - v\frac{\partial\varsigma}{\partial y} - \beta v. \qquad (8.3.2)$$

Equation 8.3.2 indicates that the movement of a vortex such as a typhoon is primarily controlled by the advection of relative vorticity and the beta effect. This is the classical two-dimensional beta-advection model (BAM).

This is also the Charney's (1948) approach, which utilized the conservative properties of potential vorticity. A special case of this model, the equivalent barotropic model, was used in 1950 to make the first successful numerical forecast. This model provided forecasts of the geopotential near 500 hPa. It did not forecast "weather" in the usual sense. It could, however, be used by forecasters as an aid in predicting the local weather associated with large-scale circulations. Later, multi-level versions of the quasi-geostrophic model provided explicit predictions of the surface pressure and temperature distributions.

Holland (1983) and Marks (1992) applied this BAM to hurricane track forecasting. The advection and the beta effect of TC motion have been described in more details by Chan (2005). There are several versions of the BAM (Velden and Leslie 1991; Marks 1992; Simpson 2003): the BAM Shallow (BAMS) to the 850–700 hPa layer, the BAM Medium (BAMM) to the 850–400 hPa layer, and the BAM Deep (BAMD) to the 850–200 hPa layer. In early days, the BAM was initialized using the vertically averaged horizontal wind predictions from the NMC (the predecessor of NCEP) spectral model (Marks 1992) and also included the beta effect (i.e., variations of the Coriolis force with latitude) (Holland 1983). Chandan and Rita (2012) found that BAM performance was poor in the NW Pacific basin compared to the North Atlantic basin, while in the North Atlantic basin the BAMD and BAMM performed slightly better than the BAMS.

Which layer (shallow, medium, or deep) to apply is an issue for the classical BAM. Qian et al. (2014b) developed a generalized BAM (GBAM), which was written at an optimal level for a BAM-type model to apply for short-term TC track predictions. This level is close to the levels of the max-VA and min-DA. The max-VA and min-DA levels can be estimated between 850 and 200 hPa over the TC center at model initialization time. The GBAM can be derived as follows.

Total wind (vorticity) can be decomposed into climatic and anomalous winds (vorticities) as follows:

$$\begin{cases} u = \tilde{u} + u' \\ v = \tilde{v} + v' \\ \varsigma = \tilde{\varsigma} + \varsigma' \end{cases}. \qquad (8.3.3)$$

The climatic field such as $\tilde{u}$ is estimated by averaging over 30 years (1981–2010), based on the reanalysis data on a calendar day d and at a diurnal time t,

$$\tilde{u}_d(\lambda, \varphi, p, t) = \sum_{y=1981}^{2010} u_{\{d,y\}}(\lambda, \varphi, p, t)/30, \tag{8.3.4}$$

where y covers 1981–2010. It is assumed that the positive and negative anomalies of atmospheric variables at a specific grid point (λ, φ, p) cancel each other out during the 30 year period to approximate the quasi-static climatic state. The climate defined by Eq. 8.3.4 varies from hour to hour and from day to day with a diurnal cycle and a seasonal cycle. We have done an experiment by removing an intense typhoon. The result shows that the "noise" is very small to influence the temporal climatology by using 30-year average with the typhoon or 29-year average without the typhoon.

By plugging Eqs. 8.3.3 into 8.3.1, the vorticity equation can be rewritten as

$$\frac{\partial(\tilde{\varsigma}+\varsigma')}{\partial t} = -(\tilde{u}+u')\frac{\partial(\tilde{\varsigma}+\varsigma')}{\partial x} - (\tilde{v}+v')\frac{\partial(\tilde{\varsigma}+\varsigma')}{\partial y} - \beta(\tilde{v}+v') + \widetilde{F} + F' \tag{8.3.5}$$

Taking the climatic average on Eq. 8.3.5, we can obtain a pure climatic vorticity Eq. 8.3.6 driven by climatic vorticity advection, climatic flow beta effect, climatic averaged forcing from anomalous weather systems (transient eddies), and other climatic averaged generic forcing,

$$\frac{\partial\tilde{\varsigma}}{\partial t} = -\tilde{u}\frac{\partial\tilde{\varsigma}}{\partial x} - \tilde{v}\frac{\partial\tilde{\varsigma}}{\partial y} - \beta\tilde{v} - \widetilde{X} - \widetilde{Y} + \widetilde{F}, \tag{8.3.6}$$

where $X = u'\frac{\partial\varsigma'}{\partial x}$ and $Y = v'\frac{\partial\varsigma'}{\partial y}$. Since the climatic wind and vorticity are estimated by Eq. 8.3.4, there is no need to predict climatic vorticity by Eq. 8.3.6. In all current weather and climate models, climatic wind is predicted so that calculated error is inevitable.

The prediction of anomalous vorticity can then be obtained by subtracting the climatic vorticity (Eq. 8.3.6) from the total vorticity (Eq. 8.3.5)

$$\begin{aligned} \frac{\partial\varsigma'}{\partial t} = & \left[-u'\frac{\partial\varsigma'}{\partial x} - v'\frac{\partial\varsigma'}{\partial y} - \beta v'\right] \\ & + \left[-\tilde{u}\frac{\partial\varsigma'}{\partial x} - \tilde{v}\frac{\partial\varsigma'}{\partial y}\right] + \left[-u'\frac{\partial\tilde{\varsigma}}{\partial x} - v'\frac{\partial\tilde{\varsigma}}{\partial y}\right] \\ & + \widetilde{X} + \widetilde{Y} + F' \end{aligned} \tag{8.3.7}$$

Based on our estimations of many TCs, it is found that

$$\left[-\tilde{u}\frac{\partial\varsigma'}{\partial x} - \tilde{v}\frac{\partial\varsigma'}{\partial y}\right] > \left[-u'\frac{\partial\tilde{\varsigma}}{\partial x} - v'\frac{\partial\tilde{\varsigma}}{\partial y}\right]. \tag{8.3.8}$$

Based on the 30-year ERA-Interim data, it is also found that $\widetilde{X}+\widetilde{Y}$ can also be neglected because it is of order $10^{-10}/\mathrm{s}^{-2}$, an order smaller than the daily anomalous vorticity advection by total wind (an order $10^{-9}/\mathrm{s}^{-2}$), i.e.,

$$\left|-u\frac{\partial\varsigma'}{\partial x}-v\frac{\partial\varsigma'}{\partial y}\right|>\left|\widetilde{X}+\widetilde{Y}\right| \tag{8.3.9}$$

Therefore, under the non-divergent, frictionless, and adiabatic condition ($F'=0$), Eq. 8.3.7 can be simplified as

$$\frac{\partial\varsigma'}{\partial t}=-u\frac{\partial\varsigma'}{\partial x}-v\frac{\partial\varsigma'}{\partial y}-\beta v'. \tag{8.3.10}$$

Equation 8.3.10 is the generalized beta-advection model (GBAM) used by Qian et al. (2014b) and Huang et al. (2015). The GBAM basically describes the advection of a TC disturbance (anomaly) by the total flow, which is the sum of the climatic and anomalous flows. To examine the different roles of climatic and anomalous flows, a climatic-flow beta-advection model (CBAM[3]) and an anomalous-flow beta-advection model (ABAM) can be analogously defined by Eqs. 8.3.11 and 8.3.12, respectively

$$\frac{\partial\varsigma'}{\partial t}=-\tilde{u}\frac{\partial\varsigma'}{\partial x}-\tilde{v}\frac{\partial\varsigma'}{\partial y}-\beta v', \tag{8.3.11}$$

$$\frac{\partial\varsigma'}{\partial t}=-u'\frac{\partial\varsigma'}{\partial x}-v'\frac{\partial\varsigma'}{\partial y}-\beta v', \tag{8.3.12}$$

where the vorticity anomaly $\varsigma'=\frac{\partial v'}{\partial x}-\frac{\partial u'}{\partial y}$. The CBAM can be viewed as the steering process of a TC vortex by the climatic flow, while the ABAM explicitly describes the interaction between a TC vortex and other surrounding anomalous weather disturbances. The separate descriptions of the impacts from large-scale background flow and the anomalous systems by the CBAM and the ABAM are the advantages over the traditional total-based BAM. In operation, the CBAM might be similar to the classical BAM with differences in the definition of flows, i.e., climatic (temporally anomalous) flow versus large-scale environment (residual) flow.

The vorticity equation includes the divergence term of horizontal wind, which can be neglected if its value is small enough, so we hope to find a layer with minimum divergence. At 1200 UTC 19 October 2010, when the Megi vortex

[3]The inclusion of β in the CBAM was not found to affect the typhoon trajectory in a number of random test cases, despite the potential to modify the vorticity anomaly that defines the shape and position of a typhoon.

started to turn right, the layer of minimum divergence anomaly (MDA) was examined. Figure 8.20a showed that two MDA layers are located between 300 and 400 hPa as well as around 700 hPa over the location of Typhoon Megi. Theoretically, maximum ascending velocity and maximum vorticity anomaly (MVA) should be observed over the location of a typhoon near its MDA layer. So, two MVA layers are found, not surprisingly, located between 300 and 400 hPa as well as around 850 hPa (Fig. 8.20b).

After finding the layer of Megi's MDA and MVA at every moment, three models including CBAM, ABAM, and GBAM from Eqs. 8.3.10 to 8.3.12 are used to predict the motion of positive vorticity center, respectively. These models are written and calculated globally, but we only focus on the typhoon of our interest and its vicinity area at the layer of MDA and MVA. The calculations are carried out at a single pressure level or the average pressure layer of MDA and MVA with fixed grid interval of 0.75° × 0.75°. In the calculation, the initial wind field is obtained from the ERA-Interim product. The vorticity anomaly is updated every 10 min and then new wind anomalies u' and v' are derived from vorticity anomaly via spherical harmonic expansion. The climatic wind components $\tilde{u}$ and $\tilde{v}$ are the mean winds averaged over 1981–2010, as calculated by Eq. 8.3.4. In order to restrain sub-grid waves, the vorticity field is smoothed by triangular truncation on spherical harmonic coefficients. This technique makes the computation stable but also causes a side effect: the amplitude of vorticity anomaly keeps decreasing and becomes smaller than the observed after two days.

All the integration Eqs. 8.3.10, 8.3.11, and 8.3.12 are written in planar coordinates for easy understanding and simplicity. In our actual calculations, these

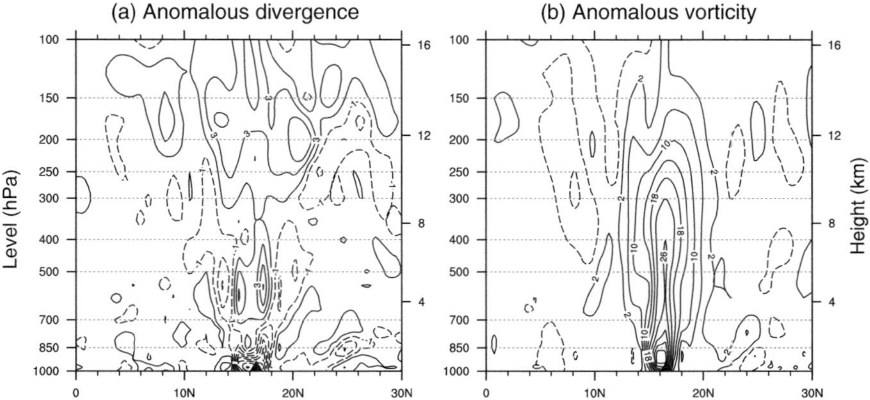

Fig. 8.20 Vertical cross-sections of **a** divergence anomaly (4 × 10^{-5} s^{-1} interval) and **b** vorticity anomaly (4 × 10^{-5} s^{-1} interval) at 1200 UTC 19 October 2010 along 117.75°E. Sourced from Qian et al. (2014b) © 2014 John Wiley & Sons, Inc., with permission

equations are written and calculated in spherical coordinates. For example, the GBAM is in the form of Eq. 8.3.13

$$\frac{\partial \varsigma'}{\partial t} = -\frac{u}{a\cos(\theta)}\frac{\partial \varsigma'}{\partial \lambda} - \frac{v}{a}\frac{\partial \varsigma'}{\partial \theta} - \beta v', \tag{8.3.13}$$

where the anomalous vorticity $\varsigma' = -\frac{1}{a\cos(\theta)}\frac{\partial v'}{\partial \lambda} - \frac{1}{a}\frac{\partial u'}{\partial \theta}$.

8.4 GBAM Predictions on Typhoon Megi (2010)[4]

This section is a brief description of the computational methods used to integrate the GBAM forward in time (Qian et al. 2014b). Model integration is done in spherical coordinates with 0.75° longitude–latitude grid at a pressure level. Initial condition is obtained from the ERA-Interim product, where the global climate is derived through Eq. 8.3.4 and anomaly is the difference between the total field and climatic component. The time step for the model integration is 10 min and the climate can be updated every 6 h (since the reanalysis is available 6 hourly), but the best way is to update the climate more frequently. At each time step, a new vorticity anomaly is calculated using Eqs. 8.3.10, 8.3.11, and 8.3.12. Then new wind anomalies are derived from the new vorticity anomaly via a spherical harmonic expansion, and climatic wind is linearly interpolated between two climatic winds that are 6 h apart (ideally climatic winds are needed at every time step, i.e., 10 min). Since the vorticity anomaly tends to grow too strong and causes numerical instability during integration, the new model-derived vorticity anomalous field is smoothed by performing a triangular truncation on the spherical harmonic coefficients. This technique makes the computation stable but also has a side effect: the amplitude of vorticity anomaly decreases steadily with time and becomes weaker than the observations after 2 days. The spherical harmonic expansion and triangular truncation technique are described at this website http://dx.doi.org/10.5065/D6WD3XH5.

To verify the model results, Typhoon Megi was chosen as an example here. Figure 8.21 showed the average of vorticity anomaly between 300 and 400 hPa, along with the Megi's best track and its central location at different times: 0000 UTC 19 October (pre-turning point), 0012 UTC 19 October (turning moment), and from 20 to 22 October (post-turning period) 2010. It is clear that the center of vorticity anomaly is consistent with the actual track of Megi. From 1200 UTC 20 to 0000 UTC 22 October, three centers of vorticity anomaly, i.e., the Megi vortex in

---

[4]Modified article originally published in *Qian W.H., Shan X.L., Liang H.Y., Huang J., Leung C.H.* (2014b) *A generalized beta advection model to improve unusual typhoon track prediction by decomposing total flow into climatic and anomalous flows. J Geophys Res Atmos 119(3):1097–1117.* Published with kind permission © 2014 John Wiley & Sons, Inc.. All Rights Reserved.

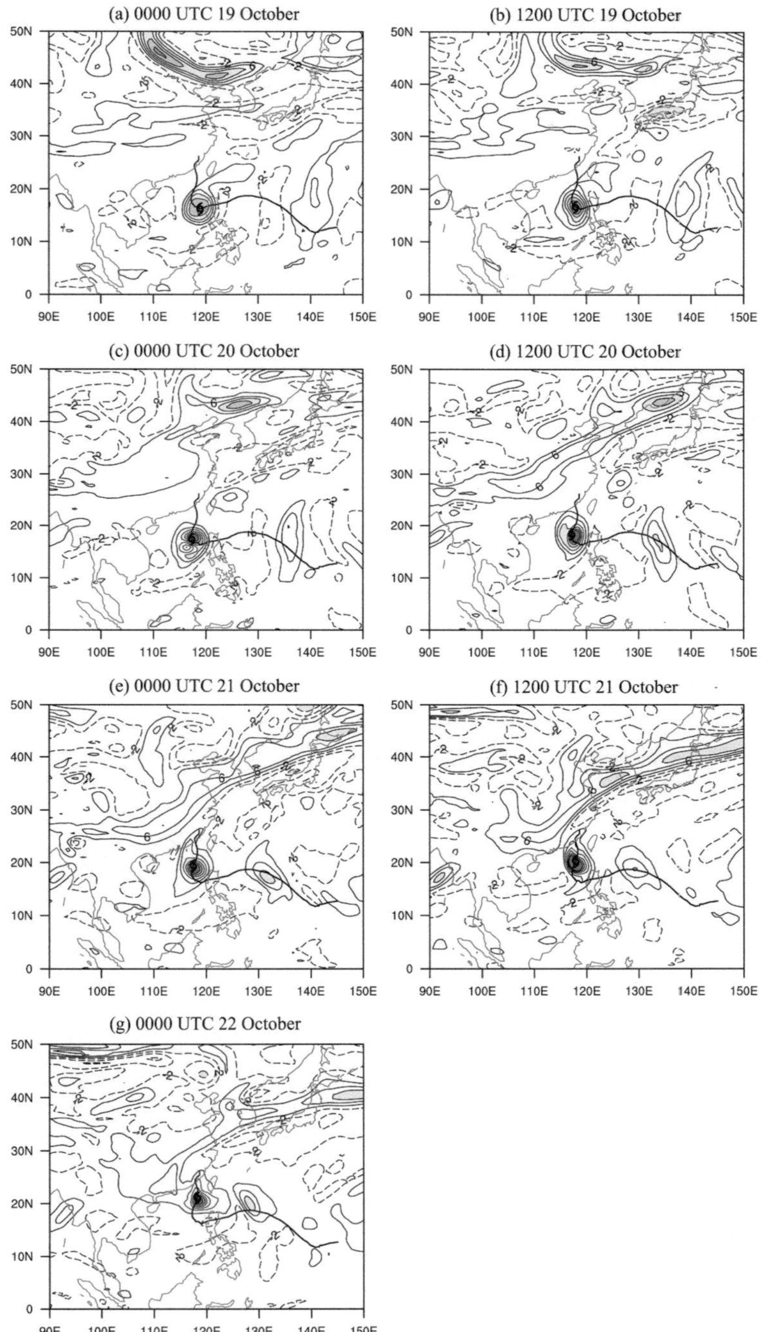

Fig. 8.21 Average vorticity anomaly (2×10^{-5} s^{-1} interval) between 300 and 400 hPa at **a** 0000 UTC 19, **b** 1200 UTC 19, **c** 0000 UTC 20, **d** 1200 UTC 20, **e** 0000 UTC 21, **f** 1200 UTC 21, and **g** 0000 UTC 22 October 2010. The Megi (2010) track and its current location are denoted by the *solid curve* and the symbol 𝟼. Sourced from Qian et al. (2014b) © 2014 John Wiley & Sons, Inc., with permission

the northeastern SCS and two vortices in Southwest China and the Philippine Sea, circled around each other and became closer to each other.

First, the prediction by GBAM was examined. This model is used to predict the sudden right turning at 1200 UTC 19 October, starting from 0000 UTC 19 October 2010. Figure 8.22 shows the prediction for the next 2.5 days (60 h). At 0000 UTC 20 and 1200 UTC 20 October 2010, the model successfully predicts the right turn, although the predicted speed is faster than Megi's actual speed.

Then the prediction by CBAM starting from 0000 UTC 19 October 2010 for the next 60 h was examined (Fig. 8.23). At 0000 UTC 20 and 1200 UTC 20 October, the predicted center of vorticity anomaly keeps moving westwards with increasing speed without turning right. On 21 October, the predicted center of vorticity anomaly moves to the Indo-China Peninsula. Therefore, Megi's right turning in the northeastern SCS cannot be predicted by the CBAM. The form of CBAM's governing equation is similar to that of the conventional BAM, differing only in climatic flow versus large-scale environmental flow, and temporal anomalous flow versus residual flow. It is also clear to see from Fig. 8.7c, e, and g that the right turning cannot be predicted using the 5-day-mean wind, the monthly mean wind, and the spatially filtered wind, because the Megi vortex is located in different large-scale environmental easterly flows. Three experiments show that: (i) using the 5-day-mean wind (Fig. 8.7c) and its residual flow (Fig. 8.7d), the model predicted a local-circling path of the typhoon in the western sea of the northern Philippines for the next 48 h; (ii) using the monthly mean wind (Fig. 8.7e) and its residual flow (Fig. 8.7f), the model predicted a slow straight-moving path of the typhoon for the next 72 h; and (iii) using the spatially filtered flow (Fig. 8.7g) and its residual flow (Fig. 8.7h), the model predicted a westward straight-moving path of the typhoon for the next 48 h.

Qian et al. (2014b) further examined whether the anomalous interaction alone (Eq. 8.3.12), the ABAM, can predict the sudden right turning at 1200 UTC 19 October starting from 0000 UTC 19 October 2010. Figure 8.24 shows the prediction for the next 2.5 days. From 0600 UTC 19 to 1200 UTC 19 and 1200 UTC 20 October, the center of vorticity anomaly turns right but the speed is faster than Megi's actual speed. During the first two days (48 h), the track prediction is comparable to the observation. On 21 October, however, the predicted center of vorticity anomaly returns southward, incorrectly. This result implies that the anomalous flow may be a dominant factor influencing the turning motion of Megi in the SCS.

As analyzed by Xu (2011), some complex general circulation models (GCMs) failed to predict the sudden right turning of Megi in the SCS when they were initialized at 0000 UTC 18 October 2010. At that time, the MDA and MVA layers of Megi were basically centered between 300 and 400 hPa in the vicinity of the typhoon (Fig. 8.25). Thus, Qian et al. (2014b) applied the GBAM, at the average layer between 300 and 400 hPa. Figure 8.26 showed the GBAM's result, where the prediction of vorticity anomaly started from 0000 UTC 18 October 2010 before Megi made landfall on the northern Philippines. The result showed that the right turning from 1200 UTC 19 October was predicted but its speed after the right

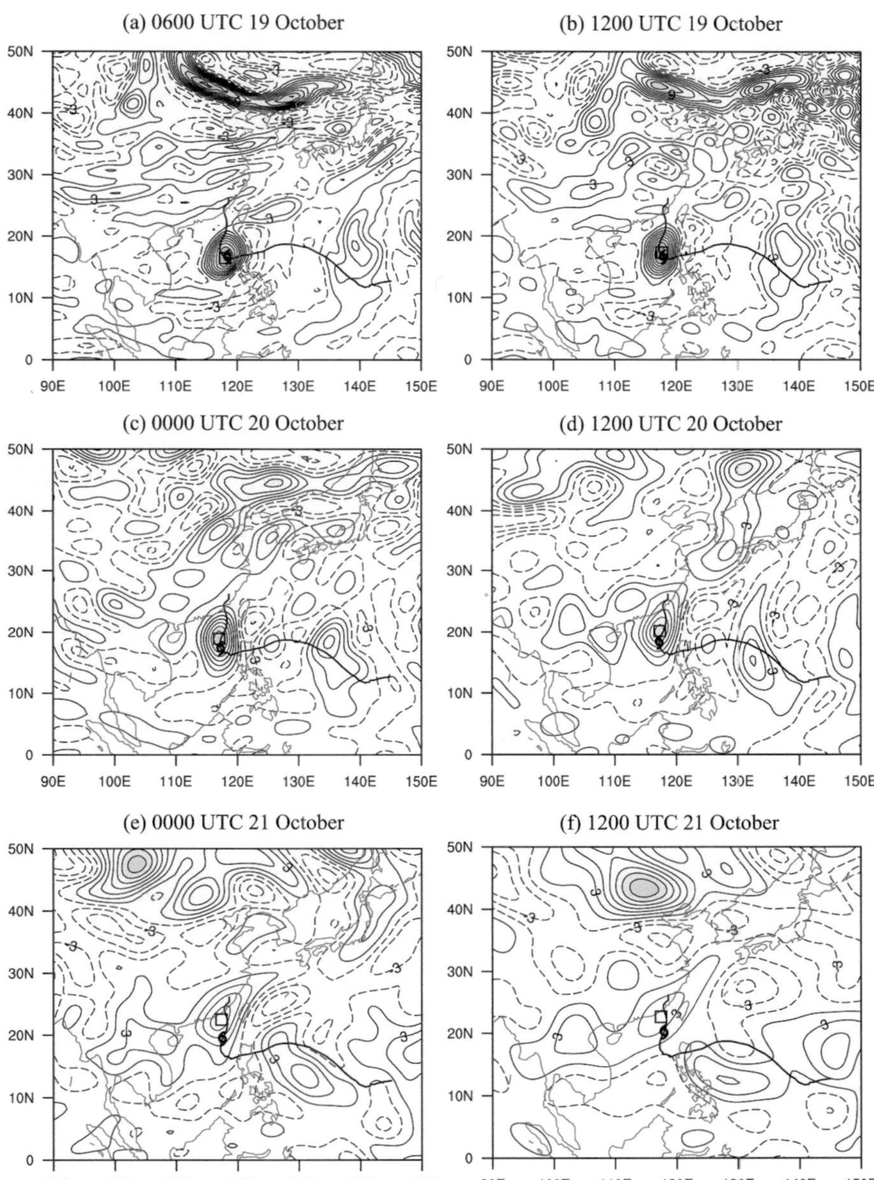

Fig. 8.22 Predictions of vorticity anomaly (2×10^{-5} s^{-1} interval) started from 0000 UTC 19 October 2010 by the GBAM at 300–400 hPa at **a** 0600 UTC 19, **b** 1200 UTC 19, **c** 0000UTC 20, **d** 1200 UTC 20, **e** 0000 UTC 21, and **f** 1200 UTC 21 October 2010. The small box "□" indicates the predicting location of typhoon vortex. The typhoon track and its current location are denoted by the *solid curve* and the symbol "໑". Sourced from Qian et al. (2014b) © 2014 John Wiley & Sons, Inc., with permission

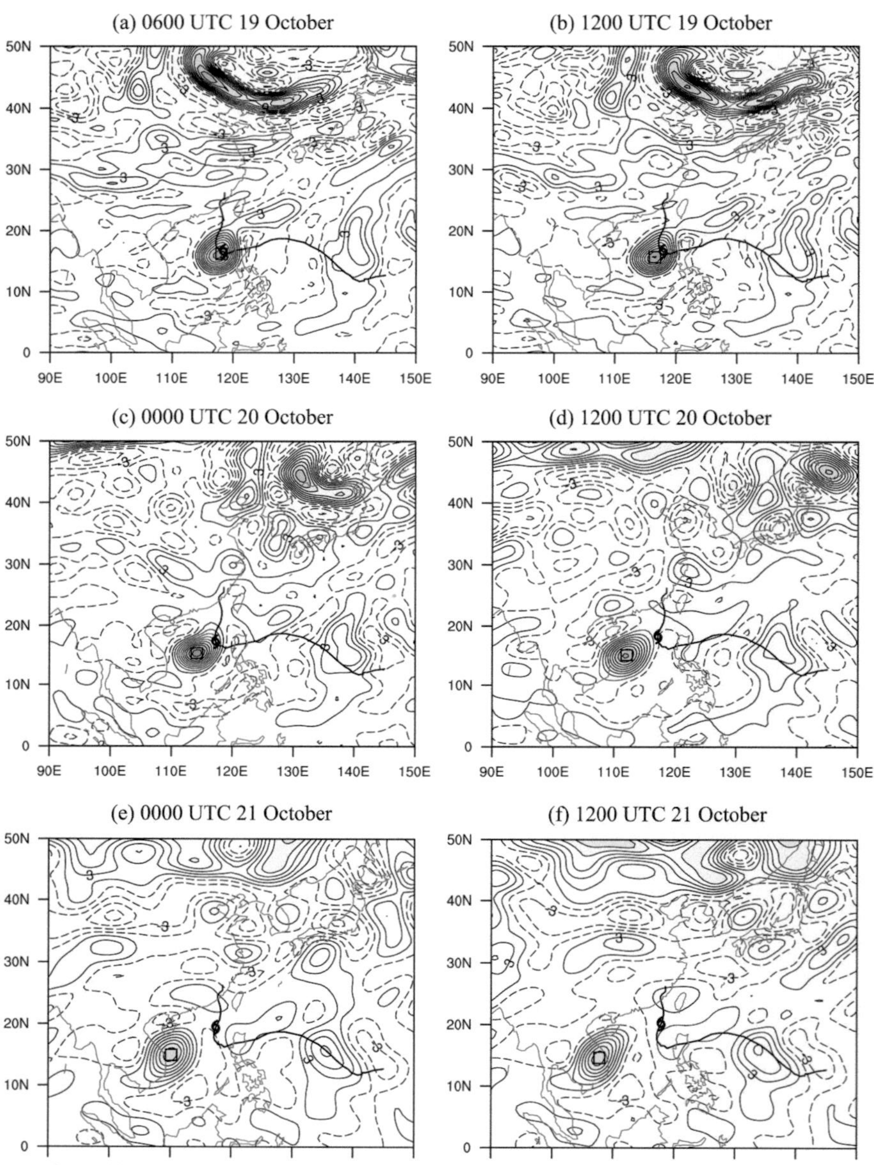

Fig. 8.23 Same as in Fig. 8.22 except using the CBAM. Sourced from Qian et al. (2014b) © 2014 John Wiley & Sons, Inc., with permission

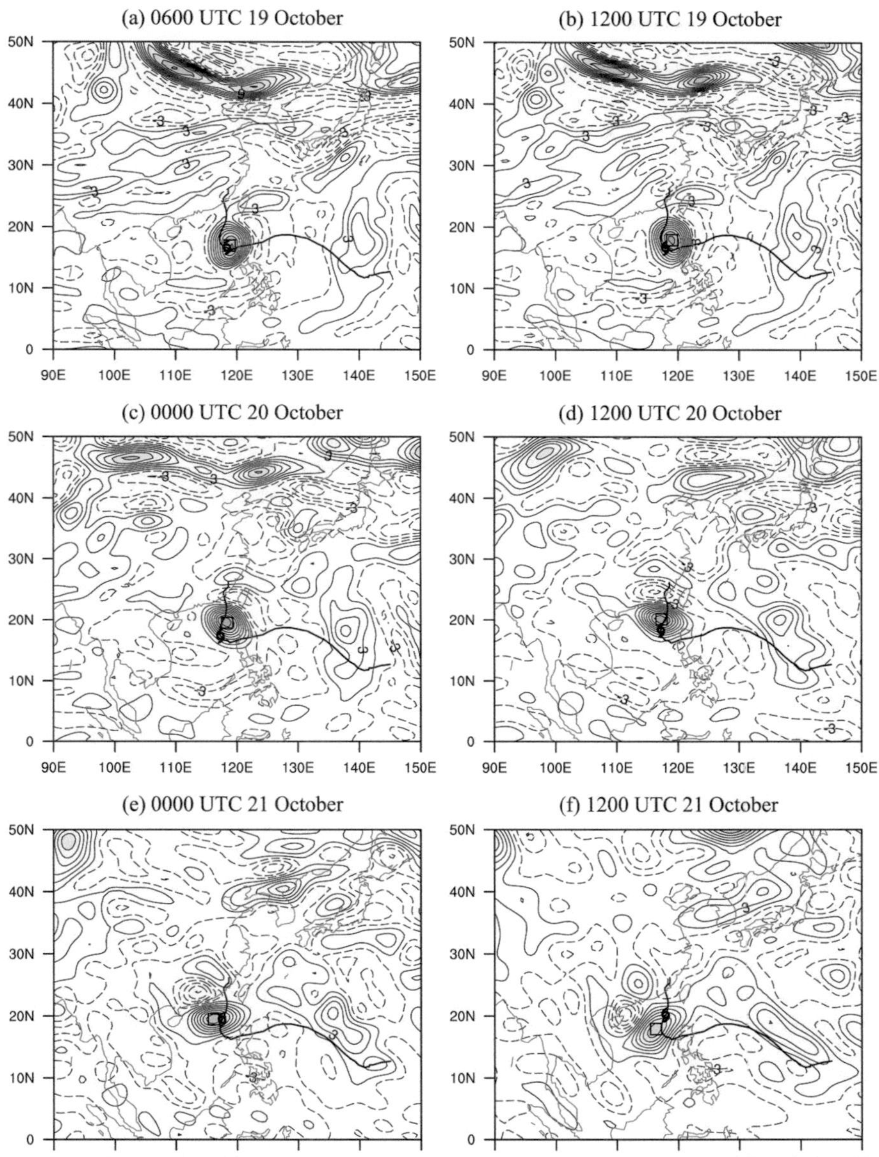

Fig. 8.24 Same as in Fig. 8.22 except using the ABAM. Sourced from Qian et al. (2014b) © 2014 John Wiley & Sons, Inc., with permission

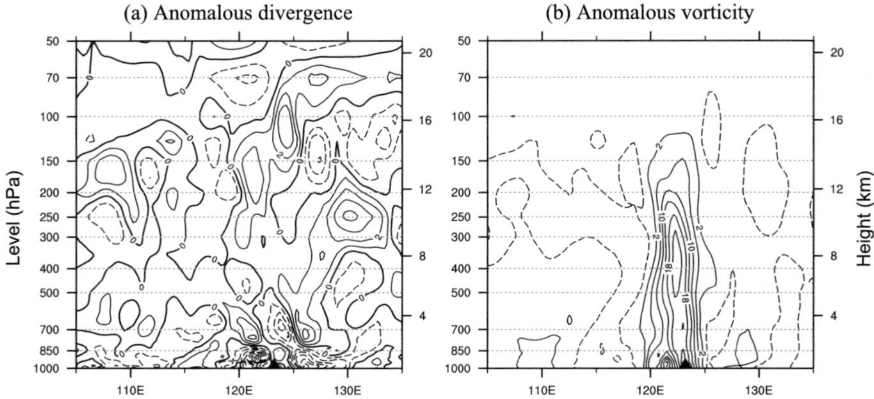

Fig. 8.25 Vertical cross-sections of **a** the divergence anomaly ($1 \times 10^{-5}\ s^{-1}$ interval) and **b** the vorticity anomaly ($4 \times 10^{-5}\ s^{-1}$ interval) at 0000 UTC 18 October 2010 along 17.25°N. Sourced from Qian et al. (2014b) © 2014 John Wiley & Sons, Inc., with permission

turning tended to be faster and the predicted intensity tended to be weaker than the observation.

From 15 to 17 October 2010, Megi experienced a left turning in the Philippine Sea. The GBAM was applied at 500 hPa, the layer of MDA and MVA. The left turning is well predicted from 0000 UTC 15 October 2010. But, again, the predicted speed was faster and intensity was weaker than the observation. The right turning of Megi that started from 0000 UTC 13 to 1800 UTC 14 October 2010 was also well predicted by the GBAM at 850 hPa.

8.5 More Applications of the GBAM[5,6]

Every year, about seven TCs hit mainland China from either the southern or eastern coasts (Xue et al. 2012). Over the past five years (2007–2011), the average typhoon track forecast errors at the NMC are 114, 190, and 287 km for 24, 48, and 72 h forecasts, respectively (Qian et al. 2012). Similar progress has been made for TCs in the North Atlantic basin at the U.S. National Hurricane Center, where TC position forecasts have gained approximately 1 day in skill, e.g., 48 h position errors during

[5]Modified article originally published in Qian W.H., Shan X.L., Liang H.Y., Huang J., Leung C.H. (2014b) A generalized beta advection model to improve unusual typhoon track prediction by decomposing total flow into climatic and anomalous flows. J Geophys Res Atmos 119(3):1097–1117. Published with kind permission © 2014 John Wiley & Sons, Inc. All Rights Reserved.

[6]Modified article originally published in *Huang J., Du J., Qian W.H.* (2015) *A comparison between Generalized Beta-Advection Model and classical Beta-Advection Model in predicting and understanding unusual typhoon tracks in eastern China seas. Wea Forecasting 30:771–792.* Published with kind permission) © American Meteorological Society. All Rights Reserved.

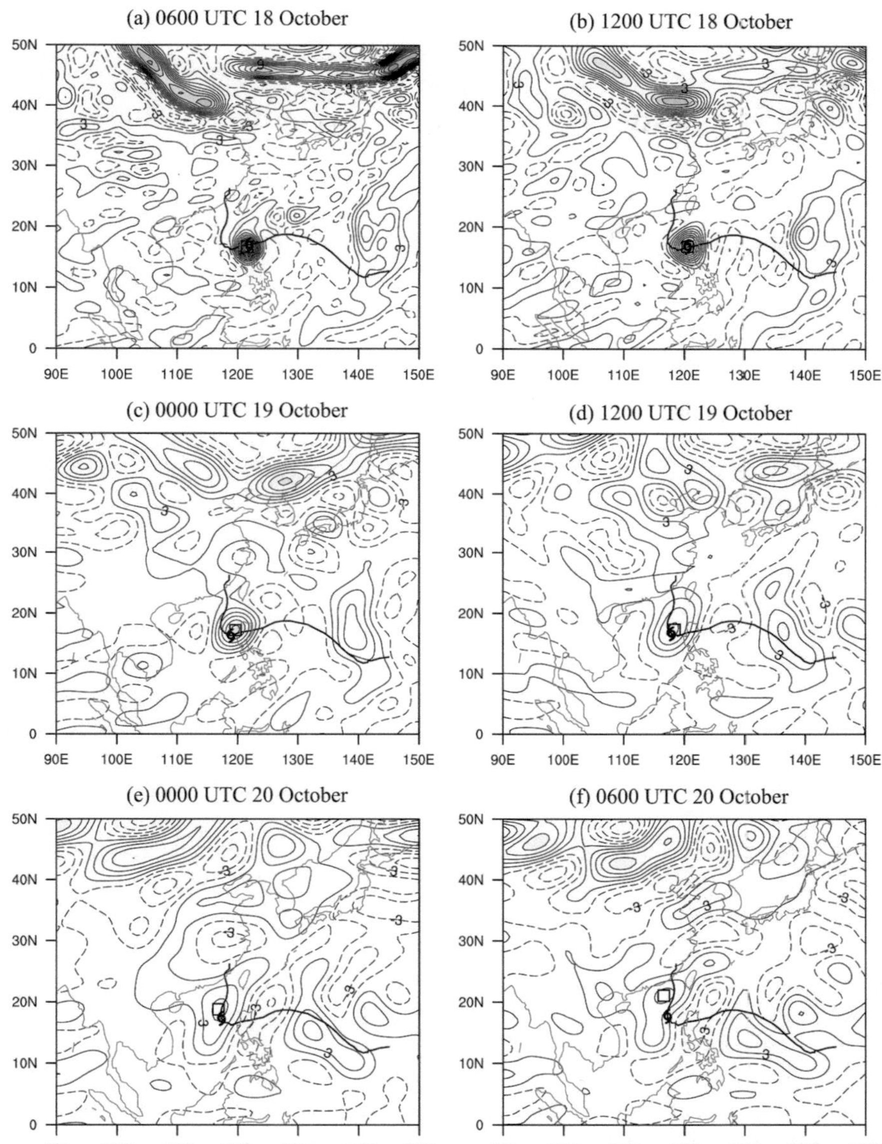

Fig. 8.26 Same as in Fig. 8.22 except the predictions of vorticity anomaly (10^{-5} s^{-1}) from 0000UTC 18 October 2010 using the GBAM at 300–400 hPa at **a** 0060 UTC 18, **b** 1200 UTC 18, **c** 0000 UTC 19, **d** 1200 UTC 19, **e** 0000 UTC 20, and **f** 0600 UTC 20 October 2010, respectively. Sourced from Qian et al. (2014b) © 2014 John Wiley & Sons, Inc., with permission

2000–2008 are comparable to 24 h position errors from 1970 to 1989, at about 100 km (Rappaport et al. 2009). Much of the advance in forecast skill can be attributed to the improvements in model physics, the increases in model resolution, the availability of aircraft and satellite observations (e.g., Aberson 2010), and new data assimilation techniques (e.g. Hamill et al. 2011). Statistically, all typhoons that make landfall in mainland China must reach the off-shore area to the north of 15°N and west of 130°E during their lifetimes. From 1979 to 2011, there have been a total of 330 typhoons over the two ocean basins (SCS and ECS) that reached or passed through this area (Qian et al. 2014a). Among the 330 TCs, most of them had smooth tracks and only a few had abrupt turnings. It is those smooth-track (normal) TCs that contribute to the high accuracy in track forecasts, while false predictions are often associated with the rapid-turning TCs. Most TCs move westward from the NW Pacific to the SCS (normal track), but a few of them swerve to the north with a sudden right-turning track (Deng et al. 2010). With such sudden right-turning typhoons, their landfall locations are difficult to be accurately predicted with current methods even at 24–48 h in advance.

During 1979–2011, there have been totally 19 typhoons that experienced right turnings in the SCS. Their tracks and the turning points in the SCS are shown in Fig. 8.27. As indicated by two arrows in each panel, the right-turning typhoon is defined when the moving direction rapidly changes at least 30° in the SCS from pre-turning period to post-turning period. The first arrow direction follows four pre-turning locations along a typhoon track 24 h before turning, and the second arrow direction follows four post-turning locations within 24 h after turning. Right turnings of the 15 typhoons are well predicted at least 24 h in advance. However, four typhoons' right turnings are not correctly predicted in leading 24–48 h (Fig. 8.27a). Their turning locations are all concentrated near the Hainan Island in the northwest SCS, where they experience stagnant motion during the turning stage. Four TCs' intensities at turning times as well as 24 and 48 h prior to the turning are listed in Table 8.1. Except the 24th TC in 1995 reaching the classification (34–63 knots of the maximum wind speed) of TS, other three cases were TD (below 33 knots) or disturbance. The durations when these four cases' intensities reaching the typhoon classification (64 knots) lasted for only a short period (6–24 h), which occurred after their turning. Other 15 typhoons had an average intensity of 75, 71, and 78 knots at their turning time, 24 and 48 h prior to the turning, respectively. Although this model correctly predicts the right turning for all the 15 typhoons, the predicted speeds tend to be relatively faster than their observations.

Fig. 8.27 Tracks of **a** four weak typhoons and **b–f** other 15 strong typhoons indicated by the symbol "ɕ" and their track predictions started from pre-turning location 24 h ahead. The location of large symbol "ɕ" indicates the turning point. In each panel, *two arrows* indicate directions of typhoon in pre-turning 24 h (4 points) and post-turning 24 h (other four points). In the *bottom left* of **a**: symbol and number (× 041981) indicate the typhoon track × and the 4th TC in 1981. In the *bottom left* of **b**: symbol and number (O 261981) indicate the predicted typhoon track O and the 26th TC in 1981. Sourced from Qian et al. (2014b) © American Meteorological Society, with permission ▶

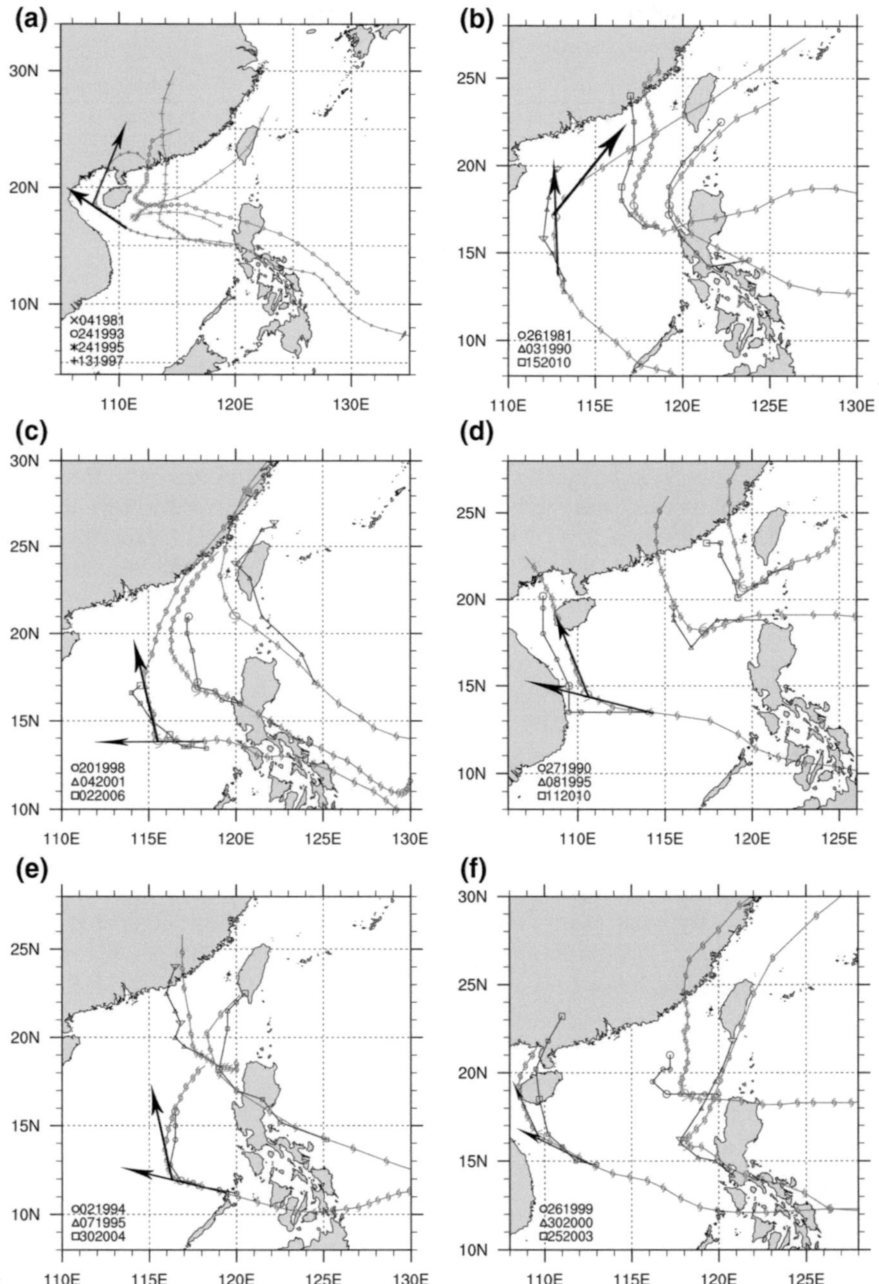
(a)
×041981
o241993
*241995
+131997
(b)
o261981
Δ031990
□152010
(c)
o201998
Δ042001
□022006
(d)
o271990
Δ081995
□112010
(e)
o021994
Δ071995
□302004
(f)
o261999
Δ302000
□252003

Table 8.1 Four TC intensities (knots) at three moments. Sourced from Qian et al. (2014b) © American Meteorological Society, with permission

Typhoon: no. (turning Time)	Pre-48 h	Pre-24 h	Turning times
1981: 04 (12 UTC 09 June)	20	25	30
1993: 24 (06 UTC 24 Sep)	27	29	30
1995: 24 (12 UTC 12 Oct)	35	55	70
1997: 13 (12 UTC 31 July)	15	20	32
Average	24	32	40.5

The typhoons whose right turnings are experimentally predicted are listed in Table 8.2. For comparison, the MVA and MDA levels as well as the best level to predict the turning track, at 24 and 48 h prior to the turning, are all listed together in Table 8.2. The best level is determined when a predicted turning track matches the best with observation. In Table 8.2, the first typhoon is the 26th TC in 1981 with its right turning time at 0000 UTC 25 November. In the pre-turning 24 h, the MVA level is well consistent with the best level (500 hPa), but there is a 100-hPa difference between the best level and the MDA level. In the pre-turning 48 h, the differences of both the MVA and the MDA levels relative to the best level are 50 hPa. Among the 15 typhoons, the durations of two typhoons (1995_No.07 and 2010_No.11) were 48 h shorter from their genesis to the right turning. On average, the difference between the best level and the MVA level is 47 hPa in the pre-turning 24 h and 77 hPa in the pre-turning 48 h; and the difference between the best level and the MDA level is 40 hPa in the pre-turning 24 h and 46 hPa in the pre-turning 48 h. The maximum difference between the best level and the MVA level is 150 hPa in the pre-turning 48 h among four cases and that between the best level and the MDA level is 100 hPa among seven cases. This statistical analysis indicates that the data with higher vertical resolution may be an important factor to obtain the best level based on the diagnostic results of MDA and MVA.

Using the 5-day-mean wind and spatially filtered wind, the canonical BAM is also applied to these 15 right-turning typhoons (Table 8.2). It turns out that none of the typhoon tracks has been correctly predicted to turn right. For 5-day-mean wind, the model shows that seven of them are expected to have local-circling motion in the next 48 h. Three of them have straight motion, four turn left, and only one typhoon (the 30th TC in 2004) turns right with distance errors of more than 500 (1000) km at leading 24 (48) hours. For spatially filtered winds, results show that nine of them are expected to move westward or southwestward in the next 48 h, five have local-circling motion, and only one typhoon (the 02nd TC in 1994) turns right rapidly with a distance error of more than 500 (800) km at leading 24 (48) hours.

The abrupt turning of typhoons is one kind of extreme tracks, while a lasting straight-moving track is another type of extremes. During 1979–2011, there have been totally 10 typhoons that experienced straight-moving tracks, lasted for at least 48 h in the SCS and then made landfall over South China and northeastern Vietnam (Fig. 8.28). For all these 10 typhoons, their straight-moving tracks are generally well predicted by the GBAM initiated from the strongest moment of each typhoon

Table 8.2 MVA and MDA levels of 15 right-turning typhoons[a]

Typhoon: no. (turning time)	Pre-turn time (h)	MVA level/ difference (hPa)	MDA level/ difference (hPa)	Best level (hPa)
1981: 26 (00 UTC 25 Nov)	24	500/0	400/100	500
	48	500/50	500/50	450
1990: 03 (12 UTC 17 May)	24	550/50	550/50	600
	48	500/100	400/0	400
1990: 27 (12 UTC 15 Nov)	24	500/0	500/0	500
	48	550/0	450/100	550
1994: 02 (06 UTC 6 Apr)	24	450/50	500/0	500
	48	550/150	700/0	700
1995: 07 (00 UTC 30 Jul)	24	850/150	600/100	700
	48	–	–	–
1995: 08 (12 UTC 10 Aug)	24	600/50	550/0	550
	48	700/150	450/100	550
1998: 20 (00 UTC 24 Oct)	24	500/0	600/100	500
	48	450/150	500/100	600
1999: 26 (12 UTC 6 Oct)	24	600/100	700/0	700
	48	750/50	700/0	700
2000: 30 (12 UTC 29 Oct)	24	550/0	600/50	550
	48	500/100	500/100	600
2001: 04 (00 UTC 23 June)	24	550/50	450/50	500
	48	500/50	450/50	500
2003: 25 (06 UTC 17 Nov)	24	600/0	600/0	600
	48	500/150	600/50	650
2004: 30 (00 UTC 3 Dec)	24	450/50	450/50	500
	48	550/50	600/0	600
2006: 02 (18 UTC 14 May)	24	400/0	450/50	400
	48	500/0	550/50	500
2010: 11 (12 UTC 8 Sep)	24	500/100	350/50	400
	48	–	–	–
2010: 15 (12 UTC 19 Oct)	24	500/100	400/0	400
	48	400/0	400/0	400
Averaged bias (hPa)	24	47	40	/
	48	77	46	/

[a]MVA level and MDA level at pre-turning locations for 24 and 48 h as well as the best level to predict the turning track at pre-turning 24 and 48 h. The difference (hPa) between the MVA (or MDA) level and the best level. Symbol "–" in column means that the storm not existing at that time

Sourced from Qian et al. (2014b) © American Meteorological Society, with permission

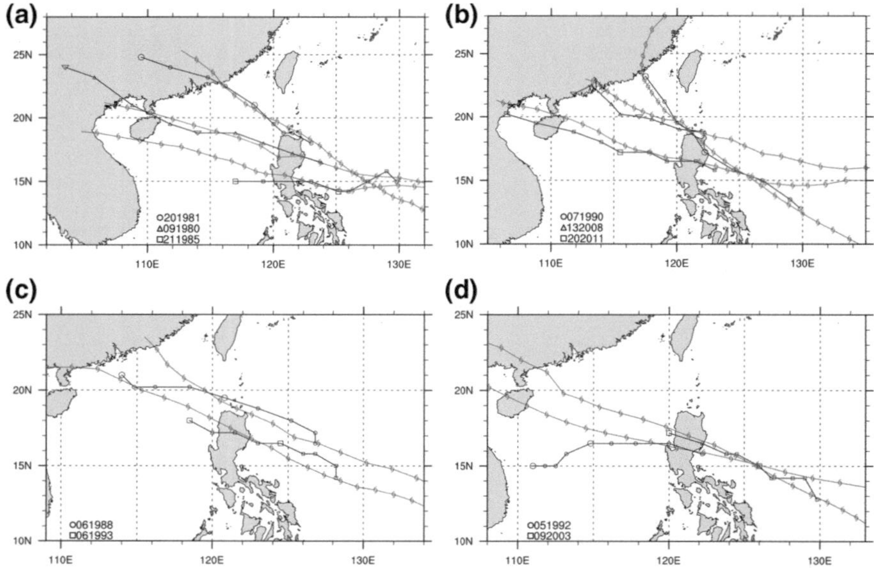

Fig. 8.28 Tracks and predictions of 10 typhoons started from the strongest moment of their lifetimes. **a** Three typhoons in 1981, 1980, and 1985; **b** three typhoons in 1990, 2008, and 2011; **c** two typhoons in 1988 and 1993; and **d** two typhoons in 1992 and 2003. The symbol ∮ indicates the typhoon location of 6-h interval. In **a**: "○ 201981" means the 20th TC in 1981 and its track prediction indicated by the symbol "○" every 6 h starting from the strongest moment at 1200 UTC 20 July 1981 for the next 48 h. Sourced from Qian et al. (2014b) © American Meteorological Society, with permission

for a 48-h period. In Table 8.3, the average difference between the best level and the MVA level is 65 hPa, and that between the best level and the MDA level is 40 hPa, at the strongest moment during a "straight-moving" typhoon lifetime.

During 1979–2011, a total of 163 TCs have occurred in the ECS and reached typhoon strength (74 mph, 64 knots, or 33 ms^{-1}). The ECS includes the Bohai Sea, the Yellow Sea, the East China Sea, and the Taiwan Island as shown in Fig. 8.29a. Their tracks are classified into four types (Table 8.4). Right-turning and straight-moving are the two most common tracks (Huang et al. 2015). 113 TCs have a right-turning track. Their initial, peak, and dissipating positions are indicated by open dots, solid dots, and crosses, respectively (Fig. 8.29a). As shown by the fitted lines, the initial positions of these right-turning TCs are located in the lower latitudes, mostly in the tropical NW Pacific, while the dissipating positions are in the middle latitudes. The peak positions are observed mostly in the subtropical NW Pacific near the ECS. 33 TCs have straight-moving tracks (Fig. 8.29b). Their initial, peak, and dissipating positions are mostly located over the tropical NW Pacific, the ECS, and inland East China, respectively. From the initial to peak and dissipating positions, their general movement directions are illustrated by the arrows in Fig. 8.29a, b.

Table 8.3 MVA and MDA levels of 10 straight-moving typhoons[a]

Typhoon: no. (strongest time)	MVA level/ difference (hPa)	MDA level/ difference (hPa)	Best level (hPa)
1980: 09 (12 UTC 20 July)	550/50	550/50	500
1981: 20 (18 UTC 19 Sep)	850/50	850/50	800
1985: 21 (18 UTC 16 Oct)	350/50	350/50	300
1988: 06 (00 UTC 17 July)	500/150	650/0	650
1990: 07 (06 UTC 25 June)	450/0	500/50	450
1992: 05 (00 UTC 11 July)	500/200	750/50	700
1993: 06 (12 UTC 24 June)	800/50	700/50	750
2003: 09 (18 UTC 20 July)	700/50	750/0	750
2008: 13 (00 UTC 20 Aug)	450/50	500/100	400
2011: 20 (18 UTC 26 Sep)	600/0	600/0	600
Averaged bias (hPa)	65	40	/

[a]MVA level and MDA level as well as the best level at the strongest moment of typhoons. The difference (hPa) between the MVA (or MDA) level and the best level
Sourced from Qian et al. (2014b) © American Meteorological Society, with permission

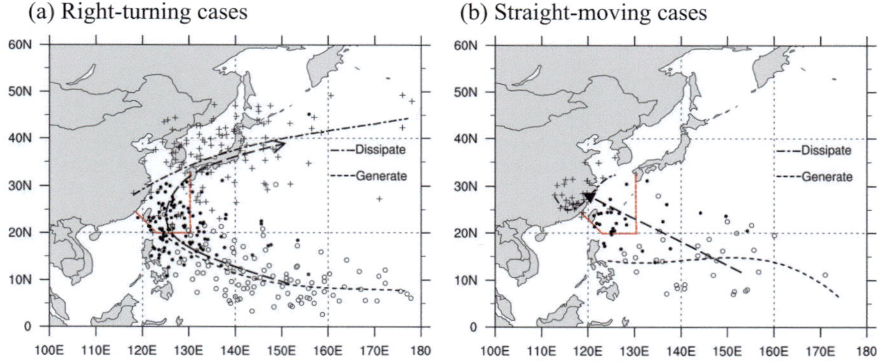

Fig. 8.29 The initial (*open dots*), peak (*solid dots*), and dissipating (*crosses*) position of each TC that belongs to **a** the 113 right-turning tracks and **b** the 33 straight-moving tracks in the ECS from 1979 to 2011. The *dashed line* at low latitudes and the *dashed-dot line* at middle latitudes are the cubic splines fitting the initial and dissipating positions, respectively. Two *dashed-line arrows* indicate the movement tendency for all TCs from initial to dissipating positions. *Red dashed lines* outline the area of the ECS. Sourced from Huang et al. (2015) © American Meteorological Society, with permission

Figure 8.30 shows the three representative groups of tracks in the ECS. The first group shows five TCs with a single right-turn point near the box (24–26°N and 124–126°E) located to the northeast of Taiwan (Fig. 8.30a). Most right-turning TCs reach their maximum intensity near the turning point (Richard and Davidson 2004; Zeng et al. 2007). The second group shows four TCs with multiple (two or more) right-turning points (Fig. 8.30b). The third group shows seven TCs with a

Table 8.4 The classification of TCs over ECS (1979–2011)

Total typhoon number (163)	Class (number)
Right-turning cases (113)	Single turning point (99)
	Multiple turning points (*14*)[a]
Straight-moving cases (33)	Landfall in East China (29)
	Landfall in Japan (1)
	Disappeared over sea (*3*)[a]
Left-turning and landfall cases (4)	Left-turning angle >40°
Complex path cases (*13*) *	Circling and serpentine paths

[a]These 30 TCs tracks are not modeled

Sourced from Huang et al. (2015) © American Meteorological Society, with permission

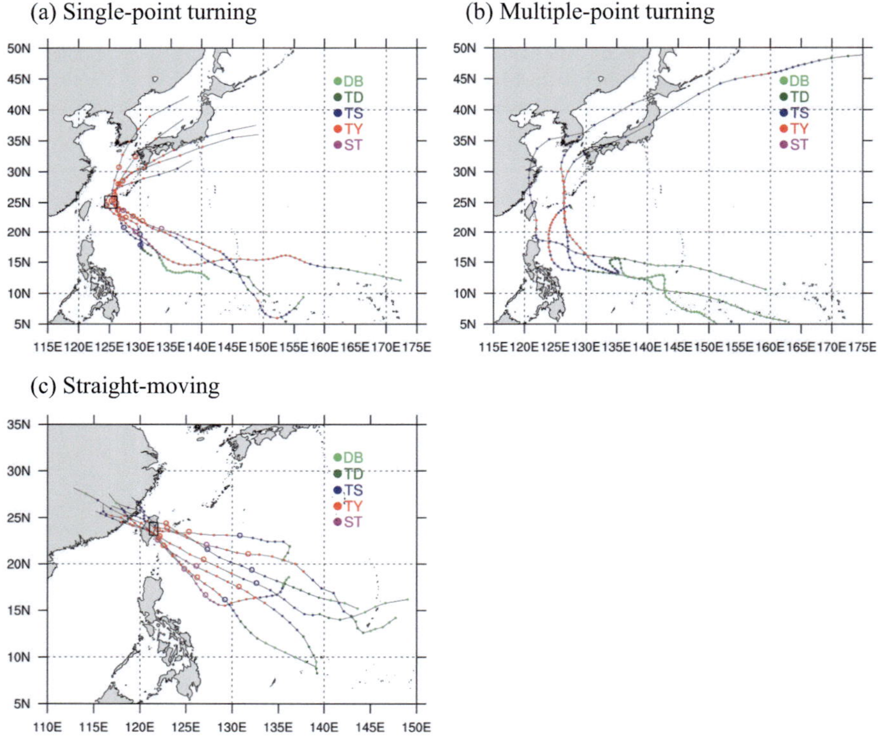

Fig. 8.30 Tracks of **a** five single-point right-turning TCs, **b** four multiple-point right-turning TCs, and **c** seven straight-moving TCs. The TC intensity is indicated by *DB*, *TD*, *TS*, *TY* and *ST*. *Big open circles* indicate positions in **a** 24 and 48 h before and **c** 24 h after the TC's turning/landing point. The *boxes* depict the areas of 24–26°N and 124–126°E in **a** and 23–24.5°N and 121–122°E in **c**. Sourced from Huang et al. (2015) © American Meteorological Society, with permission

straight-moving track and an initial landfall in Taiwan (Fig. 8.30c). In total, there are 33 straight-moving TCs with 29 of them making landfall in mainland China, one making landfall in Japan, and three dissipating over the open sea. Among the 29 land falling TCs, 20 of them made their initial landfalls in Taiwan. The seven selected cases in Fig. 8.30c moved straight toward Taiwan and made first landfall on the island within the box (23–24.5°N and 121–122°E) during the last 48 h of their life cycles.

The four typhoons with unusual left-turning tracks are shown in Fig. 8.31. The five pairs of arrows indicate a direction change exceeding 40 degrees between the post-turning point and the point with 48 h prior to the turning. After their final left turnings, all the four TCs made landfall on the eastern coast of China. Although the occurrence of left-turning TCs is rare (four out of 163 or about 2.5%), they cause severe damage after their unexpected landfalls and are woefully difficult to predict in current operations.

Historical cases show that unusual TC tracks are difficult to predict even at 24 h to 48 h in advance (Chen et al. 2002). Typhoon Aere (Fig. 8.31a) formed in the equatorial NW Pacific at 1700 UTC 17 August 2004, reached the typhoon intensity on 21 August and then moved northwestward. Its intensity peaked at 1800 UTC 24 August off the northeast shore of Taiwan and then it unusually turned to the west-southwest. It landed on the southeast coast of China (Shishi City, Fujian Province) at 1330 UTC 25 August and dissipated in Guangdong Province at 1200 UTC 26 August 2004. From 24 to 26 August, it caused severe floods in both southeast Fujian Province and northern Taiwan. As indicated in Fig. 8.31a, the left turning was quite sharp with a turning angle of about 65°. For this TC, its optimal level of max-VA was observed approximately at 350 hPa.

After obtaining its optimal level, Fig. 8.32 shows a 2-dimensional view of Aere at 350 hPa at 0600 23 UTC August 2004 (24 h before the turning). From the total fields (Fig. 8.32a), Aere was intuitively expected to continue its northwestward or

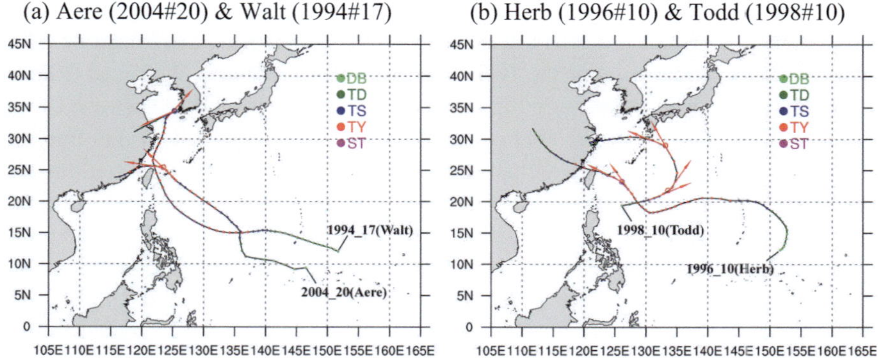

Fig. 8.31 Left-turning TC tracks: **a** Aere (2004#20) and Walt (1994#17), and **b** Herb (1996#10) and Todd (1998#10). The five pairs of *red arrows* indicate turning angles of at least 40°. *Open circles* denote the left-turn point. Sourced from Huang et al. (2015) © American Meteorological Society, with permission

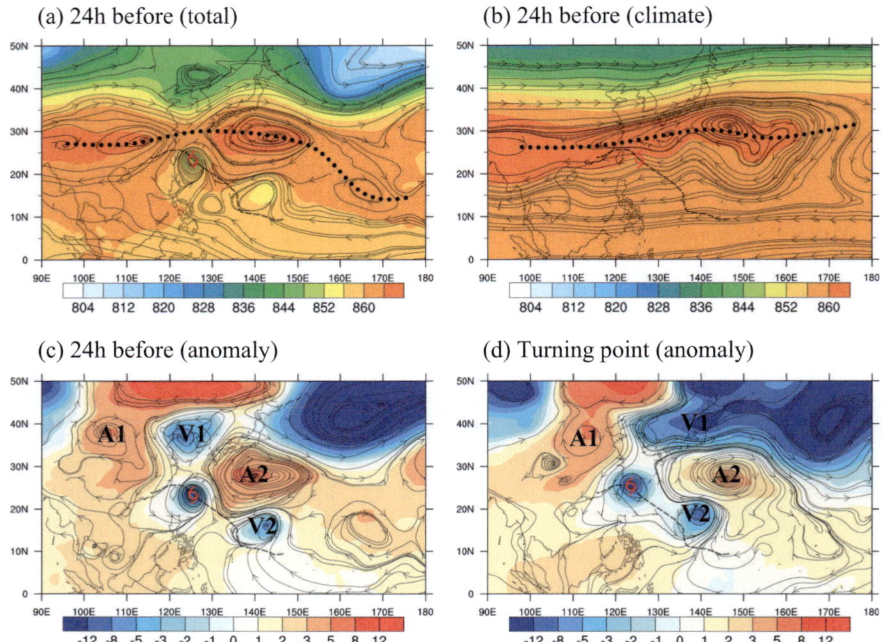

Fig. 8.32 Horizontal distributions at 350 hPa of **a** total flow (stream) and height (shaded, 40 gpm intervals), **b** climatic flow (stream) and height, and **c** anomalous flow (stream) and height (shaded, 10, 20, 30, and 40 gpm intervals) 24 h before the turning, **d** same as **c** but at the turning point. The symbol "໑" denotes Typhoon Aere's position and the *dashed line* is the best track of Aere. In **a** and **b**, the *heavy dotted line* indicates the subtropical high ridge. In **c** and **d**, the two anomalous vortices and two anomalous anticyclonic systems are indicated by "V1", "V2", "A1", and "A2", respectively. Sourced from Huang et al. (2015) © American Meteorological Society, with permission

northward path, because it was located on the western edge of a strong subtropical NW Pacific high. Since it turned toward the southwest in reality, there must be other stronger impacts acting upon the typhoon. Li et al. (2006) and Wei et al. (2006) argued that the northeasterly of a high pressure system over land to the west of the TC as well as the interaction with a nearby typhoon contributed to this unusual southwestward turning. The climatically easterly (Fig. 8.32b) suggested that it should be steered to the west. In the anomalous fields (Fig. 8.32c), two positive centers (anticyclonic, A1 and A2) and two negative centers (cyclonic, V1 and V2) of height anomaly surrounded Typhoon Aere (໑). Vortex V2 was another TC, which interacted with Typhoon Aere as reported in some early studies (Lin et al. 2005; Wei et al. 2006; Li et al. 2006). From Fig. 8.32c, both V1 and A2 seemed to pull the typhoon northward to northeastward. At the turning point (Fig. 8.32d), two positive and two negative height anomalous centers still surrounded Aere. It had approached the anticyclonic system A1 and could be directly influenced by A1 to make a southwestward turning. These speculations will be examined by the

GBAM, CBAM, and ABAM below to quantitatively assess the roles played by the total, climatic, and anomalous flows in causing this unusual left turning.

Nine experiments (Table 8.5) are carried out for the four left turning typhoons. For each experiment, an optimal level for the GBAM to predict a TC track, which ranges from 850 to 300 hPa (Table 8.5), is first detected. The corresponding max-VA and min-DA levels of the nine experiments are listed in Table 8.5. For locating max-VA and min-DA levels, 2-dimensional divergence and vorticity anomalies are first obtained at each pressure level from 850 to 300 hPa. They can then be found by manually comparing the central values at all levels over a typhoon center. Table 8.5 shows that the optimal level is in the neighborhood of the

Table 8.5 Same as in Table 8.2 except optimal, max-VA, and min-DA levels of the four left-turning TCs

Year# No.	Time	max-VA/ difference (hPa)	min-DA/ difference (hPa)	Opti. level (hPa)
1994#17 (Walt)	06 UTC 10 Aug	450/+100	550/0	550
	06 UTC 11 Aug	700/–50	700/–50	650
1996#10 (Herb)	18 UTC 28 July	800/+50	850/0	850
	18 UTC 29 July	450/–50	450/–50	400
1998#10 (Todd)	00 UTC 15 Sep	700/+50	800/–50	750
	00 UTC 16 Sep	800/–200	600/0	600
	12 UTC 18 Sep	400/0	400/0	400
	12 UTC 19 Sep	650/+150	500/0	500
2004#20 (Aere)	06 UTC 23 Aug	350/0	350/0	350

Sourced from Huang et al. (2015) © American Meteorological Society, with permission

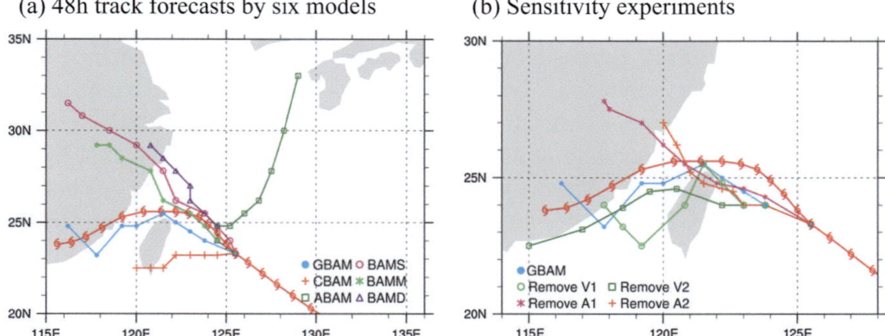

Fig. 8.33 The 48 h track forecasts of Typhoon Aere, initiated at 0600 UTC 23 August 2004 (24 h before the left turn), by **a** six models: GBAM (*blue solid dots*), CBAM (*orange pluses*), ABAM (*light green squares*), BAMS (*pink open circles*), BAMM (*dark green asterisks*), and BAMD (*purple triangles*). **b** four GBAM sensitivity experiments that each removes one of the four anomalous systems V1, V2, A1, and A2 as well as the original GBAM forecast. Symbol "6" indicates the best track. Sourced from Huang et al. (2015) © American Meteorological Society, with permission

max-VA and min-DA levels, and is closer to the min-DA level. This is probably consistent with the "zero divergence" assumption. Therefore, the GBAM could be operated at the min-DA level in a real-time prediction since the optimal level is not known beforehand. Figure 8.33a shows the 48 h predictions of Typhoon Aere by the six models initialized at 0600 UTC 23 August 2004. The GBAM, CBAM, and ABAM are integrated at the optimal level of 350 hPa. The TC turns left (westward) immediately if just the climatic flow is considered (CBAM) , while it turns right (northeastward) rapidly if just the anomalous flow is considered (ABAM). The prediction is the best and closest to the best track if both climatic and anomalous flows are considered by the GBAM. To compare with the classical BAM (using the full wind and vorticity with no decomposition), all three versions, BAMS, BAMM, and BAMD, are also employed to predict Aere's track. By definition, the mean wind and vorticity of 850–700, 850–400, and 850–200 hPa layers are, respectively, used for the BAMS, BAMM, and BAMD. Unfortunately, all three classical BAM models wrongly predict a straight-moving track (Fig. 8.33a). To better understand the above result, the differences in the three controlling factors between the GBAM and the three BAMs are compared: advection, vorticity itself and the beta effect. The largest difference is in the advection, with smaller differences in the vorticity and beta terms (figures not shown). By replacing the advections in the BAMS, BAMM, and BAMD with the GBAM's, the three BAMs, especially BAMD, yield slight improvements in track forecasts. But they are still much worse than the GBAM and far away from the best track (figures not shown). Even when initiating the BAM with the full wind and full vorticity from the optimal level of the GBAM's, the BAM still incorrectly predicts a locally circling track for Aere (figures not shown, the same is true for the other three left turning TCs). This indicates that the TC vortex structure and intensity modified by the decomposition method are also important in determining the track. The beta term plays the least role in track prediction.

To investigate the relative impacts of the four anomalous systems in Fig. 8.32c, another four 48 h track forecasts of Aere are made by the GBAM, each of them with one particular anomalous system removed. The results are compared in Fig. 8.33b. To remove an anomalous system, such as V2, total winds u, v are replaced by the corresponding climatic winds $\tilde{u}, \tilde{v}$ within the area covered by the positive vorticity anomaly. New anomalous u', v' and vorticity can then be obtained by integrating the GBAM. After removing the nearby TC vortex (V2), the GBAM prediction has an immediate deviation from the originally predicted track and then moves in a straight line to the northwest. By removing the mid-latitude vortex V1, the GBAM follows the original track closely in the first 24 h but makes a sharp left turning to noticeably deviate the original track in the second 24 h, although the direction trend is still close. However, by removing either the A1 or A2 anomalous high, the predicted Aere continues to move northwestward with no left turnings at all during the entire 48 h, similar to the tracks predicted by the BAMs. These results suggest that the left-turning motion of Aere is greatly influenced by the two anomalous highs A1 and A2 as well as the nearby TC, and is less influenced by the mid-latitude cyclone.

The same experiment is repeated for the other three left-turning TCs. Their optimal, max-VA, and min-DA levels are all listed in Table 8.5. Figure 8.34a shows the model track predictions of Typhoon Walt (1994#17) initiated at 0600 UTC 9 August 1994 (48 h prior to the left turn). The CBAM predicts a right-turning track under the influence of climatic southwesterly steering flow, the ABAM predicts an immediate left-turning track. Only the GBAM prediction is more consistent with the best track: moving northeastward initially followed by a sharp left turning, although the direction is to the west instead of the southwest. All three classical BAMs (BAMS, BAMM, and BAMD) predict tracks in wrong directions (two to the northeast and one to the northwest). When initiated at 24 h prior to the left turning (Fig. 8.34b), the GBAM correctly predicts a left-turning track after 24 h and the ABAM predicts an immediate left turning, while the CBAM track continues to move northeastward under the climatic steering flow. The other three models, BAMS, BAMM, and BAMD, all wrongly predict some kinds of circular tracks.

Typhoon Todd (1998#10) experienced a continuous left-turning loop and even a returning movement after its landfall in East China. Experiments are carried out at the two turning points: a left turning to the north near 20°N and a left turning to the west with a surprising northeastward return after a landfall near 30°N. Figure 8.35a, b are the two 48 h forecasts for the first turning, where the GBAM obviously outperforms all other models, i.e., the CBAM, ABAM, and the three classical BAMs. Initiated from 1200 UTC 18 September (24 h before the landfall, Fig. 8.35c), both GBAM and especially ABAM give the correct direction, while the CBAM goes in an opposite direction. The BAMS predicts a northwestward motion, while the BAMM and BAMD predict two stagnating motions wandering locally. Initiated at 1200 UTC 19 September (12 h before the returning movement, Fig. 8.35d), the GBAM correctly predicts the returning movement but the ABAM

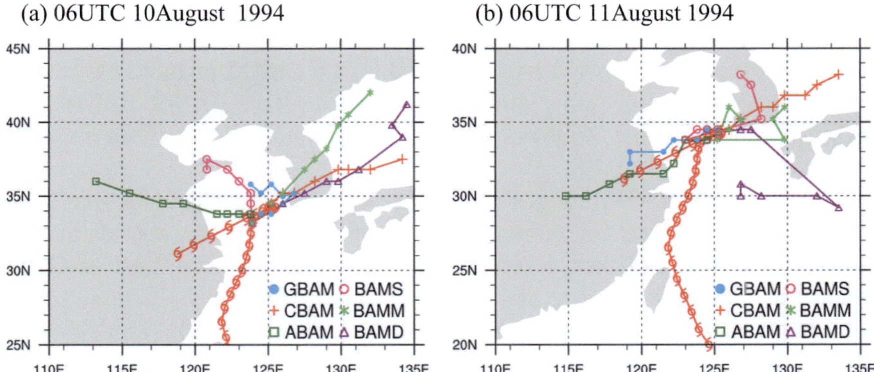

Fig. 8.34 The 48 h track forecasts of Typhoon Walt at 6 h intervals by the six models, initiated at **a** 0600 UTC 10 August (48 h before the left-turn) and **b** 0600UTC 11 August (24 h before the left turn) 1994. Symbol "ϩ" indicates the best track. Sourced from Huang et al. (2015) © American Meteorological Society, with permission

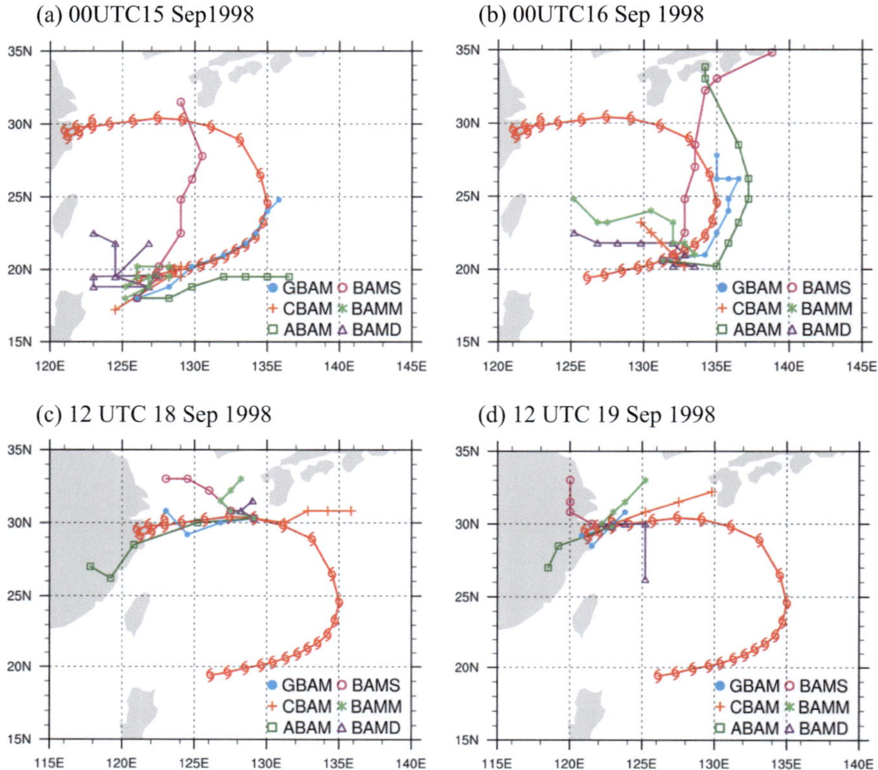

Fig. 8.35 The track forecasts of Typhoon Todd at 6-h intervals by the six models. **a** and **b** Two 48 h forecasts initiated at 0000 UTC 15 and 0000 UTC 16 September. **c** and **d** Two 24 h forecasts initiated at 1200 UTC 18 (24 h before landfall) and 1200 UTC 19 September (12 h before the return) 1998. Symbol "ɕ" indicates the best track. Sourced from Huang et al. (2015) © American Meteorological Society, with permission

still forecasts a southwestward movement, while the CBAM predicts a northeastward movement steered by the climatic flow. The classical BAMS, BAMM, and BAMD wrongly predict northwestward, northeastward, and eastward motions, respectively.

Typhoon Herb (1996#10, Fig. 8.36) experienced a 14-day path with several left and right turnings. After a left turning to the southeast of Taiwan, it made the first landfall in Taiwan and then a second landfall in the southeast coast of China. Finally, it continued to move northwestward and dissipated in Central China. Initiated at 1800 UTC 28 July 1996 (72 h before the first landfall, Fig. 8.36a), the GBAM correctly predicts a left-turning track closely following the best track and landfall in Taiwan leading 48 h, while the CBAM moves too far north in position and too slow in speed, and the ABAM departs too far south. For the three classical BAMs, the BAMS predicts a fast straight track instead of a left-turning path initially, although the overall track "accidently" matches with the best track pretty

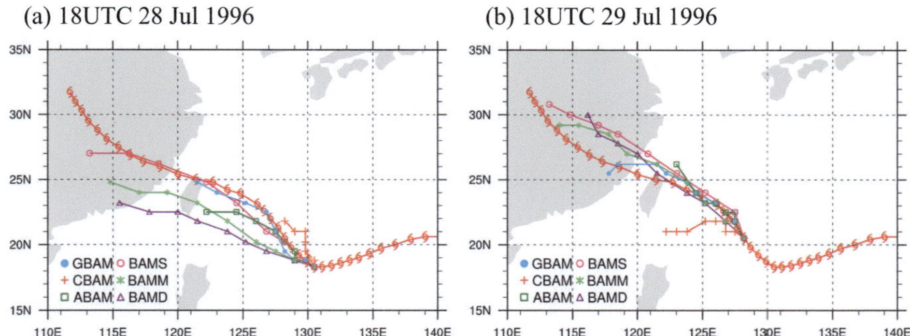

Fig. 8.36 The 48 h track forecasts of Typhoon Herb at 6 h intervals by the six models, initiated at **a** 1800 UTC 28 July (72 h before the landfall in Taiwan) and **b** 1800 UTC 29 July (48 h before the landfall on the southeast coast of China) 1996. Symbol "6" indicates the best track. Sourced from Huang et al. (2015) © American Meteorological Society, with permission

well. Both the BAMM and BAMD predict straight tracks too far to the south. Initiated at 1800 UTC 29 July 1996 (48 h before the second landfall, Fig. 8.36b), the GBAM correctly predicts landfall in the area close to the observed on the southeast coast of China 48 h in advance (unfortunately the landfall in Taiwan is missed in this prediction). The ABAM prediction is too slow in speed and too far north in position, and the CBAM predicts a straight westward track which is completely off the best track. All three classical BAMs maintain a northwestward track with a speed about 18 h faster and make their landfalls too far north from the observed.

Figure 8.37 shows the 48 h track forecasts of the five right-turning TCs by the GBAM and the three classical BAMs (the CBAM and ABAM are no longer shown for clarity) initiated at 24 h before the right turning. The superiority of the GBAM over the three classical BAMs is evident: the GBAM successfully predicts right-turning tracks closely following the best track for all five cases, while the classical BAMs fail in all cases (moving straight in the northwest direction) except for one occasion when the BAMD predicts a right-turning track for the 1980#23 TC (Fig. 8.37d).

Figure 8.38 shows 48 h track forecasts of the seven straight-moving TCs by the four models initiated at 48 h before landfall. The GBAM predicts quite well of all the seven cases. As expected, this is the best group to predict by the classical BAM models: all the three BAMs have skill similar to the GBAM for two cases (2005#13 in Fig. 8.38d and 1980#15 in Fig. 8.38e), slightly worse than the GBAM for three cases (2007#09 in Fig. 8.38b, 1990#17 in Fig. 8.38f, and 2000#18 in Fig. 8.38g), and much worse than the GBAM for two cases (1994#08 in Fig. 8.38a and 2009#09 in Fig. 8.38c). To assess the overall forecast capability, the GBAM is applied to all 30 northwestward straight-moving TCs which made landfall either in mainland China or in Japan. The 48 and 24 h track forecasts are made separately by initializing the GBAM at 48 and 24 h before landfall. Landfall location errors are

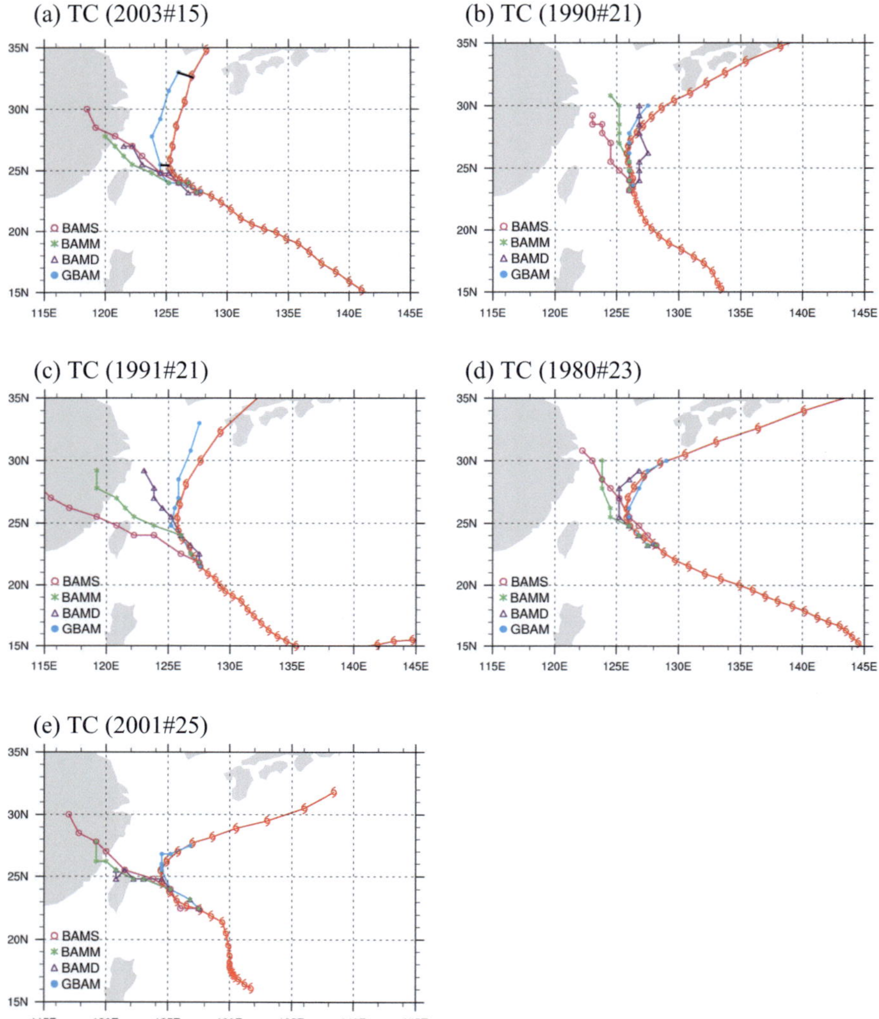

Fig. 8.37 The 48 h track forecasts of the five right-turning TCs by the four models initiated at 24 h before the right turning. Symbol "6" indicates the best track, and the blue (*dot*), pink (*open circle*), green (*asterisks*), and purple (*triangle*) curves represent the GBAM, BAMS, BAMM, and BAMD forecasts, respectively. **a** TC (2003#15), **b** TC (1990#21), **c** TC (1991#21), **d** TC (1980#23), and **e** TC (2001#25). Sourced from Huang et al. (2015) © American Meteorological Society, with permission

then also measured separately for the 24 and 48 h forecasts. Averaged over the 33 TCs, the mean landfall location errors of the GBAM are 33 and 101 km for 24 and 48 h forecasts, respectively. Again, this performance is surprisingly impressive for such a simple model compared to the current operational forecasts involving state-of-the-art complex models (Powell and Aberson 2001).

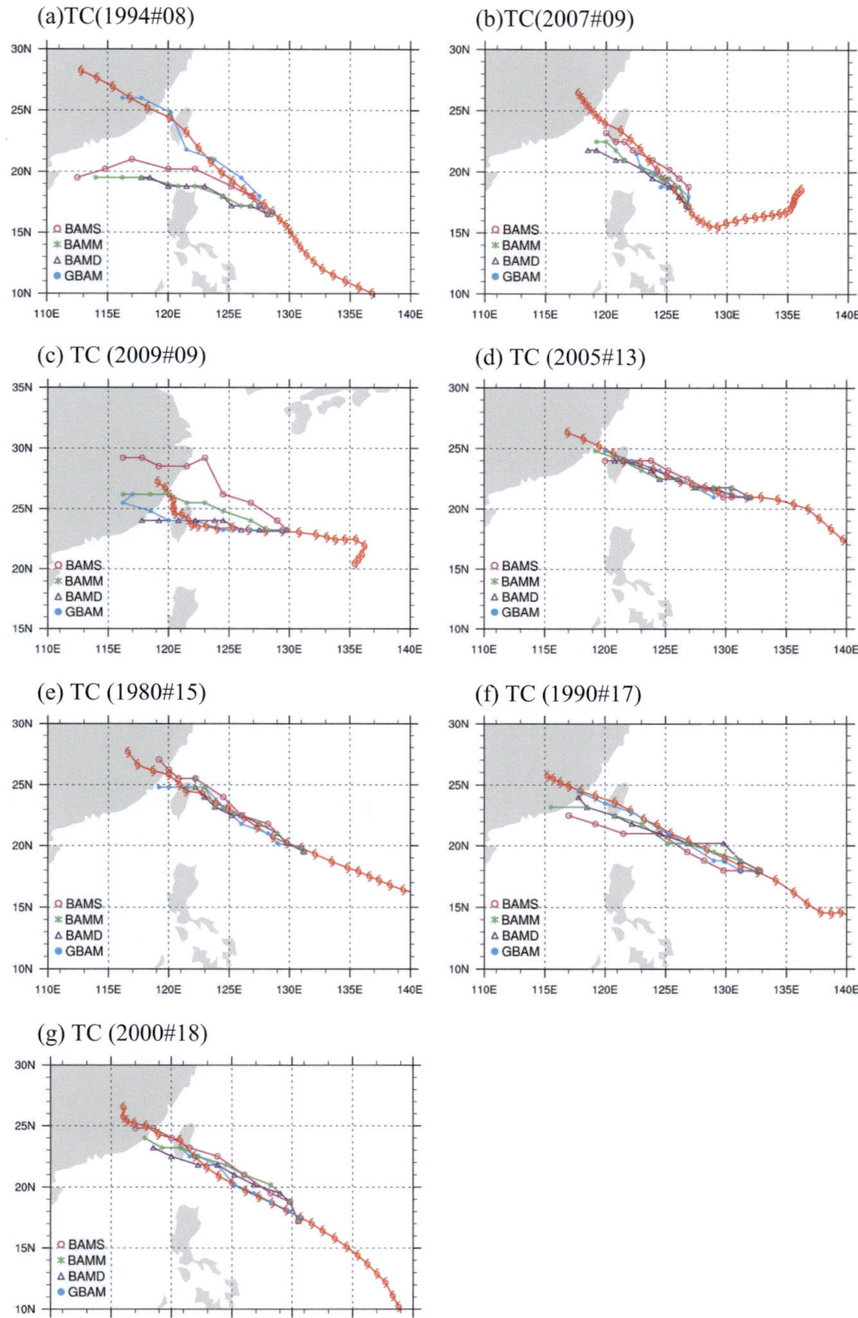

Fig. 8.38 Same as Fig. 8.37 but for the seven straight-moving TCs initiated at 48 h before landfall. Sourced from Huang et al. (2015) © American Meteorological Society, with permission

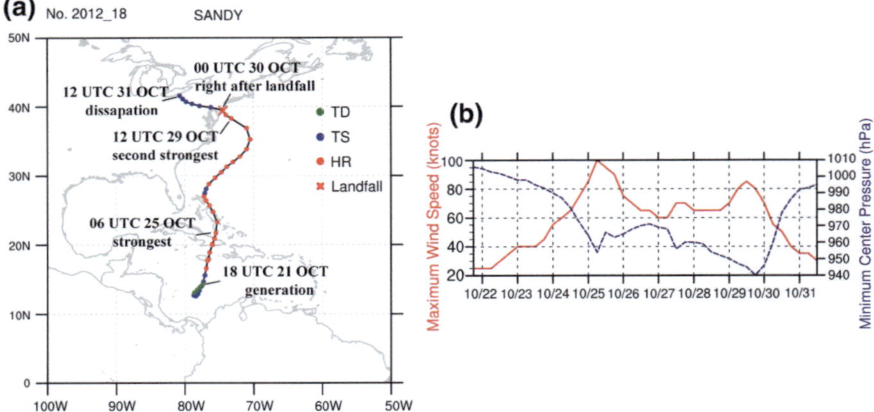

Fig. 8.39 **a** Best track (*black curve* with *color dots*) and **b** intensity evolution indicated by maximum wind speed (*red solid line*, knots) and minimum central pressure (*blue dashed line*, hPa) of Hurricane Sandy with 6-h interval from 1800 UTC 21 October to 1200 UTC 31 October 2012. It formed at 1800 UTC 21 October with the strongest wind speed at 0600 UTC 25 October, the second peak at 1200 UTC 27 October, and the third peak at 1200 UTC 29 October, landfall near 0000 UTC 30 October, and dissipation at 1200 UTC 31 October 2012. Sandy was classified as TD, TS, and hurricane (HR) during its lifetime. Sourced from Qian et al. (2016a) © 2016 W. Qian et al., with permission

8.6 Hurricane Sandy (2012) and Left-Turning Prediction[7]

Decomposition approach and the GBAM can also predict hurricanes in the North Atlantic. Qian et al. (2016a) showed an example of Hurricane Sandy (2012) using this approach. The "best track" of the cyclone's path, starting at 1800 UTC 21 October 2012, is given in Fig. 8.39a with the maximum wind speed and pressure evolution in Fig. 8.39b. Surface and satellite data suggested that a low-pressure circulation became well defined about 200 NM south of Jamaica on 21 October 2012 (http://www.nhc.noaa.gov/data/tcr/AL182012_Sandy.pdf). At the first stage, Sandy first moved southwestward slowly for about one day as a TD and then returned northward for about two days as a TS. It intensified into a hurricane (HR) quickly by late 23 October and continually moved northward. Its strongest intensity (the first peak of wind speed) was recorded at 0600 UTC 25 October. Six hours later, it turned northwestward for about one day. It weakened into a TS at 0000 UTC 27 October for a half day and turned northeastward entering the northwest Atlantic. Starting at 1200 UTC 27 October, Sandy intensified again into a HR (the second peak wind speed). After 0000 UTC 29 October, it turned left

[7]Modified article originally published in *Qian W.H., Huang J., Du J.* (2016a) *Examination of Hurricane Sandy's (2012) structure and intensity evolution from full-field and anomaly-field analyses. Tellus-A* 68:29029. Published with kind permission) © 2016 W. Qian et al. All Rights Reserved.

suddenly to the northwest and reached its third peak of wind speed of 85 knots around 1200 UTC and the lowest central pressure of 940 hPa at around 1800 UTC 29 October. Five and a half hours later (around 2330 UTC 29 October), it made landfall near Brigantine, New Jersey, just north of Atlantic City as a "post-tropical" cyclone. After landfall, it moved westward inland and dissipated at 1200 UTC 31 October 2012.

To illustrate the advantage of the anomaly based analysis over the traditional total-based analysis, a few comparisons are given. In a horizontal field (Fig. 8.40),

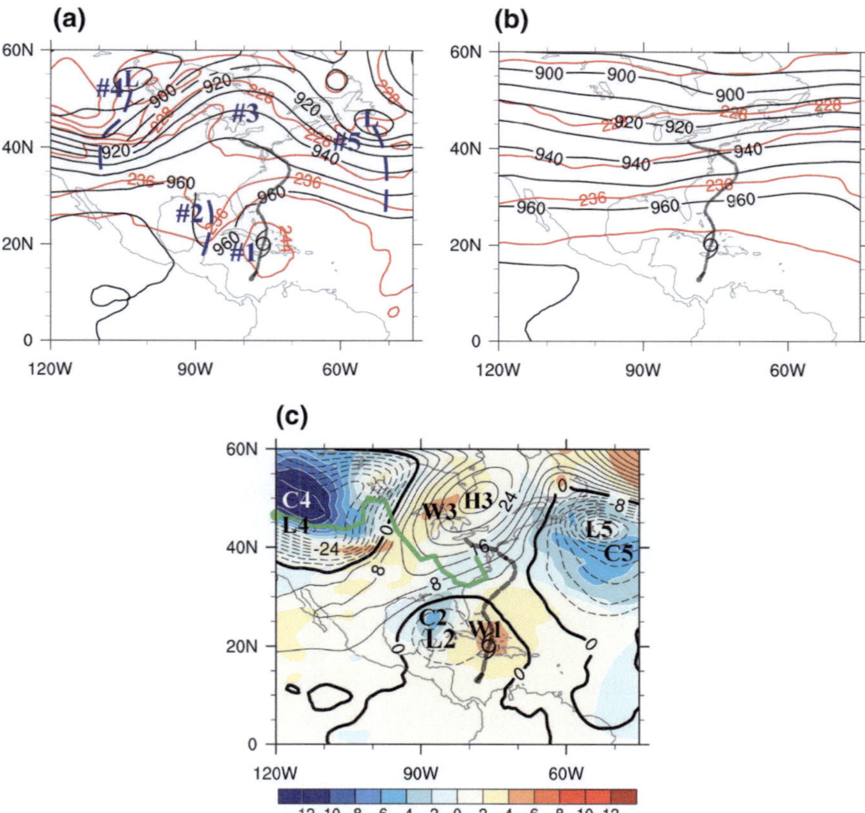

Fig. 8.40 **a** Total height (*black line*, 10 × 10 gpm interval) and total temperature (*red line*, 4 K interval), **b** climatic height (*black line*, 10 × 10 gpm interval) and climatic temperature (*red line*, 4 K interval), and **c** height anomaly (*thin solid* and *dashed black contours*, 4 × 10 gpm interval) and temperature anomaly (*shading* with *white contour*, 2 K interval) at 300 hPa at 0600 UTC 25 October 2012. The *gray-dashed line* with the symbol "୨" is the best track of Sandy. The *thick green line* in **c** is the low center track of anomalous vortex L4 with the current location (*green spot*). In **a**: the *heavy dashed-blue lines* indicate three troughs (#2, #4, and #5) while a ridge and the TC are indicated by #3 and #1, respectively. In **c**, letters "H" and "L" indicate the positive and negative centers of height anomalies, and "W" and "C" are the warm and cold centers of temperature anomalies. Sourced from Qian et al. (2016a) © 2016 W. Qian et al., with permission

three troughs (#2, #4, and #5) and a ridge (#3) in a total field (Fig. 8.40a) become easily identifiable into three anomalous low centers (L2, L4, and L5) and an anomalous high center (H3) (Fig. 8.40c) after removing climatic height (Fig. 8.40b). The advantage is even more visible in vertical cross-sections such as in Fig. 8.41. Figures 8.41a, c show the vertical–latitude cross-sections of the total and climatic height and temperature while Fig. 8.41e is height and temperature anomalies along 75.75°W (crossing the TC center) at 0600 UTC 25 October 2012. Although the differences also exist if one carefully compares the total field (Fig. 8.41a) with the climatic field (Fig. 8.41c), the detailed structures of Sandy and other nearby synoptic systems are not obviously visible in the total field. However, they are clearly revealed in the anomaly based field (Fig. 8.41e). For example, there is a negative center (L0) of height anomalies near the 1000 hPa and a positive center (W0) of temperature anomalies at 700 hPa over Sandy. Another pair of positive height anomalies (H1) at 150 hPa and positive temperature anomalies (W1) at 250 hPa is also found over Sandy. From the view of hydrostatic equilibrium, the positive temperature anomalies (W1) at 250 hPa should be accompanied by a negative height anomaly below it. From thermodynamic point of view, the merger of anomalous warm center W1 and the Sandy vortex favors the first intensification of Sandy. To the north (50°N) of Sandy, a positive center (H3) of height anomalies is located around 300 hPa, which separates a positive center (W3) of temperature anomalies below and a negative center (C3) of temperature anomalies above. Similar features of total, climatic, and anomalous fields can also be observed from a vertical-longitude cross section along 20.25°N. The vertical structures of height and temperature anomalies are clearly visible over Sandy (W0, L0, W1, H1, and C1) and to its west (L2, W2, and C2) in the anomaly based field (Fig. 8.41f), but not in the total field (Fig. 8.41b) and the climatic field (Fig. 8.41d).

When Sandy reached the strongest wind speed at 0600 UTC 25 October 2012, there were three major synoptic-scale systems (#2, #3, and #4 in Fig. 8.40a) which could potentially impact Sandy: a trough#2 to its west, a strong ridge#3 to its north, and a deep trough#4 to its northwest over the central U.S. at 300 hPa. At this level, Sandy is not clear. These systems are well observed in the anomaly-based fields (Fig. 8.41e, f). For example, L0 and W0 reflect the lower level structure of Sandy itself; L2, W2, and C2 reflect the trough#2; H3, W3, and C3 reflect the mid-latitude

Fig. 8.41 Vertical–latitude cross sections of **a** total height (*black contour*, 200 × 10 gpm interval) and total temperature (*red contour*, 10 K interval), **c** climatic height (*black contour*, 200 × 10 gpm interval), and climatic temperature (*red contour*, 10 K interval), and **e** height anomalies (*solid black contour* for positive and *dashed black contour* for negative, 2 × 10 gpm interval) and temperature anomalies (*red shading* for positive and *blue shading* for negative with *white line*, 1 K interval), along 75.75°W at 0600 UTC 25 October 2012. In **e**, the *dotted line* indicates the axis of negative height anomalies, and the long-dashed line indicates the axis of positive height anomalies, while letters "H" and "L" denote positive and negative centers of height anomalies, and letters "W" and "C" denote positive and negative centers of temperature anomalies. Symbol "*Triangle*" indicates the current position of Sandy. Panels **b**, **d**, and **f** are respectively the same as **a**, **c**, and **e**, but for vertical-longitude cross sections along 20.25°N. Sourced from Qian et al. (2016a) © 2016 W. Qian et al., with permission ▶

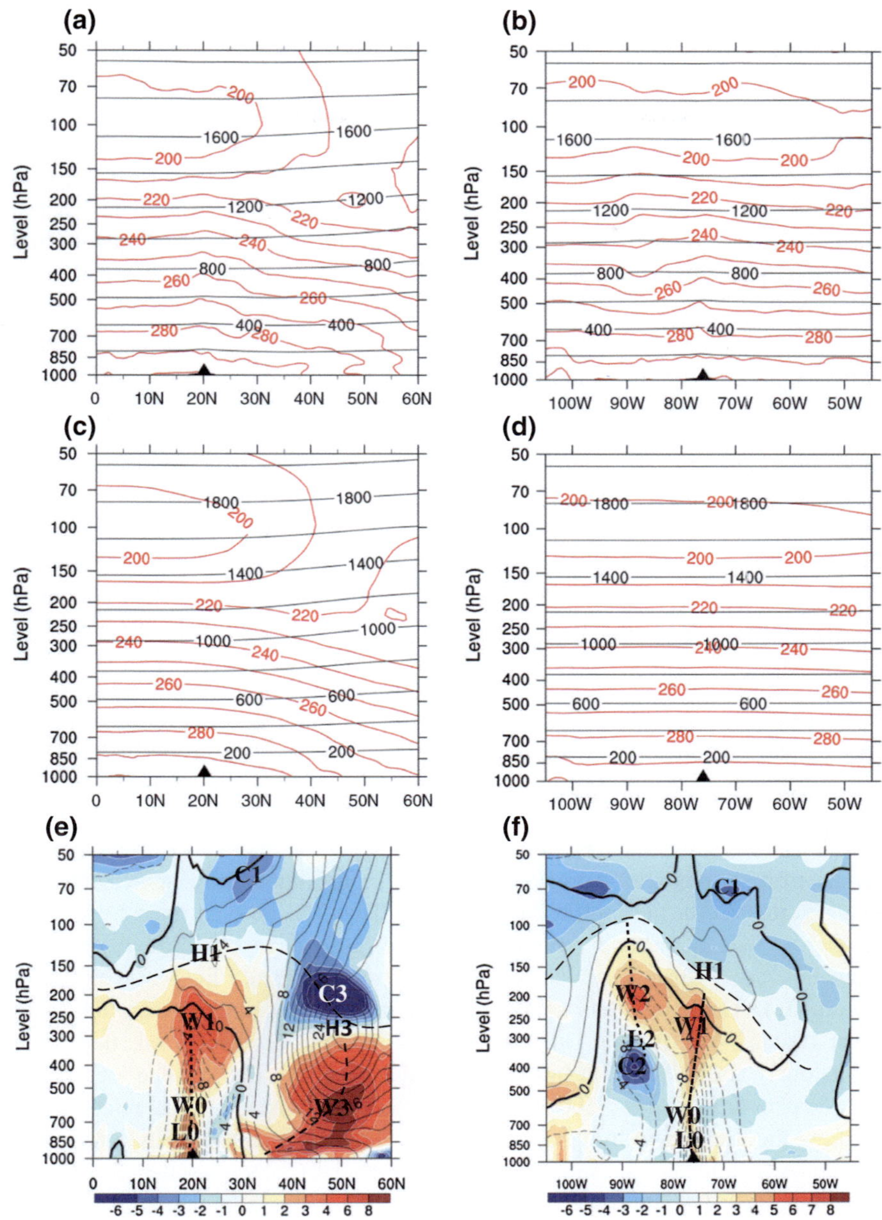
(a)
(b)
(c)
(d)
(e)
(f)
Level (hPa)
50
70
100
150
200
250
300
400
500
700
850
1000
0
10N
20N
30N
40N
50N
60N
100W
90W
80W
70W
60W
50W
C1
H1
C3
H3
W1
W2
W3
W0
L0
L2
C2
-6 -5 -4 -3 -2 -1 0 1 2 3 4 6 8
-6 -5 -4 -3 -2 -1 0 1 2 3 4 5 6 7 8

ridge#3; while L4, W4, and C4 reflect the mid-latitude trough#4. How these anomalous systems influence Sandy should be discussed.

When Sandy headed northeastward on 27 October 2012, the anomalous vortex (L2 in Fig. 8.42c) moved eastward and merged into the Sandy vortex. This anomalous vortex appeared as a trough in the total field (#2 in Fig. 8.42e). The merger of vortex L2 with Sandy deepened the TC vortex. This intensification can be observed from Sandy's maximum wind speed and central pressure in Fig. 8.39b. Meanwhile, an anomalous warm center W3 (below the mid-latitude anomalous high H3 in the upper troposphere) and an anomalous warm center W2 (above the vortex L2) moved close to each other over the Sandy's vortex (Fig. 8.42a, b). This was thermodynamically favorable to Sandy's intensification as a TC system when the warm centers merged at the upper troposphere. In this process, Sandy's warm core (W0) became indistinguishable from the two warm cores (W2 and W3) as they moved close to W0. The anomalous low L4 over the Central U.S. was also approaching but still far away from Sandy (Fig. 8.42b–d). This anomalous vortex L4 was located at the upper troposphere (Fig. 8.42c) while the Sandy vortex L0 was observed in the mid-low troposphere (Fig. 8.42d). Unlike a typical tropical cyclone, the original warm core of Hurricane Sandy was limited to the lower layer while the storm size was large (Zhu and Weng 2013). The height and temperature anomalies described above were not clear in the total field synoptic charts as shown in Fig. 8.42e, f.

Sandy started to turn left at 0000 UTC 29 October. During the following 24 h, Sandy experienced its third intensification when it was approaching the coastline. The vertical combination of the anomalous warm center W3 in the upper troposphere and the anomalous vortex L4 in the middle troposphere was the primary reason for Sandy's pre-landfall intensification. After the turning point (0000 UTC 29 October, Fig. 8.43), interactions occurred among the anomalous low L4 in the middle troposphere, the anomalous high H3 in the upper troposphere, and the Sandy vortex L0 in the mid-low troposphere. This will be confirmed by a set of BAMs. At this moment, a positive center and a negative center of temperature anomalies are respectively located to the northeast and the west of Sandy (Fig. 8.43c). The anomalous cold center C4 had not yet merged into the Sandy warm center (W0) (Fig. 8.43d). Sandy cannot move northeastward, due to an anomalously strong blocking high to its northeast (Fig. 8.43a, c). The total height and temperature fields at 300 and 700 hPa cannot illustrate these relationships although the upper cold trough was visible to the west of Sandy (Fig. 8.43e, f).

Approximately 5–6 h before landfall (1800 UTC 29 October 2012), the anomalous low L4 with cold air mass intruded into the Sandy vortex from the southwest (Fig. 8.44e). Two vortices, L0 and L4, were experiencing a counter-clockwise rotation. The Sandy vortex L0 moved northwestward and the anomalous cold vortex L4 moved southeastward, as illustrated by the two track lines in Fig. 8.44e. The anomalous vortex L4 was located in the upper troposphere (Fig. 8.44b, e) while L0 was in the mid-low troposphere (Fig. 8.44b, f). Below the 250 hPa, an anomalous cold air column and an anomalous warm air column formed a strong contrast associated with three anomalous height systems (L0, L4, and H3)

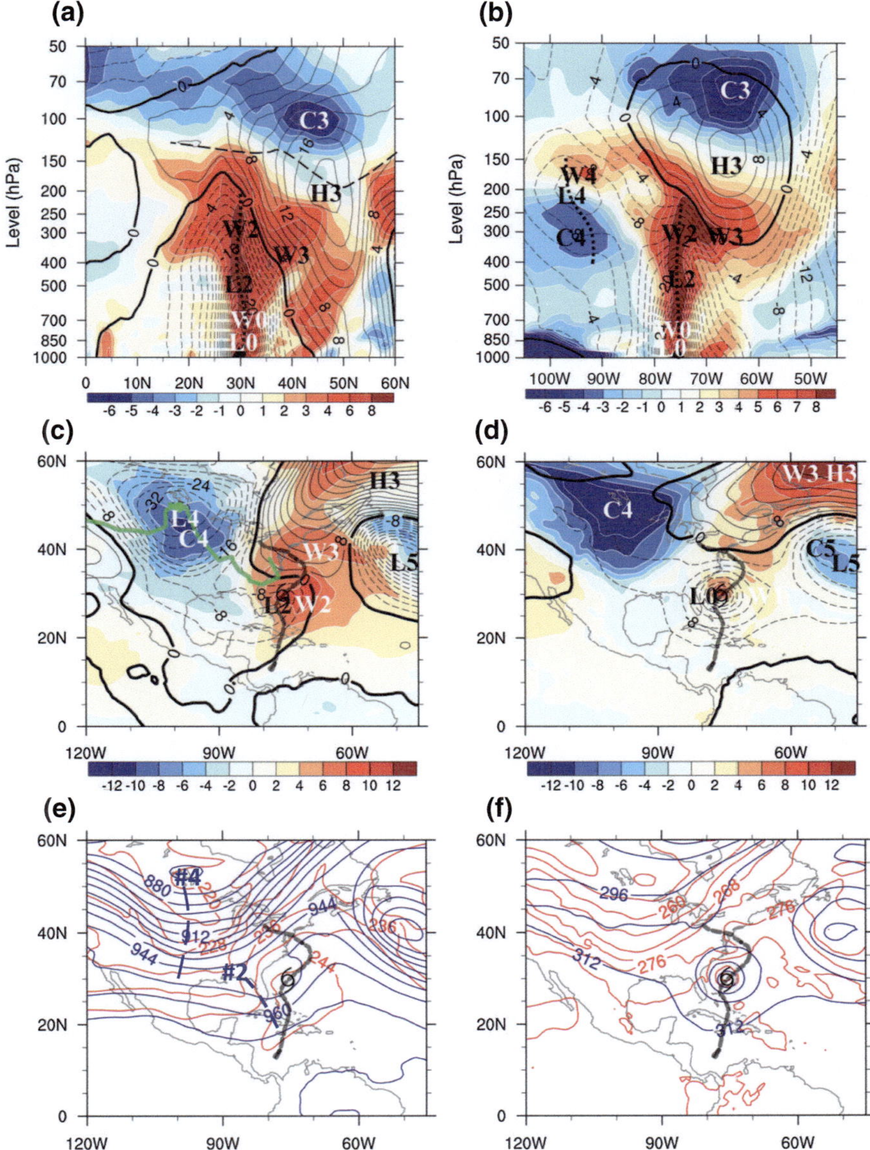

Fig. 8.42 Vertical sections of height anomalies (*contour*, 2 × 10 gpm interval) and temperature anomalies (*shading*, 1 and 2 K intervals) along **a** 75.75°W and **b** 30°N at 1800 UTC 27 October 2012, as well as the two horizontal synoptic charts of height anomalies (*contour*, 4 × 10 gpm interval) and temperature anomalies (*shading*, 2 K interval) at **c** 300 hPa and **d** 700 hPa, respectively. Panels **e** and **f** are the same as **c** and **d**, but for total height (*blue contour*, 8 × 10 gpm interval) and total temperature (*red contour*, 4 K interval) at 300 hPa and 700 hPa, respectively. In **c**: the *thick green line* is the low center track of anomalous vortex L4 and the *green spot* is its current location. In **e**: two *blue-dashed lines* indicate the height troughs. Sourced from Qian et al. (2016a) © 2016 W. Qian et al., with permission

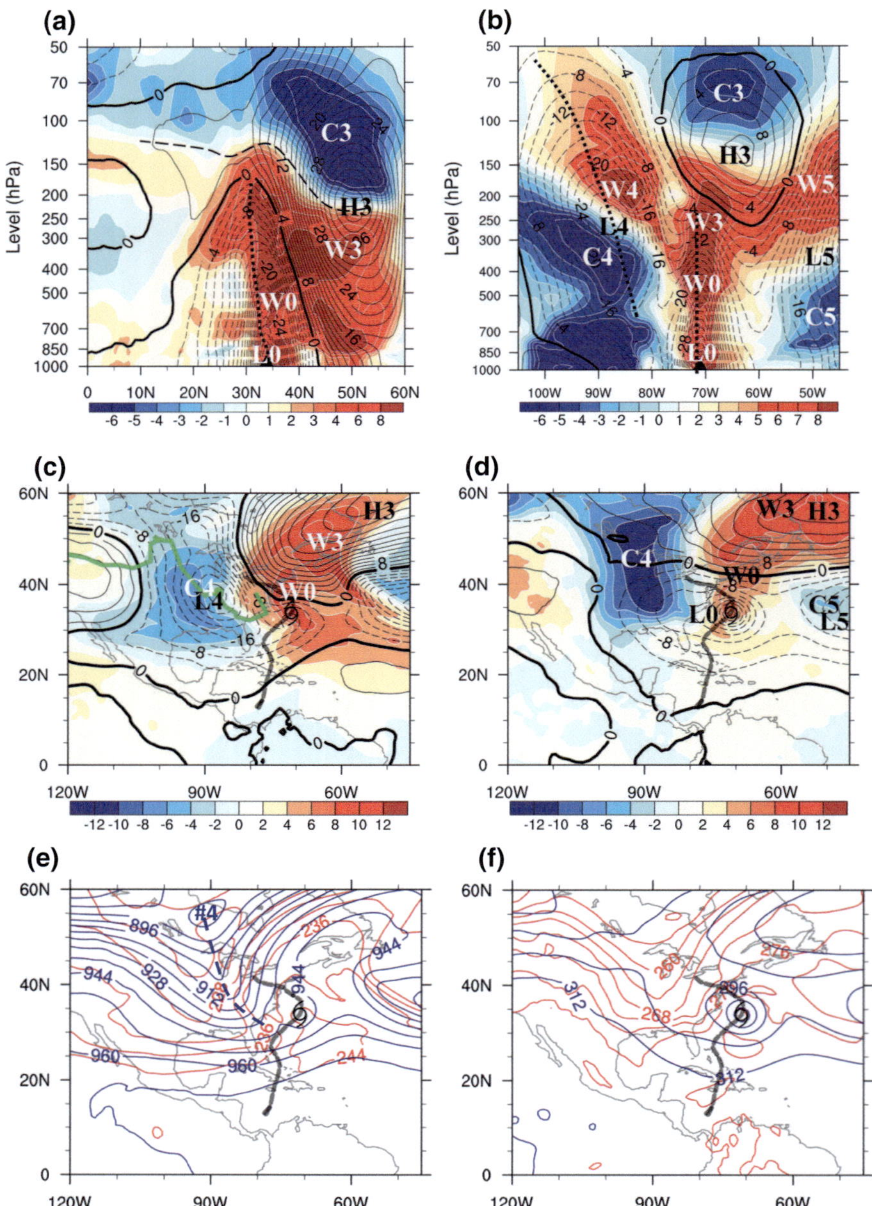

Fig. 8.43 Same as Fig. 8.42 except for the vertical sections along **a** 71.25°W and **b** 33.75°N at 0000 UTC 29 October 2012, as well as the two horizontal anomaly based synoptic charts at **c** 300 hPa and **d** 700 hPa, respectively. Total height and total temperature are shown at **e** 300 hPa and **f** 700 hPa. Sourced from Qian et al. (2016a) © 2016 W. Qian et al., with permission

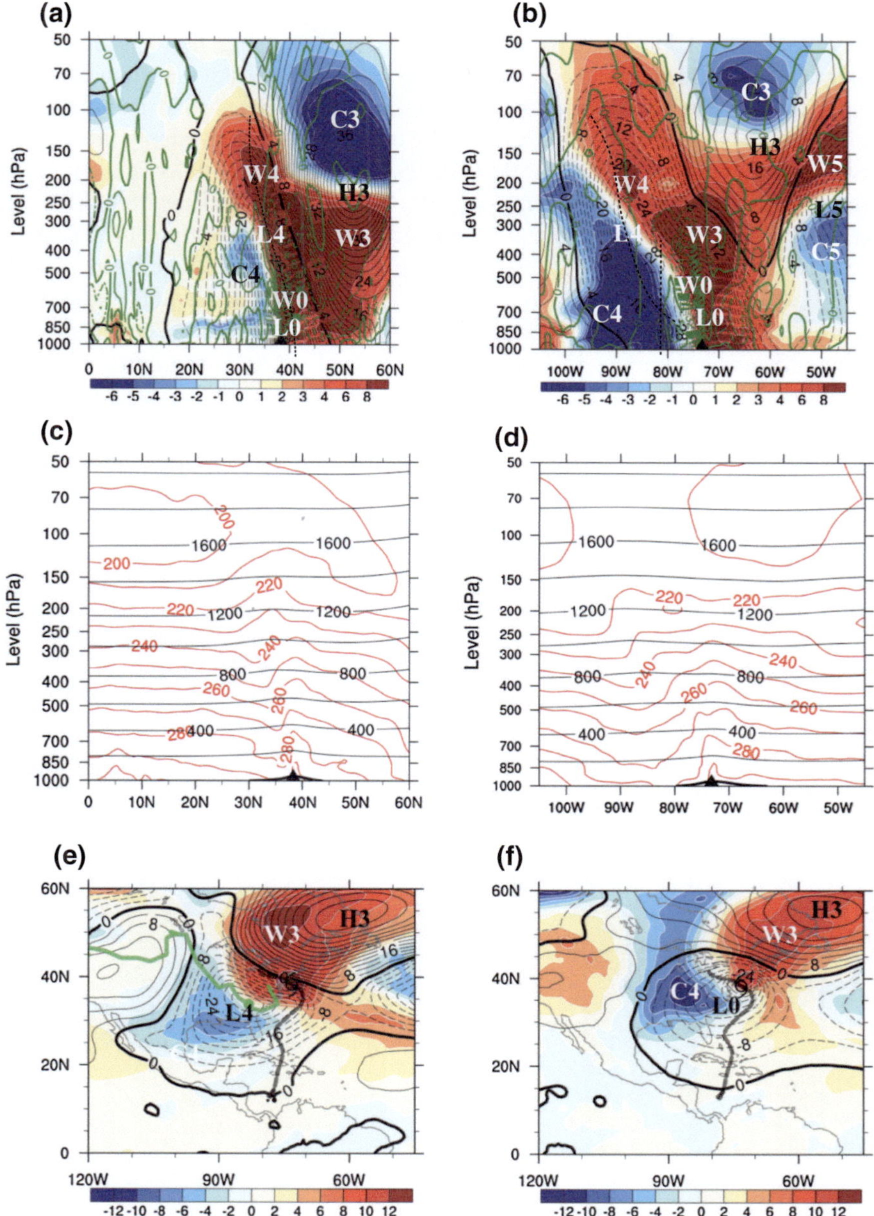

Fig. 8.44 Same as Figs. 8.41 and 8.42 except at 1800 UTC October 2012 along **a** 73.5°W and **b** 38.25°N. Panels **c** and **d** are total height (*black solid contour*, 200 × 10 gpm interval) and total temperature (*red contour*, 10 K interval) at 1800 UTC 29 October 2012 along 73.5°W and 38.25° N, respectively. Height and temperature anomalies are on **e** 300 hPa and **f** 850 hPa at 1800 UTC 29 October 2012. In **a** and **b**, the *green solid* and *dashed contours* respectively indicate anomalous descending and rising pressure velocities (10 hPa/s interval). Sourced from Qian et al. (2016a) © 2016 W. Qian et al., with permission

over the Sandy vortex (Fig. 8.44a, b). The anomalous warm air column was located over the Sandy vortex and there was an anomalous rising flow (green-dashed contour). The anomalous cold air column was observed beneath the anomalous low center L4 and there was an anomalous sinking flow (green solid contour). The rising and sinking flows should theoretically transform potential energy into kinetic energy. From 0000 UTC 29 to 1800 UTC 29 October, the central values of Sandy vorticity anomalies increased from 34×10^{-5} to 46×10^{-5} s^{-1} near 850 hPa, and the central values of Sandy anomalous kinetic energy increased from 1410 to 3100 m^2 s^{-2} at 250 hPa and from 350 to 600 m^2 s^{-2} near 700 hPa. The climatic kinetic energy also increased from 380 to 530 m^2 s^{-2} at 175 hPa over the Sandy vortex due to the northward moving of the hurricane (figures not shown). When vortex L4 captured Sandy's vortex (e.g., Figure 8.45a, b), Sandy transformed into a large and intense extratropical cyclone. The strong gradient of pressure (height) was clearly

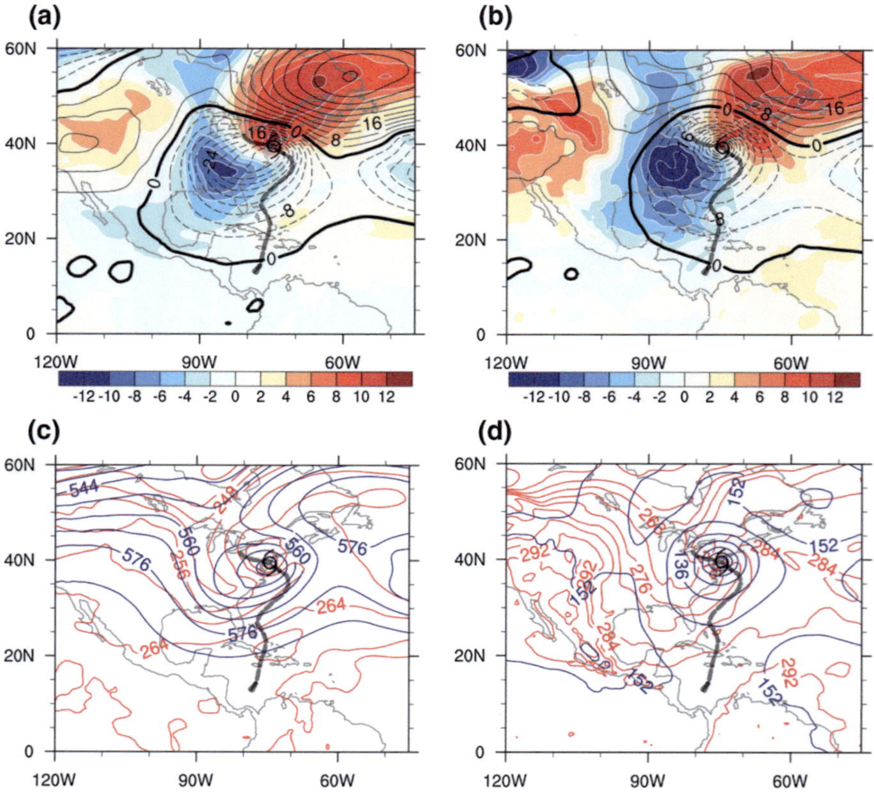

Fig. 8.45 Height anomalies (*contour*, 4 × 10 gpm interval) and temperature anomalies (*shading*, 2 K interval) at **a** 500 hPa and **b** 850 hPa at 0000 UTC 30 October 2012. Panels **c** and **d** are the same as **a** and **b** but for total height (*blue contour*, 8 × 10 gpm interval) and total temperature (*red contour*, 4 K interval) respectively. Sourced from Qian et al. (2016a) © 2016 W. Qian et al., with permission

observed in the northeast of Sandy (Fig. 8.45b). The enhanced energy resulted in an expansion of Sandy's wind field, which exacerbated the storm surge in her northeastern quadrant. In this process, the anomalous cold air mass cut off the anomalously warm and moist air underneath and forced it to ascend quickly. This resulted in heavy snowfall in its southwestern quadrant. Again, the contrasts of positive–negative height and temperature anomalies were not clearly visible in the total-based analysis (Figs. 8.44c, d and 8.45c, d).

At 0000 UTC 31 October 2012, Sandy had already moved inland. Figure 8.46 shows the post-landfall structure of height and temperature anomalies. In the vertical profile, two anomalous vortices L0 and L4 merged together into a single center around 400–500 hPa, accompanied by two anomalous warm centers (W4 and W0) above and an anomalous cold center (C4) below (Fig. 8.46a, b). This was a

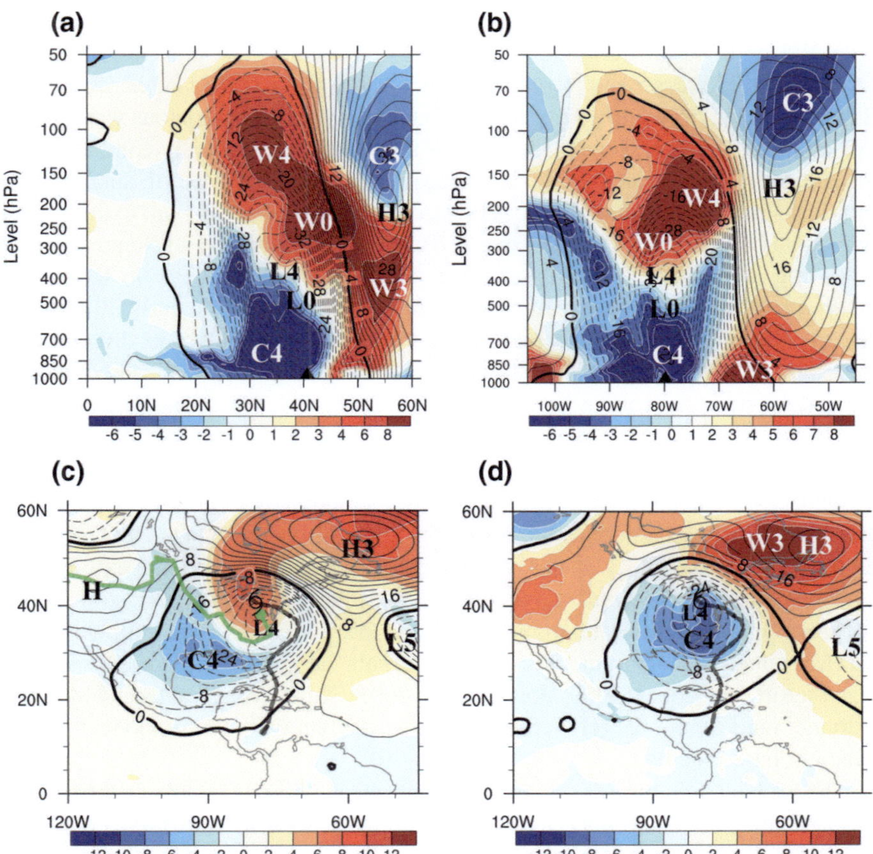

Fig. 8.46 Same as Fig. 8.42 except at 0000 UTC 31 October 2012 for the two vertical sections along **a** 79.5°W and **b** 40.5°N, as well as the two horizontal anomaly based synoptic charts at **c** 300 hPa and **d** 700 hPa, respectively. Sourced from Qian et al. (2016a) © 2016 W. Qian et al., with permission

thermally stable structure with an anomalous warm layer above and an anomalous cold layer below after Sandy's landfall as it reached the warm seclusion phase (Galarneau et al. 2013). At this moment, negative temperature anomalies were located over Sandy's vortex in the lower troposphere (Fig. 8.46d), while the upper troposphere was still covered by positive temperature anomalies (Fig. 8.46c).

Figure 8.47a summarizes the temporal evolution of the height and temperature anomalous structure of Sandy during its lifetime. Following the TC center, a time–pressure section of height and temperature anomalies can be depicted every 6 h from 1800 UTC 21 to 1200 UTC 31 October 2012. Three major time points of intensification can be identified in Fig. 8.47a. The anomalous Sandy vortex strengthened slowly from 21 October to early 24 October and its first rapid intensification occurred later on 24 October when the warm center W1 overlapped the Sandy vortex vertically (the first vertical dotted line from the left). On 25 October, the intensity of Sandy weakened. A longer period of intensification occurred from 25 to 27 October when the anomalous center L2 at 500 hPa merged with W2 and W3 at 300 hPa, overlapping the Sandy vortex at 1800 UTC 27 October 2012 (the second vertical dotted line). The last rapid intensification of Sandy occurred from late 29 to early 30 October, when the anomalous warm center W3 at 300 hPa and negative height center L4 at 350 hPa vertically combined with the anomalous warm center W0 and negative height center L0 of Sandy in the mid-low troposphere (the third vertical-dotted line). Finally, an extra-tropical transition (ET[8]), i.e., transition from a TC to an extratropical cyclone, occurred when the anomalous vortex L4 (or the mid-high latitude trough in total field) merged into the Sandy vortex L0 in the mid-low troposphere on early 30 October. The cyclone occluded on 30 October when the anomalous cold vortex driven by the mid-latitude flow invaded the lower portion of the Sandy vortex. This evolution of the Sandy structure cannot be well illustrated from the total-based analysis. In Fig. 8.47b, no detailed evolution can be observed from the total height and total temperature following the Sandy center from 1800 UTC 21 to 1200 UTC 31 October 2012, except a rapid decrease of temperature in the mid-low troposphere and a rapid increase of temperature from the upper troposphere to stratosphere after Sandy's landfall. This is also a special case because most intense extratropical cyclones after ET processes move northward or northeastward (Harr et al. 2000). As previous studies indicate, tropical cyclones that undergo ET can produce strong surface winds and extreme rainfall that result in significant societal impacts (Thorncroft and Jones 2000; Jones et al. 2003). The strong surface winds and extreme rainfall are largely contributed by the anomalous component in the anomaly based method.

The sudden left turning of Sandy occurred at 0000 UTC 29 October 2012 (Fig. 8.39a). The Sandy track in the following 24 and 48 h was examined using a set of BAMs by Qian et al. (2016a). In order to find the optimal level for the simulation of BAMs, Fig. 8.48 gives the meridional and zonal vertical distributions

[8]Defined as the transition from a warm-core TC to a cold-core extratropical cyclone.

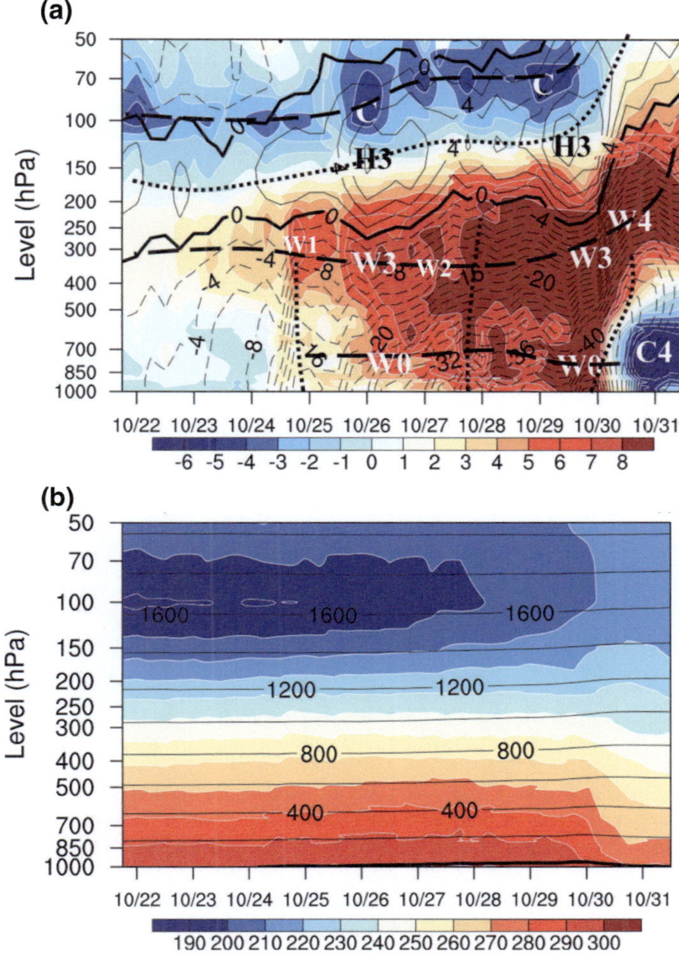

Fig. 8.47 **a** Time–pressure section of height anomalies (*contour*, 2 × 10 gpm interval) and temperature anomalies (*shading*, 1 K interval) with 6-h interval from 1800 UTC 21 to 1200 UTC 31 October 2012. The *heavy-dotted line* and the *heavy dashed line* are the axes of height anomalies and temperature anomalies, respectively. Three vertical axes of height anomalies at the mid-low troposphere are indicated by the *heavy dotted line* at 1800 UTC 24, 1800 UTC 27, and 0000 UTC 30 October, respectively. **b** Same as in **a** except for total height (*contour*, 200 × 10 gpm interval) and total temperature (*shading*, 10 K interval). Sourced from Qian et al. (2016a) © 2016 W. Qian et al., with permission

of vorticity anomalies crossing the center of Sandy at 0000 UTC 29 October 2012. As an anomalous system, the vertical structure of Sandy was well described by the positive vorticity anomalies below 200 hPa with its maximum center at about 850 hPa. In the northern side of Sandy, a negative vorticity maximum was centered at around 50°N and 250 hPa, which corresponded to the anomalous high H4 (Fig. 8.48a). In the western side of Sandy, positive vorticity anomalies existed over

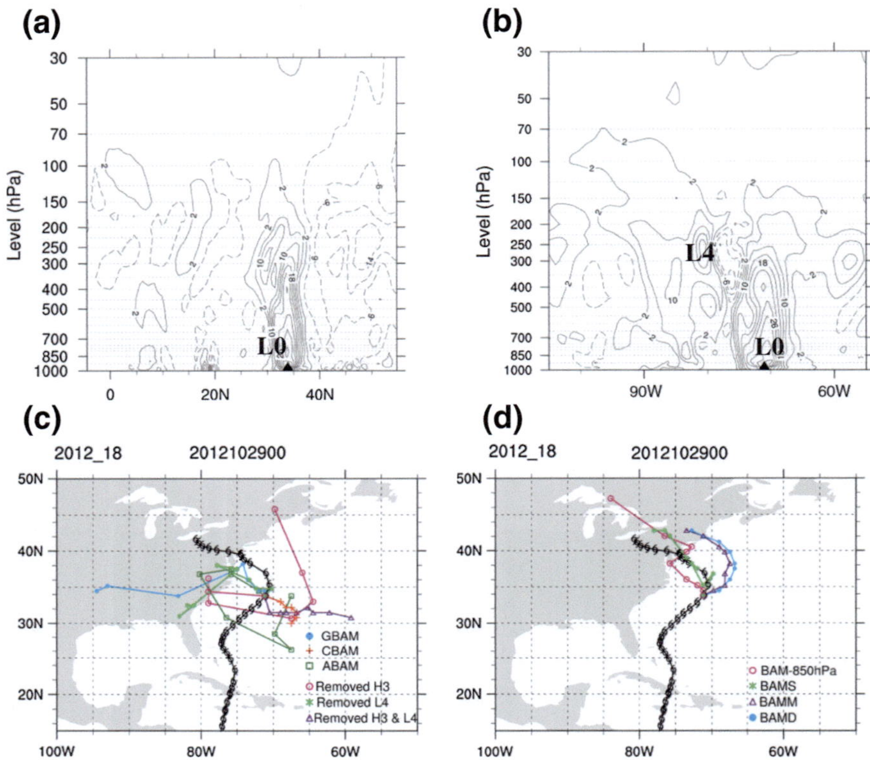

Fig. 8.48 **a** Meridional and **b** zonal sections of vorticity anomalies (4×10^{-5} s^{-1} interval) crossing the center of Sandy at 0000 UTC 29 October 2012. Letters "L0" and "L4" indicate the locations of Sandy vortex and middle-latitude vortex, respectively. The 48-h track forecasts of Sandy every 6 h by **c** the GBAM, CBAM, ABAM, and the results after removing other anomalous systems at 850 hPa and by **d** classical BAMs initiated at 0000 UTC 29 October 2012. The *black line* with symbol "6" indicates the best track of Sandy in **c** and **d**. Sourced from Qian et al. (2016a) © 2016 W. Qian et al., with permission

the US inland at the upper troposphere, which corresponded to the anomalous low L4 (Fig. 8.48b). At 1800 UTC 29 October, these two positive centers of vorticity anomalies merged together and intensified Sandy's strength (figures not shown).

Using a set of BAMs, Qian et al. (2016a) examined the influence of climatic steering flow on Sandy's track and the interaction among Sandy, H3, and L4 since 0000 UTC October 2010. The optimal level can be determined at 850 hPa with the maximum positive center of vorticity anomalies (Fig. 8.48a). In Fig. 8.48c, the CBAM predicted that Sandy would head toward southeast in the following 24 and 48 h. The ABAM predicted that Sandy will have a circular track in the coming 48 h. The GBAM showed that the 24-h prediction was close to the best track. It can be easily understood that the predicted track of CBAM follows the climatic steering flow while the predicted track of ABAM is a result of interaction among the three anomalous systems. If we only remove the anomalous low L4 from the GBAM, the predicted track is similar to that of the GBAM, which implies that the anomalous low L4 has little effect on the left turning of Sandy. The method of removing an anomalous system is simply to replace its anomalous vorticity by the climatic one. For example, a positive anomalous vortex L4 is replaced by the climatic vorticity there. If we only remove the anomalous high H3 from the GBAM, the predicted track is a circle, which implies that the anomalous high H3 has an important role in the left turning of Sandy. If we remove the two anomalous systems of high H3 and low L4 together from the GBAM, the predicted track is similar to that of the CBAM, which implies that the predicted track is basically influenced by the climatic steering flow now.

Qian et al. (2016a) further examined the predicted tracks of three classical BAMs in Fig. 8.48d. If using the BAM at 850 hPa, it predicts a reversed "z" type track (BAM-850 hPa). Among the three classical BAMs, the best prediction is made by the BAMS, because it captures the positive center of vorticity anomalies in the lower troposphere. The other two models of BAMM and BAMD have larger track deviations from the best track, but their left turnings are well predicted due to the influence of the high#3. The results of the simple BAM experiments provide insight into the dynamics of the interactive anomalies and the influence of the climatic steering flow on Sandy's track. However, the GBAM cannot predict any TC tracks beyond 3 days because the optimal level is not constant and varies with time.

On the other hand, the ECMWF model provided an excellent forecast of Sandy's landfall up to 8 days in advance at 0000 UTC 22 October 2012 (Bassill 2014). Figure 8.49 shows 6 track predictions initialized at 0000 UTC 22, 23, 24, 25, 26, and 27 October 2012, respectively. All the three turning processes, the first left turning, then a right turning and particularly the last left turning on 29 October, were successfully predicted. To compare with the analyzed result in Fig. 8.47a, two time–pressure sections of height-temperature anomalies in 6-h interval predicted by the ECMWF model initialized from 0000 TUC 22 and 24 October 2012 are given in Fig. 8.50. The three intensification processes of Sandy revealed by Fig. 8.47a were also predicted by the ECMWF model (Fig. 8.50). However, the predicted intensity was stronger than the observed. For example, during the ET process on 30 October, the predicted values at the height anomalous center were −663 gpm (Fig. 8.50a) and −736 gpm (Fig. 8.50b), which were stronger than the analyzed value −553 gpm (Fig. 8.47a).

The NCEP model has a shorter predictability than the ECMWF model for the Sandy case. The GFS initialized at 0000 UTC 24 and 0000 UTC 25 and predicted the right turning initially, but it failed to predict the left-turning afterward (Fig. 8.51a and b). It started to predict the correct left turning from the cycle of 0000 UTC 26 (Fig. 8.51c, d), with a 4-day delay than the ECMWF model. As a result, the ET process was not predicted by the GFS in Fig. 8.52a and b until the initial time of 0000 UTC 27 October 2012 (Fig. 8.52c). The detailed strengthening and weakening processes of the intensity predicted by the GFS were also not as well as the ECMWF model did.

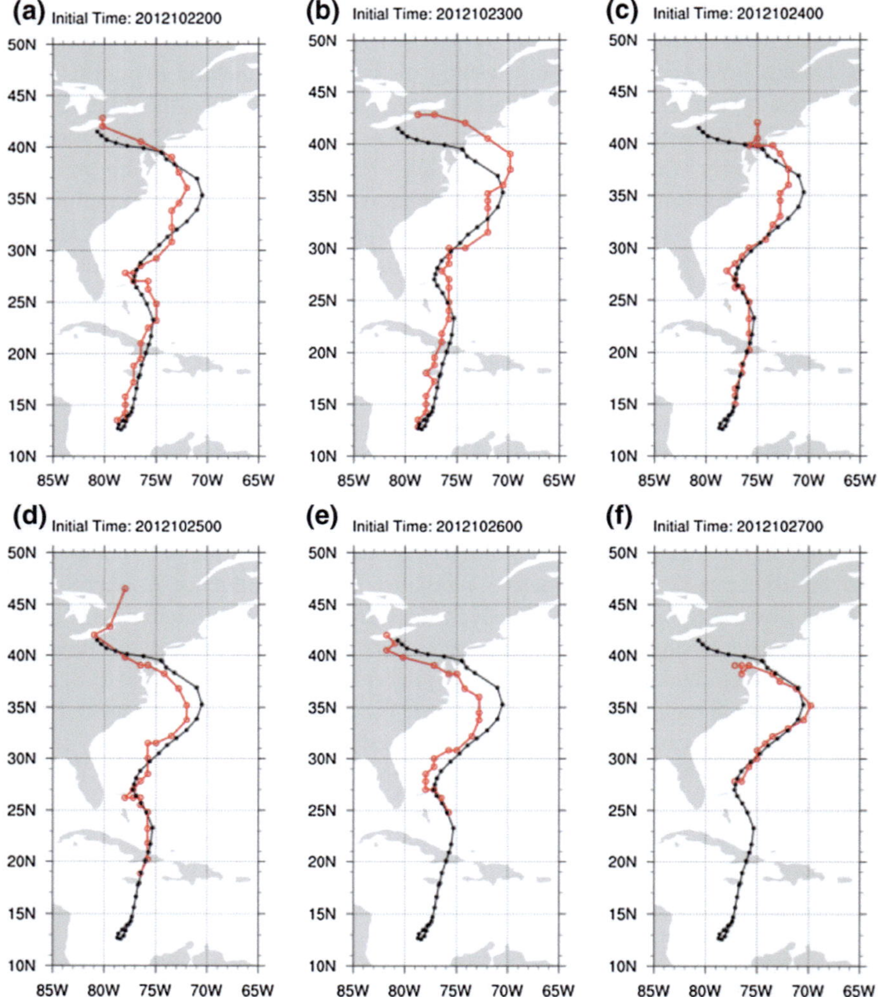

Fig. 8.49 Model track predictions of the ECMWF initialized from **a** 0000 UTC 22, **b** 0000 UTC 23, **c** 0000 UTC 24, **d** 0000 UTC 25, **e** 0000 UTC 26, and **f** 0000 UTC 27 October, 2012, respectively. The *black line* with *solid dot* and the *red line* with *open dot* are the best track and model prediction with 6 h interval, respectively

The prediction in Figs. 8.51 and 8.52 by the GFS is an ensemble-mean from 20 member initial conditions (ICs). All GFS predictions from every IC initialized from 0000 UTC 22 October 2012 have been examined. Three group tracks predicted from different ICs are shown in Fig. 8.53. The first group shows two ICs resulting a correct left-turning track (Fig. 8.53a), the second group has three ICs resulting a straight-northward track (Fig. 8.53b), and the third group is ICs resulting a right-turning track (Fig. 8.53c).

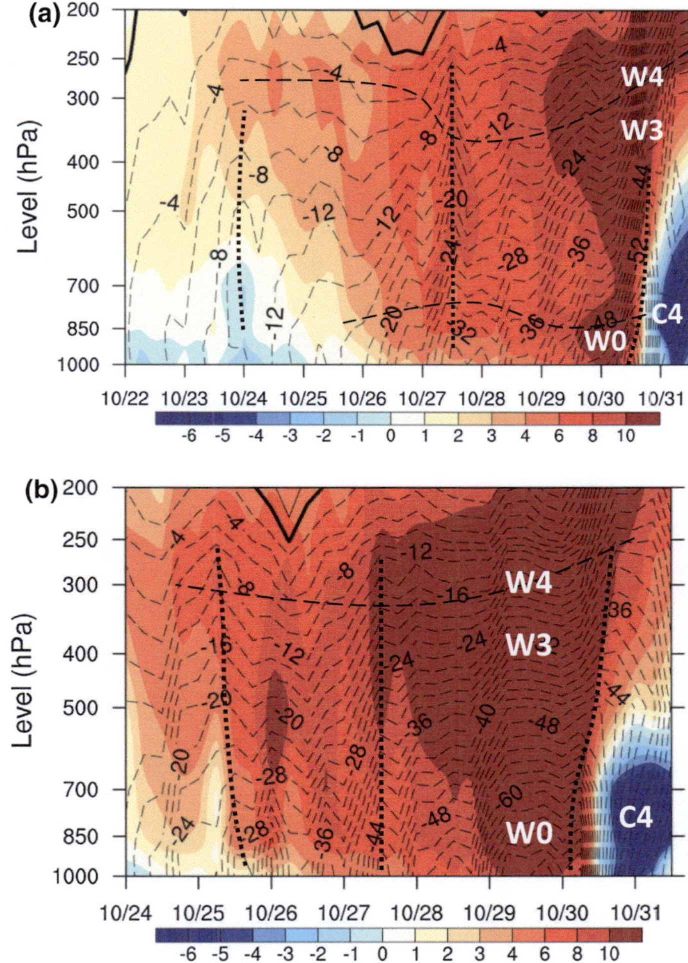

Fig. 8.50 Same as Fig. 8.47a but following the predictions of ECMWF model initialized from **a** 0000 UTC 22 and **b** 0000 UTC 24 October 2012, respectively

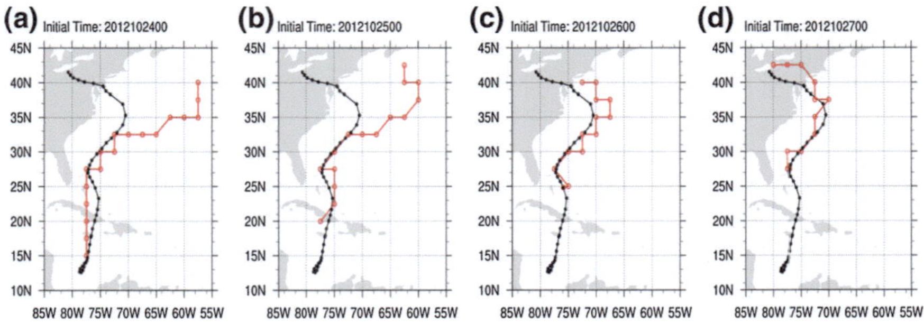

Fig. 8.51 Same as Fig. 8.49 but for the NCEP GFS model initialized from **a** 0000 UTC 24, **b** 0000 UTC 25, **c** 0000 UTC 26, and **d** 0000 UTC 27 October 2012, respectively

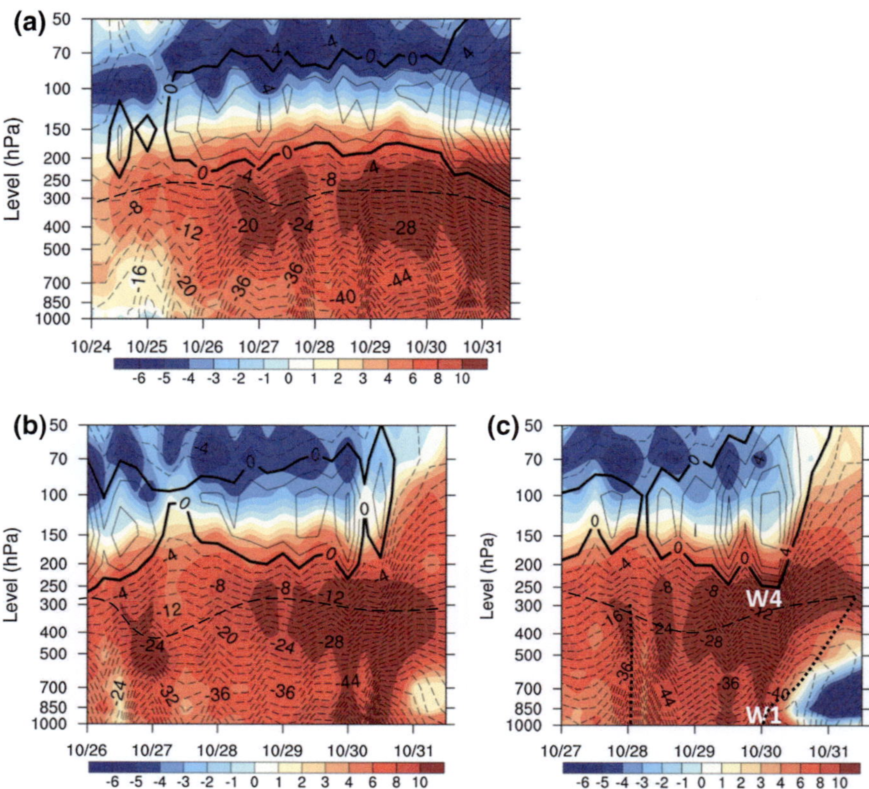

Fig. 8.52 Same as Fig. 8.50 but for the NCEP GFS model initialized from **a** 0000 UTC 24, **b** 0000 UTC 26, and **c** 0000 UTC 27 October 2012, respectively

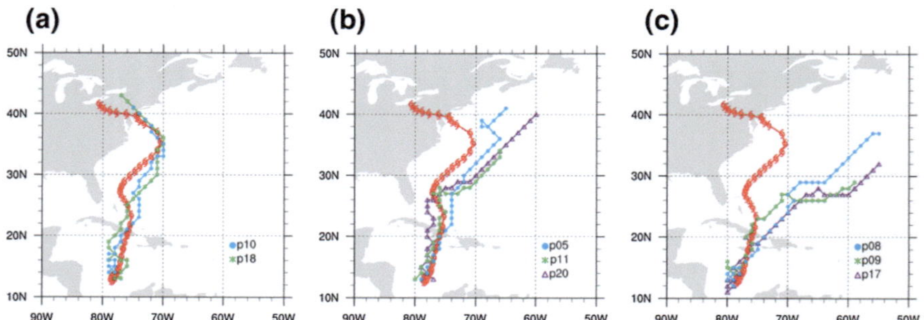

Fig. 8.53 Three group tracks predicted by the GFS model from different ICs. **a** Correct left-turning track with two members, **b** straight-northward track with three members, and **c** right-turning track with three members

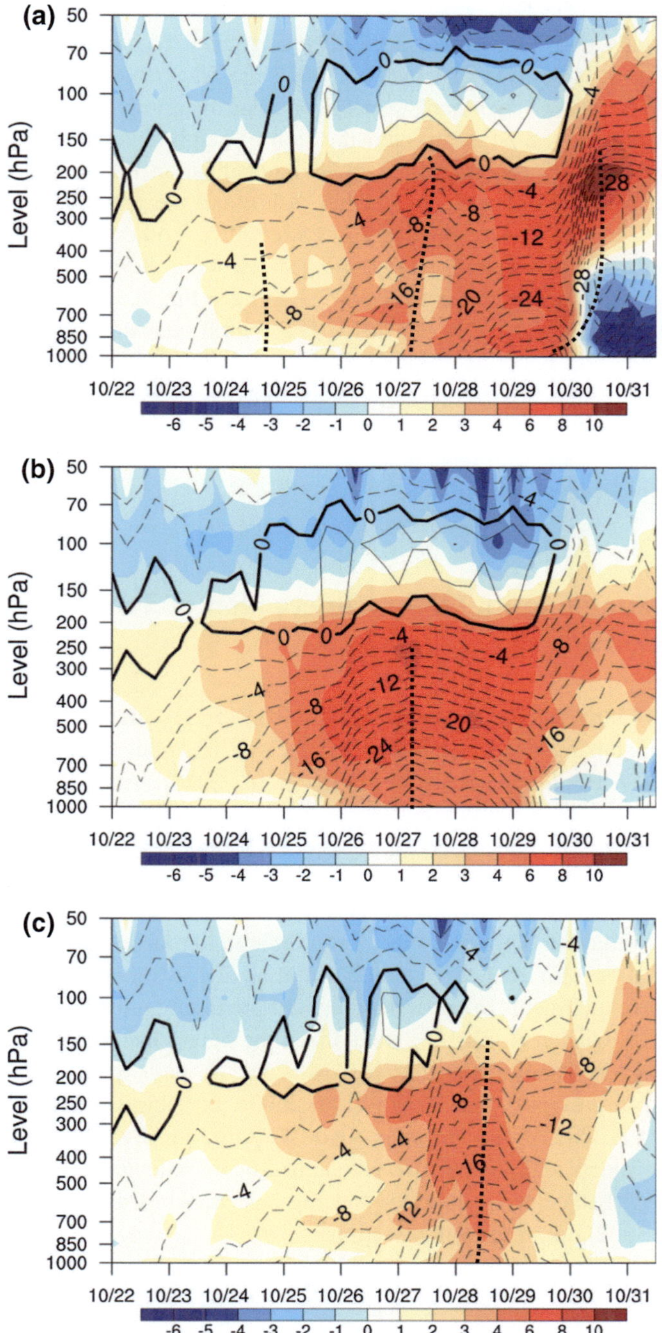

Fig. 8.54 Same as Fig. 8.52 but for the NCEP GFS model initialized at 0000 UTC 22 October 2010 from three groups of **a** correct left-turning track prediction, **b** straight-northward track prediction, and **c** right-turning track prediction, respectively

Same as shown in Figs. 8.52 and 8.47a, three group time–pressure sections of height anomaly and temperature anomaly using the GFS model initialized from different ICs at 0000 UTC 22 October 2012 are illustrated in Fig. 8.54. Three groups have 2, 3, and 3 members, respectively. For the group of correct left-turning track prediction (Fig. 8.54a), an intensification of height anomalies from 26 to 27 October and an ET process with a rapid intensification on 30 October can be comparable between the reanalysis in Fig. 8.47a and the predictions of ECMWF model in Fig. 8.50, respectively. Differently, other two groups of straight-northward track and right-turning track (Fig. 8.53b, c) predicted the strongest height anomaly respectively on 27 and 28 (Fig. 8.54b, c), showing none ET process from 30 to 31 October 2012. The central intensities of positive height anomaly (H3) and negative height anomaly (L4) have also not been correctly predicted in the latter two groups (figures not shown). This comparison tells that the GFS model has the ability to predict the unusual Sandy track correctly as the ECMWF model done for 9 days in advance. A critical effort which should be made is how to obtain those useful ICs.

8.7 Reexamining the Binary Interaction of Tropical Cyclones[9]

In mainland China, forecasters are particularly concerned with three types of unusual TC tracks. One is the sudden right-turning track of TCs while they move westward from the NW Pacific to the SCS (Deng et al. 2010). Another is the sudden left-turning (west-turning) TCs that occur in the ECS (Wu et al. 2013) while they move northward. The above two TC types can cause direct damage in China when they rapidly turn and make landfall in China. The third type of unusual TC tracks is the binary interaction of two nearby TCs in the NW Pacific. The short-term (2–3 days) track prediction of this TC type is still difficult in practice (Wu et al. 2003a, b; Yang et al. 2008; Wu et al. 2011, 2012; Xu et al. 2013; Bai et al. 2014).

Given the experiments in understanding and predicting unusual TC tracks in the SCS and ECS, Qian et al. (2016b) reexamined the binary interaction of unusual TC tracks in the NW Pacific using the GBAM. Four TC pairs in the ECS have been documented in previous studies (Wu et al. 2003a, b; Yang et al. 2008; Wu et al. 2012; Xu et al. 2013). Among them, two TC pairs are investigated by complex model experiments and another two are studied using potential vorticity diagnosis. The ECMWF ERA-Interim product and the best tracks derived from the JTWC (Chu et al. 2002) are used in this section.

---

[9]Modified article originally published in *Qian W.H., Huang J., Zhang G.W.* (2016b) *Reexamining the binary interaction of four pairs of tropical cyclones in the Northwest Pacific. J Meteorol Soc Japan.* doi:10.2151/jmsj.2016-016. Published with kind permission) © Meteorological Society of Japan. All Rights Reserved.

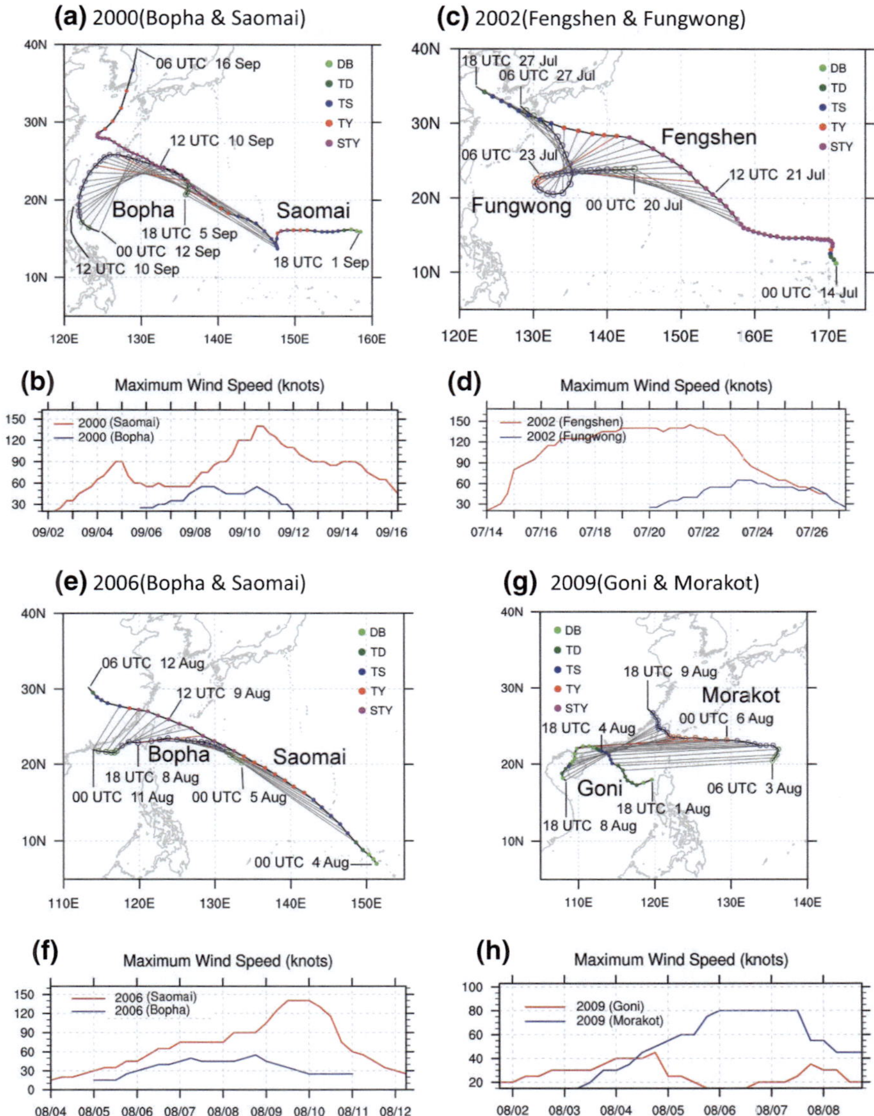

Fig. 8.55 Tracks and intensities of four TC pairs: **a** and **b** Bopha and Saomai (2000), **c** and **d** Fengshen and Fungwong (2002), **e** and **f** Bopha and Saomai (2006), and **g** and **h** Goni and Morakot (2009). Two TC tracks are drawn by *hollow* and *solid dots* respectively. The *dot color* indicates the TC scale: DB, TD, TS, TY, and STY. Maximum surface wind speeds (knots) of two TCs with 6-h interval are denoted by *red* and *blue lines* respectively in **b**, **d**, **f**, and **h**. The *gray lines* in **a**, **c**, **e**, and **g** connect a TC pair at the same time while *red lines* connect two points used in context. Sourced from Qian et al. (2016b) © Meteorological Society of Japan, with permission

Since the four TC pairs have been studied in previous works, their tracks and intensities are simply described here. In Fig. 8.55a, b, Saomai generated in the tropical NW Pacific at 1800 UTC 1 September 2000 and slowly moved westward. After a local circular movement, it moved northwestward and its intensity (140 knots) peaked at 1200 UTC 10 September 2000. Before it rapidly turned right in the ECS, its speed was very slow. Bopha occurred in the heading direction of Saomai at 1800 UTC 5 September 2000 and reached its strongest (55 knots) at 1200 UTC 10 September 2000 near the southeast of Taiwan Island. The lifetime period of Bopha was shorter and weaker than that of Saomai (Fig. 8.55b). Bopha experienced a left turning or southward movement in the east of Taiwan Island, while Saomai followed a northwestward course.

The second pair of TCs was Fengshen and Fungwong, shown in Fig. 8.55c, d. Fengshen generated in the tropical NW Pacific at 0000 UTC 14 July 2002 and moved northwestward before finally dissipating in the Yellow Sea at 0600 UTC 27 July 2002. Its intensity (145 knots) peaked at 1200 UTC 21 July 2002. Fungwong occurred in the NW of Fengshen at 0000 UTC 20 July 2002 and reached its strongest (65 knots) at 0600 UTC 23 July 2002. The lifetime period of Fungwong was shorter and weaker than that of Fengshen. After experiencing a left-turning circle movement, Fungwong dissipated in the Yellow Sea at 1800 UTC 27 July 2002.

In Fig. 8.55e, f, the two TCs, Saomai and Bopha, both moved northwestward after generating in the tropical NW Pacific at 0000 and 0000 UTC 4 and 5 August 2006, respectively. Their intensities (140 and 55 knots) peaked at 1200 and 1800 UTC 9 and 8 August 2006, respectively, when they were getting closer. Finally, Bopha dissipated near the South China coast and Saomai dissipated in Central China. The lifetime period of Bopha was shorter and its strength was weaker than that of Saomai in 2006 (Fig. 8.55f).

In recent years, the notorious and deadly TC Morakot hit Taiwan Island and caused severe damage (Wu et al. 2011, 2012; Yu and Cheng 2013; Zheng et al. 2014; Huang et al. 2014). Morakot generated at 1600 km east of Taiwan Island at 0600 UTC August 2009 and moved westward before making landfall on Taiwan Island at 1545 UTC 8 August 2009 (Fig. 8.55g). Its intensity (80 knots) peaked at 0000 UTC 6 August 2009. Goni generated in the northeastern SCS at 1800 UTC 1August 2009 before Morakot's formation. Goni experienced a slow movement in the northern SCS and the South China coast for 6 days with a very weak intensity and two intensity peaks (Fig. 8.55h). The right-turning movement of Morakot after making landfall on Taiwan Island and its interaction with Goni have been examined by Qian et al. (2016b).

Saomai experienced a long-lived track, while Bopha appeared only from 1800 UTC 5 September 2000 to 0000 UTC 12 September 2000 (Fig. 8.55a, b). The vertical structures of the two TCs at 1200 UTC 8 September 2000 and 1200 UTC 9 September 2000 are depicted in Fig. 8.56. The total height and temperature as well as climatic height and temperature are vertically shown, respectively, in Fig. 8.56a, b crossing two central positions of the two TCs (red lines in Fig. 8.55a). No significant signals can be found indicating the two TCs at 1200 UTC 8

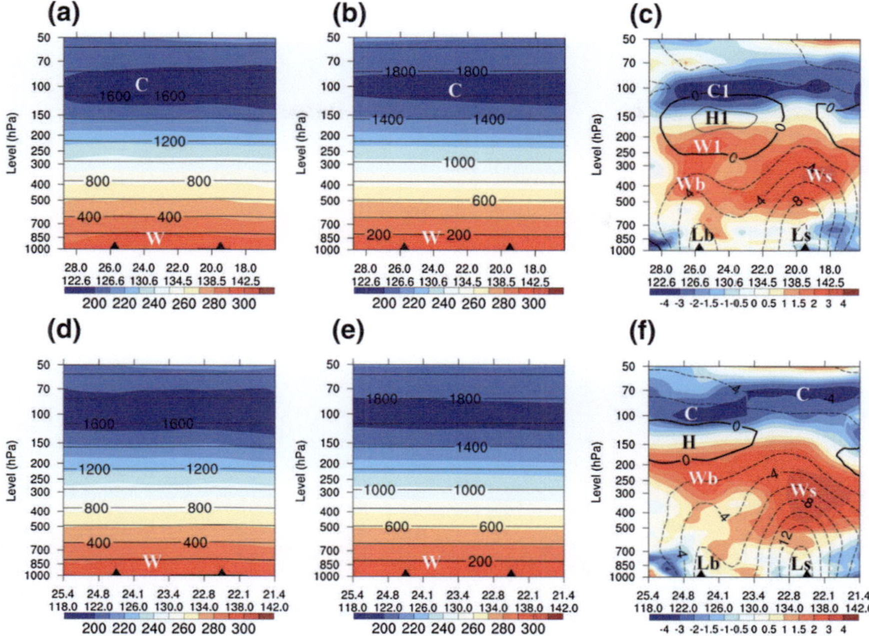

Fig. 8.56 Vertical sections of height and temperature along the connection between two TC centers of Bopha and Saomai at 1200 UTC 8 (*upper panel*) and 1200 UTC 9 (*lower panel*) September 2000. **a** and **d** Total height (*contour*, 200 × 10 gpm interval) and total temperature (*shading*, 10 K interval), **b** and **e** climatic height (*contour*, 200 × 10 gpm interval) and climatic temperature (*shading*, 10 K interval), and **c** and **f** height anomalies (*contour*, 2 × 10 gpm interval) and temperature anomalies (*shading*, 0.5 K and 1 K intervals). Letters "Lb" and "Ls" indicate the height anomalous centers of Bopha and Saomai (2000) and letters "Wb" and "Ws" are their temperature anomalous centers, respectively. Letters "H1", "W1", and "C1" denote the anomalous systems of an upper turbulence. Symbol "*Triangle*" indicates the position of each TC and longitude–latitude grids are digital below the x-axis. Sourced from Qian et al. (2016b) © Meteorological Society of Japan, with permission

September 2000. However, the differences between total and climatic fields show two significant negative centers of height anomalies in the surface and two positive centers of temperature anomalies in the upper troposphere right above the two TCs in Fig. 8.56c. From the troposphere to the stratosphere, the height and temperature anomalies satisfy the hydrostatic equilibrium vertically. It is clearly noted that a negative center (C1) and a positive center (W1) of temperature anomalies are situated above and below the positive center (H1) of height anomalies at the upper troposphere. Even in the lower troposphere, a positive center (Wb) of temperature anomalies is located above the negative center (Lb) of height anomalies at the surface. This hydrostatic equilibrium relationship with obvious and defined positive-negative signals of height and temperature anomalies is also observed in Fig. 8.56f at 1200 UTC 9 September 2000, but there is not any signal indicating the

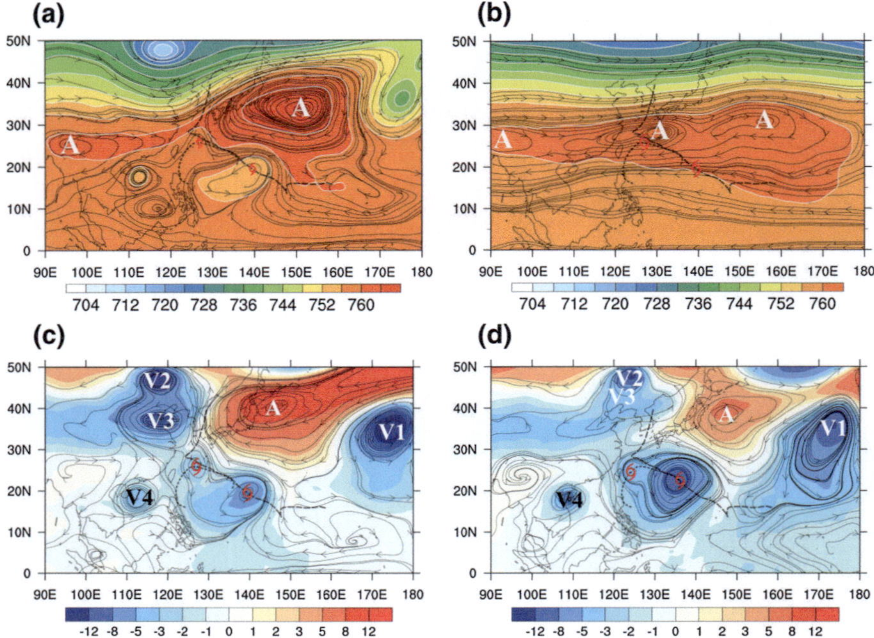

Fig. 8.57 **a** Total height (*shading*, 4 × 10 gpm interval) and total flow, **b** climatic height (*shading*, 4 × 10 gpm interval) and climatic flow, **c** height anomaly (s1 × 10, 2 × 10, 3 × 10, and 4 × 10 gpm intervals) and anomalous flow at 400 hPa at 1200 UTC 8 September 2000. **d** Same as **c** but at 450 hPa at 1200 UTC 9 September 2000. Letters "A" and "V" indicate the centers of anticyclone and vortex, respectively. The symbol "϶" denotes the position of each TC and the *dashed line* is the best track. Sourced from Qian et al. (2016b) © Meteorological Society of Japan, with permission

two TCs in Fig. 8.56d, e from total height, total temperature, climatic height, and climatic temperature.

Figure 8.57a is the total flow that can be decomposed as climatic flow (Fig. 8.57b) and anomalous flow (Fig. 8.57c). The strong TC (Saomei) was closer to the subtropical high than that of the weak TC (Bopha), so the track of Saomei was mainly influenced by total flow (Fig. 8.57a). Under the climatic flow in Fig. 8.57b, Bopha and Saoma should move westward. However, considering the interactions among the two TCs and other anomalous systems, a complex movement can be observed from Fig. 8.57c–d. At 1200 UTC 8 September 2000, an anomalous anticyclone over Japan and four anomalous vortices near the two TCs from East Asia to the NW Pacific existed in the height and wind anomalous fields at 400 hPa (Fig. 8.57c).

As shown in Fig. 8.56a, b, the total field cannot be directly distinguished from climatic field, but it is clear from the anomalous field in Fig. 8.56c. Similarly, the height and flow deviations cannot be identified in the total field (Fig. 8.57a) and the climatic field (Fig. 8.57b), but their differences or anomalies are clearly shown in the anomalous field (Fig. 8.57c), where an anomalous anticyclone and four anomalous vortices from East Asia to NW Pacific surround two TC vortices. These anomalous systems' locations and intensities change from day to day. From 1200 UTC 8 September 2000 to 1200 UTC 9 September 2000, the distance of the two anomalous TC vortices decreased and Saomai strengthened, but the anomalous anticyclone weakened in the NW Pacific while two mid-latitude vortices (V2 and V3) merged (Fig. 8.57c, d). Whether there are interactions among the two TCs and other anomalous systems needs to be examined by the GBAM.

The vorticity anomalies, derived from anomalous wind, in the vertical section and horizontal plane are shown from Fig. 8.58a–d at 1200 UTC 8 September 2000 and 1200 UTC 9 September 2000, respectively. The significant positive vorticity

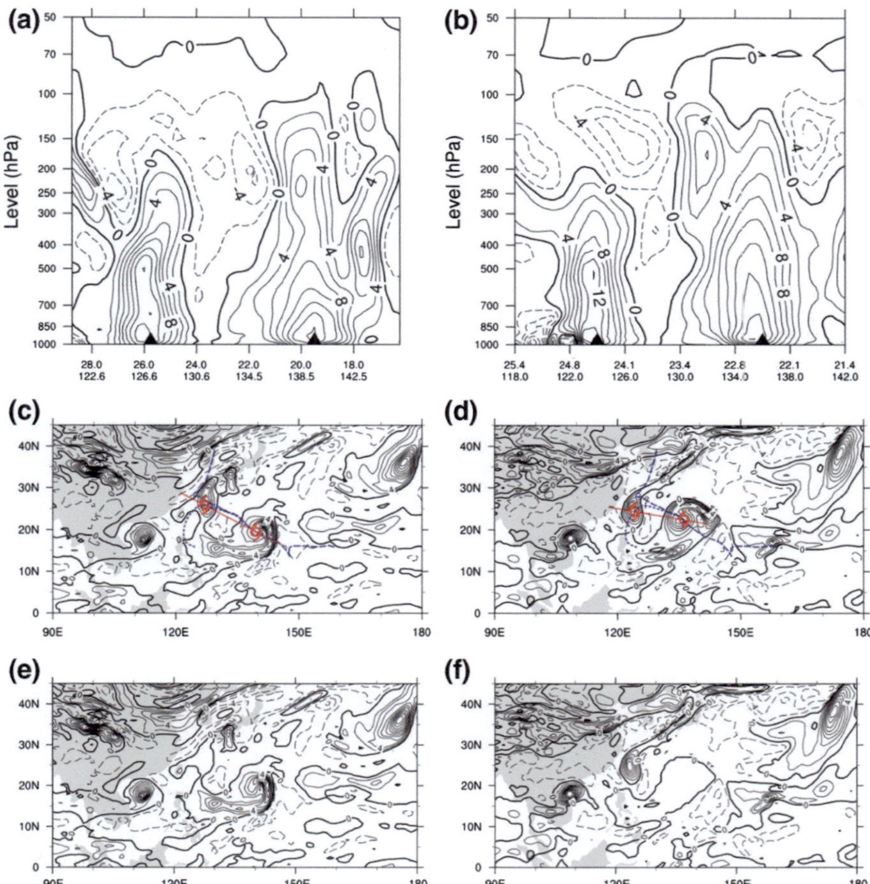

Fig. 8.58 Vorticity anomalies (10^{-5} s^{-1}) in vertical section along the connection of two TC centers at **a** 1200 UTC 8 and **b** 1200 UTC 9 September 2000. Vorticity anomalies (10^{-5} s^{-1}) at **c** 400 hPa at 1200 UTC 8 and at **d** 450 hPa at 1200 UTC 9 September 2000. Vorticity anomalies in **e** and **f** are the same as in **c** and **d**, except the positive vorticity anomalies around the Bopha and Saomei are respectively replaced by climatic vorticity. Sourced from Qian et al. (2016b) © Meteorological Society of Japan, with permission

anomalies correspond well to the two TCs extending from the surface to the upper troposphere (Fig. 8.58a, b). There is a maximum center of positive vorticity anomalies at around 400–500 hPa, and thus the horizontal distribution of vorticity anomalies at 400 and 450 hPa is plotted, respectively, at 1200 UTC 8 September 2000 and 1200 UTC 9 September 2000 (Fig. 8.58c, d). Near the two TC vortices, there is another anomalous vortex (a large center of positive vorticity anomalies) in the SCS. The two TC centers circle each other in the counterclockwise direction, as observed from the red lines in Fig. 8.58c, d.

To explain whether there is a binary interaction between two nearby TCs, comparison experiments are shown in Fig. 8.58e, f as examples. The positive vorticity anomaly around the Bopha (weak one) in the ECS was replaced by climatic vorticity at 1200 UTC 8 September 2000 (Fig. 8.58e). Similarly, the positive vorticity anomaly around the Saomei (strong one) was replaced by the climatic vorticity at 1200 UTC 9 September 2000 (Fig. 8.58f). To examine whether the two

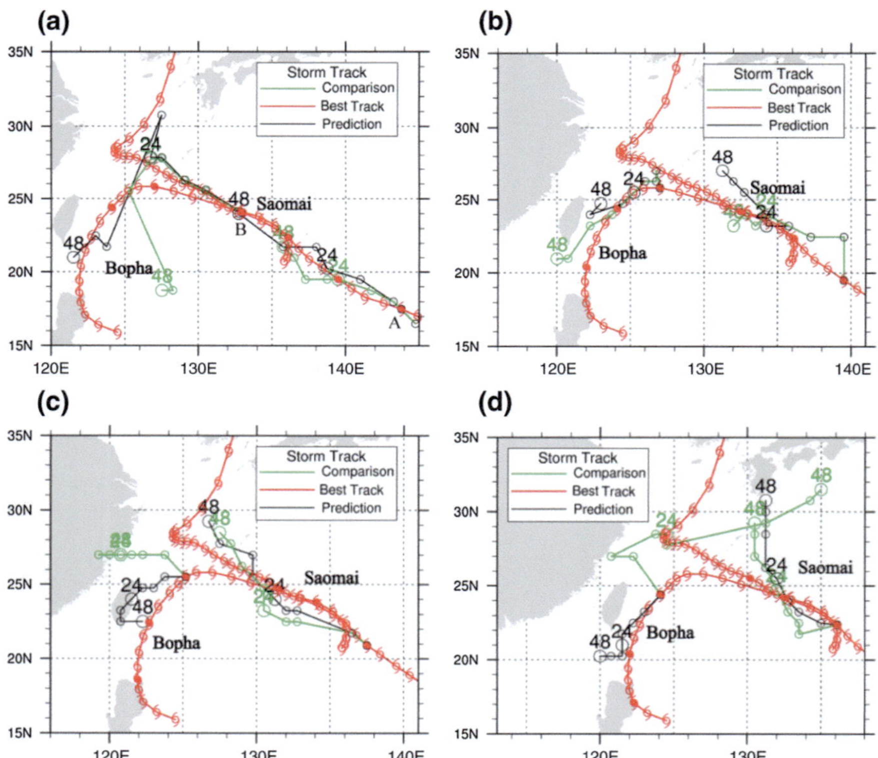

Fig. 8.59 Best track (*red line*), original GBAM prediction track (*black line*), and compared prediction track (*green line*) when another TC vortex is replaced by climatic flow initiated respectively at **a** 450 hPa at 1200 UTC 7, **b** 400 hPa at 1200 UTC 8, **c** 600 hPa at 0000 UTC 9, and **d** 450 hPa at 1200 UTC 9 September 2000. The highlighted symbol "$\mathfrak{6}$" and the numbers 24 and 48 indicate each TC following the best track and model prediction for coming 24 and 48 h. In **a**, two letters "A" and "B" are two location points of Saomei and Bopha at the model initial time. Sourced from Qian et al. (2016b) © Meteorological Society of Japan, with permission

TCs interact, the GBAM is used to compare their track predictions over the following 48 h initiated at 1200 UTC 7, 1200 UTC 8, 0000 UTC 9, and 1200 UTC 9 September 2000, respectively. Track deviation is qualitatively examined, but their speed deviation is not considered in Fig. 8.59. The black line ("Prediction") is the track predicted by the GBAM, while the green line ("Comparison") is the predicted track when another TC vortex's positive vorticity anomaly is replaced by climatic vorticity. If the black and green lines are consistent with a TC best track (red line), it means that another TC does not influence on this TC's track. If the two predicted lines show a large deviation in distance, it means that another TC directly influences this TC's track. If large deviations of the two lines are found in both TCs' predictions, it means that there is a binary interaction between two TCs.

In this case, the track of Saomai is not influenced by the Bopha vortex, but the track of Bopha is influenced by Saomai, particularly from 24 to 48 h, when their prediction is initiated from 1200 UTC 7 September 2000 (Fig. 8.59a). It is noted from point "A" that the prediction (black line) and comparison (green line) tracks of Saomai followed the best track (red line) from 24 to 48 h. From point "B" the prediction track curved to the best track of Bopha, but the comparison showed a rapid return. Starting from 1200 UTC 8 September 2000 (Fig. 8.59b), the two TCs had no significant interaction. At other two initial times of 0000 and 1200 UTC 9 September 2000, the predicted track of Saomai was accurate in the first 24 h in both cases with and without Bopha, but the predicted track of Bopha was different if Saomai's vortex was removed (Fig. 8.59c, d). This implied that Bopha was influenced by Saomai from 1200 UTC 7 September 2009 when Bopha approached Taiwan Island, but the latter was not influenced by the former. The dynamical

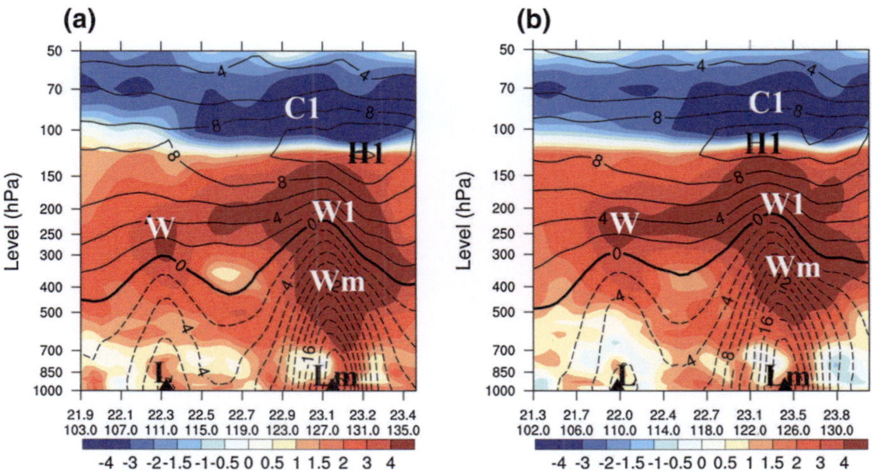

Fig. 8.60 Same as in Fig. 8.56c and f except vertical sections of height anomalies (*contour*, 2 × 10 gpm interval) and temperature anomalies (*shading*, 0.5 and 1 K intervals) along the connection of the two TCs (Goni and Morakot) at **a** 0000 UTC and **b** 1200 UTC 6 August 2009. Sourced from Qian et al. (2016b) © Meteorological Society of Japan, with permission

model showed that the southward movement of Bopha was only influenced by Saomai, but without any binary interaction.

The TC Morakot caused severe damage when it hit Taiwan Island on 7 August 2009, so we also give an analysis. Figure 8.60 shows the vertical sections of height and temperature anomalies along the two TC centers at 0000 and 1200 UTC 6 August 2009. The anomalous vortex of Morakot has a depth up to 250 hPa, while that of Goni is relatively shallower at 300 hPa. The hydrostatic equilibrium is also clearly observed in Fig. 8.60a, b for the anomalous Morakot vortex. A negative center (C1) and a positive center (W1) of temperature anomalies are situated above and below, respectively, the positive center (H1) of height anomalies at the upper troposphere. In the lower troposphere, a positive center (Wm) of temperature anomalies is located above the negative center (Lm) of height anomalies at the

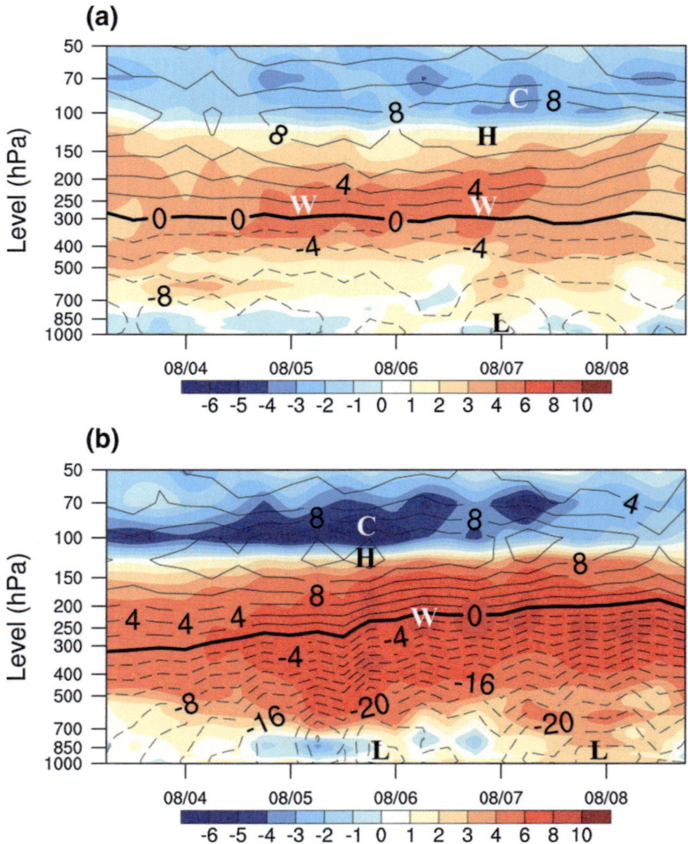

Fig. 8.61 Vertical–time sections of height anomalies (*contour*, 2 × 10 gpm interval) and temperature anomalies (*shading*, 1 and 2 K intervals) for **a** Goni and **b** Morakot in their common lifetime period from 0600 UTC 3 to 1800 UTC 8 August 2009. Sourced from Qian et al. (2016b) © Meteorological Society of Japan, with permission

surface. It means that the temperature anomalies can be derived from the height anomalies.

The intensity evolutions of the two TCs are compared from their vertical-time sections of height anomalies and temperature anomalies in Fig. 8.61. The strongest center of Goni with the value −120 gpm occurred at 0000 UTC 7 August 2009 (Fig. 8.61a), and the strongest center of Morakot with the value −240 gpm occurred at 0000 UTC 8 August 2009 (Fig. 8.61b). Negative temperature anomalies sometimes appeared during their lifetime periods in the lower layer. Positive temperature anomalies always existed in the upper troposphere centered around 250–300 hPa.

During the common lifetime period of the two TCs, anomalous flows and heights changed with time and altitude. At 450 hPa at 0000 UTC 6 August 2009, Morakot was located in the eastern sea of Taiwan Island and directly influenced by the total steering flow of the NW Pacific subtropical high, while Goni made landfall on the South China coast without obvious total steering flow (Fig. 8.62a). The two TCs were situated in a climatic easterly (Fig. 8.62b). An anomalous anticyclone was located in the NW Pacific and another weak anomalous anticyclone was in the central SCS, while at the same time there was an anomalous vortex (V1) in the southeast of Morakot (Fig. 8.62c). At 450 hPa at 1200 UTC 6 August 2009,

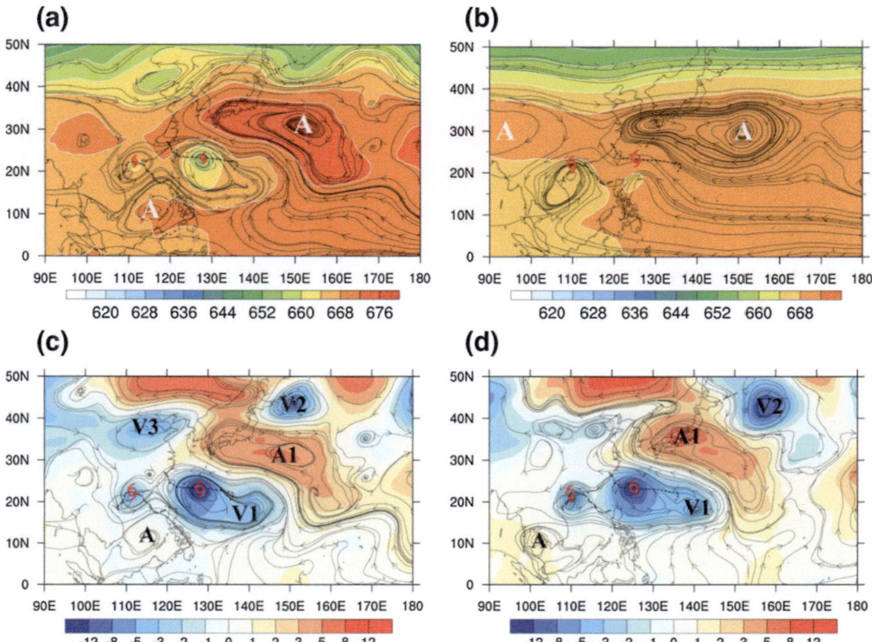

Fig. 8.62 Same as Fig. 8.57 but at 450 hPa **a** and **c** at 0000 UTC 6 August 2009, and **b** and **d** at 1200 UTC 6 August 2009. Sourced from Qian et al. (2016b) © Meteorological Society of Japan, with permission

Morakot propagated closer to Taiwan Island in the ECS and Goni remained at the same position. The anomalous anticyclone in the NW Pacific strengthened and extended westward, but the anomalous anticyclone in the central SCS had moved to the Indo-China Peninsula, while the two TCs and another anomalous vortex (V1) oriented from the South China coast to the NW Pacific (Fig. 8.62d).

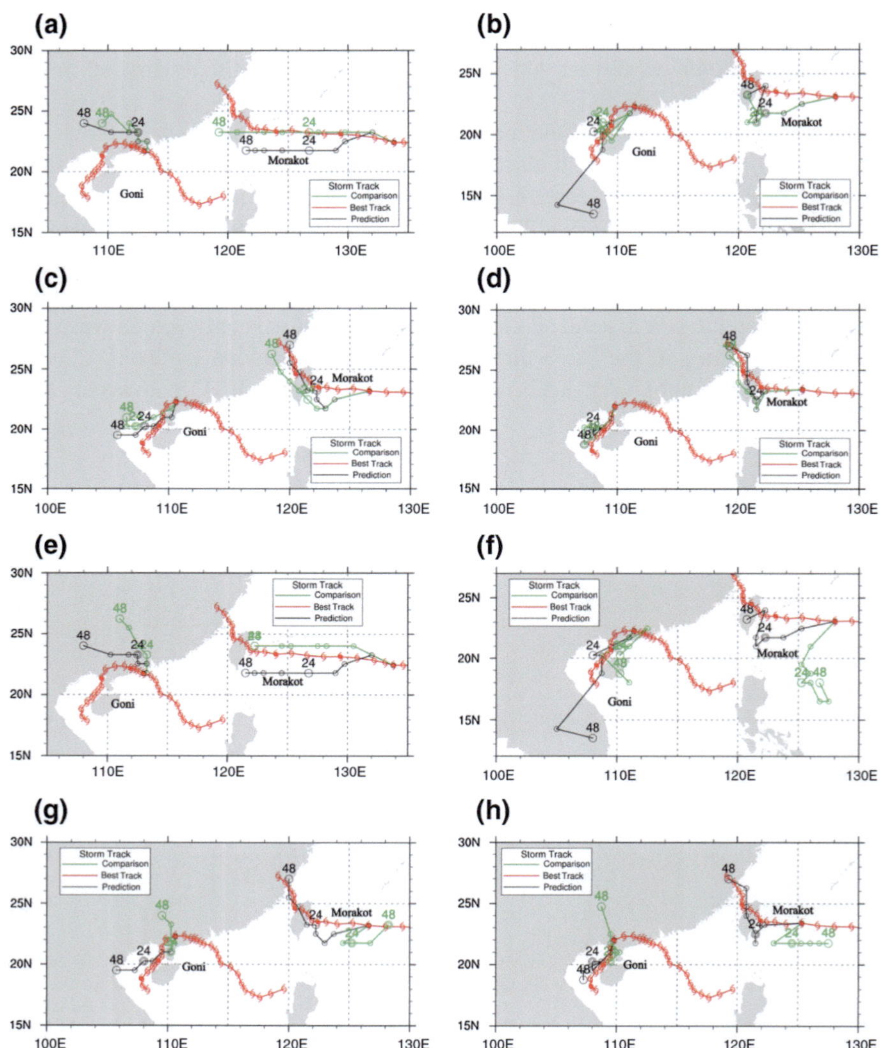

Fig. 8.63 Same as in Fig. 8.59 but for the two TCs (Goni and Morakot) initiated respectively at **a** 1800 UTC 4, **b** 0000 UTC 6, **c** 0600 UTC 6, and **d** 1200 UTC 6 August 2009 at 400, 400, 450, and 450 hPa as well as the movement of two TCs (Goni and Morakot) after replacing the anomalous anticyclone by the climatic flow initiated respectively at **e** 1800 UTC 4, **f** 0000 UTC 6, **g** 0600 UTC 6, and **h** 1200 UTC 6 August 2009. Sourced from Qian et al. (2016b) © Meteorological Society of Japan, with permission

The GBAM showed that there was no binary interaction between the two TCs before Morakot made landfall on Taiwan Island. As illustrated in Fig. 8.63a–d, the black and green lines showed consistency for the model prediction that initiated from 1800 UTC 4, 0000 UTC 6, 0600 UTC 6, and 1200 UTC 6 August 2009. However, deviations became large after replacing the anomalous anticyclone (A1) with the climatic flow at 1800 UTC 4, 0000 UTC 6, 0600 UTC 6, and 1200 UTC 6 August 2009 (Fig. 8.63e–h). The green line turned southward in Fig. 8.63f, but it returned eastward in Fig. 8.63g, h. The model experiment showed that the right-turning track of Morakot was caused by the anomalous anticyclone in the NW Pacific when it approached Taiwan Island.

Similarly, another two TC pairs, Fengshen and Fungwong (2002), as well as Bopha and Saomai (2006) , are analyzed (figures not shown) and modeled. The large center of vorticity anomalies was vertically observed around 500 hPa for Fengshen and Fungwong. The GBAM showed that Fengshen had an influence on the movement of Fungwong, but the latter did not directly influence the former's movement (Fig. 8.64). The GBAM can accurately predict the straight track of Fengshen whether Fungwong exists or not. The circular movement of Fungwong is also accurately predicted by the GBAM, but it is influenced by Fengshen's

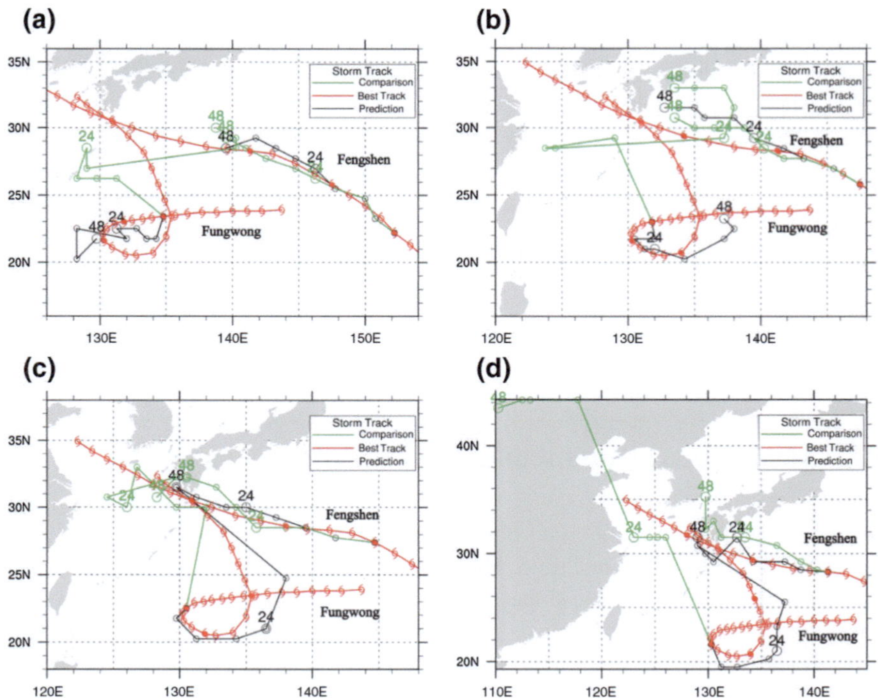

Fig. 8.64 Same as in Fig. 8.59 but for the two TCs of Fengshen and Fungwong at **a** 0000 UTC 22, **b** 0000 UTC 23, **c** 1200 UTC 23, and **d** 0000 UTC 24 July 2002 at 500 hPa. Sourced from Qian et al. (2016b) © Meteorological Society of Japan, with permission

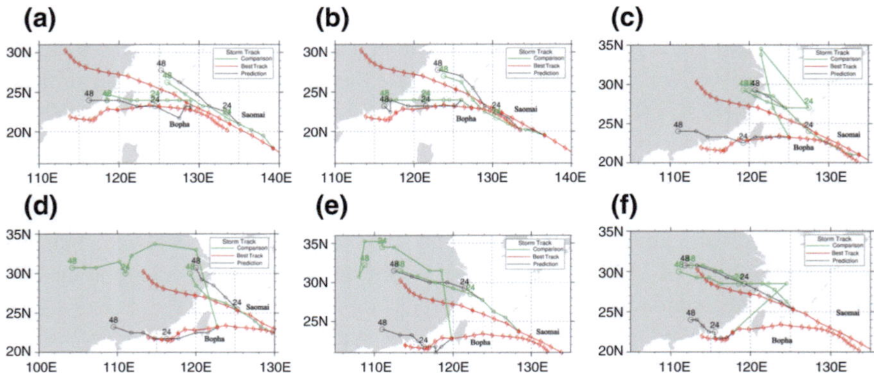

Fig. 8.65 Same as in Fig. 8.59 but for the two TCs of Bopha and Saomai initiated respectively at **a** 0000 UTC 7, **b** 1200 UTC 7, **c** 0000 UTC 8, **d** 1200 UTC 8, **e** 0000 UTC 9, and **f** 1200 UTC 9 August 2006 at 500, 450, 550, 550, 850, and 850 hPa, respectively. Sourced from Qian et al. (2016b) © Meteorological Society of Japan, with permission

circulation. It is noted that if the Fengshen vortex is replaced by a climatic steering flow, the predicted track (green line) of Fungwong marches northward instead of the circular movement. This result from the GBAM shows that it cannot be referred to as a binary interaction, because only the Fengshen affects the movement of Fungwong but the reverse is not true.

The GBAM showed that there was no interaction at 0000 and 1200 UTC 7 August 2006 when the intensity of Saomai was relatively weaker than that of Bopha (Fig. 8.65a, b). The GBAM could accurately predict the movement in terms of direction from six initial times. The last four panels showed that Saomai had a direct influence on the movement of Bopha, but the reverse was not true. From Fig. 8.65c–f, the model predicted that Bopha moved northward (green line) instead of the best track (red line) and the original GBAM prediction (black line) for westward movement. This was the result by replacing the Saomai vortex with the climatic flow. The above results show that the strong TC had a direct influence on the track of the weak TC, but the reverse was not true.

8.8 Ideal Binary Interaction of Tropical Cyclones

In summer and autumn, we often observe two or more TCs or typhoons appearing in the NW Pacific at the same time. Their tracks have attracted many scientists since the last century. This is because they can move in irregular or unusual tracks, which are difficult to forecast in practice. The steering flow cannot be used in their track predictions because there is binary interaction between them. The binary interaction was first studied by Sakuhei Fujiwhara, the Japanese meteorologist who initially described the effect. So, the binary interaction is also known as the Fujiwhara effect. The concept of binary interaction was first demonstrated in tank experiments (Fujiwhara 1921, 1923, 1931). Then many observations (Brand 1970; Dong and

Neumann 1983; Smith et al. 1990; Wang and Zhu 1992a, b; Lander and Holland 1993; Wang and Holland 1995; Carr et al. 1997; Carr and Elsberry 1998), and modeling studies (Chang 1983, 1984; DeMaria and Chan 1984; Ritchie and Holland 1993; Holland and Dietachmayer 1993; Wang and Holland 1995) indicated that two TCs can interact and develop mutual orbiting and possible approaching, merging, and escaping processes (Lander and Holland 1993). In the last decade, dynamical diagnostic method and complex numerical models have been used to study the binary interaction between two TCs (Kuo et al. 2000; Wu et al. 2003a, b; Prieto et al. 2003; Shin et al. 2006; Yang et al. 2008).

The binary interaction is not only an interesting and fundamental problem in fluid mechanics and vortex dynamics, but also a practical concern in improving the forecast accuracy of multiple TCs in close vicinity. In practice, the situation of two TCs located within a large environmental flow is complex. A basic problem of how to separate a TC vortex from its environmental flow is still unsolved. It is important to study the relationship between a TC vortex and its environmental flow.

Using the decomposition method and the GBAM, Qian et al. (2016b) reexamined four TC pairs in the ECS, which have been documented in previous studies (Wu et al. 2003a, b; Yang et al. 2008; Wu et al. 2012; Xu et al. 2013). Among them, two TC pairs were investigated by complex model experiments and other two were studied using potential vorticity diagnosis. The intensities of the TC pairs were not exactly equal to each other during their common lifetime period, so the result of GBAM revealed that the strong TC had a direct influence on the track of the weak TC, but the reverse was not true. The binary interaction between two adjacent TCs should occur when they have equal intensity during their common lifetime period.

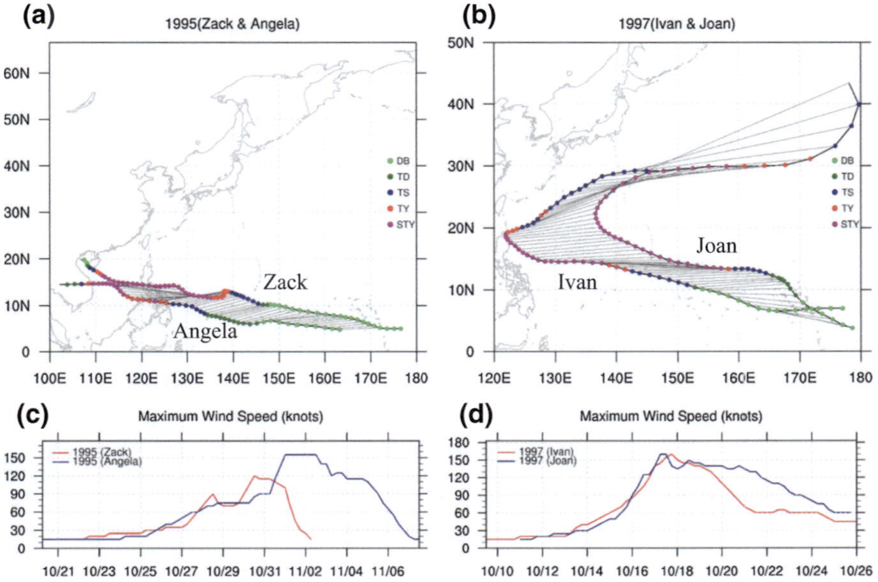

Fig. 8.66 Same as in Fig. 8.55 except for Two pair TCs with their intensities (knot) exceeding 120 knots in **a** and **c** 1995 (Zack and Angela) and **b** and **d** 1997 (Ivan and Joan)

After that we can well understand the concept of binary interaction proposed by Fujiwhara (1921, 1923, 1931) in his tank experiments.

The interaction between two pair TCs with their intensities exceeding 120 knots in 1995 (Zack and Angela) and 1997 (Ivan and Joan) are examined here as examples. Their tracks and intensities are given in Fig. 8.66. Zack and Angela experienced straight-moving tracks from the equatorial western Pacific to the SCS, while Ivan and Joan showed right-turning movement in the NW Pacific.

Figure 8.67 shows the vertical sections of height and temperature along the connection between two typhoon centers of Angela and Zack at 0000 UTC 30 (upper panel) and 0000 UTC 31 (lower panel) October 1995. Although the differences also exist if one carefully compares the total field (Fig. 8.67a) with the climatic field (Fig. 8.67b), the detailed structures of the two TCs are not obviously visible in the total field. However, they are clearly revealed in the anomaly based field (Fig. 8.67c). For example, there are two negative centers (La and Lz) of height

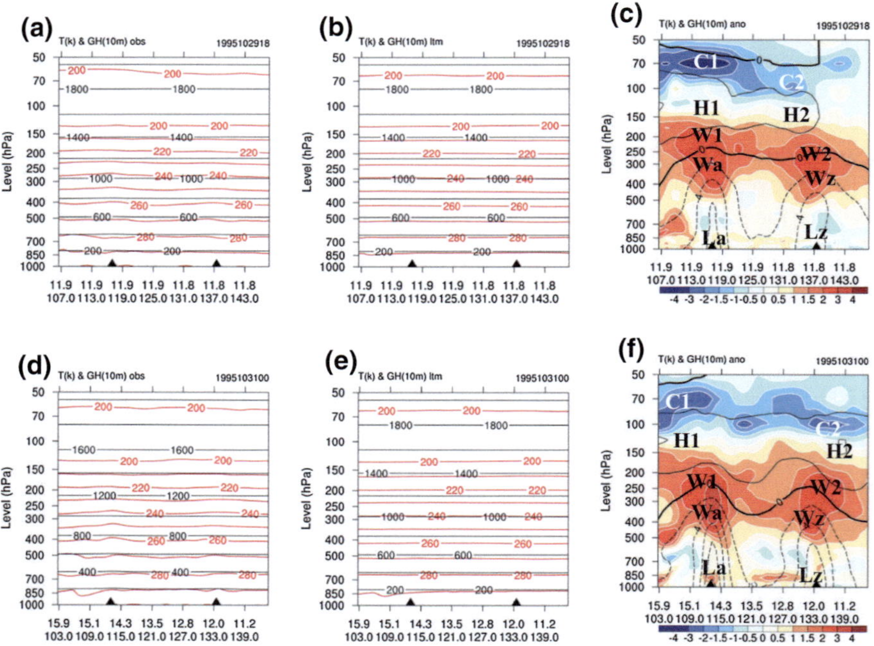

Fig. 8.67 Vertical sections of height and temperature along the connection between two typhoon centers of Angela and Zack at 0000 UTC 30 (*upper panel*) and 0000 UTC 31 (*lower panel*) October 1995. **a** and **d** Total height (*contour*, 2000 gpm interval) and total temperature (*shading*, 10 K and 20 K interval), **b** and **e** climatic height (*contour*, 2000 gpm interval) and climatic temperature (*shading*, 10 and 20 K interval), and **c** and **f** height anomalies (*contour*, 2 × 10 gpm interval) and temperature anomalies (*shading*, 0.5 K and 1 K interval). Letters "La" and "Lz" indicate the negative centers of height anomalies of Angela and Zack and letters "Wa/Ca" and "Wz/Cz" are their positive (warm) and negative (cold) centers of temperature anomalies, respectively. Letters "H1/H2", "W1/W2", and "C1/C2" are other anomalous systems at upper troposphere. Symbol "*Triangle*" indicates the position of typhoon

anomalies near the 1000 hPa and two positive centers (Wa and Wz) of temperature anomalies at 350 hPa right over the two TCs. Other two positive centers (H1 and H2) of height anomalies at 125 hPa and two positive centers (W1 and W2) of temperature anomalies at 250 hPa are also found over the two TCs. From the view of hydrostatic equilibrium, the positive temperature anomalies (Wa, Wz, W1, and W2) between 200 and 300 hPa should have negative height anomalies (La and Lz) below and positive height anomalies (H1 and H2) above. At 0000 UTC 31 October 1995, the two TCs (La and Lz) and the two positive centers of height anomalies strengthened (Fig. 8.67f). The two negative centers (C1 and C2) and the two positive

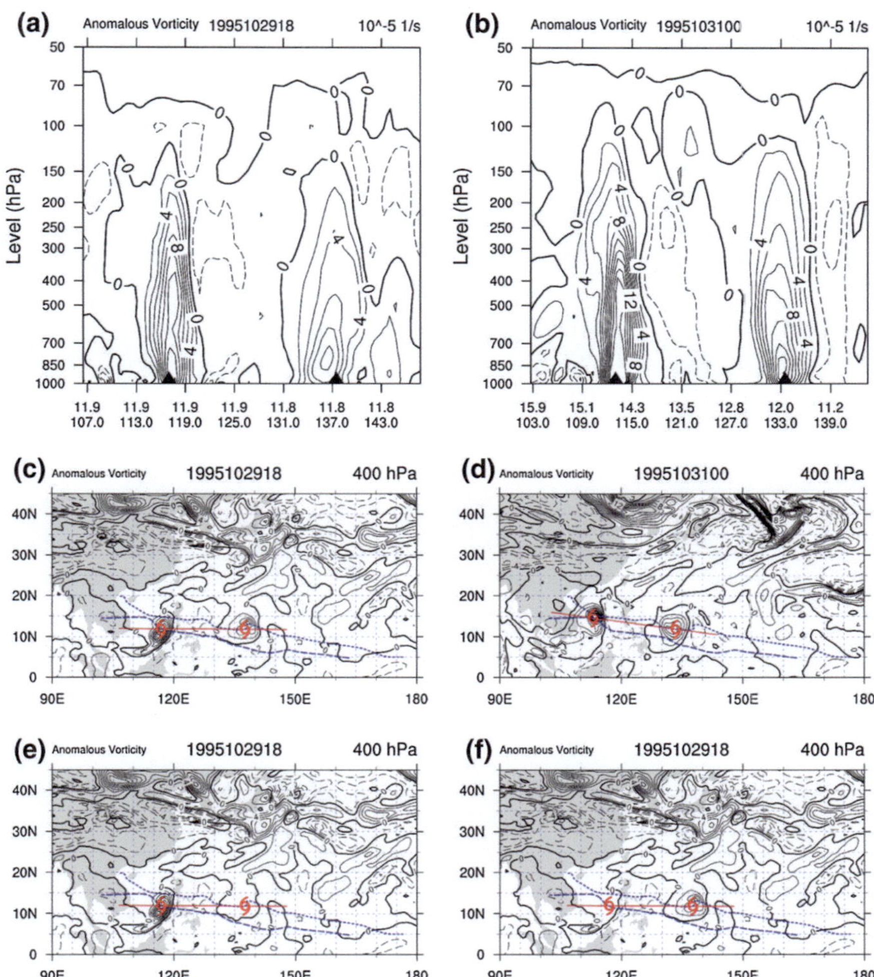

Fig. 8.68 Vertical sections of vorticity anomalies (10^{-5} s^{-1}) along the connection of two typhoon centers at **a** 1800 UTC 29 and **b** 0000 UTC 31 October 1995. Vorticity anomalies (10^{-5} s^{-1}) at **c** 400 hPa at 1800 UTC 29 and at **d** 400 hPa at 0000 UTC 31 October 1995. Vorticity anomalies in **e** and **f** are the same as in **c** and **d**, except the Zack and Angela vorticities are respectively replaced by the climatic vorticities

centers (W1 and W2) of temperature anomalies have intensified with the two positive centers (H1 and H2) of height anomalies at this moment. These basic features of two TCs cannot be directly identified from the total fields in Fig. 8.67a, d, because they contain the climatic component which overwhelms the anomalies superimposed on it.

Using the wind anomalies, Fig. 8.68a, b displays the vertical sections of vorticity anomalies along the connection of two typhoon centers at 1800 UTC 29 and 0000 UTC 31 October 1995. The two positive columns of vorticity anomalies can well represent the two TCs spatially. They extend from the surface to the upper troposphere near 100 hPa, but the upper strongest center is located at 400 hPa except a center at 850 hPa. Their horizontal distributions of vorticity anomalies at 400 hPa are illustrated in Fig. 8.68c, d, where the anomalous intensity of Angela is stronger than that of Zack. A counterclockwise rotation can be noted from their connection of the two times. But the existence of binary interaction needs to be examined using the GBAM.

The total, climatic, and anomalous fields of height and wind anomalies at 400 hPa are examined in Fig. 8.69. In the total height and flow fields (Fig. 8.69a),

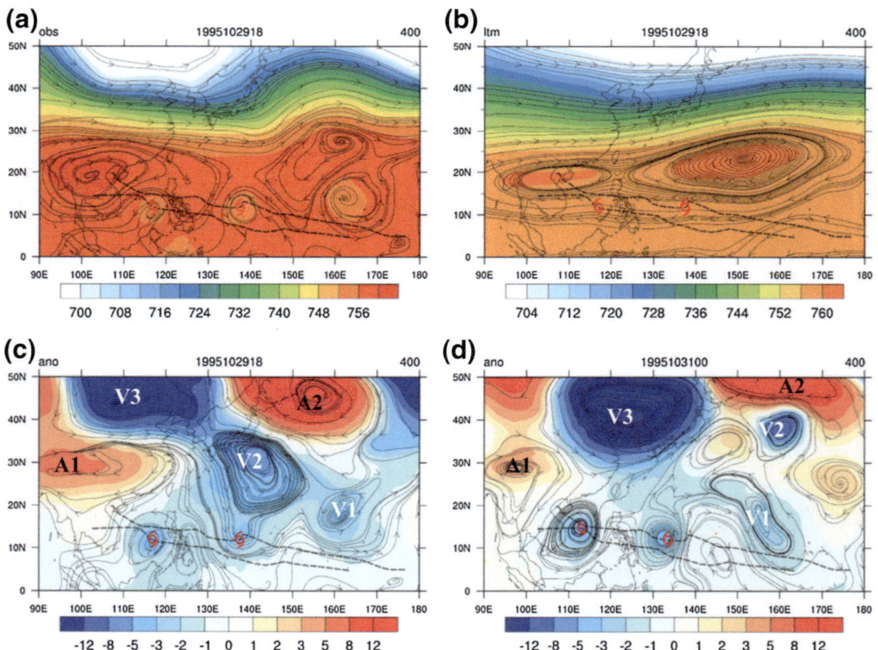

Fig. 8.69 **a** Total height (*shading*, 40 gpm interval) and total flow, **b** climatic height (*shading*, 4 × 10 gpm interval) and climatic flow, **c** height anomaly (*shading*, 1 × 10, 2 × 10, and 4 × 10 gpm intervals) and anomalous flow at 400 hPa at 1800 UTC 29 October 1995. **d** is the same as in **c** except height anomaly and anomalous flow at 0000 UTC 31 October 1995. Letters "A" and "V" indicate the centers of anticyclone and vortex. The symbol "6" denotes the positions of two typhoons and the *dashed line* is the best track

there were many cyclones and anticyclones surrounding the two TCs and a strong westerly in higher latitudes at 1800 UTC 29 October 1995. It is difficult to separate the two TCs from the total flow. Using the decomposition approach, the climatic height and flow were shown in Fig. 8.69b, where the two TCs were simply located within the climatic easterly, so they should move westward under the climatic steering flow. The complexity of total height and flow in Fig. 8.69a is contributed by those anomalous systems of height and wind anomalies in Fig. 8.69c. Indeed, there were three TCs (two typhoons and V1) respectively in the SCS, the east of the Philippines, and the central NW Pacific. In the middle latitudes, there were two anomalous vortices (V2 and V3) and two anomalous anticyclones (A1 and A2). All these anomalous systems changed their locations and intensities from 1800 UTC 29 to 0000 UTC 31 October 1995, while the climatic height and flow had no obviously changes in the two days. The movements of the two TCs should be influenced by the climatic steering flow and interact with other anomalous systems in the vicinity.

The GBAM is used to explore whether there was interaction between the two TCs. The GBAM is run at 400 hPa because it is the level of maximum vorticity anomalies at 1800 UTC 29 October 1995 (Fig. 8.68a, b). One of the two anomalous TC vortices to examine their tracks is removed respectively. As shown in Fig. 8.68c, e, the anomalous vortex of Zack is replaced by the local climatic flow. Similarly, the anomalous vortex of Angela is also replaced by the local climatic flow (Fig. 8.68d, f). A comparison is shown in Fig. 8.70 using the GBAM. In Fig. 8.70a, the control prediction with the two TC vortices are indicated by two blue lines, while compared predictions are respectively made after removing the Angela vortex and the Zack vortex. Their tracks are indicated by a green line and a yellow line. The results imply that there is no any interaction between the two TCs during their peak period. If we remove all other anomalous systems in Fig. 8.69c, the predicted tracks are shown by the two green lines in Fig. 8.70b. Their differences between the blue and green lines in Fig. 8.70b indicate that there is a little influence caused by other anomalous systems.

The two TCs of Ivan and Joan at 1800 UTC 19 and 0000 UTC 21 October 1997 were stronger than that of Zack and Angela at 1800 UTC 29 and 0000 UTC 31 October 1995. The vertical structure of height and temperature anomalies is clearly observed in Fig. 8.71c, f, but cannot be found in their total and climatic fields (Fig. 8.71a, d and b, e). The negative Joan center (Lj) of height anomalies is located at the surface, but the negative Ivan center (Li) is found at around 500 hPa. That is why there is a negative center (Cj) of temperature anomalies at 700 hPa.

In the vertical section along the connection of two typhoon centers of Ivan and Joan at 1800 UTC 19 and 0000 UTC 21 October 1997, the two positive centers of vorticity anomalies are respectively located at 500 and 850 hPa (Fig. 8.72a, b). The two positive columns of vorticity anomalies extend upward to 100 hPa over the two TCs. The horizontal distributions of vorticity anomalies at 500 hPa are depicted in Fig. 8.72c, d, where the anomalous strength of Joan is stronger than that of Ivan.

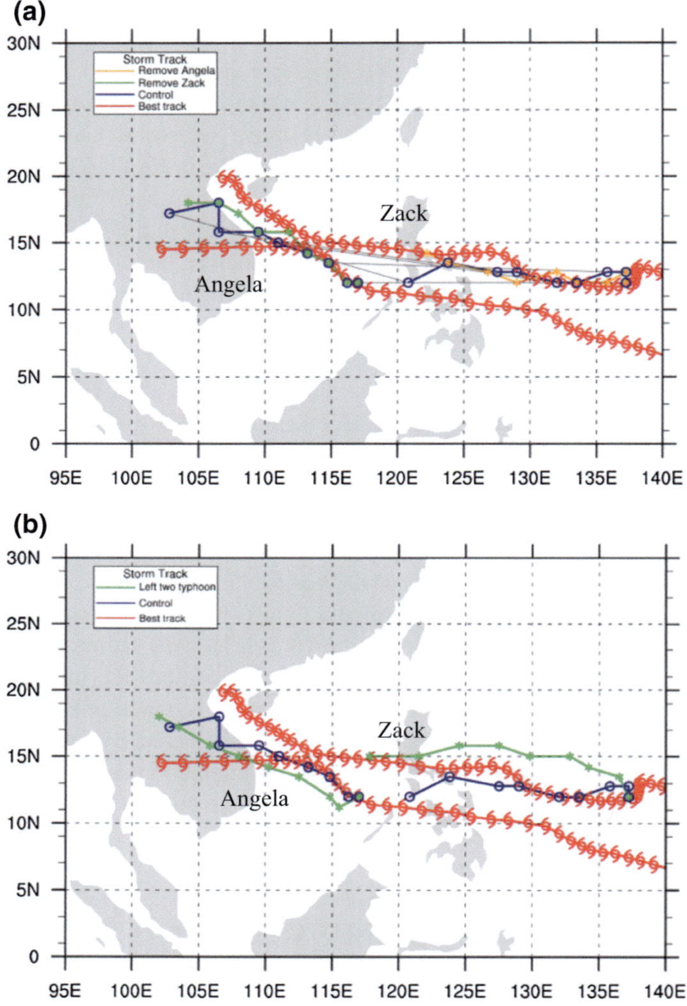

Fig. 8.70 **a** The best tracks (*red line*) of the two TCs, the GBAM prediction tracks (*blue line*) and compared tracks (*yellow and green lines*) when another TC is removed at 400 hPa at 1800 UTC 29 October 1995, **b** is the same as **a** but removing all other anomalous systems with only the two anomalous TC vortices left under the climatic flow

A counterclockwise rotation manner is clearly noted from their connection in the two times. The existence of binary interaction needs to be examined using the GBAM by removing one of the two TC vortices (Fig. 8.72e, f).

The two TCs seem to be located in simple total height and flow fields as shown in Fig. 8.73a, because Ivan and Joan have arrived in the westerly flow. Therefore, their tracks can be expected to turn northeastward. The two TCs are located in the

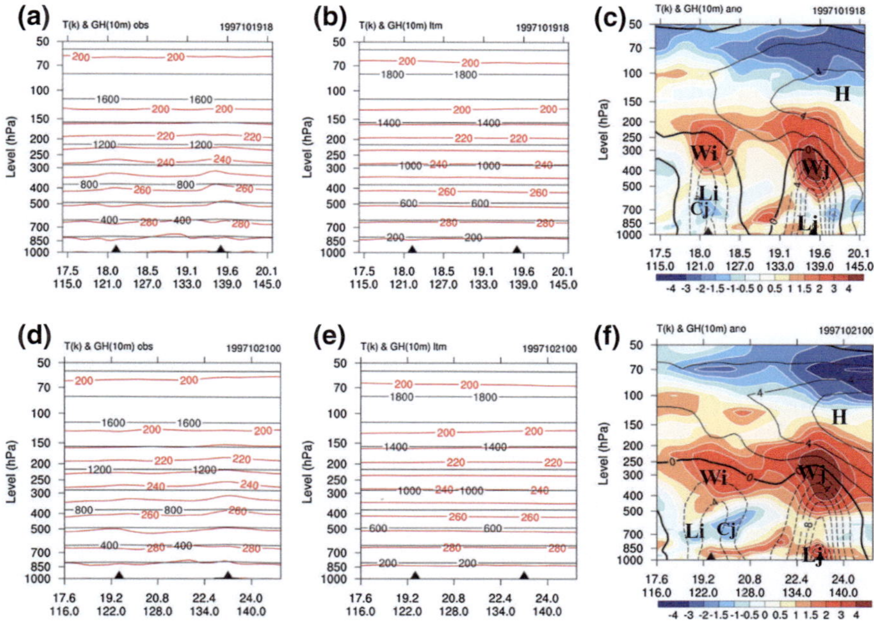

Fig. 8.71 Same as in Fig. 8.67 but for Ivan and Joan at 1800 UTC 19 (*upper panel*) and 0000 UTC 21 (*lower panel*) October 1997

climatic easterly steering flow, so they are expected to move westward (Fig. 8.73b). The right-turning movement of the two TCs is due to the interaction with other anomalous systems (Fig. 8.73c, d). From 1800 UTC 19 to 0000 UTC 21 October 1997, the vortex V1 strengthened in the China continent.

A comparison is also shown in Fig. 8.74 for another pair of TCs using the GBAM. In Fig. 8.74a, the control prediction of the two TC vortices are indicated by the two blue lines, while compared predictions by removing the Ivan vortex and the Joan vortex are indicated by a green line and a yellow line. The results indicate that the track of Joan vortex is directly influenced by the Ivan vortex during the right-turning period. If we remove all other anomalous systems and only retain the two TC vortices (Fig. 8.74b), the predicted tracks are shown by the two green lines. The differences between the blue and green lines are clearly observed in Fig. 8.74b, and the result indicates that there is an influence caused by other anomalous systems. If we retain the two TC vortices and the vortex V1 (Fig. 8.74c), the right-turning track of Joan vortex is influenced by the Ivan vortex, while the right-turning track of the Ivan vortex is influenced by the vortex V1.

During the era without computers, the concept of binary interaction was firstly described by tank experiments (Fujiwhara 1921, 1923, 1931). Since the computer era, many numerical experiments of two TC interactions have been carried out.

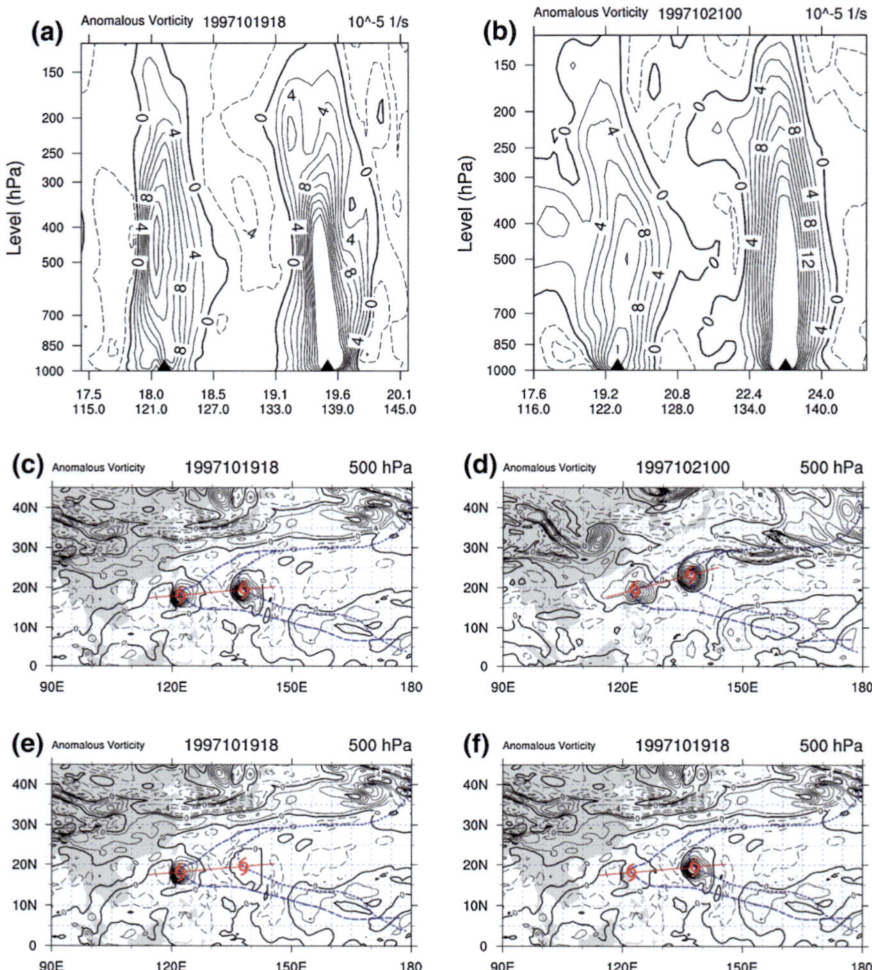

Fig. 8.72 Same as in Fig. 8.68 but for Ivan and Joan at 1800 UTC 19 and 0000 UTC 21 October 1997

These experiments can be categorized into two types. One is based on two observed TCs and environmental fields, but it is difficult to separate the two TC vortices from the environmental flow. Another is based on two ideal TCs and environmental fields, but their results cannot be compared with reality. We have made efforts but it is hard to find the binary interaction of two TCs in observation. In this section, some results of ideal ABAM experiment based on the two strongest TC pairs only comparing from different distances and intensities are given.

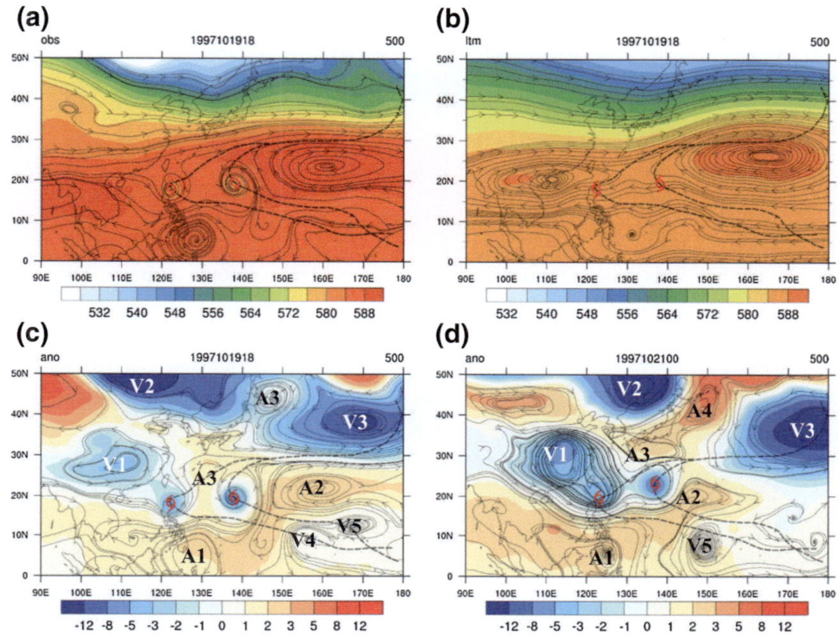

Fig. 8.73 Same as in Fig. 8.69 but for Ivan and Joan at 1800 UTC 19 and 0000 UTC 21 October 1997

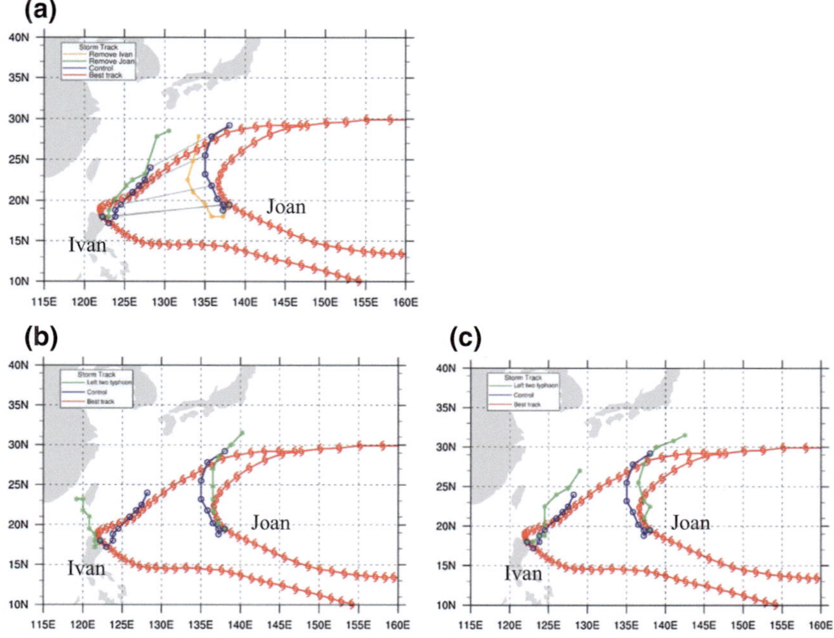

Fig. 8.74 **a**, **b** are the same as in Fig. 8.70 but for Ivan and Joan at 1800 UTC 19 October 1997. In **c**, all other anomalous systems are removed, with only the two anomalous TC vortices and the V1 vortex left under the climatic flow

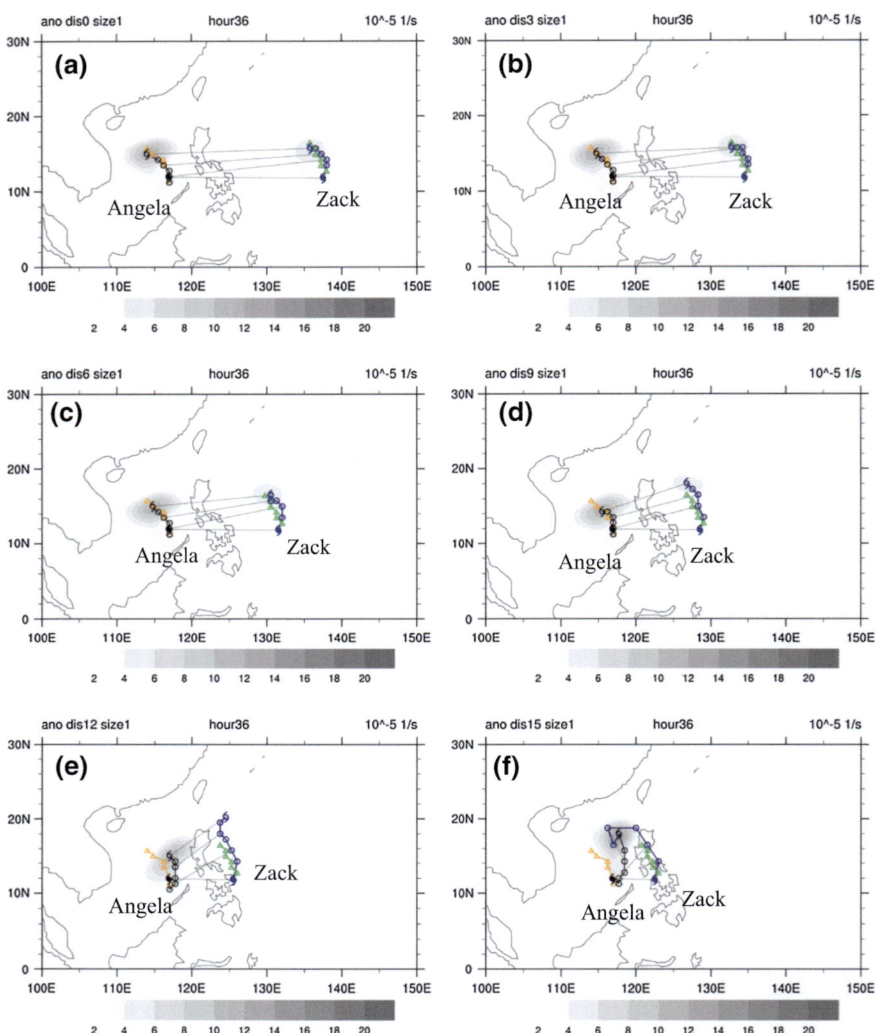

Fig. 8.75 Two TC track predictions (*black* and *blue lines*) of Angela and Zack using the ABAM for the following 36 h when the central location of Zack vortex moves **a** 0, **b** 3, **c** 6, **d** 9, **e** 12, and **f** 15 longitudes westward respectively at the initial time. *Yellow* and *green lines* are the tracks predicted by the ABAM using a signal TC vortex. The shadings are positive values of vorticity anomalies ($2 \times 10^{-5}\ s^{-1}$ interval)

Figure 8.75 shows two TC track predictions only with Angela and Zack anomalous vortices using the ABAM for the following 36 h when the central location of Zack vortex moves 0, 3, 6, 9, 12, and 15 longitudes westward, respectively. The left-turning tracks of two vortices are comparable to their single TC track in Fig. 8.75a. It implies that there is no any interaction between them, because the two tracks of black (blue) and yellow (green) lines are consistent. When

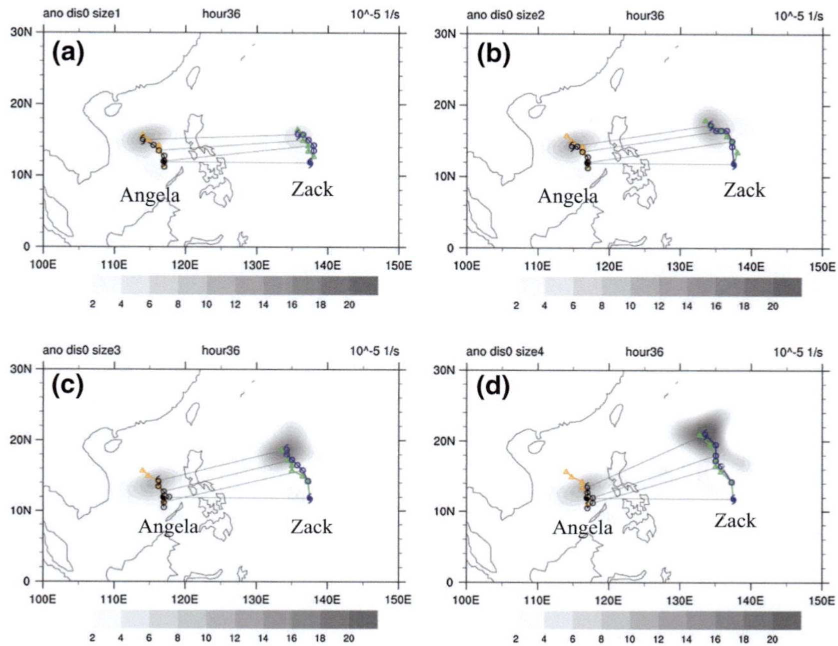

Fig. 8.76 Same as in Fig. 8.75 except amplifying the intensity of Zack vorticity anomalies to **a** 1, **b** 2, **c** 3, and **d** 4 times respectively

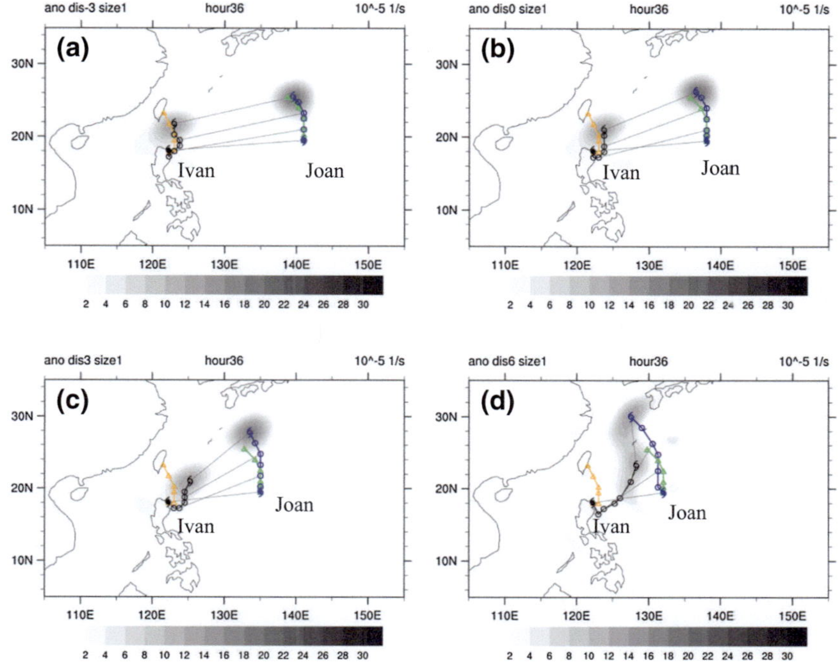

Fig. 8.77 Same as in Fig. 8.75 except moving the central location of Joan vorticity **a** 3 longitudes eastward as well as **b** 0, **c** 3, and **d** 6 longitudes westward, respectively

the central location of Zack vortex moves 3, 6, and 9 longitudes westward respectively, their tracks show an increasing trend of deviations between two TCs and single vortex (Fig. 8.75b–d). The binary interaction is clearly observed in Fig. 8.75e when their distance is 9 latitudes. The two TCs finally combine or merge together when their initial distance is shorter than 600 km (Fig. 8.75f).

We further examine how the Zack intensity influences their track. The central intensity of Zack is amplified from 6 to 24 × 10^{-5} s^{-1}. Figure 8.76 shows the results of ABAM experiments. When amplifying the intensity of Zack vorticity anomalies, Zack moves fast and has an influence on the slowdown of Angela. It means that a stronger TC should have an influence on a weaker TC's track.

Using the ABAM, the binary interaction between Ivan and Joan is clearly observed when the central location of Joan moves westward or eastward (Fig. 8.77). A counterclockwise rotation manner with a merging tendency is significant when their distance reduces to 1300 km in Fig. 8.77d. The binary interaction between Ivan and Joan is also found when the intensity of Joan is amplified from 10 to 44 × 10^{-5} s^{-1} (Fig. 8.78). A counterclockwise rotation manner with a merging trend is also observed during this intensity amplification (Fig. 8.78c, d).

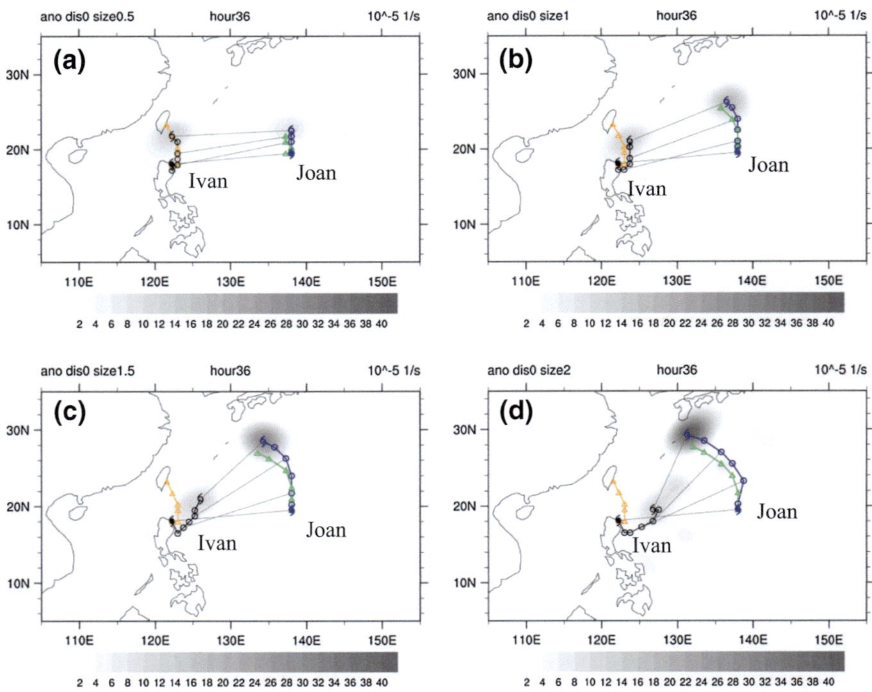

Fig. 8.78 Same as in Fig. 8.75 except amplifying the intensity of Joan vorticity anomalies to **a** 0.5, **b** 1, **c** 1.5, and **d** 2 times respectively

Questions

1. How to separate the TC vortex from the surrounding flow? How many methods are usually used in the TC studies?
2. How many types of unusual and complex TC tracks can be classified from observations?
3. Why do current complex dynamical models sometimes fail to predict unusual TC tracks?
4. What are the differences between the Fujiwhara effect and the binary interaction of TC in observation?
5. How to improve the forecasting skill of TC track and intensity?

References

Aberson SD (2010) 10 years of hurricane synoptic surveillance (1997–2006). Mon Wea Rev 138:1536–1549

Bai LN, Ma LM, Zeng ZH, Huang W, Wang DL (2014) Numerical analysis on the role of tropical storm Namtheum in the unusual tracks of three tropical cyclones in 2010. J. Tropical Meteorology 20(4):297–307

Bassill NP (2014) Accuracy of early GFS and ECMWF Sandy (2012) track forecasts: Evidence for a dependence on cumulus parameterization. Geophys Res Lett 41:3274–3281

Brand S (1970) Interaction of binary tropical cyclones of the western North Pacific Ocean. J Appl Meteor 9:433–441

Carr III LE, Elsberry RL (1998) Objective diagnosis of binary tropical cyclone interactions for the western North Pacific Basin. Mon Wea Rev 126:1734–1740

Carr III LE, Mark AB, Elsberry RL (1997) Observational evidence for alternate modes of track-altering binary tropical cyclone scenarios. Mon Wea Rev 125:2094–2111

Chan JCL, Gray WM (1982) Tropical cyclone movement and surrounding flow relationships. Mon Wea Rev 110:1354–1374

Chan JCL (2005) The physics of tropical cyclone motion. Annu. Rev. Fluid Mech. 37:99–128

Chang SW (1983) A numerical study of the interaction between two tropical cyclones. Mon Wea Rev 111:1806–1817

Chang SW (1984) Reply. Mon We. Rev, 112:1646–1647

Chandan R, Rita K (2012) Tropical cyclone track forecasting techniques—A review. Atmospheric Research 104–105:40–69

Charney JG (1948) On the scale of atmospheric motions. Geofys Publ 17(2):1–17

Chen LS, Luo ZX, Xu XD (2002) An introduction to tropical cyclone unusual motion (in Chinese). 25th annual Conf., Beijing, China, Chin Meteor Soc 195–199

Chen LS, Ding YH (1979) An Introduction to the Western Pacific Typhoons (in Chinese). Science Press, Beijing

Chu JH, Sampson CR, Levine AS, Fukada E (2002) The Joint Typhoon Warning Center tropical cyclone best tracks 1945–2000. Joint Typhoon Warning Cent., Pearl Harbor, Hawaii

Davis C, Snyder C, Didlake AC (2008) A vortex-based perspective of eastern Pacific tropical cyclone formation. Mon Wea Rev 136:2461–2477

Dee DP, and Coauthors (2011) The ERA-Interim reanalysis: Configuration and performance of the data assimilation system. Quart J Roy Meteor Soc 137:553–597

DeMaria M, Chan J C L. (1984) Comments on "A numerical study of the interactions between two tropical cyclones". Mon Wea Rev 112(8):1643–1645

Ding T, Qian WH (2012) Statistical characteristics of heat wave precursors in China and model prediction (in Chinese). Chin J Geophys 55:1472–1486

Dong KQ, Neumann CJ (1986) The relationship between tropical cyclone motion and environmental geostrophic flows. Mon Wea Rev 114:115–122

Dong KQ, Neumann CJ (1983) On the relative motion of binary tropical cyclones. Mon Wea Rev 111:945–953

Deng G, Zhou YS, Liu LP (2010) Use of a new steering flow method to predict tropical cyclone motion. J Tropical Meteor 16(2):154–159

Du J, Grumm RH, Deng G (2014) Ensemble anomaly forecasting approach to predicting extreme weather demonstrated by extremely heavy rain event in Beijing (in Chinese). Chin J Atmos Sci 38:685–699

Dvorak V (1975) Tropical cyclone intensity analysis and forecasting from satellite imagery. Mon Wea Rev 103(5):420–430

Franklin JL (1990) Dropwindsonde observations of the environmental flow of hurricane Josephine (1984): Relationships to vortex motion. Mon Wea Rev 118:2732–2744

Fujiwhara S (1921) The mutual tendency towards symmetry of motion and its application as a principle in meteorology. Quart J Roy Meteor Soc 47:287–293

Fujiwhara S (1923) On the growth and decay of vortical systems. Quart J Roy Meteor Soc 49:75–104

Fujiwhara S (1931) Short note on the behavior of two vortices. Proc Phys Math Soc Japan Ser 3 13:106–110

Galarneau TJ, Davis CA (2013) Diagnosing forecast errors in tropical cyclone motion. Mon Wea Rev 141:405–430

Galarneau TJ, Davis CA, Shapiro MA (2013) Intensification of Hurricane Sandy (2012) through extratropical warm core seclusion. Mon Weather Rev 141:4296–4321

Graham RA, Grumm RH (2010) Utilizing normalized anomalies to assess synoptic-scale weather events in the western United States. Wea Forecasting 25:428–445

Grumm RH (2011a) The Central European and Russian heat event of July–August 2010. Bull Amer Meteor Soc 92:1285–1296

Grumm RH (2011b) New England record maker rain event of 29–30 March 2010. Electronic J Operational Meteor 12(4):1–31

George JE, Gray WM (1976) Tropical cyclone motion and surrounding parameter relationships. J Appl Meteorol 15:1252–1264

Grumm RH, Hart R (2001) Standardized anomalies applied to significant cold season weather events: preliminary findings. Wea Forecasting 16:736–754

Hamill TM, Whitaker JS, Kleist DT, Fiorino M, Benjamin SG (2011) Predictions of 2010's tropical cyclones using the GFS and ensemble-based data assimilation methods. Mon Wea Rev 139:3243–3247

Harr PA, Elsberry RL, Hogan TF (2000) Extratropical transition of tropical cyclones over the Western North Pacific. Part II: The impact of midlatitude circulation characteristics. Mon Rea Rev 128:2634–2653

Hart RE, Grumm RH (2001) Using normalized climatological anomalies to rank synoptic scale events objectively. Mon Wea Rev 129:2426–2442.

Holland GJ (1983) Tropical cyclone motion: Environmental interaction plus a beta effect. J Atmos Sci 40:328–342

Holland GJ (1984) Tropical cyclone motion: A comparison of theory and observation. J Atmos Sci 41:68–75

Holland GJ, Dietachmayer GS (1993) On the interaction of tropical-cyclone-scale vortices. III: Continuous barotropic vortices. Quart J Roy Meteorol Soc 119(514):1381–1398

Huang HL, Yang MJ, Sui CH (2014) Water budget and precipitation efficiency of Typhoon Morakot (2009). J Atmos Sci 71(10):112–129

Huang J, Du J, Qian WH (2015) A comparison between Generalized Beta-Advection Model and classical Beta-Advection Model in predicting and understanding unusual typhoon tracks in eastern China seas. Wea Forecasting 30:771–792

Jones SC and coauthors (2003) The extratropical transition of tropical cyclones: Forecast challenges, current understanding, and future directions. Wea Forecasting 18:1052–1092

Junker NW, Grumm RH, Hart R, Bosart LF, Bell KM, Pereira FJ (2008) Use of standardized anomaly fields to anticipate extreme rainfall in the mountains of northern California. Wea Forecasting 23:336–356

Junker NW, Brennan MJ, Pereira F, Bodner MJ, Grumm RH (2009) Assessing the potential for rare precipitation events with standardized anomalies and ensemble guidance at the Hydrometeorological Prediction Center. Bull Amer Meteor Soc 90:445–453

Kasahara A (1957) The numerical prediction of hurricane movement with the barotropic model. J Atmos Sci 14:386–402

Kalnay E and Coauthors (1996) The NCEP/NCAR 40-Year Reanalysis Project. Bull Amer Meteor Soc 77:437–471

Kuo HC, Chen GTJ, Lin CH (2000) Merger of tropical cyclones Zeb and Alex. Mon Wea Rev 128:2967–2975

Lalaurette F (2003) Early detection of abnormal weather conditions using a probabilistic extreme forecast index. Q J R Meteorol Soc 129:3037–3057

Lander M, Holland GJ (1993) On the interaction of tropical-cyclone-scale vortices. I:observations. Q J R Meteorol Soc 119:1347–1361

Li P, Lin LN, Lei F (2006) The characteristic analysis of typhoon 0418 (Aere) track (in Chinese). Marine Forecasts 23(4):47–50

Lin Y, Jiang XN, Lin XH (2005) The motion path analysis to the typhoon Aere left folded twice (in Chinese). J Guangxi Meteorol 26(A01):174–176

Marks DG (1992) The beta and advection model for hurricane track forecasting. National Oceanic and Atmospheric Administration. National Weather Service, National Meteorological Center (U.S.), Washington D.C.

Miller BI, Moore PL (1960) A comparison of hurricane steering levels. Bull. Am. Meteorol. Soc. 41:59–63

Peng XD, Che YZ, Chang J (2013) A novel approach to improve numerical weather prediction skills by using anomaly integration and historical data. J Geophy Res 118:1–13

Peixoto JP, Oort AH (1992) Physics of Climate. American Institute of Physics Press, New York

Prieto R, McNoldy BD, Fulton SR, Schubert WH (2003) A classification of binary tropical-cyclone-like vortex interactions. Mon Wea Rev 131: 2656–2666

Powell MD, Aberson SD (2001) Accuracy of United States tropical cyclone landfall forecasts in the Atlantic basin (1976–2000). Bull Amer Meteor Soc 82:2749–2767

Qian CH, Duan YH, Ma SH, Xu YL (2012) The current status and future development of China operational typhoon forecasting and its key technologies (in Chinese). Adv Meteor Sci Technol 2(5):36–43

Qian WH, Huang J, Du J (2016a) Examination of Hurricane Sandy's (2012) structure and intensity evolution from full-field and anomaly-field analyses. Tellus-A 68:29029

Qian WH, Huang J, Zhang GW (2016b) Reexamining the binary interaction of four pairs of tropical cyclones in the Northwest Pacific. J Meteorol Soc Japan doi:10.2151/jmsj.2016-016

Qian WH, Shan XL, Liang HY (2014a) Typhoon turning atlas. World Scientific Publishing Co. Pte. Ltd., Singapore

Qian WH, Shan XL, Liang HY, Huang J, Leung CH (2014b) A generalized beta advection model to improve unusual typhoon track prediction by decomposing total flow into climatic and anomalous flows. J Geophys Res Atmos 119(3):1097–1117

Qian WH, Zhang ZJ (2012) Precursors to predict low-temperature freezing-rain events in southern China (in Chinese). Chin J Geophys 55:1501–1512

Qian WH, Li J, Shan XL (2013) Application of synoptic-scale anomalous winds predicted by medium-range weather forecast models on the regional heavy rainfall in China in 2010. Sci China Earth Sci 56:1059–1070

Qian WH, Zhang GW, Huang J (2015) Intensity evolution of Typhoon Megi (2010) revealed from anomaly-based atmospheric variables. Mon Meteorol 41(7): 818–827

Randy G, Alcott T, Hosenfeld N, Grumm R (2013) Anticipating a rare event utilizing forecast anomalies and a situational awareness display: The western U.S. Storms of 18–23 January 2010. Bull Amer Meteor Soc 94:1827–1836

Rappaport EN, and Coauthors (2009) Advances and challenges at the National Hurricane Center. Wea Forecasting 24:395–419

Richard AD, Davidson NE (2004) Characteristics of tropical cyclones in the Australian region. Mon Wea Rev 132(12):3049–3065

Ritchie EA, Holland GJ (1993) On the interaction of tropical-cyclone-scale vortices. II: Discrete vortex patches. Q J R Meteorol Soc 119(514):1363–1379

Roy C, Kovordanyi R (2012) Tropical cyclone track forecasting techniques—A review. Atmos. Res. 104–105:40–69

Sanders F, Adams AL, Gordon NJB, Jensen WD (1980) Further development of a barotropic operational model for predicting paths of tropical storms. Mon Wea Rev 108:642–654

Smith RK, Ulrich W, Dietachmayer G (1990) A numerical study of tropical cyclone motion using a barotropic model. 1. The role of vortex asymmetries. Q J R Meteorol Soc 116:337−362

Simpson RH (2003) Hurricane: Coping with Disaster: Progress and Challenges since Galveston, 1900. American Geophysical Union

Singh SK, Kishtawal CM, Jaiswal N, Singh R, Pal PK (2012) Impact of vortex-removal from environmental flow in cyclone track prediction using Lagrangian advection model. Meteorol Atmos Phys 117:103–120

Shin SE, Han JY, Baik JJ (2006) On the critical separation distabance of binary vortices in a nondivergent barotropic atmosphere. J Meteor Soc Japan 84:853–869

Trinh VT, Krishnamurti TN(1992) Vortex initialization for typhoon track prediction. Meteorol Atmos Phys 47:117–126

Thorncroft C, Jones SC (2000) The extratropical transitions of hurricanes Felix and Iris in 1995. Mon Wea Rev 128:947–972

Van den Dool (2007) Empirical methods in short-term climate prediction. Oxford University Press, Oxford, UK

Velden CS, Leslie LM (1991) The basic relationship between tropical cyclone intensity and the depth of the environmental steering layer in the Australian region. Wea Forecasting 6:244–253

Wang B, Elsberry RL, Wang YQ, Wu LG (1998) Dynamics in tropical cyclone motion: A review. Chin J Atmos Sci 22:535–547

Wang Y, Holland GJ (1995) On the interaction of tropicalcyclone-scale vortices. IV: Baroclinic vortices. Q J R Meteor Soc 121:95–126

Wang YQ, Zhu YQ (1992a, 1992b) Analysis and numerical study of the interactions of binary tropical cyclones. Part I (Analysis of physical mechanism) and Part II (Numerical simulation). Scientia Atmospheric Sinica, 16: 573–582(659–668)

Wei YZ, Zhang CG, Lin XB, Wu CF (2006) Analyses of abnormal path and landing site of Typhoon Aere (2004) (in Chinese). J Natural Disasters 15(5):54–60.

Wu CC, Huang TS, Huang WP, Chou KH (2003) A new look at the binary interaction: Potential vorticity diagnosis of the unusual southward movement of Tropical Storm Bopha (2000) and its interaction with Supertyphoon Saomai (2000). Mon Wea Rev 131: 1289–1300

Wu X, Fei JF, Huang XG, Cheng XP, Ren JQ (2011) Statistical classification and characteristics analysis of binary tropical cyclones over the western North Pacific Ocean. J Tropical Meteorology 17(4):335–344

Wu X, Fei JF, Huang XG, Zhang X, Cheng XP, Ren JQ (2012) A numerical study of the interaction between two simultaneous storms: Goni and Morakot in September 2009. Adv Atmos Sci 29(3):561–574

Wu LG, Ni ZP, Duan JJ, Zong HJ (2013) Sudden tropical cyclone track changes over the western North Pacific: A composite study. Mon Wea Rev 141:2597–2610

Wu CC, Huang TS, Huang WP, Chou KH (2003) A new look at the binary interaction: Potential vorticity diagnosis of the unusual southward movement of tropical storm Bopha (2000) and its interaction with Supertyphoon Saomai (2000). Mon Wea Rev 131(7):1289–1300

Xu YL (2011) Forecast analysis on the abrupt northward recurvature of super typhoon Megi (1013). Meteorol Monthly 37(7):821–826

Xu HX, Zhang XJ, Xu XD (2013) Impact of tropical storm Bopha on the intensity change of super Typhoon Saomai in the 2006 typhoon season. Adv Meteorol doi:10.1155/2013/487010

Xue JJ, Li JY, Zhang LS, Wang XR, Xu YL (2012) Characteristics of typhoon disasters in China and risk prevention strategies. Meteorology and Disaster Reduction Research 35(1):59–64

Yang CC, Wu CC, Chou KH, Lee CK (2008) Binary interaction between typhoons Fengshen (2002) and Fungwong (2002) based on the potential vorticity diagnosis. Mon Wea Rev 136 (12):4593–4611

Yu CK, Cheng LW (2013) Distribution and mechanisms of orographic precipitation associated with Typhoon Morakot (2009). J Atmos Sci 70(9):2894–2915

Zebiak SE, Cane MA (1987) A model El Niño-Southern Oscillation. Mon Wea Rev 115:2262–2278

Zeng ZH, Wang YQ, Wu CC (2007) Environmental dynamical control of tropical cyclone intensity—An observational study. Mon Wea Rev 135:38–59, doi:10.1175/MWR3278.1

Zheng ZW, Zheng QA, Lee CY, Gopalakrishnan G (2014) Transient modulation of Kuroshio upper layer flow by directly impinging Typhoon Morakot in east of Taiwan in 2009. J Geophy Res Ocean 119(7):4462–4473

Zhu T, Weng FZ (2013) Hurricane Sandy warm-core structure observed from advanced Technology Microwave Sounder. Geophys Res Lett 40:3325–3330

Zsoter E (2006) Recent developments in extreme weather forecasting. ECMWF Newsletter 107:8–17

Chapter 9
Circulation and Climatic Anomalies

The Arctic is warming two to four times faster than the global average and the Arctic sea ice has undergone a substantial long-term decline with superimposed interannual sea ice minimum (SIM) events over the past decade. Eurasian and tropical climatic events frequently occur with severe drought/flood and cold/warm episodes. What are the relationships among anomalous activities of those general circulations and regional surface climatic anomalies? This chapter will reveal their relationships. First, the seasonal, interannual, and interdecadal variabilities of the Arctic and Polar cell strengths are used to explore their connections with the long-term trends and interannual events of Arctic sea ice as well as precipitation and surface air temperature (SAT) in the Northern Hemisphere (NH). Then, regional and persistent anomalous events are also paid particular attention associated with geopotential anomalies of the troposphere and even the stratosphere. As two examples, the anomalous activities of the South Asian High (SAH), the largest circulation system in the boreal summer over upper troposphere, and the western Pacific subtropical high, are investigated using anomaly-based approach.

9.1 Meridional Circulation Anomalies and Arctic Sea-Ice Events[1]

The Arctic sea ice has undergone a substantial long-term decline (Serreze et al. 2007; Graversen et al. 2008; Screen and Simmonds 2010a, b; Ding et al. 2014) with superimposed interannual SIM events (Kay et al. 2008; Graversen et al. 2011; Bennartz et al. 2013; Parkinson and Comiso 2013) over the last decade. The meridional circulation is considered as an important part in climate change, such as the intensity and coverage of the Hadley cell (Hu and Fu 2007; Chen et al. 2014),

[1]Modified article originally published in Qian et al. (2015a). Published with kind permission © 2015 W. Qian et al. All Rights Reserved.

© Springer Nature Singapore Pte Ltd. 2017

W. Qian, *Temporal Climatology and Anomalous Weather Analysis*,
Springer Atmospheric Sciences, DOI 10.1007/978-981-10-3641-5_9

but it cannot directly explain the fast climate warming and interannual SIM events in the Arctic (Kumar et al. 2010; Graversen et al. 2011; Devasthale et al. 2013; Dong et al. 2014). In the Arctic, one of the most concerned topics is the interaction between sea ice and atmospheric circulation (Qian et al. 2015a). Various feedback processes are considered to exist in the Arctic atmosphere–ocean–ice interaction (Hansen and Nazarenko 2004; Shindell and Faluvegi 2009; Vihma 2014) including the role of meridional heat transport via the atmosphere (Jeong et al. 2014). But there is no consensus on the relative importance of various factors responsible for the Arctic amplification and sea ice decline.

Beside feedback mechanisms, storm tracks, jet stream, and planetary waves also play important roles in the Arctic sea ice change (Cohen et al. 2014). Honda et al. (2009) and Sato et al. (2014) showed that late autumn–early winter atmospheric anomalies can force an immediate, local, and direct atmospheric response to the warm anomalies over the Barents and Kara seas locally, which in turn change the baroclinicity and affect large-scale planetary or Rossby waves in the atmosphere. Jaiser et al. (2012) found that sea ice anomalies in early autumn can force a delayed, remote, and indirect atmospheric response through altered baroclinicity and synoptic-planetary wave interaction, or through increased Eurasian snow cover and high pressure over the Barents and Kara seas that force upward planetary waves into the stratosphere. The "summertime" Arctic sea ice concentration's influence on the following "wintertime" climate and circulation variability suggests a potential application in operational seasonal climate and circulation forecasts, although it is unclear whether the sea-ice anomalies in late summer and early autumn, or late autumn and early winter are more skillful at predicting the winter weather patterns. As summarized by Cohen et al. (2014) as well as Deser and Teng (2008a), observational analyses have shown significant correlations between reduced Arctic sea-ice cover and the negative phase of the winter North Atlantic Oscillation/Arctic Oscillation (NAO/AO) and Northern Annular Mode (NAM). On the other hand, numerous studies indicate that the atmospheric circulation plays an important role in driving Arctic sea ice declines (Rigor et al. 2002; Hu et al. 2002; Serreze and Francis 2006; Deser and Teng 2008b; Jeong et al. 2014), particularly showing that summer sea ice variations appear to be initiated by atmospheric circulation anomalies over the Arctic in late spring (Deser et al. 2000). The consensus of the observational studies suggests that interannual variability in sea ice conditions is caused by the variability in the large-scale atmospheric circulation, which locally manifests as surface air temperature and wind anomalies (Prinsenberg et al. 1997). In order to confirm the effect of the Arctic and Polar cells on the Arctic sea ice variation, we explore the possible relationship between the Arctic sea ice change and meridional circulation anomalies in the NH, particularly in high latitudes.

Reanalysis datasets, one ensemble mean CESM1 (CAM5) model simulation, and sea ice concentration (SIC) dataset are used. They include the NCEP reanalysis 1 (R1) (Kalnayet al. 1996), the NCEP reanalysis 2 (R2) (Kanamitsu et al. 2002), the ERA-Interim product (Dee et al. 2011), the Japanese 55-year Reanalysis (JRA-55) product (Ebita et al. 2011), the ensemble mean CAM5 model dataset

(Neale et al. 2010), and the monthly mean gridded SIC and sea ice extent (SIE) from the HadISST1 version for 1979–2013 (Rayner et al. 2003).

The monthly mean and seasonal-mean climatology of variables were used by Qian et al. (2015a). An monthly mean atmospheric variable field $F_{m,y}(\lambda,\varphi,p,t)$ is decomposed into its climatology $\tilde{F}_m(\lambda,\varphi,p)$ and temporal deviation $\tilde{F}'_{m,y}(\lambda,\varphi,p,t)$ on a calendar month t in a year y at a spatial point of pressure level p, longitude λ, and latitude φ, i.e.,

$$F_{m,y}(\lambda,\varphi,p,t) = \tilde{F}_m(\lambda,\varphi,p) + \tilde{F}'_{m,y}(\lambda,\varphi,p,t), \tag{9.1.1}$$

In Eq. 9.1.1, the subscript m indicates monthly mean, which is the 30-year (1981–2010) average of an observed or modeled variable field on the calendar month m in this case. For example, when m is set to May, t is the months from May 1981, May 1982, …, May 2010 for totally 30 years.

$$\tilde{F}_m(\lambda,\varphi,p) = \sum_{y=1981}^{2010} F_{t=m,y}(\lambda,\varphi,p,t)/30, \tag{9.1.2}$$

The climatic mean field or climatology defined in Eq. 9.1.2 varies from month to month, depicting a seasonal cycle. Climatic seasonal-mean field $\tilde{F}_s(\lambda,\varphi,p)$ and climatic annual mean field $\tilde{F}_a(\lambda,\varphi,p)$ can be calculated by monthly mean fields. The meridional stream function, $\psi(\varphi,p)$, satisfying the zonally averaged continuity equation in spherical coordinates, can be calculated at each pressure level p, and latitude φ using the meridional wind integration (Stachnik and Schumacher 2011).

Using four reanalysis products and one model simulation, Qian et al. (2015a) revealed that there are four meridional cells in the NH from individual monthly mean and climatic seasonal mean flow as well as climatic seasonal mean and annual mean meridional mass stream functions (MSFs). If the Arctic cell is real and climatologically active in the high latitudes, it should have a significant impact on Arctic climate change. In the Arctic, the formation of new sea ice is followed by a strong cold air outbreak, when a cold air mass with a lower temperature (below 0°C) than sea surface temperature passes by. The cold air outbreak is a consequence of strong descending flow in high latitudes and northerly temperature advection near the surface, which are dependent upon the seasonal migration of the Polar and Arctic cells. The adiabatic warming during the descending process is weaker than the cooling of the cold advections, which can be observed from the following analysis.

As shown in Fig. 2.10, the descending flow zone is concentrated from 69°N to 82°N while the ascending flow zone is located beyond 83°N. The strongest descending flow is climatologically observed in the boreal winter. They defined two areas with the descending flow intensity (DFI) averaged over 500–850 hPa from 75 to 83°N and the ascending flow intensity (AFI) averaged over 500–850 hPa from 85 to 90°N based on the ERA and JRA-55 reanalysis products.

Figure 9.1a, b shows the climatic DFI and AFI. The DFI derived from the two reanalysis products is consistent, with the maximal DFI in February and the minimal DFI in July (Fig. 9.1a), while the negative AFI peaks in January and July (Fig. 9.1b). The total SIE and mean SIC, which indicate the amount of sea ice in the NH, reach the maximum in March and minimum in September (Serreze et al. 2007; Budikova 2009) (Fig. 9.1c). In summer and autumn, the minimal SIE and SIC lag the minimal DFI and the maximal AFI for about 2 months. This lag relationship implies that the weakening DFI and the strengthening AFI in July would reduce cold air transport vertically and horizontally from 80°N to 70°N, resulting in the minimal SIE or SIC between 80°N and 70°N usually in late autumn (Figs. 2.10 and 9.2). It is also noted that the climatic temperature is always low in higher latitudes but varies with season (Fig. 9.2). In winter, the strongest descending flow (>4 × 10^{-2} Pa s^{-1}) is

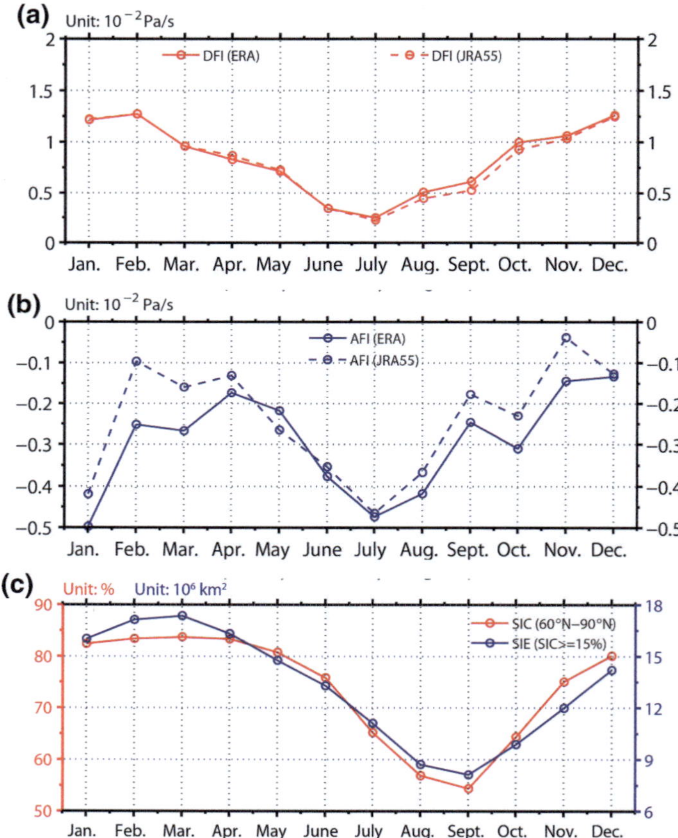

Fig. 9.1 Seasonal series of **a** the monthly mean DFI (10^{-2} Pa s^{-1}) averaged over 500–850 hPa from 75 to 83°N based on ERA-Interim and JRA-55 reanalysis products, **b** the monthly mean AFI averaged over 500–850 hPa from 85 to 90°N, and **c** SIC (%) and SIE (the area of sea ice with SIC > 15%, 10^6 km^2) averaged over 60–90°N based on the HadISST1 dataset. Sourced from Qian et al. (2015a) © 2015 W. Qian et al., with permission

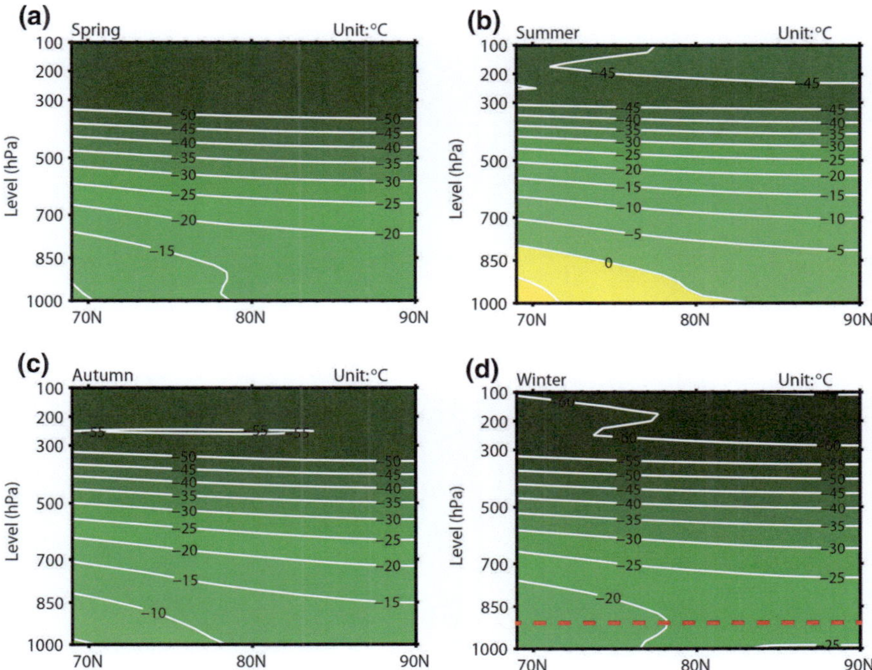

Fig. 9.2 Vertical and meridional sections of zonally averaged climatic seasonal-mean temperature (5 °C interval) between 69°N and 90°N and from 100 to 1000 hPa in the boreal **a** spring, **b** summer, **c** autumn, and **d** winter using a 30-year average (1981–2010) of the ERA-Interim product. In **d**, below the *red-dashed line* (925 hPa) is the temperature inversion layer. Sourced from Qian et al. (2015a) © 2015 W. Qian et al., with permission

climatologically located in the lower troposphere near the 80°N (Fig. 2.10d), while there is a layer of temperature inversion below 925 hPa due to the effects of radiation and temperature advection in higher latitudes (Fig. 9.2d).

The climatic DFI is about two times stronger than the AFI, which motivates us to focus on the relationship between the DFI and the SIC. After removing the climatic seasonal cycle and long-term linear trends of SIE (-0.272×10^6 km^2 per decade), SIC (-1.33% per decade), and the DFI (0.012×10^{-2} Pa s^{-1} per decade), the 15-month running means of the three indices are calculated. A significant correlation ($r = 0.79$) is detected between SIC and SIE. A lagged correlation analysis shows that the DFI anomaly statistically leads the SIC anomaly by about 10 months with a maximal positive correlation coefficient of 0.54, reaching the 99% confidence level using the Student's t-test. This implies the Polar–Arctic cell interaction is a strong factor inducing large-scale sea-ice change (Fig. 9.3a). This significant correlation provides a potential for operational seasonal sea ice forecasts. Figure 9.3a displays a total of 12 DFI peaks and a total of 8 SIC peaks exceeding two thresholds (dashed lines, $\pm 0.1 \times 10^{-2}$ Pa/s or $\pm 1\%$) since 1980. The DFI contains a clear precursory signal of SIC. In fact, most of the SIC peaks, i.e., 66.7%

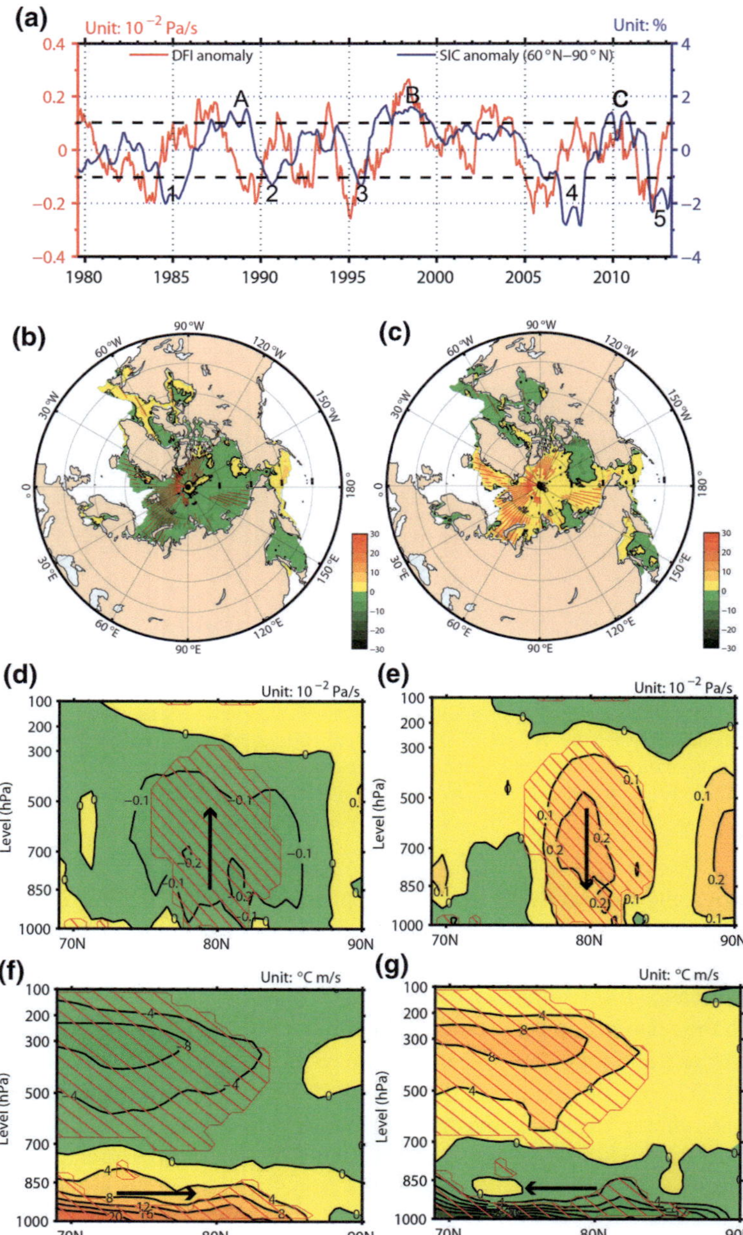
(a)
Unit: 10 −2 Pa/s
Unit: %
DFI anomaly
SIC anomaly (60 °N–90 °N)
A
B
C
1
2
3
4
5
(b)
(c)
(d)
Unit: 10 −2 Pa/s
(e)
Unit: 10 −2 Pa/s
Level (hPa)
(f)
Unit: °C m/s
(g)
Unit: °C m/s

◀ **Fig. 9.3** DFI and SIC anomalies from time series to spatial distributions. **a** Monthly mean time series of the DFI anomalies (*red line*, 10^{-2} Pa s^{-1}) and the SIC anomalies (*blue line*, %) averaged over 60–90°N after removing their linear trends from 1979 to 2013 and calculating the 15-month running mean; **b** composite average of SIC anomalies (*yellow-red shading* for positive and *green* for negative, 5% interval) over five warm cases with 15 consecutive months centered in January 1985, September 1990, September 1995, September 2007, and September 2012; Same as **b** but for **c** three cold cases centered in October 1988, June 1998, and April 2010; **d** composite average of DFI anomalies (*yellow* as descending, *green* as ascending, 0.1×10^{-2} Pa s^{-1} interval) with 15 consecutive months centered in July 1983, June 1989, December 1994, December 2005, and January 2012; Same as **d** but for **e** three descending flow anomalies centered in January 1987, April 1998, and February 2010; **f** same as **d** but for composite average of meridional temperature advection anomaly (*yellow* as warm advection, *green* as cold advection, 4 °C m s^{-1} interval); **g** same as **f** but for three periods in **e**. In **a**, the numbers denote five warm events while the *capital letters* (A, B, and C) are three clod events. In **d** and **e**, the *arrows* indicate the strengthened descending and ascending flows, respectively. In **g** and **f**, the *arrows* indicate the strengthened warm and cold advection, respectively. The *red-lined* area in the last six panels covers the region reaching the 95% confidence level for the composite anomaly. Sourced from Qian et al. (2015a) © 2015 W. Qian et al., with permission

(8/12), are associated with prior DFI peaks, indicating that the DFI is a useful early signal for seasonal SIC prediction.

When a negative SIC anomaly less than −1% persists for at least half a year, it is regarded as a warm Arctic event. Similarly, a cold Arctic event is defined here as SIC anomaly is greater than +1% for at least half a year. Among the five warm events indicated in Fig. 9.3a, two recorded September SIM events in 2007 (e.g., Kay et al. 2008; L'Heureux et al. 2008; Graversen et al. 2011) and 2012 (Bennartz et al. 2013; Parkinson and Comiso 2013) are particularly studied. The strongest warm event with the SIC anomaly <−2% persisting for 15 months was centered in September 2007 and the last warm event with the SIC anomaly <−1.5% was centered in September 2012. Similarly, the most recent four SIM events have their negative SIC anomaly peaked or centered in September.

Previous studies only focused on the last two Arctic SIM events mentioned above. Qian et al. (2015a) reported a composite result of the five warm or SIM events in the past three decades. The central month of warm events is easily determined at its negative SIC peak or the middle month of each period with SIC less than −1%. For each of the five warm events, its early signal can be found from corresponding persistent negative DFI anomalies (Fig. 9.3a). Five central months and their 15 consecutive months of anomalous DFI episodes are also determined from the anomalies DFI series using the same method (negative DFI peak or the middle month of each period with DFI less than -0.1×10^{-2} Pa/s). The five anomalous DFI episodes lead the five Arctic warm events by about 14 months in average. The composite averages of five negative SIC and DFI events are shown in Fig. 9.3b, d, respectively.

Three major cold events are also identified in Fig. 9.3a when using SIC anomaly >+1% as a threshold. A signal of positive DFI anomalies can be found before the cold event "A," while another positive DFI anomaly was strong and simultaneously coincided with the cold event "B." The composite averages of the three positive SIC and DFI anomalies are shown in Fig. 9.3c, e, respectively.

The two composites of SIC anomalies show a significant contrast, particularly in the Eastern Hemisphere (EH; Fig. 9.3b vs. Fig. 9.3c). Other significant contrasts are from the composites of DFI anomalies illustrated in Fig. 9.3d, e as well as those horizontal temperature advections in Fig. 9.3f, g. In the mid-low troposphere between 78 and 85°N, a negative DFI anomaly (Fig. 9.3d) is associated with an Arctic warm (negative SIC anomaly) pattern (Fig. 9.3b), while a positive DFI anomaly (Fig. 9.3e) is linked with an Arctic cold (positive SIC anomaly) pattern (Fig. 9.3c). The warm (Fig. 9.3f) and cold (Fig. 9.3g) horizontal advections below 850 hPa can well explain the Arctic warm (negative SIC anomaly) (Fig. 9.3b) and cold (positive SIC anomaly) (Fig. 9.3c) patterns. This finding strongly supports the notion that the Polar and Arctic cells have a strong impact on the Arctic sea ice variation through the descending flow and the horizontal temperature advection between them in the interannual timescale.

Two interesting interaction mechanisms can be summarized from previous eddy momentum and heat fluxes and Fig. 9.3. The strengthening meridional heat and momentum fluxes in high latitudes will intensify the Polar cell and weaken the Arctic cell, causing a positive DFI anomaly (Fig. 9.3e) and a cold horizontal advection (Fig. 9.3g). These lead to a positive SIC anomaly (Fig. 9.3c) in the Arctic. The reverse influence can be observed from a negative DFI (Fig. 9.3d) and a warm horizontal advection (Fig. 9.3f), resulting in a negative SIC anomaly (Fig. 9.3b). Therefore, we find that the spread of SIE is followed by a cold air outbreak with strong descending flow and surface cold advection from 80°N to the subpolar region. The result shows that the change in Arctic SIC lags the intensity variations of descending flow associated with the Polar and Arctic cells by about 2 months in the climatic seasonal cycle and by about 10 months for the interannual anomaly. Five Arctic SIM events during the past 3 decades support this finding. This work provides a new perspective in understanding the relationship between atmospheric circulation and sea-ice variations, and predicting the Arctic sea ice changes.

9.2 Seasonal Variation of the Arctic and Polar Cells[2]

The Arctic cell as a reversed and closed loop next to the Polar cell has been recently revealed in the NH (Qian et al. 2015a). Qian et al. (2016a) demonstrated that the interannual variability of SAT, precipitation, and SIC at mid- and high latitudes of the NH is associated with the Polar and Arctic meridional cell strengths during 1979–2012. They showed that there is a significant negative correlation between the Polar and Arctic cell strengths, which can well describe the recurring climatic anomalies of SAT, precipitation, and SIC in their extreme winters. The surface large-scale interannual cold-warm and dry-wet patterns across the entire 34 years

[2]Modified article originally published in Qian et al. (2016a). Published with kind permission © Springer-Verlag Berlin Heidelberg 2016. All Rights Reserved.

are directly linked with the vertical structure of height and temperature anomalies in the troposphere.

Changes in the location, strength, and width (or coverage) of any meridional circulations should influence the climate variability in circular bands or local regions. Over the past decade, studies concerned the decadal changes in the Hadley cell strength. Chen et al. (2002) and Wielicki et al. (2002) suggested that the Hadley cell had intensified in the 1990s by using outgoing longwave radiation (OLR) datasets. Quan et al. (2002), Hu et al. (2005), and Mitas and Clement (2005) showed that the Hadley cell in winter experienced a large intensification using reanalysis datasets. Although such meridional circulation variation is considered as an important part in climate change, the Hadley cell cannot directly explain the rapid Arctic warming (Screen et al. 2012) and interannual SIM events in the Arctic (Kumar et al. 2010; Graversen et al. 2011; Devasthale et al. 2013; Dong et al. 2014). As summarized by Cohen et al. (2014) and Deser and Teng (2008a, b), observational analyses have shown significant correlation between the reduced trend of Arctic sea-ice cover and the negative phase of the winter AO/NAO. Thus, climate change in the high latitudes or the NH Polar region should be directly associated with the Polar cell strength variation. However, how the Polar cell strength variation influences the cold and warm events in the Polar areas is still not clear. This is partly because the sparse observations of long-term atmospheric radiosonde and satellite in the Arctic before 1979 (Kahl et al. 1991, 1992). In the Arctic, the qualities of existing reanalysis products have improved due to the assimilated satellite and radiosonde datasets since 1979. Bromwich et al. (2007) has made a tropospheric assessment of the ERA-40, NCEP, and JRA-25 global reanalysis products in the Polar region. In particular, Bromwich et al. (2015) has reported that the biases between the Arctic system reanalysis (ASR) , the ERA-Interim, and observations are comparable throughout the troposphere in the mid- and high latitudes of the NH.

Using the NCEP R1 and NCEP R2 as well as the ERA-Interim products since 1979, Qian et al. (2015a) found that there is a fourth meridional circulation in the Arctic, named the Arctic cell. The Arctic cell is a thermally indirect and eddy-driven circulation. In the four-cell model, there is a southwesterly band showing a cyclonic circulation beyond 80°N while the northeasterly is centered at around 70°N, which is different from the three-cell model. Therefore, the four-cell model in the NH should have entirely different influence on climate change and extreme climatic events in the Polar region, compared with the traditional three-cell model. Qian et al. (2015b, 2016c) noted that the surface cold and warm extremes in the synoptic scale directly correspond to negative and positive centers of geopotential height anomalies, respectively, in the upper troposphere over China. This section examines whether similar relationships or robust patterns are also valid to indicate climatic anomalies in the NH Polar region on the interannual timescale since 1979.

Five reanalysis products, a precipitation dataset, and a gridded SIC dataset, as well as the AO/NAO indices are used. The first two reanalysis products are the monthly mean NCEP R1 and the monthly mean NCEP R2 (Kalnay et al. 1996; Kanamitsu et al. 2002). The third is the monthly mean ERA-Interim product (Dee et al. 2011). The fourth is the monthly mean JRA-55 product (Ebita et al. 2011).

The fifth is the interim version of monthly mean ASR dataset from 2000 to 2010, a 3-DVAR assimilation product which has been designed and optimized for Polar region obtained from the website (http://rda.ucar.edu/datasets/ds631.4/#!description). The ASR dataset has a higher horizontal resolution of 30 km by 30 km and 34 vertical levels beyond entire 50°N (Bromwich et al. 2015), which blends modeling and observations to provide a high-resolution description in space and time of the sea-air-ice-land surface system of the great Arctic since 2000.

The Climate Prediction Center (CPC) Merged Analysis of Precipitation (CMAP) (Xie and Arkin 1997) is used to examine the climatic precipitation rate and its anomalous distribution in the NH. The seventh dataset is the monthly mean gridded SIC from the HadISST1 version during 1979–2012 (Rayner et al. 2003). The normalized time series of AO/NAO indices are derived from the website (http://www.cpc.ncep.noaa.gov/products/precip/CWlink/daily_ao_index/teleconnections.shtml). The monthly mean AO index series is constructed by projecting the monthly mean 1000-hPa height anomalies onto the leading empirical orthogonal function (EOF) mode. To obtain the teleconnection patterns of NAO, the rotated principle component analysis (PCA) technique (Barnston and Livezey 1987) is applied to monthly mean standardized 500-hPa height anomalies from the Climate Data Assimilation System (CDAS) in the analysis region 20°N–90°N between January 1950 and December 2000. The monthly mean NAO indices are calculated using the least squares regression approach to represent the combination of teleconnection patterns that account for the most spatial variance of the observed monthly mean anomalous map.

Global tropospheric- and stratospheric-monthly mean geopotential height, temperature, and wind relative anomalies (RAs) can be extracted by Eq. 9.1.1. The monthly mean wind vector $\vec{V}_m(\lambda, \varphi, p, y) = \vec{i}u_m(\lambda, \varphi, p, y) + \vec{j}v_m(\lambda, \varphi, p, y) + \vec{k}\omega_m(\lambda, \varphi, p, y)$ consists of three components: zonal wind velocity u, meridional wind velocity v, and vertical pressure velocity ω (positive downward). A zonally averaged monthly mean climatic wind vector in a specific month m of a year y is calculated by

$$\left[\tilde{\vec{V}}_m(\varphi, p)\right] = \sum_{y=1}^{N} \sum_{\lambda=1}^{K} \vec{V}_m(\lambda, \varphi, p, y)/(K \times N) \tag{9.2.1}$$

where N = 30 years from 1981 to 2010, K is the total grid number along a zonal circle. Similarly, the zonally averaged seasonal mean and annual mean climatic wind vectors can also be calculated from the monthly mean climatic wind vector. The climatic meridional cells such as the Polar and Arctic cells can be illustrated by monthly mean wind $\left[\tilde{\vec{V}}_m(\varphi, p)\right]$ or seasonal mean wind $\left[\tilde{\vec{V}}_m(\varphi, p)\right]$.

Meridional cells can be visualized by wind vectors, but the zonally averaged monthly mean vertical pressure velocity is too small when comparing with the zonally averaged monthly mean meridional velocity. Therefore, the meridional-vertical section is used to depict the meridional cell structures. It is simply drawn using the meridional velocity and vertical pressure velocity magnified by 1000 times, so there is no boundary effect near the North Pole. However, this structure is not able to quantitatively illustrate the strength of each cell. Usually, zonally averaged meridional cells are considered as non-divergent flow so its strength can be calculated using the meridional MSF, which has been used in many previous studies (Lindzen and Hou 1988; Mitas and Clement 2005; Hu and Fu 2007; Chen et al. 2014). The meridional MSFs $\psi(\varphi, p)$, satisfying the zonally averaged continuity equation in spherical coordinates, can be calculated at each pressure level p, and latitude φ using the meridional wind integration (Stachnik and Schumacher 2011). The detailed calculation can be referred to Qian et al. (2015a, 2016a, b).

The meridional MSFs and spatial circulation structures of the four meridional cells in the NH have been illustrated by Qian et al. (2016a). The three meridional cells beyond 30°N are shown in Fig. 9.4 using a 30-year (1981–2010) average of the monthly mean ERA-Interim data. It is noted that the Ferrel cell covers from 30°N to 60°N, and the Polar and Arctic cells are climatologically located beyond 60°N in the subpolar zone and near the North Pole, respectively. A westerly core is observed at the upper troposphere and its location shifts seasonally between 30°N and 40°N. Weak easterlies are observed mostly below 850 hPa in the high latitudes in spring and winter (Fig. 9.4a, d). Only a shallow layer of weak easterly is centered around 70°N in summer and autumn (Fig. 9.4b, c). The upper layer strong westerly and the lower layer weak easterly in Fig. 9.4e is well summarized in Fig. 2.13b, which describes the four-cell model. A pair of meridional cells with opposite directions from 60°N to 90°N (Fig. 9.4a–e) and opposite signs in MSF values (Fig. 9.4f–j), respectively, indicate the existence of the Polar cell and the Arctic cell. It is noted that when the Polar cell is stronger in winter and spring, the easterly can extend northward from around 70°N to the North Pole in the lower troposphere. Similar spatial distributions of the Polar and Arctic cells can also be derived from other four reanalysis products (figures not shown).

From Fig. 9.4, the central altitude of the Arctic cell is at approximately 850 hPa, which is lower than that of the Polar cell at around 600 hPa. Under the Arctic cell center, the zonally averaged meridional flow converges near the North Pole, so there should be a climatic southerly in the lower troposphere. This feature has been revealed in Fig. 2.12. Figure 9.4 shows that the climatic strength and shape of the Arctic cell vary seasonally with the Polar cell. The strength anomaly of a cell is defined as the deviation of the central MSFs relative to its climatology. The Arctic cell strength and the MSF value are opposite in sign, so the Arctic cell strength anomaly is the MSF anomaly multiplied by −1 for convenience in the following analysis. Figure 9.5 shows the normalized anomalies (NAs) of the two cells' strengths in terms of seasonal-mean and annual mean values. The NA is a ratio of the RA to its standard deviation (SD). If the SD is defined as 1, the NA becomes the RA. There are significant negative correlations ($R = -0.64$, -0.75, and -0.72 in spring, summer, and winter, respectively) between seasonal-mean strength anomalies of the Polar and Arctic cells from 1979 to 2012 as well as their annual mean anomalies ($R = -0.59$), except a weak correlation ($R = -0.32$) in autumn.

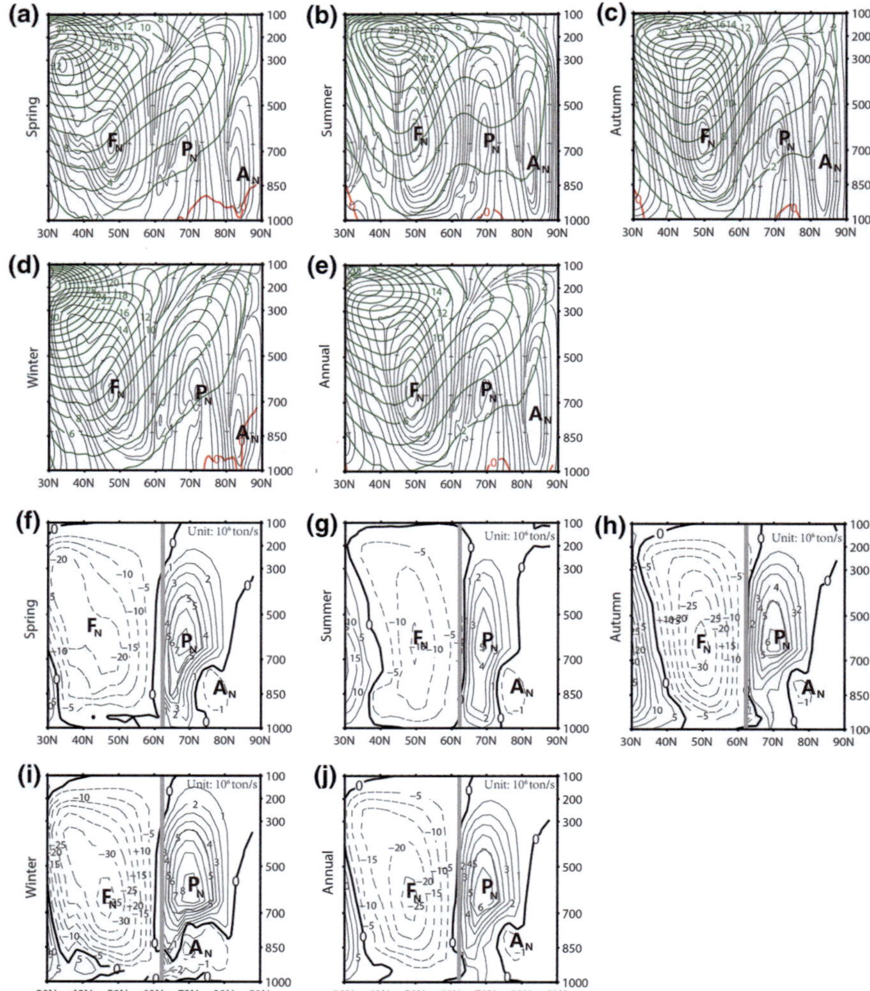

Fig. 9.4 Zonally averaged climatic component: vertical-meridional streamline (*black lines* with *arrows*) and westerly velocity (*green* contour for the westerly velocity, 2 m s^{-1} interval; the easterly velocity is very weak and located below the *red zero line*) using a 30-year (1981–2010) average of the monthly mean ERA-Interim data in boreal **a** spring, **b** summer, **c** autumn, **d** winter, and **e** annual mean; as well as meridional MSFs (10^6 ton s^{-1}) in boreal **f** spring, **g** summer, **h** autumn, **i** winter, and **j** annual mean. "F_N," "P_N," and "A_N" represent the Ferrel, Polar, and Arctic cells. The MSF interval is 5 × 10^6 $tons^{-1}$ between 30°N–62.25°N and 1 × 10^6 ton s^{-1} between 62.25°N and 90°N, separated by the *thick gray line* near 62.25°N. Sourced from Qian et al. (2016a) © Springer-Verlag Berlin Heidelberg 2016, with permission

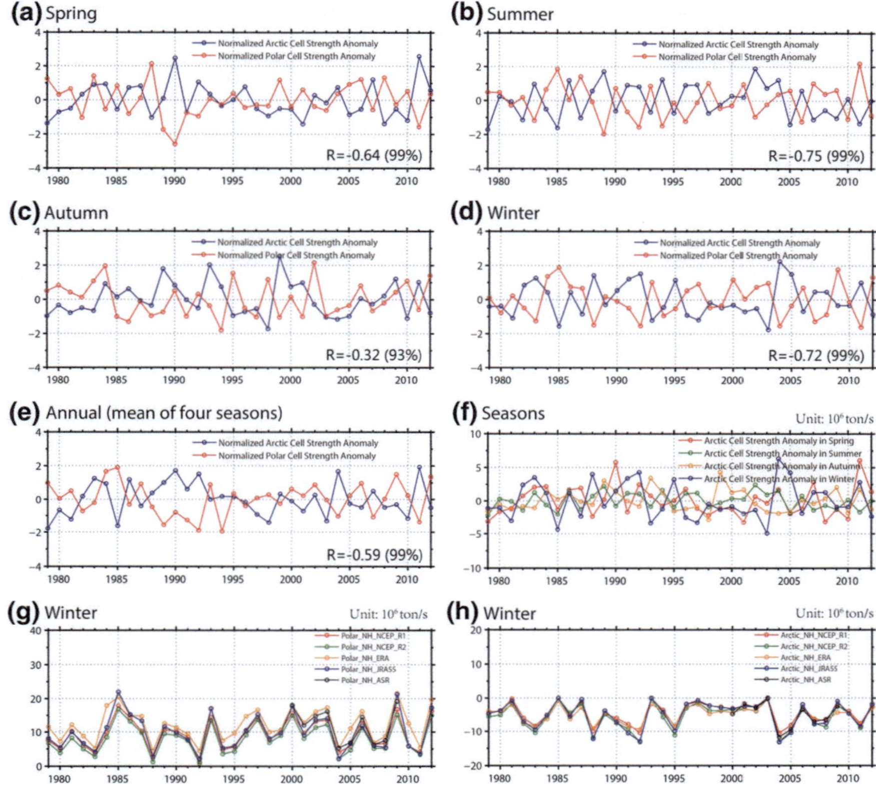

Fig. 9.5 Strength NA series (unit: SD) of the Polar cell (*red line*) and the Arctic cell (*blue line*) in boreal **a** spring, **b** summer, **c** autumn, **d** winter, and **e** annual mean after removing the linear trends from 1979 to 2012 using the monthly mean ERA-Interim dataset. Panel **f** is the same as panels **a**–**d** except for the Arctic cell strength RA series in four boreal seasons. Panel **g** is the non-detrended central MSF series (unit: 10^6 ton/s) of the Polar cell in boreal winter using five different reanalysis datasets. Panel **h** is the same as **g** except for the Arctic cell. In each panel from **a** to **e**, *R* is the correlation coefficient of the two series with corresponding confidence level. Sourced from Qian et al. (2016a) © Springer-Verlag Berlin Heidelberg 2016, with permission

9.3 Meridional Cell Anomalies and Climatic Anomalous Patterns[3]

Ideally in an aqua-planet, the seasonal mean and annual mean strength variability of the Arctic and Polar cells should produce circular patterns of SAT and precipitation anomalies. However, when sea–land contrast is taken into account, patterns will become complicated. A comparison of the climatic anomalies at mid- and high

[3]Modified article originally published in Qian et al. (2016a). Published with kind permission © Springer-Verlag Berlin Heidelberg 2016. All Rights Reserved.

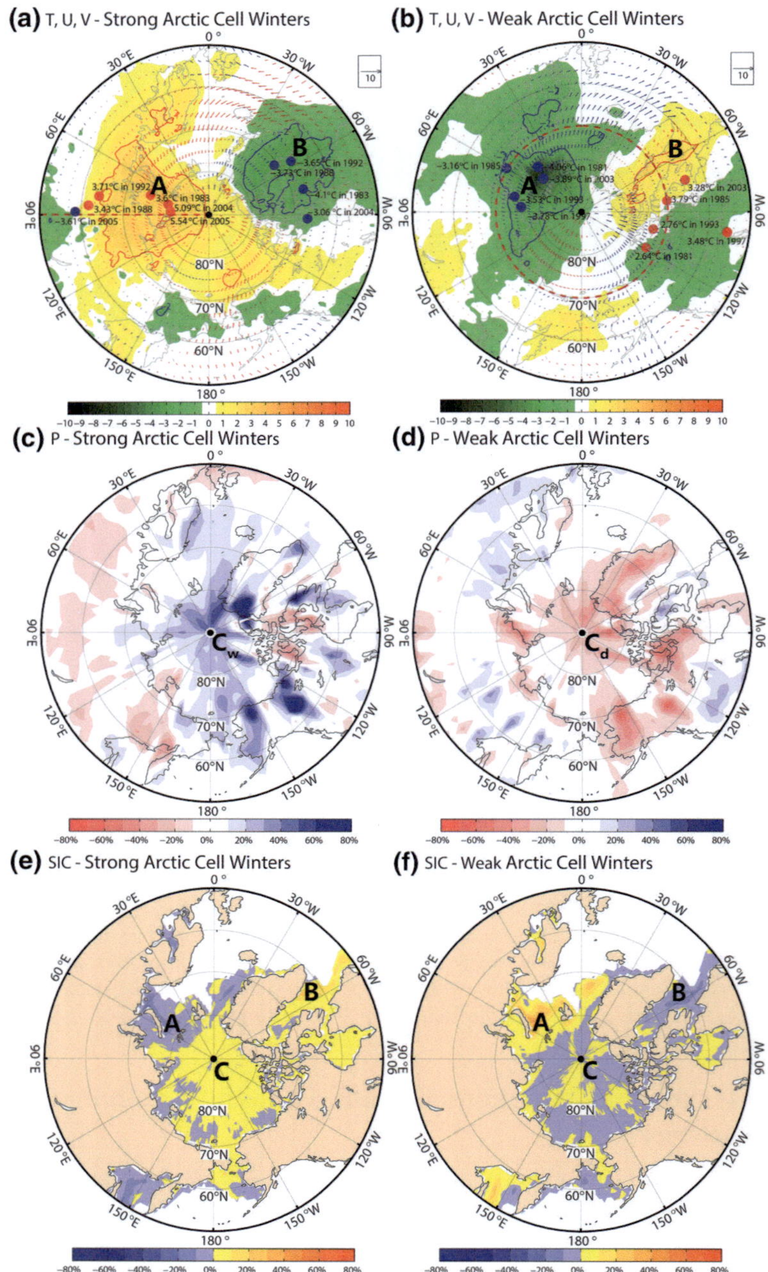
(a) T, U, V - Strong Arctic Cell Winters
(b) T, U, V - Weak Arctic Cell Winters
A
B
80°N
70°N
60°N
(c) P - Strong Arctic Cell Winters
(d) P - Weak Arctic Cell Winters
C_w
C_d
(e) SIC - Strong Arctic Cell Winters
(f) SIC - Weak Arctic Cell Winters
C

◀ **Fig. 9.6** **a** Composite SAT RAs beyond 50°N at the 2 m level (*shading*, 2 °C interval) and surface wind RAs at the 10 m level (*blue* and *red arrows* as northerly and southerly anomalies respectively, m s^{-1}) in the five strongest Arctic cell winters using the ERA-Interim dataset. Panel **b** is the same as panel **a** except for the five weakest Arctic cell winters. Panels **c** and **d** are the same as panels **a** and **b** but for precipitation RAs (percentage, 10% interval). Panels **e** and **f** are the same as panels **a** and **b** but for the SIC RAs (percentage, 10% interval). The *red-dashed line* in **a** is along 90°E and in **b** along 70°N. In **a** and **b**, the *red* and *blue dots* denote the central locations of regions with SAT RAs respectively above 2 °C and below −2 °C for each extreme winter. *Capital letters* "A" and "B" indicate regions of composite SAT RAs above 2 °C (*red line*) or below −2 °C (*blue line*). Regional-mean SAT RA value and the corresponding year are marked beside each *dot*, such as 3.71 °C in 1992. Letters "C_w" and "C_d" indicate the centers with wet and dry anomalies in **c** and **d**. Letters "A," "B," and "C" indicate the centers with negative and positive SIC anomalies in **e** and **f**. Sourced from Qian et al. (2016a) © Springer-Verlag Berlin Heidelberg 2016, with permission

latitudes of the NH among years of extreme Arctic and Polar cell strengths was made by Qian et al. (2016a). The largest relative variation of the Arctic cell strength is in winter, indicated by the MSF RAs in Fig. 9.5f. Thus, extreme winters are selected according to the Arctic cell strength NAs in Fig. 9.5d and RAs in Fig. 9.5f. The five strongest Arctic cell winters are 1983, 1988, 1992, 2004, and 2005 while the five weakest Arctic cell winters are 1981, 1985, 1993, 1997, and 2003. The five strongest Polar cell winters are 1984, 1985, 2000, 2009, and 2012 while the five weakest Polar cell winters are 1988, 1992, 2004, 2007, and 2011. These extreme winters and their interannual variability of Arctic and Polar cell strengths are also confirmed by other four reanalysis products as shown in Fig. 9.5g, h. The winterly central MSF series of both the Polar and Arctic cells using all the four reanalysis products have the same amplitude and phase basically in the annual timescale during 1979–2012. The winterly central MSF series of the Arctic and Polar cells (black lines in Fig. 9.5g, h) derived from the ASR dataset are comparable with those from other four reanalysis products during 2000–2009. This confirms the recent comparison of the regional ASR and the global ERA-Interim product for the Arctic (Bromwich et al. 2015).

Figure 9.6 shows the composite spatial distributions of SAT and precipitation RAs in the extreme winters of Arctic cell strength. Figure 9.6a illustrates the composite SAT RAs of the five strongest Arctic cell winters. The two capital letters "A" and "B" indicate a large positive (warm) area and a large negative (cold) area of composite SAT RAs above 2 °C and below −2 °C beyond 50°N, respectively. The central locations of these regions for each extreme winter are calculated using Eqs. 7.1.1 and 7.7.2 and marked by red (positive SAT RAs) and blue (negative SAT RAs) dots. All the five positive central locations are near the area "A" over northern Siberia in Fig. 9.6a. Four negative central locations are located between northeastern Canada and the Greenland. An exception is in winter 2005, showing a central location of larger negative SAT RA region around 90°E and 60°N, with a small center near "B" (figure not shown). In the composite SAT RAs of the five weakest Arctic cell winters (Fig. 9.6b), the five negative central locations are located within the green area "A," but the five positive central locations are scattered from northeastern Canada to the Greenland around "B." In other words, all the positive and negative central locations are stable in the subpolar region near Russia but scattered in the Greenland side. It seems interesting that the anomalies of southerly and northerly, respectively, contribute to the positive and negative SAT RAs in high latitudes (Fig. 9.6a, b). Separately, southerly anomalous flow and warm anomalies over northern Eurasia are directly associated with strong Arctic cell in the winters, whereas opposite wind and SAT anomalies are observed in the winters of weak Arctic cell.

A wet center "C_w" and a dry center "C_d" over the Arctic are, respectively, observed in the composites of the strongest and the weakest Arctic cell winters (Fig. 9.6c, d). Spatially, positive precipitation RAs in the Arctic and negative precipitation RAs in Eurasia are observed in the strongest Arctic cell winters (Fig. 9.6c). An opposite pattern of precipitation RAs is shown in the weakest Arctic cell winters (Fig. 9.6d). The positive precipitation RAs in the Arctic (Fig. 9.6c) correspond well with the surface anomalous convergence of southerlies in the subpolar regions (Fig. 9.6a). The relationship between the negative precipitation RAs in the Arctic and anomalous northerly winds can be also found in Fig. 9.6b, d.

Opposite spatial distributions of SIC RAs are also clearly observed in the composites of the strongest and the weakest Arctic cell winters, respectively. In Fig. 9.6e, "A" denotes a negative area of SIC RAs while positive SIC RA areas are indicted by "B" and "C." The opposite spatial distribution is illustrated in Fig. 9.6f. Three major areas "A," "B," and "C" with opposite spatial distributions can be explained by the comprehensive effects of SAT and precipitation anomalies. The SIC RAs in areas "A" and "B" are sensitive to the SAT RAs in the subpolar regions such as in the Barents Sea and the Labrador Sea. The SIC RA area "C," where climatic SATs are lower than 250 K (or lower than −23 °C in Fig. 2.12d), corresponds well with the precipitation RAs in the central Arctic Ocean, implying that snowfall would help increase SIC.

It is noted that the central Arctic precipitation anomaly is affected by the SIC RAs in the open ice areas over 70°N–80°N. In the strongest Arctic cell winters, the negative SIC RAs in the open ice areas ("A" in Fig. 9.6e) enhance evaporation in that area. Then, moisture is transported from the open ice areas to the Arctic region by the anomalous southerly (Fig. 9.6a), and thus increases the precipitation in the central Arctic Ocean (Fig. 9.6c). Reversed SIC and wind anomalies in "A" are observed in the weakest Arctic cell winters, and thus decrease precipitation in the central Arctic Ocean (Fig. 9.6b, d, f). This shows that the evaporation and moisture transport in the open ice areas contribute to the precipitation in the Arctic region.

Surface variable anomalies should be theoretically opposite between the Arctic and the subpolar zone without considering sea–land difference in the NH high latitudes. In reality, the sea–land difference leads to the uneven distribution of warm and cold anomalies in Fig. 9.6a, b. Particularly, two negative (green) areas of SAT RAs are separated by positive (yellow) areas of SAT RAs along the 70°N (Fig. 9.6b). A contrast of positive (yellow) and negative (green) SAT anomalies is observed along the 90°E (Fig. 9.6a). Therefore, we should explore the vertical features of geopotential height and temperature RAs along these two lines (Fig. 9.7). In Fig. 9.7a, a positive center (H_N) and a negative center (L_S) of height

RAs around 300 hPa are detected, respectively, in the subpolar region (70°N) and the middle latitudes (50°N). A negative center "C_N" and a positive center "W_N" of temperature RAs are, respectively, located above and below the positive center "H_N" of height RAs at the upper troposphere in the vertical structure near 70°N. A reversed pattern can be observed in the vertical structure near 50°N in Fig. 9.7a. The temperature and pressure (or height) RAs vertically satisfy the hydrostatic equilibrium. Therefore, the warm center "W_N" of temperature RA (2.89 °C) in the north is dynamically correlated to the positive center "H_N" of height RAs (4.03 × 10 gpm) near 70°N, while the cold center "C_S" of temperature RA (−1.08 °C) in the south is dynamically related to the negative center "L_S" of height RAs (−2.24 × 10 gpm) near 50°N. A connection is clear from the SAT RAs in Fig. 9.6a and the vertical structure of height and temperature RAs in Fig. 9.7a.

The spatial connections among these variables (SAT, precipitation, SIC, height, and temperature anomalies, as well as strength variations of the Arctic and Polar cells) correlate with each other from the surface, the troposphere, and the stratosphere (Figs. 9.6 and 9.7). In daily synoptic-scale analysis, anomalous systems

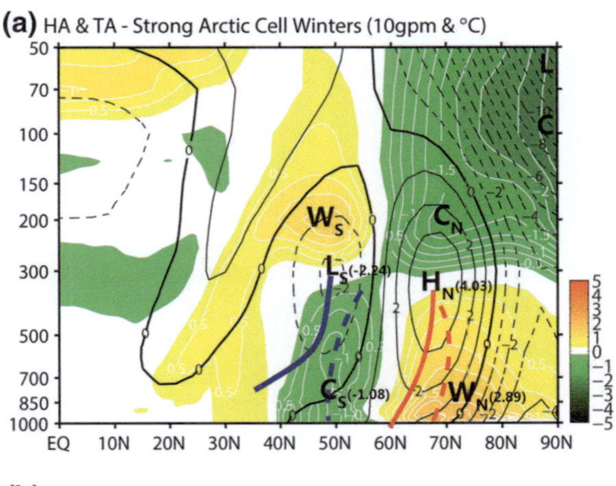

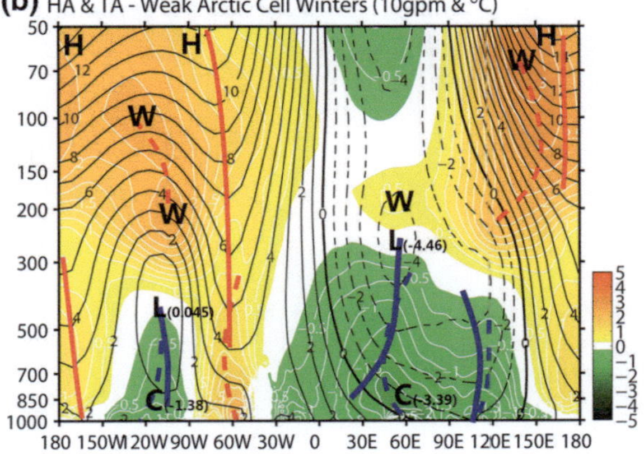

Fig. 9.7 Vertical cross sections of composite height RAs (*solid* and *dashed lines*, 1 × 10 gpm interval) and temperature RAs (*shading*, 0.5 °C interval) in **a** the five strongest Arctic cell winters along 90°E and **b** the five weakest winters along 70°N using the ERA-Interim dataset. Letters "H/L" and "W/C" indicate the "high/low" centers of height RAs and the "warm/cold" centers of temperature RAs, respectively, with their central values marked beside. Subscripts "S" and "N" in **a** indicate south and north. The *thick solid-* and *thick-dashed blue lines* are the axes of negative height and temperature RAs respectively, while the *thick solid-* and *thick-dashed red lines* are the axes of positive height and temperature RAs. Sourced from Qian et al. (2016a) © Springer-Verlag Berlin Heidelberg 2016, with permission

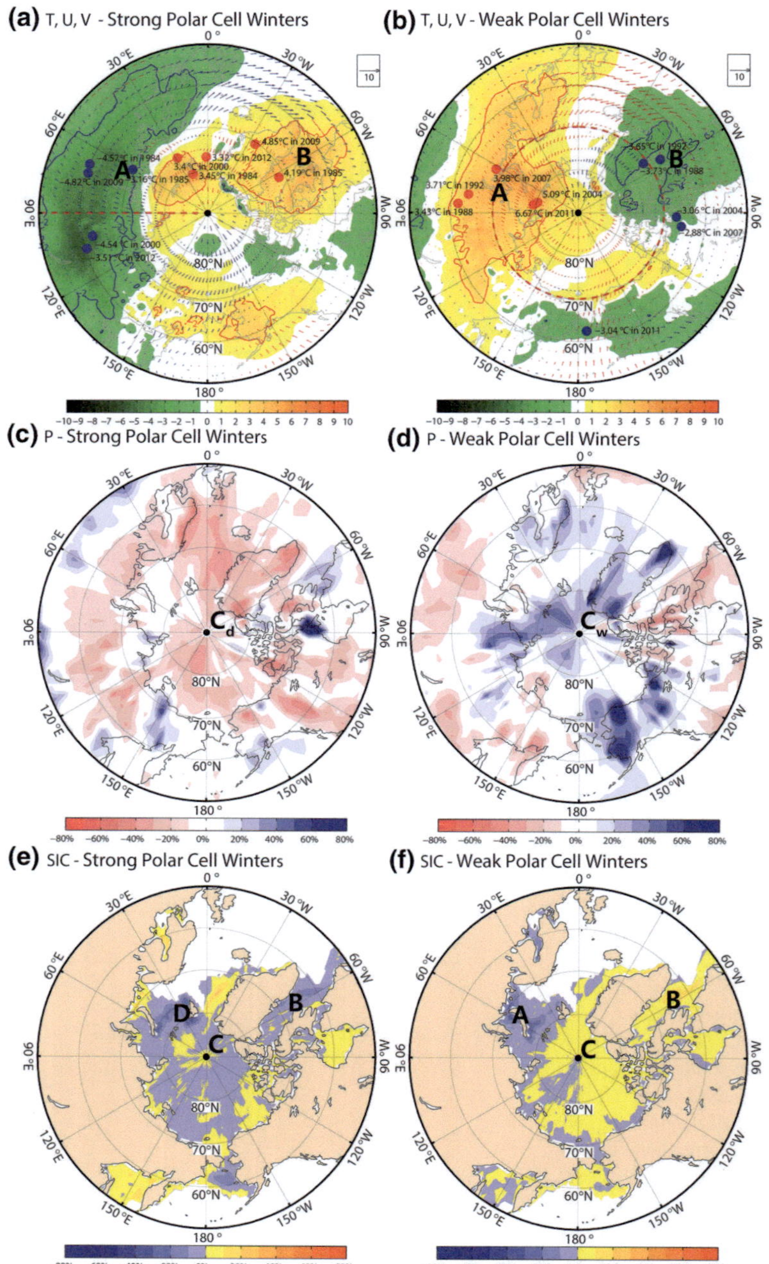

Fig. 9.8 Same as Fig. 9.6 but for the five strongest Polar cell winters and the five weakest Polar cell winters. Sourced from Qian et al. (2016a) © Springer-Verlag Berlin Heidelberg 2016, with permission

depicted by height and temperature anomalies are some internal turbulence in the atmosphere, which can be traced several days ago to indicate local surface heat wave and cold surge events (Qian et al. 2015b, 2016c). The horizontal distribution of original SATs and the original vertical structure can be distinguished from daily height and temperature RAs, which satisfy hydrostatic relationship. Similarly in the seasonal timescale, the two negative areas of SAT RAs separated by positive areas of SAT RAs in Fig. 9.6b can be explained by the two cold air columns below the two negative centers of height RAs in Fig. 9.7b. For each case, a positive (or negative) center of height RAs at the upper troposphere and a positive (or negative) center of temperature RAs in the lower troposphere are always found above the surface regional positive (or negative) center of SAT RAs.

Composite analyses on SAT, surface wind, precipitation, and SIC RAs in the five strongest Polar cell winters as well as the five weakest Polar cell winters are shown in Fig. 9.8. The opposite spatial distribution of SAT RAs is found between northern Eurasia ("A") and the region surrounding the Greenland ("B") with opposite wind directions (Fig. 9.8a, b). The influences of the Polar and Arctic cell

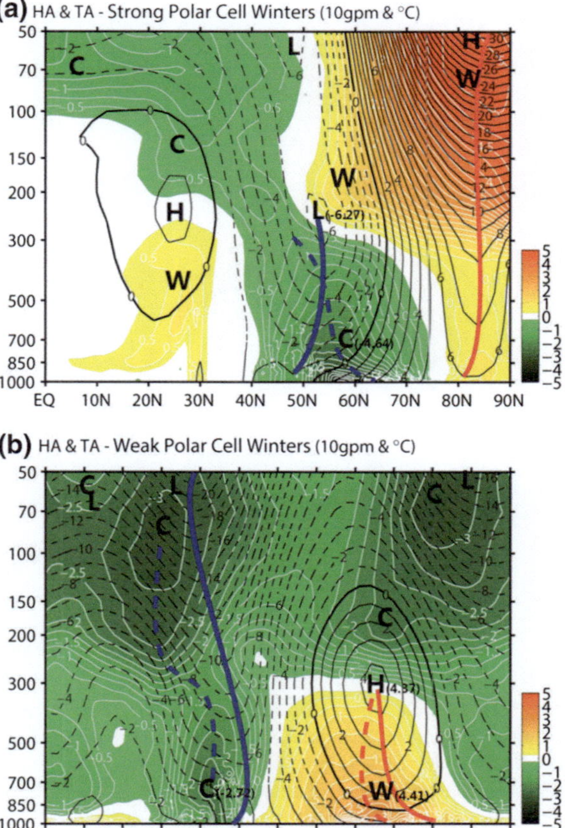

Fig. 9.9 Same as Fig. 9.7 but for **a** the five strongest Polar cell winters and **b** the five weakest Polar cell winters. Sourced from Qian et al. (2016a) © Springer-Verlag Berlin Heidelberg 2016, with permission

strengths on SAT, wind, and precipitation RAs are mostly opposite. Comparisons are found from Fig. 9.8a–d and from Fig. 9.6a–d. An opposite distribution is clearly observed between the precipitation RAs in Fig. 9.8c, d. In Fig. 9.8e, SIC RAs of three negative areas, "B," "C," and "D," are located within the warm domains in Fig. 9.8a and the negative precipitation RAs (Fig. 9.8c). In Fig. 9.8f, negative area "A" and positive area "B" of SIC RAs are directly influenced by

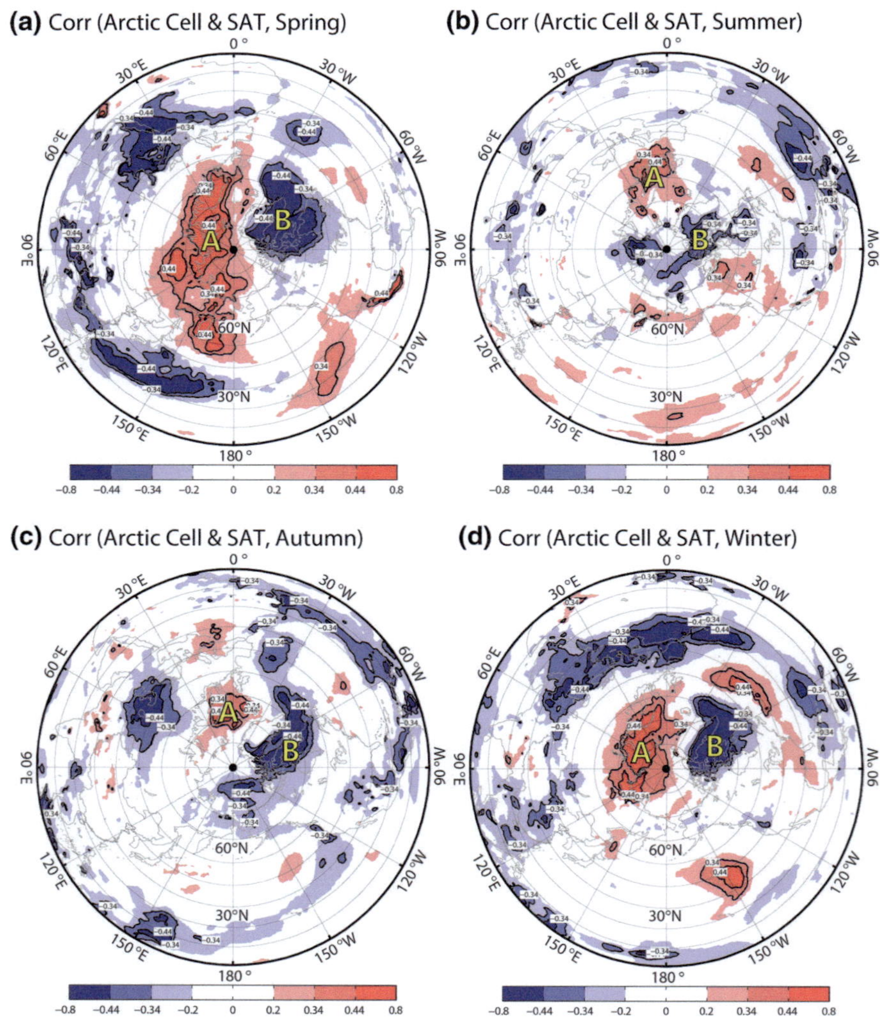

Fig. 9.10 Correlation coefficients between the Arctic cell strength and SAT RAs in boreal **a** spring, **b** summer, **c** autumn, and **d** winter in the NH after removing their trends from 1979 to 2012 using the ERA-Interim dataset. The isolines of 0.34/−0.34 and 0.44/−0.44 show their correlations reaching the 95 and 99% confidence levels. Letters "A" and "B" indicate two major areas of significant correlation. Sourced from Qian et al. (2016a) © Springer-Verlag Berlin Heidelberg 2016, with permission

warm area "A" and cold area "B" in the subpolar regions (Fig. 9.8b), while positive area "C" corresponds well with positive precipitation RAs in the central Arctic (Fig. 9.8d), implying that snowfall helps to increase SIC in the central Arctic. As in Qian et al. (2016a), the strong Polar cell strengthens descending flow and northerly, so there is a continental-scale cold region in northern Eurasia (Fig. 9.8a). A reversed distribution is identified in Fig. 9.8b.

As in Fig. 9.7, vertical-section composites of height and temperature RAs along 90°E and 70°N in the five strongest and five weakest Polar cell winters are,

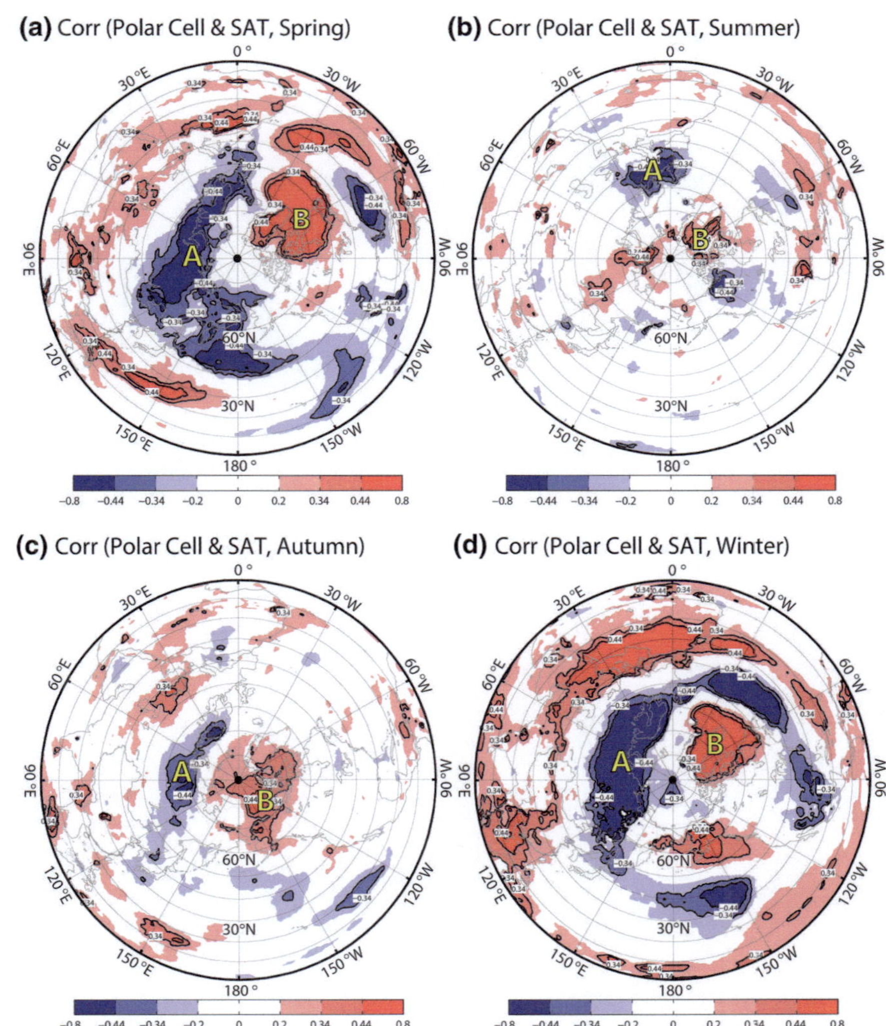

Fig. 9.11 Same as Fig. 9.10 but for the correlation coefficients between the Polar cell strength and SAT RAs in boreal **a** spring, **b** summer, **c** autumn, and **d** winter. Sourced from Qian et al. (2016a) © Springer-Verlag Berlin Heidelberg 2016, with permission

respectively, shown in Fig. 9.9a, b. Along 90°E, an axis of negative temperature RAs extends downward below the center (-6.27×10 gpm) of negative height RAs at 50°N and 250 hPa in the strongest Polar cell winters (Fig. 9.9a). The negative area and the positive area of SAT RAs along 90°E in Fig. 9.8a are well correlated with the downward extension of a cold air column and a warm air column, respectively, in middle and high latitudes (Fig. 9.9a). Along 70°N, a warm (cold) center of SAT RAs (Fig. 9.8b) corresponds well with a warm (cold) air column below a positive (negative) center of height RAs (Fig. 9.9b). The

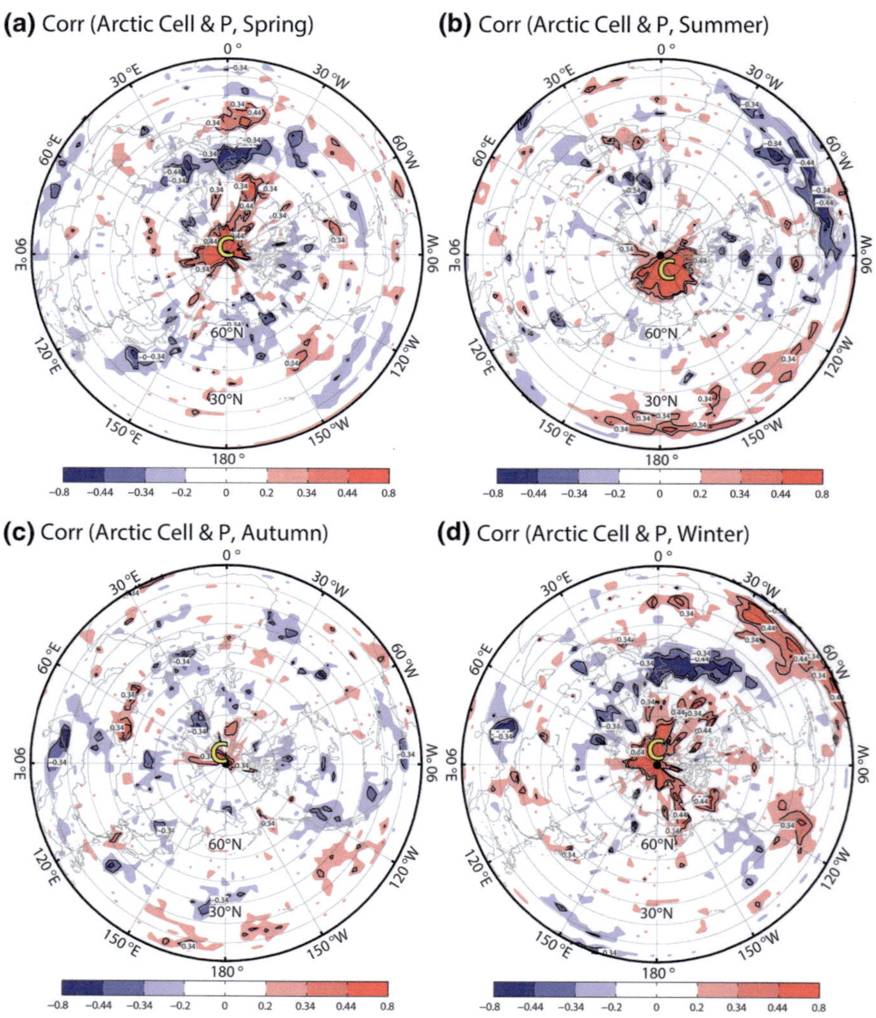

Fig. 9.12 Same as Fig. 9.10 but for the correlation coefficients between the Arctic cell strength and precipitation RAs in boreal **a** spring, **b** summer, **c** autumn, and **d** winter. Sourced from Qian et al. (2016a) © Springer-Verlag Berlin Heidelberg 2016, with permission

meridional contrast of height and temperature RAs associated with the Arctic and Polar cell strengths are illustrated in Fig. 9.6a (Fig. 9.8a) and Fig. 9.7a (Fig. 9.9a), while the zonal deviations of height and temperature anomalies associated with sea–land distribution are observed in Fig. 9.6b (Fig. 9.8b) and Fig. 9.7b (Fig. 9.9b). Since the Polar cell is stronger than the Arctic cell, it is expected that the temperature anomalies associated with the extreme Polar cell winters (Fig. 9.9) are much stronger than the corresponding anomalies in the extreme Arctic cell winters (Fig. 9.7).

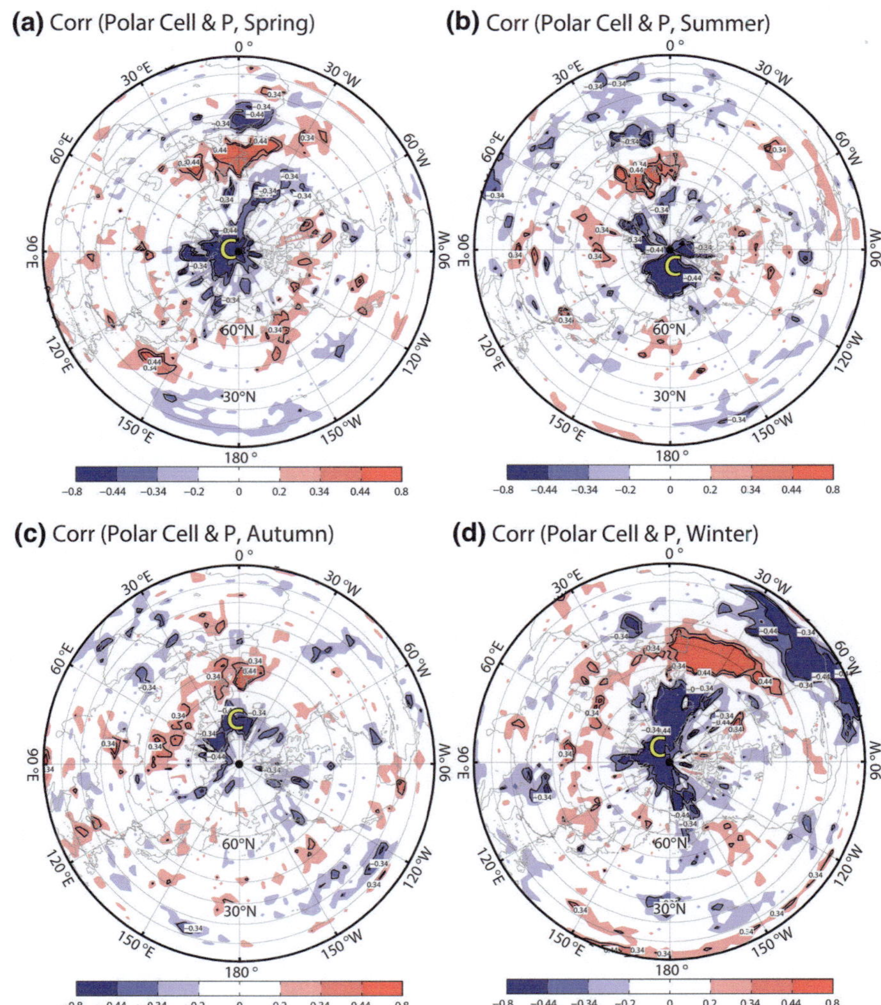

Fig. 9.13 Same as Fig. 9.10 but for the correlation coefficients between the Polar cell strength and precipitation RAs in boreal **a** spring, **b** summer, **c** autumn, and **d** winter. Sourced from Qian et al. (2016a) © Springer-Verlag Berlin Heidelberg 2016, with permission

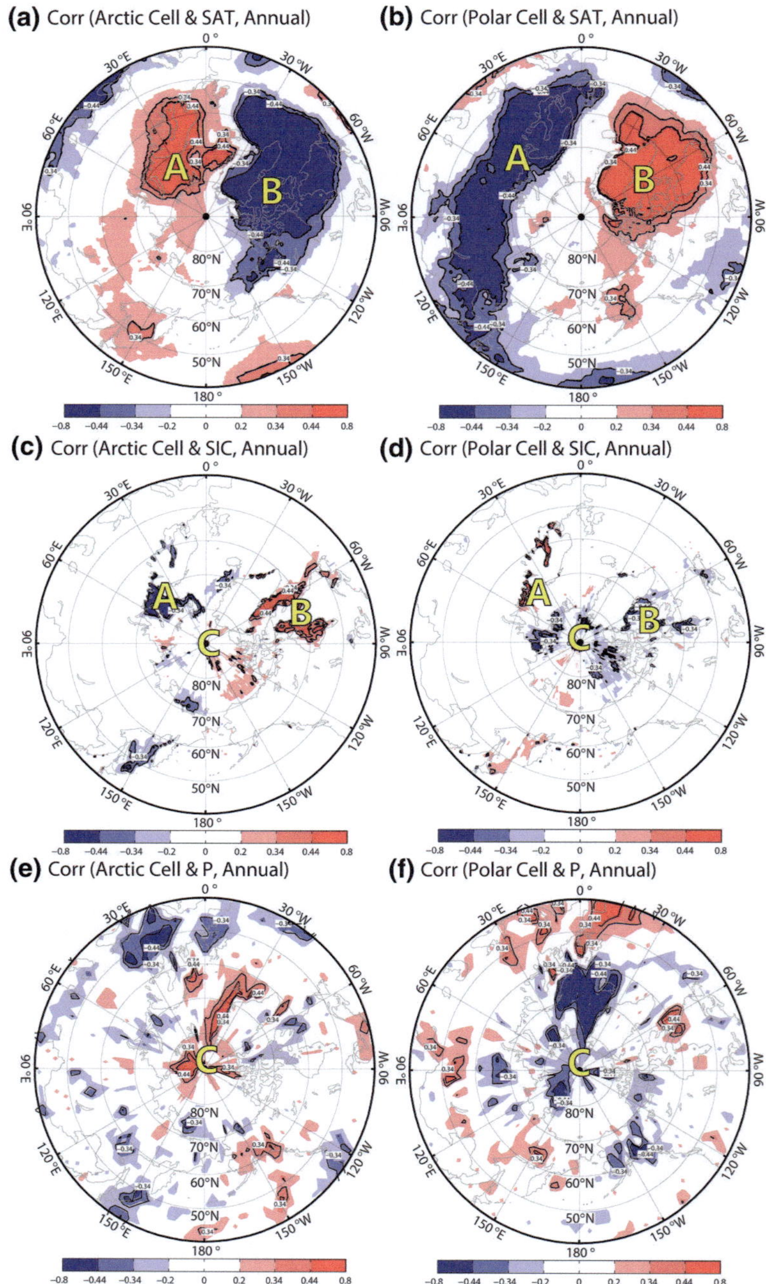
(a) Corr (Arctic Cell & SAT, Annual)
(b) Corr (Polar Cell & SAT, Annual)
(c) Corr (Arctic Cell & SIC, Annual)
(d) Corr (Polar Cell & SIC, Annual)
(e) Corr (Arctic Cell & P, Annual)
(f) Corr (Polar Cell & P, Annual)
A
B
C
0°
30°E
30°W
60°E
60°W
90°E
90°W
120°E
120°W
150°E
150°W
180°
80°N
70°N
60°N
50°N
−0.8 −0.44 −0.34 −0.2 0 0.2 0.34 0.44 0.8

◀ **Fig. 9.14** The annual mean correlation coefficients between **a** the Arctic cell strength and SAT RAs, **b** the Polar cell strength and SAT RAs, **c** the Arctic cell strength and SIC RAs, **d** the Polar cell strength and SIC RAs, **e** the Arctic cell strength and precipitation RAs, and **f** the Polar cell strength and precipitation RAs beyond 40°N. Letters "A," "B," and "C" indicate major areas of significant correlation. The isolines of 0.34/−0.34 and 0.44/−0.44 show their correlations reaching the 95 and 99% confidence levels. Sourced from Qian et al. (2016a) © Springer-Verlag Berlin Heidelberg 2016, with permission

To examine the relationships of the two meridional cells in the NH high latitudes with surface climatic anomalies at the interannual timescale, long-term linear trends of all variables and indices from 1979 to 2012 are removed. In Figs. 9.6 and 9.8, comparable patterns of SAT and precipitation RAs associated with the strongest and weakest Arctic and Polar cells are only given beyond 50°N. Figure 9.10 illustrates the distributions of correlation coefficients between the Arctic cell strength and SAT RAs in four seasons. The spatial distributions in spring and winter are similar compared to those in other two seasons, except of their correlation significant levels. In the NH, a positive region of correlation over 0.34, reaching the 95% confidence level in the high-latitude EH, separates a negative region in the high-latitude Western Hemisphere (WH) and a scattered negative zone in the EH mid-low latitudes (Fig. 9.10a, d). This pattern of correlation between the Arctic cell strength and SAT RAs in spring and winter changes in summer and autumn, because thermal contrast varies with season. In all four seasons, a negative correlation center "B" always stands over the Greenland in the WH and a positive zone "A" near northern Eurasia.

Opposite distributions of correlation coefficients between the Polar cell strength and SAT RAs are also found in spring and winter. A positive area centered over the Greenland is surrounded by a scattered negative circular band and a scattered positive circular band, respectively, in high and middle-low latitudes (Fig. 9.11a, d). Correlation distributions between the Polar cell strength and SAT RAs in summer and autumn are also similar to that in spring, but with weaker scattered positive and negative correlation regions (Fig. 9.11b, c). These significant positive (negative) correlation areas "A" in Fig. 9.10 (Fig. 9.11) surround a center "B" located over the Greenland.

Correlation between the Arctic cell strength and precipitation RAs in all seasons is shown in Fig. 9.12. A positive correlation area is always centered in the Arctic (indicated by letter "C") and surrounded by negative correlation spots in mid-high latitudes in all seasons. For the Polar cell, a negative correlation area indicated by the letter "C" is also centered in the Arctic and surrounded by positive correlation spots in mid-high latitudes in all seasons (Fig. 9.13). These positive correlation spots, which are associated with ascending flow over the southern boundary of the Polar cell, move from 60°N in summer to 50°N in winter.

The annual mean correlation coefficients between the Arctic cell strength and SAT RAs as well as between the Polar cell strength and SAT RAs were also calculated by Qian et al. (2016a). Evidently, a negative correlation area is centered over the Greenland ("B") in the WH and a positive correlation ("A") in the EH between the Arctic cell strength and SAT RAs in the interannual timescale (Fig. 9.14a). A reversed relationship of the Polar cell strength and SAT RAs is well observed in Fig. 9.14b. This correlation pattern between the two cell strengths and the SAT anomalies can well explain the correlation pattern of SIC anomalies in the Arctic region as shown in Fig. 9.14c, d. The positive area "A" and the negative area "B" of correlation in Fig. 9.14a can well explain the negative area "A" and the positive area "B" of correlation in Fig. 9.14c. For details, when the Arctic cell is strong (weak), the SAT RAs are above (below) normal in the Barents Sea and below (above) normal from the Hudson Bay to the Davis Strait (Fig. 9.14a), so the SIC RAs in the two areas are, respectively, below (above) normal and above (below) normal (Fig. 9.14c). A reversed relationship is observed in Fig. 9.14b, d. Figure 9.14e, f show the correlation distributions of the precipitation RAs associated with annual mean Arctic and Polar cell strength RAs. Positive and negative correlations centered in the Arctic ("C"), respectively, in Fig. 9.14e, f correspond well with that in Fig. 9.14c, d.

This section demonstrates an interesting phenomenon that the negative (positive) correlation center between Arctic (Polar) cell strength and SAT RAs is always found over the Greenland in all four seasons, which is surrounded by positive (negative) correlation. If there is no sea–land contrast in the NH high latitudes, the climatic cold center and the climatic ascending flow should symmetrically exist over the North Pole, driven by the climatic Arctic cell. During strong Arctic cell year (when the Polar cell is weak), a weakening subsidence flow appears between

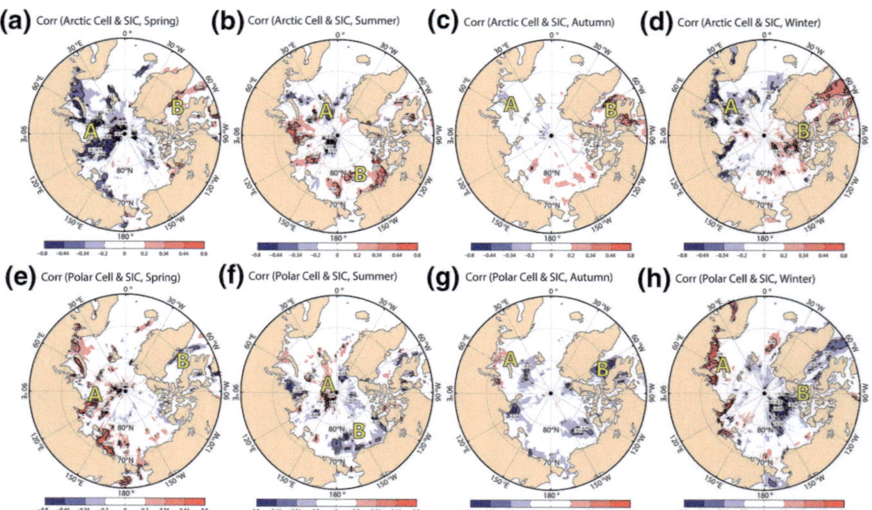

Fig. 9.15 Same as Fig. 9.13c, d except of the seasonal-mean correlation coefficients between the Arctic cell strength and SIC RAs in **a** spring, **b** summer, **c** autumn, and **d** winter, as well as between the Polar cell strength and SIC RAs in **e** spring, **f** summer, **g** autumn, and **h** winter. Sourced from Qian et al. (2016a) © Springer-Verlag Berlin Heidelberg 2016, with permission

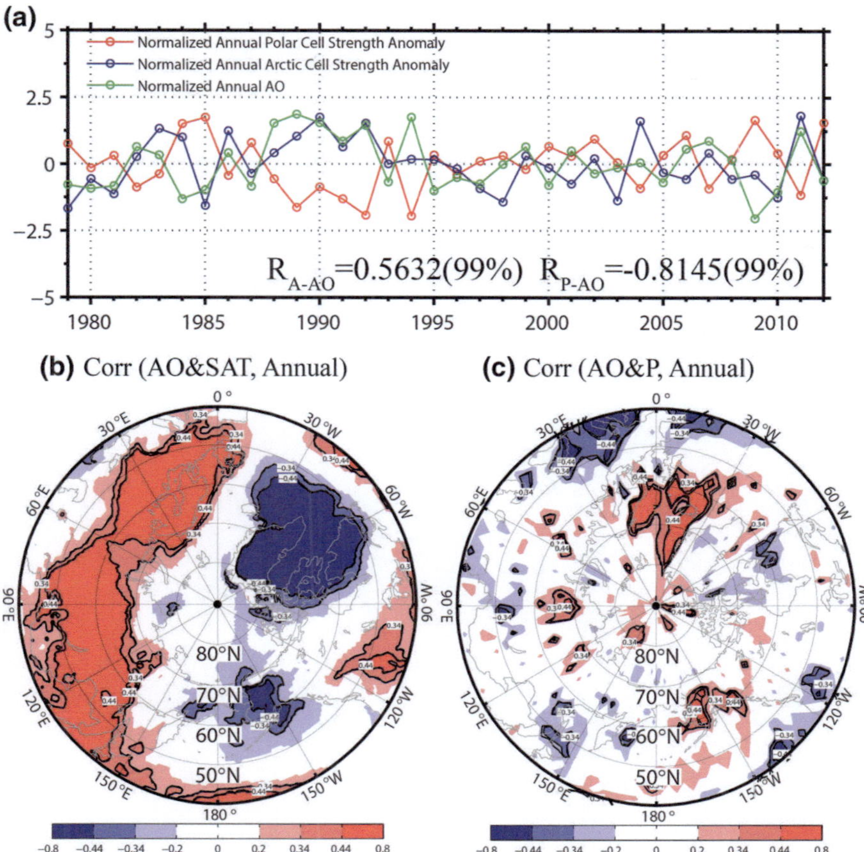

Fig. 9.16 **a** Series of the Polar cell strength NAs (*red line*) and Arctic cell strength NAs (*blue line*) as well as the normalized annual mean AO index series (*green line*) during 1979–2012. The spatial distributions of correlation coefficient between **b** the normalized annual mean AO index series and SAT RAs using the ERA-Interim dataset, as well as **c** the normalized annual mean AO index series and precipitation RAs beyond 40°N. The isolines of 0.34/−0.34 and 0.44/−0.44 show their correlations reaching the 95 and 99% confidence levels, respectively. The correlation coefficient between the Arctic (Polar) cell strength NAs and the normalized AO index series is 0.56 (0.81) at the 99% confidence level

the two cells (Qian et al. 2015a, 2016a). This weakens the cold subsidence flow and forms anomalous southerly warm-wet advection in the subpolar zone, and thus results in warming subpolar zone and more snowfall near the North Pole. Under this process, SIC should decline in the subpolar zone and increase near the North Pole. The reversed process and distributions of SAT, precipitation, and SIC RAs should similarly form when the Arctic cell is weak and the Polar cell is strong. However, the sea–land distribution is not symmetric, so negative (positive) center of correlation between Arctic (Polar) cell strength and SAT (precipitation) RAs systematically shift to the Greenland.

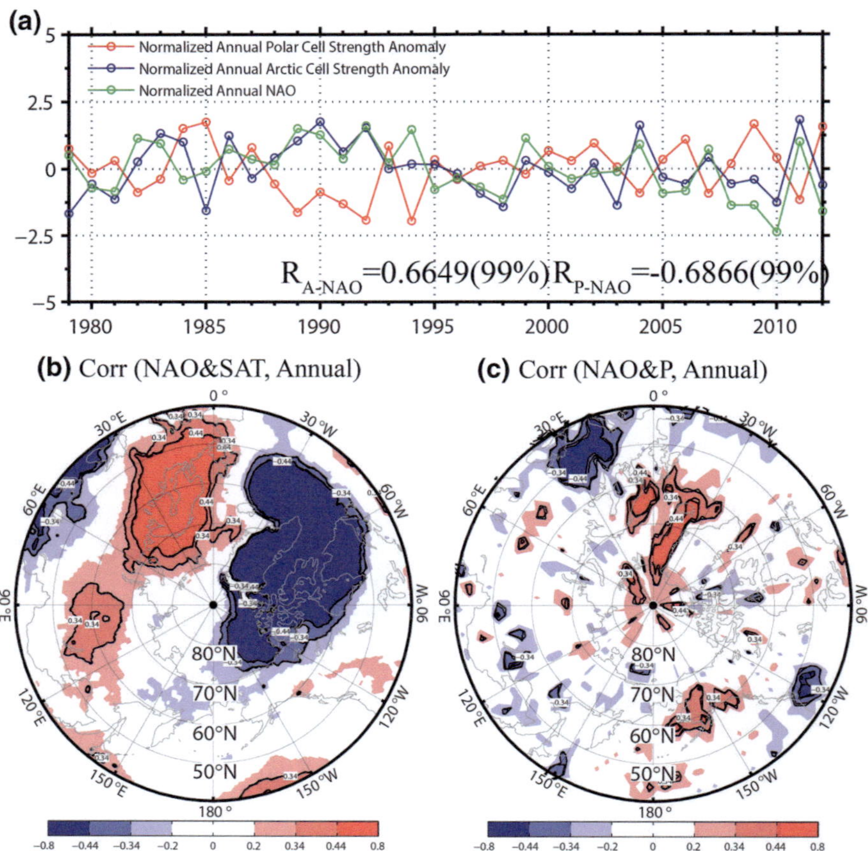

Fig. 9.17 Same as in Fig. 9.16 except for the normalized annual mean NAO index series (*green line*) during 1979–2012

Under this asymmetrical sea–land distribution, Qian et al. (2016a) further examined the seasonal-mean correlation coefficients between the two cell strengths and SIC RAs. Opposite spatial distributions of correlation coefficients between the Arctic cell strength and SIC RAs as well as between the Polar cell strength and SIC RAs are illustrated in Fig. 9.15 for the four seasons. In spring (Fig. 9.15a, e), opposite correlations are clearly comparable from the Eurasian margin seas to the Sea of Okhotsk in the EH. In summer, scattered but opposite correlations can be observed in Fig. 9.15b, f. In autumn, major negative correlation between the Polar cell strength and SIC RAs is concentrated over the Arctic Ocean (Fig. 9.15g). In winter (Fig. 9.15d, h), correlation distributions between the two cell strengths and SIC RAs are similar but stronger than those in spring.

Climatic indices such as the AO index and the NAO index have been widely applied in the studies of climatic extreme events in the NH. The AO index is a planetary-scale circulation index, which is derived from the 1000-hPa height.

The NAO index is a regional circulation index over the North Atlantic and its surrounding continents. The correlation coefficients (0.56 and -0.81) between the Arctic/Polar cell strengths and the AO index series from 1979 to 2012 are significant at the 90% confidence level (Fig. 9.16a). This means that the AO index is significantly correlated with the Arctic and Polar cell strengths. This result also shows that the strength variation of the two cells is dynamically connected with the AO horizontal change. At the same time, the spatial distribution of correlation between the AO index and SAT (precipitation) RAs in Fig. 9.16b (9.16c) is similar to the correlation pattern between the Arctic cell strength and SAT (or precipitation) RAs.

Similar to the AO index, the correlations between the Arctic/Polar cell strengths and the NAO index series from 1979 to 2012 reach the 99% confidence level with their correlation coefficients 0.66 and 0.69 (Fig. 9.17a). The spatial distribution (Fig. 9.17b, c) of correlation between the NAO index and SAT/precipitation RAs in the NH is also similar to that between the AO index and SAT/precipitation RAs. This implies that the Arctic/Polar cells and the AO/NAO occupy internal dynamical connection in the NH high latitude.

9.4 Annual Mean Meridional Cell Trends and the Arctic Climatic Anomalies[4]

The strengthening and broadening trends of the Hadley cell have been revealed, while the existence of the Arctic cell has also been confirmed in previous studies. Qian et al. (2016b) extended previous strengthening trend analysis of the Hadley cell to the Polar and Arctic cells in the NH and explored their climate influences. The results showed that the Polar cell experienced an abrupt change from a slow to a rapid strengthening trend in 1989, while the Arctic cell showed an insignificant strengthening trend and a significant weakening trend successively. The strengthening subsidence associated with the Polar and Arctic cells can partly explain the warming SAT and declining SIC, through the increasing tropospheric height and temperature trends.

The Hadley cell is fundamentally important in the Earth's climate system and has an important impact on the global climate (Diaz and Bradley 2004). In recent decades, many studies have determined that climate changes, such as the mid-latitude tropospheric warming and the poleward shift of the subtropical dry zone, are associated with the strengthening trend and poleward expansion of the Hadley cell (Chen et al. 2002; Mitas and Clement 2005; Fu et al. 2006; Hu and Fu 2007; Song and Zhang 2007; Stachnik and Schumacher 2011; Liu et al. 2012; Su et al. 2014). However, these studies were mostly limited between 60°S and 60°N, with the widest range between 80°S and 80°N, used by Peixoto and Oort (1992).

[4]Modified article originally published in Qian et al. (2016a). Published with kind permission © 2016 John Wiley & Sons, Inc. All Rights Reserved.

The climate influences of other meridional cells, such as the Ferrel and Polar cells, have been paid less attention.

Sea ice is a sensitive component of the climate system, influenced by both atmosphere and ocean. The physical mechanisms underlying the decline of Arctic sea ice are still not fully understood. It is believed that the sea ice is related to changes in winds, ocean currents, and thermodynamic processes involving air temperature, radiative or turbulent energy fluxes, ocean heat storage, and ice-albedo feedback (Serreze et al. 2007; Francis and Hunter 2007; Overland and Wang 2010; Screen and Simmonds 2010a, b; Vihma et al. 2012; Ogi and Rigor 2013; Timmermans 2015). Based on such climate change, Qian et al. (2016b) extended previous studies from the Hadley cell to the Polar and Arctic cells, and further explored their influence on the long-term sea ice decline in the Arctic region. This is because that the Arctic is warming two to four times faster than the global average (Screen et al. 2012).

Beside previous reanalysis products, another dataset used is the NOAA-CIRES Twentieth Century Reanalysis (20CR) dataset (Compo et al. 2011). Considering the comparability, five reanalysis products for the same period 1981–2010 are used to climatologically depict meridional cells. The climatic annual mean MSFs averaged over 30 years (1981–2010) using the NCEP R1, NCEP R2, ERA-Interim, JRA-55, and 20CR reanalysis products and their composite are displayed in Fig. 9.18. Four meridional cells are commonly observed in the NH, with comparable central intensities and locations from all five datasets. The intensity ratios of the Ferrel, Polar, and Arctic cells to the Hadley cell are approximately 1:3, 1:13, and 1:80, respectively, while the intensity ratio of the Arctic cell to the Polar cell is approximately 1:6 (Qian et al. 2015a). The Arctic cell exists next to the Polar cell as a weak opposite circulation and is centered at 850 hPa and 80°N. The sinking flow is seasonally located on the boundary between the Polar and Arctic cells, while the rising flow is to the north of the Arctic cell center. They revealed that the subpolar high and the polar low climatologically exist in the middle-low troposphere over the NH high latitudes. In the composite result of the five datasets (Fig. 9.18f), the Hadley cell strength (101.47×10^6 ton s^{-1}) in the Southern Hemisphere (SH) is climatologically stronger than that (74.52×10^6 ton s^{-1}) in the NH. However, differences in cell strengths are found between different datasets (Table 9.1). The maximum difference (33.65×10^6 ton s^{-1}) of the Hadley cell strength in the SH is found between the JRA-55 and NCEP R1 datasets, while that (15.65×10^6 ton s^{-1}) of the Hadley cell strength in the NH is found between the ERA-Interim and NCEP R1 datasets. The ratios of maximum differences to the composite MSF values are close between the Hadley cell in SH and Arctic cell in NH (last column in Table 9.1). The ratios for other cells, especially the Ferrel and Polar cells, are smaller. On the other hand, the SDs of four NH cells' central MSFs are calculated using the five reanalysis datasets. It is shown that the annual mean Hadley cell strength has the largest variation but the smallest ratio relative to the corresponding climatic value. In contrast, annual mean strength of the Arctic cell has the smallest variation but the largest ratio relative to the corresponding climatic value for the five datasets and their composite (Table 9.2). The similar results of SDs for Polar and

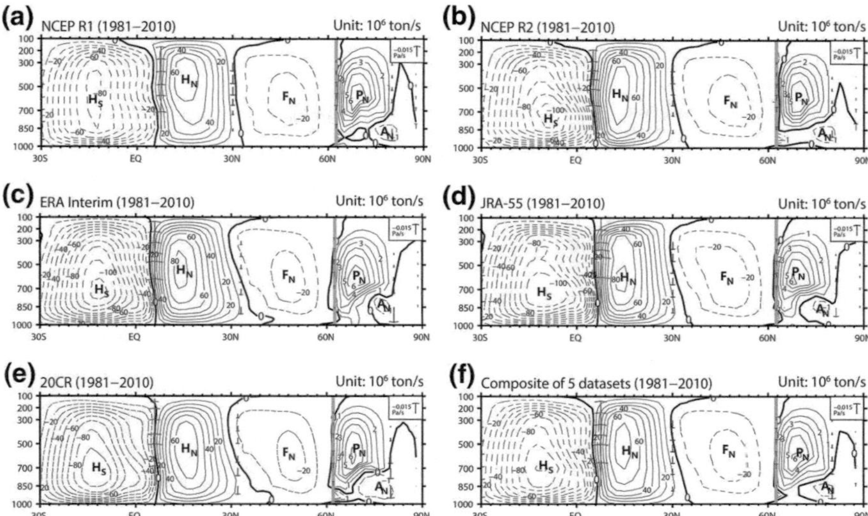

Fig. 9.18 Five zonally averaged meridional cells described by the pressure-latitude section of climatic annual mean MSFs (10^6 ton s^{-1}, *solid contour* for positive and *dashed contour* for negative) averaged over 30 years (1981–2010) using the five reanalysis products: **a** NCEP R1, **b** NCEP R2, **c** ERA-Interim, **d** JRA-55, **e** 20CR (V2c), and **f** their composite. *Red* and *blue arrows* indicate, respectively, ascending and descending pressure velocities (−0.015 Pa s^{-1}). The MSF interval is 10 × 10^6 $tons^{-1}$ between 30°S and 62.25°N and 1 × 10^6 ton s^{-1} between 62.25°N and 90°N, separated by the *thin grey line* near 62.25°N. Sourced from Qian et al. (2016b) © 2016 John Wiley & Sons, Inc., with permission

Table 9.1 Climatic strength values (absolute values of central MSF, 10^6 ton s^{-1}) of five meridional cells based from five reanalysis products and their composite

Strength	NCEP R1	NCEP R2	ERA-interim	JRA-55	20CR (V2c)	Composite	Max difference	Ratio (%)
H_S	82.83	108.67	114.03	116.48	98.54	101.47	33.65	33.16
H_N	68.20	79.49	83.85	83.54	69.40	74.52	15.65	21.00
F_N	26.81	27.94	27.03	30.22	26.56	27.21	3.66	13.45
P_N	6.61	6.01	6.92	6.49	6.20	6.17	0.91	14.75
A_N	1.89	1.78	1.59	1.36	1.90	1.62	0.54	33.33

The letter "H_S" indicates the Hadley cell in the SH and the letters "H_N," "F_N," "P_N," and "A_N" denote the Hadley, Ferrel, Polar, and Arctic cells in the NH. The last three columns are the composite results, maximum differences of five reanalysis products, and the ratios of maximum differences to the composite strength values, respectively. Sourced from Qian et al. (2016b) © 2016 John Wiley & Sons, Inc., with permission

Arctic cells imply that all the reanalysis products can consistently describe the annual mean strength variations of the two cells.

Figure 9.19 depicts the central MSF interannual variations and long-term trends of the four cells. All time series start from the first available year of each dataset.

Table 9.2 Central MSF SDs (10^6 ton s^{-1}), calculated from the five reanalysis datasets and their composites from 1979 to 2011, for the four meridional cells in the NH

Standard deviation	Hadley	Ferrel	Polar	Arctic
NCEP R1 (1979–2011)	4.50 (6.60%)	2.85 (10.63%)	1.65 (24.96%)	1.20 (63.49%)
NCEP R2 (1979–2011)	4.61 (5.60%)	3.12 (11.17%)	1.70 (28.29%)	1.24 (69.66%)
ERA-Interim (1979–2011)	5.05 (6.05%)	2.79 (10.32%)	1.69 (24.42%)	1.18 (74.21%)
JRA-55 (1979–2011)	7.03 (8.42%)	2.93 (9.73%)	1.93 (29.74%)	1.34 (98.53%)
20CR (1979–2011)	3.02 (4.35%)	2.74 (10.31%)	1.83 (29.51%)	1.37 (72.11%)
Composites of 5 reanalysis datasets	3.77 (5.06%)	2.85 (9.47%)	1.70 (27.55%)	1.22 (75.31%)

The percentages in the brackets are the ratios of the SDs of cell strength variations to their climatic values. Sourced from Qian et al. (2016b) © 2016 John Wiley & Sons, Inc., with permission

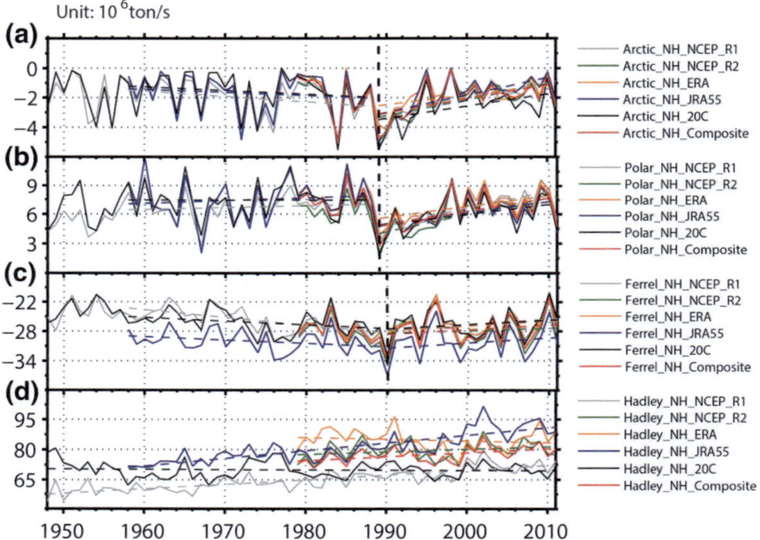

Fig. 9.19 Annual mean central MSF (10^6 ton s^{-1}) series of **a** the Arctic cell, **b** the Polar cell, **c** the Ferrel cell, and **d** the Hadley cell from 1948 to 2011. For each cell, the strength series are calculated by using the annual mean MSF from the five reanalysis products of NCEP R1, NCEP R2, ERA-Interim, JRA-55, 20CR (V2c) and their composites. *Dashed lines* denote the linear trends of different periods and various products. The *heavy black-dashed lines* denote 1989 in **a** and **b**, and 1990 in **c**. Qian et al. (2016b) © 2016 John Wiley & Sons, Inc., with permission

Many previous studies have revealed that the Hadley cell experienced a strengthening trend in the last three decades (Liu et al. 2012; Su et al. 2014). The long-term strengthening trend of the Hadley cell is evidently observed from the NCEP R1 and JRA-55 reanalysis products, with the values of 3.00×10^6 and 5.53×10^6 ton s^{-1} decade^{-1}, respectively, from 1979 to 2011 (Table 9.3). However, neither strengthening nor weakening trends are statistically significant in the other three

Table 9.3 Long-term central MSF trends (10^6 ton s^{-1} $decade^{-1}$) of the four cells in different periods and with different confidence levels (e.g., 99%)

Trend	NCEP R1	NCEP R2	ERA- Interim	JRA-55	20CR (V2c)	Composite
Hadley (1979–2011)	3.00 (99%)	1.23 (85%)	−0.89 (65%)	5.53 (99%)	0.58 (70%)	1.89 (99%)
Hadley (1958–2011)	/	/	/	3.79 (99%)	/	/
Hadley (1948–2011)	2.35 (99%)	/	/	/	−0.16 (43%)	/
Ferrel (1990–2011)	1.17 (72%)	0.51 (32%)	1.03 (66%)	1.02 (63%)	0.91 (61%)	0.93 (60%)
Ferrel (1958–1989)	−1.58 (99%)	/	/	−0.69 (89%)	−0.76 (90%)	/
Polar (1989–2011)	1.78 (99%)	1.75 (99%)	1.15 (97%)	1.16 (96%)	1.06 (93%)	1.38 (99%)
Polar (1979–1988)	0.20 (11%)	0.02 (1%)	0.75 (33%)	0.58 (24%)	−0.65 (28%)	0.18 (9%)
Polar (1958–1988)	0.20 (48%)	/	/	0.30 (50%)	0.01 (2%)	/
Arctic (1989–2011)	1.06 (99%)	1.08 (99%)	0.66 (95%)	1.19 (99%)	0.83 (96%)	0.96 (99%)
Arctic (1979–1988)	−1.53 (67%)	−1.52 (68%)	−1.88 (71%)	−1.89 (70%)	−1.44 (58%)	−1.65 (68%)
Arctic (1958–1988)	−0.37 (82%)	/	/	−0.21 (52%)	−0.25 (62%)	/

The trends are also shown in Fig. 9.19. The positive (negative) MSF trends of the Hadley and Polar cells represent strengthening (weakening), while the positive (negative) MSF trends of the Ferrel and Arctic cells represent weakening (strengthening). Sourced from Qian et al. (2016b) © 2016 John Wiley & Sons, Inc., with permission datasets (Table 9.3 and Fig. 9.19d). The strengthening trend 1.89×10^6 ton s^{-1} $decade^{-1}$ reaches the 99% confidence level for the Hadley cell in NH from the composite of 5 datasets in the period 1979–2011. For the Ferrel cell (Fig. 9.19c), a significant strengthening trend is detected from the NCEP R1 dataset during 1958–1989, while insignificant strengthening trends are detected from both the JRA-55 and 20CR MSF series (Table 9.3). The weakest MSF value of the Ferrel cell is observed in 1990, and an insignificant weakening trend is detected from all datasets and their composite after 1990.

For the Arctic and Polar cells, an extreme year of MSF values is observed in 1989, and the interannual variability is consistent for all 5 datasets (Fig. 9.19a, b). The maximum weakening trend of the Arctic cell and strengthening trend of the Polar cells are detected from 1989 to 2011 (Fig. 9.19a, b; Table 9.3) using the Mann-Kendall trend test. From 1958 to 1988, insignificant strengthening trends of the Polar and Arctic cells are detected from three datasets, as listed in Table 9.3. From 1979 to 1988, the Arctic cell's strengthening trend (-1.65×10^6 ton s^{-1} $decade^{-1}$ from the composite) and the Polar cell's strengthening cell (0.18×10^6 ton s^{-1} $decade^{-1}$ from the composite) are also insignificant (Table 9.3). Therefore, the strength trends of the Arctic and Polar cells are considered since 1989. After removing their long-term trends, the strengths of the Arctic (Arctic central MSF multiplied by -1) and Polar cells show opposite interannual variations with a significant negative correlation (-0.53, exceeding the 99% confidence level from their composites) from 1989 to 2011. The composite series also show a significant strengthening trend of the Polar cell (1.38 ton s^{-1} $decade^{-1}$, exceeding the 99% confidence level) since 1989, while the Arctic cell has a weakening trend (0.96 ton s^{-1} $decade^{-1}$, exceeding the 99% confidence level) in Table 9.3.

Arctic sea ice has undergone a substantial long-term decline (Serreze et al. 2007; Graversen et al. 2008; Screen and Simmonds 2010a, b; Ding et al. 2014), so the relationship between the Arctic sea ice's long-term decline and the strengthening trend of the Polar cell and the weakening trend of the Arctic cell in the recent two decades is revealed. The SIE and SIC change mostly in the subpolar zone between 70° and 80°N. Thus, the Arctic Ocean is divided into four regions: A (0°–90°E, 70°–80°N) over the Barents Sea and the Kara Sea, B (0°–90°W, 70°–80°N) over the Laptev Sea and the East Siberian Sea, C (90°E–180°, 70°–80°N) over the Beaufort Sea, and D (90°W–180°, 70°–80°N) near Baffin Bay. Figure 9.20 depicts four regional-averaged monthly mean time series of SIC anomalies from 1979 to 2011. Declining trends of the four SIC anomalous series from regions A to D are -5.50, -0.68, -3.77, and -2.75% $decade^{-1}$, respectively, reaching the 99, 96, 99, and 96% confidence levels. An interesting phenomenon is noted from regions C and D (Fig. 9.20b): the SIC, which usually showed positive anomalies in September of 1979–1988, started to show negative anomalies in September since 2005. During the transition period of 1989–2005, irregular positive and negative SIC anomalies occurred in September. Regions C and D are entirely covered by ice in spring and partly melted in other seasons, particularly in autumn. Thus, the interannual variation and long-term trends are apparent in September. In contrast, the other two regions, A and B, are open to and influenced by the North Atlantic.

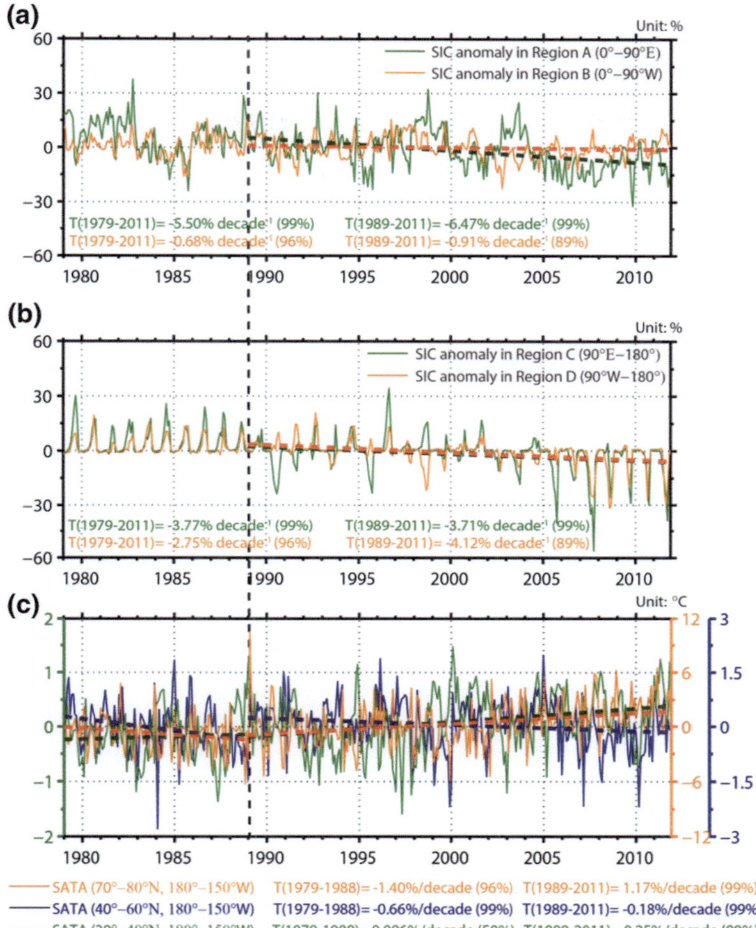

Fig. 9.20 **a** Time series of monthly mean SIC anomalies averaged over two regions: *A* (0°–90°E, 70°–80°N; *green line*) and *B* (0°–90°W, 70°–80°N; *orange line*) using the HadISST1 dataset. **b** Time series of monthly mean SIC anomalies averaged over two regions: *C* (90°E–180°, 70°–80°N; *green line*) and *D* (90°W–180°, 70°–80°N; *orange line*). **c** Time series of monthly mean SAT anomalies (°C) at the 2-m level averaged over three areas, indicated by the *orange line* (180°–150°W, 70°–80°N), *blue line* (180°–150°W, 40°–60°N), and *green line* (180°–150°W, 20°–40°N) using the ERA-Interim data. The letter "T" indicates the trends of different series and the corresponding confidence levels. Sourced from Qian et al. (2016b) © 2016 John Wiley & Sons, Inc., with permission

Thus, their SIC anomalies can be observed in all four seasons. The SIC trends in regions A to D are −6.47, −0.91, −3.71, and −4.12% decade^{-1} during 1989–2011, respectively, reaching 99, 89, 99, and 89% confidence levels. Declining trends became larger for the period 1989–2011 than those for the whole period 1979–2011 in all four regions (Fig. 9.20a, b). The maximum declining trend (−6.47%

decade^{-1}) of the SIC anomalies is noted in the marginal sea of northern Europe (region A), i.e., over the Barents and Kara sea, after an extreme year in 1989. The interdecadal trend transition of the SIC variation is basically consistent with that of the Polar and Arctic cells in their strengths (Fig. 9.20a, b), which has the maximum trends since 1989.

The above results show that the SICs and the strengths of Arctic and Polar cells have internal connections in the interannual and interdecadal timescales. The trend standard deviation (SD_T) and the interannual standard deviation (SD_I) are calculated for the last period 1989–2011 using three series. The percentages of SD_T to the total variations ($SD_T + SD_I$) are 18, 29%, and 48%, respectively, for the composite central MSF series of the Polar and Arctic cells, as well as the SIC anomalous series. This confirms that the long-term trends are not negligible in the variations of SIC and the two cell intensities. Then, the contributions of the Polar and Arctic cells toward the sea ice change are examined. Binary linear regression analysis is done by fitting the long-term trend of SIC to the long-term trends of Arctic and Polar cell central MSFs. The regression coefficients of Arctic and Polar cell MSF values are −1.05 and −2.49, respectively, which implies that the Polar cell has a larger negative influence on the SIC long-term trend than the Arctic cell does. Other portions might be influenced by ocean currents, sunlight, water, radiation, turbulent energy fluxes, ocean heat storage, and ice-albedo feedback (Perovich et al. 2008; Toole et al. 2010; Timmermans 2015).

The Arctic sea ice decline is closely related to the change in SAT. To investigate the long-term trends in SAT variability, the optimum domain should be determined. As mentioned by Wang et al. (2005) and Qian et al. (2015a), the North Pacific is an ideal place without obvious terrain effect. Three SAT series, i.e., for high latitudes (70°–80°N, 180°–150°W), middle latitudes (40°–60°N, 180°–150°W), and subtropical latitudes (20°–40°N, 180°–150°W) over the central North Pacific, are given in Fig. 9.20c. Significant warming trends (1.17 and 0.25 °C decade^{-1}) in high and subtropical latitudes and a significant cooling trend (−0.18 °C decade^{-1}) in middle latitudes are observed from 1989 to 2011 (all reaching the 99% confidence level). Their trends in the period 1979–1988 are −1.40 °C decade^{-1} (96%), −0.66 °C decade^{-1} (99%), and 0.096 °C decade^{-1} (50%), respectively, for high, middle, and subtropical latitudes. The high latitudes have experienced a shift from a cooling period (1979–1988) to a warming period (1989–2011), which is consistent with the sea ice declining trend (1989–2011).

Fu et al. (2006) suggested that the subtropical warming and tropospheric height strengthening have been associated with Hadley cell strengthening and widening since 1979. To examine whether a similar relationship exists between SAT and the two meridional cells in high latitudes, the strengthening of the Hadley and Polar cells and the weakening of the Ferrel and Arctic cells can be observed by subtracting the average annual mean MSFs during 1989–2000 from those during 2001–2011 (Fig. 9.21a–f). The Hadley cell has the maximum strengthening trend, with a difference of more than 6×10^6 ton s^{-1} compared with all datasets and their composite. The maximum MSF difference of the Hadley cell in the NH is located in the upper troposphere (300 hPa) based on the NCEP R1 and NCEP R2 datasets, but

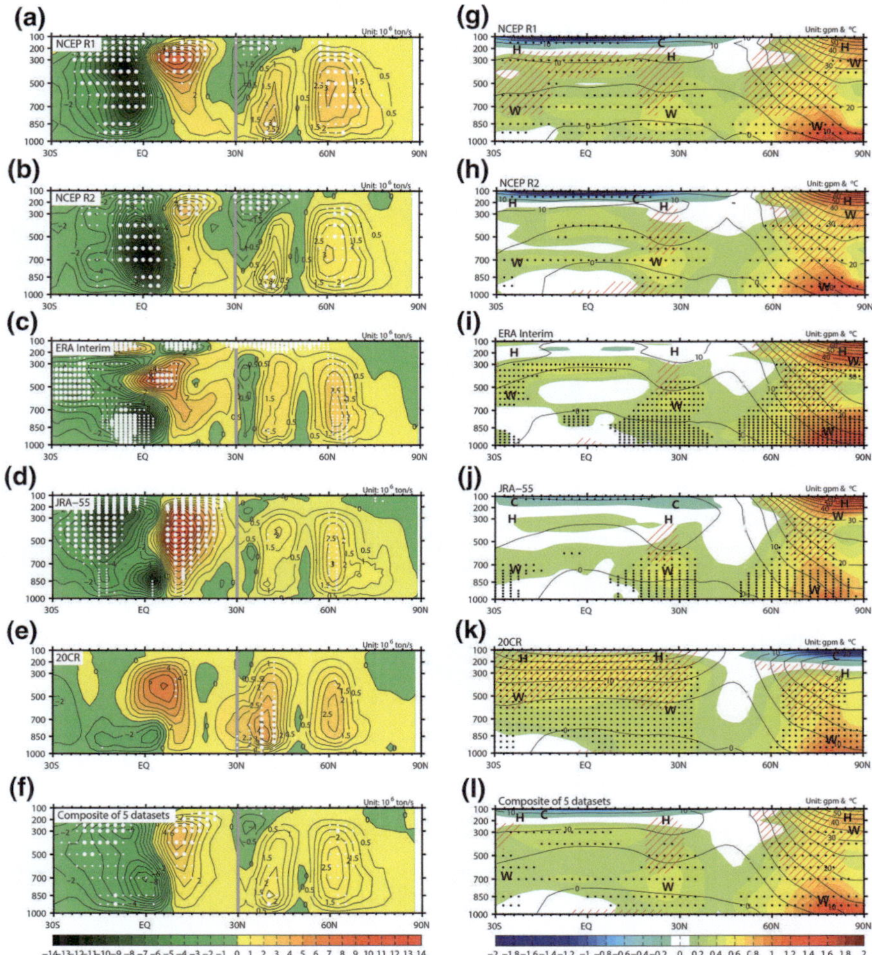

Fig. 9.21 Differences of climatic annual mean MSFs (10^6 ton s^{-1}) between the period 2001–2011 and the period 1989–2000 using the five reanalysis products **a** NCEP R1, **b** NCEP R2, **c** ERA-Interim, **d** JRA-55, **e** 20CR (V2c), and **f** their composite. Panels **g–l** are the corresponding differences in zonally averaged geopotential height (*solid line*, 5 gpm interval) and temperature (*shading*, 0.1 °C interval) between the two periods for the five reanalysis products and composite, respectively. In **a–f**, the *small white dots* denote that the difference in the MSF between the two periods reaches the 95% confidence level, while *big white dots* denote reaching the 99% level. In **g–l**, the *red-slant line* and *black-dot* areas denote that the difference in height and temperature between the two periods, respectively, reaches the 99% confidence level. The letters "H" and "W" in **g–l** are the centers of enhanced height and temperature, respectively. Sourced from Qian et al. (2016b) © 2016 John Wiley & Sons, Inc., with permission

in the middle troposphere (400–500 hPa) based on the ERA-Interim, JRA-55, and 20CR reanalysis products. Particular attention is the influence of the Polar and Arctic cell intensity change on the climate changes in the NH middle-high latitudes.

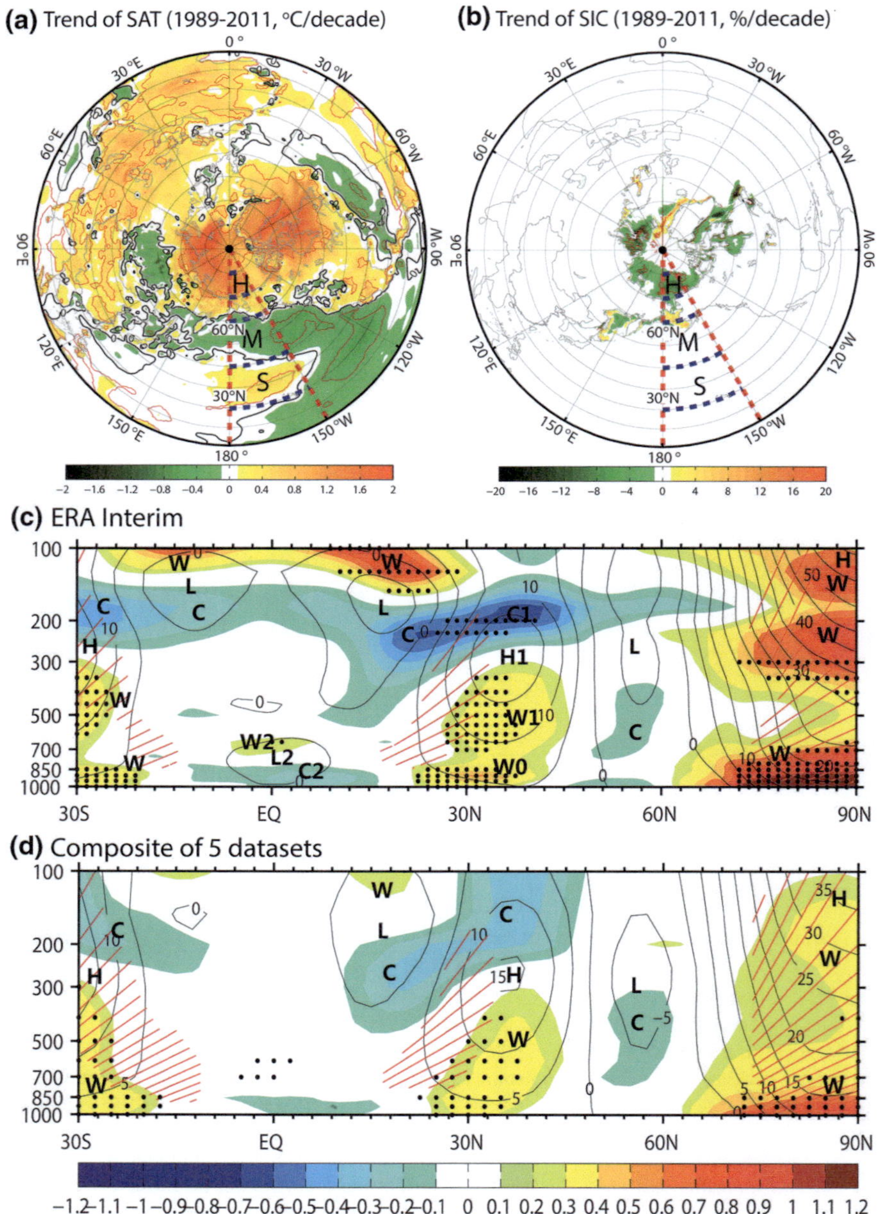
(a) Trend of SAT (1989-2011, °C/decade)
(b) Trend of SIC (1989-2011, %/decade)
0°
30°E
30°W
60°E
60°W
90°E
90°W
120°E
120°W
150°E
150°W
180°
60°N
30°N
H
M
S
-2 -1.6 -1.2 -0.8 -0.4 0 0.4 0.8 1.2 1.6 2
-20 -16 -12 -8 -4 0 4 8 12 16 20
(c) ERA Interim
(d) Composite of 5 datasets
100
200
300
500
700
850
1000
30S
EQ
30N
60N
90N
W1
W2
W0
C1
C2
H1
L2
-1.2 -1.1 -1 -0.9 -0.8 -0.7 -0.6 -0.5 -0.4 -0.3 -0.2 -0.1 0 0.1 0.2 0.3 0.4 0.5 0.6 0.7 0.8 0.9 1 1.1 1.2

◀ **Fig. 9.22** **a** SAT trend (*shading*, 0.1 °C decade^{-1} interval) derived from the ERA-Interim product at the 2-m level and **b** SIC trend (*shading*, 1% decade^{-1} interval) in the NH during 1989–2011. *Red contours* in **a** and **b** cover the area with trends reaching the 99% confidence level. Two *red-dashed lines* in **a** and **b** shape a fan domain with three areas: "H" in the high latitude, "M" in the middle latitude and "S" in the subtropical area over the central North Pacific between 150°W and 180°W, of which the monthly mean SAT anomalies are drawn in **a**. **c** The vertical sections of height (*contour*, 5 gpm decade^{-1} interval) and temperature (*shading*, 1 K decade^{-1} interval) trends are derived from the ERA-Interim product and averaged along the central North Pacific. **d** is the same as **c**, except it is derived from the composite of the five datasets. In **c** and **d**, the *red-slant line* and *black-dot* areas denote that the height and temperature trends, respectively, reach the 99% confidence level. Letters "H" and "L," as well as "W" and "C" in **c** and **d** are the centers of increasing and decreasing height trends, as well as increasing and decreasing temperature trends, respectively. Sourced from Qian et al. (2016b) © 2016 John Wiley & Sons, Inc., with permission

Beyond 30°N, the five reanalysis products consistently show strengthening of the Polar cell and weakening of the Arctic cell in the lower troposphere. At the levels of 500–700 hPa and at approximately 63°N, the difference in MSFs between two decades reaches approximately 3×10^6 ton s^{-1}. In the NH, almost all latitude bands are occupied by positive MSF differences in the middle-low troposphere from all five reanalysis products and their composite, except for two narrow bands of negative differences at 30°N and 50°N.

The changes of zonally averaged geopotential height and temperature in the NH also have increasing trends during 1989–2000 compared with those during 2001–2011. Figures 9.21g–l depicts the height and temperature differences between the two periods based on the five reanalysis products and their composite. The increasing trend of height is located mainly in the middle-upper troposphere and particularly in high latitudes beyond 60°N. Three positive centers ("H") of height difference are found in the upper troposphere above 30°S, 30°N, and the North Pole. Below each positive height difference center, positive centers ("W") of temperature difference are apparent vertically. Three low-level warming centers exist in the two subtropical zones and the NH high latitudes. In Fig. 9.22, two enhanced subtropical highs are associated with the strengthening Hadley cell, warming the mid-low troposphere by enhancing subsidence near 30°N and 30°S, respectively. Moreover, the enhanced subpolar high, associated with the strengthening Polar cell, forces warming by enhancing subsidence at the NH high latitudes. Spatial patterns of SAT trends derived from the ERA-Interim reanalysis at the 2-m level and SIC trends for the period 1989–2011 are compared in Fig. 9.22a, b. Two regions with significant warming from the Mediterranean Sea to East Asia and the high latitudes beyond 70°N are separated by a large cooling area in the eastern and northern Pacific, extending westward to central Russia (Fig. 9.22a). The warming at high latitudes is consistent with the declining SIC trend in the Arctic, as shown in Fig. 9.22a, b.

In fact, the warming SAT and the declining SIC, as well as the height and temperature variations, are not entirely oriented along any zonal bands due to the influence of sea–land contrast. The fan-shaped domain crossing the central North Pacific between 150°W and 180° (Fig. 9.22a, b) is less influenced by the large-scale sea–land contrast. In this domain, two negative (green shading) SAT areas, located at the tropic and middle latitudes (40–60°N), are separated by two positive (yellow-orange shading) SAT areas in the subtropical North Pacific (20–40°N) and the Arctic (Fig. 9.22a). A declining SIC trend in the Chukchi Sea (the Arctic) and an increasing SIC trend in the Bering Sea correspond well with the increasing and declining SAT trends in the two seas, respectively. Regional SAT and SIC trend patterns also correspond well with the height and temperature trend patterns in the vertical atmosphere. Figure 9.22c shows the vertical height and temperature trends averaged over the central North Pacific. The trend has a pattern similar to hydrostatic relationship between daily height and temperature anomalies vertically. In the vertical subtropical North Pacific, a positive trend center "H1" at 300 hPa separates a negative temperature trend center "C1" at 200 hPa and a positive temperature trend center "W1" at 500 hPa. Another positive temperature trend center "W0" at 925 hPa is a downward extension of "W1" below the height center "H1." Above the equatorial band, a negative height trend center "L2" at 750 hPa separates a positive temperature trend center "W2" at 600 hPa and a negative temperature trend center "C2" at 925 hPa. The strongest positive centers of height and temperature trends in the high latitudes correspond well with the warming trend of SAT and the declining trend of SIC in the Arctic. The three positive centers of height trend in the upper troposphere and the three positive centers of temperature trend in the lower troposphere over the fan-band-mean analysis (Fig. 9.22c) are comparable to difference centers from the zonally averaged band analysis in Fig. 9.21i based on the ERA-Interim dataset. These spatial distributions of significant height and temperature trends are also evidently observed in Fig. 9.22d from the composite of the five datasets.

9.5 Boreal Winter–Spring Arctic Amplification and Meridional Cells

This section demonstrates that winter–spring strength trends of the Polar and Arctic cells can be used to explain the possible linkage of Arctic amplification (AA) and the mid-latitude temperature reduction during the past two decades. Results show that the winter (spring) Polar cell has experienced an insignificant (significant) long-term strengthening trend since 1988 (1990), while the winter (spring) Arctic cell has undergone an insignificant weakening trend since 1988 (1990), respectively. The meridional SAT trend pattern of the AA and the mid-latitude temperature reduction, respectively, corresponds to the positive geopotential height trends in the Arctic troposphere and negative geopotential height trends in the mid-latitude troposphere. The positive tropospheric geopotential height trends averaged in the Arctic and subtropics over the central North Pacific are connected with the strengthening Polar and Hadley cells in the NH similarly. A vertical structure of a positive (negative) geopotential height trend center in the middle troposphere with a cooling (warming) trend above and a warming (cooling) trend below can indicate the warming (cooling) trend in the surface very well.

The global mean surface temperature has increased 0.6–0.7 °C since the mid-1960s (Kennedy et al. 2007). The temperature over the Arctic region during the same period has risen by 1.9–2.0 °C, two to four times faster than the global average (Screen et al. 2012), and even faster during winter and spring (Richter-Menge et al. 2008; Graversen et al. 2008). The phenomenon is called the AA. Therefore, the winter–spring AA phenomenon becomes a hot topic and attracts a great interest. Whether the winter–spring AA is linked with the long-term intensity trends of Arctic and Polar cells along with the atmospheric height anomalies is further examined in this section.

Figure 9.23 depicts the boreal winter- and spring-mean central MSF variations and long-term trends of the Arctic and Polar cells in the NH using five reanalysis datasets and their composite. It should particularly pay attention to the Polar cell variation because its intensity and spatial coverage are stronger and larger than

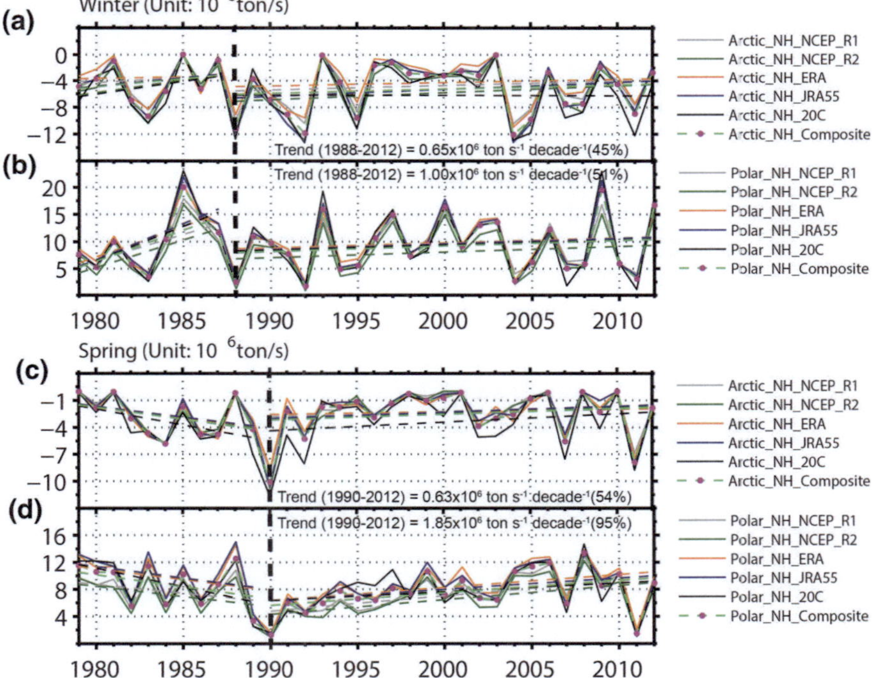

Fig. 9.23 Boreal winter central MSF (10^6 ton s^{-1}) series of **a** the Arctic cell and **b** the Polar cell from 1979 to 2012 calculated by NCEP R1, NCEP R2, ERA-Interim, JRA-55, 20CR datasets and their composite. The boreal winter refers to December in current year and January, February in next year. Panels **c** and **d** are boreal spring series. The boreal spring refers to March, April, and May in current year. *Straight dashed lines* denote the linear trends of Arctic cell and Polar cell central MSFs in different periods and various datasets. The *heavy black-dashed lines* indicate the extreme years of the Polar cell and Arctic cell strengths

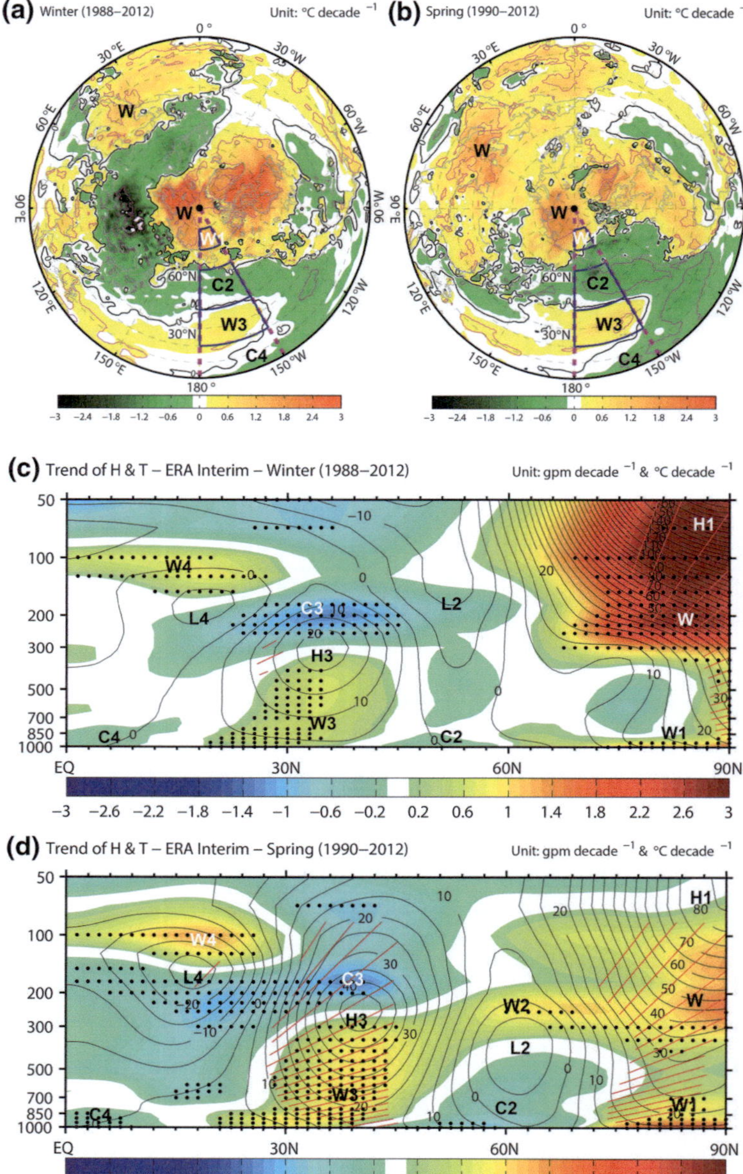
(a) Winter (1988–2012)
Unit: °C decade −1
(b) Spring (1990–2012)
Unit: °C decade −1
W
W1
C2
W3
C4
60°N
30°N
0°
30°E
30°W
60°E
60°W
90°E
90°W
120°E
120°W
150°E
150°W
180°
−3 −2.4 −1.8 −1.2 −0.6 0 0.6 1.2 1.8 2.4 3
(c) Trend of H & T – ERA Interim – Winter (1988–2012)
Unit: gpm decade −1 & °C decade −1
H1
W4
L4
C3
L2
H3
W3
C2
C4
W1
EQ
30N
60N
90N
50
100
200
300
500
700
850
1000
−3 −2.6 −2.2 −1.8 −1.4 −1 −0.6 −0.2 0.2 0.6 1 1.4 1.8 2.2 2.6 3
(d) Trend of H & T – ERA Interim – Spring (1990–2012)
Unit: gpm decade −1 & °C decade −1
W2
L2

◀ **Fig. 9.24** **a** Seasonal-mean boreal winter SAT trends (*shading*, 0.15 °C decade^{-1} interval) derived from the ERA-Interim product at the 2-m level in the NH during 1988–2012. **b** Same as **a** but of boreal spring SAT trends during 1990–2012. In **a** and **b**, *red contours* cover the areas with SAT trends reaching the 99% confidence level. **c** The vertical section of boreal winter height (*contour*, 5 gpm decade^{-1} interval) and temperature (*shading*, 0.1 °C decade^{-1} interval) trends averaged along the sector area (180°–150°W) of central North Pacific during 1988–2012, derived from the ERA-Interim product. **d** Same as **c** but of boreal spring height and temperature trends during 1990–2012. In **c** and **d**, the *red-slant line* and *black-dotted* areas denote that the trend in height and temperature, respectively, reaches the 95% confidence level. "H/L" and "W/C" are the centers of increasing/decreasing height and temperature trends, respectively

those of the Arctic cell. The composite results make the following analysis. It is found that there was an extreme boreal winter in 1988, when the Polar cell's MSF value reached the lowest (Fig. 9.23b). From 1988 to 2012, an insignificant strengthening trend of the winter Polar cell (1.00×10^6 ton s^{-1} decade^{-1} reaching the 51% confidence level) and an insignificant weakening trend of winter Arctic cell (0.65×10^6 ton s^{-1} decade^{-1} reaching the 45% confidence level) are detected in Fig. 9.23a, b. After removing their long-term trends, the strengths of the winter Arctic cell (Arctic cell central MSF multiplied by −1) and strengths of the winter Polar cell (Polar cell central MSF) show opposite interannual variations with a significant negative correlation (−0.80, exceeding the 99% confidence level from their composites) from 1988 to 2012.

An extreme year of the boreal spring Polar and Arctic cell strengths is noted in 1990 (Fig. 9.23c, d). From 1979 to 1989, there were a strengthening trend (-2.66×10^6 ton s^{-1} decade^{-1} reaching the 78% confidence level) of the Arctic cell and a weakening trend (-3.54×10^6 ton s^{-1} decade^{-1} reaching the 75% confidence level) of the Polar cell in spring. Since 1990, a significant strengthening Polar cell (1.85×10^6 ton s^{-1} decade^{-1}, reaching the 95% confidence level) and an insignificant weakening Arctic cell (0.63×10^6 ton s^{-1} decade^{-1} reaching the 54% confidence level) are detected in spring. The interannual correlations between the Polar and Arctic cell strengths are −0.84 after removing their long-term trends since 1990 from the composite series. It is noted that the transition time of the two cell strength trends in winter (December in 1988, January and February in 1989) is one year earlier than that in spring (March, April, and May in 1990). For the five composite series in Fig. 9.23, only the central MSF of spring Polar cell series has exceeded the 95% confidence level. The maximum negative trend before 1990 and the significant positive trend after 1990 can be detected by the Mann-Kendall test.

Spatial distributions of boreal winter and spring SAT trends derived from the ERA-Interim product at the 2-m level during 1988–2012 and 1990–2012 are shown in Fig. 9.24a, b, respectively. It can be noted that a significant warming trend is mainly concentrated beyond 60°N over the Arctic Ocean in both seasons. In the mid-latitude band between 40°N and 60°N, a significant cooling trend is separated by two warming trends in the Arctic Ocean and the subtropical areas, especially over the subtropical EH and the subtropical North Pacific. Opposite SAT trends can be identified between the Arctic Ocean and the mid-latitude band since 1988 or 1990. Therefore, the recent AA is not only an Arctic phenomenon, but also related to the mid-low latitudes. On the other hand, the fact shows that the warming or cooling SAT is not entirely oriented along any latitude zones due to the influence of sea–land contrast. In the central North Pacific, there is fewer influence of the large-scale sea–land contrast.

To explain the regional SAT trend, Fig. 9.24c, d shows the vertical sections of boreal winter (spring) height and temperature trends averaged over the central North Pacific during 1988–2012 (1990–2012). The central North Pacific is an ideal place to examine the influence from the strength trends of the four cells in the NH without obvious sea–land contrast. The height and temperature trend displays a vertical pattern similar to hydrostatic relationship between daily height and temperature anomalies in the analysis of synoptic extreme events (Chen et al. 2016). Vertically above the subtropical North Pacific, a positive height trend center "H3" at 300 hPa associated with the intensified Hadley cell locally separates a negative temperature trend center "C3" at 200 hPa and a positive temperature trend center "W3" at 600 hPa. Another positive temperature trend center "W1" at 925 hPa is a downward extension of warming trend below the height center "H1" at the stratosphere over the North Pole, which is associated with the intensified descending flow of the Polar cell extending from the lower stratosphere to the upper troposphere near the North Pole locally (Fig. 9.4). Above the tropical North Pacific, a negative height trend center "L4" at 200 hPa separates a positive temperature trend center "W4" at 100 hPa and a negative temperature trend center "C4" at 925 hPa near the equator. Another cooling SAT trend between 40° and 60°N is associated with a center "L2" of negative height trends. The strongest positive centers of height and temperature trends in the high latitudes correspond well with the warming trend of SAT and the declining trend of SIC in the Arctic. It is clear that the two positive height trend centers "H1" and "H3" at the North Pole and the subtropical area as well as the two negative height trend centers "L2" and "L4" at the mid-latitude area and the tropical North Pacific are directly connected with the SAT trends below. These spatial distributions of significant height and temperature trends in the troposphere are also evidently observed from the composite of the five datasets (Fig. 9.25).

As noted in Fig. 9.24a, b, due to sea–land contrast, SAT trends are different along the same zonal band, such as the band between 40° and 60°N. In Fig. 9.24a, the two positive SAT trends are centered at around 50°E and 60°W respectively, with the widespread negative SAT trends in other areas during boreal winter. This contrast of tropospheric temperature trends along the zonal band is clearly observed in Fig. 9.26a. The positive (negative) SAT trend at the lower troposphere is the result of local positive (negative) height trend center at the upper troposphere. According to the hydrostatic relationship, an upper tropospheric positive height anomaly may induce a positive (negative) temperature anomaly below (above). Therefore, the negative temperature trend center at the stratosphere can be explained by the positive height trend center at the upper troposphere. The similar relationship between the height and temperature trends is also observed in spring (Fig. 9.26b), except that there are more positive and negative trend centers during 1990–2012.

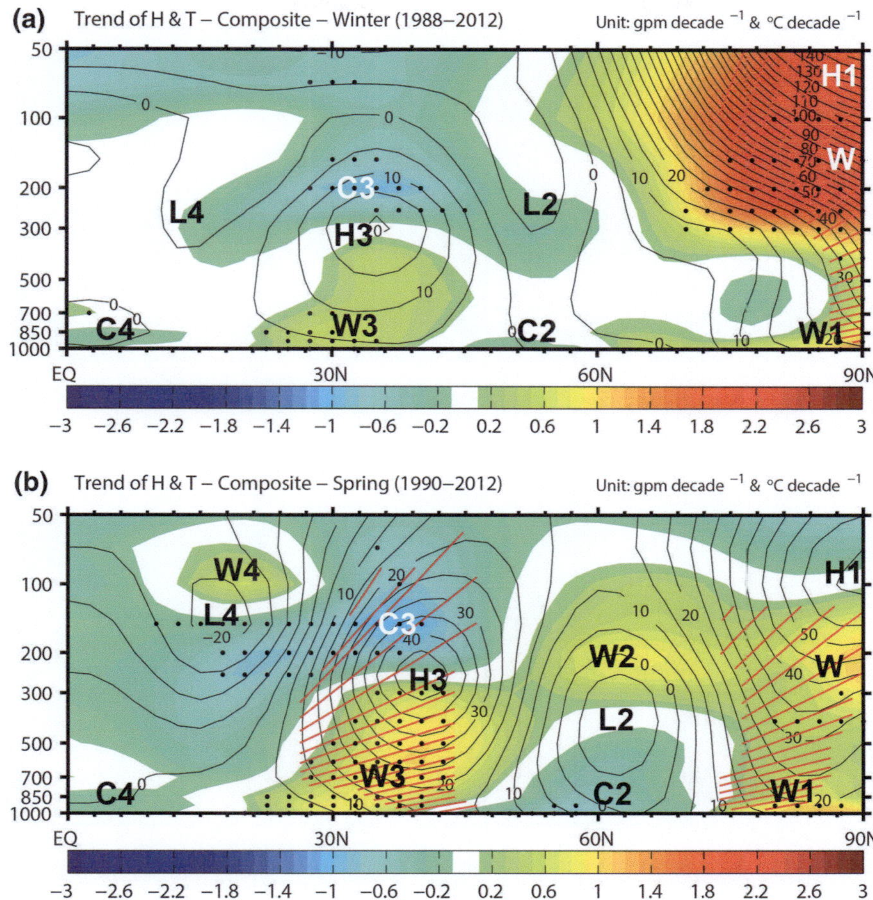

Fig. 9.25 Same as in Fig. 9.24c, d, except the composite of five reanalysis products

The meridional pattern of SAT trends in Fig. 9.24c, d can be further confirmed by the trends of zonally averaged height and temperature in Fig. 9.27a, b, respectively. The AA is partly the result of significant positive height trend centered at the stratosphere, which induces a significant positive temperature trend extending downward from the stratosphere to the troposphere and the surface Arctic area (Fig. 9.27a). The subtropical warming in the mid-low troposphere is the direct result of the enhancing subtropical high and the strengthening Hadley cell. Between the AA and the subtropical warming, the SAT cooling in the mid-latitude is associated with a negative height trend in the atmosphere. Similar result and relationship between the zonally averaged height and temperature trends can also be found from the composite of five reanalysis products (Fig. 9.27b).

The boreal winter and spring anomalous SAT series (Fig. 9.28a, b) correspond well to the meridional pattern of SAT trends with a warming trend in high latitude

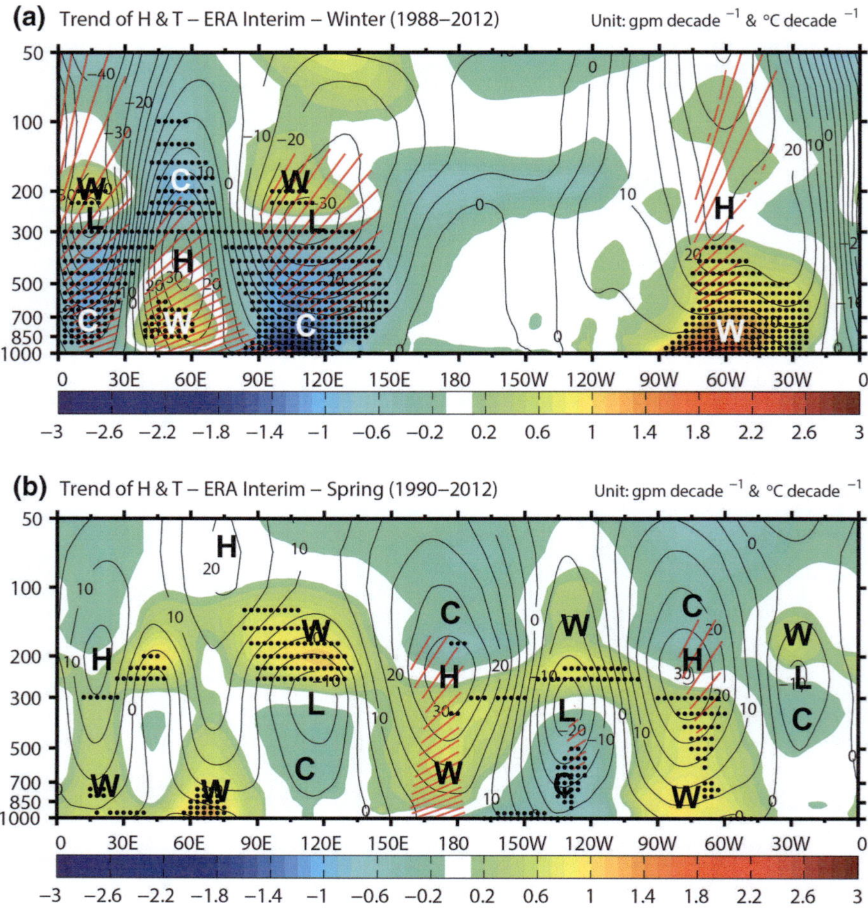

Fig. 9.26 Same as in Fig. 9.24c, d except along the zonal band between 40° and 60°N

and a cooling trend in the middle latitude since 1988 and 1990, respectively. The boreal winter SAT has the maximum warming trend after 1988 detected by the Mann-Kendall test. The boreal winter warming trend (1.31 °C decade^{-1}) in high latitude reaches the 98% confidence level, while the boreal spring cooling trend (−0.55 °C decade^{-1}) in the middle latitude reaches the 99% confidence level. Figure 9.27a, b shows that the maximum winter and spring warming trends are observed in the polar stratosphere and upper troposphere. The positive SAT trend is associated with positive height trend from the stratosphere to the troposphere. We believe that under this vertical structure of tropospheric height and temperature trends, the AA and SIC declining have a positive feedback. Figure 9.29a shows the time series of Arctic SIC anomaly in winter half year (DJFMAM) from 1979 to 2012. Before 1988, there was no obvious SIC anomalous trend but an abrupt decrease significantly since 1988 detected by the moving *t*-test. The stronger

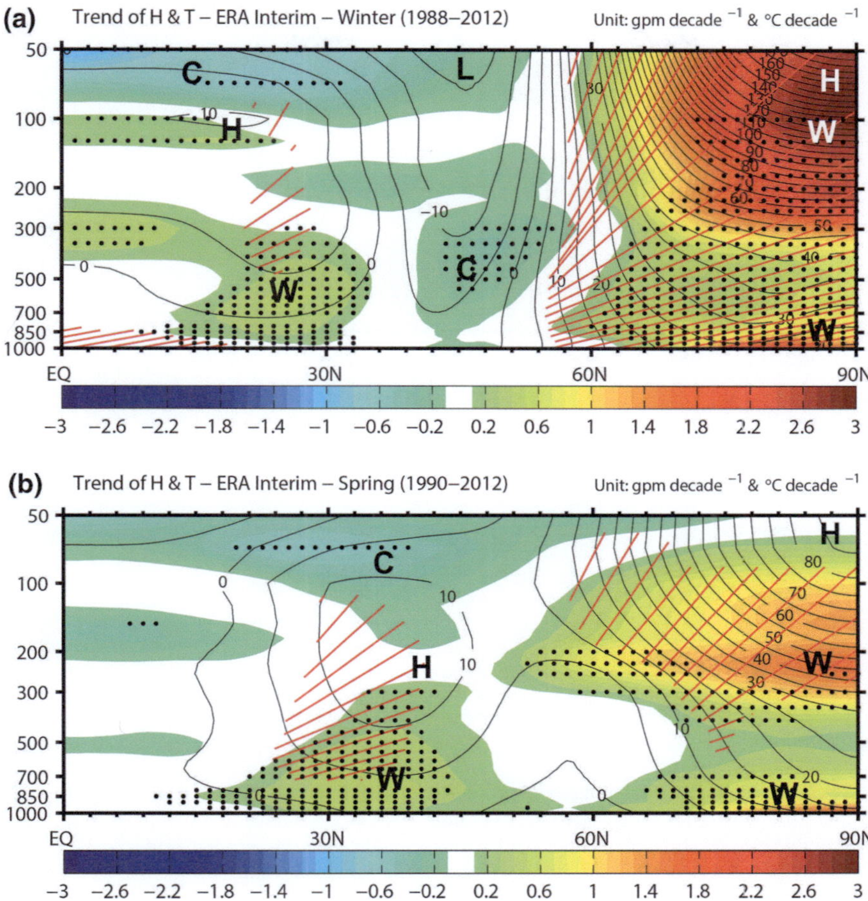

Fig. 9.27 **a** Trends of zonally averaged boreal winter height (*solid line*, 5 gpm decade^{-1} interval) and temperature (*shading*, 0.1 °C decade^{-1} interval) based on the ERA-Interim product during 1988–2012. **b** Same as **a** but of boreal spring height and temperature trends during 1990–2012. "H/L" and "W/C" are the centers of increasing/decreasing height and temperature trends, respectively. The *red-slant line* and *black-dotted* areas denote that the trends in height and temperature respectively reach the 95% confidence level

declining trend of Arctic SIC series is also found in summer half year (JJASON) in Fig. 9.29b.

A weak strengthening trend of the winter Polar cell and a weak weakening trend of the winter Arctic cell have started since 1988, while a significant strengthening trend of the spring Polar cell and a weak weakening trend of the spring Arctic cell are observed since 1990. The strength trends of the Polar and Arctic cells are indicated by the zonally averaged MSFs, which are significantly and seasonally influenced by land–sea thermal contrast caused by the topography of the Greenland. It is also found that the AA is not an isolated Arctic phenomenon in the NH. The

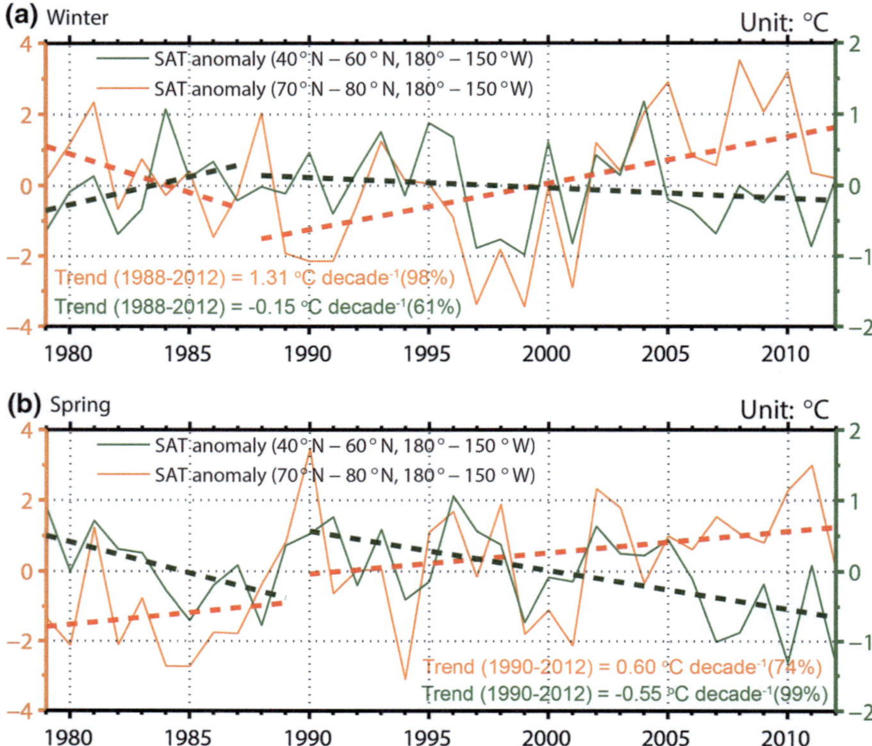

Fig. 9.28 The boreal **a** winter and **b** spring anomalous SAT series (°C) averaged over (70°–80°N, 180°–150°W) and (40°–60°N, 180°–150°W) from 1979 to 2012. The corresponding linear trends and significance levels of anomalous SAT series are indicated

meridional pattern of opposite SAT trends shows that the AA is mainly concentrated beyond 60°N. The zonal band between 40°N and 60°N has experienced a cooling SAT trend over the past two decades in winter and spring. The five reanalysis products show that the meridional pattern of SAT trends corresponds to a positive height trend in the whole polar atmosphere and a negative height trend in the mid-latitude troposphere. Below and above the positive (negative) height trend center, there are a warming (cooling) air column and a cooling (warming) air column vertically. The surface warming and cooling trends can be dynamically diagnosed from the height trend in the troposphere by the hydrostatic relationship.

The most significant declining trend of Arctic SIC is found in autumn, while weak declining trends of SAT are observed in winter and spring. This is because that the Arctic is iced in winter and spring over most areas of the Arctic Ocean. This may imply that the significant warming in the polar atmosphere during winter and spring can influence the SIC declining in late summer and autumn, just like the subsidence anomaly ahead of the SIC anomaly (Qian et al. 2015a).

The above descriptions show that the strengthening Polar cell and the weakening Arctic cell in the past two decades lead to the positive geopotential height and

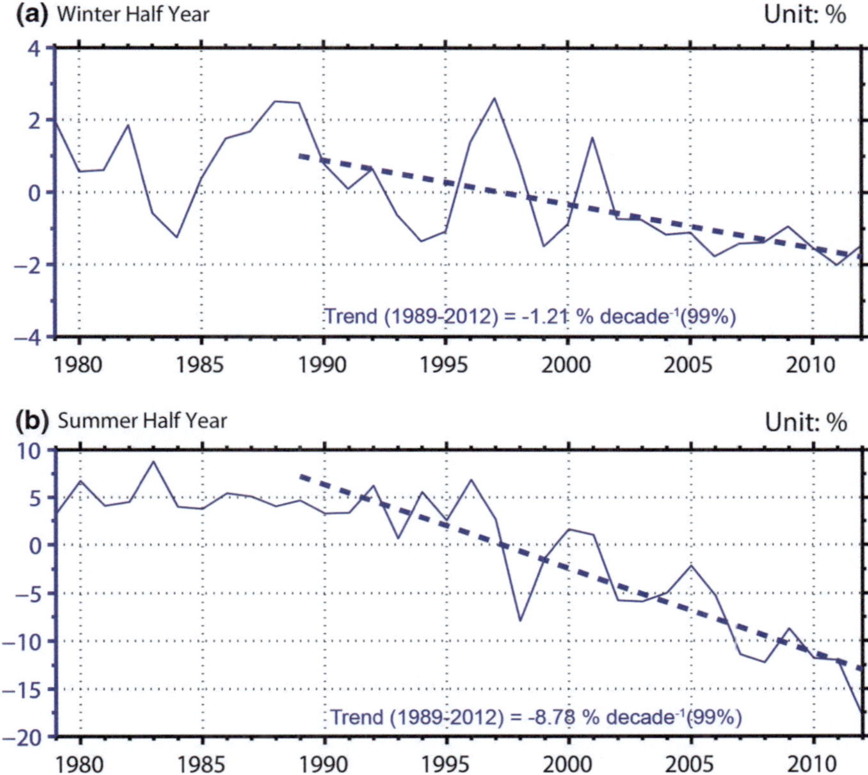

Fig. 9.29 Anomalous SIC series (%) of **a** winter half year (DJFMAM) and **b** summer half year (JJASON) from 1979 to 2012 along the zonal band 70°–80°N with 99% confidence level

temperature trends in the polar atmosphere and the SAT warming beyond 60°N. The SAT warming and the Arctic SIC declining have a positive feedback. Other causes of the AA need to be linked with the vertical and horizontal structures of height and temperature trends from the surface to the troposphere and the stratosphere in the Arctic.

9.6 Regional Relationship Between Arctic Sea Ice and Atmospheric Anomalies

Two record Arctic SIM events in 2007 and 2012 have been studied widely in recent years (Giles et al. 2008; Stroeve et al. 2008; Devasthale et al. 2013; Dong et al. 2014). Using the method of ensemble empirical mode decomposition (EEMD), there were four interannual Arctic SIM events since 1990, respectively in 1990, 1995, 2007, and 2012, and two decadal opposite phases of SIC anomalies during

1979–1989 and 2005–2013. The four SIM events are mainly contributed by the interannual variability, but the interdecadal variation also has non-negligible impacts. The opposite phases of decadal SIC anomalies are mainly characterized by regional positive and negative SIC anomalies along the Barents Sea, the East Siberian Sea to the Beaufort Sea in summer and autumn. The SIC anomalies with different timescales are associated with the local geopotential height, temperature, and wind anomalies in the upper and lower troposphere respectively, which dynamically satisfy the hydrostatic equilibrium. This spatial configuration of atmospheric height and temperature anomalies provides a new perspective for understanding the Arctic sea ice variations and the AA physically.

The AA has raised great concern in recent decade (Graversen et al. 2008; Screen and Simmonds 2010a, b). There are different explanations for the warming Arctic and reducing Arctic SIE, including oceanic and atmospheric causes. To discover the different regional and timescale features of Arctic sea ice variation, it is useful to expand the previous studies and explore the relationship between the regional sea ice variation in Arctic marginal seas and the atmospheric variable anomalies from intraseasonal, interannual to interdecadal timescales in the high latitudes of NH since 1979.

After monthly mean variables are decomposed as monthly mean climatology and anomaly, the ensemble empirical mode decomposition (EEMD) is used to decompose the 35-year (1979–2013) monthly mean series into multi-timescale oscillation components and a secular nonlinear trend. The EEMD was developed by Wu and Huang (2009) from the original empirical mode decomposition (EMD) by Huang et al. (1998). The EMD provides an adaptive time-frequency data analysis method. The generic principle of EMD is to decompose any time series into a set of Intrinsic Mode Function (IMF) components of the data itself, based on scale separation. The EEMD is further developed to eliminate mode mixing by adding white noise. The EEMD method has been used in climate change study in recent years (e.g., Qian et al. 2010; Wu and Qian 2015). The multidimensional EEMD (MEEMD) as done by Feng et al. (2014) is also employed, namely the EEMD process is applied on variables of each grid points, instead of the areal average of variables. For the MEEMD process, the ratio of the SD of the added white noise to that of the targeted time series is 1, and the ensemble size is 100 members.

The total SIE (the area of sea ice with SIC > 15%) in the NH and mean SIC over 60°–90°N are the two indices that indicate the amount of Arctic sea ice. Figure 9.30 shows the monthly mean series of the SIE anomalies in the NH and the Arctic SIC anomalies (60°–90°N) from 1979 to 2013 using the HadISST1 dataset. The historical SIC and SIE series can be divided into three decadal periods of 1979–1989, 1990–2004, and 2005–2013 according to the significant differences among their mean values and SDs (Fig. 9.30). For the two decadal periods of 1979–1989 and 2005–2013, their average SICs (SDs) are 1.37% (1.66%) and −2.51% (3.19%), respectively. The linear trends of SIE and SIC from 1979 to 2013 are -0.27×10^6 km^2 decade^{-1} (99% confidence level) and −1.33% decade^{-1} (99% confidence level), respectively. It is found that the SDs of the intraseasonal and interannual variability are large in the third decadal period, so the SIC anomalies

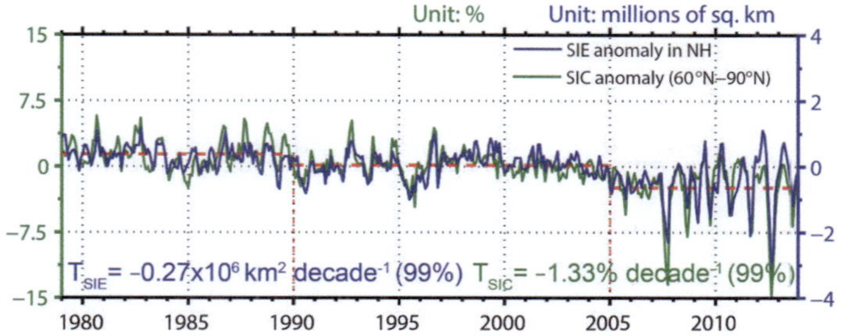

Fig. 9.30 Monthly mean series of the Arctic SIE anomalies (*blue line*, unit: 10^6 km^2) in the NH and the Arctic SIC anomalies (*green line*, unit: %) averaged over 60°–90°N from 1979 to 2013 using the HadISST1 dataset. The periodic mean (SD) of SIC anomalies over 1979–1989, 1990–2004, and 2005–2013 are 1.37% (1.66%), 0.14% (1.38%), and −2.51% (3.19%), respectively

from 1979 to 2013 vary on at least three timescales, including intraseasonal, interannual, and decadal timescales. There is a significant positive correlation (0.76, 99% confidence level) between the two series of SIE and SIC anomalies, so only the SIC anomalies are analyzed.

The spatial SD distributions of the SIC anomalies during three periods of 1979–1989, 1990–2004, and 2005–2013 are illustrated in Fig. 9.31. The largest area with SD beyond 15% is observed during the last period (Fig. 9.31c). In all the three periods, the high SD zone exceeding 15% is mainly located between 70°N and 80°N. Other high SD areas are observed in the Davis Strait, the Bering Sea, and the Sea of Okhotsk. The major high SD zone is divided into four regions indicated by letters "A" (0°–90°W, 70°–80°N), "B" (0°–90°E, 70°–80°N), "C" (90°E–180°, 70°–80°N), and "D" (90°W–180°, 70°–80°N) in Fig. 9.31. Regions A and B are partly open to the North Atlantic over the Greenland Sea, Norwegian Sea, and the Barents Sea, so their regional SIC anomalies should be influenced by the atmospheric flow and ocean current from the Atlantic. Differently, other two regions C and D, over the East Siberian Sea and the Beaufort Sea, are almost closed to the North Pacific.

Long-term linear trends of seasonal-mean SIC anomalies during 1979–2013 are different among four regions and four seasons (Table 9.4). The most significant declining trend (−9.88% decade^{-1} reaching 99% confidence level) is the autumn SIC anomalies in region C (the East Siberian Sea), then in region D (the Beaufort Sea; −7.44% decade^{-1}) and region B (the Barents Sea; −6.84% decade^{-1}) in autumn. Relatively weak trends are found in winter–spring and region A (near the Greenland). In regions C and D, increasing SIC trends are observed in winter and spring. For annual mean SIC anomalies, the maximum declining trend is found in region B (the Barents Sea) with the value −6.09% decade^{-1} and the minimum declining trend −0.73% decade^{-1} in region A (near the Greenland).

To quantitatively identify the secular nonlinear trend and multi-timescale oscillations of SIC, the four regional SIC series are decomposed by the EEMD method. In Fig. 9.32, each regional SIC series is decomposed into 8 IMF

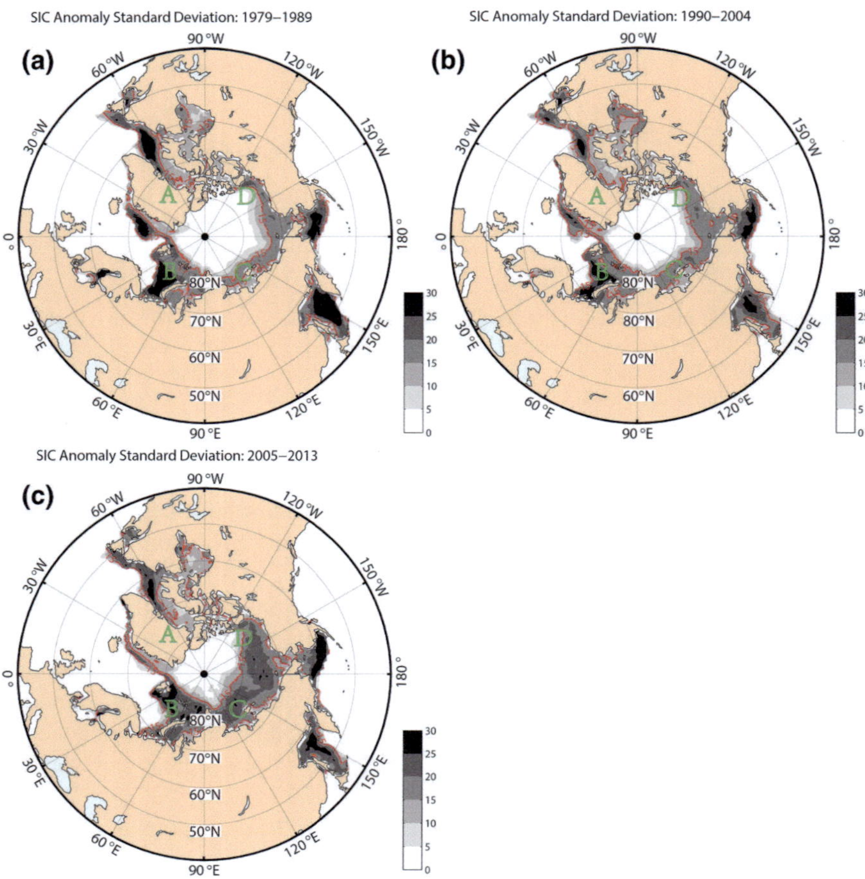

Fig. 9.31 SD (*gray shading*, 5% interval) of monthly mean SIC anomalies during **a** 1979–1989, **b** 1990–2004, and **c** 2005–2013 using the HadISST1 dataset. Symbols "A" (0°–90°W), "B" (0°–90°E), "C" (90°E–180°), and "D" (90°W–180°) show four divided regions in Arctic marginal seas over 70°–80°N. The *red line* covers the area with SD exceeding 15%

Table 9.4 Long-term linear trends (% decade^{-1}) and their confidence levels (%) of the seasonally mean SIC anomalies during 1979–2013 in four regions respectively

Trends	Regions A	Regions B	Regions C	Regions D
Spring	−0.17 (21%)	−4.34 (99%)	−0.23 (86%)	0.18 (93%)
Summer	−1.24 (81%)	−6.51 (99%)	−5.71 (99%)	−4.30 (99%)
Autumn	−1.34 (78%)	−6.84 (99%)	−9.88 (99%)	−7.44 (99%)
Winter	−0.61 (59%)	−5.69 (99%)	0.12 (89%)	0.35 (99%)
Annual mean	−0.73 (98%)	−6.09 (99%)	−3.91 (99%)	−2.80 (99%)

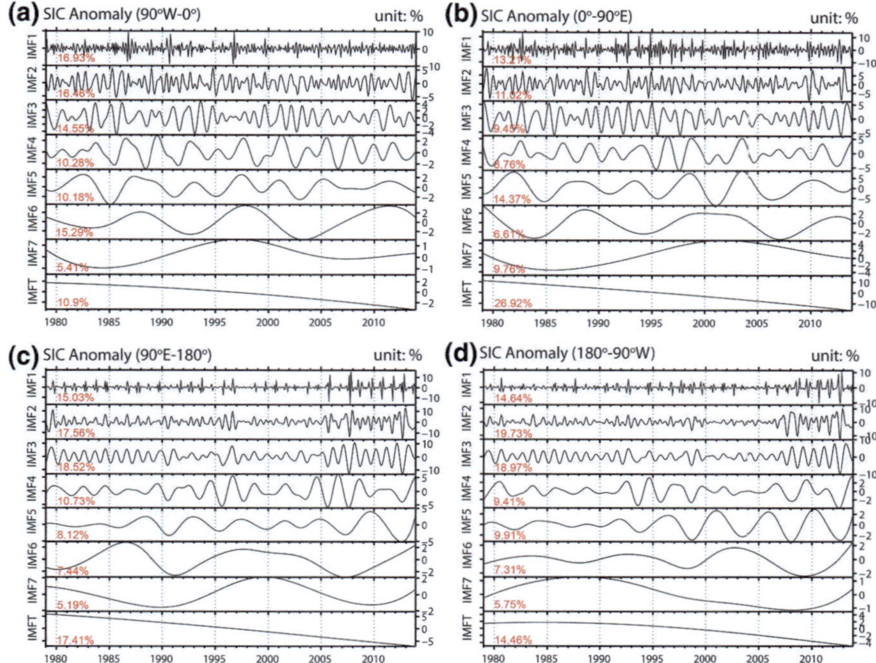

Fig. 9.32 The 8 IMF components of the monthly mean SIC anomaly (*black line*, unit: %) averaged in **a** region A (0°–90°W, 70°–80°N), **b** region B (0°–90°E, 70°–80°N), **c** region C (90°E–180°, 70°–80°N), and **d** region D (90°W–180°, 70°–80°N) from 1979 to 2013. The percentages (%) are the contribution of each component's SD to the sum of all SDs. IMF1–7 are the multi-timescale oscillations while IMFT is the nonlinear trend

components. The last component IMFT depicts a secular nonlinear trend. Each component from IMF1 to IMF7 is characterized as neither a regular oscillation nor an irregular oscillation, referring them as quasi-periodic oscillations. The first two components, IMF1 and IMF2, describe the intraseasonal oscillation. The interdecadal variation is contributed by the IMF6, IMF7, and IMFT components. So the interannual SIC events should be observed from the IMF3, IMF4, and IMF5 components. For the SIC variation in region A, the contributions of intraseasonal, interannual, and interdecadal timescale variations are 33, 35, and 32%, respectively (Fig. 9.32a). This implies that the three timescales are of similar importance to sea ice variability. Similarly, the three timescales of SIC variation are also detected in regions B, C, and D from Fig. 9.32b–d.

Using the EEMD method, the interannual SIC events can be quantitatively identified by summing up the three (IMF3 to IMF5) components (Fig. 9.33). A positive (negative) SIC event is defined when the interannual SIC anomalies exceed 10% (−10%) for 5 consecutive months. Neither positive nor negative anomalous SIC events can be identified in region A (Fig. 9.33a). In region B (Fig. 9.33b), three positive SIC events (1982, 1998, and 2003) and two negative

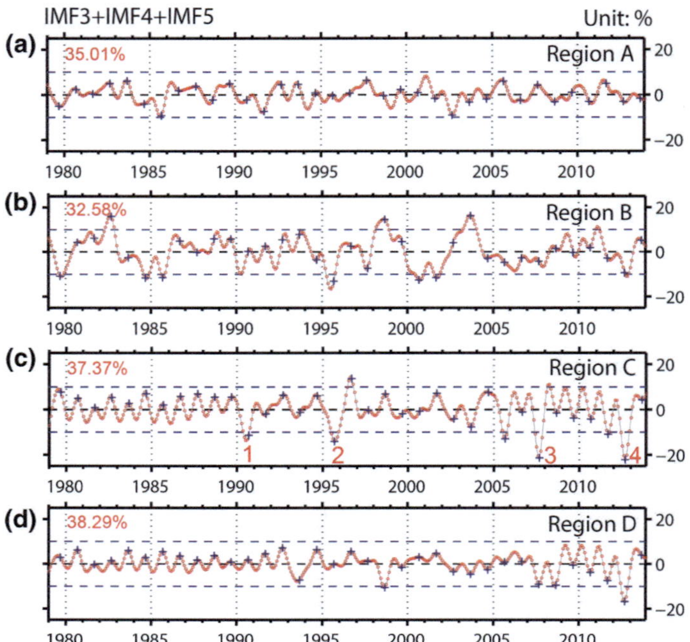

Fig. 9.33 The sum of IMF3 to IMF5 components of SIC anomalies (*grey line* with *red circle*, unit: %) averaged in **a** region A (0°–90°W, 70°–80°N), **b** region B (0°–90°E, 70°–80°N), **c** region C (90°E–180°, 70°–80°N), and **d** region D (90°W–180°, 70°–80°N) during 1979–2013. The percentages (%) are the contribution of interannual SDs to the sum of all SDs. The *red numbers* show the four negative SIC events respectively in 1990, 1995, 2007, and 2012. The *blue crosses* denote the values in September

SIC events (1995, 2000) reach the threshold. Four negative SIC events (1990, 1995, 2007, and 2012) and one positive SIC event (1996) can be observed from region C (Fig. 9.33c). In region D (Fig. 9.33d), only one negative SIC event (2012) reaches the threshold. Among all SIC events, only the two negative SIC events (2007, 2012) in region C both reach the thresholds of −20%. This is why they have been paid more attentions by scientists. The negative SIC event in 2012 (marked by "4") covered two regions (C and D). The negative SIC event in 1995 (marked by "2") influenced regions B and C. The negative SIC events in 2007 and 2012 (marked by "3" and "4") peaked in September while the event in 1990 (marked by "1") peaked in July. The event in 1995 peaked in July in region B, but in September in region C.

The interdecadal variation of SIC anomalies is a combination of the last three (IMF6, IMF7, and IMFT) components. The nonlinear long-term trend is not clear in regions A (Fig. 9.34a) and D (Fig. 9.34d) after adding the IMF6 and IMF7 to IMFT. The decline interdecadal variation (43.29%) is the largest in region B (Fig. 9.34b), implying that the long-term sea ice decline mainly occurs in region B. The first decadal transition (1989–1990) is confirmed by the two curves of regions

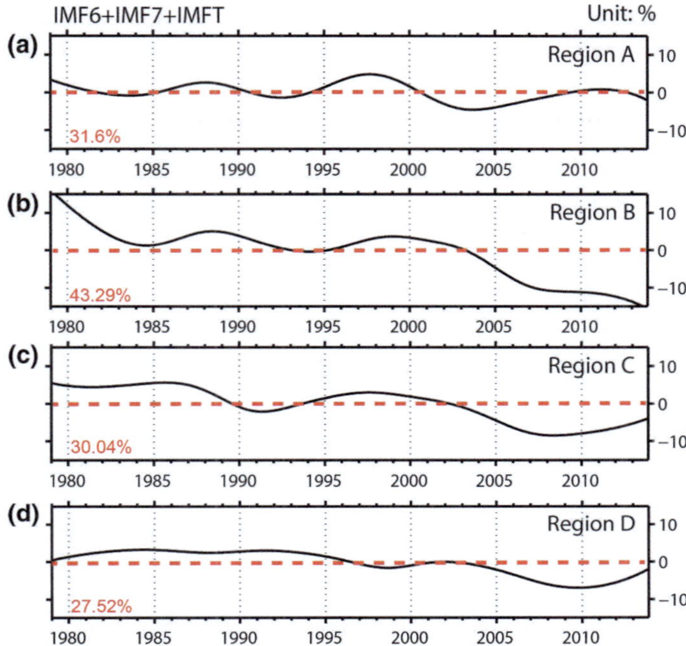

Fig. 9.34 Same as in Fig. 9.33 except for the sum of last three IMF components (IMF6, IMF7, and IMFT) of SIC anomalies (*black line*, unit: %)

Table 9.5 Four SIM events contributed by intraseasonal (IMF1 + 2), interannual (IMF3 + 4 + 5), interdecadal (IMF6 + 7 + T) and overall components of five-month mean SIC anomalies (%) averaged over region C and all regions, respectively

Region C	IMF1 + 2 (%)	IMF3 + 4 + 5 (%)	IMF6 + 7 + T (%)	Overall (%)
1990	−2.76	−13.96	−2.71	−19.43
1995	−0.99	−12.97	+2.22	−11.75
2007	−3.63	−17.69	−8.51	−29.84
2012	−3.62	−17.14	−6.35	−27.11
All regions	IMF1 + 2 (%)	IMF3 + 4 + 5 (%)	IMF6 + 7 + T (%)	Overall (%)
1990	−0.59	−4.23	+0.83	−4.00
1995	−0.04	−5.58	+0.99	−4.55
2007	−1.27	−7.54	−6.23	−15.0
2012	−1.89	−10.83	−5.40	−18.12

A (Fig. 9.34a) and C (Fig. 9.34c), while the second decadal transition (2004–2005) is obvious in the three curves of the regions B, C, and D (Fig. 9.34b–d).

Table 9.5 compares the contributions of SIC anomalies at each timescale to the four SIM events averaged over region C and all regions, respectively. For the four SIM events in region C, the sea ice decline is mainly contributed from the interannual components (−12.97 to −17.69%), while the intraseasonal components only account for small amounts (−0.99 to −3.63%). The interdecadal components also play a role in the 2007 (−8.51%) and 2012 (−6.35%) SIM events, but have an opposite impact (increasing sea ice) to the 1995 (+2.22%) SIM event. For all regions, the largest contribution of negative SIC anomalies also came from the interannual components (−4.23 to −10.83%), while the intraseasonal components (−0.04 to −1.89%) contribute much less. Negative contributions of interdecadal SIC anomalies to the SIM events are observed in 1990 and 1995. The above results suggest that the four SIM events are generally the impacts of interannual variability, while the interdecadal variability has opposite effects to the events in different decades. The intraseasonal variability is negligible, therefore, the intraseasonal contribution is not considered in the following analysis.

The four negative SIC or SIM events (1990, 1995, 2007, and 2012) identified above (Table 9.5) are mainly contributed from the interannual variability of SIC anomalies, so the horizontal distribution of interannual SIC anomalies is examined. Considering the peak months of the four SIM events, Fig. 9.35 shows the composite interannual (IMF3 + 4 + 5) monthly mean SIC anomalies averaged over five consecutive months centered in July 1990, August 1995, September 2007, and September 2012, respectively. The four center months are chosen with comprehensive consideration of their negative SIC peaks in regions B, C, and D. The strongest negative SIC anomalies occurred in 2007 with its center in regions C and D (Fig. 9.35c), while the largest coverage of negative SIC anomalies occurred in 2012 covering regions B, C, and D (Fig. 9.35d). For the two SIM events in 1990 and 1995, the negative SIC anomalies were mainly contributed from regions B and C. At the same time, there were large areas of positive SIC anomalies in region D (Fig. 9.35a–b). The distributions of SIC anomalies in the four SIM events indicate that the negative SIC anomalies are mainly concentrated in the EH along 70°–80°N. This distribution of SIC anomalies is consistent with the first mode of EOF with the contribution rate of 20.2% using annual mean detrended SIC anomalies from 1979 to 2013 (figure not shown). Figure 9.36 clearly shows that the two SIM events in 2007 and 2012 are also contributed by the interdecadal and long-term trend components over almost the four regions, while the two SIM events in 1990 and 1995 show opposite interdecadal contributions. Therefore, there is a new feature that Arctic sea ice has undergone a substantial long-term decline with superimposed interannual SIM events in past three decades.

The asymmetric distribution of the interannual SIC anomalies is clearly observed along 70°–80°N even during the strongest SIM event. In Fig. 9.35c, a positive area "P2" and a negative area "N1" of SIC anomalies respectively appear in the Greenland Sea and the East Siberian Sea (the dateline). The connection between the SIC anomalies and the tropospheric height-temperature anomalies is examined in Fig. 9.37. All the variables in Fig. 9.37 illustrate the interannual combination (IMF3 + IMF4 + IMF5) of SIC, height, and temperature anomalies. As an example of the 2007 SIM event, Fig. 9.37c shows a positive center ("P2" within green shading) and a negative center ("N1" within red shading) of SIC anomalies in the lower panel, which are respectively associated with the warm (W1, red shading)

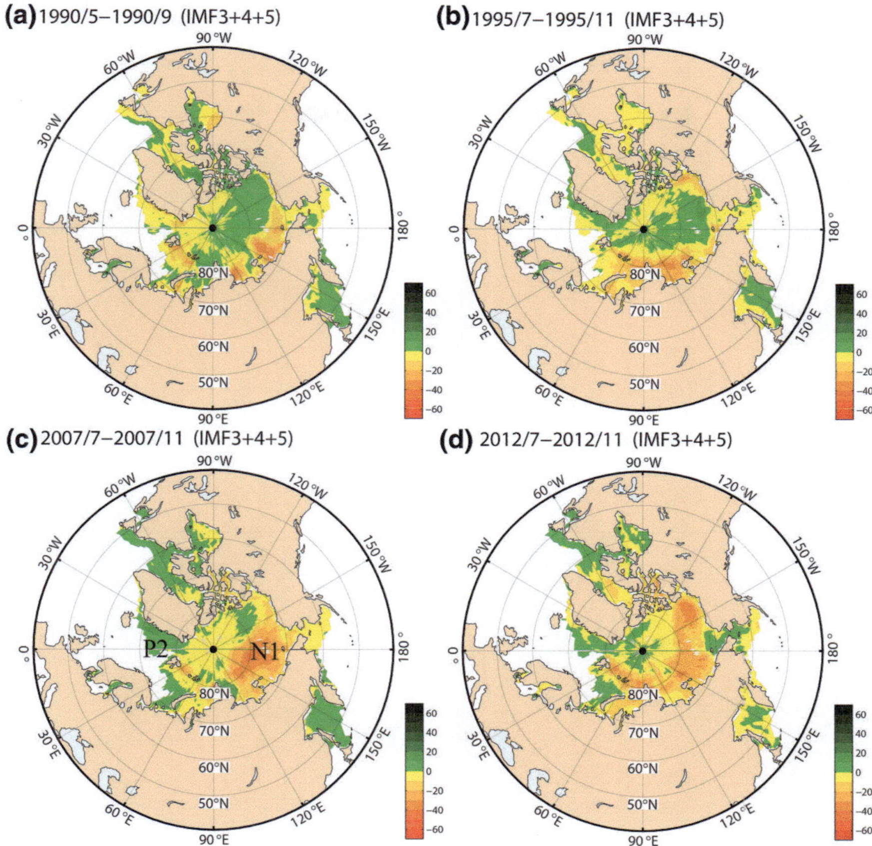

Fig. 9.35 Composite interannual (IMF3 + 4 + 5) SIC anomalies (*shading*, 10% interval) of the SIM events in **a** 1990, **b** 1995, **c** 2007, and **d** 2012, respectively

and cold (C2, green shading) temperature anomalies in the upper panel. It is observed that the warm center "W1" is located below the positive center ("H1," 500 gpm) of height anomalies at 400 hPa with another cold center of temperature anomalies at 200 hPa near the dateline. Another negative center "L2" of height anomalies at 200 hPa separates the cold center "C2" below and a warm center "W2" above. It is clear that the two centers of temperature anomalies "W1" and "C2" vertically correspond to the positive "H1" and negative "L2" centers of height anomalies, respectively, at the upper troposphere. Similar vertical configurations between height and temperature anomalies can also be observed in other three vertical sections in Fig. 9.37. In Fig. 9.37a, four warm air columns of temperature anomalies at the surface and the lower troposphere extended from the four positive centers (200 gpm) of height anomalies at the upper troposphere in 1990. Similarly in 1995, two warm air columns were located below two positive centers of height anomalies (Fig. 9.37b). In 2012, the largest positive center (700 gpm) of height

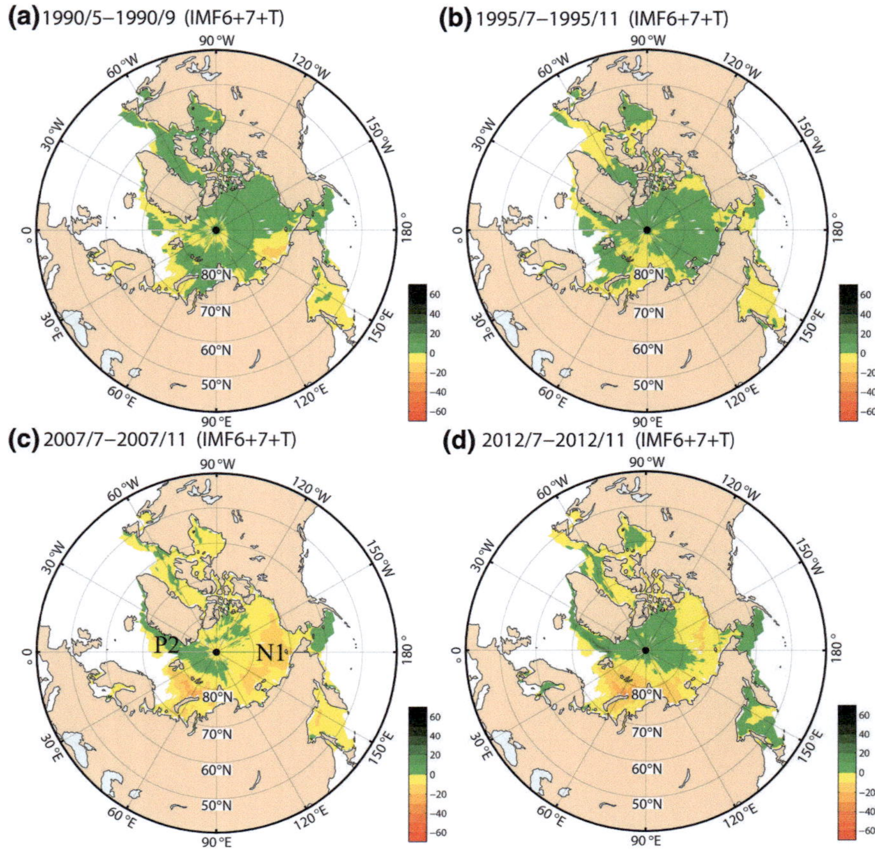

Fig. 9.36 Composite interdecadal (IMF6 + 7) and long-term trend (IMFT) SIC anomalies (*shading*, 10% interval) of the SIM events in **a** 1990, **b** 1995, **c** 2007, and **d** 2012, respectively

anomalies was located at the stratosphere and extended downward to the troposphere, so that there was the widest warm air column in the lower troposphere (Fig. 9.37d). This large coverage of warm air column can perfectly explain the surface warming event and the SIM event in 2012 (Figs. 9.35d and 9.37d).

This vertical configuration of height-temperature anomalies in Fig. 9.37 is entirely the same as previous studies of daily surface heat waves and cold events in China and western Eurasia. In these studies of surface warm (cold) events, the centers of temperature anomalies at the mid-low troposphere and the centers of height anomalies at the upper troposphere can be traced for 2–3 weeks in advance. The same vertical pattern is clear that a positive center "H" of height anomalies at the upper troposphere separates a positive center "W" below and a negative center "C" of temperature anomalies above. A reversed pattern is that a negative center "L" of height anomalies separates a negative center "C" below and a positive center "W" of

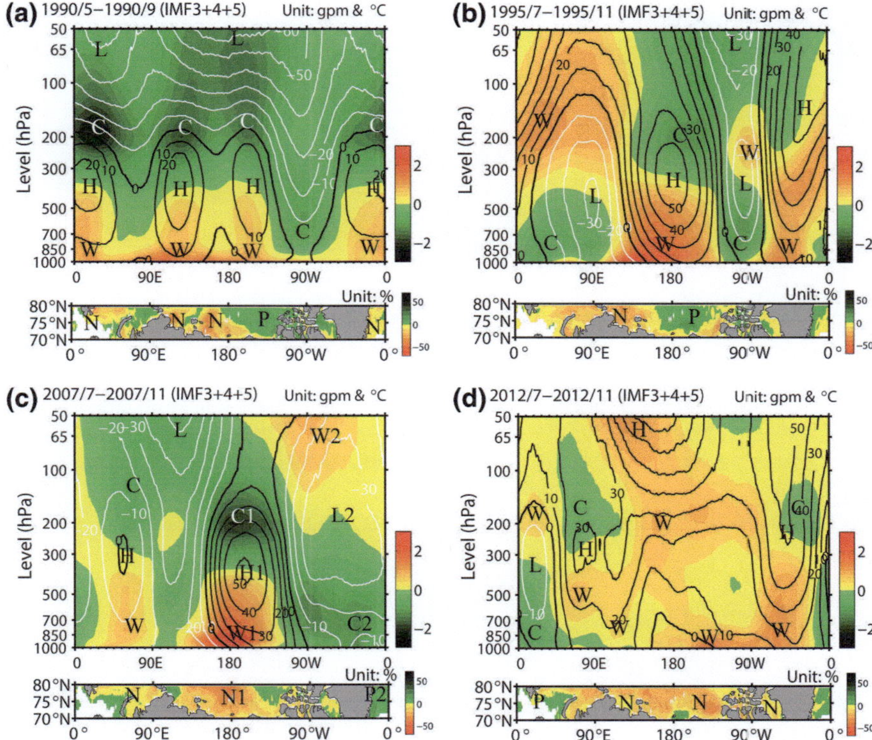

Fig. 9.37 Vertical (50–1000 hPa) and zonal sections of composite interannual (IMF3 + 4 + 5) height anomalies (*black contour* for positive and *white contour* for negative, 10 gpm interval) and temperature anomalies (*red shading* for positive and *green shading* for negative, 0.5 °C interval) averaged over 70°–80°N in the four SIM events in **a** 1990, **b** 1995, **c** 2007, and **d** 2012, respectively, using the monthly mean ERA-Interim product. The corresponding SIC anomalies (*red shading* for negative and *green shading* for positive, 10% interval) along the band of 70°N and 80°N in the four SIM events are plotted in the *lower part* of each panel

temperature anomalies above. This vertical pattern of height and temperature anomalies satisfies the hydrostatic equilibrium, i.e., the temperature anomalies can be derived from the vertical gradient of height anomalies. Therefore, a strong warm center "W1" is expected near the surface and it can explain the strongest SIM event.

The above description indicates that the interannual SIC events are directly associated with the lower tropospheric temperature anomalies. Then, the relationship between the surface wind and temperature anomalies is examined. Figure 9.38 is the composite SAT anomalies at the 2-m level and surface wind anomalies at the 10-m level over the spans of the four SIM events. For the strongest SIM event in 2007, the strongest southerly anomalies coincided with the strongest surface warm center of temperature anomalies (W1 in Fig. 9.38c) and the strongest negative center of SIC anomalies over the East Siberia Sea (N1 in Fig. 9.35c). In the same period, a positive center "P2" of SIC anomalies (Fig. 9.35c) and a negative center

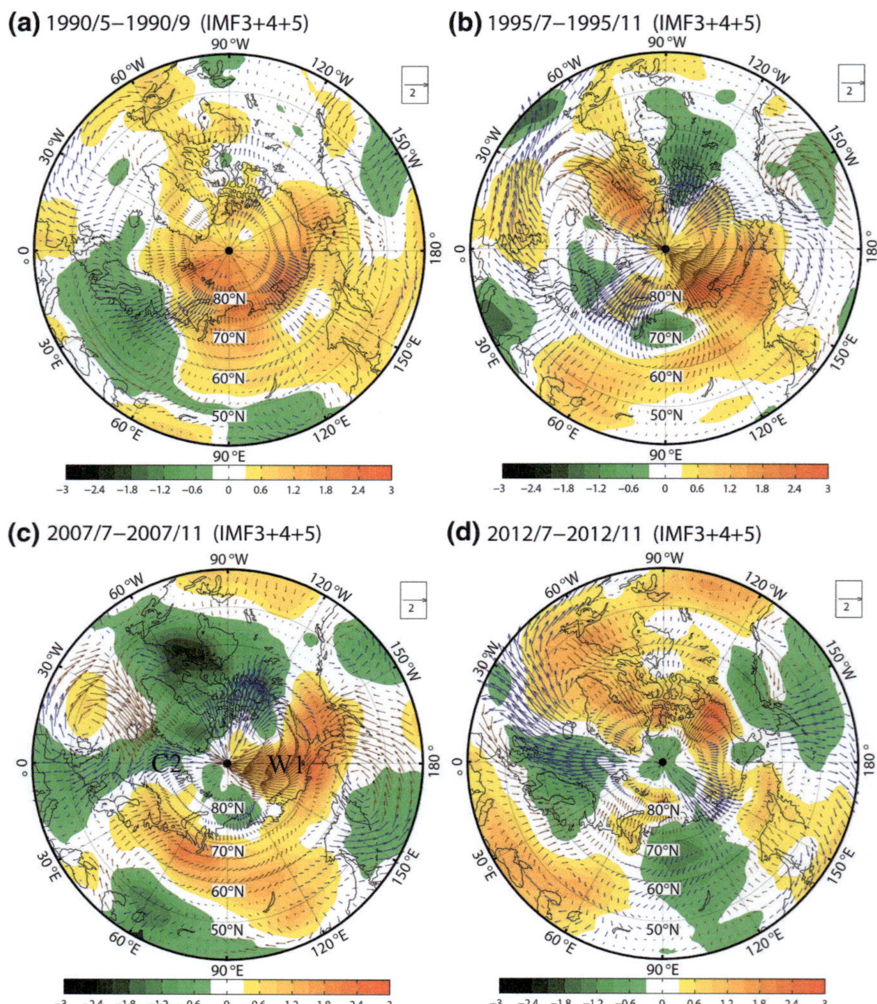

Fig. 9.38 Composite interannual (IMF3 + 4 + 5) SAT anomalies (*shading*, 0.3 °C interval) at the 2-m level and surface wind anomalies (*blue* vectors as northerly anomalies and *brown* vectors as southerly anomalies respectively, ms^{-1}) at the 10-m level in the four SIM events in **a** 1990, **b** 1995, **c** 2007, and **d** 2012, respectively, using the monthly mean ERA-Interim dataset. The time spans of the four SIM events are same as those in Fig. 9.35

"C2" of temperature anomalies (Figs. 9.37c and 9.38c) were directly linked to the northerly anomalies near the Greenland Sea. The opposite correlation can be observed from "N1" in Figs. 9.35c and 9.37c, with the strongest surface southerly in Fig. 9.38c. In 2012, only one positive area (Fig. 9.35d) of SIC anomalies in the Greenland Sea corresponded well with the negative SAT and northerly anomalies in Fig. 9.38d. Similar relationship among the SIC, SAT, and wind anomalies can also be found from the first two SIM events.

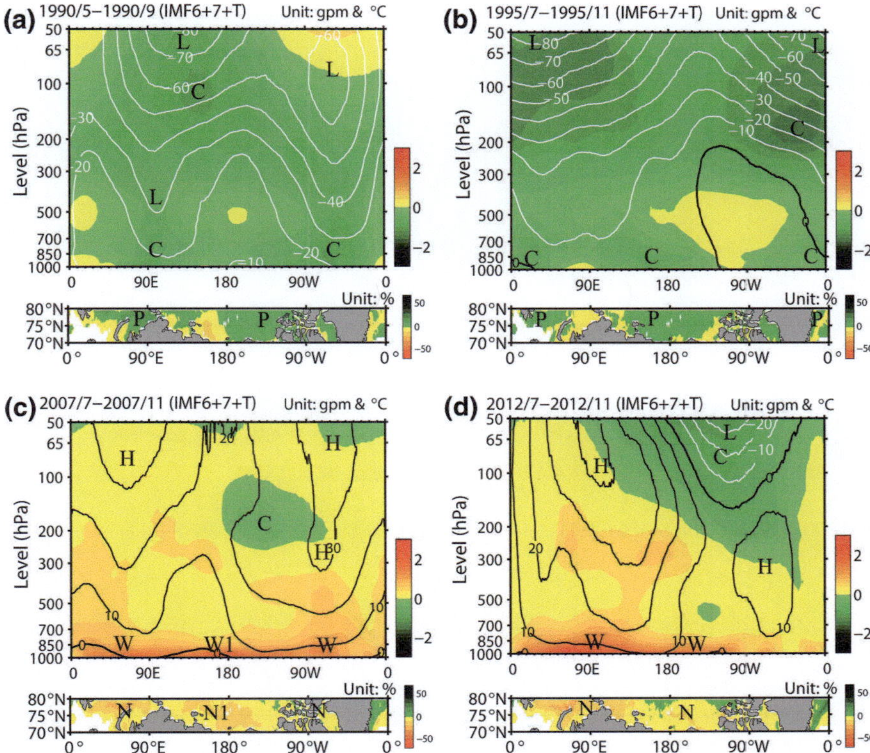

Fig. 9.39 Same as Fig. 9.37 but for the composite interdecadal and long-term trend (IMF6 + 7 + T) height anomalies (*black contour* for positive and *white contour* for negative, 10 gpm interval) and temperature anomalies (*red shading* for positive and *green shading* for negative, 0.5 °C interval)

The decadal air-ice connection can be further recognized from the last three IMFs components (IMF6 + IMF7 + IMFT) of monthly mean SIC anomalies and height-temperature anomalies (Fig. 9.39). For the four events, contributions from interdecadal and long-term trends are different between the last two events and the first two. For the last two events, decadal component of positive temperature anomalies in the mid-low troposphere is responsible for the decadal surface warm and the SIC negative anomalies, which are the results of decadal positive height anomalies at the upper troposphere (Figs. 9.36c–d and 9.39c–d). Reversely, the decadal component of negative temperature anomalies in the atmosphere is responsible for the decadal surface cold and the SIC positive anomalies, which are also the results of decadal negative height anomalies at the atmosphere (Figs. 9.36a–b and 9.39a–b).

The AA since this century and its contribution to the SIC anomalies can also be observed from the interdecadal component of SAT anomalies in Fig. 9.40. The decadal components of positive SAT anomalies in 2007 and 2012 covered over

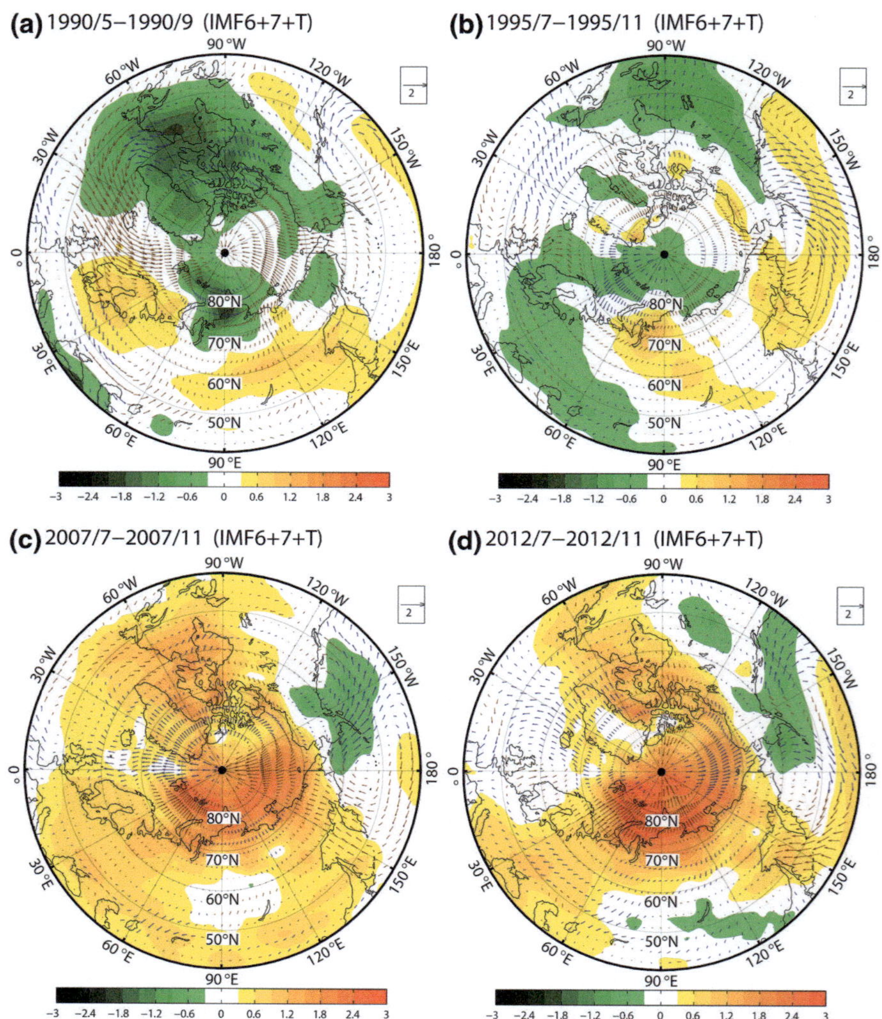

Fig. 9.40 Same as Fig. 9.38 but of composite interdecadal (IMF6 + 7 + T) SAT anomalies (*shading*, 0.5 °C interval) at the 2-m level and surface wind anomalies (*blue* and *red* vectors as northerly and southerly anomalies respectively, ms^{-1}) at the 10-m level

the whole NH, only except a small part in the Northeast Pacific (Fig. 9.40c–d). The warm center beyond 70°N in the EH was stronger in 2012 than that in 2007. This anomalous warm area in the EH was coincident with southerly anomalies at the interdecadal timescale. The interdecadal component of negative SAT anomalies in 1990 was mainly situated in the WH (Fig. 9.40a), while the decadal SAT anomalies was weak in 1995 (Fig. 9.40b). A strong negative center of SAT anomalies in eastern Canada and the Greenland was coincident with northerly anomalies at the interdecadal timescale (Fig. 9.40a). These positive and negative SAT anomalies can also be explained by the height anomalies because they also satisfy the hydrostatic equilibrium relationship at the interdecadal timescale.

9.7 The South Asian High and Eurasian Summer Climatic Anomalies[5]

Using the daily mean anomaly of atmospheric variables from the NCEP R1, Shi and Qian (2016) revealed the connection between anomalous zonal activities of the SAH and Eurasian climatic anomalies in boreal summer. An analysis of variance identifies two major domains with larger geopotential height variability, located, respectively, in the eastern and western flanks of the SAH at around 100 and 150 hPa. For both eastern and western domains, extreme events are selected during 1981–2014 with their normalized height anomalies greater than 1.0 (less than −1.0) SD for at least 10 consecutive days. Based on these events, four SAH modes including strong and weak Tibetan modes (STM and WTM) as well as strong and weak Iranian modes (SIM and WIM) are defined to depict the zonal SAH features. Positive composites show a robust wave-like pattern including a low center at 150 hPa denoting surface cold-wet anomalies over Mongolia and northern China (Kazakhstan and the Russia north of it) surrounded by three high centers at 150 hPa indicating surface warm-dry anomalies in the eastern (western) domain. Opposite distributions can be found in the negative composites of both domains (WTM and WIM). The SAT anomalies are the downward extension of an anomalous air column aloft while the precipitation is directly associated with the height anomalies above the air column.

As illustrated in Chap. 3, the SAH is a planetary-scale anticyclonic climatic system stably located in the upper troposphere and lower stratosphere in the NH during boreal summer. Many studies described its intensity variations and location shifts at 100 hPa (Mason and Anderson 1963; Zhang et al. 2002). Differently, Wei et al. (2014, 2015) studied the interannual variation of the SAH at 200 hPa. Only a few studies used the variables at 150 hPa to describe the SAH activities (Liu et al. 2009, 2013; Li et al. 2011). All these studies employed a fixed level mainly based on experience.

In zonal direction, the longitudinal shift of the SAH was first pointed out by Mason and Anderson (1963) and has been treated as the most significant feature of the SAH ever since then. More precisely, the SAH was classified into the east pattern, the west pattern, and the zonal (transitional) pattern according to its central location in total geopotential height (Luo et al. 1982). The three patterns have been widely used in previous studies from then on. Zhang et al. (2002) documented the longitudinal bimodality of the SAH including the Tibetan mode (TM) and the

[5]Modified article originally published in Shi and Qian (2016). Published with kind permission © American Meteorological Society. All Rights Reserved.

Iranian mode (IM) in a climatological perspective and its differences from the east–west shift in synoptic scale. Recently, Wei et al. (2014, 2015) constructed three SAH indices to elaborate the movements in zonal and meridional directions as well as the combined southeast–northwest movement. Relationships between these indices and summer rainfall anomalies in China and India have also been revealed in their studies.

In early years, Tao and Zhu (1964) pointed out that the east–west shift of the SAH has a quasi-biweekly timescale. The maintaining period for one mode is about 10–13 days (Luo et al. 1982), which is close to the 13–15 day oscillation period of the stream function over the Tibetan Plateau (TP) revealed by Krishnamurti et al. (1973). In recent decades, however, many studies illustrated the east–west shift of the SAH using datasets with coarse temporal resolution. For example, monthly mean data were employed in some studies (e.g., Huang and Qian 2003; Peng et al. 2010), while seasonal-mean results were also common choice (e.g., Tan et al. 2005; Wei et al. 2014, 2015). These studies treated the SAH seasonal (or monthly) features as a whole and did not consider the specific activities of the SAH within a month. Recently, on the intraseasonal timescales, Ren et al. (2015) investigated the subseasonal eastward extension of the SAH based on an index defined near its eastern flank. They attributed this movement to the joint role of diabatic heating feedback over eastern Asia and mid-latitude wave train. Further, they diagnosed the evolution of the SAH movement.

Based on previous studies, several issues need to be clarified. First, which level is the best to describe the SAH zonal activities? Second, how many patterns or modes are there to describe the SAH zonal variabilities and how long are their durations? Third, is there connection between the SAH and other crucial systems nearby at the same levels and what is the relationship between these systems and the anomalous climate pattern in Eurasia? To illustrate some background information of SAH, it is clear that the SAH acts with great seasonality. In May, it moves onto the Indo-China Peninsula from the western equatorial Pacific, and then moves northwestward in June and July (Fig. 9.41a). The location of the SAH in August is similar to that in July. It shrinks southeastward in September. So, the SAH is predominant from May to September, with the strongest intensity and widest coverage in boreal summer (June–July–August; JJA).

In order to find the best domain to reveal the SAH zonal activities, SDs of height anomalies are obtained in the summers during 1981–2010. Higher SDs indicate larger variability of the SAH. As the SAH is prominent in the upper troposphere and lower stratosphere, Fig. 9.41b plots the spatial distribution of gridded maximal SDs of height anomalies between 400 and 70 hPa over the whole summer (JJA). All maximal SDs are located between 300 and 100 hPa depending mainly on latitudes. The white-shading area in lower latitudes denotes that the SD peak does not exist between 400 and 70 hPa. In fact, over the white-shading area, SDs increase with height (figures not shown). In other words, there is no SD peak but monotonic increase within those certain pressure levels. As shown in Fig. 9.41b, the oval-shaped SAH as outlined by the 14,340-gpm contour bestrides above most of the southern part of Asia, with the ridge line (red line) climatologically along

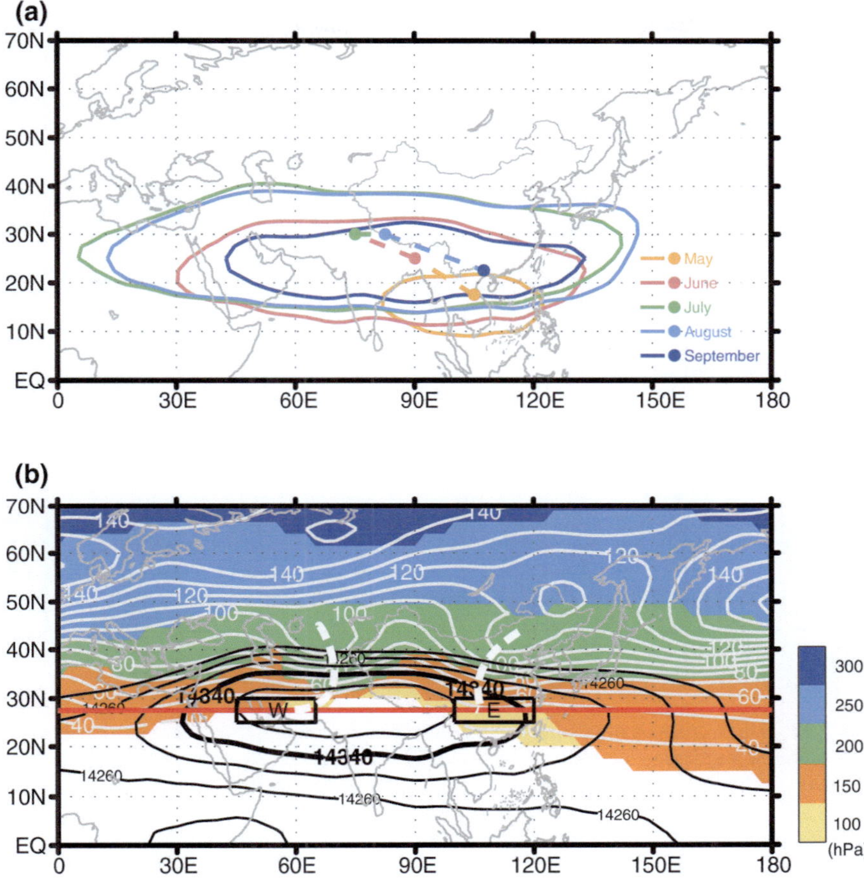

Fig. 9.41 **a** Climatic 14,300-gpm contour at 150 hPa from May to September (*solid lines*) and the *colored-dashed lines* indicate the seasonal track of maximum centers (*colored dots*) of the SAH. **b** Vertical height SD peaks (*white contour*, 10 gpm interval) between 400 and 70 hPa and corresponding pressure levels (*shading*) over JJA from 1981 to 2010. In **b**, climatic heights (*black lines*, 40 gpm interval) at 150 hPa are averaged over JJA; the *white shading* denotes the area with no SD peaks between 400 and 70 hPa; the *red line* denotes the mean ridge of the SAH in JJA; the two *white-dashed lines* denote the extension lines of maximal SDs from higher latitudes. The two *black boxes* are for domain-E (100°–120°E, 25°–30°N) and domain-W (45°–65°E, 25°–30°N). Sourced from Shi and Qian (2016) © American Meteorological Society, with permission

27.5°N. Therefore, three standard latitudes including 25°N, 27.5°N, and 30°N are chosen to reveal the zonal activities of the SAH. The body of the SAH mainly covers the longitudes from 30°E to 120°E, with its center at around 75°E (Fig. 9.41b). The primary domain of the SAH is mainly covered by orange and yellow shadings, denoting the peak levels of height SDs at 150 and 100 hPa, respectively.

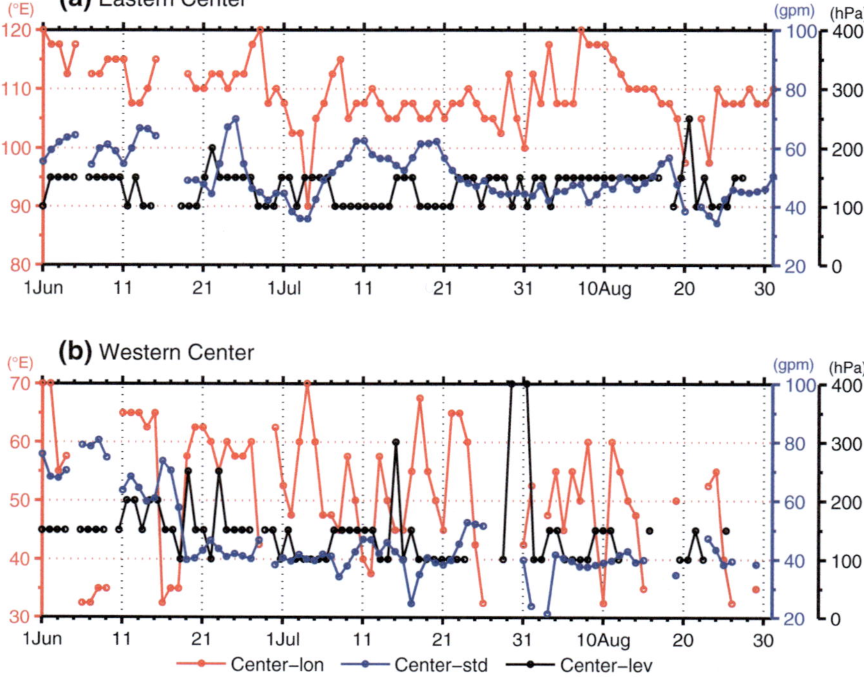

Fig. 9.42 Daily variations of maximal height SDs (*blue lines*, gpm) over 1981–2010 along 27.5° N with the corresponding longitudes (*red lines*, °E) and pressure levels (*black lines*, hPa) for the **a** eastern and **b** western centers of the SAH from 1 June to 31 August. Sourced from Shi and Qian (2016) © American Meteorological Society, with permission

Taking the standard latitudes as a reference, the SD peaks only exist in the two flanks of the SAH (Fig. 9.41b), with no SD peak near its central location. In addition, two SD extension lines (white-dashed lines in Fig. 9.41b) extend to the two SAH flanks from middle latitudes. So, the two SAH flanks are the emphasis. To determine the exact locations of the two disturbance centers with relatively higher SDs, daily variations of maximal height SDs over 1981–2010 and the corresponding longitudes and pressure levels along 27.5°N are shown in Fig. 9.42. The SDs of height anomalies in summer are smaller than those in other seasons (figures not shown). Most daily SD centers of height anomalies (black lines with black dots) are stably located at 150 hPa or 100 hPa (Fig. 9.42a, b).

For the eastern center (Fig. 9.42a), the maximal SD values (blue line) are stable with small oscillation and the longitudinal shift of its central location (red line) is relatively small. Similar results are obtained at 25°N and 30°N (figures not shown). In order to cover more positional information, horizontal grids (100°–120°E, 25°–30°N) are chosen for the following study. For vertical pressure levels, 100 hPa and 150 hPa are nearly of the same importance along 27.5°N (Fig. 9.42a), so they are both considered with the same weight. Moreover, 100 hPa (150 hPa) is the main level of maximal SDs at 25°N (30°N) (figures not shown), which is also consistent with Fig. 9.41b. In summary, the proportion of different longitudes and latitudes is equal while the levels are chosen at 100 hPa (150 hPa) at 25°N (30°N), both 100 and 150 hPa at 27.5°N for all the 9 zonal grids from 100°E to 120°E. The zone above composes the eastern crucial domain (domain-E) to describe the SAH zonal activities.

The height SD peaks near the western flank mainly range from 20 to 80 gpm with a larger east-west shift of its peak longitudes (Fig. 9.42b). Some missing points exist in Fig. 9.42b because the SD extrema near the SAH western flank disappear occasionally. Most height SD centers are located within 45°E and 65°E (9 zonal grids) in longitude and within 150 and 100 hPa vertically. The results along 25°N and 30°N (figures not shown) for the western flank are nearly the same, except the most important level at 30°N being only 150 hPa. Based on the above, horizontal grids (45°–65°E, 25°–30°N) are chosen. The proportion of different longitudes and latitudes is equal. In the vertical direction, pressure levels are chosen at both 100 and 150 hPa for 25°N and 27.5°N, while only 150 hPa is chosen for 30°N. So, domain-W (the western box in Fig. 9.41b) is determined to measure the SAH zonal activities in its western part.

The two 3D domains defined above are, respectively, referred to as domain-E and domain-W, hereafter. They are optimum to capture the SAH zonal variability and not fixed on a certain level. The zonal range in domain-W is larger than that of the IM defined by Zhang et al. (2002) which lies between 55° and 65°E, while domain-E is located to the east of their TM ranging over 82.5°–92.5°E. Recently, two indices occupying the longitudes (55°–75°E) and (85°–105°E) as well as (50°–80°E) and (85°–115°E) were defined by Wei et al. (2014, 2015). The zonal ranges defined by Wei et al. (2014, 2015) were based on the SAH locations from total height while our study determines the two domains by height SDs, which indicate the SAH variability. Similarly, the zonal range in domain-E is the same as that of the SAH eastward extension index defined by Ren et al. (2015), only with a smaller meridional range compared with theirs (22.5°–32.5°N).

Two daily series of height NAs averaged over domain-E (Eq. 9.7.1) and domain-W (Eq. 9.7.2) are, respectively, constructed in summers from 1981 to 2014 and they can well capture the SAH zonal activities day by day.

$$\begin{aligned} H_E = \Bigg[& \sum_{\mathrm{lon}=100}^{\mathrm{lon}=120} H(\mathrm{lon}, \mathrm{lat} = 25°\mathrm{N},\ \mathrm{lev} = 100\,\mathrm{hPa}) \\ & + \sum_{\mathrm{lon}=100}^{\mathrm{lon}=120} H(\mathrm{lon}, \mathrm{lat} = 27.5°\mathrm{N},\ \mathrm{lev} = 100\,\mathrm{hPa}) \\ & + \sum_{\mathrm{lon}=100}^{\mathrm{lon}=120} H(\mathrm{lon}, \mathrm{lat} = 27.5°\mathrm{N},\ \mathrm{lev} = 150\,\mathrm{hPa}) \\ & + \sum_{lon=100}^{lon=120} H(\mathrm{lon}, \mathrm{lat} = 30°\mathrm{N},\ \mathrm{lev} = 150\,\mathrm{hPa}) \Bigg] / (9 \times 4) \end{aligned} \tag{9.7.1}$$

$$
\begin{aligned}
H_W = & \left[\sum_{\mathrm{lon}=45}^{\mathrm{lon}=65} H(\mathrm{lon}, \mathrm{lat}) = 25^{\circ}\mathrm{N}, \ \mathrm{lev} = 100\,\mathrm{hPa}) \right. \\
& + \sum_{lon=45}^{lon=65} H(\mathrm{lon}, \mathrm{lat}) = 27^{\circ}\mathrm{N}, \ \mathrm{lev} = 150\,\mathrm{hPa}) \\
& + \sum_{\mathrm{lon}=45}^{\mathrm{lon}=65} H(\mathrm{lon}, \mathrm{lat}) = 27.5^{\circ}\mathrm{N}, \ lev = 100\,\mathrm{hPa}) \\
& + \sum_{lon=45}^{lon=65} H(\mathrm{lon}, \mathrm{lat}) = 27.5^{\circ}\mathrm{N}, \ \mathrm{lev} = 150\,\mathrm{hPa}) \\
& \left. + \sum_{\mathrm{lon}=45}^{\mathrm{lon}=65} H(\mathrm{lon}, \mathrm{lat}) = 30^{\circ}\mathrm{N}, \ \mathrm{lev} = 150\,\mathrm{hPa}) \right] / (9 \times 5)
\end{aligned} \tag{9.7.2}
$$

The extreme SAH events can be determined according to the above two series, respectively. A positive event occurs when the height NAs exceed 1.0 SD for at least 10 consecutive days with at most one-day interruption. The definition for a negative event is similar, but with the criterion for less than −1.0 SD. The absolute value of average NAs indicates the intensity of an event. All persistent events in domain-E and domain-W together with their average intensities and orders are, respectively, listed in Tables 9.6 and 9.7. It should be noted that no filter or running mean is used here, in order to examine the extreme events contributed by oscillations from different timescales.

Table 9.6 Persistent (equal to or longer than 10 days) and extreme [larger (smaller) than 1.0 (−1.0) SD] events including their years, start dates, durations, average intensities and orders in summers (June 1 –August 31, JJA) from 1981 to 2014 in domain-E

Year	Start date	Duration	Average NA	Order	Year	Start date	Duration	Average NA	Order
1981	**14 August**	**18**	**1.79**	**1**	1992	06 July	14	−1.62	5
1983	04 July	20	1.69	4	**1997**	**10 June**	**20**	**−1.98**	**1**
1987	12 June	23	1.77	2	1999	19 August	13	−1.88	2
1988	05 June	13	1.54	5	2002	09 August	15	−1.88	2
1988	15 July	12	1.41	9	2004	07 June	13	−1.40	7
1990	15 August	10	1.49	7	2004	04 July	36	−1.72	4
1991	20 July	16	1.74	3	2008	03 August	13	−1.48	6
1998	17 July	10	1.29	12					
2003	06 August	11	1.26	13					
2006	10 July	15	1.40	10					
2009	13 July	13	1.39	11					
2009	30 July	14	1.50	6					
2010	28 June	13	1.47	8					

The orders are based on their average NAs. The strongest events are in bold fonts. Sourced from Shi and Qian (2016) © American Meteorological Society, with permission

Table 9.7 Same as Table 9.1, but for domain-W

Year	Start date	Duration	Average NA	Order	Year	Start date	Duration	Average NA	Order
1983	22 July	20	1.69	3	1982	25 June	23	−1.33	11
1983	18 August	14	1.38	9	1991	24 June	10	−1.64	9
1984	09 June	10	1.21	10	1992	01 June	17	−1.88	4
1988	12 July	15	1.73	2	1992	08 July	14	−1.73	7
1988	05 August	13	1.58	6	1992	10 August	12	−1.50	10
1991	30 July	12	1.39	8	1996	19 August	13	−1.76	6
1998	**24 June**	**22**	**1.98**	**1**	1997	01 June	15	−2.34	3
1998	09 August	12	1.63	4	1999	14 August	10	−1.86	5
2006	03 July	12	1.49	7	**2000**	**18 July**	**14**	**−2.44**	**1**
2009	05 August	13	1.60	5	2004	10 July	19	−1.69	8
					2009	03 June	24	−2.42	2

Sourced from Shi and Qian (2016) © American Meteorological Society, with permission

For domain-E (Table 9.6), 13 positive events lasted for 10–23 days with average intensities of 1.26–1.79 SDs, while 7 negative events occurred with average intensities of −1.40 to −1.98 SDs. The longest event lasted for 36 days. In the western domain, 10 positive events persisted for 10–22 days with average intensities of 1.21–1.98 SDs (Table 9.7). The durations of the 11 negative events ranged from 10 to 24 days with higher intensities than those of the positive events.

In domain-E (Table 9.6), positive events were less common in the 1990s, with the same number in the 1980s and 2000s. There were more negative cases in the 2000s while no negative event occurred in the 1980s. In domain-W (Table 9.7), 5 positive events but only one negative event occurred in the 1980s. In the 1990s, however, there were 3 positive events and 7 negative events. The number of events in the 2000s was similar between positive and negative ones. The distribution for different months is uniform. In addition, most SAH events lasted for less than one month, leading to their intraseasonal feature of circulation anomalies (Tables 9.6 and 9.7). Therefore, these extreme events are random phenomena and it is reasonable to use the daily mean variables instead of monthly mean or seasonal-mean variables.

The strongest positive and negative events in each domain are highlighted in Tables 9.6 and 9.7. Their time series are plotted in Fig. 9.43. In domain-E, the strongest positive case (3.03 SDs; Fig. 9.43b) and the strongest negative case (−2.62 SDs; Fig. 9.43d) peaked on 25 August 1981 and 24 June 1997, respectively. In domain-W, the strongest positive case (Fig. 9.43a) and the strongest negative case (Fig. 9.43c) occurred with their peaks (2.85 and −3.32 SDs) on 29 June 1998 and 27 July 2000, respectively.

Composite total, climatic and anomalous heights for all the positive and negative cases in the two domains are presented in Fig. 9.44. For the positive composite in domain-W (Fig. 9.44a), a positive center (white H) of height anomalies is identified

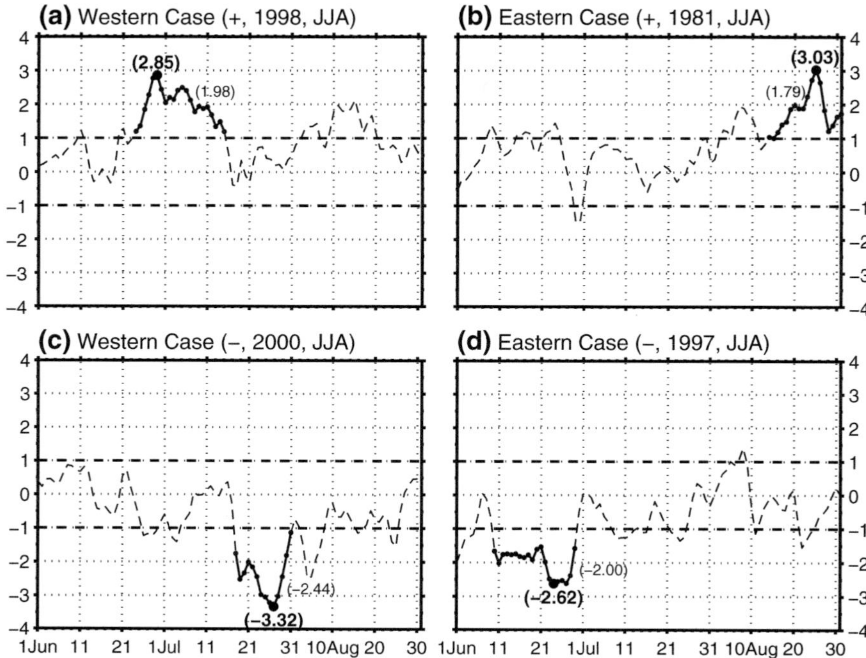

Fig. 9.43 Normalized series of daily height anomalies in JJA for the four strongest events. **a** The strongest positive case in domain-W in 1998. **c** The strongest negative case in domain-W in 2000. Panels **b** and **d** are the corresponding cases in domain-E in 1981 and 1997, respectively. The *thick solid lines with dots* indicate the event processes. The numbers in *brackets with bold* and *normal fonts* denote the strongest and average intensities of the events, respectively. In each panel, two *thick-dashed lines* denote ±1 SD as the criteria for event identification. Sourced from Shi and Qian (2016) © American Meteorological Society, with permission

near this domain and the SAH strengthens with its center (black H) moving westward over the Iranian Plateau (IP) compared with its climatic location (red H). Moreover, most of the plotted area is covered by positive anomalies over the subtropical area, indicating the consistent variations of geopotential heights in planetary scale. The SAH center shows a preference for the largest positive anomalies. In the negative composite for domain-W (Fig. 9.44c), the SAH center (black H) located over the TP is much weaker in intensity, compared to its climatic center (red H). The SAH shrinks and strides over the TP due to the negative center (white L) near domain-W (Fig. 9.44c). Similarly in domain-E, an anomalous positive center (white H) superimposes the climatic height and forms a strong SAH with its center over the TP for the positive composite (Fig. 9.44b). The SAH (black H) shifts eastward and is located near the anomalous high center (white H). Reversely, a negative center (white L) near domain-E weakens the climatic SAH (red H) and forms a weak SAH (black H) over IP in the negative composite for domain-E (Fig. 9.44d). In fact, the disturbances in the two domains take the most important responsibility. In summary, four SAH modes are detected to reveal zonal

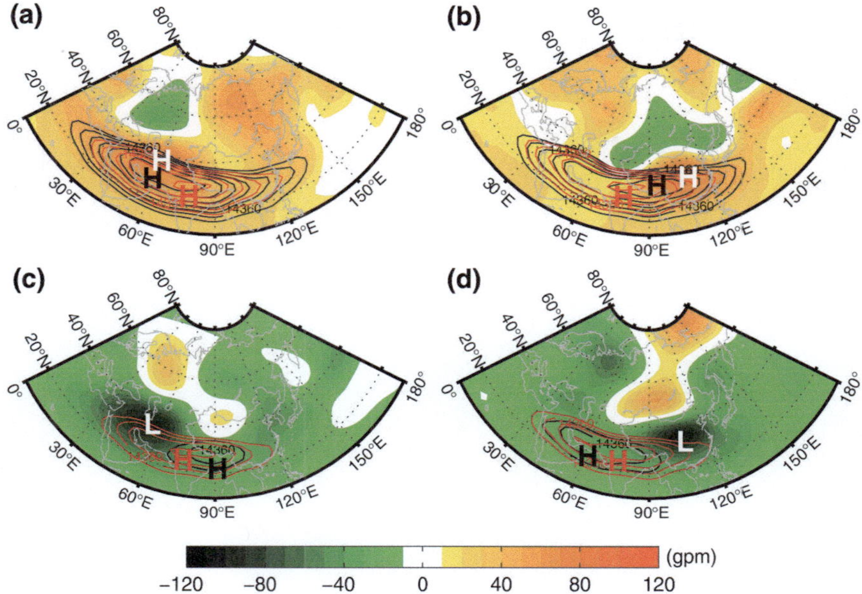

Fig. 9.44 Composites of height anomalies (*shading*, 10 gpm interval), total heights (*black contour*, 20 gpm interval, 14,360 gpm marked), and corresponding climatic heights (*red contour*, 20 gpm interval, 14,360 gpm marked) denoting the SAH at 150 hPa. Panels **a** and **c** are the composites of positive and negative events in domain-W, respectively. Panels **b** and **d** are the same but in domain-E. *Black* letter "H" indicates the center of the SAH in original (total) field, while the *red* one is for the climatic field. *White* letters "H/L" indicate the anomalous high/low centers. Sourced from Shi and Qian (2016) © American Meteorological Society, with permission

variabilities of the SAH, showing different characteristics in location and intensity. They can be named as strong IM (SIM; Fig. 9.44a), strong TM (STM; Fig. 9.44b), weak TM (WTM; Fig. 9.44c) and weak IM (WIM; Fig. 9.44d), respectively. The four SAH modes are persistent zonal modes, which are similar to the east and west patterns of Luo et al. (1982) in location and duration.

It is conspicuous that the four SAH modes are actually associated with the four anomalous height centers at the upper troposphere near domain-E and domain-W (Fig. 9.44). Further examination of the composite vertical structures of height and temperature anomalies in domain-E, from the troposphere to the lower stratosphere is given in Fig. 9.45. It should be noted that all the composite analyses in the following are the results averaged over the whole duration of each case, with no emphasis on their evolutions. Vertical structures along 27.5°N for the positive and negative composites in domain-E are shown in Fig. 9.45a, c, respectively. When STM occurs, a warm center (W1) below and a cold center (C1) above are separated by a positive height center (H1) at the upper troposphere around 150–100 hPa (Fig. 9.45a). The zonal range of H1 resembles that in domain-E. Positive height anomalies are prevalent over the whole hemisphere in the planetary

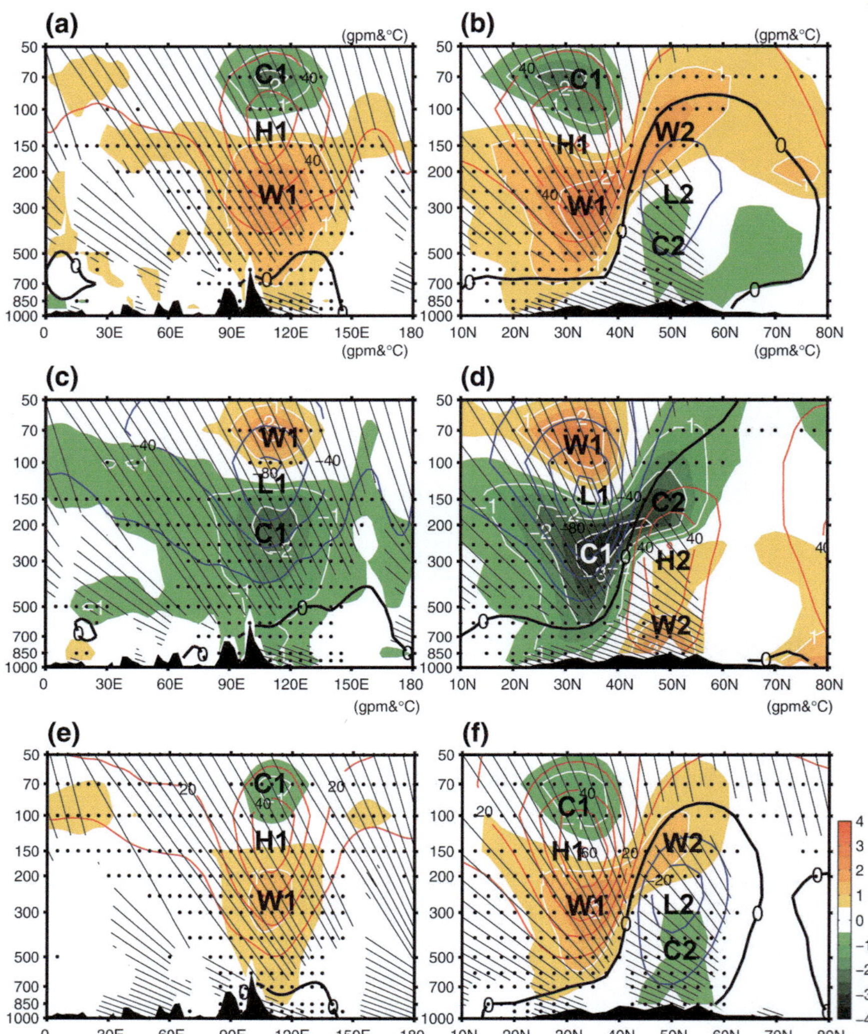

Fig. 9.45 Composite vertical sections of height anomalies (*red contour* for positive and *blue contour* for negative, 20 gpm interval) and temperature anomalies (*white solid contour with shadings*, 1 °C interval) along **a** 27.5°N and **b** 110°E for positive events in domain-E. Panels **c** and **d** are the same as **a** and **b**, respectively, but the composites for negative events. *Black slashes* for height and *dots* for temperature in **a**–**d** indicate the composite differences exceeding 95% confidence level. **e** Regressed height anomalies (*red contour* for positive and *blue contour* for negative, 10 gpm interval) and temperature anomalies (*shadings with white contour*, 1 °C interval) along 27.5°N against daily height NA series in domain-E. **f** The same as **e**, but along 110°E. *Black slashes* for height and *dots* for temperature indicate the areas exceeding 95% confidence level. Letters "H/L" and "W/C" indicate the high/low centers of height anomalies and the warm/cold centers of temperature anomalies, respectively. Shi and Qian (2016) © American Meteorological Society, with permission

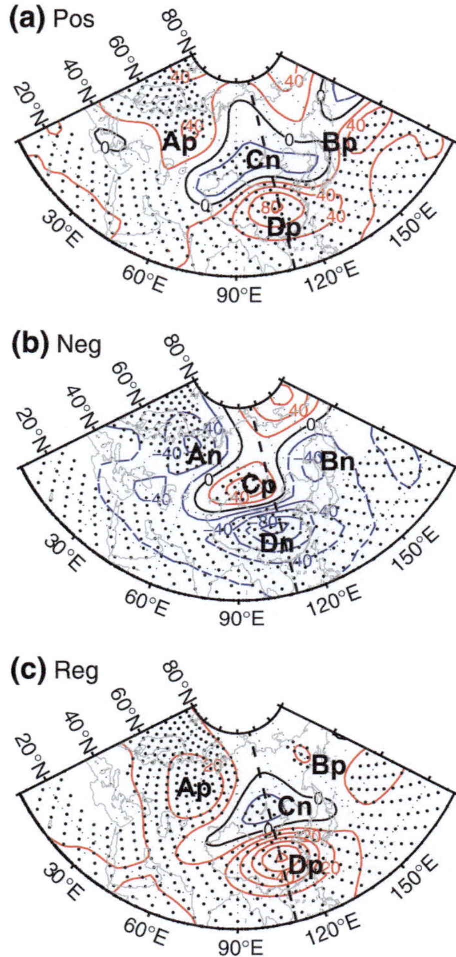

Fig. 9.46 **a** Positive composite of height anomalies (*red contour* for positive and *blue contour* for negative, 20 gpm interval) at 150 hPa in domain-E. **b** The same as **a**, but for negative composite. *Dots* in **a** and **b** denote the composite differences exceeding 95% confidence level. **c** Regressed height anomalies (*red contour* for positive and *blue contour* for negative, 10 gpm interval) at 150 hPa against daily height NA series in domain-E. *Dots* indicate the areas exceeding 95% confidence level. Letters "A," "B," "C," and "D" denote four areas with positive "p" and negative "n" for height anomalies. The *dashed lines* are along 110°E. Sourced from Shi and Qian (2016) © American Meteorological Society, with permission

scale. Opposite features for both height and temperature anomalies can be observed in the negative composite (WIM; Fig. 9.45c). Composite results for height and temperature anomalies along 110°E are also calculated in Fig. 9.45b, d. A dipolar height pattern including a positive center (H1) near 30°N and a negative center (L2) near 50°N is noted in the meridional direction (Fig. 9.45b). The center H1 separates a warm center (W1) below and a cold center (C1) above, while the center L2 separates a cold center (C2) below and a warm center (W2) above. The vertical location of H1 is around 150 hPa while L2 is lower and located near 300 hPa. A reversed and significant dipole is evident in the negative composite in domain-E (Fig. 9.45d). The height and temperature anomalies satisfy the hydrostatic relationship. It means that the temperature anomalies from the troposphere to stratosphere can be derived from the height anomalies. It can be concluded that the SAT anomalies are the downward extension of temperature anomalies below the height anomalies at the upper troposphere. Generally, regression patterns against daily height NA series in domain-E reconfirm the results above in the vertical sections (Fig. 9.45e, f). In addition, height and temperature anomalies in the negative composite are stronger than those in the positive composite in domain-E, which is consistent with the intensities of extreme events (Table 9.6).

Composite horizontal distribution of height anomalies consisting of all the positive and negative cases in domain-E is drawn at 150 hPa (Fig. 9.46a, b), which corresponds to the STM (Fig. 9.44b) and the WIM (Fig. 9.44d), respectively. As shown in the positive composite (Fig. 9.46a), three positive centers (Ap, Bp, and Dp) of height anomalies are separated by a negative center (Cn). The positive centers Ap, Bp, and Dp are located mainly over western Russia and Kazakhstan, the far eastern Russia and southern China, respectively, while the negative center (Cn) mainly covers Mongolia and northern China. The center Dp well captures the center H1 in Fig. 9.45a, b, leading to the STM of the SAH. Inverse distribution of height anomalous centers appears in the negative composite (Fig. 9.46b). Three negative centers (An, Bn, and Dn) separated by a positive center (Cp) are clearly observed in Fig. 9.46b, and Dn is the crucial reason for WIM of the SAH. The regression pattern (Fig. 9.46c) shows great consistency with the composite results (Fig. 9.46a, b). The wave train in the middle latitudes and the dipolar pattern in the meridional direction over East Asia are clear in Fig. 9.46, connecting the SAH with systems in higher latitudes. The wave train pattern was also revealed in Ren et al. (2015). Differently, the composite propagation of the wave train is not plotted because the three anomalous systems in the wave train vary largely among different cases.

Figure 9.47a, b shows the composite SAT distributions of all the positive and negative cases in domain-E using the NCEP R1, respectively. When STM occurs (Fig. 9.47a), three warm centers (Aw, Bw, and Dw) separated by a cold center (Cc) on the surface correspond well to the anomalous height pattern in Fig. 9.46a. In the negative composite (Fig. 9.47b), a warm center (Cw) is surrounded by three cold centers (Ac, Bc, and Dc), which also matches the horizontal pattern in Fig. 9.46b. In China, a dipolar pattern with a warm center in the south and a cold center in the north is observed in the STM (Fig. 9.47a) and is reversed in the WIM (Fig. 9.47b). The opposite patterns result from the opposite extensions of temperature anomalies in the troposphere (Figs. 9.45 and 9.46). The SAT anomalies in southern China are directly influenced by the SAH variations in intensity and location. Other anomalous SAT centers are directly affected by the corresponding local temperature anomalies aloft in the troposphere. Figure 9.47c plots the SAT differences between the composites of positive and negative cases. The four centers (Aw, Bw, Cc, and Dw) all reach the 95% confidence level in Fig. 9.47c, showing significant SAT differences between the positive (Fig. 9.47a) and negative (Fig. 9.47b) composites. To further verify the spatial patterns of SAT anomalies, correlations between daily height NA series in domain-E and daily SAT anomalies in summers from 1981 to 2010 are calculated (Fig. 9.47d). The above four centers (Ap, Bp, Cn, and Dp) are robust and significant exceeding 95% confidence level.

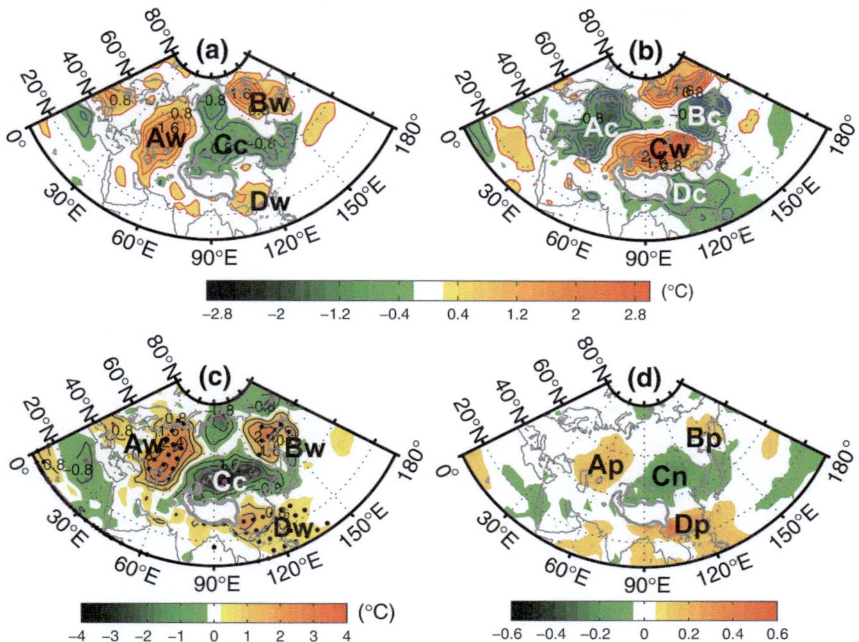

Fig. 9.47 Composites of SAT anomalies (*red contour* with *yellow-orange shadings* for positive and *blue contour* with *green shadings for negative*, 0.4 °C interval) at the 995-sigma level for **a** positive events and **b** negative events in domain-E. **c** The differences (*shadings* and *contours*, 0.8 °C interval) between **a** and **b**. *Red* and *blue dots* denote the areas exceeding 95% confidence level of *t*-test. **d** Correlation coefficients (*shading*) between daily height NA series in domain-E and SAT anomalies in JJA from 1981 to 2010. *Shading* areas indicate correlations exceeding the 95% confidence level. Letters "A," "B," "C," and "D" denote four areas with warm "w" or cold "c" and positive "p" or negative "n" correlations. *Thick grey line* indicates the Tibetan Plateau region with topography exceeding 3000 m. Sourced from Shi and Qian (2016) © American Meteorological Society, with permission

Composite distributions of precipitation anomalies for positive and negative cases in domain-E are shown in Fig. 9.48a, b, respectively. For the positive composite (STM of the SAH), three dry centers (Ad, Bd, and Dd) isolated by a rainy center (Cr) are observed in Fig. 9.48a, in the same places as the ones in Figs. 9.46a and 9.47a. In addition, another dry center (Ed) appears over the southern foot of the TP. For the WIM, a reversed pattern of precipitation anomalies can be found (Fig. 9.48b). Further, differences of the precipitation anomalies between Fig. 9.48a, b are calculated in Fig. 9.48c. Five centers (Ad, Bd, Cr, Dd, and Ed) can be identified again by those significant points exceeding the 95% confidence level. Again, correlation map reconfirms the spatial pattern of precipitation anomalies along with the SAH variations simultaneously (Fig. 9.48d). The correlation with precipitation is weaker than that of temperature. An important reason is that the former is affected by other factors, such as moisture.

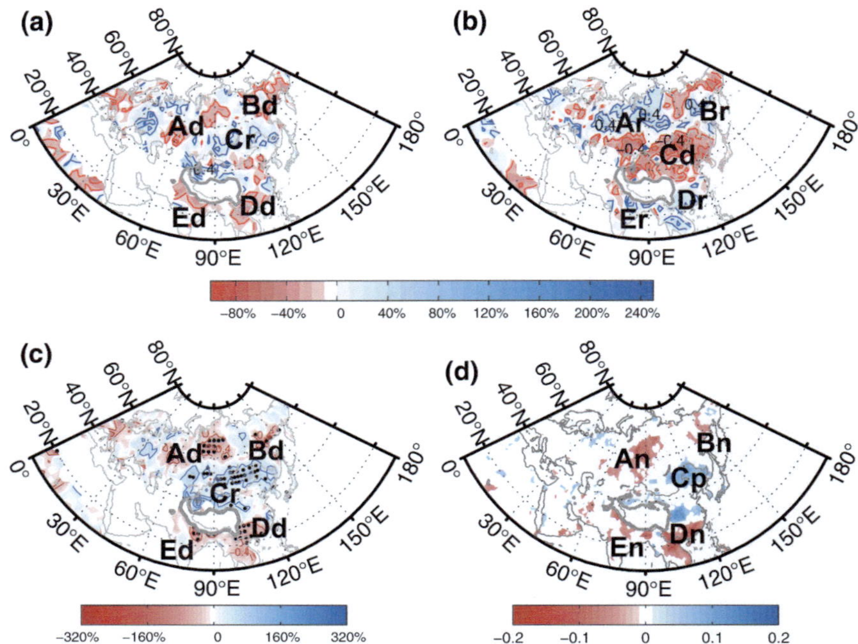

Fig. 9.48 Composites of precipitation percentage anomalies (*red contour* with *red shadings* for negative and *blue contour* with *blue shadings* for positive, 20% interval) for **a** positive events and **b** negative events in domain-E. **c** The differences (*shadings* and *contours*, 40% interval) between **a** and **b**. *Red* and *blue dots* denote the areas exceeding 95% confidence level of *t*-test. **d** Correlation coefficients (*shading*) between daily height NA series in domain-E and precipitation percentage anomalies in JJA from 1981 to 2010. *Shading* areas indicate correlations exceeding the 95% confidence level. Letters "A," "B," "C," "D," and "E" denote five areas with dry "d" or rainy "r" and positive "p" or negative "n" correlations. *Thick grey lines* indicate the Tibetan Plateau region with topography exceeding 3000 m. Sourced from Shi and Qian (2016) © American Meteorological Society, with permission

Similar composite analyses are conducted for domain-W. Figure 9.49a, c display the vertical structures along 27.5°N for the positive and negative composites in domain-W, respectively. For the SIM, a warm center (W1) and a cold center (C1) of temperature anomalies are located vertically below and above the positive center of height anomalies (H1) at the upper troposphere around 100 hPa (Fig. 9.49a). Positive height anomalies are prominent between 500 and 50 hPa, with only one positive height center over IP at 100 hPa. Reversely, opposite configuration is observed from the negative composite of height and temperature anomalies (Fig. 9.49c). Moreover, the amplitude of height and temperature anomalies is stronger with an anomalous low center (L1) near 150 hPa.

In latitude-pressure section along 55°E (Fig. 9.49b), the positive center (H1) and the negative center (L2) form a dipolar pattern in the positive composite in domain-W. The positive center (H1) of height anomalies near 32.5°N separates the

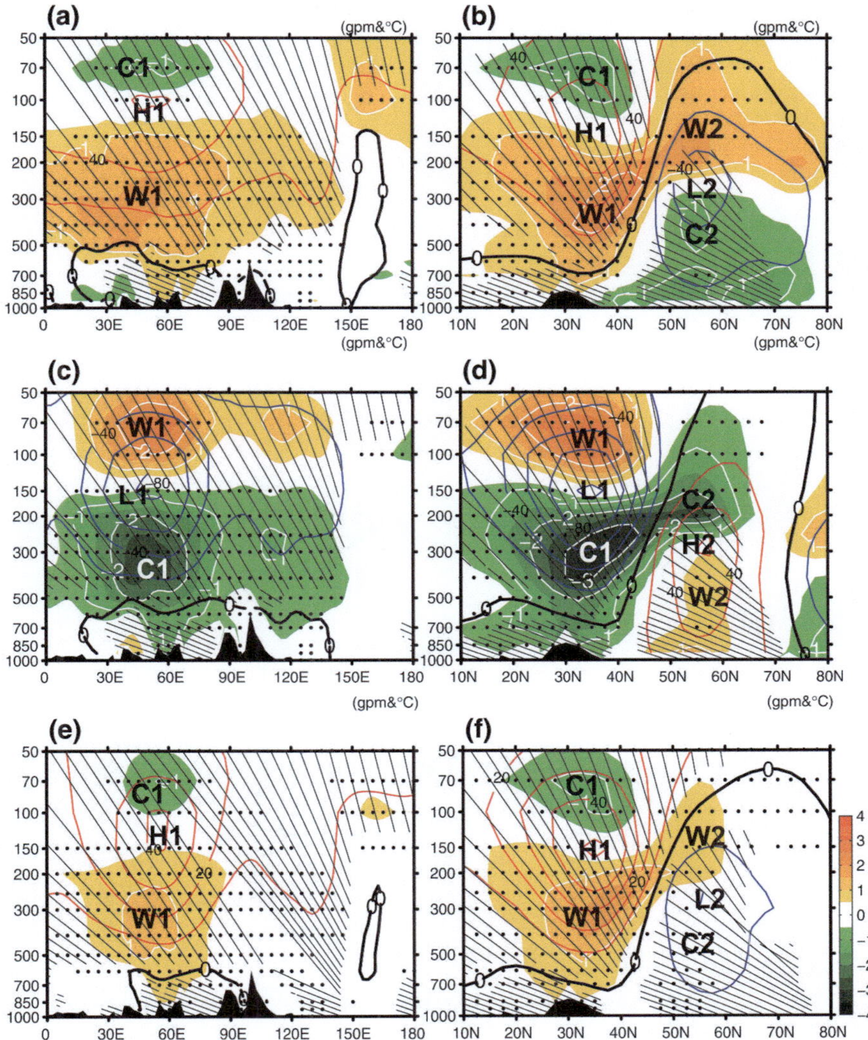

Fig. 9.49 Same as Fig. 9.45, but for domain-W. Panels **b**, **d**, and **f** are along 55°E. Sourced from Shi and Qian (2016) © American Meteorological Society, with permission

warm center (W1) below and the cold center (C1) above. In contrary, anomalous low center (L2) exists around 55°N with opposite temperature anomalies. The vertical location of H1 is around 150 hPa while L2 is near 300 hPa. Reversed but comparable configuration is clear in the negative composite in domain-W (Fig. 9.49d). All these centers of height and temperature anomalies satisfy the hydrostatic relationship. Figure 9.49a–d clearly show that the SAT anomalies are the downward extension of temperature anomalies below the height anomalous centers at the upper troposphere. To further confirm this vertical structure,

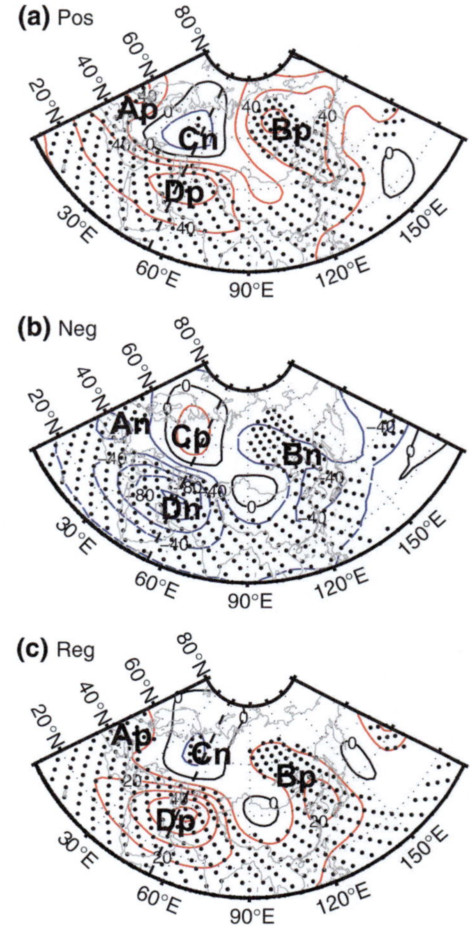

Fig. 9.50 Same as Fig. 9.46, but for domain-W. The *dashed lines* are along 55°E. Sourced from Shi and Qian (2016) © American Meteorological Society, with permission

regression patterns against daily height NA series in domain-W are calculated (Fig. 9.49e, f). It is clear that most areas near domain-W are statistically significant in the troposphere and lower stratosphere, showing a similar structure to the extreme events demonstrated.

Figure 9.50 depicts the composite horizontal structures at 150 hPa over domain-W. When the SIM occurs, four large-scale systems of height anomalies (Ap, Bp, Cn, and Dp) can be found (Fig. 9.50a). Three positive centers with Ap located in Western Europe, Bp mainly over middle to eastern Russia and Dp over the vicinity in domain-W surround a negative center Cn over Kazakhstan and Russia north of it. They display a wave train over Eurasia in the middle latitudes and a dipolar pattern near 55°E in the meridional direction, indicating the simultaneous relationship between the height anomalies in the subtropics and middle latitudes. Again, opposite pattern can be discovered in the negative composite for domain-W (Fig. 9.50b), which matches the WTM (Fig. 9.44c) above. The four centers (An, Bn, Cp, and Dn) with opposite signs exist at similar locations. In addition, the regression pattern against the daily height NA series in domain-W is significant over the above areas (Fig. 9.50c). So, the anomalous pattern at 150 hPa is robust, including the wave train in the middle latitudes and a dipolar pattern near 55°E in the meridional direction.

Again, to identify the possible connection between the above height-temperature anomalies and the surface temperature extremes, the anomalous SAT patterns corresponding to the positive and negative composites in domain-W are illustrated (Fig. 9.51a, b). For the positive composite (SIM; Fig. 9.51a), three warm centers (Aw, Bw, and Dw) and one cold center (Cc) can be found easily. The four temperature centers are consistent with the four anomalous height centers in Fig. 9.50. As indicated by Fig. 9.49a, b, the warm surface center Dw (Fig. 9.51a) is the downward extension of the warm center W1 in the upper troposphere, which matches well with H1. The anomalous height-temperature configurations of other three temperature centers in Fig. 9.51a also accord with this vertical structure (figures not shown). In contrast, three cold centers (Ac, Bc, and Dc) and one warm center (Cw) exist in the negative composite in domain-W (WTM; Fig. 9.51b). The differences (Fig. 9.51c) confirm the robustness of the four centers by the statistical *t*-test, in consistency with the wave train and dipole in anomalous height field at

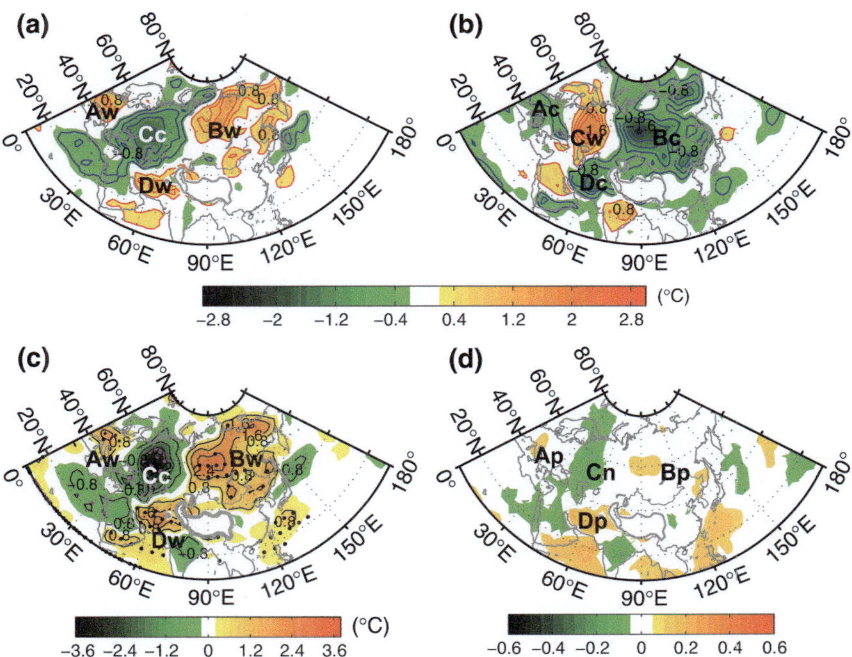

Fig. 9.51 Same as Fig. 9.47, but for domain-W with SAT anomalies. Sourced from Shi and Qian (2016) © American Meteorological Society, with permission

150 hPa (Fig. 9.50). Generally, three positive centers (Ap, Bp, and Dp) and one negative center (Cn) stand out significantly in the correlation map between daily height NAs in domain-W and daily SAT series over Eurasia (Fig. 9.51d).

Further, the relationship between the SAH zonal variation in domain-W and precipitation over Eurasia is illustrated through composite and correlation analyses (Fig. 9.52). Positive composite shows three centers (Ad, Bd, and Cr) in the middle latitudes (Fig. 9.52a), which correspond well with anomalous height centers (Ap, Bp, and Cn in Fig. 9.50a) and SAT centers (Aw, Bw, and Cc in Fig. 9.51a). Moreover, another rainy center (Er) is obvious over the southern foot of TP due to the negative height anomalies in this region (Fig. 9.49a). For negative composite (Fig. 9.52b), opposite centers (Ar, Br, Cd, and Ed) appear in the similar places to those in the positive composite (Fig. 9.52a). These four precipitation centers are statistically significant under the *t*-test of the differences between positive and negative composites (Fig. 9.52c). For further detection, the four centers above (An, Bn, Cp, and Ep) can also be distinguished in the correlation map (Fig. 9.52d). However, the centers in the middle latitudes (An, Bn, and Cp) are much weaker, while the Ep over TP becomes dominant. It is interesting that the negative center (Dn) only appears in the correlation map. Dry climatic feature over this region may lead to the disappearance of this center in Fig. 9.52a–c, although it has a significant negative correlation with height anomalies.

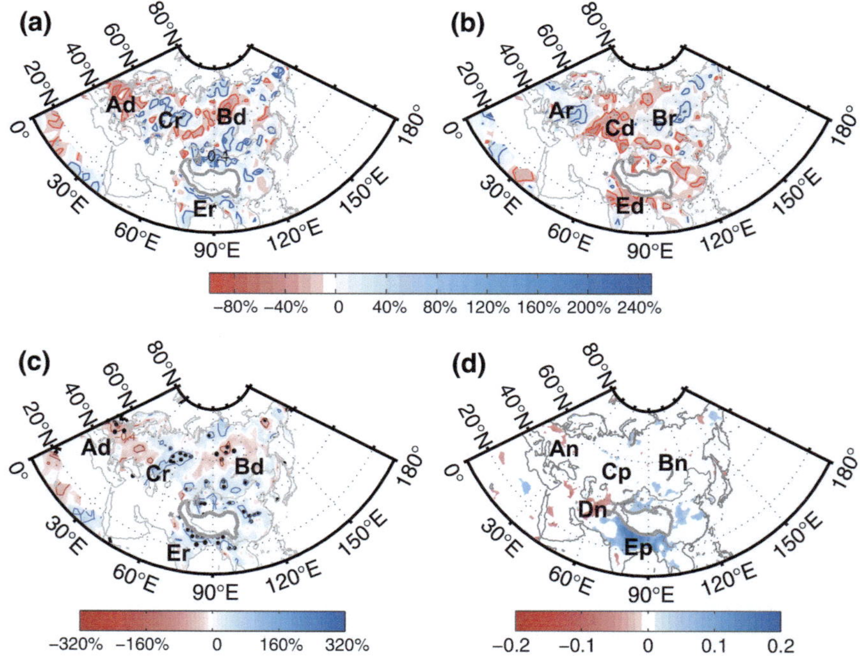

Fig. 9.52 Same as Fig. 9.48, but for domain-W with precipitation percentage anomalies. Sourced from Shi and Qian (2016) © American Meteorological Society, with permission

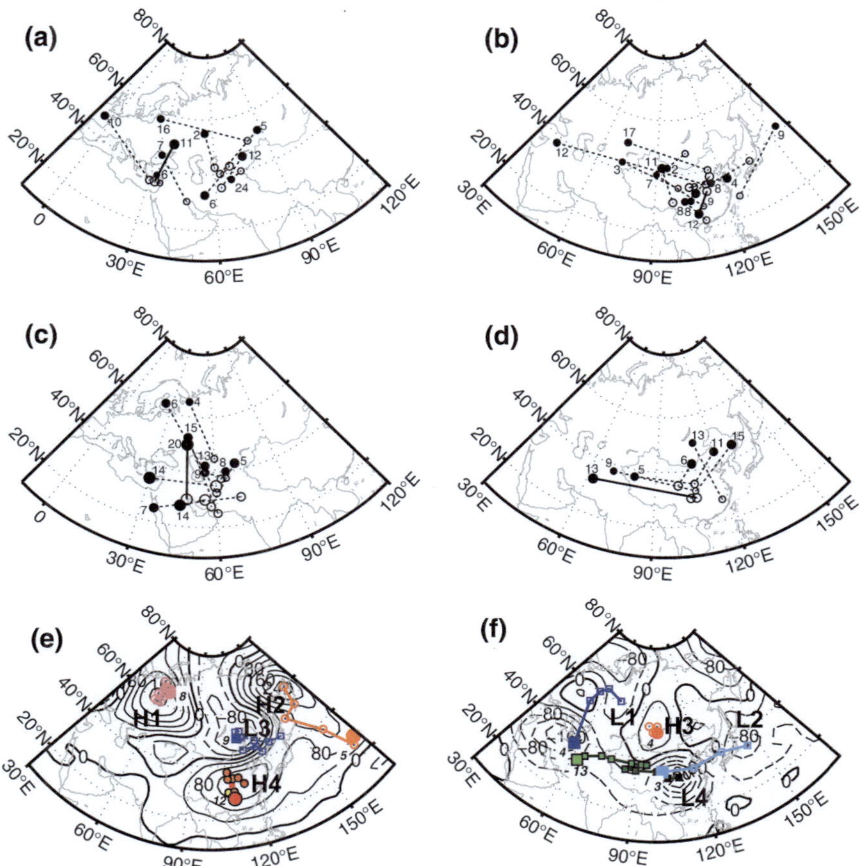

Fig. 9.53 The *lines* (*solid* for the strongest case and *dashed* for other cases) connecting the original central locations (*solid circles*) of anomalous height centers with leading days (*numbers around solid circles*) and their central locations on the peak days (*hollow circles*) for all the events, respectively. Panels **a** and **c** include the positive and negative centers related to the SAH near domain-W, respectively. Panels **b** and **d** are the same as **a** and **c**, but for domain-E. The height anomalies (*solid contour* for positive and *dashed contour* for negative, 40 gpm interval) at 150 hPa on **e** August 25, 1981 and **f** June 24, 1997, the peak days during the strongest positive and negative events in domain-E, respectively. The *biggest solid circles* (*squares*) indicate the original locations of positive (negative) height centers. The tracks of anomalous centers near domain-E are denoted by the *lines* with *solid circles* in **e** and *solid squares* in **f** with *black* edges, respectively. The *lines* with *hollow circles* (*squares*) indicate the central tracks of other positive (negative) centers. The *small numbers* beside each *biggest solid circles* or *squares* indicate the leading days. Letters "H/L" indicate the high/low anomalous centers. Sourced from Shi and Qian (2016) © American Meteorological Society, with permission

The anomalous centers of height anomalies can be traced through anomalous height maps day by day until they cannot be identified. Based on this, original locations and leading days of anomalous height centers as well as their locations on the peak days for each SAH mode are illustrated in Fig. 9.53. For each positive (negative) event, the peak day is defined by the day with the biggest (smallest) height NAs. For the 10 positive events in the domain-W, both original locations and locations on peak days were scattered (Fig. 9.53a). Two distant cases originated near Spain and the Baltic Sea at leading 10 and 16 days, respectively. However, most negative centers of height anomalies finally concentrated near the domain-W (Fig. 9.53c). They emerged mainly from the west and the north. Two remote cases initiated disturbances from northern Europe 6 and 4 days before the peak days of each case, respectively. For cases in domain-E, most anomalous height centers concentrated in central and eastern China on the peak days (Fig. 9.53b, d). Most signals came from the west or the north and they originated from China or the countries nearby. One exception aroused east of Japan at leading 9 days and matured over the East China Sea while the other one originated near Turkey at leading 12 days and moved to China (Fig. 9.53b). Further, in order to clarify the anomalous systems related to the SAH in detail, case enumeration is used on the strongest positive and negative cases in domain-E (Fig. 9.53e, f). It is observed that all the anomalous systems showed complexity in their tracks. Some were stable and motionless, while others originated from distant areas and moved in different ways. Their leading days spanned 3–13 days. The moving tracks of anomalous systems in other cases were also varied (figures not shown). So, based on the particularity of each case and each anomalous system, composite event evolutions are not carried out here.

9.8 The Western Pacific Subtropical High and Precipitation Pattern[6]

The SAH and the western Pacific Subtropical High (WPSH) are the two planetary-scale circulation systems, which are active and influence regional climate. Using the daily variables of NCEP R1 from 1981 to 2014, Qian and Shi (2016) identified the zonal extreme episodes of WPSH and then examined their climate impacts. All the ten positive and six negative episodes range from 10 to 17 days based on a new WPSH zonal index calculated by normalized relative vorticity (RV) averaged over a core domain (112.5°–142.5°E, 10°–30°N, 925–500 hPa). Composite results demonstrate a tripole pattern with a wet (dry) zone over the tropical western Pacific (TWP) separating two dry (wet) zones in the meridional direction. The northern one covers the Meiyu belt while the southern one extends from the tropical Indian Ocean (TIO) to the Maritime Continent (MC). The WPSH activity and monsoon trough form the height anomalies over the core domain, while its northern height anomalies are due to the westerly trough in middle latitudes.

[6]Modified article originally published in Qian and Shi (2016). Published with kind permission © 2016 John Wiley & Sons, Inc. All Rights Reserved.

The WPSH is a crucial component of the East Asian summer monsoon system, dominant over North Pacific in both lower and middle troposphere. As shown in many previous studies, the WPSH variations in intensity, location, and shape exert significant influence on the climate in East Asian monsoon region (Tao and Chen 1987; Wu et al. 2003; Yang and Sun 2005; Ding and Chan 2005). In boreal summer, the northward and then southward migration of the WPSH is the best-known feature and is closely related to the variation of rainbelt in eastern China. A typical northward jump occurs in mid-July, which indicates the end of Meiyu (rainy) period along the Yangtze River valley (YRV) and the beginning of rainy season in North China (Tao and Chen 1987; Ding and Chan 2005). However, the above migrations in summer for both the WPSH and rainbelt are climatic features (Qian et al. 2002; Wang and Ding 2008). The anomalous WPSH activities are the real causes of drought or flood disasters and should be paid more attention.

In the zonal direction, the east–west shift of the WPSH is also a predominant characteristic of the WPSH activities, and it plays a vital part in the climate over Asian-Pacific region (e.g. Kawatani et al. 2008; Ren et al. 2013). The circulation pattern in lower middle troposphere and the consequent water vapor condition over the South China Sea (SCS) and the tropical western Pacific (TWP) can be different due to the zonal variation of the WPSH, which further contributes to different rainfall distributions (Wang et al. 2000; Liu et al. 2008). The relationship between the rainfall over the YRV and zonal activities of the WPSH has been manifested in many studies (e.g. Mao et al. 2010; Yang et al. 2010). To a larger extent, Nitta (1986, 1987) and Huang (1992) used correlation maps and a case study to reveal dry–wet distributions over Asian-Pacific areas by limited data. The Pacific-Japan (PJ) pattern and East Asia/Pacific (EAP) pattern were respectively named by them.

To better understand and depict WPSH variability, many quantitative indices have been established, and they differ mainly in the selection of atmospheric variables, vertical levels, and horizontal domains. For the determination of variables, geopotential height (GH) and relative vorticity (RV) are often used for the definition of the indices. Using GH, Lu and Dong (2001) introduced an index, defined as the mean GH at 850 hPa in June-July-August (JJA) averaged over the western flank of the WPSH (110°–140°E, 10°–30°N). Lu (2002) made some modifications for the horizontal area with the eastern most longitude to be 150°E. The GHs at 500 hPa have also been widely used to depict the WPSH changes. However, using RV to define the WPSH zonal indices has been proven to have advantages over the GH in recent years (Yang and Sun 2005; Lu et al. 2008). The above definitions also differ in vertical levels and horizontal domains. In short, due to the complexity of the WPSH, there has been no consensus on the optimum index to reveal its zonal variations.

The WPSH acts with variability in intraseasonal, seasonal, interannual, and interdecadal timescales. Many previous studies paid more attention to its interannual variability (Lu and Dong 2001; Yang and Sun 2003; Lu et al. 2008). In the intraseasonal timescales, although some studies (Ren et al. 2013; Yang et al. 2010; Mao et al. 2010) pointed out some features of the WPSH activities, they focused more on the regional rainfall over the YRV.

Based on above descriptions, several issues need to be reconfirmed and clarified including (1) the optimum and objective index to indicate the WPSH zonal activities, (2) the temporal characteristics of the extreme WPSH zonal episodes, (3) the connection between the WPSH zonal activities and other large-scale circulation systems, and (4) the dry–wet pattern in the Asian-Pacific region associated with the WPSH zonal episodes.

To illustrate the above issues, the optimum three-dimensional (3D) domain or core area (112.5°–142.5°E, 10°–30°N, 925–500 hPa) to represent the most active domain of the WPSH is firstly determined by the daily analysis of variance. Based on the defined core domain, a new WPSH zonal index ($WPSHI_P$) has been constructed using the daily series of RV NA averaged over the above core domain in summers from 1981 to 2014 by Qian and Shi (2016). This index can reflect not only the zonal variation of the WPSH, but also the interaction between WPSH and the monsoon trough. To make a comparison with previous WPSH zonal indices, their correlation coefficients are listed in Table 9.8. The $WPSHI_P$, $WPSHI_{L1}$, $WPSHI_{L2}$, $WPSHI_{Y1}$, $WPSHI_{Y2}$, $WPSHI_{L3}$, and $WPSHI_X$ denote the indices proposed in the present study, by Lu and Dong (2001), Lu (2002), Yang and Sun (2003, 2005), Lu et al. (2008), and Xie et al. (2009), respectively. The correlations of the new index with the other representative indices (the first column in Table 9.8) all exceed 99% confidence level, indicating the consistency and reasonability of this new index ($WPSHI_P$). Moreover, in terms of variability, the $WPSHI_P$ based on daily RV variance (or SD) has advantages over the selection of extreme episodes.

According to the new WPSH zonal index ($WPSHI_P$) defined by daily RV NAs, extreme WPSH episodes can be determined. A positive (negative) episode occurs when the RV NAs are greater (less) than 1.0 (−1.0) SD for at least 10 days with at most one-day interruption. The positive and negative WPSH episodes together with their average intensities and orders are, respectively, listed in Table 9.9. The absolute value of average RV NAs indicates the average intensity of an episode. When selecting the extreme WPSH episodes, no filter or running mean is conducted on the daily time series ($WPSHI_P$), because the extreme episodes combine oscillations from different timescales.

Table 9.8 The correlation coefficients among the new-defined WPSH zonal index and other previous indices (All the values exceed the 99% confidence level based on the student's *t*-test, except the one between $WPSHI_{Y1}$ and $WPSHI_{Y2}$)

Indices	$WPSHI_P$	$WPSHI_{L1}$	$WPSHI_{L2}$	$WPSHI_{Y1}$	$WPSHI_{Y2}$	$WPSHI_{L3}$	$WPSHI_X$
$WPSHI_P$	1.000	−0.712	−0.666	0.584	0.533	0.701	−0.611
$WPSHI_{L1}$	−0.712	1.000	0.975	−0.456	−0.402	−0.585	0.517
$WPSHI_{L2}$	−0.666	0.975	1.000	−0.406	−0.347	−0.617	0.576
$WPSHI_{Y1}$	0.584	−0.456	−0.406	1.000	−0.034	0.410	−0.364
$WPSHI_{Y2}$	0.533	−0.402	−0.347	−0.034	1.000	0.415	−0.276
$WPSHI_{L3}$	0.701	−0.585	−0.617	0.410	0.415	1.000	−0.822
$WPSHI_X$	−0.611	0.517	0.576	−0.364	−0.276	−0.822	1.000

Sourced from Qian and Shi 2016) © 2016 John Wiley & Sons, Inc., with permission

Table 9.9 Persistent (equal to or exceeding 10 days) and extreme (larger than 1.0 SD) RV episodes including their years, start dates, durations, average intensities and orders in summer (1 June–31 August, JJA) from 1981 to 2014

Year	Start date	Duration	Average NAs	Order	Year	Start date	Duration	Average NAs	Order
1981	05 June	12	1.83	3	1983	17 Jul	16	−1.49	3
1982	**21 June**	**14**	**2.08**	**1**	1987	28 Jul	12	−1.47	4
1984	15 June	10	1.16	10	**1988**	**15 Aug**	**17**	**−1.57**	**1**
1985	12 June	10	1.79	4	1998	03 Aug	17	−1.54	2
1987	09 June	10	1.37	9	2003	21 Jun	11	−1.26	6
1990	14 August	15	1.51	8	2007	15 Jul	14	−1.45	5
2002	28 June	15	2.02	2	Ave	22 Jul	14.5	−1.46	
2004	15 June	17	1.68	6					
2004	06 August	10	1.57	7					
2011	08 July	11	1.79	4					
Ave	28 June	12.4	1.68						

The orders are based on the average NAs. The strongest RV episodes are in bold fonts. The last row denotes the average start date, duration and average NAs of the positive and negative RV episodes, respectively. Sourced from Qian and Shi (2016) © 2016 John Wiley & Sons, Inc., with permission

Statistically, 10 positive WPSH episodes are detected during 1981–2014, persisting from 10 to 17 days with average intensities from 1.16 to 2.08 SDs (left column in Table 9.9). Among these episodes, five episodes occurred in the 1980s, with the largest proportion. Only one occurred in the 1990s while others were after 2000. For the negative episodes, the durations of the six negative episodes ranged from 11 to 17 days with their intensities from −1.26 to −1.57 SDs (right column in Table 9.9). Three negative episodes were found in the 1980s, and other three were in 1998, 2003, and 2007 respectively. It implies that the WPSH was more active in the 1980s in terms of zonal direction. In view of the months, 8 out of 10 positive episodes occurred before mid-July, while 5 out of 6 negative episodes occurred after mid-July. It can be inferred that positive RV anomalies tend to occur more frequently in early summer due to the impact of monsoon trough, whereas negative RV anomalies are more active in late summer because of the WPSH activities. Moreover, the positive RV episodes had a larger intensity span but the two types of episodes shared similar durations.

In detail, the $WPSHI_P$ time series of the strongest positive and negative episodes over the summers are respectively given in Fig. 9.54a–d. All the four extreme cases lasted for less than a month, presenting extreme WPSH activities within a season. The two positive cases went through their extreme variabilities before mid-July (Fig. 9.54a, b), while the experiences of the two negative cases were after mid-July (Fig. 9.54c, d). The extreme episodes reflect the interactions between the WPSH and monsoon trough. During the residual days of the same summer, the RV NAs fluctuate in the normal range (between −1.0 and 1.0 SD) or become abnormal within 10 days. Further, to detect the relative contribution from different timescales,

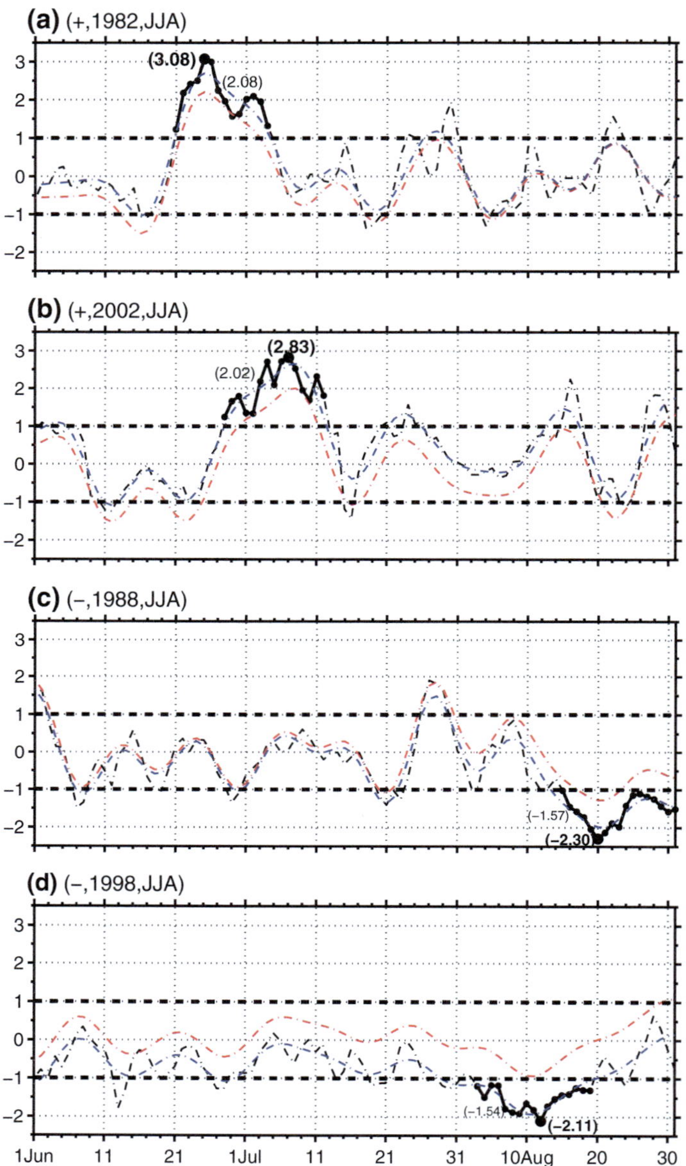

Fig. 9.54 Normalized series of daily $WPSHI_P$ (*black-dashed line*, with the *solid* section for episode lifetime) for **a** the strongest positive episode in 1982, **b** the second strongest positive episode in 2002, **c** the strongest negative episode in 1988, and **d** the second strongest negative episode in 1998. The *numbers* in *brackets* with *bigger* and *normal fonts* denote the strongest and average intensities of the episodes, respectively. The $WPSHI_P$ series with 10–90-day bandpass filter and 10-day lowpass filter are denoted by *red* and *blue lines*, respectively. In each panel, two *thick-dashed lines* denote ± 1 SD as the criteria of episode identification. Sourced from Qian and Shi (2016) © 2016 John Wiley & Sons, Inc., with permission

filter is used here. Firstly, the 10-day lowpass series (blue-dashed lines in Fig. 9.54) are illustrated here, which exclude the anomalies in synoptic timescale and contain the signals in intraseasonal, interannual, and decadal timescales. Their results show a smoother series than the original $WPSHI_P$ but their amplitudes are nearly the same. Second, the 10–90-day bandpass results are computed and displayed by red-dashed lines in Fig. 9.54. They demonstrate the anomalies in intraseasonal timescale with a relatively smaller amplitude than the original series. Thus, the differences between the 10-day lowpass series and 10–90-day bandpass series denote the oscillation in interannual and decadal timescales. The 10–90-day oscillation during extreme episode is only about 50–70% out of the original $WPSHI_P$ anomalies, indicating that these natural extreme episodes cannot be selected by bandpass series. For example, the second strongest negative episode occurred in the summer of 1998 (Fig. 9.54d), which was following the strongest El Niño in record. As El Niño dominates the oscillation in interannual timescale, the interannual anomalies are nearly as large as the intraseasonal ones so the episode is missed from the red-dashed line. To sum up, the synoptic anomalies are not requisite for the identification of extreme episodes, while the intraseasonal, interannual, and decadal oscillations are all necessary. These results also validate the procedure without filter.

Based on the selected episodes, the composite vertical structures of height and temperature anomalies, from the troposphere to lower stratosphere, are given in Fig. 9.55. It should be noted that the composite results in Figs. 9.55, 9.56, 9.57, 9.58, and 9.59 are averaged over the whole duration of each case. Composite vertical cross sections of height and temperature anomalies along 20°N and 127.5°E for positive episodes are respectively depicted in Fig. 9.55a, b. A negative height center (L0) is located near 130°E and 22.5°N in the lower troposphere accompanied with a cold air mass (C0) below and a warm one above (W0), which indicates the eastward retreat of WPSH and the enlargement of monsoon trough (Fig. 9.55a, b). Above L0, a positive height center (H1) is located at about 125 hPa, separating a warm center (W1) below and a cold center (C1) above. As reported by Tao and Chen (1987), the upper level easterly flow serves as the main component of the East Asian summer monsoon system, and it varies due to the positive height center H1 (Fig. 9.55a, b). The negative composite shows an opposite configuration of height and temperature anomalies within the whole troposphere and lower stratosphere above the core domain (Fig. 9.55c, d). The differences in height and temperature anomalies between positive and negative composites are significant below than 100 hPa aloft the core region (112.5°–142.5°E, 10°–30°N; Fig. 9.55a–d). Also, a robust dipole of height anomalies can be observed meridionally, but the high and low anomalous centers are located at different pressure levels (Fig. 9.55b, d). The centers in the middle latitudes (H2 in Fig. 9.55b) and L2 in Fig. 9.55d) signify the variations of westerly trough. The height and temperature anomalies satisfy the hydrostatic relationship vertically in these two composites. Further, to verify the robust linkage between the $WPSHI_P$ and circulation anomalies, regressed vertical pattern against $WPSHI_P$ is illustrated along 20°N (Fig. 9.55e) and 127.5°E (Fig. 9.55f). Overall, the configuration of height-temperature anomalies holds true

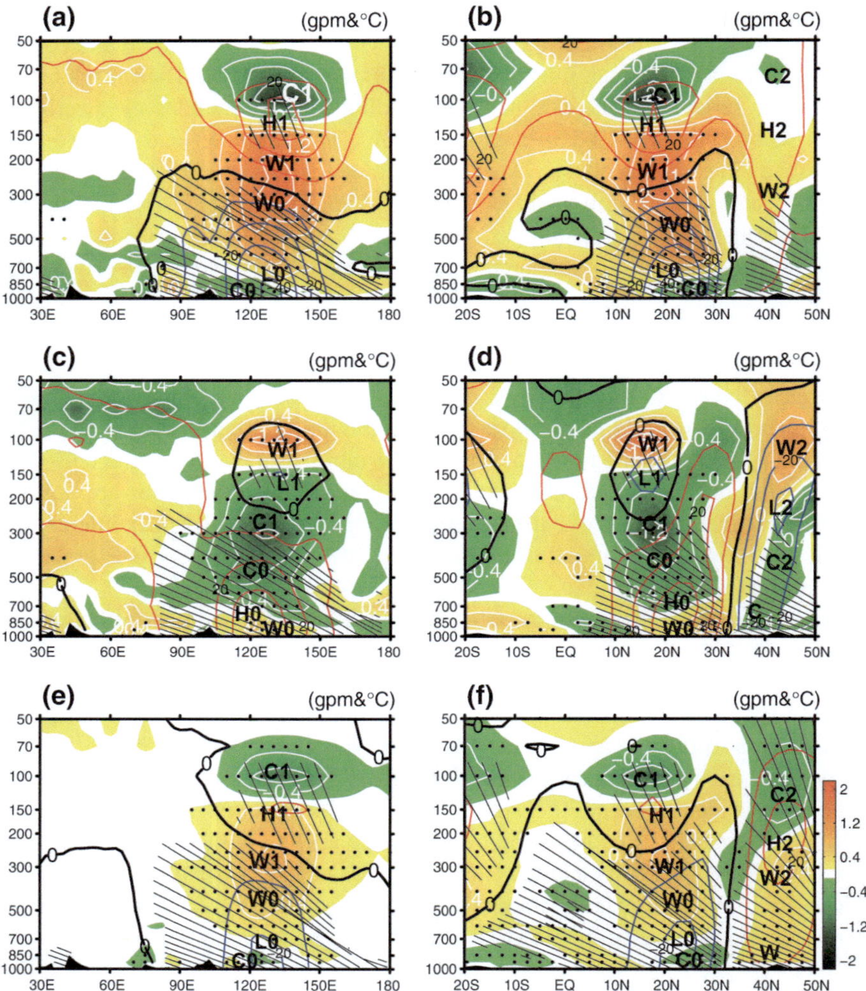

Fig. 9.55 Composite vertical sections of height anomalies (*red contour* for positive and *blue contour* for negative, 10 gpm interval) and temperature anomalies (*white solid* contour with *shadings*, 0.4 °C interval) along **a** 20°N and **b** 127.5°E for positive episodes. Panels **c** and **d** are the same as **a** and **b**, respectively, but for negative composite. *Black slashes* for height anomalies and *dots* for temperature anomalies in **a**–**d** indicate the composite differences exceeding 95% confidence level. **e** Regressed height anomalies (*red contour* for positive and *blue contour* for negative, 10 gpm interval) and temperature anomalies (*shadings* with *white lines*, 0.4 °C interval) along 20°N against daily $WPSHI_P$. **f** The same as **e**, but along 127.5°E. *Black slashes* for height anomalies and *dots* for temperature anomalies indicate the areas exceeding 95% confidence level. Letters "H/L" and "W/C" indicate the high/low centers of height anomalies and the warm/cold centers of temperature anomalies, respectively. Sourced from Qian and Shi (2016) © 2016 John Wiley & Sons, Inc., with permission

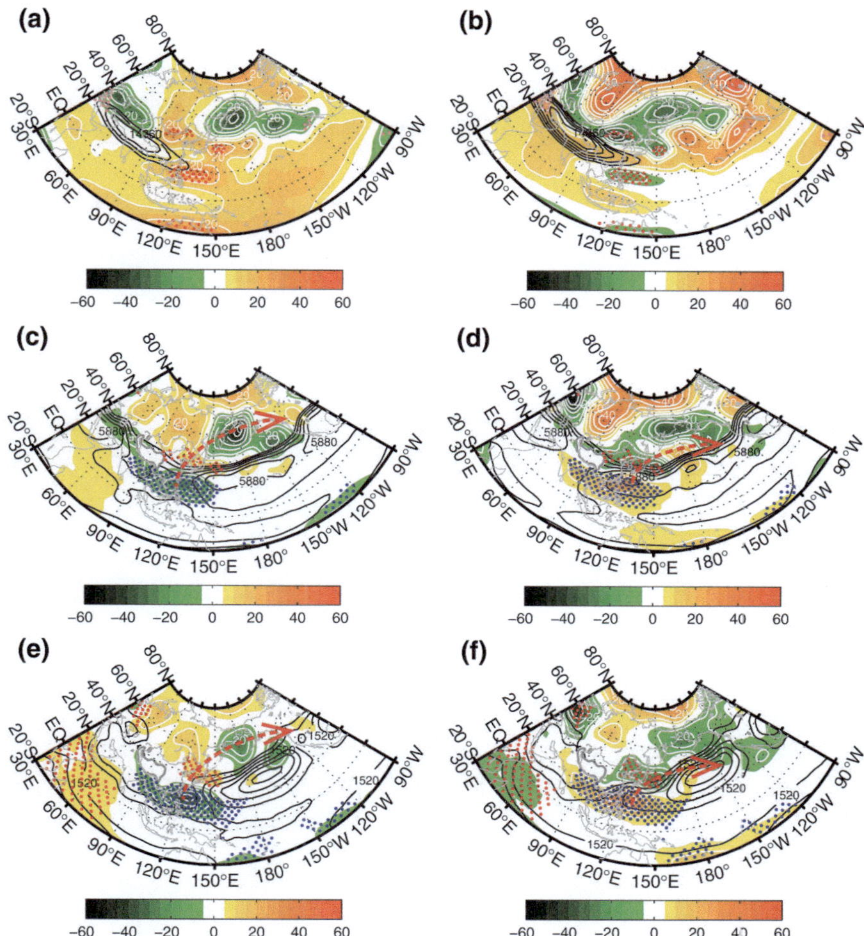

Fig. 9.56 Composites of height anomalies (*shadings* and *white contour*, 10 gpm interval) and total heights (*black contour*, 20 gpm interval) for the composite of positive RV episodes in panels **a**, **c**, and **e** as well as the negative RV composite in panels **b**, **d** and **f** indicating 150, 500, and 850 hPa, respectively. *Red* (*blue*) *dots* denote the positive (negative) differences between positive and negative composites exceeding 95% confidence level of *t*-test. *Thick grey lines* indicate the Tibetan Plateau region with topography exceeding 3000 m. The *red arrows* indicate the wave train. Sourced from Qian and Shi (2016) © 2016 John Wiley & Sons, Inc., with permission

significantly in a general sense. So, the $WPSHI_P$ is able to depict the circulation and temperature features related to the WPSH variation.

Based on the extreme episodes, the horizontal composites of both total and anomalous heights are displayed in Fig. 9.56. For the positive (negative) RV composite, negative (positive) centers of height anomalies are both observed at 850 and 500 hPa over the core domain in Fig. 9.56e, c (Fig. 9.56f, d), while there is a positive (negative) height center at 150 hPa aloft in Fig. 9.56a (Fig. 9.56b). So, the

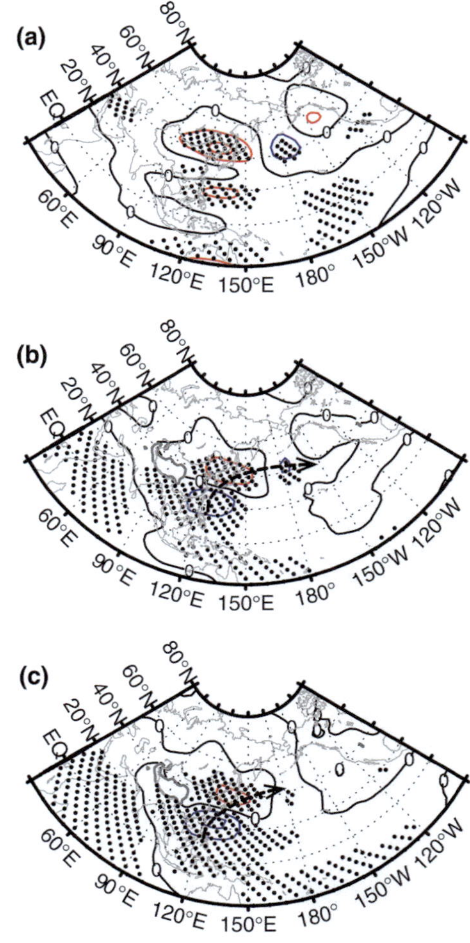

Fig. 9.57 Regressed height anomalies (*red contour* for positive and *blue contour* for negative, 10 gpm interval) at **a** 150 hPa, **b** 500 hPa, and **c** 850 hPa against daily $WPSHI_P$ series, respectively. *Dots* indicate the areas exceeding 95% confidence level. The *black arrows* indicate the wave train. Sourced from Qian and Shi (2016) © 2016 John Wiley & Sons, Inc., with permission

anomalous centers above the core domain show opposite features in the lower-middle troposphere and the upper troposphere vertically. In addition, the height anomalies at 850 hPa are stronger than those at 500 hPa. The monsoon trough exhibits different features in the positive (Fig. 9.56e) and negative (Fig. 9.56f) composites, respectively. In the positive RV composite (Fig. 9.56e), the monsoon trough extends eastward to the TWP with a stronger intensity, accompanied by the eastward retreat of the WPSH. Meanwhile, the westerly trough as East Asian Trough (EAT) at 500 hPa is situated westward due to positive height anomalies north to the core region (Fig. 9.56c). Oppositely, the monsoon trough becomes weak and recesses northwestward in the negative RV composite (Fig. 9.56f), while the WPSH extends westward and intensifies. In middle latitudes, the EAT deepens and moves southeastward as a result of the negative height anomalies over the Northeast China and Japan (Fig. 9.56d, f). At 150 hPa, the SAH is weaker and retreats westward for the positive RV composite (Fig. 9.56a), while the SAH shifts eastward with a stronger

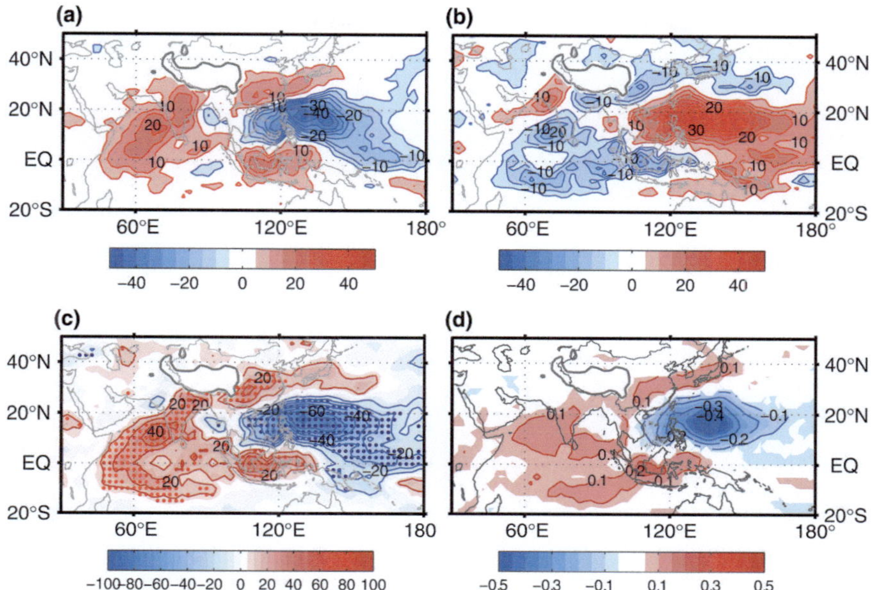

Fig. 9.58 Composite OLR anomalies (*red contour* with *red shadings* for positive and *blue contour* with *blue shadings* for negative, 5 W m^{-2} interval) for **a** positive and **b** negative RV episodes. Panel **c** is the difference (10 W m^{-2} interval) between **a** and **b**. The *red* and *blue dots* in **c** denote the areas exceeding 95% confidence level of *t*-test. Panel **d** displays the correlation coefficients (*shadings* and *contours*, 0.1 interval) between daily $WPSHI_P$ series and OLR anomalies in summers from 1981 to 2013. *Shading areas* in **d** indicate correlations exceeding the 95% confidence level. *Thick grey line* in each panel indicates the Tibetan Plateau region with topography exceeding 3000 m. Sourced from Qian and Shi (2016) © 2016 John Wiley & Sons, Inc., with permission

intensity for the negative RV composite (Fig. 9.56b). It coincides well with the WPSH locations in Fig. 9.56c–f, indicating opposite directions of the zonal movements of the two systems. In other words, the SAH moves eastward (westward) when the WPSH moves westward (eastward), as revealed by Tao and Zhu (1964). Moreover, the variation of the SAH can trigger the modification of easterly flow in upper troposphere.

In the lower and middle troposphere, there is a planetary-scale wave train of height anomalies covering three main colored centers, respectively, over the SCS and the Philippine Sea, the eastern China and Japan and the North Pacific (Fig. 9.56c–f). The meridional dipole pattern of height anomalies over East Asia and its nearby seas is clear at 500 and 850 hPa. Statistically, this dipole pattern is also significant under the difference *t*-test of positive and negative RV composites. Similar results were proposed as the emanation and propagation of Rossby waves, called the PJ pattern (Nitta 1986, 1987) and the EAP teleconnection (Huang 1992). The PJ pattern can be obtained in both interannual and intraseasonal timescales (Nitta 1987), while the EAP pattern is mainly contributed by the intraseasonal

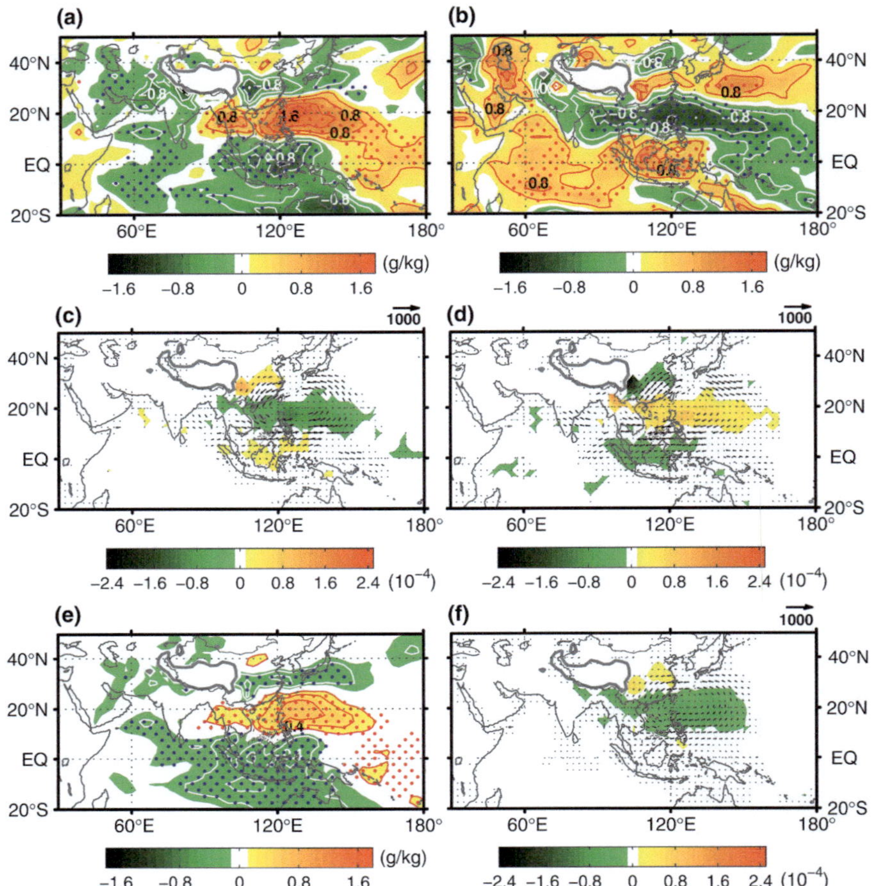

Fig. 9.59 Spatial distributions of specific humidity anomalies (*shadings* and *contours*, 0.4 g/kg interval) at 850 hPa of the **a** positive composite and **b** negative composite, respectively. *Dots* indicate the area exceeding 95% confidence level of difference *t*-test. Panels **c** and **d** are the same as **a** and **b**, but for the integrated water vapor flux anomalies (vectors, unit: kg s^{-1} m^{-1}) from 1000 to 300 hPa and their divergence anomalies (*shadings*, unit: kg s^{-1} m^{-2}). All the *shaded* area exceeds 95% confidence level of difference *t*-test. **e** Same as **a**, but for regressed anomalies against $WPSHI_P$. *Dots* indicate the area exceeding 95% confidence level. **f** Same as **c**, but for regressed anomalies against $WPSHI_P$. *Thick grey lines* indicate the Tibetan Plateau region with topography exceeding 3000 m. Sourced from Qian and Shi (2016) © 2016 John Wiley & Sons, Inc., with permission

variations. The extreme WPSH cases determined above contain both the two timescales in Fig. 9.54.

In order to examine the general ability of the $WPSHI_P$ to reflect the circulation anomalies in troposphere and lower stratosphere. Regressed height anomalies against the daily $WPSHI_P$ series at 150 hPa (Fig. 9.57a), 500 hPa (Fig. 9.57b) and 850 hPa (Fig. 9.57c) are respectively plotted. Obviously, the height anomalous pattern similar to the PJ pattern and EAP pattern still stands out significantly in lower-middle troposphere (Fig. 9.57b, c). Also, at 150 hPa (Fig. 9.57a), the reversed center over the core domain exists vertically compared to that in lower-middle troposphere. Moreover, another wave train is identified zonally in the middle latitudes, with one center near Northeast China and one opposite center south of Aleutian Islands (Fig. 9.57a). In short, the $WPSHI_P$ not only depict the circulation anomalies in the lower and middle troposphere but also their related features in the upper troposphere.

Precipitation anomalies are direct responses of the height and temperature anomalies. The daily precipitation anomalies are denoted by the outgoing longwave radiation (OLR) anomalies. Composite OLR anomalies for positive and negative RV episodes are respectively displayed in Fig. 9.58a, b. Obviously, a tripole dry–wet pattern can be identified. In the positive composite (Fig. 9.58a), a wet zone indicated by negative OLR anomalies ranges from the SCS to the TWP, which is directly related to the negative height anomalies (or stronger SCS monsoon trough) over the core domain (Fig. 9.55a–c, e). In addition, the OLR anomalies here indicate the convective activities in this region, and are revealed to be crucial in the generation of Rossby waves and further the formation of PJ pattern (Nitta 1987) or EAP teleconnection (Huang and Sun 1992). One dry zone with positive OLR anomalies covers the Meiyu belt from eastern China to Japan, in the region of East Asian summer monsoon. Another dry zone extends from the TIO to the MC and is also very important for the global heat balance, which is wider than the one in Nitta (1987). All the above three meridional OLR centers play a vital part in the regional or even global climate variations. Further, a reversed spatial distribution of OLR anomalies is displayed in the negative RV composite (Fig. 9.58b). This tripole dry–wet pattern is statistically detected again by the difference using *t*-test (Fig. 9.58c), derived from the composites for positive and negative episodes. Finally, the distribution of tripole OLR anomalies or dry–wet pattern is reconfirmed by the correlation between daily $WPSHI_P$ series and simultaneous series of OLR anomalies (Fig. 9.58d). It means that the $WPSHI_P$ not only can reflect the dry–wet pattern during extreme episodes but also general conditions.

These dry–wet zones are largely affected by the WPSH activities, in terms of both circulation and water vapor. In order to reveal the underlying dynamical conditions of water vapor responsible for the anomalous dry–wet pattern, composite analyses of the specific humidity anomalies at 850 hPa as well as the water vapor flux integrated from 1000 to 300 hPa and their divergence anomalies are given in Fig. 9.59. Three zones of specific humidity anomalies situated over Asian-Pacific region (Fig. 9.59a, b) are similar to the anomalous dry–wet pattern shown in Fig. 9.58. The cyclonic circulation and convergent zone over SCS and TWP (Fig. 9.59c) favor more precipitation, while the two zones in its northern and southern sides experience dry condition with divergent flow. The opposite pattern for water vapor flux anomalies is easily detected in Fig. 9.59c, d. Generally, the regressed pattern of specific humidity anomalies at 850 hPa is displayed in Fig. 9.59e. The three centers including one center with more water vapor surrounded by two centers with less water vapor are statistically significant and robust,

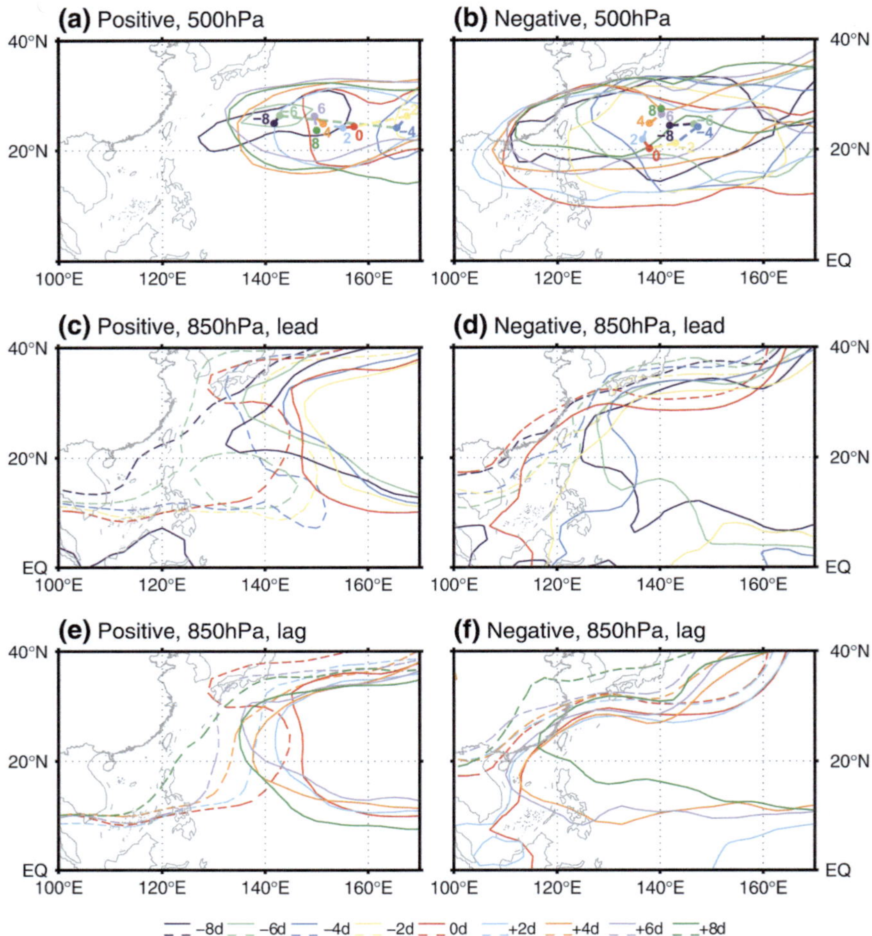

Fig. 9.60 Composite evolutions of the 5880-gpm line at 500 hPa for **a** positive and **b** negative RV episodes. The 5880-gpm lines lead (or lag) the peak day from 0 to 8 days, with an interval of 2 days. The *colored lines* and *dots* in **a** and **b** denote the central track of the WPSH with lead (negative number) or lag (positive number) days. Panels **c** and **d** are the same as **a** and **b**, but for the lead composites of 1490-gpm (*dashed lines*) and 1510-gpm (*solid lines*) lines at 850 hPa. Panels **e** and **f** are the same as **c** and **d**, but for the lagged composites. Sourced from Qian and Shi (2016) © 2016 John Wiley & Sons, Inc., with permission

which favor the dry–wet tripole pattern. Also, the regressed water vapor flux integrated from 1000 to 300 hPa and its divergence is calculated (Fig. 9.59f). Only two centers are detected significantly, with one over the core domain, and the other in eastern China. So, under extreme episodes, the $WPSHI_P$ can simultaneously identify the tripole pattern for specific humidity anomalies and integrated water vapor flux anomalies. However, in general case, it is less qualified to reveal a center over the MC for the divergence of integrated water vapor flux anomalies.

To further examine the time evolutions of the extreme WPSH episodes, Fig. 9.60 plots the composite evolutions of GHs at 500 and 850 hPa from lead = 8 days to lag = 8 days, with an interval of 2 days. The lead (lag) days are relative to the peak days of extreme episodes. The peak day refers to the day with maximum (minimum) $WPSHI_P$ of a positive (negative) RV episode. In the positive RV composite at 500 hPa (Fig. 9.60a), the WPSH is weak as represented by the 5880-gpm line. This intensity feature is consistent with the results in the overall composites (Fig. 9.56c, d). Within the 16 days, the WPSH first shrinks and retreats eastward, reaching its easternmost location at lead = 2 days (first stage). Then, the WPSH strengthens and extends westward (lead = 2 days to lag = 8 days; second stage). On the westernmost day (lead = 8 days and lag = 6, 8 days), the 5880-gpm lines are located at around 130°E, far away from the Eurasian Continent. In the negative RV composite at 500 hPa (Fig. 9.60b), the WPSH is much stronger and stretches more westward than those in the positive RV composite (Fig. 9.60a). Similar to the positive RV composite, the WPSH firstly shrinks and withdraws eastward, reaching its easternmost location at lead = 4 days (first stage; Fig. 9.60b) in the negative RV composite. Afterwards, the WPSH stretches westward with three big strides (lead = 4 to lag = 2 days; second stage), and then moves northeastward (lag = 2 days to lag = 8 days; third stage). Over the whole process, the WPSH can reach Indo-China Peninsula in its westernmost point, more westward around 20° compared to the positive RV composite.

At 850 hPa, similar to 500 hPa, the WPSH weakens with eastward retreat from lead = 8 days to lead = 2 days, and then extends westward gradually in the positive RV composite (1510-gpm lines in Fig. 9.60c, e). For the negative RV composite at 850 hPa (Fig. 9.60d, f), the WPSH tracks are also similar to those at 500 hPa. In addition, the zonal span of the WPSH is smaller at 850 hPa compared to that at 500 hPa. Further, as another crucial system at 850 hPa, the monsoon trough indicated by the 1490-gpm line also displays large variability over the plotted 16 days (Fig. 9.60c–f). When the WPSH is weak and withdraws eastward, the monsoon trough intensifies and penetrates southeastward into the TWP (Fig. 9.60c, e). However, the monsoon trough is limited over the SCS and Indo-China Peninsula when the WPSH extends westward (Fig. 9.60d, f).

Questions

(1) How to explain that the Arctic sea ice has undergone a substantial long-term decline with superimposed interannual sea ice minimum (SIM) events while the Antarctic sea ice has increased over the last decades?

(2) What are the causes of the Arctic amplification (AA) and the mid-latitude temperature reduction during last two decades?

(3) What the pattern of climatic and general circulation anomalies looks like if there is no sea–land contrast in the Northern Hemisphere?

(4) Why can't we simulate the interannual and interdecadal variability of the four cell intensities in the Northern Hemisphere using the current climatic models?
(5) How to present the anomalous activities of the South Asian High and other planetary-scale high and low systems?

References

Barnston AG, Livezey RE (1987) Classification, seasonality and persistence of low-frequency atmospheric circulation patterns. Mon Wea Rev 115:1083–1126. doi:10.1175/1520-0493(1987)115<1083:CSAPOL>2.0.CO;2

Bennartz R, co-authors (2013) July 2012 Greenland melt extent enhanced by low-level liquid clouds. Nature 496(7443):83–86. doi:10.1038/nature12002

Bromwich DH, Fogt RL, Hodges KI, Walsh JE (2007) A tropospheric assessment of the ERA-40, NCEP, and JRA-25 global reanalyses in the polar regions. J Geophys Res Atmos 112 (D10):185–194. doi:10.1029/2006JD007859

Bromwich DH, Wilson AB, Bai LS, Moore GW, Bauer P (2015) A comparison of the regional Arctic System Reanalysis and the global ERA-Interim Reanalysis for the Arctic. Q J R Meteor Soc. doi:10.1002/qj.2527

Budikova D (2009) Role of Arctic sea ice in global atmospheric circulation: A review. Glob Planet Change 68(3):149–163. doi:10.1016/j.gloplacha.2009.04.001

Chen JY, Carlson BE, Del Genio AD (2002) Evidence for strengthening of the tropical general circulation in the 1990s. Science 295:838–841. doi:10.1126/science.1065835

Cohen J, Screen JA, Furtado JC, Barlow M, Whittleston D, Coumou D, Francis J, Dethloff K, Entekhabi D, Overland J, Jones J (2014) Recent Arctic amplification and extreme mid-latitude weather. Nature Geosci 7:627–637. doi:10.1038/NGEo2234

Chen Y, Hu Q, Yang YM, Qian WH (2016) Anomaly based analysis of extreme heat waves in Eastern China during 1981–2013. Inter J Climatol doi:10.1002/joc.4724

Chen S, Wei K, Chen W, Song L (2014)Regional changes in the annual mean Hadley circulation in recent decades. J Geophys Res119(13):7815–7832. doi:10.1002/2014JD021540

Compo GP, et al. (2011) The Twentieth Century Reanalysis Project. Q J R Meteor Soc 137: 1–28, doi:dx.doi.org/10.1002/qj.776

Dee DP, Uppala SM, Simmons AJ et al. (2011) The ERA-Interim reanalysis: Configuration and performance of the data assimilation system. Q J R Meteor Soc 137(656):553–597. doi:10.1002/qj.828

Deser C, Teng H (2008a) Recent trends in Arctic sea ice and the evolving role of atmospheric circulation forcing, 1979–2007. Arctic Sea Ice Decline: Observations, Projections, Mechanisms, and Implications 7–26. doi:10.1029/180GM03

Deser C, Teng H (2008b) Evolution of Arctic sea ice concentration trends and the role of atmospheric circulation forcing, 1979–2007. Geophys Res Lett 35:L02504. doi:10.1029/2007GL032023

Deser C, Walsh JE, Timlin MS (2000) Arctic sea ice variability in the context of recent atmospheric circulation trends. J Climate 13(3):617–633. doi:10.1175/1520-0442(2000)013<0617:ASIVIT>2.0.CO;2

Devasthale A, Sedlar J, Koenigk T, Fetzer EJ (2013) The thermodynamic state of the Arctic atmosphere observed by AIRS: comparisons during the record minimum sea ice extents of 2007 and 2012. Atmos Chem Phys 13(15):7441–7450

Diaz HF, Bradley RS (2004) The Hadley Circulation: present, past and future, vol. 21, Springer, pp 85–120

Ding QH, et al. (2014) Tropical forcing of the recent rapid Arctic warming in northeastern Canada and Greenland. Nature 509(7499):209–212. doi:10.1038/nature13260

Ding YH, Chan JCL (2005) The East Asian summer monsoons: An overview. Meteorol. Atmos. Phys. 89:117–142, doi:10.1007/s00703-005-0125-z

Dong X, Zib B, Xi B, Stanfield R, Deng Y, Zhang X, Lin B, Long C (2014) Critical mechanisms for the formation of extreme Arctic sea-ice extent in the summers of 2007 and 1996. Clim Dyn 43(1–2):53–70

Ebita A, Kobayashi S, Ota Y et al. (2011) The Japanese 55-year Reanalysis "JRA-55": An interim report. J Meteor Soc Japan 7:149–152. doi:10.2151/sola.2011-038

Feng J, Wu Z, Zou X (2014) Sea Surface Temperature Anomalies off Baja California: A Possible Precursor of ENSO. Journal of the Atmospheric Sciences 71:5 1529–1537

Francis JA, Hunter E (2007) Drivers of declining sea ice in the Arctic winter: A tale of two seas. Geophys Res Lett 34:L17503. doi:10.1029/2007GL030995

Fu Q, Johanson CM, Wallace JM, Reichler T (2006) Enhanced mid-latitude tropospheric warming in satellite measurements. Science, 312:1179

Giles KA, Laxon SW, Ridout AL (2008) Circumpolar thinning of Arctic sea ice following the 2007 record ice extent minimum. Geophys Res Lett 35:L22502

Graversen RG, Mauritsen, T, Drijfhout S, Tjernström M, Mårtensson S (2011) Warm winds from the Pacific caused extensive Arctic sea-ice melt in summer 2007. Clim Dyn 36(11–12):2103–2112. doi:10.1007/s00382-010-0809-z

Graversen RG, Mauritsen T, Tjernstrom M, Kallen E, Svensson G (2008) Vertical structure of recent Arctic warming. Nature 451:53–56

Honda M, Inoue J, Yamane S (2009) Influence of low Arctic sea-ice minima on anomalously cold Eurasian winters. Geophys Res Lett 36:L08707. doi:10.1029/2008GL037079

Hansen J, Nazarenko L (2004) Soot climate forcing via snow and ice albedo. Proc Natl Acad Sci USA 101:423–428. doi:10.1073/pnas.2237157100

Hu YY, Fu Q (2007) Observed poleward expansion of the Hadley circulation since 1979.Atmos Chem Physics 7:5229–5236. doi:10.5194/acp-7-5229-2007

Hu A, Rooth C, Bleck R, Deser C (2002) NAO influence on sea ice extent in the Eurasian coastal region. Geophys Res Lett 29(22):2053. doi:10.1029/2001GL014293

Hu YY, Tung KK, Liu J (2005) A closer comparison of early and late-winter atmospheric trends in the Northern-Hemisphere. J Clim 18:3204–3216. doi:10.1175/JCLI3468.1

Huang Y, Qian YF (2003) Relationships between South Asian high and summer rainfall in North China (in Chinese). Plateau Meteor 22:602–607

Huang NE, Shen Z, Long SR, Wu ML, Shih HH, Zheng QN, Yen NC, Tung CC, Liu HH (1998) The empirical mode decomposition and the Hilbert spectrum for nonlinear and non-stationary time series analysis. Proc Roy Soc London Ser A 454(1971):903–995

Huang RH (1992) The East Asia/Pacific pattern teleconnection of summer circulation and climate anomaly in East Asia. Acta Meteor Sin 6: 25–37

Huang RH, Sun FY. (1992) Impacts of the tropical western Pacific on the East Asian summer monsoon. J. Meteorol. Soc. Japan 70:243–256

Jaiser R, Dethloff K, Handorf D, Rinke A, Cohen (2012) Impact of sea ice cover changes on the Northern Hemisphere atmospheric winter circulation. Tellus A64: 11595. doi:10.3402/tellusa.v64i0.11595

Jeong JH, Kug JS, Linderholm HW, Chen D, Kim BM, Jun SY (2014) Intensified Arctic warming under greenhouse warming by vegetation-atmosphere-sea ice interaction. Environ Res Lett 9 (9):094007. doi:10.1088/1748-9326/9/9/094007

Kalnay E, Kanamitsu M, Kistler R et al. (1996) The NCEP/NCAR 40-year reanalysis project. Bull Amer Meteor Soc 77:437–471

Kahl JD, Serreze MC, Schnell RC (1991) The historical Arctic rawinsonde archive. In: Climate Monitoring and Diagnostics Laboratory No. 19 Summary report 1990 (eds. E. E. Fergusson and R. M. Rosson), National Oceanic and Atmospheric Administration, Boulder, CO, pp. 95–98

Kahl JD, Skony SM, Serreze MC, Shiotani S, Schnell RC (1992) In situ meteorological sounding archives for Arctic studies. Bull Amer Meteor Soc 73:1824–1830

Kanamitsu M, Ebisuzaki W, Woollen J et al. (2002) NCEP–DOE AMIP-II Reanalysis (R-2). Bull Amer Meteor Soc 83:1631–1643. doi:10.1175/BAMS-83-11-1631

Kawatani Y, Ninomiya K, Tokioka T. (2008) The North Pacific subtropical high characterized separately for June, July, and August: zonal displacement associated with submonthly variability. J. Meteorol. Soc. Japan 86:505–530

Kay JE, co-authors (2008) The contribution of cloud and radiation anomalies to the 2007 Arctic sea ice extent minimum. Geophys Res Lett 35:L08503. doi:10.1029/2008GL033451

Kennedy JJ, Titchner HA, Hardwick J, et al. (2007) Global and regional climate in 2007. Weather 62:232–242

Krishnamurti T N, Daggupaty SM, Fein J, Kanamitsu M, Lee JD (1973) Tibetan high and upper tropospheric tropical circulation during northern summer. Bull Amer Meteor Soc54:1234–1239

Kumar A, Perlwitz J, Eischeid J, et al. (2010) Contribution of sea ice loss to Arctic amplification. Geophys Res Lett 37(21):L21701. doi:10.1029/2010GL045022

L'Heureux ML, Kumar A, Bell GD, Halpert MS, Higgins RW (2008) Role of the Pacific-North American (PNA) pattern in the 2007 Arctic sea ice decline. Geophys Res Lett 35:L20701. doi:10.1029/2008GL035205

Li CY, Li L, Tan YK (2011) Structure of South Asia high in the stratosphere and influence of ENSO. J Trop Meteor 17:293–301

Lindzen RS, Hou AY (1988) Hadley circulations for zonally averaged heating centered off the equator. J Atmos Sci 45(17):2416–2427. doi:10.1175/1520-0469(1988)045<2416:HCFZAH>2.0.CO;2

Liu BQ, He JH, Wang LJ (2009) Characteristics of the South Asia high establishment processes above the Indo-China Peninsula from April to May and their possible mechanism (in Chinese). Chin J Atmos Sci 33:1319–1332

Liu J, Song M, Hu Y, Ren X (2012) Changes in the strength and width of the Hadley Circulation since 1871. Clim Past 8:1169–1175, doi:10.5194/cp-8-1169-2012

Liu B, Wu G, Mao J, He J (2013) Genesis of the South Asian high and its impact on the Asian summer monsoon onset. J Climate 26:2976–2991. doi:10.1175/JCLI-D-12-00286.1

Liu H, Zhang DL, Wang B (2008) Daily to submonthly weather and climate characteristics of the summer 1998 extreme rainfall over the Yangtze River basin. J. Geophys. Res. 113:D22101, doi:10.1029/2008JD010072

Lu RY (2002) Indices of the summertime western North Pacific subtropical high. Adv Atmos Sci19: 1004–1028

Lu RY, Dong BW (2001) Westward extension of North Pacific subtropical high in summer. J. Meteorol. Soc. Japan 79:1229–1241

Lu RY, Li Y, Ryu CS (2008) Relationship between the zonal displacement of the western Pacific subtropical high and the dominant modes of low-tropospheric circulation in summer. Progress Nat. Sci. 18:161–165

Luo SW, Qian ZA, Wang QQ (1982) The climatic and synoptical study about the relation between the Qinghai-Xizang high pressure on the 100 mb surface and the flood and drought in east China in summer (in Chinese). Plateau Meteor 1:1–10

Mao JY, Sun Z, Wu GX (2010) 20–50-day oscillation of summer Yangtze rainfall in response to intraseasonal variations in the subtropical high over the western North Pacific and South China Sea. Clim. Dyn. 34:747–761. doi:10.1007/s00382-009-0628-2

Mason RB, Anderson CE (1963) The development and decay of the 100 mb summertime anticyclone over southern Asia. Mon Wea Rev 91:3–12. doi:10.1175/1520-0493(1963) 091,0003:TDADOT.2.3.CO;2

Mitas CM, Clement A (2005) Has the Hadley cell been strengthening in recent decades? Geophys Res Lett 32(3):L03809. doi:10.1029/2004GL021765

Neale RB, co-authors (2010) Description of the NCAR Community Atmosphere Model (CAM5.0). NCAR Tech Rep NCAR/TN-4861STR, 268 pp

Nitta T (1986) Long-term variations of cloud amount in the western Pacific region. J Meteorol Soc Japan 64:373–390

Nitta T (1987) Convective activities in the tropical western Pacific and their impact on the Northern Hemisphere summer circulation. J. Meteorol. Soc. Japan 65:373–390

Ogi M, Rigor IG (2013) Trends in Arctic sea ice and the role of atmospheric circulation. Atmos Sci Lett 14(2):97–101. doi:10.1002/asl2.423

Overland JE, Wang M (2010) Large-scale atmospheric circulation changes are associated with the recent loss of Arctic sea ice. Tellus 62(1):1–9. doi:10.1111/j.1600-0870.2009.00421.x

Parkinson CL, Comiso JC (2013) On the 2012 record low Arctic sea ice cover: combined impact of preconditioning and an August storm. Geophys Res Lett 40(7):1356–61. doi:10.1002/grl.50349

Peixoto JP, Oort AH (1992) Physics of climate. Springer-Verlag, American Institute of Physics Press, New York

Peng LX, Sun ZB, Chen HS, Ni DH (2010) The persistent anomaly of the south Asia high and its association with ENSO events (in Chinese). Acta Meteor Sin 68:855–864

Perovich DK, Richter-Menge JA, Jones KF, Light B (2008) Sunlight, water, and ice: Extreme Arctic sea ice melt during the summer of 2007. Geophys Res Lett 35:L11501, doi:10.1029/2008GL034007

Prinsenberg SJ, Peterson IK, Narayanan S, Umoh JU (1997) Interaction between atmosphere, ice cover, and ocean off Labrador and Newfoundland from 1962–1992. Can J Aquat Sci 54(S1):30–39. doi:10.1139/f96-150

Qian C, Wu ZH, Fu CB, Zhou TJ (2010) On multi-time scale variability of temperature in China in modulated annualcycle reference frame. Adv Atmos Sci 27:1169 − 1182

Qian WH, Chen Y, Jing M, Hu Q (2015b) An anomaly-based method for identifying signals of spring and autumn low temperature events in the Yangtze River Valley, China. J Appl Meteorol Climatol 54:1216–1233. doi:10.1175/JAMC-D-14-0240.1

Qian WH, Kang H-S, Lee D-K (2002) Distribution of seasonal rainfall in the East Asian monsoon region. Theor. Appl. Climatol. 73:151–168

Qian WH, Shi J (2016) Tripole precipitation pattern and SST variations linked with extreme episodes of the western Pacific subtropical high. Int J Climatol

Qian WH, Wu KJ, Chen DL (2015a) The Arctic and Polar cells act on the Arctic sea ice variation. Tellus A 67:27692

Qian WH, Wu KJ, Leung JCH, Shi J (2016b) Long-term trends of the Polar and Arctic cells influence the Arcticclimate since 1989. J Geophys Res Atmos doi:10.1002/2015JD024252

Qian WH, Wu KJ, Leung JCH (2016a) Climatic anomalous patterns associated with the Arctic and Polar cell strength variations. Clim Dyn doi:10.1007/s00382-016-3067-x

Qian WH, Yu TT, Du J (2016c) A unified approach to trace surface heat and cold events by using height anomaly. Clim Dyn. doi:10.1007/s00382-015-2666-2

Quan X, Diaz HF, Hoerling MP (2002) Changes of the Hadley circulation since 1950. The conference on the Hadley circulation: Present past and future 12–15 November, Honolulu, Hawaii

Rayner NA, Parker DE, Horton EB et al. (2003) Global analyses of sea surface temperature, sea ice, and night marine air temperature since the late nineteenth century. J Geophys Res 108 (D14):1–29. doi:10.1029/2002JD002670

Ren XJ, Yang XQ, Sun XG(2013) Zonal oscillation of western Pacific subtropical high and subseasonal SST variations during Yangtze persistent heavy rainfall events. J Clim 26:8929–8946

Ren XJ, Yang DJ, Yang XQ (2015) Characteristics and mechanisms of the subseasonal eastward extension of the South Asian high. J Clim 28:6799–6822

Richter-Menge J, Comiso J, Meier W, et al. (2008) Arctic report card 2008: Sea Ice Cover. http://www.arctic.noaa.gov/report08/seaice.html

Rigor IG, Wallace JM, Colony RL (2002) Response of sea ice to the Arctic oscillation. J Climate 15:2648–2663. doi:10.1175/1520-0442(2002)015<2648:ROSITT>2.0.CO;2

Sato K, Inoue J, Watanabe M (2014) Influence of the Gulf Stream on the Barents Sea ice retreat and Eurasian coldness during early winter. Environ Res Lett 9:084009. doi:10.1088/1748-9326/9/8/084009

Screen JA, Deser C, Simmonds I (2012) Local and remote controls on observed Arctic warming, Geophys Res Lett 39:L10709, doi:10.1029/2012GL051598

Serreze MC, Francis JA (2006) The Arctic amplification debate. Clim Change 76:241–264. doi:10.1007/s10584-005-9017-y

Screen JA, Simmonds I (2010a) The central role of diminishing sea ice in recent Arctic temperature amplification. Nature 464(7293):1334–1337

Screen JA, Simmonds I (2010b) Increasing fall-winter energy loss from the Arctic Ocean and its role in Arctic temperature amplification. Geophys Res Lett 37:L16707

Serreze MC, Holland MM, Stroeve J (2007) Perspectives on the Arctic's shrinking sea-ice cover. Science 315:1533–1536, doi:10.1126/science.1139426

Shi J, Qian WH (2016) Connection between anomalous zonal activities of the South Asian High and Eurasian summer climate anomalies. J Clim

Shindell D, Faluvegi G (2009) Climate response to regional radiative forcing during the twentieth century. Nature Geosci 2:294–300. doi:10.1038/ngeo473

Song H, Zhang M (2007) Changes of the boreal winter Hadley circulation in the NCEP-NCAR and ECMWF reanalyses: A comparative study, J Climate 20:5191–5200

Stachnik JP, Schumacher C (2011) A comparison of the Hadley circulation in modern reanalyses. J Geophys Res 116:D22. doi:10.1029/2011jd016677

Stroeve J, Serreze M, Drobot S, Gearheard S, Holland M, Maslanik J, Meier W, Scambos T (2008) Arctic sea ice extent plummets in 2007. Eos Trans AGU 89(2):13–14

Su H, Jiang JH, Zhai C, Shen TJ, Neelin JD, Stephens GL, Yung YL (2014) Weakening and strengthening structures in the Hadley circulation change under global warming and implications for cloud response and climate sensitivity. J Geophys Res 119(10):5787–5805

Tan J, Yang H, Sun SQ, Wang PX (2005) Characteristics of the longitudinal oscillation of South Asia high during summer (in Chinese). J Nanjing Institute Meteor 28:452–460

Tao SY, Chen LX (1987) A review of recent research on the East Asian summer monsoon in China. In Monsoon Meteorology, Chang CP, Krishnamurti TN (eds). Oxford University Press: Oxford; pp 60–92

Tao SY, Zhu FK (1964) The 100-mb flow patterns in southern Asia in summer and its relation to the advance and retreat of the west-Pacific subtropical anticyclone over the Far East (in Chinese). Acta Meteor Sin 34:385–396

Timmermans ML (2015) The impact of stored solar heat on Arctic sea ice growth. Geophys Res Lett 42:6399–6406. doi:10.1002/2015GL064541

Toole J, Timmermans ML, Perovich D, Krishfield R, Proshutinsky A, Richter-Menge J (2010) Influences of the ocean surface mixed layer and thermohaline stratification on Arctic sea ice in the Central Canada Basin. J Geophys Res 115:C10018. doi:10.1029/2009JC005660

Vihma T (2014) Effects of Arctic sea ice decline on weather and climate: A review. Surv Geophy 35:1175–1214. doi:10.1007/s10712-014-9284-0

Vihma T, Tisler P, Uotila P (2012) Atmospheric forcing on the drift of Arctic sea ice in 1989–2009. Geophys Res Lett 39(2):L02501, doi:10.1029/2011GL050118

Wang ZY, Ding YH (2008) Climatic characteristics of rainy seasons in China. Chin J Atmos Sci 32:1–13 (in Chinese)

Wang WC, Gong W, Wei H (2000) A regional model simulation of the 1991 severe precipitation event over the Yangtze-Huai River valley. Part I: Precipitation and circulation statistics. J Clim 13:74–92

Wang WL, Wang Y, Wu RS (2005) A new view on the Ferrel cell (in Chinese). Chin J Geophys 48(3):539–545

Wei W, Zhang RH, Wen M, Kim BJ, Nam JC (2015) Interannual variation of the South Asian high and its relation with Indian and East Asian summer monsoon rainfall. J Climate 28:2623–2634

Wei W, Zhang RH, Wen M, Rong XY, Li T (2014) Impact of Indian summer monsoon on the South Asian high and its influence on summer rainfall over China. Clim Dyn 43:1257–1269. doi:10.1007/s00382-013-1938-y

Wielicki BA, Wong T, Allan RP, Slingo A, Kiehl JT, Soden BJ, Gordon CT, Miller AJ, Yang SK, Randall DA, Robertson F, Susskind J, Jacobowitz H (2002) Evidence for large decadal variability in the tropical mean radiative energy budget. Science 295:841–843. doi:10.1126/science.1065837

Wu GX, Chou JF, Liu YM, Zhang QY, Sun SQ (2003) Review and prospect of the study on the subtropical anticyclone. Chin J Atmos Sci 27:503–517 (in Chinese)

Wu ZH, Huang NE (2009) Ensemble Empirical Mode Decomposition: A noise-assisted data analysis method. Adv Adapt Data Anal 1(01):1 − 41

Wu KJ, Qian WH (2015) Secular non-linear trends and multi-timescale oscillations of regional surface air temperature in eastern China. Climate Res 63:19 − 30

Xie P, Arkin PA (1997) Global precipitation: A 17-year monthly analysis based on gauge observations, satellite estimates, and numerical model outputs. Bull Am Meteorol Soc 78:539–2558. doi:10.1175/1520-0477(1997)078<2539:GPAYMA>2.0.CO;2

Xie SP, Hu KM, Hafner J, Tokinaga H, Du Y, Huang G, Sampe T (2009) Indian Ocean capacitor effect on Indo-western Pacific climate during the summer following El Niño. J Clim 22:730–747, doi:10.1175/ 2008JCLI2544.1

Yang H, Sun SQ (2003) Longitudinal displacement of the subtropical high in the western Pacific in summer and its influence. Adv Atmos Sci 20:921–933

Yang H, Sun SQ(2005) The characteristics of longitudinal movement of the subtropical high in the western Pacific in the pre-rainy season in South China. Adv Atmos Sci 22:392–400

Yang J, Wang B, Wang B, Bao Q (2010) Biweekly and 21–30-day variations of the subtropical summer monsoon rainfall over the lower reach of the Yangtze River basin. J Clim 23:1146–1159

Zhang Q, Wu GX, Qian YF (2002) The bimodality of the 100 hPa South Asia high and its relationship to the climate anomaly over East Asia in summer. J Meteor Soc Japan 80:733–744. doi:10.2151/jmsj.80.733

Chapter 10
Low-Frequency Oscillations

Since the early 1970s and the early 1980s, the Madden-Julian Oscillation (MJO) in the tropics and some teleconnections in worldwide have respectively become two important topics in climate study. The former is moving low-frequency fluctuations while the latter is standing oscillations. The goal of this chapter is to re-identify the tropical MJO and worldwide teleconnections using anomaly-based analysis method. For examining teleconnections, the rotated empirical orthogonal Function (REOF) is used to analyze the grid maximum height anomalies (MHAs) respectively in the tropics and the two hemispheres. The height anomalies are filtered and applied to represent MJOs, while eastward and westward propagations of tropical oscillations in the three tropical ocean basins are also introduced using the height anomalies. The interaction between the MJO in the tropics and moving low-frequency fluctuations such as intraseasonal oscillations in extra-tropical regions is most important for improving extended-range weather forecasts. The height anomalies can be commonly used in the tropics and extra-tropical regions to investigate this interaction in the future.

10.1 Maximum Anomalies of Atmospheric Variables

The biosphere is mainly concentrated near the Earth's surface. The atmosphere and the lithosphere provide resource and environment of biological growth. Thus, anomalous variations of the atmosphere and the lithosphere can form disasters of biosphere. Many disasters such as persistent heat waves, cold surges, and heavy rains have been examined in previous chapters using the anomaly-based method. These surface weather extremes are directly linked with the height anomalies in the atmosphere. Standing and moving systems of height anomalies have also been identified in Chap. 7. The South Asian High, the largest circulation system, is active in the boreal summer. The duration of its anomalous activity is usually shorter than one month. Sometimes, anomalous activity of one system can accompany or

© Springer Nature Singapore Pte Ltd. 2017

W. Qian, *Temporal Climatology and Anomalous Weather Analysis*,
Springer Atmospheric Sciences, DOI 10.1007/978-981-10-3641-5_10

influence other systems nearby, even at a large distance. This phenomenon is called as teleconnections in the atmosphere.

Teleconnections in atmospheric science refer to climate anomalies related to each other at large distances (typically thousands of kilometers). The most emblematic teleconnection is the linkage of sea-level pressure (SLP) at Tahiti and Darwin, which is defined as the Southern Oscillation (SO). Teleconnections were first noted by the British meteorologist Sir Gilbert Walker in the late nineteenth century, through computation of the correlations between time series of atmospheric pressure, temperature, and rainfall. They served as a building block for the understanding of climate variability, showing that the latter was not purely random.

Indeed, the El Niño-Southern Oscillation (ENSO) is a phenomenon underlying variability in several locations at once. It is noticed that associated teleconnections occurred all over North America, as embodied by the Pacific-North American (PNA) teleconnection pattern. In the 1980s, improved observations allowed the detection of teleconnections at larger distances throughout the troposphere (Wallace and Gutzler 1981). Concomitantly, the theory emerged, indicating such patterns could be understood through the dispersion of Rossby waves due to the spherical geometry of the Earth (Hoskins and Karoly 1981). This is sometimes called the "proto-model" (Trenberth et al. 1998).

Most teleconnections have been revealed at 500 hPa (Wallace and Gutzler 1981; Barnston and Livezey 1987). Basic methods are one-point correlation map, composite analysis, and eigenvector analysis expansion (Kutzbach 1970; Dickson and Namias (1976); Wallace and Gutzler 1981; Ding and Wang 2005; Hannachi et al. 2007). The goal of studying teleconnections is to predict regional climate anomalies in different areas. However, Chap. 7 has documented that the MHA is not situated near 500 hPa or in the lower troposphere. Thus, we first examine which level of the MHA is mostly located. Figure 10.1 shows the vertical-latitude distribution of daily mean height standard deviation (SD) along 120°E on 10 February (boreal winter) and 9 August (boreal summer). In boreal winter, two axes of MHA SDs are located at the upper troposphere near 300 hPa in the mid-high latitudes, except that there is a vertical axis near the Arctic in the stratosphere (Fig. 10.1a). In boreal summer, three centers of MHA SDs in the Northern Hemisphere (NH) are observed near 45° N, 70°N, and 90°N at the upper troposphere (Fig. 10.1b). An axis of MHA SDs is horizontally found at the upper troposphere near 250 hPa in the mid-high latitudes of the Southern Hemisphere (SH), except that there is a vertical axis near 60°S in the stratosphere (Fig. 10.1b). The two vertical axes imply that the large variation of height anomalies is mostly concentrated in local winter season, in the Arctic and near 60°S.

To examine the spatial feature of height anomalies, Fig. 10.2 shows the distribution of height SDs at 300 hPa on 10 February and 9 August in the two hemispheres. In boreal winter, three maximum centers of height SDs larger than 200 gpm are respectively located in the northern Atlantic, the northeastern Pacific, and western Russia in the NH. In boreal summer, six maximum centers of height SDs larger than 160 gpm are respectively located in the northern Atlantic, the Svalbard Island, western-central Russia, northern Pacific, and northeastern Canada.

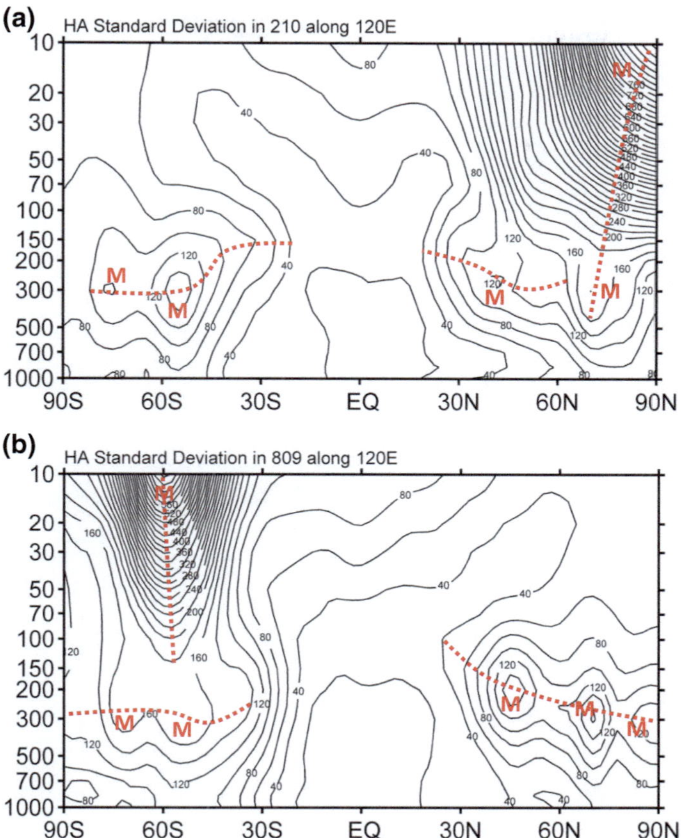

Fig. 10.1 Vertical-latitude distribution of daily mean height SD (20 gpm interval) along 120°E on **a** 10 February and **b** 9 August using the NCEP R1 data from 1981 to 2010. The *red dotted line* is the maximum axis of height SD. Letter "M" indicates the maximum SDs

Relative low SDs are located in the Arctic Ocean in winter and over the Greenland in summer. In the SH, seven maximum centers of height SDs larger than 180 gpm have been detected on 10 February and 9 August mostly along middle latitudes.

Similar to that in Fig. 10.1, the vertical-latitude distribution of temperature SD is given in Fig. 10.3. In boreal winter, there are three horizontal axes of temperature SD in mid-high latitudes. The largest temperature SD is at the lower stratosphere (Fig. 10.3a). From Figs. 10.1a and 10.3a, it is noted that the horizontal height SD axis near 300 hPa separates a temperature SD axis above near the tropopause and another temperature SD axis near 850 hPa. The hydrostatic relationship between the height and temperature anomalies has been revealed in many previous cases. It has been illustrated that a positive (negative) center of height anomalies near 300 hPa separates a negative (positive) center of temperature anomalies above and a positive (negative) center below. This relationship is also found in boreal summer

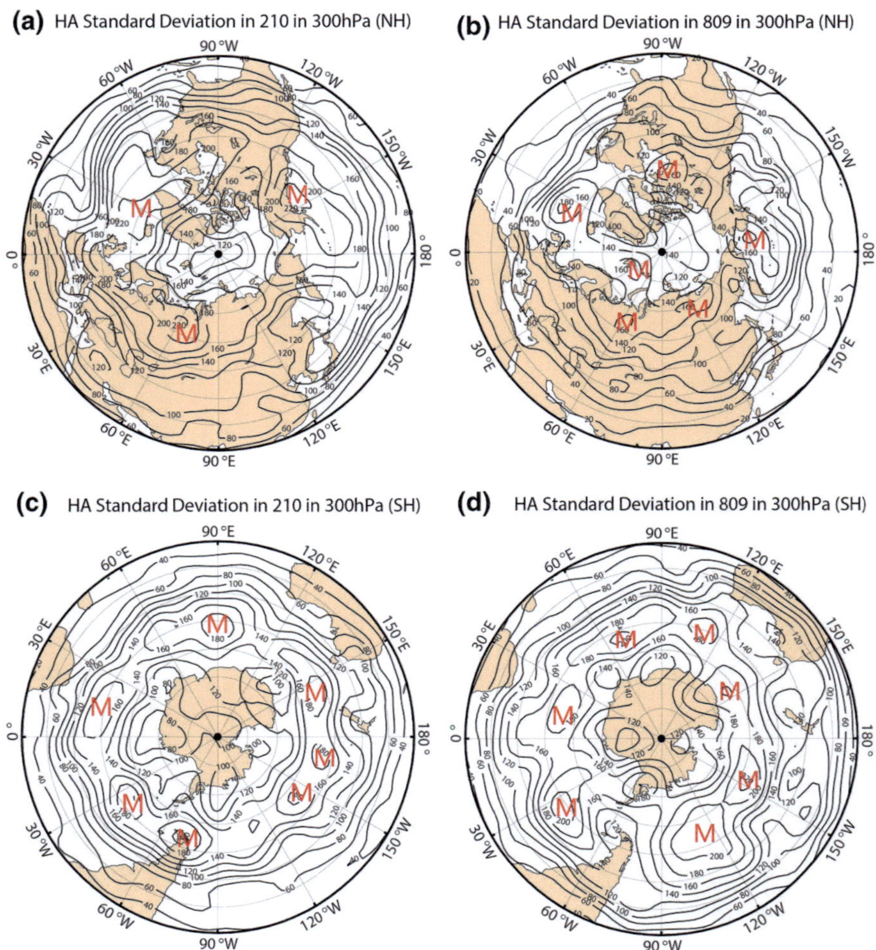

Fig. 10.2 Spatial distribution of height SD (20 gpm interval) at 300 hPa on 10 February and 9 August. Letter "M" indicates maximum centers of height SDs

(Fig. 10.3b). Figure 10.3 also implies that the sudden stratospheric warming occurs in the Arctic and in 60°S in local winter.

Figure 10.4 depicts the distribution of westerly SD along 120°E on 10 February and 9 August. A major axis of westerly SD with many centers is located at the upper troposphere. The highest altitude is at 150 hPa near the equator, while the lowest altitude is at 300 hPa in high latitudes. In the stratosphere, large westerly SD is concentrated in their winter hemisphere.

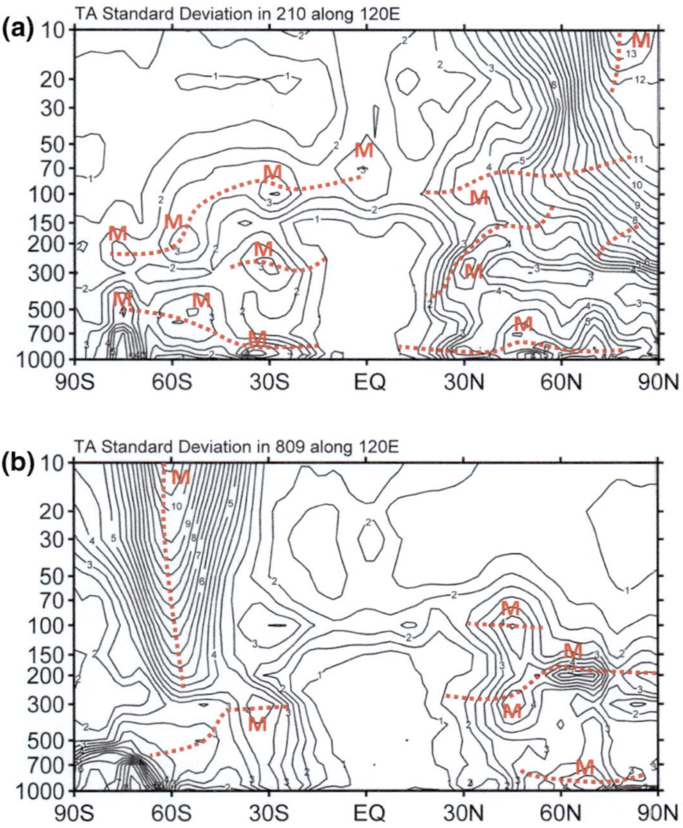

Fig. 10.3 Same as Fig. 10.1 except of temperature SD (1 K interval) along 120°E on **a** 10 February and **b** 9 August

10.2 Low-Frequency Standing Oscillations in the Northern Hemisphere

Standing oscillations, such as the North Atlantic Oscillation (NAO) and the North Pacific Oscillation (NPO), have been identified by Walker and Bliss (1932). Using the 500-hPa height in boreal winter, five patterns and their indices were identified and defined by Wallace and Gutzler (1981).

The eastern Atlantic (EA) pattern is characterized by three centers: one located southwest of the Canary Island (25°N, 25°W), one west of Great Britain (55°N, 20°W), and a third one near the Black Sea (50°N, 40°E). The EA pattern index is defined as:

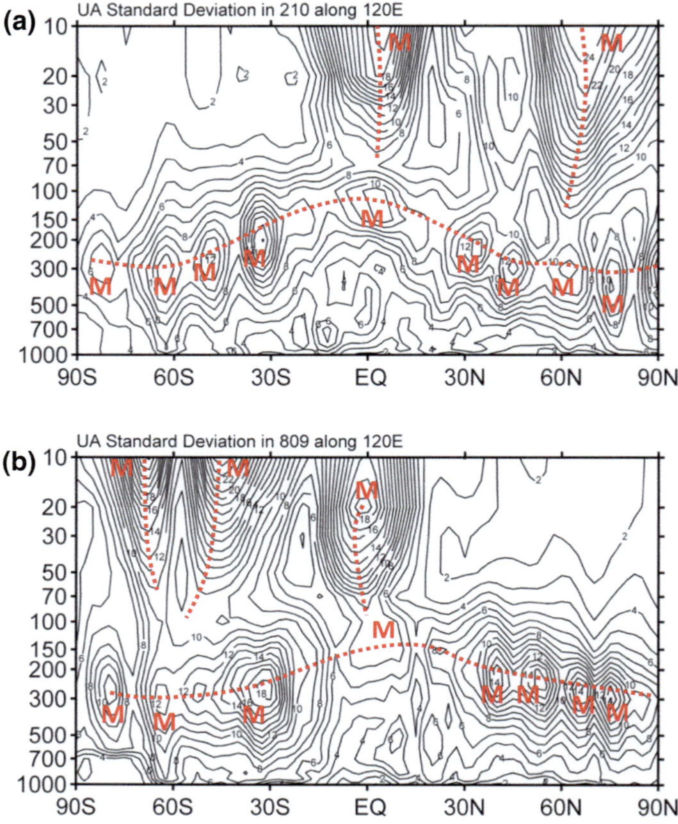

Fig. 10.4 Same as Fig. 10.1 except of westerly SD (1 m/s interval) along 120°E on **a** 10 February and **b** 9 August

$$\begin{aligned}\mathrm{EA} = {} & 0.5Z(55^\circ\mathrm{N}, 20^\circ\mathrm{W}) \\ & - 0.25Z(25^\circ\mathrm{N}, 25^\circ\mathrm{W}) - 0.25Z(50^\circ\mathrm{N}, 40^\circ\mathrm{E})\end{aligned} \tag{10.2.1}$$

The PNA pattern has been noted by Lorenz (1951). It is comprised of four centers: one located near Hawaii (20°N, 160°W), second one over the North Pacific (45°N, 165°W), third one over Alberta (55°N, 115°W), and fourth one over the Gulf Coast of the United States (30°N, 85°W). The PNA pattern index of each winter month is defined as a linear combination of the normalized height anomalies at the four pattern centers:

$$\begin{aligned}\mathrm{PNA} = {} & 0.25[Z(20^\circ\mathrm{N}, 160^\circ\mathrm{W}) - Z(45^\circ\mathrm{N}, 165^\circ\mathrm{W}) \\ & + Z(55^\circ\mathrm{N}, 115^\circ\mathrm{W}) - Z(30^\circ\mathrm{N}, 85^\circ\mathrm{W})]\end{aligned} \tag{10.2.2}$$

The western Atlantic (WA) pattern is based on two centers over the Atlantic Ocean, (55°N, 55°W) and (30°N, 55°W). The WA pattern index is defined as

$$\mathrm{WA} = 0.5[Z(55^\circ\mathrm{N}, 55^\circ\mathrm{W}) - Z(30^\circ\mathrm{N}, 55^\circ\mathrm{W})] \tag{10.2.3}$$

The western Pacific (WP) pattern is characterized by a weak Aleutian low (particularly west of the date line) (60°N, 155°E) and a weak jet-stream over Japan (30°N, 155°W), and vice versa. The WP pattern index is defined as

$$\mathrm{WP} = 0.5[Z(60^\circ\mathrm{N}, 155^\circ\mathrm{E}) - Z(30^\circ\mathrm{N}, 155^\circ\mathrm{W})] \tag{10.2.4}$$

Three active centers respectively over Scandinavia and Poland (55°N, 20°E), over Siberia (55°N, 75°E), and over Japan (40°N, 145°E) are used to construct the Eurasian (EU) pattern. The EU pattern index is defined as

$$\begin{aligned} \mathrm{EU} = & -0.25Z(55^\circ\mathrm{N}, 20^\circ\mathrm{E}) \\ & + 0.5Z(55^\circ\mathrm{N}, 75^\circ\mathrm{E}) - 0.25Z(40^\circ\mathrm{N}, 145^\circ\mathrm{E}) \end{aligned} \tag{10.2.5}$$

In the NH, another planetary-scale pattern is the Arctic Oscillation (AO). The AO or Northern Annular Mode (NAM)/Northern Hemisphere Annular Mode is an index (which varies over time with no particular periodicity) of the dominant pattern of non-seasonal SLP variations north of 20°N. It is characterized by pressure anomalies of one center in the Arctic with the opposite center over 37–45°N (Thompson and Wallace 1998). The AO, NAO, and PNA indices are derived from the NCEP website http://www.cpc.ncep.noaa.gov/products/precip/CWlink/daily_ao_index/teleconnections.shtml.

In previous chapters, the persistence of anomalous MHA activity is usually shorter than one month. In a calendar month, the monthly mean MHA will be zero if the first-half MHA is the same as that in the second-half only with opposite signs. We assume that there are many months with only positive or negative daily mean MHAs without opposite signs. Statistically, we can obtain a database of monthly mean MHA at all grids from 1981 to 2010. Using the monthly mean MHA database in the NH, we calculate the REOF. In statistics and signal processing, the method of EOF analysis is a decomposition of a signal or dataset, in terms of orthogonal basic functions which are determined from the data. It is the same as performing a principal component analysis on the data, except that the EOF method gives both time series and spatial patterns. The REOF analysis is useful to detect localized behaviors, whereas EOF analysis tends to detect behaviors spanning the whole domain, which are sometimes difficult to interpret. Thus, REOF analysis is useful in the identifying teleconnection.

The first eight rotated eigenvectors and their monthly mean time series are derived using the REOF analysis. Spatial distributions of the first rotated eigenvector with its variance 32.3%, as well as its monthly mean time series from 1981 to 2010, are shown in Fig. 10.5. The first mode is characterized by the negative MHAs in the Arctic, with the opposite anomalies in about 30–60°N. This REOF mode is

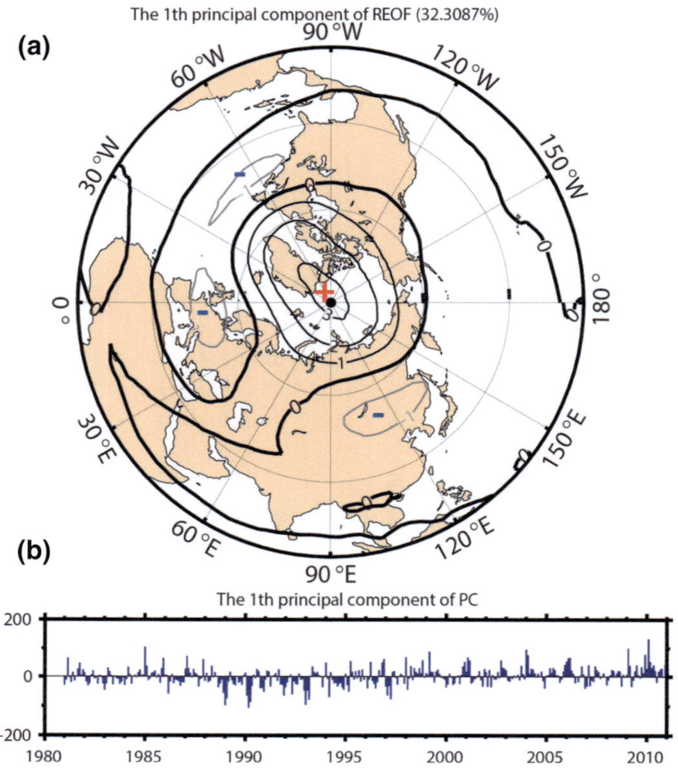

Fig. 10.5 **a** Spatial distributions of the first rotated eigenvector of MHAs with its variance of 32.3% and **b** its monthly mean time series (*blue bars*) from 1981 to 2010

Table 10.1 Correlations of the first eight MHA REOF mode series against other indices over 1981–2010

REOF (%)	AO	NAO	PNA1	PNA	WA	EA	WP	EUP
1 (32.31)	***−0.770***	***−0.454***	0.154	0.115	0.366	−0.043	−0.044	0.176
2 (6.21)	−0.047	−0.197	***0.400***	*0.471*	0.146	0.092	−0.169	0.004
3 (5.18)	−0.138	−0.174	0.352	***0.581***	0.261	***0.402***	0.108	***0.537***
4 (4.67)	−0.171	−0.128	−0.028	0.061	−0.335	−0.163	***0.558***	−0.311
5 (4.42)	0.223	0.138	−0.154	−0.297	*−0.332*	*−0.357*	−0.052	0.033
6 (4.12)	0.051	0.266	0.086	0.207	−0.135	−0.292	0.281	*0.376*
7 (4.00)	0.116	0.387	0.372	0.292	***−0.404***	−0.113	*−0.410*	−0.144
8 (3.68)	0.114	*0.332*	0.074	0.005	0.075	−0.229	−0.245	*−0.329*

The correlation coefficient 0.136 reaches the 99% confidence level and the italic (bold) number indicates higher (maximum) correlation between mode series and indices. The PNA1 index is derived from NOAA, while the PNA index and other indices are calculated using Eqs. 10.2.1–10.2.5. Symbol % indicates the variance of REOF modes the same as the dominant pattern of non-seasonal SLP variations north of 20°N, namely the AO or NAM. The correlation between the monthly mean AO series and the time series of the first mode is −0.77, reaching the 99% confidence level. The first mode also has a high correlation (−0.454) with NAO (Table 10.1).

The second rotated eigenvector is shown with its variance 6.2% (Fig. 10.6). Two wave trains are clearly observed in the Western Hemisphere (WH) and the Eastern Hemisphere (EH). The wave train in the WH is similar to that of the PNA pattern. Their correlation coefficient is 0.40 for PNA index and 0.47 for PNA1 index (Table 10.1). The EH wave train with four centers is standing from South Europe to western Russia and East Asia. The two opposite centers at South Europe and western Russia should be a standing oscillation which has not been named.

The third rotated eigenvector with three pairs of centers shows a circle-zonal wave train around 40–70°N with its variance 5.2% (Fig. 10.7). Two centers are located in northern Pacific and northern Atlantic, respectively. Two pairs of MHA

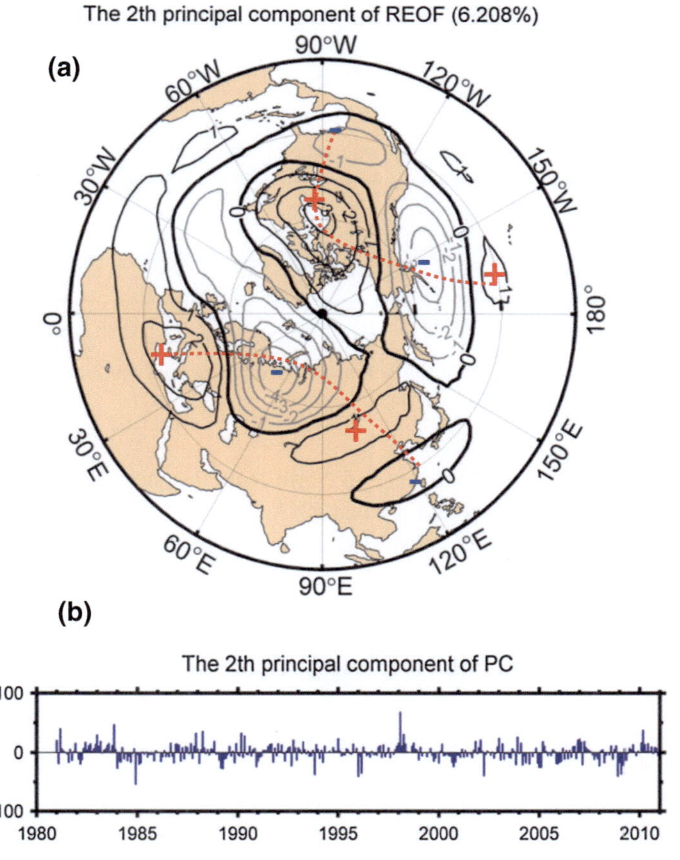

Fig. 10.6 Same as Fig. 10.5 except for the second rotated eigenvector with its variance 6.2%. The *red dashed line* connects positive and negative centers

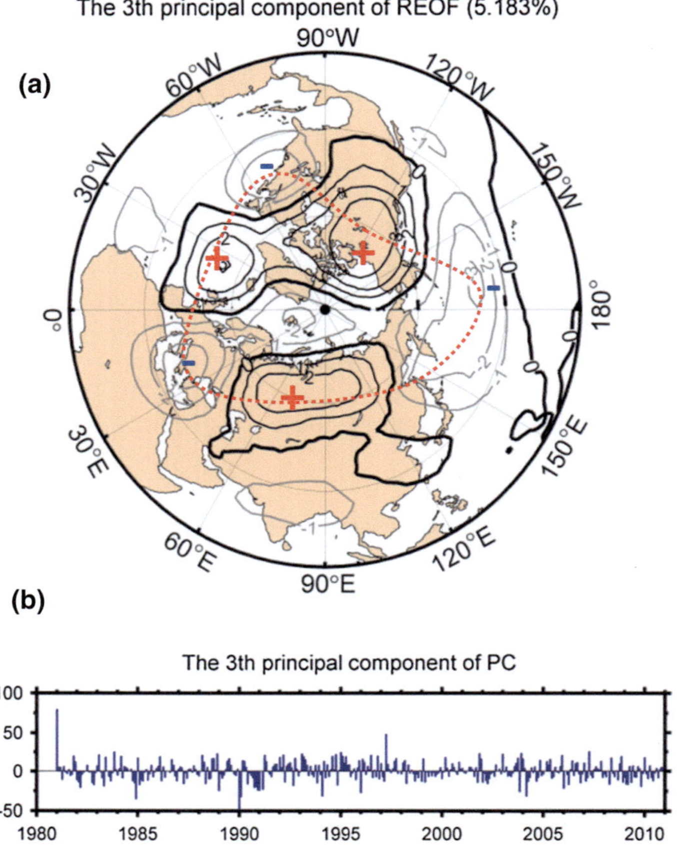

Fig. 10.7 Same as Fig. 10.5 except for the third rotated eigenvector with its variance 5.2%. The *red dashed line* connects positive and negative centers

centers are separated in the Eurasian and North American continents. In Table 10.1, three high correlations are found against PNA, EA, and EUP indices. Indeed, the three patterns are statistically connected along a zonal circle in the NH mid-high latitudes.

The fourth rotated eigenvector shows a wave train from the northern Atlantic crossing the Eurasian continent in middle latitudes to the northern Pacific with its variance 4.7% (Fig. 10.8). In the Eurasian continent, there are three centers of MHA. In Table 10.1, the maximum correlation between the fourth mode and WP pattern index is 0.56. It means that there is a significant oscillation with the two centers in Northeast Asia and Northwest Pacific.

The fifth rotated eigenvector is shown with its variance 4.4% (Fig. 10.9a). Three large centers are connected by the red dashed line from Alaska to the Iceland and Europe. In Table 10.1, two higher correlations are −0.332 and −0.357 against the WA and EA pattern series. It is found that there is only a pair of centers in central

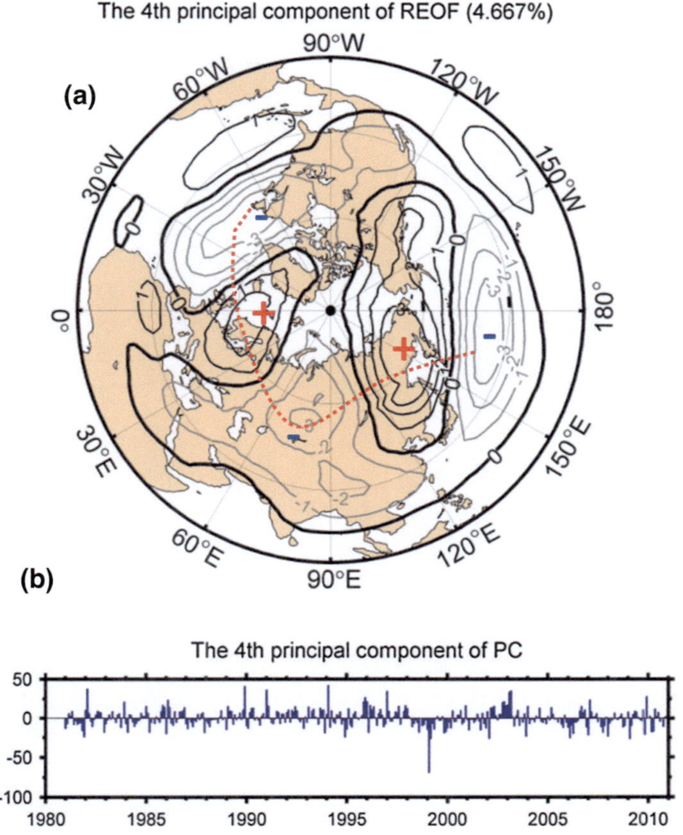

Fig. 10.8 Same as Fig. 10.5 except for the fourth rotated eigenvector with its variance 4.7%

northern Atlantic (Fig. 10.9a). The sixth rotated eigenvector shows two crossing lines centered over the north of England by two wave trains with its variance 4.1% (Fig. 10.9b). In Table 10.1, the highest correlation is 0.376 between the sixth mode and the EU index series. The seventh rotated eigenvector depicts a major wave train from the central North Pacific to northeastern Pacific and the Greenland with its variance 4.0% (Fig. 10.9c). In Table 10.1, the two highest correlations are −0.40 and −0.41 against the WA and WP pattern indices. The eighth rotated eigenvector shows a major wave train from North America to the Greenland, Europe, central Russia, and East Asia with its variance 3.7%. From the eighth mode, two higher correlations (0.33 and −0.33) are detected against the NAO and the EU indices (Table 10.1). The EU pattern is clear in Fig. 10.9d.

Table 10.1 summarizes that the first MHA mode represents the AO and NAO in the NH mid-high latitudes. The PNA pattern can be found from the second and third MHA modes. The WA and EA patterns are located within the seventh and the third MHA modes, respectively. The WP and EU patterns are respectively found in

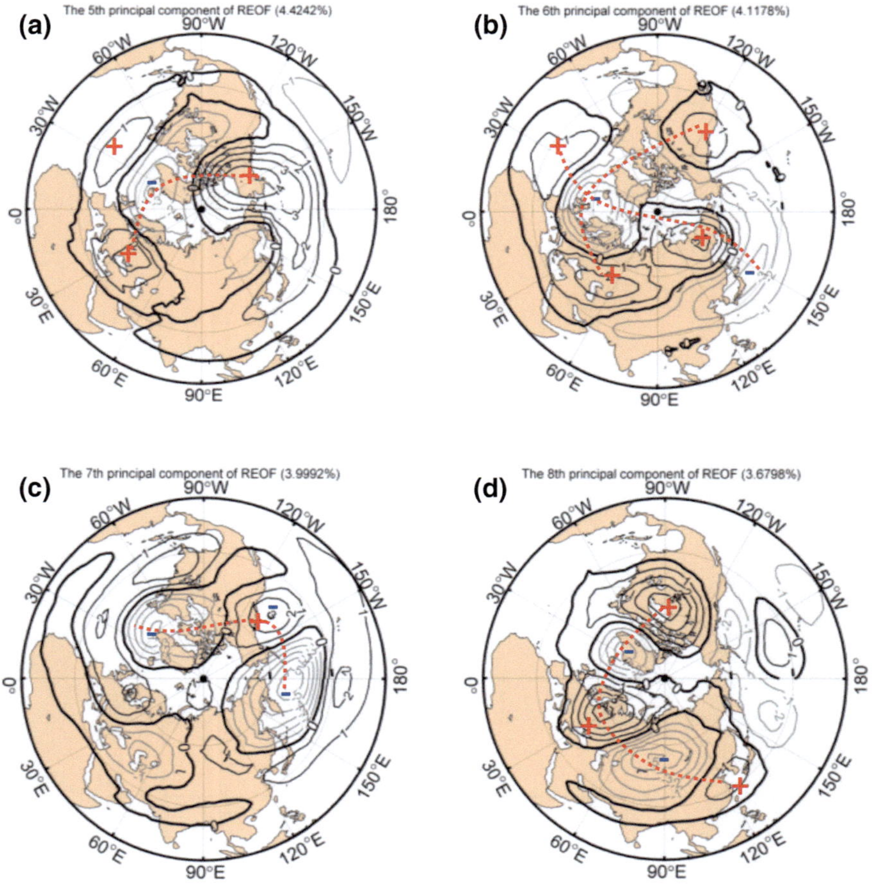

Fig. 10.9 Same as Fig. 10.5 except for **a** the fifth, **b** the sixth, **c** the seventh, and **d** the eighth rotated eigenvectors with their variances 4.4, 4.1, 4.0, and 3.7%, respectively

the fourth and third MHA modes. The third MHA mode includes the zonal wave train, so it represents three well-known (PNA, EA, and EU) patterns revealed by Wallace and Gutzler (1981).

10.3 Low-Frequency Standing Oscillations in the Southern Hemisphere

The Southern Annular Mode (SAM) is the major oscillation of the extratropical circulation in the SH and is characterized by zonally symmetric or annular structures. Its geopotential height perturbations with opposing signs in Antarctica and the surrounding zonal ring are centered near 45°S (Thompson and Wallace 2000a, b). The SAM is also called the Antarctic Oscillation (AAO, Gong and Wang 1999) and High-Latitude Mode (Rogers and Van Loon 1982). Similar to that in the NH, this section shows the spatial distributions and temporal evolutions of rotated eigenvectors based on the MHAs in the SH. The first rotated eigenvector of MHA shows a typical SAM pattern, with MHAs of opposite signs in Antarctica and the surrounding zonal ring centered near 45°S (Fig. 10.10a) The correlation between the first mode series and the AAO index series is −0.89. The second rotated eigenvector is shown with its variance 8.3% (Fig. 10.10b). The red dashed line connects three centers in higher latitudes over the Pacific and Atlantic. Indeed, the largest positive center is located near the Graham Land. A negative zonal ring is centered near 45°S while a positive zonal ring is in the tropics. This mode is closely correlated with the Nino3.4 index, i.e., the sea surface temperature (SST) anomaly over equatorial central-eastern Pacific. Their correlation coefficient is 0.26, reaching the 99% confidence level.

As shown in Fig. 10.2c, d, the maximum centers of MHA in the SH are mainly located within westerly zone outside of the Antarctic continent. From the third mode to the sixth mode (Figs. 10.11 and 10.12), all positive and negative centers of MHA are located over the SH oceans beyond 30°S. It means that there are several sets of zonal wave trains in the SH oceans. It is hard to name them in the SH.

The PNA pattern is detected in the second mode of MHA in the NH. Similarly, Fig. 10.13 shows the Pacific-South America (PSA) pattern with its variance 3.6%. If connecting the PNA and PSA, it may be the Circum-Pacific Teleconnection pattern (Wang et al. 2005).

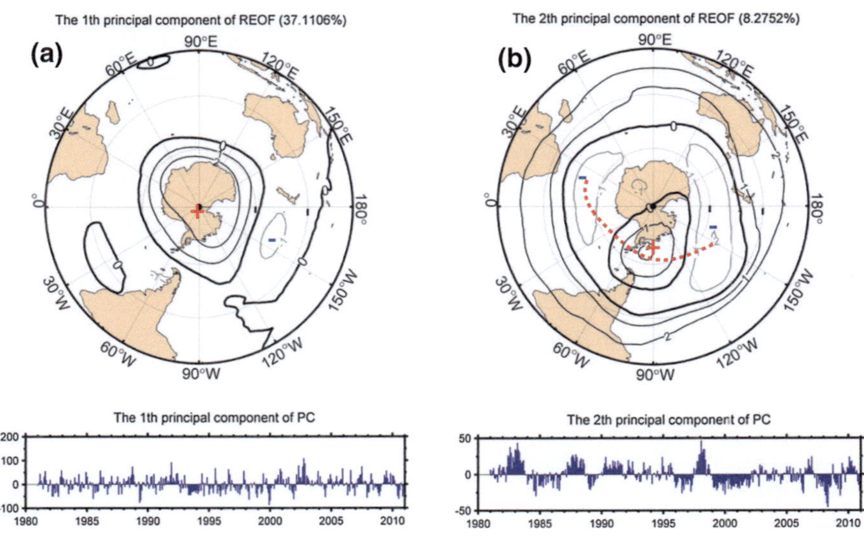

Fig. 10.10 Spatial distributions and monthly mean series (*blue bars*) of **a** the first rotated eigenvector of MHA with its variance of 37.1% and **b** the second rotated eigenvector with its variance of 8.3% from 1981 to 2010. The *red dashed line* connects positive and negative centers

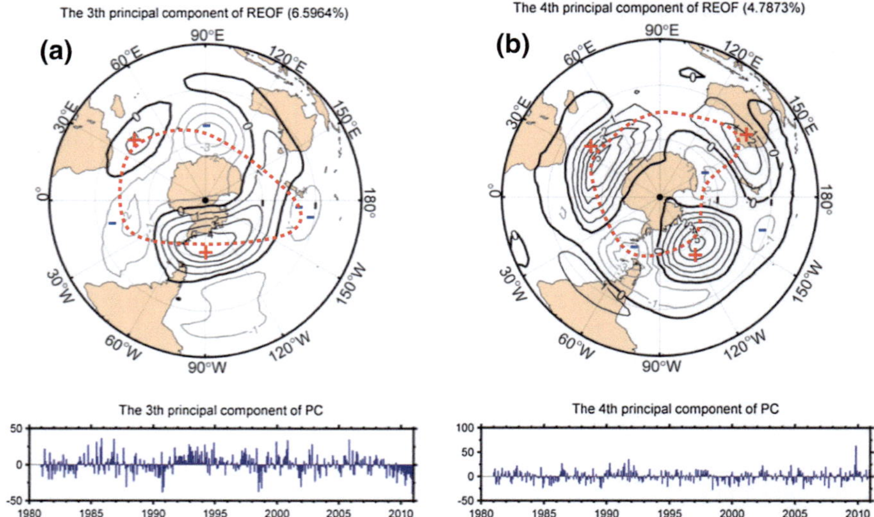

Fig. 10.11 Same as Fig. 10.10 except for **a** the third and **b** the fourth rotated eigenvectors with their variances 6.6 and 4.8%, respectively

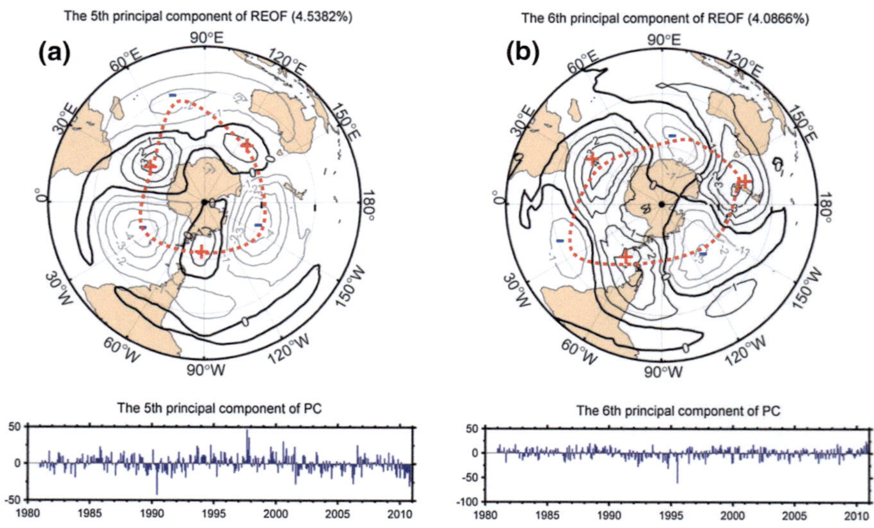

Fig. 10.12 Same as Fig. 10.10 except for **a** the fifth and **b** the sixth rotated eigenvectors with their variances 4.5 and 4.1%, respectively

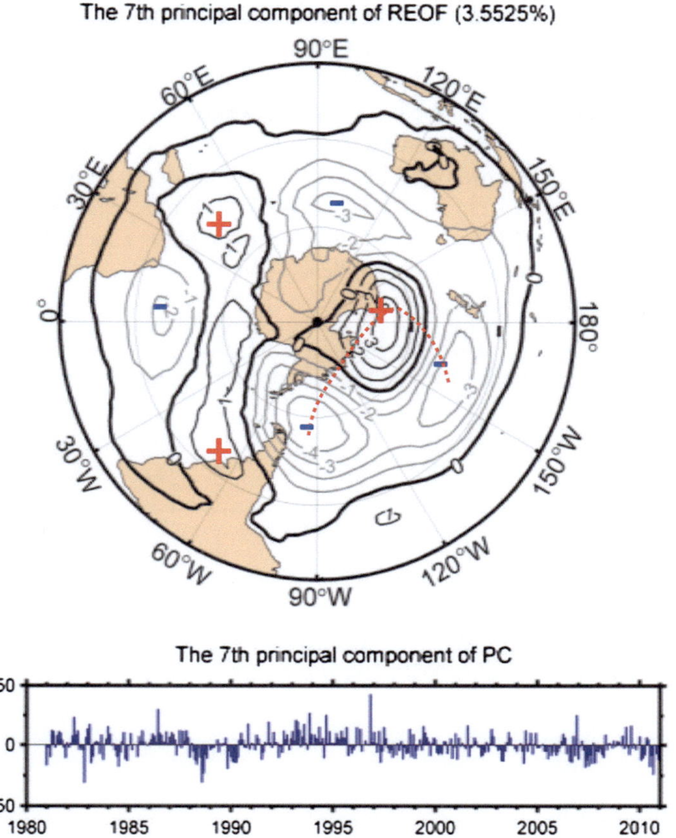

Fig. 10.13 Same as Fig. 10.10 except for the seventh rotated eigenvector with its variance 3.6%

10.4 Low-Frequency Standing Oscillations in the Tropics

Similar to those in the NH, the monthly mean MHAs in the tropical zone between 30°S and 30°N are analyzed by the REOF. Figure 10.14 illustrates the first rotated eigenvector with its variance of 61.6% and the monthly mean time series from 1981 to 2010. The correlation between the first mode time series and Niño3.4 index series is −0.56, showing a significant interannual variability. This mode is coupled with SST anomalies at interannual timescale. The tendency of MHAs is consistent in the tropics though there are several negative centers in three tropical oceans. The second rotated eigenvector is shown with its variance 5.3% (Fig. 10.15a). One standing oscillation with opposite signs in the WH and EH shows decadal variability. A predominant decadal negative series is found in Fig. 10.15b. The third rotated eigenvector is shown with its variance 4.1% (Fig. 10.16). Its special mode is the same as the first one, except there is long-term trend. Therefore, interannual, decadal, and long-term trend are the three timescales of tropical MHA variation that has been detected from Figs. 10.14, 10.15 and 10.16, respectively.

If the long-term trend and zonally averaged height anomaly are removed from the height anomalous series in the last 30 years, we can obtain another set of height anomalies. The new monthly mean MHAs in the tropics are analyzed by the REOF. A standing oscillation shows opposite signs separated along the dateline with two centers in western and eastern Pacific (Fig. 10.17). The time series of the first mode is closely correlated with the Niño3.4 index at interannual timescale. Their correlation is −0.77, reaching the 99% confidence level.

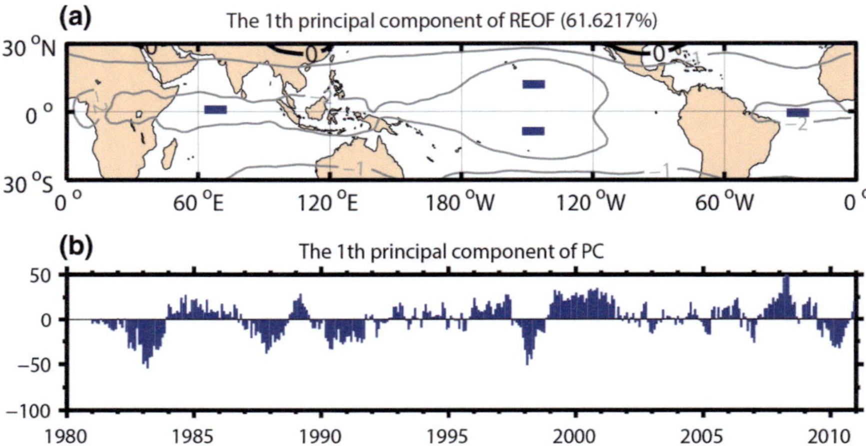

Fig. 10.14 **a** Spatial displays of the first rotated eigenvector of MHA with its variance of 61.6% and **b** its monthly mean time series (*blue bars*) from 1981 to 2010

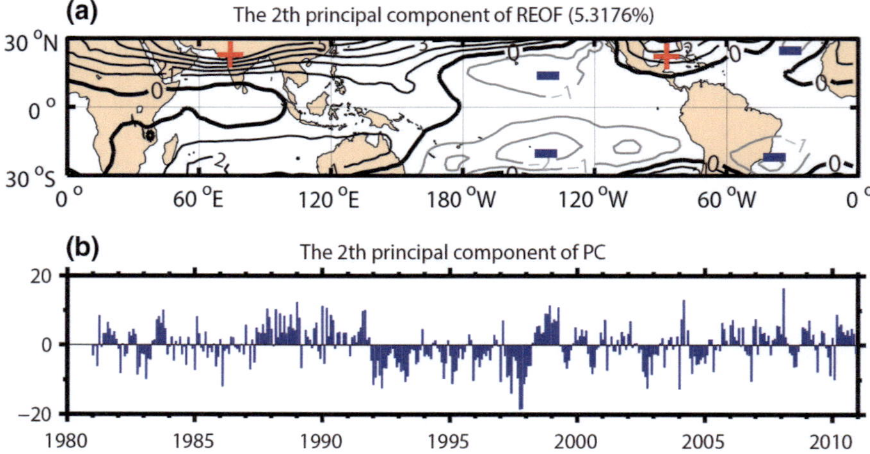

Fig. 10.15 Same as Fig. 10.14 except for the second rotated eigenvector with its variance 5.3%

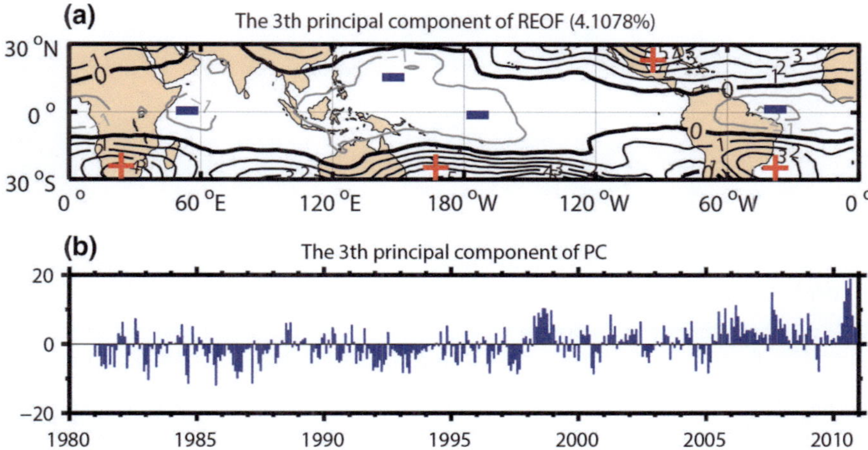

Fig. 10.16 Same as Fig. 10.14 except for the third rotated eigenvector with its variance 4.1%

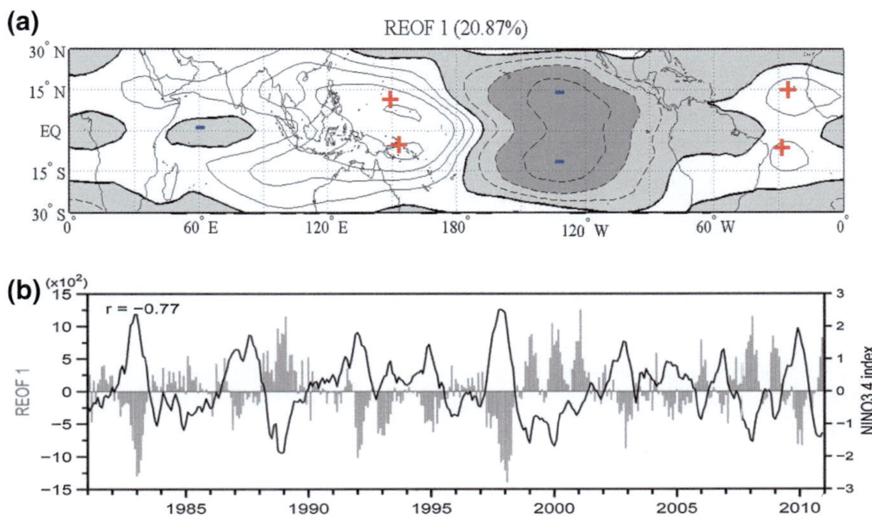

Fig. 10.17 **a** Spatial distributions of the first rotated eigenvector of MHA with its variance of 20.87% and **b** its monthly mean time series (*gray bars*) from 1981 to 2010. The new monthly mean MHAs after removing the long-term trend and zonally averaged height anomaly are used. The *black curve* is the monthly mean SST anomalous series at the Niño3.4 region

The second mode shows an oscillation with two opposite signs in the eastern Pacific and the tropical Atlantic separating along the American continent (Fig. 10.18). The correlation between the second mode and the series of Niño3.4 region SST anomaly is 0.75. The third mode depicts two opposite signs centered in the tropical Indian Ocean and the central equatorial Pacific (Fig. 10.19).

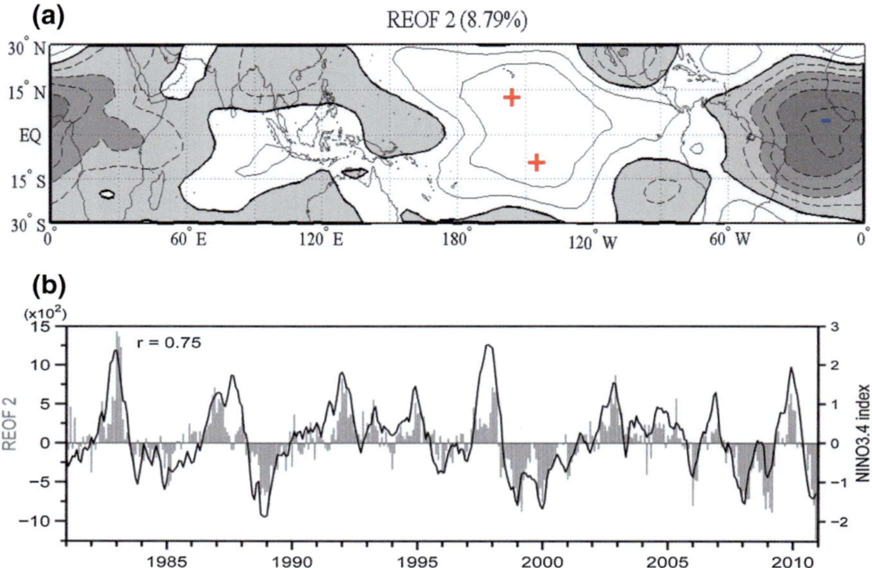

Fig. 10.18 Same as Fig. 10.17 except for the second rotated eigenvector of MHA with its variance 8.8%

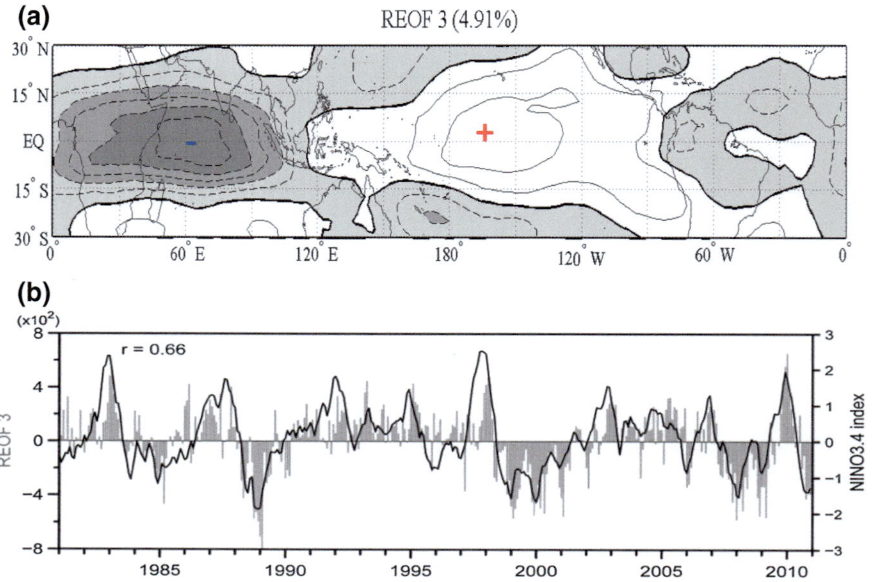

Fig. 10.19 Same as Fig. 10.17 except for the third rotated eigenvector of MHA with its variance 4.9%

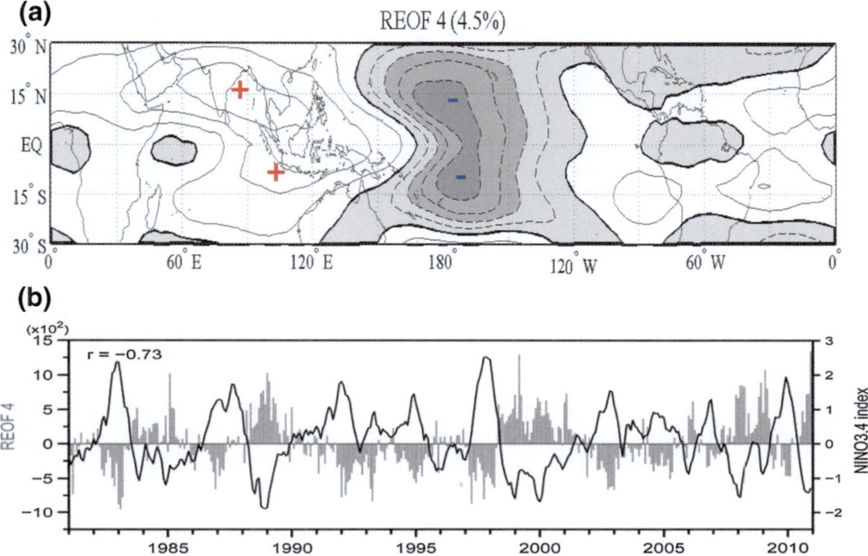

Fig. 10.20 Same as Fig. 10.17 except for the fourth rotated eigenvector of MHA with its variance 4.5%

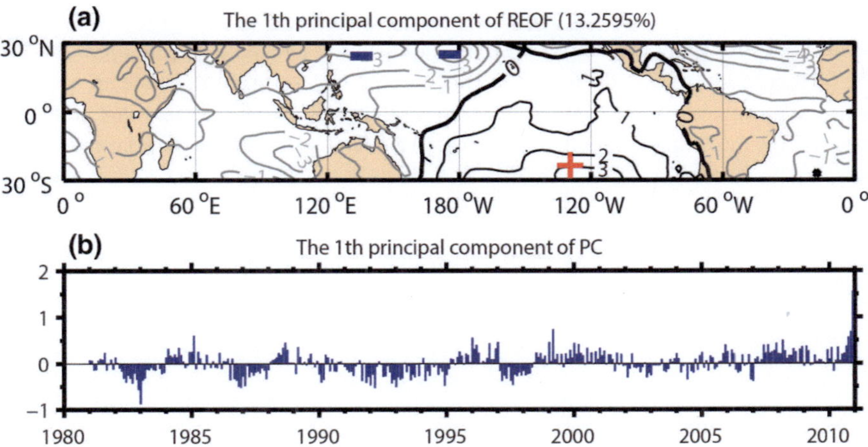

Fig. 10.21 Same as Fig. 10.14 except for the first rotated eigenvector of height anomalies at 1000 hPa with its variance 13.3%

The correlation between the third mode and the Niño3.4 region SSTA series is 0.66. The fourth mode shows two opposite signs in the subtropical Pacific and southern Asia (Fig. 10.20). The correlation between the fourth mode and the Niño3.4 region SSTA series is −0.73. It is interesting to note that the standing MHA oscillations in

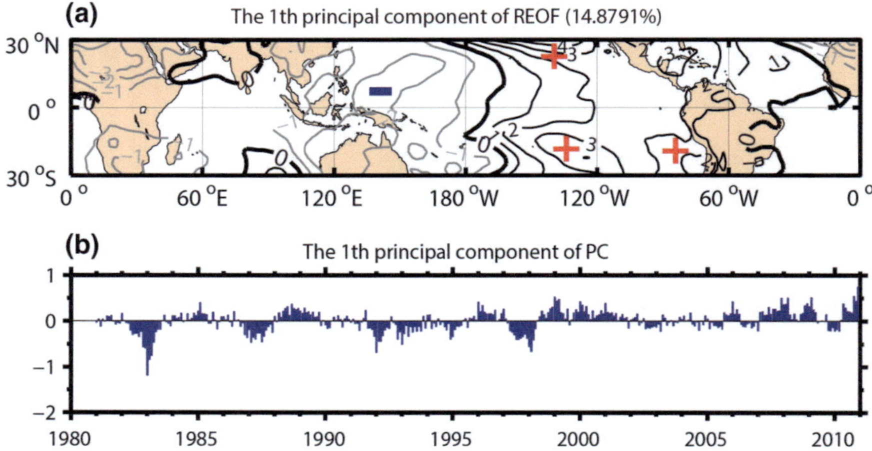

Fig. 10.22 Same as Fig. 10.21 except for removing zonally averaged height anomalies with its variance 14.9%

the tropics are associated with the tropical Pacific SSTAs and separated by oceanic-land distributions.

A comparison is made between Figs. 10.21 and 10.22, with different height anomalies at 1000 hPa. Figure 10.21 shows the spatial distribution of the first rotated eigenvector based on height anomalies at 1000 hPa, while Fig. 10.22 is the result after removing the long-term trend and zonally averaged height anomaly. For the former, the opposite centers are located in the two subtropics, while an opposite pattern is situated in the tropical western and eastern Pacific, separating along the dateline for the latter. The latter is mostly like the SO pattern. Its correlation against the Niño3.4 index is −0.79, which is higher than the former (−0.61).

10.5 The Madden-Julian Oscillation and Height Anomalies[1]

The Madden-Julian Oscillation (MJO; Madden and Julian 1971, 1972) is an intraseasonal oscillation (ISO) characterized with a large-scale convection system, moving eastward with an average speed of 5 m s^{-1} along the equatorial regions. Since its discovery by Madden and Julian (1972) in the 1970s, the MJO has been widely observed and studied over the past four decades (e.g., Weickmann 1983; Majda and Biello 2004; Zhang 2005; Adames and Wallace 2015). Leung and Qian

[1]Modified article originally published in Leung JCH, Qian WH (2016) Monitoring the Madden-Julian Oscillation with Geopotential Height. **Clim Dyn**. doi:10.1007/s00382-016-3431-x. Published with kind permission © Springer-Verlag Berlin Heidelberg 2016. All Rights Reserved.

(2016) examined the three-dimensional structure of the MJO using geopotential height and proposed that the MJO convection signals can be well reflected by upper tropospheric zonal anomalous height gradient (ZAHG, $\nabla_x z'$). A case study in 2011 and composite analyses have been done by applying the wavenumber-frequency filtering method on daily anomalies, obtained from the ERA-Interim data. Results show that the eastward-moving intraseasonal geopotential height anomalies display a clear Rossby–Kelvin wave structure and are well associated with the temperature and zonal wind anomalies. The notable tilted upper tropospheric temperature structure can be derived from the height anomalies by the hydrostatic relationship. The first baroclinic mode vertical structure of zonal wind anomalies can be approximated from the height anomalies by the geostrophic balance. It is shown that the MJO convective center can be traced by the upper tropospheric ZAHG, $\nabla_x z'$. Composite analyses of all MJO events during extended winter (NDJFM) from 1979 to 2013 suggested that positive upper tropospheric ZAHG is exactly in phase with negative outgoing longwave radiation (OLR) anomalies in the eastern Indian Ocean and Maritime Continent (MC) during a MJO event. On the other hand, positive upper tropospheric ZAHG leads (lags) negative OLR anomalies by 1 (2) days in the western Indian Ocean (western Pacific Ocean), where MJO events generally initiate (dissipate).

The MJO is characterized by planetary-scale convection in a spatial scale of 5000–10,000 km within its active phase (Nakazawa 1988; Lau and Waliser 2011). Given the dominant convection and circulation features of the MJO, its observation relies mostly on the measurements of convection such as OLR, or precipitation and circulation-based measurement such as zonal wind, velocity potential, and stream function. A number of indices have been designed to identify the MJO signals in observation, and most of them are derived based on the above two types of measurements. The most widely used MJO index is the Real-time Multivariate MJO (RMM) index developed by Wheeler and Hendon (2004). The RMM index is constructed from the combined empirical orthogonal function (CEOF) of latitudinal averaged OLR, 850- and 200-hPa zonal winds. A typical advantage of the RMM index comes from the inclusion of large-scale circulation, which helps smooth the signal and effectively discriminate the first baroclinic mode of the MJO structure. Thus, it is not necessary to perform prefiltering, such as bandpass filtering, on the observed variables and the index calculation can be done in real-time observation (Kiladis et al. 2014). Despite its advantages, the RMM index has numerous limitations. As mentioned by Liu et al. (2016), although the RMM index can effectively reflect the first baroclinic mode of the MJO structure, it has difficulty capturing the MJO's convective signal. Straub (2013) has demonstrated the problem of the RMM index in detecting the starting dates of MJO initiation.

Focusing on the weakness of the RMM index, Ventrice et al. (2013) introduced the velocity potential MJO (VPM) index, a multivariate index combining 200-hPa velocity potential, 850- and 200-hPa zonal wind, which enhances the MJO signals over the WH and during boreal summer. It was also reported that the RMM index underestimates the WH MJO signals. Kiladis et al. (2014) proposed the use of two univariate indices, the OLR-based MJO (OMI) and the filtered MJO OLR (FMO) indices, to improve the detection of MJO initiations. Liu et al. (2016) have recently revised the RMM index by rescaling the proportion of OLR and winds to solve the problem of the RMM index in capturing the MJO's convective signals. Despite the success made by previous studies, different indices may have different sensitivities to the MJO features. As indicated by Ventrice et al. (2013), exploration of more alternative indices could promote a better understanding of the MJO and its global impacts.

It has been documented in many studies that the most noteworthy characteristic, apart from the planetary-scale convective system, of the MJO structure is the horizontal pattern of combined Rossby–Kelvin wave couplet (e.g., Wang and Rui 1990). The slow MJO moving speed and its horizontal structure are similar to the wave response to an equatorial heat source (Matsuno 1966; Gill 1980). As reviewed by Zhang (2005), the low-level easterlies and upper level westerlies resemble the classical equatorial Kelvin wave to east of the MJO convective center; and to the west, the low-level westerlies and upper level easterlies are associated with the equatorial Rossby wave gyres.

In general, a Kelvin wave is featured with a pair of high and low centers along the equator. It tends to have convection over the enhanced lower tropospheric westerlies in the EH while over the low-level convergence in the WH. The Rossby wave has a pair of low (high) centers symmetrically located on both sides of the equator at the lower (upper) atmosphere. In the EH, the Rossby wave tends to have convection in the region of enhanced lower level westerlies but over lower tropospheric convergence in the WH (Yang et al. 2007). Although the geopotential height structures of Kelvin and Rossby waves have been widely studied, the MJO and its Rossby–Kelvin wave structure are seldom studied or observed in terms of geopotential height.

In a recent research, Adames and Wallace (2014) have documented some characteristics of MJO-related geopotential height by regressing it upon the two leading principal components of the daily 850- minus 150-hPa global velocity potential. This section will show that the upper tropospheric geopotential height could well reflect the Rossby–Kelvin wave structure as the zonal winds do. Therefore, the possibility of monitoring the MJO by geopotential height is further explored.

The MJO is characterized by large-scale convections, which is well reflected by OLR. The daily interpolated OLR product obtained from the National Oceanic and Atmospheric Administration (NOAA) polar-orbiting satellites (Liebmann and Smith 1996) is used. To examine the three-dimensional structure of the MJO, the geopotential height (z), air temperature (T), horizontal wind (u, v), and vertical velocity (ω) data are obtained from the European Centre for Medium-Range Weather Forecasts (ECMWF) Re-Analysis Interim (ERA-Interim) product (Dee et al. 2011).

The daily anomalies derived are comprised of, apart from ISOs, interannual variability, and synoptic signals (Qian et al. 2015a). Meanwhile, ISOs can be simply divided into eastward-propagating (ISOe) and westward-propagating

(ISOw) modes according to their behaviors and structures (Roundy and Frank 2004). The ISOe mode represents the MJO, which propagates eastward. The ISOw mode include the westward propagations, such as the meridional mode $n = 1$ equatorial Rossby wave (Wheeler and Kiladis 1999). Thus, three types of filters are further applied to the daily anomalies. The first is the Lanczos bandpass filter designed by Duchon (1979), which is used to obtain the 15–100-day anomalies including both the ISOe and ISOw. The other two filters are the wavenumber-frequency filter developed by Wheeler and Kiladis (1999), labeled ISOe filter and ISOw filter. The ISOe filter extracts the MJO signals, including wavenumbers 0–10 and periods of 15–100 days. The ISOw filter describes the westward-moving ISOs, including wavenumber 1–10 and periods of 15–100 days. The inclusion of 15–100 day period in the filter is based on a recent climatology study of the RMM index by Lafleur et al. (2015). They have documented that weak MJO events could have lifetimes longer than 60 days and mostly shorter than 100 days while the shortest complete MJO cycle is 17 days. To simplify the discussion, the above three types of filtered anomalies are named as bandpass-filtered, ISOe-filtered and ISOw-filtered anomalies, respectively.

The MJO is well known to be the dominant mode of the ISOs in the winter (Jiang et al. 2004), thus Leung and Qian (2016) focused on the extended winter period (NDJFM). Given that the ERA-Interim product is available since 1979 while the NOAA interpolated OLR product provides data up to 2013, the study period is set from 1979 to 2013, covering 35 years. Figure 10.23a shows the SD map of bandpass-filtered OLR anomalies in NDJFM of 1979–2013. Large SDs of intraseasonal OLR anomalies are observed in tropical regions to the south of the equator and are concentrated from the western Indian Ocean (WIO) to the WP. The largest SD appears near the eastern Indian Ocean (EIO) and MC. This is mainly contributed by the MJO activities, as shown in the SD map of ISOe-filtered OLR anomalies (Fig. 10.23b). The most active region of the MJO is concentrated in the EIO and MC, which is similar to that in Fig. 10.23a. On the other hand, the ISOw activities generally cover the off-equatorial regions (Fig. 10.23c).

The four active regions of MJO are respectively the WIO (0°–10°S, 65°–75°E), EIO (0°–10°S, 90°–100°E), MC (5°–15°S, 115°–125°E), and WP (5°–15°S, 160°–170°E) regions (green boxes in Fig. 10.23). The above four regions are similar to that defined by Hsu and Li (2012) when diagnosing the moisture budget associated with the eastward propagation of the MJO.

Of each region, MJO events are identified from the ISOe-filtered OLR normalized anomalies (ISOe OLR NAs). Daily mean NAs have been employed to diagnose high impact weather events and are effective in ranking the events' intensities. For each box, a MJO event is defined when the daily ISOe OLR NAs are smaller than −1 SD. Based on the above definition, 109, 104, 90, and 95 MJO events are identified respectively in the WIO, EIO, MC, and WP regions in NDJFM of the entire 35-year period (1979–2013). The events are not independent of each other, due to the propagating feature of the MJO. This corresponds to average 2.6–3.1 MJO events per winter (NDJFM).

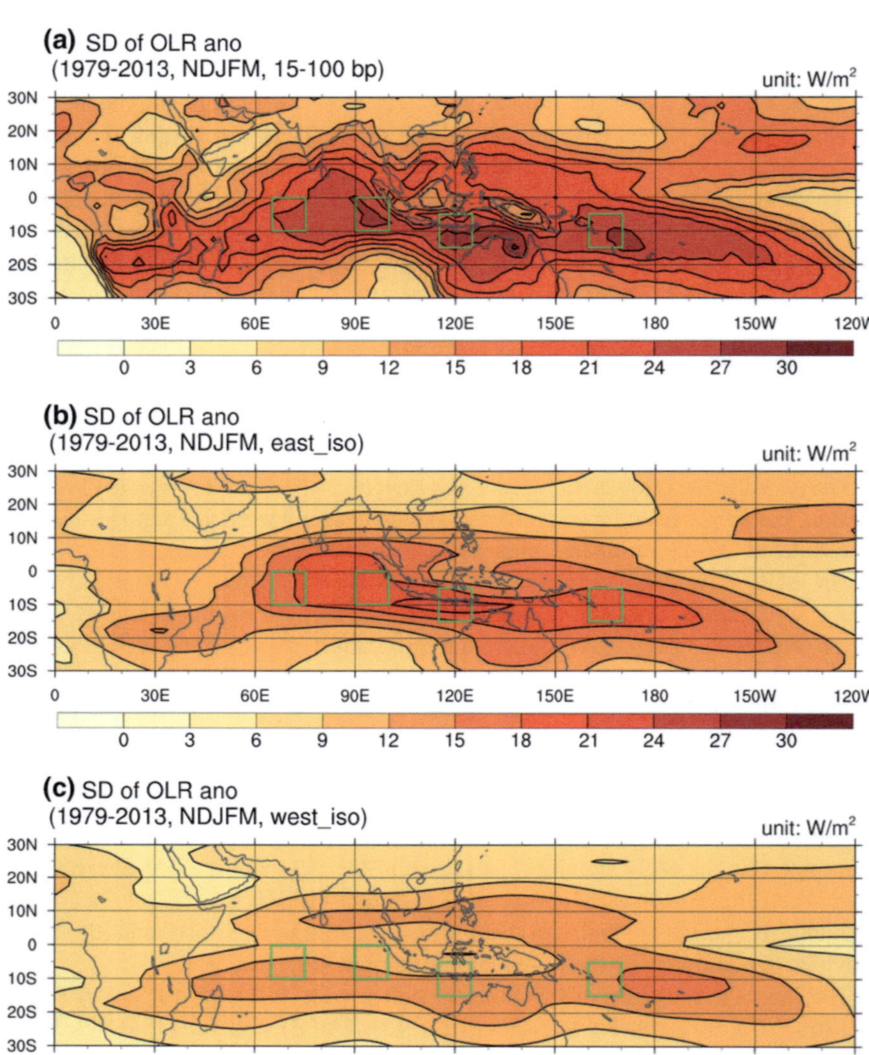

Fig. 10.23 SDs of daily mean **a** bandpass-filtered, **b** ISOe-filtered, and **c** ISOw-filtered OLR anomalies (*shadings and contours*, 3 W m^{-2} interval) in the extended winter (NDJFM) of 1979–2013. *Green boxes* are the four selected regions for composite analyses: WIO (0°–10°S, 65°–75°E), EIO (0°–10°S, 90°–100°E), MC (5°–15°S, 115°–125°E) and WP (5°–15°S, 160°–170°E). Sourced from Leung and Qian (2016) © Springer-Verlag Berlin Heidelberg 2016, with permission

The three-dimensional MJO structure of geopotential height and its relationship with other variables are examined during the 2011–2012 DYNAMO field campaign. The DYNAMO field campaign is an international campaign that collects observation for the MJO in the Indian Ocean from October 2011 to March 2012. During this period, two successive MJO events were detected from October to November 2011, and a strong MJO event occurred from late February to March 2012 (Gottschalck et al. 2013; Yoneyama et al. 2013). In this section, the second MJO event, which occurred during late November 2011, is selected as a case example to be studied in detail.

Figure 10.24a displays the Hovmöller diagram of 15–100-day bandpass-filtered OLR anomalies. It shows that the selected MJO case started in the WIO and propagated to the MC from mid-November to early December 2011, while the ISOe-filtered MJO signal extended to the WP (shadings in Fig. 10.24b). However, both filtered OLR anomalies reached their strongest values at approximately 80°E on 28 November 2010. It is noted that ISOw signals were also observed during the selected MJO event (contours in Fig. 10.24b), but they were too weak to be visible in the bandpass-filtered Hovmöller diagram (Fig. 10.24a).

Figure 10.25 plots the zonal cross sections of geopotential, temperature, zonal wind, and vertical wind anomalies along 10°S–10°N on 28 November 2010. The vertical structure of temperature anomalies is a notable characteristic of the MJO. By examining the regressed temperature, Kiladis et al. (2005) have shown a tilted upper tropospheric temperature structure with height for the MJO. The vertical tilting is a response to the eastward-moving MJO heat source. The zonal cross section of ISOe-filtered height and temperature anomalies on 28 November

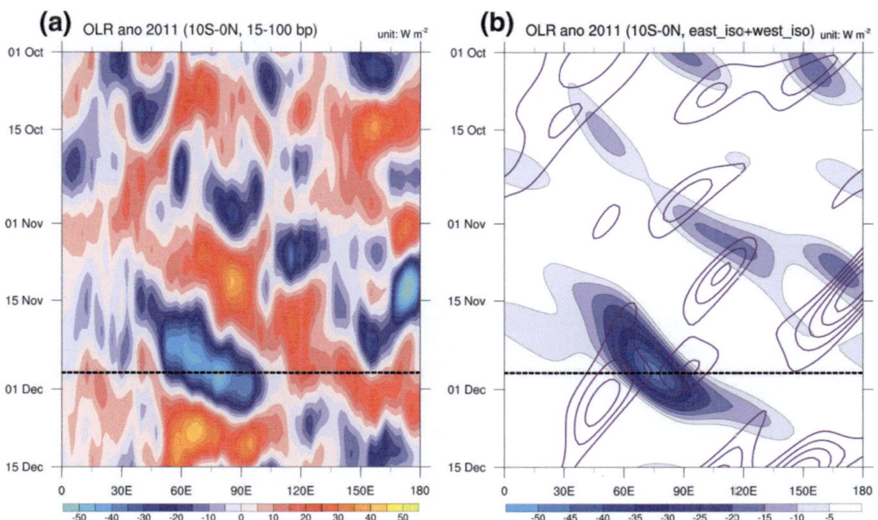

Fig. 10.24 **a** Hovmöller diagram of 15–100 days bandpass-filtered OLR anomalies (*shadings*, 5 W m^{-2} interval) averaged over 0°–10°S from October to December 2010. **b** Same as **a** but of ISOe-filtered (*shadings*, 5 W m^{-2} interval, only negative values are plotted) and ISOw-filtered (*contours*, 5 W m^{-2} interval, only negative values are plotted) OLR anomalies. *Black dashed lines* denote 28 November 2010. Sourced from Leung and Qian (2016) © Springer-Verlag Berlin Heidelberg 2016, with permission

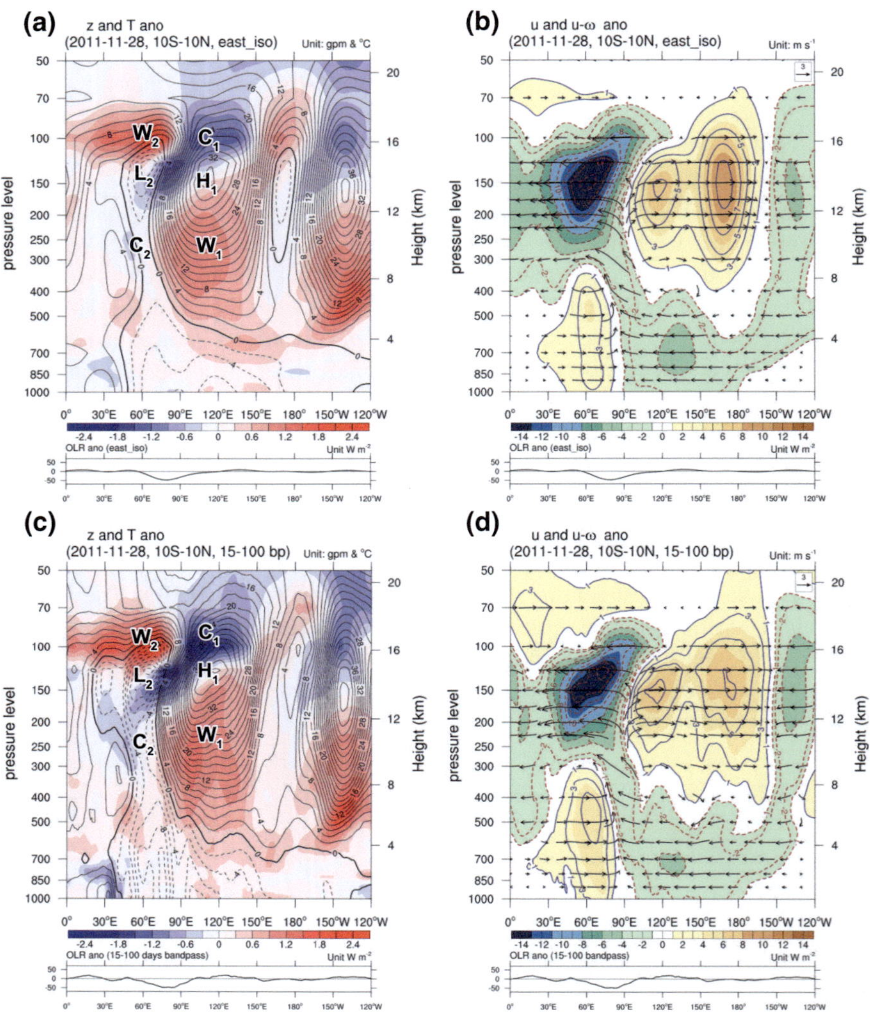

Fig. 10.25 **a** Zonal cross sections of ISOe-filtered geopotential height (*contours*, 2 gpm interval) and temperature (*shadings*, 0.3 °C interval) anomalies along the band 10°S–10°N on 28 November 2010. **b** Same as **a** but of ISOe-filtered $u - \omega$ wind (vectors, ω multiplied by 100 times) and zonal wind (*shadings*, 1 m s^{-1} interval) anomalies. **c** and **d** are the same as **a** and **b** but of bandpass-filtered anomalies. The corresponding filtered OLR anomalies (unit: W m^{-2}) are plotted at the bottom of each panel. Letters "H/L" and "W/C" denote high/low and warm/cold centers, respectively. Sourced from Leung and Qian (2016) © Springer-Verlag Berlin Heidelberg 2016, with permission

(Fig. 10.25a) gives a similar vertical pattern of temperature anomalies. Strong convection (negative OLR anomalies) is found between 75°E and 85°E, where large convergence and vertical velocity are found in the middle troposphere (Fig. 10.25b). A warm center ("W_1") is observed between 200 and 400 hPa to the east of the convective center, and another warm center ("W_2") is located above 125 hPa to the west of the convective center. A strong cold center ("C_1") is found between 70 and 150 hPa, and a weak cold perturbation ("C_2") is observed at 300 hPa to the west of the convective center (Fig. 10.25a).

Notice that the ISOe-filtered height anomalies correspond well with the temperature anomalies vertically (Fig. 10.25a). To the east of the convective center, a positive center ("H_1") of height anomalies appears at 150 hPa above "W_1" and below "C_1," while a negative center ("L_2") of height anomalies to its west is located above "C_2" and below "W_2" (Fig. 10.25a). In Fig. 10.25c, the vertical structures of bandpass-filtered temperature and height anomalies have almost the same pattern as Fig. 10.25a. It has been reported that similar patterns can also be observed in cold spells and heat waves (Qian et al. 2015b). The vertical configuration of ISOe-filtered height and temperature anomalies satisfies the hydrostatic equilibrium.

Another remarkable MJO feature, as reviewed by Li (2014), is the westerlies in the low-to-middle troposphere at and to the west of the convective center, and the easterly winds above 300 hPa. The same is observed in this MJO case as shown in the zonal vertical cross section of ISOe-filtered zonal and vertical wind anomalies on 28 November 2011 (Fig. 10.25b). Strong westerly (easterly) anomalies are situated respectively below 400 hPa (above 300 hPa) in the west of the MJO convective center. To the east of the MJO convective center, the centers of westerly and easterly anomalies are respectively above 300 hPa and below 400 hPa, respectively. The zonal wind configuration induces convergence and ascending motion in the mid-low troposphere, and divergence in the upper troposphere above the convective center.

From the lower to upper troposphere, the zonal wind anomalies tilt westward with height, which is consistent with the findings of Sperber (2003) and Kiladis et al. (2005). The upper tropospheric ISOe-filtered center of easterly anomalies (Fig. 10.25b) matches the upper tropospheric low center "L_2" in Fig. 10.25a. Similarly, the upper tropospheric center of westerly anomalies (Fig. 10.25b) matches the upper tropospheric high center "H_1" (Fig. 10.25a). This is similar to the theoretical Kelvin wave structure derived by Matsuno (1966). In addition, the upper level zonal wind anomalies are featured by a vertical tilt as the temperature and height anomalies do (Fig. 10.25a). For all the zonal cross sections of ISOe anomalies mentioned above, the bandpass-filtered anomalies do show very similar patterns (Fig. 10.25a vs. c and Fig. 10.25b vs. d). This implies that during the MJO event, ISOe-filtered anomalies are the dominant components while ISOw-filtered anomalies account for only a small proportion. As reported by Matsuno (1966) and Adames and Wallace (2014), the zonal wind anomalies near the equator satisfy the quasi-geostrophic approximation. It will be shown below that the zonal wind anomalies can be approximately derived from the geopotential height anomalies off the equator.

Indeed, the vertical structures of geopotential height and zonal wind anomalies are associated with the Rossby–Kelvin wave response to convection as documented by Wang and Rui (1990), Hsu and Li (2012), and Adames and Wallace (2014).

They have reported in previous studies that an equatorial Kelvin wave-like structure and a pair of equatorial Rossby wave gyres are respectively observed to the east and west of the MJO convective center. During a MJO event, low-level easterlies (westerlies) and upper level westerlies (easterlies) to the east (west) of the convective center resemble the equatorial Kelvin (Rossby) wave.

The Rossby–Kelvin wave packet is clearly visible in the horizontal map of ISOe-filtered height and wind anomalies at 150 hPa (Fig. 10.26a–e), the level where both height and zonal wind anomalies have the greatest magnitudes (Fig. 10.25). On 19 November 2011 (Fig. 10.26a), the MJO event was in its early stage. Upper tropospheric high (low) center of height anomalies was situated to the east (west) of the convective center along the equator, associated with weak upper tropospheric westerly (easterly) anomalies. However, the Rossby wave gyres were not observed at this stage. As the MJO strengthened (Fig. 10.26c, d), the Rossby wave gyres became visible as two upper tropospheric high centers ("H_R") and anticyclones in the off-equatorial regions. The equatorial high ("H_K") and low ("L_K") centers also strengthened with zonal wind anomalies. While the two "H_R" centers represent the Rossby wave gyres, "H_K" and "L_K" can be regarded as responses to the Kelvin wave structure. When the MJO event weakened and dissipated (Fig. 10.26e), the Rossby–Kelvin wave structure started to deform. Figure 10.26f–j illustrate that bandpass-filtered height and wind anomalies can also visualize the Rossby–Kelvin wave packet of the MJO, but it is not as clear as the ISOe-filtered anomalies, especially when the MJO is weak. This is because the ISOw signals are also included in the bandpass-filtered anomalies. When the MJO is not strong enough, the ISOw signals will distort the bandpass-filtered anomalies to some extent.

The three-dimensional structure of MJO-related geopotential height anomalies has been examined and is shown to be linked with temperature and zonal wind anomalies. In particular, the upper tropospheric ISOe-filtered height anomalies are able to depict the Rossby–Kelvin wave packet of the MJO. A new diagnostic parameter derived from height anomalies for monitoring the MJO signals is introduced here.

It has been illustrated that as the MJO convective center propagates eastward, it stays between the upper tropospheric high ("H_K") and low ("L_K") centers along the equator (Fig. 10.26). In addition, the convective center is at a position where the zonal gradient of the upper tropospheric ISOe-filtered height anomalies is the greatest (Fig. 10.25a, c). Therefore, the zonal gradient of the filtered geopotential height anomalies is introduced to monitor the MJO, namely the zonal anomalous height gradient (ZAHG) $\nabla_x z'$ (unit: gpm m^{-1})

$$\nabla_x z' = \frac{\partial z'}{\partial x}, \tag{10.5.1}$$

where z' is the (ISOe or bandpass) filtered geopotential height anomalies and x is the distance (unit: m) in the zonal direction. One should be aware that the Rossby–Kelvin wave packet is trapped along the equator, i.e., the MJO-related height

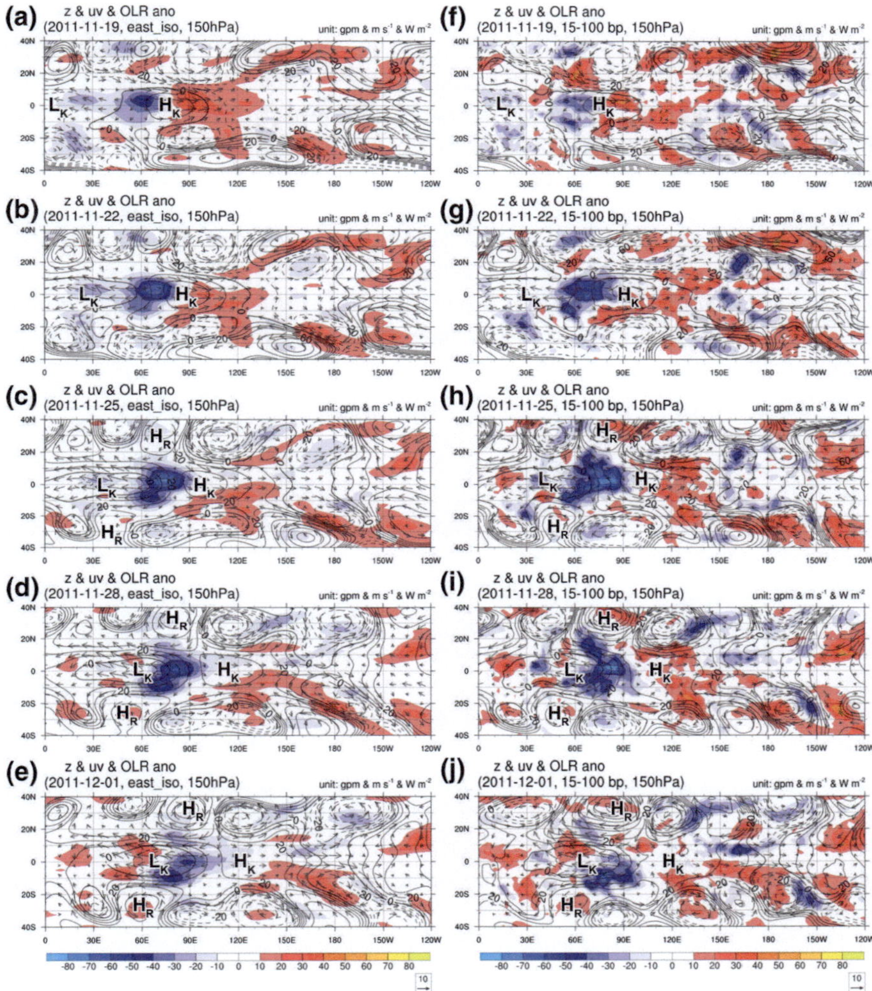

Fig. 10.26 Horizontal distributions of ISOe-filtered 150-hPa height (*contours*, 10 gpm interval for $-30 \le z \le 30$ gpm, 30 gpm for $z < -30$ gpm or $z > 30$ gpm), 150-hPa wind (vectors, m s^{-1}) and OLR (*shadings*, 10 W m^{-2} interval) anomalies on **a** 19 November, **b** 22 November, **c** 25 November, **d** 28 November, and **e** 1 December 2011, respectively. **f–j** are the same as **a–e** but of bandpass-filtered anomalies. "H"/"L" denote high/low centers. Subscripts "K"/"R" represent Kelvin/Rossby wave. Sourced from Leung and Qian (2016) © Springer-Verlag Berlin Heidelberg 2016, with permission

anomalies $\nabla_x z'$ is in between 10°S and 10°N (Fig. 10.26). Thus, $\nabla_x z'$ is averaged over 10°S–10°N below.

It has been revealed that the selected MJO case reached the strongest stage at approximately 80°E on 28 November 2010. Figure 10.27a displays the evolution of vertical profiles of the ISOe-filtered $\nabla_x z'$ and zonal wind anomalies averaged over

75°–85°E. It is obvious that both the ISOe-filtered $\nabla_x z'$ and zonal wind anomalous centers are stronger in the upper troposphere ($\sim$150 hPa). However, the ISOe-filtered $\nabla_x z'$ is insignificant in the lower troposphere, suggesting that the geopotential height response to the Rossby–Kelvin wave packet is better reflected in the upper troposphere. The time series of 150-hPa ISOe-filtered $\nabla_x z'$ and zonal wind anomalies are plotted, together with the ISOe-filtered OLR anomalies, at the bottom of Fig. 10.27a. Noted that since the MJO activity is concentrated in the south of the equator (Fig. 10.26), the ISOe-filtered OLR anomalies plotted are averaged over 0°–10°S. As shown in the time series, the 150-hPa $\nabla_x z'$ is exactly out of phase with the OLR anomalies. When the 150-hPa $\nabla_x z'$ reaches its maximum value (12×10^{-6} gpm m^{-1}), the OLR anomalies are at a minimum (-45 W m^{-2}). On the other hand, the 150-hPa easterly anomalies lag the convective center by about 5 days (Fig. 10.27a). Similar results are obtained from the bandpass-filtered anomalies (Fig. 10.27c).

From the Hovmöller diagram (Fig. 10.27b), the 150-hPa ISOe-filtered $\nabla_x z'$ (shading) can well capture the negative ISOe-filtered OLR anomalies (purple contour). The 150-hPa $\nabla_x z'$ signal appeared in the WIO in mid-November. It, then, intensified and maximized on 28 November 2011, as the OLR anomalies did. After the MJO mature stage, both the OLR anomalies and 150-hPa $\nabla_x z'$ weakened and disappeared in the WP. Signal from the 150-hPa easterly anomalies (green dashed contour) also appeared with the MJO event, and it lagged both the OLR anomalies and 150-hPa $\nabla_x z'$ by several days. Furthermore, it also showed signals of another MJO event in late October 2010. The positive $\nabla_x z'$ signals were in phase with the negative OLR anomalies, as mentioned earlier. The Hovmöller diagram of bandpass-filtered anomalies (Fig. 10.27d) also depicts the ability of 150-hPa $\nabla_x z'$ in capturing the MJO convective centers. Meanwhile, there is some westward-moving 150-hPa $\nabla_x z'$ which is not associated with negative OLR anomalies. It is believed that these anomalies belong to the equatorial Rossby waves (ISOw mode).

The three-dimensional structure of MJO-related geopotential height anomalies in a MJO event has been given a detailed description. The MJO event can be traced by ZAHG ($\nabla_x z'$). Here, composite analyses are made to evidence on all MJO events in four particular regions: (1) the WIO region, where the MJOs often begin; (2) the EIO region and (3) the MC region, the most active regions of the MJOs; and (4) the WP region, in which the MJOs weaken and dissipate (green boxes in Fig. 10.23). There are 109, 104, 90, and 95 MJO events detected in the WIO, EIO, MC, and WP regions, respectively, in the extended winter (NDJFM) during 1979–2013. The selected events are not independent of each other due to the propagating feature of the MJO. This corresponds to average 2.6–3.1 MJO events per winter (NDJFM).

Figure 10.28 displays the zonal cross sections of composite ISOe-filtered geopotential height and temperature anomalies in the four regions. A high center ("H_1") and a low center ("L_2") are obviously observed at the upper troposphere to the east and west, respectively, of the convective center (negative OLR anomalies). This is consistent with that obtained in Fig. 10.25a. The convective center is located at where the 150-hPa $\nabla_x z'$ is large. Also, a vertical tilt structure appears in the upper

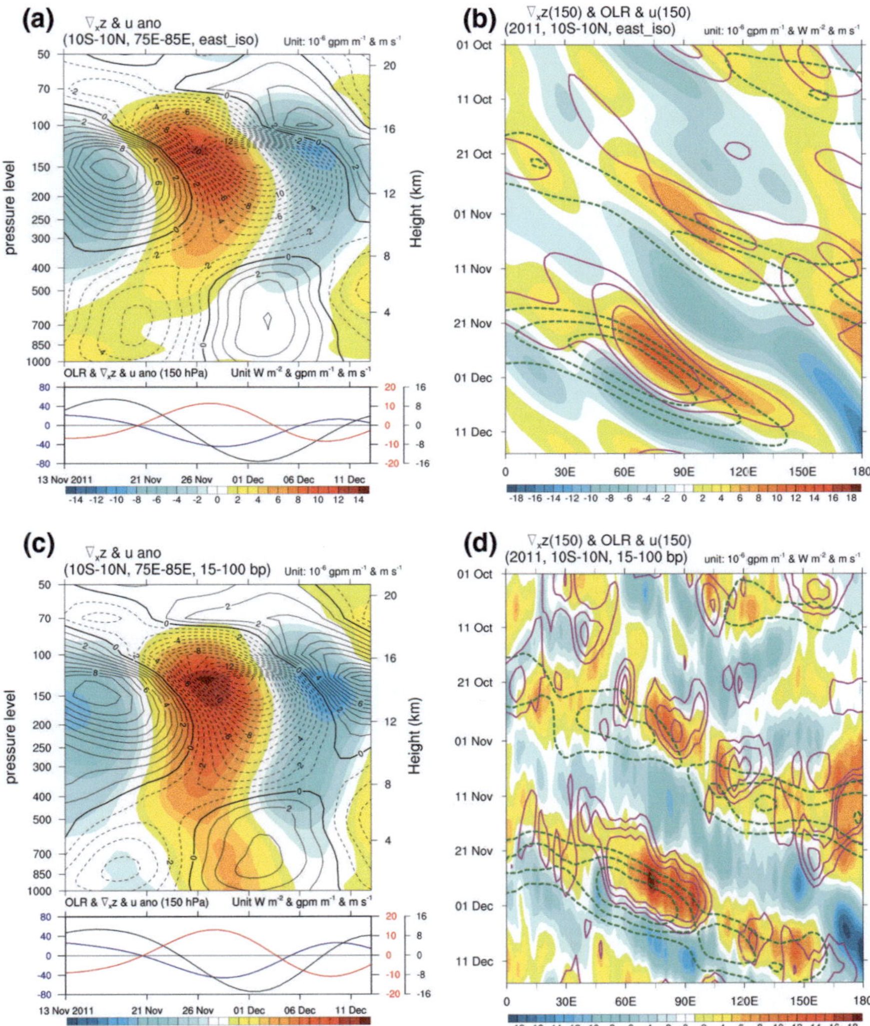

Fig. 10.27 **a** Evolution of vertical profiles of ISOe-filtered $\nabla_x z'$ (*shadings*, 1 gpm m^{-1} interval) and zonal wind anomalies (*contours*, 1 m s^{-1} interval) averaged over 10°S–10°N and 75°–85°E from 13 November to 13 December 2010. Time series of ISOe-filtered OLR anomalies (*blue*), 150-hPa $\nabla_x z'$ (*red*) and 150-hPa zonal wind (*black*) anomalies are plotted at the bottom. **b** Hovmöller diagram of ISOe-filtered 150-hPa $\nabla_x z'$ (*shading*, 1 gpm m^{-1} interval), 150-hPa zonal wind (*green dashed contours*, −4, −8 and −12 m s^{-1}) and OLR (*purple contours*, −5, −15, and −25 W m^{-2}) anomalies averaged over 10°S–10°N from October to December 2010. **c** and **d** are the same as **a** and **b** but of bandpass-filtered anomalies. Note that the all OLR anomalies are averaged over 0°–10°S. Sourced from Leung and Qian (2016) © Springer-Verlag Berlin Heidelberg 2016, with permission

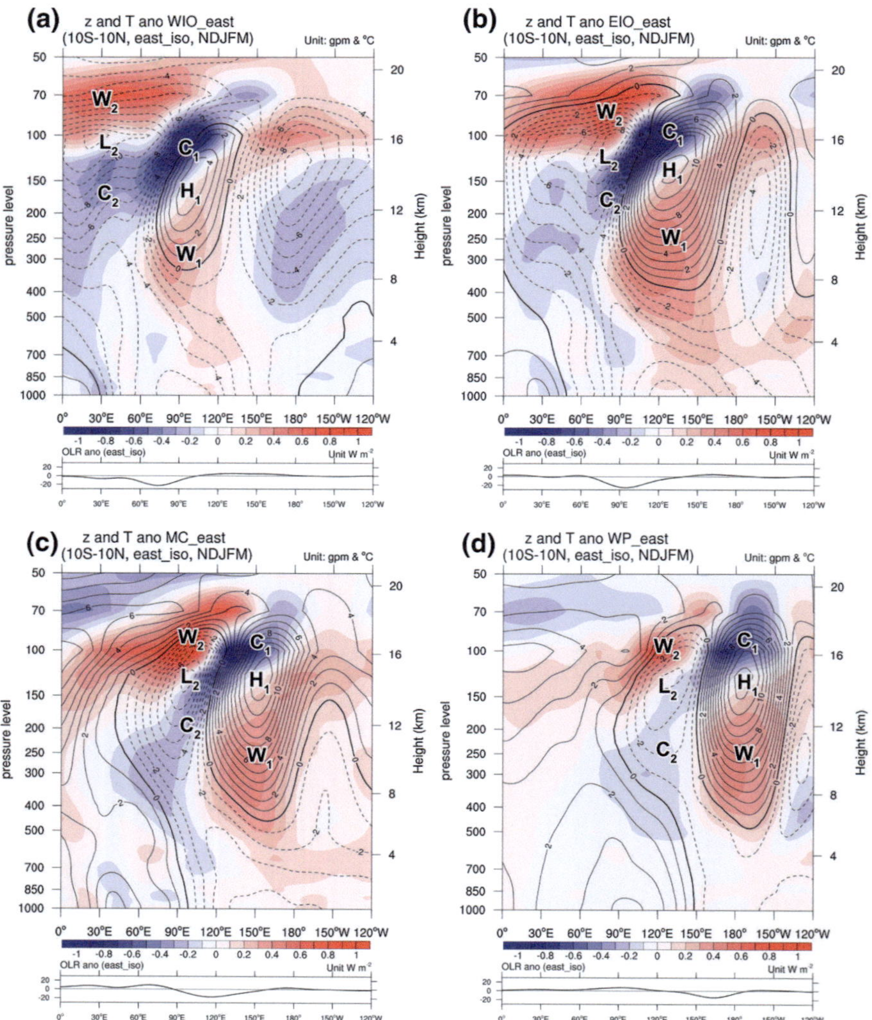

Fig. 10.28 Zonal cross sections (10°S–10°N) of composite ISOe-filtered geopotential height (*contours*, 1 gpm interval) and temperature (*shadings*, 0.1 °C interval) anomalies of all MJO events in the **a** WIO, **b** EIO, **c** MC, and **d** WP regions, respectively. The composite ISOe-filtered OLR anomalies (W m^{-2}) along the band 10°S–10°N are plotted at the bottom of each panel. Sourced from Leung and Qian (2016) © Springer-Verlag Berlin Heidelberg 2016, with permission

tropospheric ISOe-filtered temperature anomalies. As mentioned above, the MJO-related temperature and height anomalies satisfy the hydrostatic relationship, i.e., the temperature anomalies T' can be derived from the vertical gradient of height anomalies z'. Figure 10.29 plots the zonal cross sections of derived temperature anomalies by the equilibrium relationship. It displays the same pattern with Fig. 10.28 except that the derived results underestimate the temperature anomalies a bit at the upper troposphere. This may be due to the coarse vertical resolution of data at the upper level. Nevertheless, the upper tropospheric vertically tilted temperature anomalies are well reflected. This does not affect the notion that the vertical anomalous temperature patterns of the MJOs can be derived from height anomalies by the hydrostatic relationship. It is noted that positive ISOe-filtered OLR anomalies are detected to the east of "H_1" (Fig. 10.28a, b) and to the west of "L_2" (Fig. 10.28c, d), implying that negative ISOe-filtered $\nabla_x z'$ tend to inhibit

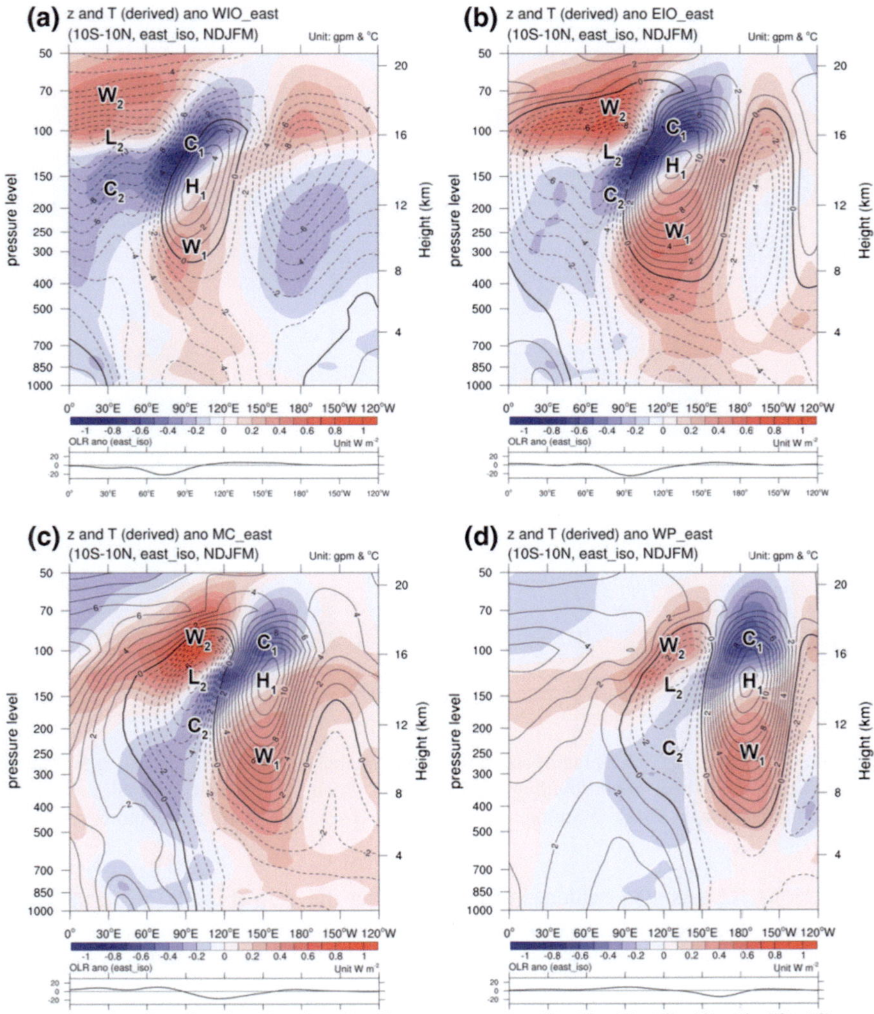

Fig. 10.29 Same as Fig. 10.28 but of observed ISOe-filtered geopotential height anomalies (*contours*, 1 gpm interval) and temperature anomalies derived by the hydrostatic equilibrium (*shadings*, 0.1 °C interval). Sourced from Leung and Qian (2016) © Springer-Verlag Berlin Heidelberg 2016, with permission

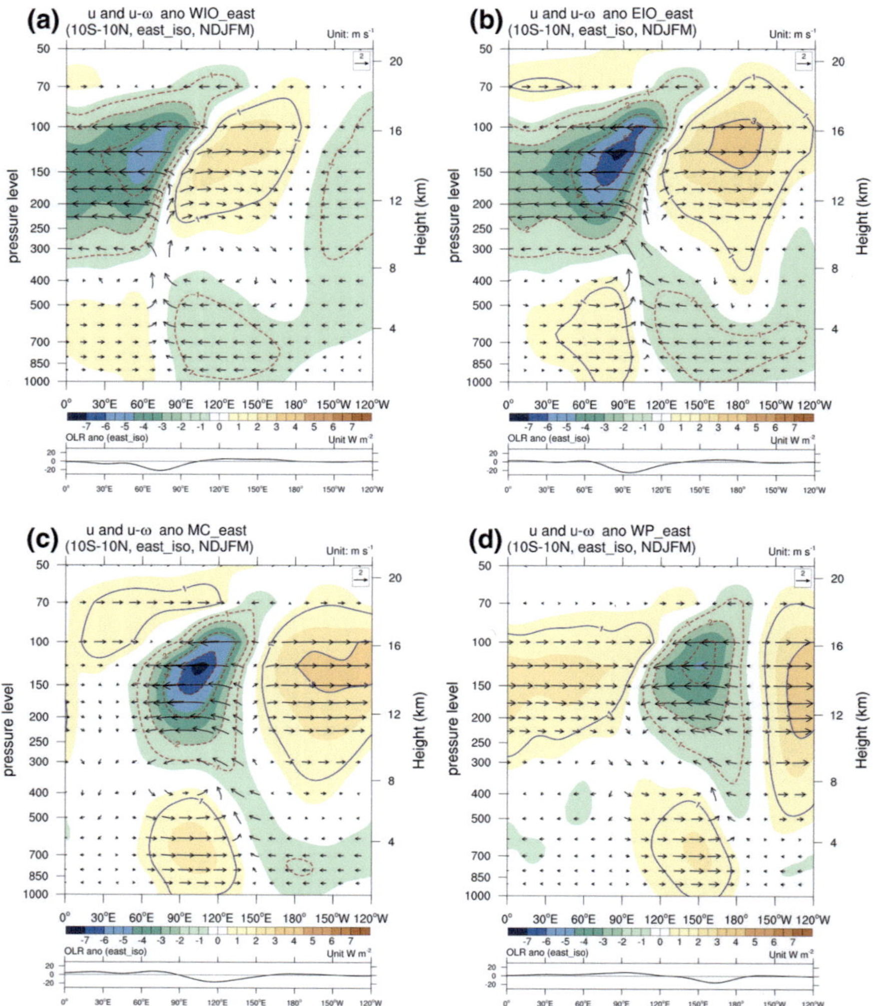

Fig. 10.30 Same as Fig. 10.28 but of composite ISOe-filtered $u - \omega$ wind (vectors, ω multiplied by 100 times) and zonal wind (*shadings and contours*, 0.5 m s^{-1} interval) anomalies. Sourced from Leung and Qian (2016) © Springer-Verlag Berlin Heidelberg 2016, with permission

convections. This is also evident in the Hovmöller diagram (Fig. 10.27b), in which negative OLR anomalies exist only when $\nabla_x z'$ is positive.

In the zonal cross sections of composite ISOe-filtered zonal and vertical wind anomalies (Fig. 10.30), the largest zonal wind anomalies are observed in the troposphere (Fig. 10.25b). While westerly (easterly) anomalous center is observed in the upper troposphere, easterly (westerly) anomalous center is at the mid-low troposphere to the east (west) of the convective center. This induces mid-low-level (500–1000 hPa) convergences and upper level (100–300 hPa) divergences, enhancing convections. As in Fig. 10.25, the upper tropospheric ISOe-filtered easterly (westerly) anomalous center (Fig. 10.30) matches the upper tropospheric low (high) center (Fig. 10.28). In addition, the upper level zonal wind anomalies are featured by a vertical tilt as the temperature and height anomalies do (Fig. 10.28). However, the vertical tilt structure of upper tropospheric zonal wind anomalies is only seen in the WIO, EIO, and MC regions (Fig. 10.30a–c), but not in the WP region (Fig. 10.30d), where the MJOs weaken and dissipate.

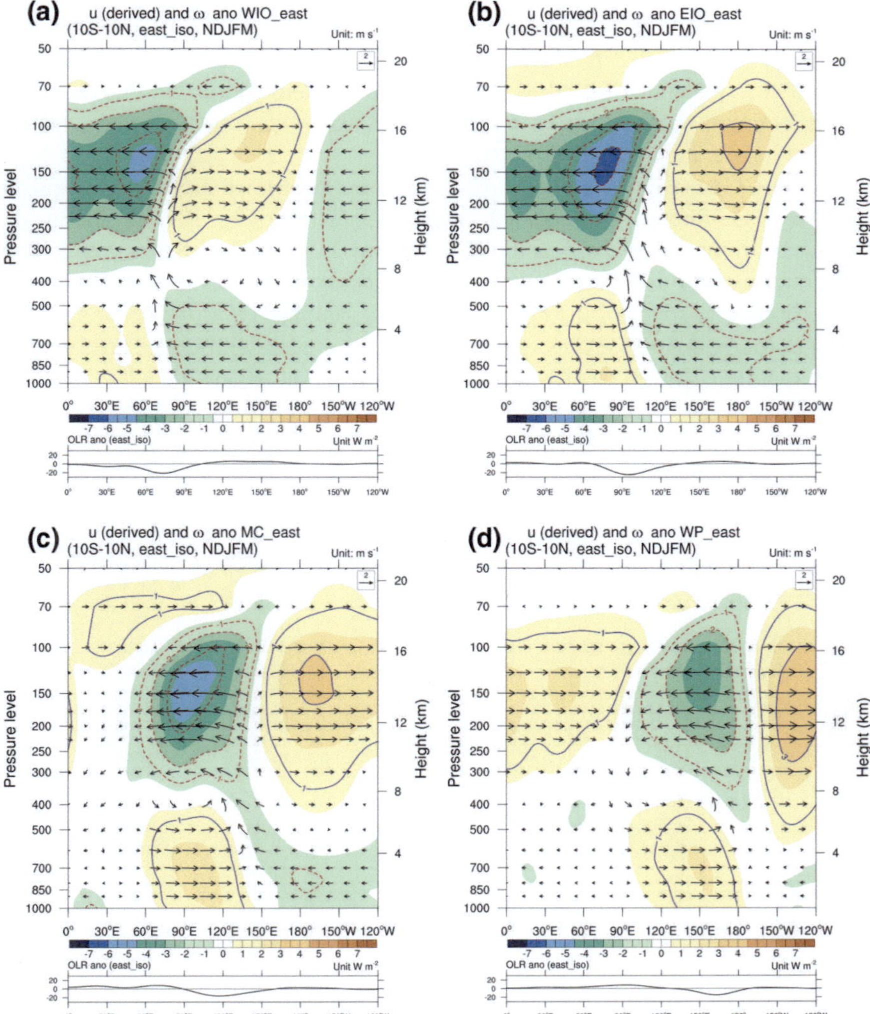

Fig. 10.31 Same as Fig. 10.30 but of observed ISOe-filtered ω-wind anomalies and u-wind anomalies derived by the geostrophic balance. Sourced from Leung and Qian (2016) © Springer-Verlag Berlin Heidelberg 2016, with permission

The close relationship between zonal wind and height anomalies is a result of quasi-geostrophic relationship between them. Figure 10.31 gives the zonal cross sections of zonal wind anomalies u' (shading and vectors) approximated from the meridional gradient of height anomalies z' by the geostrophic balance (Eq. 10.5.2):

$$u' \approx -\frac{g}{f}\left(\frac{\partial z'}{\partial y}\right)_p, \quad \text{where } |\varphi| \geq 5^\circ \tag{10.5.2}$$

where $f = 2\Omega \sin \varphi$ is the Coriolis parameter, Ω is the Earth's rotation speed and φ is the latitude. Since the Coriolis parameter f goes to zero when it is close to the equator and will result in an infinitely large u', which is physically unacceptable. Thus, only u' beyond 5°N/S are derived by the geostrophic balance while that within 5°N/S are approximated with a cubic spline interpolation. Results show that the location and vertical pattern of zonal wind anomalies can be exactly derived from the height anomalies (Fig. 10.31).

The MJO Rossby–Kelvin wave packet is also confirmed from the composite 150-hPa ISOe-filtered geopotential height anomalies (Fig. 10.32). Along the equator, a high center ("H_K") and a low center ("L_K") exist to the east and west, respectively, of the convective center, corresponding to the Kelvin wave response. To the north and south of the convective center, another two high centers ("H_R") represent the two Rossby wave gyres and are associated with upper level anticyclonic motion. The two "H_R" centers are located at about 28°N/S, which is

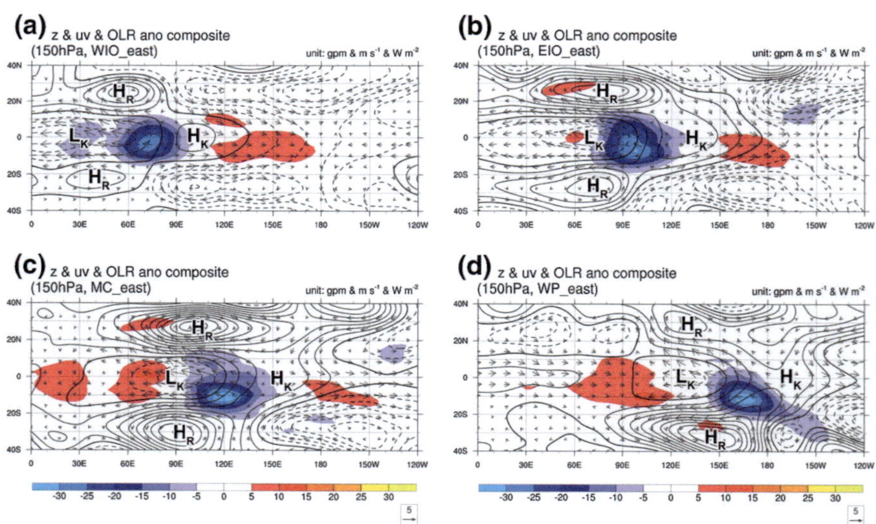

Fig. 10.32 Horizontal distribution of composite 150-hPa ISOe-filtered geopotential height (*contours*, 4 gpm interval), 150-hPa wind (vectors, m s^{-1}) and OLR (*shadings*, 5 W m^{-2} interval) anomalies of all MJO events in the **a** WIO, **b** EIO, **c** MC, and **d** WP regions, respectively. Sourced from Leung and Qian (2016) © Springer-Verlag Berlin Heidelberg 2016, with permission

consistent with the results obtained by Adames and Wallace (2014). The Rossby–Kelvin wave packet is clear in the WIO, EIO, and MC regions (Fig. 10.32a–c) and starts to deform in the WP region (Fig. 10.32d). As in Fig. 10.28, positive OLR anomalies are observed to the east of "H_K" (Fig. 10.32a, b) and to the west of "L_K" (Fig. 10.32c, d), suggesting that negative ISOe-filtered $\nabla_x z'$ tend to inhibit convections. It is also important to mark that the Rossby–Kelvin wave structure of

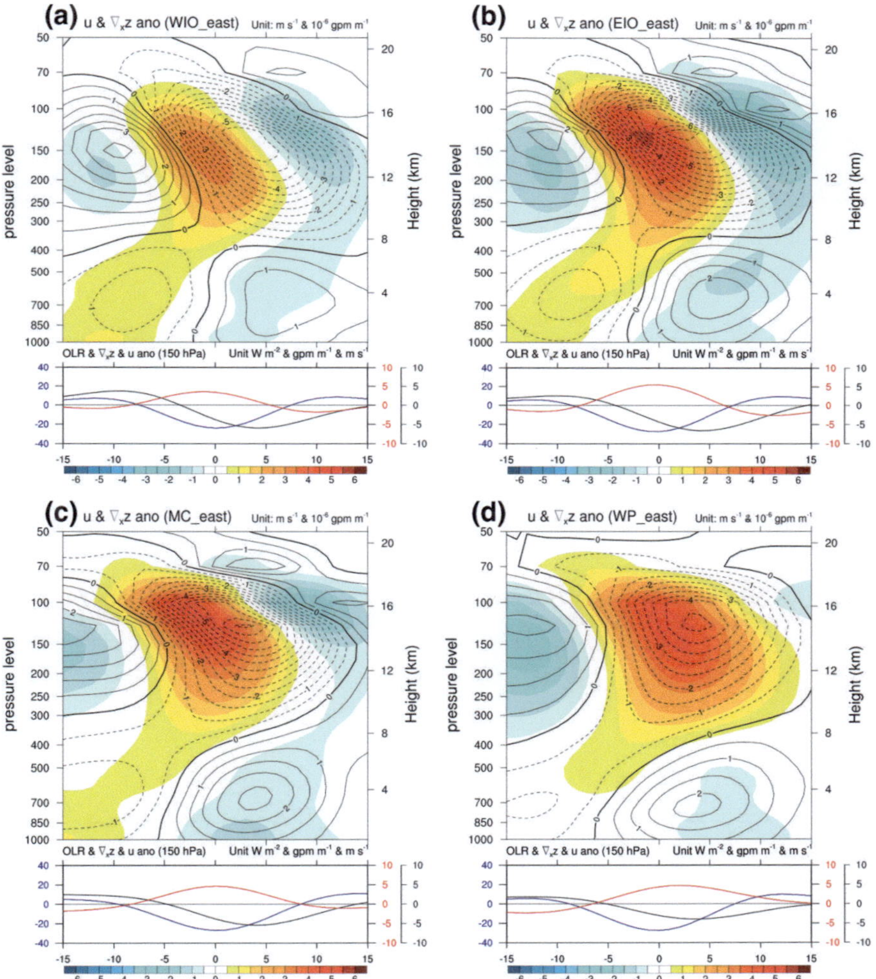

Fig. 10.33 Evolution of vertical profiles of composite ISOe-filtered $\nabla_x z'$ (*shadings*, 0.5 × 10^{-6} gpm m^{-1} interval) and zonal wind (*contours*, 0.5 m s^{-1} interval) anomalies averaged over 10°S–10°N in the **a** WIO, **b** EIO, **c** MC, and **d** WP regions, respectively. Lead-lag time series (days) of composite ISOe-filtered OLR anomalies (*blue*), 150-hPa $\nabla_x z'$ (*red*) and 150-hPa zonal wind anomalies (*black*) are plotted at the bottom of each panel. Note that the OLR anomalies are averaged over 0°–10°S in **a** and **b**, and over 5°–15°S in **c** and **d**, consistent with Fig. 10.24. Sourced from Leung and Qian (2016) © Springer-Verlag Berlin Heidelberg 2016, with permission

height anomalies are always trapped along the equator even if the convective center appears in the SH. Therefore, it is reasonable to trace the MJO by $\nabla_x z'$ averaged over 10°S–10°N.

Figure 10.33 illustrates the evolution of vertical profiles of composite ISOe-filtered $\nabla_x z'$ and zonal wind anomalies averaged over 10°S–10°N in the four regions. The largest signals of $\nabla_x z'$ and zonal wind anomalies both appear at the upper troposphere ($\sim$150 hPa), consistent with the MJO case example (Fig. 10.27a). The lead-lag time series plotted at the bottom of each panel depict that 150-hPa $\nabla_x z'$ are out of phase with OLR anomalies. The 150-hPa easterly

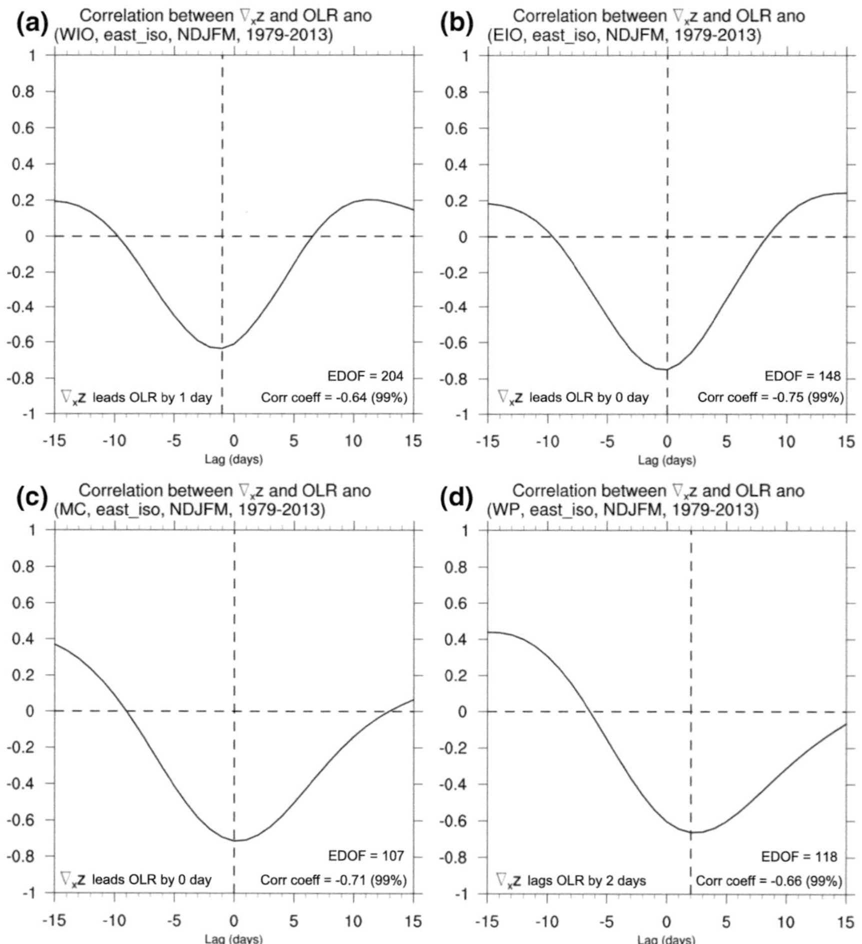

Fig. 10.34 Cross correlation between ISOe-filtered 150-hPa $\nabla_x z'$ averaged over 10°S–10°N and OLR anomalies in the **a** WIO, **b** EIO, **c** MC, and **d** WP regions. Note that the OLR anomalies are averaged over 0°–10°S in **a** and **b**, and over 5°–15°S in **c** and **d**. Sourced from Leung and Qian (2016) © Springer-Verlag Berlin Heidelberg 2016, with permission

anomalies lag OLR anomalies by about 5 days (Fig. 10.33). We have shown that negative 150-hPa $\nabla_x z'$ tends to inhibit convections (Figs. 10.28 and 10.32). This is also true from the composite result in Fig. 10.33, where positive OLR anomalies peak when the 150-hPa $\nabla_x z'$ is at a minimum.

If looking into details of the time series of 150-hPa $\nabla_x z'$ and OLR anomalies, one can easily observe that the negative OLR anomalies are in phase with positive 150-hPa $\nabla_x z'$ only in the EIO and MC regions. In the WIO and WP regions, there is a lagged relationship between 150-hPa $\nabla_x z'$ and OLR anomalies. Thus, the cross correlation between them in the four regions is examined in Fig. 10.34. Not surprisingly, ISOe-filtered 150-hPa $\nabla_x z'$ and OLR anomalies are exactly out of phase (lag = 0 days) in the EIO and MC regions (Fig. 10.34b, c), with correlation coefficients of −0.75 and −0.71 (reaching the 99% confidence level), respectively. In the WIO region (Fig. 10.34a), positive 150-hPa $\nabla_x z'$ lead negative OLR anomalies by 1 day (lag = −1 day), with a correlation coefficient of −0.64 (reaching the 99% confidence level). In the WP region (Fig. 10.34d), positive 150-hPa $\nabla_x z'$ lag negative OLR anomalies by 2 days (lag = +2 days), with a correlation coefficient of −0.66 (reaching the 99% confidence level). This implies that positive 150-hPa $\nabla_x z'$ lead (lag) negative OLR anomalies in the MJO developing (dissipating) stage, and they are in phase in the MJO mature stage.

The lagged relationship between $\nabla_x z'$ and OLR anomalies suggests that when a MJO event is mature, the positive 150-hPa ZAHG center is in phase with the negative OLR anomaly center. The convective center is located in the middle of the Rossby–Kelvin wave packet (Fig. 10.32b, c). During the early stage of a MJO event, $\nabla_x z'$ leads the convective center and this implies that the Rossby–Kelvin wave packet helps develop the deep convection. When a MJO event comes to the dissipating stage, the convective center enters the WP to the south of the equator. At this stage, the convection is affected by other systems such as the Southern Pacific convergence zone (SPCZ; Vincent 1994). One can observe that the Rossby–Kelvin wave packet starts disappearing and less influence on the OLR anomalies in Fig. 10.32d and the OLR anomalies (precipitation) are influenced by the SPCZ. Therefore, the negative OLR anomalies lag the positive $\nabla_x z'$.

10.6 Eastward and Westward Propagations of Tropical Oscillations

The MJO is an ISO characterized with a large-scale convective system, moving eastward in the tropical Indian Ocean and WP. We further examine whether there are other ISOs with different moving directions in the tropical eastern Pacific and the tropical Atlantic. Figure 10.35 shows the wavelet power spectrum of OLR anomalies at the box area (10°S–10°N, 70°E–90°E) from 1979 to 2013. Two peaks of variance are detected at the periods of 5 days and 40 days. We take the ISO period from 15 to 100 days as indicated by the two thick dashed lines in Fig. 10.35.

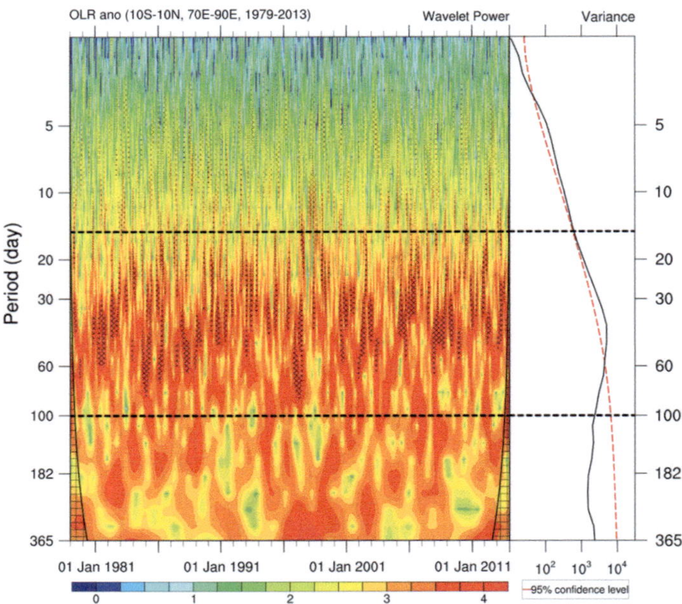

Fig. 10.35 Wavelet power spectrum of OLR anomalies at the box area (10°S–10°N, 70°E–90°E) from 1979 to 2013. The *red dashed line* denotes the 95% confidence level

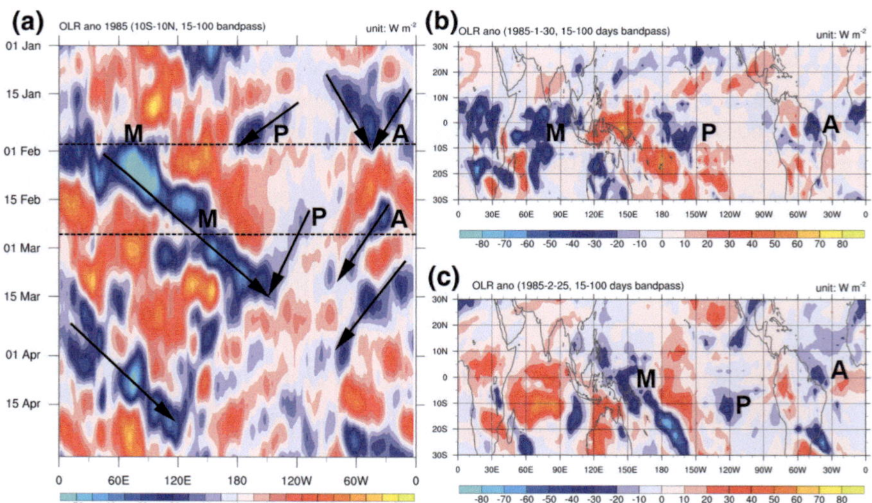

Fig. 10.36 **a** Hovmöller diagram of intraseasonal OLR anomalies (*shadings*, 5 W m^{-2} interval) in 1985. *Dashed lines* denote 30 January and 25 February 1985. **b** Horizontal distribution of intraseasonal OLR anomalies (*shadings*, 10 W m^{-2} interval) on 30 January 1985. **c** Same as **b** but on 25 February 1985. In **a** *arrows* indicate the propagation of negative OLR anomalies. Letter "M" denotes the signal of MJO from the tropical Indian Ocean to the tropical western Pacific while letters "P" and "A" denote negative OLR anomalies in the tropical eastern Pacific and Atlantic, respectively

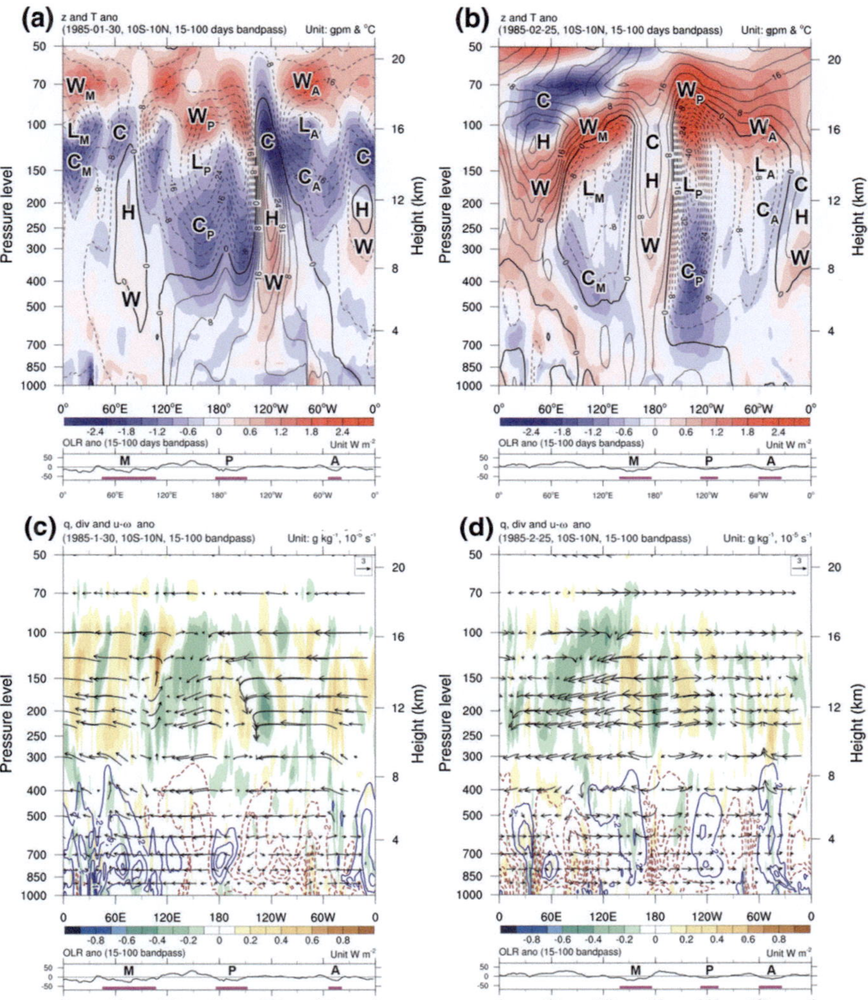

Fig. 10.37 Zonal cross sections of intraseasonal geopotential height anomaly (*contours*, 4 gpm intervals) and temperature anomaly (*shadings*, 0.3 °C intervals) along 10°S–10°N on **a** 30 January 1985 and **b** 25 February 1985, respectively. **c** and **d** are the same as **a** and **b** but of intraseasonal specific humidity anomaly (*contours*, *blue* for positive, 0.4 g kg^{-1} intervals), divergence anomaly (*shadings*, 0.05 × 10^{-5} s^{-1} intervals), $u - \omega$ wind anomaly (*arrows*, vertical wind is multiplied by 100 times). The intraseasonal OLR anomalies (W m^{-2}) along the same latitudes are plotted at the bottom

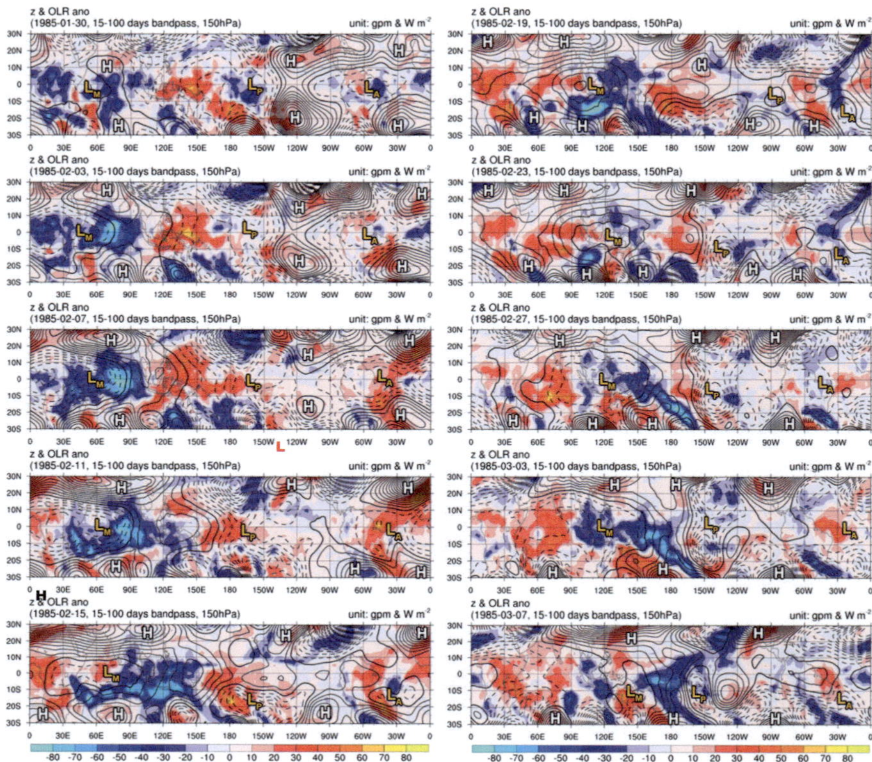

Fig. 10.38 Horizontal distribution of intraseasonal 150-hPa height anomaly (*contours*, 10 gpm intervals), and OLR anomaly (*contours*, 10 W m^{-2} intervals) from 30 January to 7 March 2012

Figure 10.36 illustrates the Hovmöller diagram of intraseasonal OLR anomalies in 1985 as an example. In Fig. 10.36a, a signal of MJO indicated by the letter "M" is originated from the tropical western Indian Ocean on 30 January 1985 and moved eastward to central tropical Pacific in middle March. Another MJO was observed from late March to April and moved eastward in the tropical Indian Ocean. On the other hand, there were ISO signals in the tropical eastern Pacific and the tropical Atlantic. These ISOs moved westward and eastward respectively in the three tropical oceans. As indicated by Roundy and Frank (2004), ISOs can be simply divided into eastward-propagating (ISOe) and westward-propagating (ISOw) modes according to their behaviors and structures. Two horizontal distributions of intraseasonal OLR anomalies are depicted in Fig. 10.36b, c. On 30 January 1985, A negative OLR center (M) was located in the tropical central Indian Ocean while other two centers (P and A) were respectively found in the equatorial central Pacific and the equatorial South America. On 25 February 1985, the center M in the

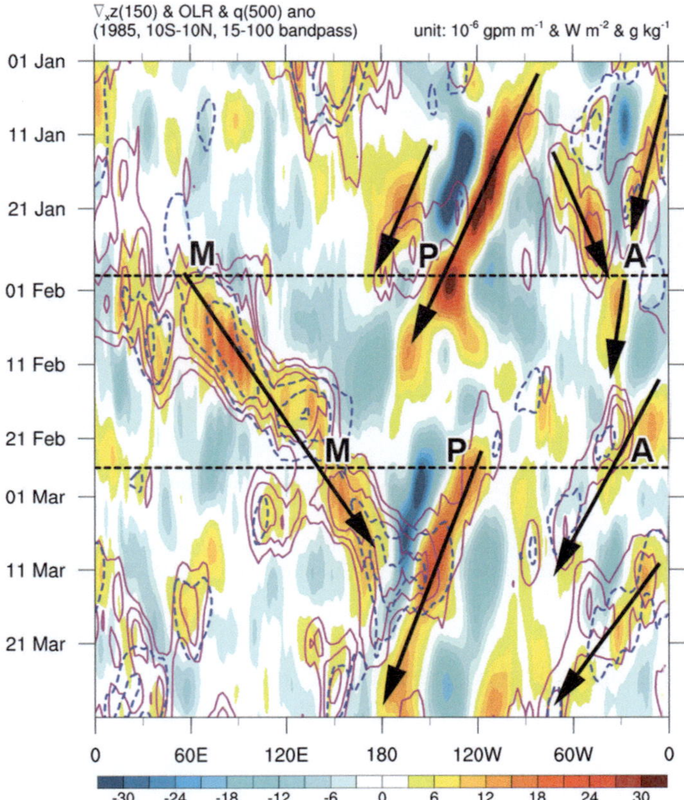

Fig. 10.39 Hovmöller diagram of intraseasonal 150-hPa $\nabla_x z'$ (*shading*, 3 gpm m^{-1} intervals), OLR anomaly (*purple contours*, −5, −15 and −25 W m^{-2}) and 500-hPa specific humidity anomaly (*blue dashed contours*, −0.5, −1.0 and −1.5 m s^{-1}) from January to March 1985. *Black dashed lines* denote 30 January and 25 February 1985

equatorial western Pacific is the MJO signal propagated from the tropical Indian Ocean while the other two centers P and A were newly generated.

As studied in Sect. 10.5, Fig. 10.37 shows the zonal cross sections of intraseasonal geopotential height and temperature as well as other variables along 10°S–10°N. On 30 January 1985, there were three centers of height anomalies respectively in the central Indian Ocean, the eastern Pacific and Atlantic (Fig. 10.37a). Below (above) the anomalous height center shows a positive (negative) center of temperature anomalies vertically. This structure satisfies the hydrostatic equilibrium. Three negative OLR signals indicated by letters M, P, and A are located in the west of the three height centers in Fig. 10.37a. This horizontal and vertical structure of height, temperature anomalies associated with three

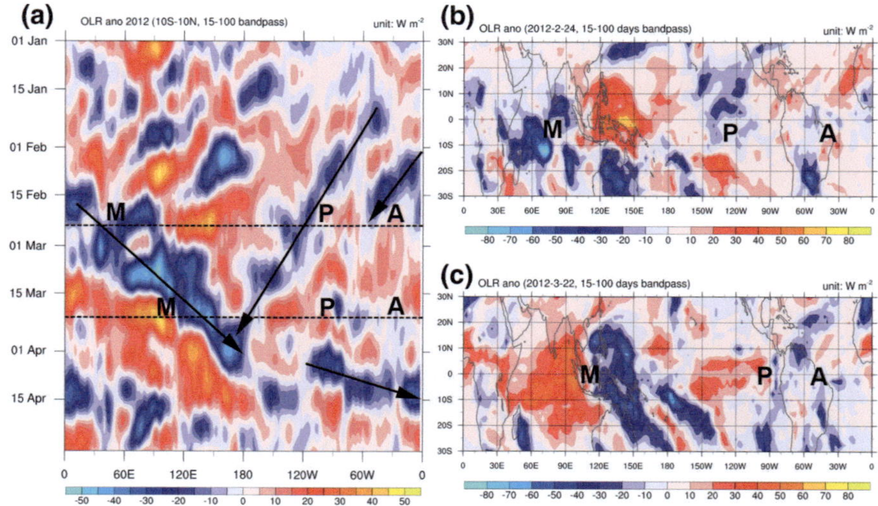

Fig. 10.40 **a** Same as Fig. 10.36a but in 2012. *Black dashed lines* denote 24 February and 22 March 2012. **b** and **c** are the same as Fig. 10.36b but on 24 February 2012 and 22 March 2012, respectively

negative OLR signals is also observed on 25 February 1985 (Fig. 10.37b). These negative OLR signals are also associated with intraseasonal specific humidity and divergence anomalies in Fig. 10.37c, d. Positive specific humidity in the lower troposphere as well as convergence anomaly and ascending flow in the mid-low troposphere are favorable to the negative OLR signal.

Figure 10.38 gives the horizontal distributions of height and OLR anomalies in tropical zone. From 30 January to 7 March 2012, three low centers of height anomalies at 150 hPa can be observed respectively in the three tropical oceans. The low "L_m" propagated eastward from the Indian Ocean to the WP with negative OLR anomalies in its east side and positive one in its west side. The low "L_p" moved southward from 30 January to 15 February and moved westward from 19 February to 7 March 1985. The low "L_A" basically moved eastward.

Using (10.5.1), Fig. 10.39 depicts the Hovmöller diagram of intraseasonal 150-hPa $\nabla_x z'$, OLR and 500-hPa specific humidity anomalies from January to March 1985. This diagram can be compared with Fig. 10.36a. This indicates that the upper tropospheric zonal anomalous height gradient (ZAHG, $\nabla_x z'$) can be used to represent OLR and specific humidity anomalies in the three tropical oceans.

Similar to that shown in Figs. 10.36, 10.37, 10.38 and 10.39, another example is illustrated from Figs. 10.40, 10.41, 10.42 and 10.43. Propagation of negative OLR anomalies from January to April and their spatial distributions on 24 February and on 22 March 2012 can also be clearly observed in Fig. 10.40. Spatial configuration

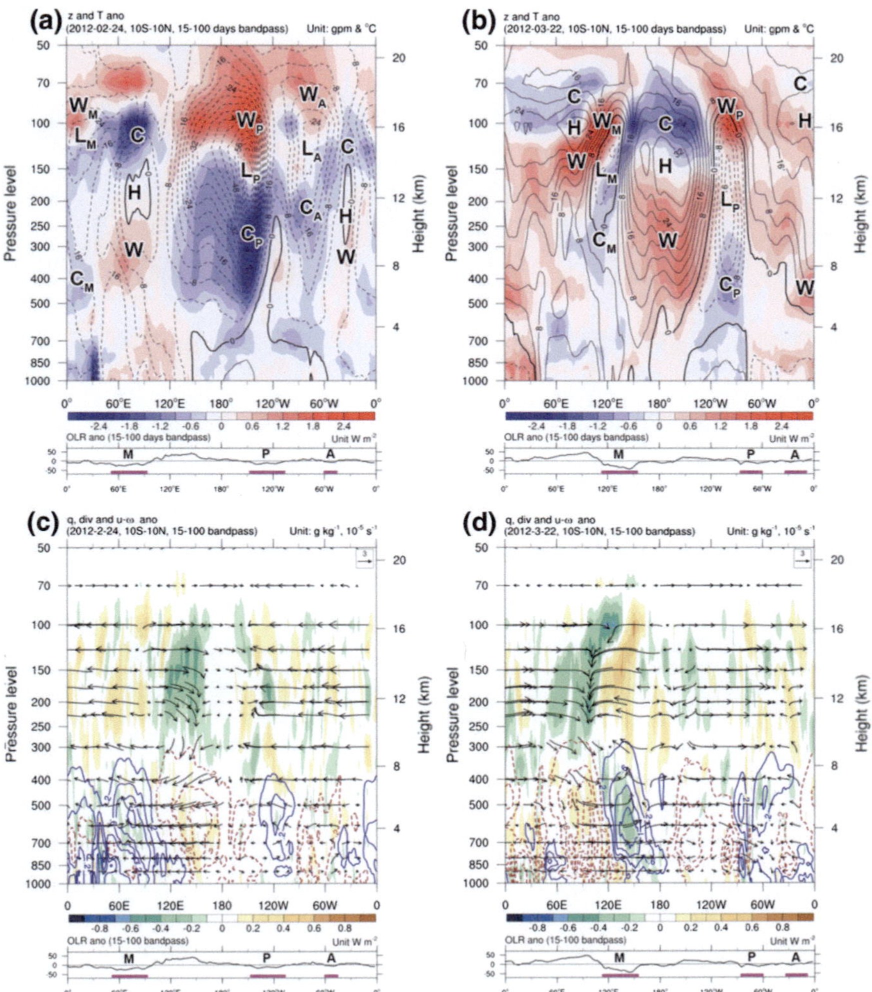

Fig. 10.41 **a** and **c** are the same as Fig. 10.37a, c but on 24 February 2012; **b** and **d** are the same as Fig. 10.37b, d but on 22 March 2012

of height, temperature, specific humidity, and wind anomalies associated with the OLR anomalies in Fig. 10.41 is very similar to that in Fig. 10.37. Three low centers of height anomalies at 150 hPa in Fig. 10.42 can be traced from day to day. Finally, Fig. 10.43 shows that the upper tropospheric zonal anomalous height gradient (ZAHG, $\nabla_x z'$) can be used to represent OLR and specific humidity anomalies in the three tropical oceans.

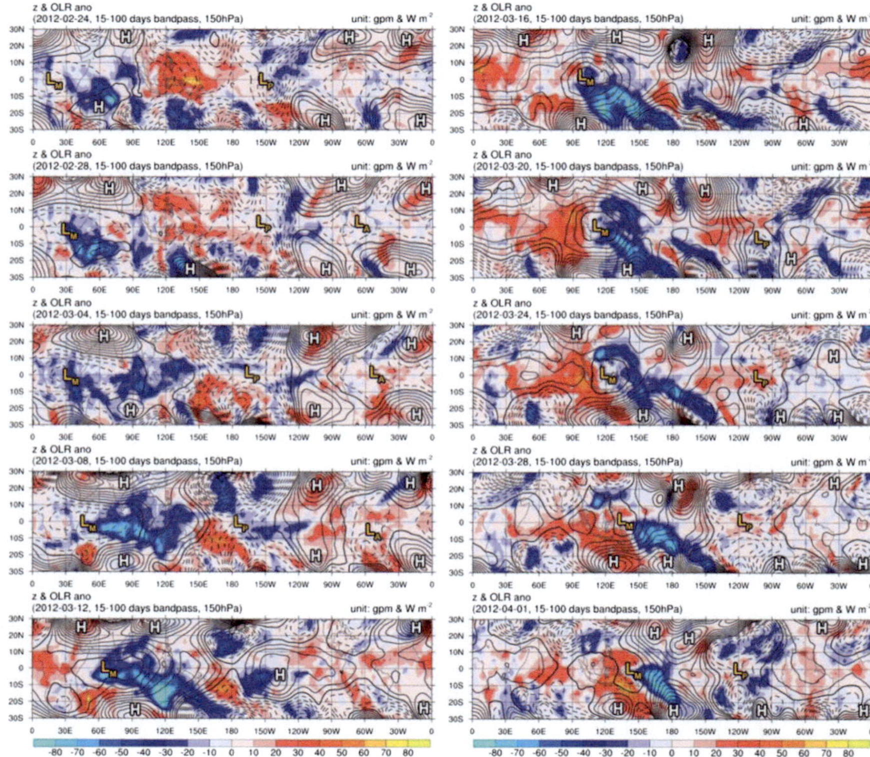

Fig. 10.42 Same as Fig. 10.38 but from 24 February to 1 April 2012

It is clear that tropical ISOs can be identified by using not only OLR anomalies but also 150-hPa height anomalies and 500-hPa specific humidity anomalies as illustrated in Figs. 10.39 and 10.43, respectively. In each region, there are eastward and westward-propagating ISOs. When the intraseasonal OLR anomalies remained lower than -5 W m^{-2} propagate eastward (westward) for more than 30° longitudes, it is defined as an eastward (westward) propagating ISO event. Figure 10.44 summarizes annual and seasonal occurring frequencies of eastward-moving ISOs in the three regions. In the Indian Ocean and western Pacific region, the annual frequency is from 7 to 15. In the boreal spring, the maximum frequency is 5 in 2001, 2006, and 2013. In boreal summer, the maximum frequency is also 5 in 1990, 2002, 2010, 2011, and 2012. In autumn and winter, the maximum frequency is 5 respectively in 2011 and 1986, respectively. In the eastern Pacific region, the annual

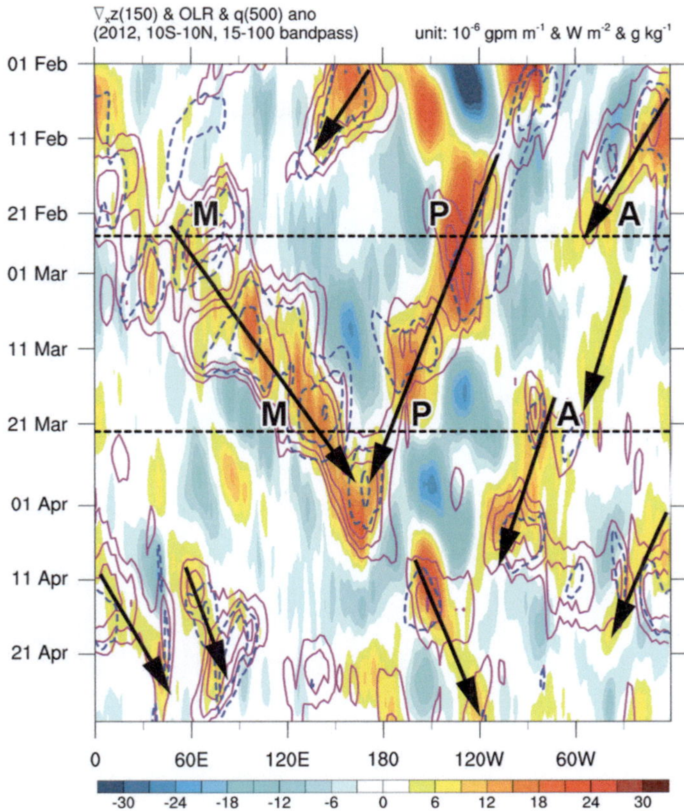

Fig. 10.43 Same as Fig. 10.39 but from February to April 2012. *Black dashed lines* denote 24 February and 22 March 2012

ISO frequency is from 1 to 13. In the tropical Atlantic region, the annual ISO frequency is from 1 to 8. The highest winter ISO frequency (4) is found in 2012.

The annual ISO frequency of westward propagation in the tropical Indian Ocean to the tropical western Pacific region is from 5 to 18 (Fig. 10.45a). The annual maximum frequency (18) occurred in 2012 and the highest winter frequency (6) was in 1990. In the eastern Pacific region, annual ISO frequency is from 3 to 10. The highest summer ISO frequency (5) is observed in 1983. In the tropical Atlantic region, annual ISO frequency is from 1 to 9. The highest winter ISO frequency (4) is found in 1997, 2010, and 2012.

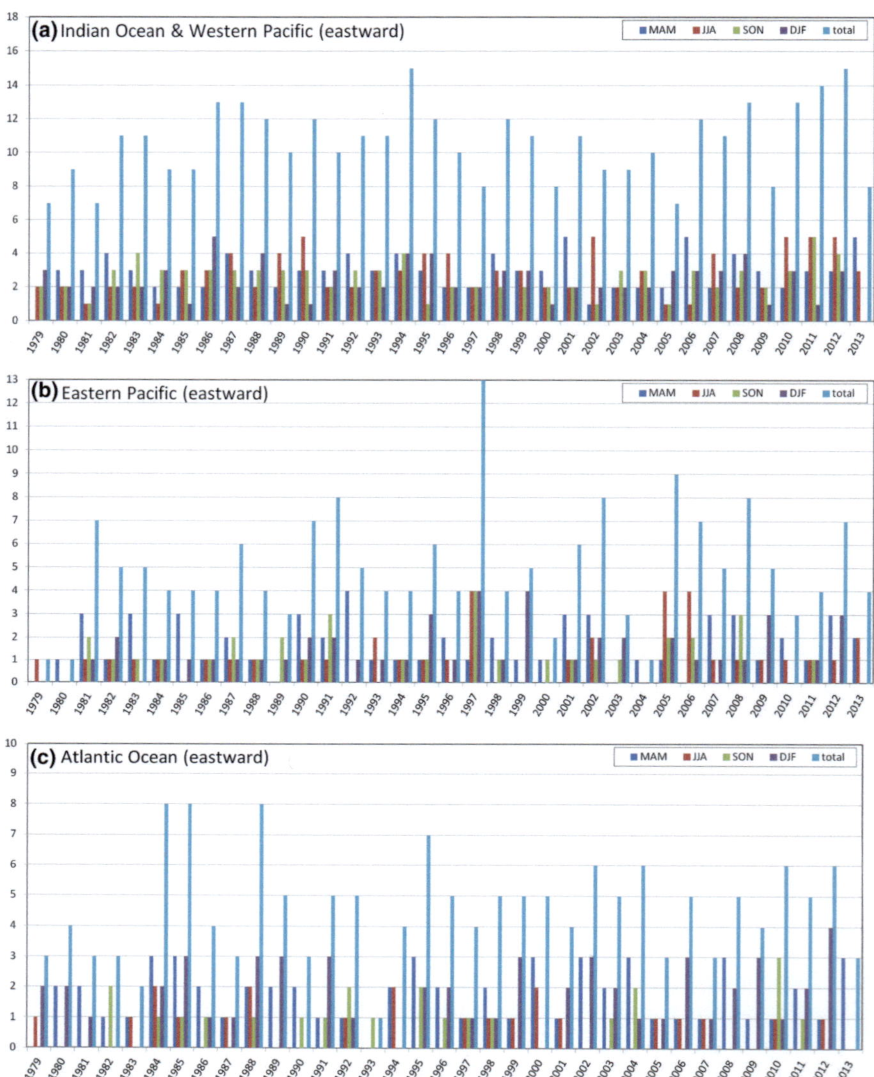

Fig. 10.44 Annual and seasonal occurring frequencies of eastward-moving ISOs in the **a** Indian Ocean and western Pacific, **b** eastern Pacific and **c** Atlantic Ocean regions. Noted that data of the first 100 days in 1979 and last 100 days in 2013 are neglected due to bandpass filtering

Fig. 10.45 Same as Fig. 10.44 but of westward-moving ISOs

Questions

1. There are many methods to reveal teleconnections, such as one-point correlation map, composite analysis, and eigenvector analysis expansion. Which one is the best?
2. How to predict these spatial teleconnections or wave trains or patterns? Are there any cases of correct prediction for leading several months?
3. What are the differences between the MJO and teleconnections?

4. How to predict MJO? Why is the bandpass filter usually used to extract MJO-related information?
5. How to physically and conveniently extract MJO and extra-tropical intraseasonal oscillation as well as teleconnections in the atmosphere?
6. Why is the hydrostatic balance relationship between height and temperature anomalies as well as geostrophic balance relationship between zonal wind and height anomalies satisfied in the tropical and extra-tropical atmospheres?

References

Adames ÁF, Wallace JM (2014) Three-dimensional structure and evolution of the MJO and its relation to the mean flow. J Atmos Sci 71:2007–2026. doi: 10.1175/JAS-D-13-0254.1

Adames ÁF, Wallace JM (2015) Three-dimensional structure and evolution of the moisture field in the MJO. J Atmos Sci 72:3733–3754

Barnston AG, Livezey RE (1987) Classification, seasonality and persistence of low frequency atmospheric circulation patterns. Mon Wea Rev 115:1083–1162

Dee DP, Coauthors (2011) The ERA-Interim reanalysis: configuration and performance of the data assimilation system. Q J Roy Meteor Soc 137:553–597

Dickson RR, Namias J (1976) North American influences on the circulation and climate of the North Atlantic sector. Mon Wea Rev 104:1255–1265

Ding QH, Wang B (2005) Circumgbobal teleconnection in the Northern Hemisphere summer. J Clim 18:3483–3505

Duchon CE (1979) Lanczos Filtering in One and Two Dimensions. J Appl Meteorol 18:1016–1022

Gill AE (1980) Some simple solutions for heat-induced tropical circulation. Q J Roy Meteor Soc 106:447–462

Gong DY, Wang SW (1999) Definition of Antarctic Oscillation index. Geophy Res Lett 26 (4):459–462

Gottschalck J, Roundy PE, Schreck III CJ,Vintzileos A, Zhang C (2013) Large-Scale Atmospheric and Oceanic Conditions during the 2011–12 DYNAMO Field Campaign. Mon Weather Rev 141:4173–4196

Hannachi A, Jolliffe IT, Stephenson DB (2007) Empirical orthogonal functions and related techniques in atmospheric science: a review. Int J Climatol 27(9):1119–1152

Hoskins BJ, Karoly DJ (1981) The steady linear response of a spherical atmosphere to thermal and orographic forcing. J Atmos Sci 38(6):1179. doi:10.1175/1520-0469(1981)038<1179:TSLROA>2.0.CO;2

Hsu PC, Li T (2012) Role of the boundary layer moisture asymmetry in causing the eastward propagation of the Madden–Julian Oscillation. J Clim 25:4914–4931

Jiang X, Li T, Wang B (2004) Structures and mechanisms of the northward propagating boreal summer Intraseasonal Oscillation. J Clim 17:1022–1039

Kiladis GN, Straub KH, Haertel PT (2005) Zonal and Vertical Structure of the Madden–Julian Oscillation. J Atmos Sci 62:2790–2809

Kiladis GN, Coauthors (2014) A Comparison of OLR and circulation-based indices for tracking the MJO. Mon Weather Rev 142:1697–1715

Kutzbach JE (1970) Large-scale features of monthly mean Northern Hemisphere anomaly maps of sea-level pressure. Mon Wea Rev 98:708–716

Lafleur DM, Barrett BS, Henderson GR (2015) Some climatological aspects of the Madden–Julian Oscillation (MJO). J Clim 28:6039–6053

Lau WKM, Waliser DE (2011) Intraseasonal Variability of the Atmosphere-Ocean Climate System (2nd E.) Springer

Leung JCH, Qian WH (2016) Monitoring the Madden-Julian Oscillation with Geopotential Height. Climate Dyn. doi:10.1007/s00382-016-3431-x

Liu P, Zhang Q, Zhang C, Zhu Y, Khairoutdinov M, Kim H-M, Schumacher C, Zhang M (2016) Arevised real-time multivariate MJO index. Mon Weather Rev 144:627–642

Li T (2014) Recent advance in understanding the dynamics of the Madden-Julian oscillation. J Meteorol Res-Prc 28:1–33

Liebmann B, Smith CA (1996) Description of a complete (interpolated) outgoing longwave radiation dataset. Bull Amer Meteor Soc 77:1275–1277

Lorenz EN (1951) Seasonal and irregular variations of the Northern Hemisphere sea-level pressure profile. J Meteor 8:52–59

Madden RA, Julian PR (1971) Detection of a 40–50 day oscillation in the zonal wind in the tropical Pacific. J Atmos Sci 28:702–708

Madden RA, Julian PR (1972) Description of global-scale circulation cells in the tropics with a 40–50 day period. J Atmos Sci 29:1109–1123

Majda AJ, Biello JA (2004) A multiscale model for tropical intraseasonal oscillations. P Natl Acad Sci 101:4736–4741

Matsuno T (1966) Quasi-Geostrophic Motions in the Equatorial Area. J Meteorol Soc Jpn 44: 25–43

Nakazawa T (1988) Tropical super clusters within intraseasonal variations over the western Pacific. J Meteorol Soc Jpn 66:823–839

Qian WH, Wu KJ, Chen D (2015a) The Arctic and Polar cells act on the Arctic sea ice variation. Tellus 67(1):27692. doi:10.3402/tellusa.v67.27692

Qian W, Yu T, Du J (2015b) A unified approach to trace surface heat and cold events by using height anomaly. Clim Dyn 46:1647–1664

Rogers JC, Van Loon H (1982) Spatial variability of sea level pressure and 500-mb height anomalies over the Southern Hemisphere. Mon Wea Rev 110:1375–1392

Roundy PE, Frank WM (2004) Effects of low-frequency wave interactions on intraseasonal oscillations. J Atmos Sci 61:3025–3040

Sperber KR (2003) Propagation and the vertical structure of the Madden–Julian Oscillation. Mon Weather Rev 131:3018–3037

Straub KH (2013) MJO initiation in the real-time multivariate MJO Index. J Clim 26:1130–1151

Thompson DWJ, Wallace JM (1998) The Arctic oscillation signature in the wintertime geopotential height and temperature fields. Geophy Res Lett 25(9):1297–1300

Thompson DWJ, Wallace JM (2000a) Annular modes in the extratropical circulation. Part I: Month-to-month variability. J Clim 13:1000–1016

Thompson DWJ, Wallace JM (2000b) Annular modes in the extratropical circulation. Part II: Trends. J Clim 13:1018–1036

Trenberth KE, Branstator GW, Karoly D, Kumar A, Lau NC, Ropelewski C (1998) Progress during TOGA in understanding and modeling global teleconnections associated with tropical sea surface temperatures. J Geophy Res 103:14291. doi:10.1029/97JC01444

Vincent DG (1994) The South Pacific Convergence Zone (SPCZ): A Review. Mon Weather Rev 122:1949–1970

Ventrice MJ, Wheeler MC, Hendon HH, Schreck III CJ, Thorncroft CD, Kiladis GN (2013) A modified multivariate Madden–Julian Oscillation index using velocity potential. Mon Weather Rev 141:4197–4210

Walker GT, Bliss EW (1932) World weather V, Memoires Roy Meteorol Soc London 4(36):53–84

Wallace JM, Gutzler DS (1981) Teleconnections in the geopotential height field during the Northern Hemisphere winter. Mon Wea Rev 109:784–812

Wang HJ, Fan K, Sun JQ (2005) The Circum-Pacific Teleconnection Pattern in meridional wind in the high troposphere. Adv Atm Sic 22(3):463–466

Wang B, Rui H (1990) Dynamics of the coupled moist Kelvin–Rossby wave on an equatorial β-plane. J Atmos Sci 47:397–413

Weickmann KM (1983) Intraseasonal circulation and outgoing longwave radiation modes during Northern Hemisphere winter. Mon Weather Rev 111:1838–1858

Wheeler M, Kiladis GN (1999) Convectively coupled equatorial waves: Analysis of clouds and temperature in the wavenumber–frequency domain. J Atmos Sci 56:374–399

Wheeler MC, Hendon HH (2004) An all-season real-time multivariate MJO index: Development of an index for monitoring and prediction. Mon Weather Rev 132:1917–1932

Yang G-Y, Hoskins B, Slingo J (2007) Convectively coupled equatorial waves. Part I: horizontal and vertical structures. J Atmos Sci 64:3406–3423 doi:10.1175/JAS4017.1

Yoneyama K, Zhang C, Long CN (2013) Tracking Pulses of the Madden–Julian Oscillation. B Am Meteorol Soc 94:1871–1891. doi:10.1175/BAMS-D-12-00157.1

Zhang C (2005) Madden-Julian Oscillation. Rev Geophy 43:1–36

Data Sources

1. The monthly mean and daily mean NCEP/NCAR Reanalysis 1 (NCEP R1) is obtained from http://www.esrl.noaa.gov/psd/data/gridded/data.ncep.reanalysis.html, with a 2.5° × 2.5° horizontal resolution and 17 vertical levels from 1948 to present (Kalnay et al. 1996).
2. The monthly mean and daily mean NCEP/DOE AMIP Reanalysis 2 (NCEP R2) is obtained from http://www.esrl.noaa.gov/psd/data/gridded/data.ncep.reanalysis2.html, with a 2.5° × 2.5° horizontal resolution and 17 vertical levels from 1979 to present (Kanamitsu et al. 2002).
3. The ECMWF Reanalysis Interim (ERA-Interim) product since 1979 is obtained from http://apps.ecmwf.int/datasets/, with a horizontal resolution of 0.75° × 0.75° and 37 vertical levels, including monthly mean and 6-hourly mean data (Dee et al. 2011).
4. The medium-range ensemble forecasts of ECMWF with 51 ensemble members and NCEP with 20 ensemble members are extracted from the THORPEX (The Observing System Research and Predictability Experiment) Interactive Grand Global Ensemble (TIGGE) archive at http://tigge.ecmwf.int (Swinbank et al. 2016). Six-hourly interval forecasts are generated from 0000 UTC initial conditions including 60 time steps from +1 to +15 days for the ECMWF while the NCEP provides 16-day forecasts every 6 h (0000, 0600, 1200, and 1800 UTC).
5. The Japanese 55-year Reanalysis (JRA-55) product since 1958 is obtained from http://rda.ucar.edu/datasets/ds628.1/, with a horizontal resolution of 1.25° × 1.25° and 37 vertical levels (Ebita et al. 2011).
6. The NOAA-CIRES Twentieth Century Reanalysis (20CR) dataset (Compo et al. 2011) is the version V2c from 1851 to 2011 obtained from http://www.esrl.noaa.gov/psd/data/gridded/data.20thC_ReanV2c.html, with a 2° × 2° horizontal resolution and 24 vertical levels.
7. The monthly mean ASR dataset is a 3-DVAR assimilation product which has been designed and optimized for Polar region from 2000 to 2010 (obtained from the website http://rda.ucar.edu/datasets/ds631.4/#!description). The ASR dataset has a high horizontal resolution of 30 km × 30 km spatial grid interval and 34 vertical levels beyond 50°N (Bromwich et al. 2010, 2015).

© Springer Nature Singapore Pte Ltd. 2017

W. Qian, *Temporal Climatology and Anomalous Weather Analysis*,
Springer Atmospheric Sciences, DOI 10.1007/978-981-10-3641-5

8. The historical run of the Community Earth System Model version 1 (CESM1) includes the Community Atmospheric Model version 5 (CAM5) referred to as CESM1 (CAM5) (Neale et al. 2010). The ensemble mean CAM5 model dataset with three historical runs from 1850 to 2005 has a horizontal 0.9375 × 1.25-degree latitude–longitude resolution and 17 vertical levels, retrieved from the website (https://www.earthsystemgrid.org/home.htm).
9. The aqua-planet model control run is from the Community Climate System Model version 4 referred to CCSM4 (Gent et al. 2011). The duration of the dataset is 5 years with a horizontal 0.9375 × 1.25-degree latitude–longitude resolution and 17 vertical levels, retrieved from the website (https://www.earthsystemgrid.org/home.htm).
10. The monthly mean gridded sea ice concentration (SIC) is from the HadISST1 version since 1979 (Rayner et al. 2003). It is obtained from http://www.metoffice.gov.uk/hadobs/hadisst/, with a 1° × 1° horizontal resolution.
11. The AO/NAO indices are obtained from the website (http://www.cpc.ncep.noaa.gov/products/precip/CWlink/daily_ao_index/teleconnections.shtml).
12. The Nino 3.4 indices are downloaded from the website (http://www.cpc.ncep.noaa.gov/products/analysis_monitoring/ensostuff/detrend.nino34.ascii.txt).
13. The Climate Prediction Center (CPC) Merged Analysis of Precipitation (CMAP) (Xie and Arkin 1997) is obtained from five kinds of satellite estimates, while the enhanced version includes also blended NCEP/NCAR Reanalysis Precipitation values. The spatial resolution of CMAP dataset is 2.5° × 2.5° latitude–longitude grid interval and the temporal coverage is from 1979 to present with monthly mean and pentad mean values at the website (http://www.esrl.noaa.gov/psd/data/gridded/data.cmap.html#detail).
14. The Climate Prediction Center (CPC) Unified Gauge-Based Analysis of Global Daily Precipitation (Xie et al. 2007; Chen et al. 2008) provides daily mean precipitation dataset. This product combines all information sources available at CPC and takes advantage of the optimal interpolation (OI) objective analysis technique. Its horizontal resolution is 0.5° × 0.5° and it covers the global land. The retrospective version ranges from 1979 to 2005, which uses about 30,000 stations. The real-time version spans 2006 to present, which uses about 17,000 stations.
15. China Meteorological Administration provides station observed precipitation. The number of observational stations has increased from fewer than 200 in 1951 to more than 650 since 1960. The average number of stations has been basically stable at 680 since 1960 except that in 2007 and 2008 went over 700 stations. Totally, 754 stations are obtained excluding two stations in the South China Sea, but only 460 stations have complete precipitation records. The distribution of stations is relatively uniform in the eastern China.
16. The grid precipitation data provided by Chen et al. (2010) used the ordinary Kriging method to interpolate station precipitations in China into 0.5° latitude–longitude interval.

17. The CPC MORPHing technique (CMORPH; Joyce et al. 2004) precipitation dataset has a 3-hourly temporal and 0.25° spatial resolution. These estimates are derived from satellites as described by Joyce et al. (2004).
18. The daily interpolated OLR product obtained from the National Oceanic and Atmospheric Administration (NOAA) polar-orbiting satellites is gridded at a horizontal resolution of 2.5° × 2.5° and available from June 1974 to December 2013 (Liebmann 1996). Due to satellite failure, no OLR data were collected from 17 March 1978 to 31 December 1978.
19. The FNL (Final Operational Global Analysis) product is obtained from NCEP (http://www.esrl.noaa.gov/psd), with a horizontal resolution of 1° × 1° and 26 vertical levels.
20. The best track data of tropical cyclones is obtained from the Joint Typhoon Warning Center (JTWC) (http://www.usno.navy.mil/NOOC/nmfc-ph/RSS/jtwc/best_tracks/) (Chu et al. 2002).
21. The Atlantic hurricane database HURDAT2 is provided by the National Hurricane Center (NCEP/NHC), which are based on the traditionally disseminated tropical cyclone historical database in the format known as HURDAT (Jarvinen et al. 1984).
22. The daily mean, maximum, and minimum surface air temperature observations in mainland China are taken from 549 observational stations across China and cover the period from 1960 to 2008 (Li and Yan 2009). Daily (24 h) precipitation observations (from 1200 UTC to the next 1200 UTC) at the same 549 stations are obtained from the China Meteorological Administration.
23. The daily maximum surface air temperatures used in defining heat waves are from China Daily Surface Temperature Dataset (V2.0), with a horizontal resolution of 0.5° × 0.5° in latitude and longitude. This dataset is based on observations from 2472 surface meteorological stations in China, most of which started taking observations since 1 January 1961.
24. The CRU TS 3.1 monthly precipitation data with the spatial 0.5° × 0.5° latitude–longitude grid for 1901–2008 is obtained from http://catalogue.ceda.ac.uk/uuid/ac3e6be017970639a9278e64d3fd5508.

References

Bromwich DH, Kuo Y-H, Serreze M, Walsh J, Bai L-S, Barlage M, Hines K, Slater A (2010) Arctic System Reanalysis: Call for community involvement. Eos Trans Amer Geophys Union 91(2):13–14.

Bromwich DH, Wilson AB, Bai LS, Moore GW, Bauer P (2015) A comparison of the regional Arctic system reanalysis and the global ERA-Interim reanalysis for the Arctic. Q J R Meteorol Soc. doi:10.1002/qj.2527.

Compo GP, Whitaker JS, Sardeshmukh PD et al (2011) The Twentieth Century Reanalysis Project. Quart J Roy Meteor Soc 137:1–28.

Chen D, Ou T, Gong L, Xu CY, Li W, Ho CH, Qian W (2010) Spatial interpolation of daily precipitation in China: 1951–2005. Adv Atmos Sci 27(6):1221–1232.

Chen MY, Shi W, Xie PP et al (2008) Assessing objective techniques for gauge-based analyses of global daily precipitation. J Geophys Res 113:D04110.

Chu JH, Sampson CR, Levine AS et al (2002) The Joint Typhoon Warning Center tropical cyclone best tracks 1945–2000. Joint Typhoon Warning Center, Pearl Harbor, Hawaii, NRL/MR/7540-02-1602.

Dee DP, Uppala SM, Simmons AJ et al (2011) The ERA-Interim reanalysis: Configuration and performance of the data assimilation system. Quart J Roy Meteor Soc 137(656):553–597. doi:10.1002/qj.828.

Ebita A, Kobayashi S, Ota Y et al (2011) The Japanese 55-year Reanalysis "JRA-55": An interim report. J Meteor Soc Japan 7:149–152. doi:10.2151/sola.2011-038.

Gent PR, Danabasoglu G, Donner LJ et al (2011) The Community Climate System Model Version 4. J Climate 24(11):3666–3682.

Harris I, Jones PD, Osborn TJ, Lister D (2014) Updated High-Resolution Grids of Monthly Climatic Observations—The CRU TS3.10 Dataset. Int J Climatol 34(3):623–642.

Jarvinen BR, Neumann CJ, Davis MAS (1984) A tropical cyclone data tape for the North Atlantic Basin, 1886–1983: Contents, limitations, and uses. NOAA Tech. Memo, NWS NHC 22, Coral Gables, FL, 21 pp.

Joyce RJ, Janowiak JE, Arkin PA, Xie PP (2004) CMORPH: A method that produces global precipitation estimates from passive microwave and infrared data at high spatial and temporal resolution. J Hydrometeor 5(3):487–503.

Kalnay E, Kanamitsu M, Kistler R et al (1996) The NCEP/NCAR 40-year reanalysis project. Bull Amer Meteor Soc 77:437–471.

Kanamitsu M, Ebisuzaki W, Woollen J et al (2002) NCEP–DOE AMIP-II Reanalysis (R-2). Bull Amer Meteor Soc 83:1631–1643. doi:10.1175/BAMS-83-11-1631.

Li Z, Yan ZW (2009) Homogenized daily mean/maximum/minimum temperature series for China from 1960 to 2008. Atmos Ocean Sci Lett 2:237–243.

Liebmann B (1996) Description of a complete (interpolated) outgoing longwave radiation dataset. Bull Amer Meteor Soc 77:1275–1277.

© Springer Nature Singapore Pte Ltd. 2017

W. Qian, *Temporal Climatology and Anomalous Weather Analysis*,
Springer Atmospheric Sciences, DOI 10.1007/978-981-10-3641-5

Neale RB, Chen C-C, Gettelman A et al (2010) Description of the NCAR Community Atmosphere Model (CAM5.0). NCAR Tech Rep NCAR/TN-4861STR, 268 pp.

Rayner NA, Parker DE, Horton EB et al (2003) Global analyses of sea surface temperature, sea ice, and night marine air temperature since the late nineteenth century. J Geophys Res 108 (D14):1–29. doi:10.1029/2002JD002670.

Swinbank R, Kyouda M, Buchanan P et al (2016) The TIGGE project and its achievements. Bull Amer Meteor Soc 97(1):49–67.

Xie PP, Arkin PA (1997) Global precipitation: A 17-year monthly analysis based on gauge observations, satellite estimates, and numerical model outputs. Bull Am Meteor Soc 78:539–2558. doi:10.1175/1520-0477(1997)078<2539:GPAYMA>2.0.CO;2.

Xie PP, Yatagai A, Chen MY et al (2007) A gauge-based analysis of daily precipitation over East Asia. J Hydrometeor 8:607–626.

Index

A
African monsoon, 171, 185, 190
Aleutian Low, 86, 128, 129, 132, 631
American monsoon, 171, 185, 190, 191
Annual cycle, 13, 18, 69, 103–106, 121, 122, 125, 127–130, 132, 137, 140, 146, 148, 150, 166, 169, 177, 427, 447, 530
Anomalous-flow beta-advection model (ABAM), 448, 449, 452, 455, 467–470, 486, 512, 515, 518
Anomalous synoptic chart, 215, 216, 222, 223, 242, 283, 285, 296, 330, 420, 421
Anomalous weather chart, 409
Antarctic cell, 31, 55, 59, 61, 63, 79
Antarctic Oscillation (AAO), 637
AO/NAO index, 524, 531, 532, 550, 551, 631
Aqua planet, 26
Arctic amplification, 524, 562, 617
Arctic cell, 31, 41, 43, 45–47, 49, 51, 53–55, 59, 61, 64, 77, 79, 525, 527, 530–535, 537–539, 542, 545, 547
Arctic Oscillation (AO), 524, 631
Arctic system reanalysis (ASR), 531
Ascending flow intensity, 525–527
Asian-Australian monsoon, 171
Atmospheric delayed operator, 72
Atmospheric–ocean–ice interaction, 524
Australian monsoon, 185, 190
Axial tilt, 15, 16

B
Beta-advection model (BAM), 446, 448, 452, 460, 468, 471, 487
Beijing extreme rainfall, 269, 328
Binary interaction, 425, 492, 498, 499, 503–505, 508, 510, 511, 513, 518
Biweekly cycle, 69, 103, 104, 112, 114, 117–122, 125–128, 148
Bjerknes, J., 39, 41
Bjerknes, V., 31, 39, 40, 49

C
Circum-Pacific teleconnection pattern, 637
Classical monsoon region, 165, 170
Climate change, 1, 13, 15, 17, 525, 531, 551, 552, 559, 572
Climate Prediction Center (CPC) Merged Analysis of Precipitation (CMAP), 532, 678
Climate Prediction Center (CPC) Unified Gauge-Based Analysis of Global Daily Precipitation, 678
Climatic clock, 17, 22
Climatic convergence line, 86
Climatic cycle, 14, 26, 70, 161, 177, 178, 184, 185
Climatic-flow beta-advection model (CBAM), 448, 449, 452, 454, 467–471, 486, 487
Climatic prediction, 20
Climatic state, 1, 12, 13, 17, 18, 20, 21, 31, 64, 69, 70, 77, 80, 101, 123, 149, 353
Climatic synoptic chart, 216, 272
Climatic trough, 86, 87, 89, 149, 222, 389
Cluster distance variation, 178
CMORPH, 269, 327
Cold dew, 22, 370
Cold event, 347, 348, 350, 352, 370, 373, 375, 386, 388, 390, 401, 403, 405, 406, 409, 529, 580
Cold semi-arid, 194, 195
Cold semi-arid climate, 194, 195
Cold surge, 228, 347, 348, 369, 420, 541, 625
Combined empirical orthogonal function (CEOF), 645
Community Atmospheric Model (CAM), 678
Community Climate System Model (CCSM), 62, 678

© Springer Nature Singapore Pte Ltd. 2017
W. Qian, *Temporal Climatology and Anomalous Weather Analysis*, Springer Atmospheric Sciences, DOI 10.1007/978-981-10-3641-5

Community Earth System Model (CESM), 678
Continent-ocean contrast, 33, 216
Convective band, 8, 61
Convective system, 300, 327, 646, 667
Coriolis force, 38, 157, 446
Cross-equatorial flow, 101, 153, 154, 158, 159, 166, 168, 170, 173, 176, 177, 186, 187, 189–191, 194, 208, 226, 229, 240

D

Descending flow intensity (DFI), 525–527, 529, 530
Diurnal cycle, 17, 18, 23, 26, 31, 33, 80, 271, 281, 352, 427, 438, 447
Dove, 37, 38
Dry–wet alteration, 154, 156, 158, 159, 162, 163, 165, 166, 168, 170–172, 174–176, 181, 182, 185, 186, 190, 208
Dry–wet alterative regions, 153
Dry–wet mode, 200, 202–204, 207

E

East Asia/Pacific pattern, 159, 292, 403, 406, 551, 585, 596, 604, 606, 613, 629
East Asian monsoon, 153, 196, 605
East Asian subtropical monsoon, 92, 196
East Asian Trough, 83, 95, 140, 612
eastern Atlantic (EA) pattern, 629
Eccentricity, 15, 16
ECMWF Reanalysis Interim (ERA-Interim), 24, 437, 677
Eddy heat flux, 51–54, 64, 65
Eddy momentum flux, 51, 53, 54, 64, 65
Edmund Halley, 36, 157
El Niño, 14, 20, 609, 626
El Niño-Southern Oscillation (ENSO), 626
Empirical mode decomposition (EMD), 572
Empirical orthogonal function (EOF), 207, 532
Ensemble empirical mode decomposition (EEMD), 571
Environmental flow, 4, 425–427, 452, 505, 512
Eurasian (EU) pattern, 631
European Centre for Medium-Range Weather Forecast (ECMWF), 19, 24, 215, 247, 248, 269, 339, 340, 365–367, 369, 427, 428, 487, 489, 492, 646
Extended-range weather forecast, 625
Extra-tropical cyclone, 5, 303, 484
Extratropical transition, 636

F

Ferrel cell, 26, 41, 45, 46, 49, 52, 54, 64, 77, 554
FNL (final) Operational Global Analysis, 437, 679
Four-cell model, 31, 50, 51, 66, 531, 533
Four-Dimensional Variational (4Dvar) assimilation, 441
Freezing-rain, 384
Fujiwhara effect, 504, 518

G

General circulation, 31, 33, 37, 39, 50, 176, 452, 523, 617
Generalized beta-advection model (GBAM), 425
Generalized moist potential vorticity, 298
Geostrophic balance, 645, 659, 660, 673
Geostrophic wind, 38, 112, 129, 216
Global Forecast System (GFS), 20, 244, 427
Global monsoon, 153, 154
Global monsoon domain, 165
Global monsoon region, 98, 101
Global Producing Centers of Long-Range Forecasts (GPCLRFs), 20

H

HadISST1, 525, 526, 532, 557, 572–574, 678
Hadley, 26, 31, 36–38, 41, 43, 45–47, 53–55, 57, 77, 79, 523, 551–555, 558, 562, 566, 567
Hadley cell, 41, 43, 45, 46, 52, 53, 77
Heat wave, 228, 273, 347, 348, 350, 352, 353, 355, 357, 360, 364–366, 388, 397, 408, 409, 411, 412, 415, 418, 420, 541, 625
Hot semi-arid, 194, 195
Hot semi-arid climate, 194, 195
HURDAT2, 679
Hurricane, 4, 425, 446, 474, 679
Hurricane Sandy (2012), 478
Hydrosphere, 10, 11
Hydrostatic equilibrium, 112, 231, 235, 347, 441, 476, 495, 500, 581, 657

I

International Research Institute (IRI) for Climate and Society, 69, 109, 110, 112, 120, 231, 244, 273, 277, 332, 357, 377, 388, 420, 437, 495, 507, 539, 669

Inter-tropical convergence zone (ITCZ), 5
Intraseasonal oscillation, 575, 625, 673
Intrinsic mode function (IMF), 572
Iranian mode, 586

J

Japanese 55-year Reanalysis (JRA-55), 524, 677
Jianghuai cyclones, 5
Joint Typhoon Warning Center (JTWC), 427, 679

K

Kelvin wave, 646, 651, 652, 660, 666
Klima, 16
K-means clustering, 153, 177, 178, 184

L

Lanczos bandpass filter, 647
Land breeze, 23, 32–34, 80
La Niña, 14, 20
Low-frequency oscillation, 625–672
Low-level jet, 23, 30, 242, 257
Low-level shear, 257
Low temperature event, 295, 369

M

Madden-Julian Oscillation (MJO), 625, 644
Mann-Kendall test, 565, 568
Mass stream function, 41, 45, 525
Maury, Matthew, 38
Maximal pentad precipitation rate, 161
Maximum height anomaly, 272, 347, 387
Maximum vorticity anomaly, 425, 449
Medium-range weather forecast, 19, 26, 269, 365, 427, 437, 646
Mei-yu front, 5
Meridional cell, 26, 36, 39–41, 43, 45–47, 53, 55, 69, 530, 532, 533, 547, 552, 553
Mesoscale convective system, 269, 328
Mesoscale vortex, 8, 61
Mexican high, 83, 129, 132
Mid-oceanic ridge, 12
Milankovitch cycle, 16
Milankovitch, Milutin, 15
Minimal pentad precipitation rate, 161, 163, 168, 175
Minimum divergence anomaly, 425, 449
Moist divergence, 298
Moist potential vorticity, 298, 303
Moist vorticity, 247, 248, 298, 323
Monsoon convergence line, 23
Monsoon onset, 23, 70, 86, 92, 173, 184, 189, 196
Monsoon retreat, 70, 190, 612
Monsoon trough, 23, 24, 87, 90, 91, 98, 105, 175, 227, 606, 607, 609, 612, 615, 617
Moving-type heat wave, 347

N

NCEP/DOE AMIP Reanalysis 2, 677
NCEP/NCAR Reanalysis 1, 677
Newtonian mechanics, 2
Newton, Isaac, 1
Newton's law, 1, 2
Niño 3.4 index, 640
NOAA-CIRES Twentieth Century Reanalysis (20CR), 552, 677
Non-classical monsoon region, 153, 170, 171, 176
Nonlinear interaction, 14
Non-monsoon region, 153, 170, 171, 176, 192, 200
Normalized anomaly, 269
North America monsoon, 183
North American high, 216
North Atlantic Oscillation (NAO), 629
Northeast Pacific High, 132
Northern Annular Mode (NAM), 524, 631
Northern Hemisphere Annular Mode, 631
North Pacific high, 86, 129, 175
North Pacific low, 93
North Pacific Oscillation (NPO), 629
Northwestern Africa monsoon, 183
Now-casting weather forecast, 19
Numerical weather prediction (NWP) model, 19, 20, 365, 370

O

Occluded cyclone, 5, 295
Offshore breeze, 32
Onset date, 180, 359, 363, 366, 367, 372
Outer edge active zones, 153, 193–195

P

Pacific-Japan pattern, 605
Pacific-North American (PNA) teleconnection pattern, 626
Pacific-South American (PSA) pattern, 637
Pierre Simon Laplace, 2
Polar cell, 41, 43, 46, 47, 49, 55, 77, 79, 523, 530, 533, 537, 541, 544, 547, 550, 551, 556, 563, 570
Polar cyclone, 40

Polar frontal convergence zone, 158
Potential vorticity, 298, 303, 446, 505
Precession, 15, 16
Prevailing wind reversal, 153, 155, 156, 158, 166, 168, 176, 191, 208
Primordial Calendar, 22

Q

Quasi-linear convective system, 8, 328
Quasi-periodic oscillation, 575
Quasi-static climatic state, 23, 271, 353, 447
Quasi-stationary Rossby wave, 415

R

Raw anomaly, 409, 427
Real-time Multivariate MJO (RMM) index, 645
Regional convective event, 342
Regional heavy rain, 248, 251, 280
Relative anomaly, 271
Returning-cold, 370
Rossby-Kelvin wave, 645, 646, 651, 652, 654, 660, 661, 666
Rossby wave, 19, 226, 388, 415, 524, 613, 615, 646, 652–654
Rotated empirical orthogonal function (REOF), 625
Russian heat wave, 271, 408, 409

S

Sakuhei Fujiwhara, 504
Sea breeze, 23, 32–34, 80
Sea ice concentration (SIC), 524, 678
Sea ice extent (SIE), 525
Sea ice minimum event, 523, 617
Sea–land contrast, 23, 40, 59, 69, 86, 175, 535, 562, 566
Seasonal cycle, 17, 21–23, 26, 33, 47, 70, 72, 73, 75, 181, 281, 525, 527
Semiannual cycle, 69, 104, 108, 112, 123, 126, 140, 145, 166, 169, 181
Short-term weather forecast, 19
Siberia High, 128, 129, 132, 140, 145
Single-cell model, 31
Site heavy rain, 251
Solar radiation, 1, 13, 16, 26, 32, 33, 69, 74, 79, 148, 157, 225, 427
South African monsoon system, 189
South American monsoon system, 185, 189
South Asian high, 73, 216, 618, 625
South Asian low, 129
South Asian monsoon, 171
Southern Annular Mode (SAM), 636
Southern Oscillation, 80, 626
Spiral rain, 4, 6
Spiral vorticity band, 61
Spring begins, 22
Spring equinox, 22, 180
Standardized anomaly, 427
Standing-type heat wave, 347, 421
Steering flow, 388, 425, 426, 431, 433, 469, 486, 487, 501, 504, 509, 510
Stream function, 32, 41, 426, 525, 586, 645
Strong Iranian mode, 585
Strong Tibetan modes, 585
Super tropical cyclone / super typhoon, 5, 427, 437

T

Temporal climate, 12, 23, 69, 101, 149
Temporal climatology, 427
Thomson, 37, 39
THORPEX Interactive Grand Global Ensemble (TIGGE) archive, 677
Threat score, 298
Three-cell model, 41, 49, 50, 66, 531
Three-Dimensional Variational (3DVar) assimilation, 441
Tibetan mode, 585
Tornado, 3, 8, 9, 251, 326, 327, 330, 331, 335, 337
Tornado outbreak, 326, 327, 331, 342
Tornado-producing system, 327, 340, 342
Total distance variation, 178, 179
Trade circulation, 35
Trade wind, 35, 36, 38, 43, 171, 173, 175, 187, 190, 192
Traditional synoptic chart, 20, 215, 225, 238, 249, 251, 263, 266, 273, 281, 283, 295, 297, 326, 337, 415, 441
Traditional weather chart, 441
Tropical cyclone, 3, 5, 9, 61, 253, 255, 300, 352, 374, 376, 425, 478, 482, 484
Tropical cyclone vortex, 4
Tropic of Cancer, 72–74, 76
Tropic of Capricorn, 74
Typhoon Angela (1995), 505, 506
Typhoon Bopha (2000), 493, 499
Typhoon Bopha (2006), 493, 503
Typhoon Fengshen (2002), 493, 494, 503
Typhoon Fungwong (2002), 493, 494, 503
Typhoon Goni (2009), 493, 494, 500, 501
Typhoon Intensification, 437
Typhoon Ivan (1997), 506

Typhoon Joan (1997), 506
Typhoon Megi (2010), 425, 427, 428, 431, 433, 437, 440, 450, 451
Typhoon Morakot (2009), 493
Typhoon Saomai (2000), 494, 504
Typhoon Saomai (2006), 493, 494, 503, 504
Typhoon Zack (1995), 506

U
Unusual tropical cyclone, 425

V
Vertical cold anomalous pattern, 413, 601
Vertical warm anomalous pattern, 413

W
Walker circulation, 69, 80
Weak Iranian mode, 585
Weak Tibetan mode, 585
Weather, 15, 18, 20, 29, 199, 238, 251, 272, 347, 374, 415, 427, 441, 625
Weather anomaly, 18, 21
Weather forecast, 15, 19, 104, 420
Walker, Gilbert, 80
Western Atlantic (WA) pattern, 631
Western Pacific subtropical high, 523, 604
Western Pacific (WP) pattern, 631
William Ferrel, 38
Withdrawal date, 180
World Meteorological Organization (WMO), 20

Z
Zonal anomalous height gradient, 645, 652, 670

MIX
Papier aus verantwortungsvollen Quellen
Paper from responsible sources
FSC® C105338

If you have any concerns about our products,
you can contact us on
ProductSafety@springernature.com

In case Publisher is established outside the EU,
the EU authorized representative is:
Springer Nature Customer Service Center GmbH
Europaplatz 3, 69115 Heidelberg, Germany

Printed by Libri Plureos GmbH
in Hamburg, Germany